Emerging Trends in
Mycology
Plant Pathology &
Microbial Biotechnology

Emerging Trends in

Mycology
Plant Pathology &
Microbial Biotechnology

Editors

Prof. G. Bagyanarayana *M.Sc., Ph.D*

Prof. B. Bhadraiah *M.Sc., Ph.D*

Dr. Mrs. I. K. Kunwar *D.Phil.*

Department of Botany, Osmania University,
Hyderabad - 500 007.

BSP **BS Publications**

4-4-309, Giriraj Lane, Sultan Bazar,
Hyderabad - 500 095 A.P.
Phone : 040 - 23445601, 23445688
e-mail : contactus@bspublications.net

Published by

 **BS Publications**

4-4-309, Giriraj Lane, Sultan Bazar,
Hyderabad - 500 095 A.P.
Phone : 040 - 23445601, 23445688, 23445600
e-mail : contactus@bspublications.net

ISBN : 978-93-89974-84-3

The Proceedings Volume of the
National Conference on Mycology,
Plant Pathology & Microbial Biotechnology is
dedicated to
Prof. C. Manoharachary
Department of Botany, Osmania University, Hyderabad.

Preface

Microorganisms are omnipresent and cosmopolitan in nature. They represent bacteria, fungi, actinomycetes, algae, protozoans etc. There exists a great diversity among various microorganisms and even within a group of microorganisms. The microbial diversity depends on the habitat and environmental factors. It represents an interdisciplinary approach of taxonomists, ecologists, molecular biologists and biotechnologists.

The microorganisms are either beneficial or detrimental in nature. The later effects are by way of causing serious diseases to plants, animals and mankind. The direct beneficial effects on plants are by way of promoting their growth, through the production of phytohormones/auxins and metabolites which stimulate the microbial community in the rhizosphere. The indirect effects are by way of producing antibiotics or toxins or volatile compounds which act as biocontrol agents and thereby reducing the damage against phytopathogens.

Recent developments in molecular biology and biotechnology have thrown light in developing suitable techniques for identification and characterization of microorganisms and developing diagnostic probes for studying their relationship and variability. This has a great impact on their application for environmental sustainability and biodiversity, apart from understanding the mechanisms involved in microbe-microbe interaction in general and soil-plant–microbe interactions in particular.

The various aspects dealing with the biodiversity and taxonomy of fungi, their asssociation with plants, molecular approaches for studying the phylogenetic relationships of microbes and their application have been covered in the ***National Conference on "Emerging Trends in Mycology, Plant Pathology and Microbial Biotechnology"*** organised by the Department of Botany, Osmania University, Hyderabad in December 2004.

The editors are grateful to Prof. K.V.B.R. Tilak and Prof. Vasant Rao for helping us in various ways during the preparation of this volume.

<table>
<tr><td>Hyderabad
15th October, 2005</td><td align="right">G. Bagyanarayana
B. Bhadraiah
I.K. Kunwar</td></tr>
</table>

Contents

Unit - I

Biodiversity and Taxonomy

Unit - II

Plant Pathology

Unit - III

Molecular Biology

Unit - IV
Microbial Biotechnology

Unit – I

Biodiversity and Taxonomy

Bahusutrabeeja manoharacharii sp. nov., a Foliicolous Hyphomycete from the Forests of Western Ghats, India

Pratibha, J. and D. J. Bhat

Dept. of Botany, Goa University, Goa - 403 206.
e-mail : bhatdj@rediffmail.com

ABSTRACT

Bahusutrabeeja manoharacharii sp. nov., a foliicolous hyphomycete isolated from live leaves of *Bridelia scandens* Roxb (Euphorbiaceae) gathered from Chorla Ghat, Goa, India, is illustrated and described. In addition, hitherto described species of the genus are compared in this paper.

KEY WORDS : Biodiversity; conidial fungi; *Bahusutrabeeja manoharacharii*.

We deem it an honour to have been invited to present a paper to the Volume dedicated to Professor C. Manoharachary, Osmania University, Hyderabad, who made significant contribution to the development of mycology in India.

The hyphomycete genus *Bahusutrabeeja* Subram. & Bhat was established with *B. dwaya* Subram. & Bhat as type species (Subramanian and Bhat, 1977). The genus is characterized by mononematous, percurrently prolifereating conidiophores with simple, phialidic conidiogenous cells producing setulate conidia which are usually aggregated into a slimy mass at the tip of the phialide. To date, a further four species namely, *B. angularis* Rao & de Hoog, (Vasant Rao and de Hoog, 1986), *B. globosa* Bhat & Kendrick (Bhat and Kendrick, 1993), *B. dubhashii* Bhat (Bhat, 1994) and *B. bunyensis* Mckenzie (Mckenzie, 1997) have been described in the genus.

During the course of our studies on biodiversity and taxonomy of foliicolous fungi of the Western Ghat forests, we recovered an isolate of *Bahusutrabeeja*, from live leaves of *Bridelia scandens* Roxb (Euphorbiaceae) gathered from Chorla Ghat, Goa, India, which differed markedly from all known species in the genus in its conidiophore and conidial dimensions. The fungus is disposed in a new species.

Materials and Methods

Fresh, green leaves with leaf spots gathered from the plant were placed in polythene bags and brought to the laboratory. The samples were first examined under a stereoscope to locate the fungal colony. On spotting, the fungus was lifted by a fine-tipped needle, placed on a drop of lactophenol on a clean slide and examined under a light microscope.

Taxonomic Part

Bahusutrabeeja manoharacharii Pratibha et Bhat sp. nov. (Figs. 1 & 2) (Etym., Specific epithet: in honour of Prof. C. Manoharachary, Osmania University, Hyderabad, India).

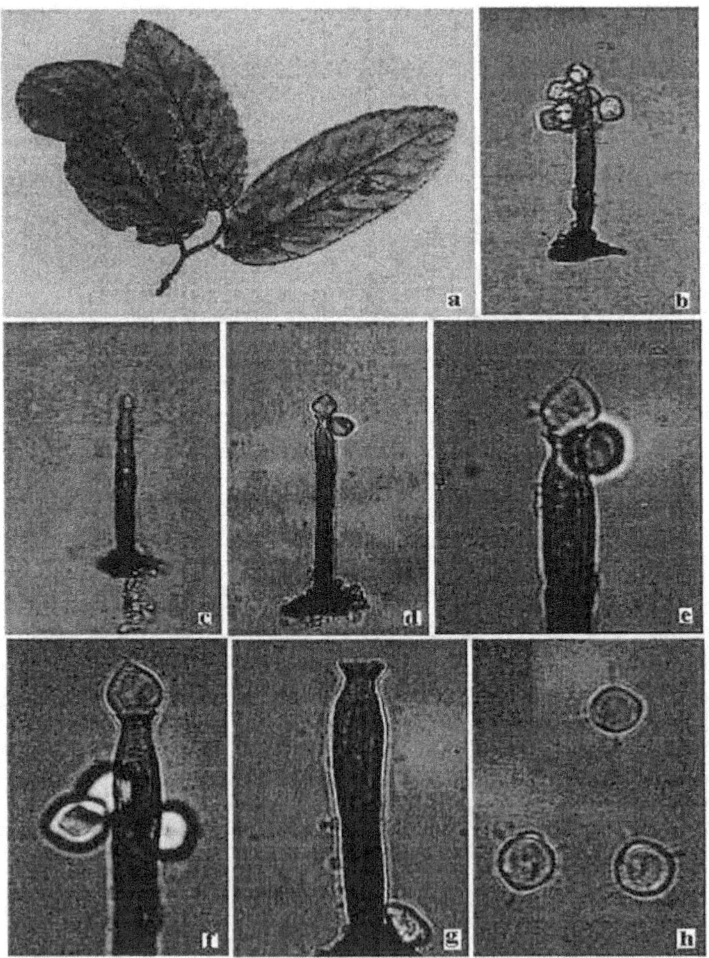

Figure 1 *B. manoharacharii*, a: leaf spots; b-f: conidiophores with conidia; g: phialide; h: conidia

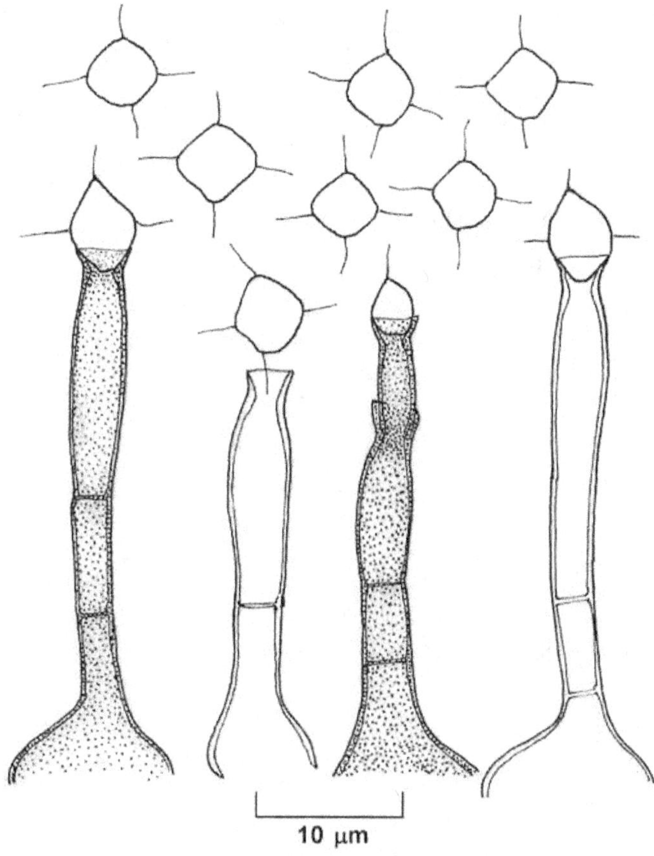

Figure 2 *B. manoharacharii*, Conidiophores and conidia

Ad fungos conidiales, hyphomycetes. *Leaf spots* amphigenae, dissitae, sphaericae, atrobrunneae, 4-6 mm dia. *Coloniae* effusae, atrobrunneae, velutinae. *Mycelium* immersum, laevis, subhyalinis, septatis, ex hyphis ramosis 2-2.5 μm lat. compositum. *Conidiophora* mononematica, erecta, recta vel flexuosa, laevia, brunnea, crassitunicata, 1-2-septata, non-ramosa, percurrenter proliferations, 40-87.5 × 5.5-8.5 μm. *Cellulae conidiogenae* terminales, integratae, monophialidicae, subcylindricalis, laeviae, 25.5-50 × 5.8-8.5 μm, cum collarulo conspicuo. *Conidia* solitaria, hyalina, crassitunicata, angulata, aseptata, 9.5-11.5 × 7.5-9 μm, ad setulis hyalina up to 4.5 μm longa, ad apicem cellulae conidiogenae in massa mucosa aggregata.

HOLOTYPUS in foliis viva *Bridelia scandens* Roxb. (Euphorbiaceae), Chorle Ghat, Goa, India, leg. Pratibha J., 2.7.2004, Herb. GUBH No. 1143.

Conidial fungi, hyphomycetes. *Leaf spots* amphigenous, scattered, irregular to circular, dark brown with a black spot in the center, 4-6 mm diam. *Colonies* on natural substrate effuse, dark brown, velvety. *Mycelium* immersed, composed of smooth, subhyaline, septate, branched hyphae 2-2.5 µm wide. *Conidiophores* mononematous, erect, straight or flexuous, smooth, brown, thick-walled, 1-2-septate, unbranched, sometimes percurrently proliferating 1-2 times, 40-87.5 × 5.5-8.5 µm. *Conidiogenous cells* terminal, integrated, monophialidic, cylindrical, smooth, 25.5-50 × 5.8-8.5 µm, with flared conspicuous collarette. *Conidia* hyaline, thick-walled, angular, 0-septate, 9.5-11.5 × 7.5-9 µm, with fine hyaline setulae up to 4.5 µm long at each angulate end, aggregating into a slimy mass at the tip of the phialide.

The fungus was brought into culture by sowing conidia on antibiotic-embedded 2% malt extract agar medium. The colony was slow growing, attaining a diam of 1-2 cm in 10 days, reddish brown at the center and pale brown towards the periphery, with superficial, floccose mycelium. The fungus did not produce any spore producing structures in culture.

Discussion

The diagnostic features of the species of *Bahusutrabeeja* are given in Table 1. *B. manoharacharii* can be compared with *B. angularis* in view of its angulate conidia. However, the conidia of *B. angularis* are smaller and 7-8 µm diam., whereas in the former it is 9.5-11.5 × 7.5-9 µm. The conidiophores in *B. manoharacharii* are shortest of all species so far described in the genus. Further, *B. manoharacharii* is the first species recognized as foliicolous in the genus *Bahusutrabeeja*.

Table 1 Conidia and conidia producing structures of species of *Bahusutrabeeja*

Species	*Conidiophore length*	*Phialide collaratte*	*Conidia : shape size*	*Conidial appendages Number and length*
B. angularis	200-500	Well developed	Angular, 7-8 µm	3-5; up to 4 µm
B. bunyensis	Up to 210 µm	Well developed	Spherical, 7.5-10.5 µm	3; up to 5-9 µm
B. dubhashii	Up to 170 µm	Well developed	Obpyriform, 15-20 × 5-8 µm	6; up to 18 µm
B. dwaya	170-290 µm	Well developed	Spherical, 12-14 µm	8-16; 4.5-12 µm
B. globosa	Up to 350 µm	inconspicuous	Spherical, 18-22 µm	9-12; 6.5-12.5 µm
B. manoharacharii	40 - 90 µm	Well developed	Rounded to angular, 9.5-11.5 × 7.5-9 µm	4; up to 4.5 µm

Acknowledgement

We thank the Council of Scientific & Industrial Research, New Delhi and Ministry of Environment & Forests, Government of India, for financial support in the form of research grants during the tenure of which this work was carried out.

References

Bhat D. J. 1994. Two undescribed species of conidial fungi from forests of Western Ghats in Sourthern India. *Indian Journal of Forestry.* 17 (2): 129-133.

Bhat D. J. and **Kendrick, B.** 1993. Twenty-five new conidial fungi from the Western Ghats and the Andaman Islands (India). *Mycotaxon.* 49: 19-90.

McKenzie, E. H. C. 1997. *Bahusutrabeeja bunyensis* sp. nov. (Hyphomycetes) from Queensland, Australia, and new name for *Chalara australis* McKenzie. *Mycotaxon.* 61 : 303-306.

Rao, V. and **de Hoog, G.S.** 1986. New or critical Hyphomycetes from India. *Studies in Mycology.* 28: 1-84.

Subramanian, C. V. and **Bhat D. J.** 1977. *Bahusutrabeeja,* a new genus of the Hyphomycetes. *Can. J. Bot.* 55 (16): 2202-2206.

Rhizobial Diversity : Changing Perspectives

Manvika Sahgal and Bhavdish N. Johri*

Department of Microbiology, G.B. Pant University of Agriculture & Technology, Pantnagar - 263 145, Uttaranchal, India.

e-mail : bhavdishnjohri@rediffmail.com

ABSTRACT :

Rhizobial diversity is fast changing. This changed scenario has opened new vistas on survey of many legumes including wild ones since forms other than Rhizobium are involved in legumes. In recent years, several molecular techniques have been adopted in expressing the phylogeneic relationship among various members. The present review highlights the strategy and importance of rhizobial diversities

KEY WORDS :

Rhizobium, diversity, legumes, nitrogen fixation, phylogeny.

Soil health is critical for production of high quality and high yielding crop in order to feed the rising population of estimated five billion people. Increase in population is associated with decrease in land under cultivation and thus the dependence is on new management strategies and scientific interventions.

1. Legumes and Sustainable Development

Leguminous plants are excellent tool for improving soil health through regeneration. Legumes also enter into symbiotic association with soil bacteria, the rhizobia that form characteristic nodules on plant roots. Nodules are site of biological nitrogen fixation where atmospheric dinitrogen (N_2) is converted into ammonia ($-NH_3$), a form which is readily assimilated by plants. This phenomenon is of considerable environmental and agricultural importance since it contributes about 120×10^6 metric tonnes year^{-1} towards the global nitrogen economy and is estimated to represent more than 65% of the nitrogen used in agriculture (Graham, 1988). Usually indigenous legumes nodulate with rhizobia present in soil. However, if a legume is introduced into an ecosystem for the first time, compatible rhizobia may not be available and there is need for artificial inoculation. Inherent with the use of bio-inoculants is the problem of variability in field performance and successful establishment of introduced strain(s), on account of competition with the indigenous rhizobacterial population. To achieve maximum efficiency in a legume-rhizobium association it is essential to characterize and identify rhizobia before they are used for field application.

In the present scenario of sustainable agricultural development, rhizobial inoculants have gained considerable significance but there is concern as well on account of recovery of pathogenic isolates from the nodules and rhizosphere (Chen *et al.*, 2001; Tripathi *et al.*, 2002). Such reports point towards an urgent need to properly characterize and identify rhizobacteria before they are made available for field applications.

2. Taxonomy of Rhizobiaceae : The Changing Scenario

Rhizobia are nitrogen fixing bacterial symbionts of legumes that are of economic importance in low input sustainable agriculture, agroforestry and land reclamation. In view of their role in sustainable development, descriptions of phenotypic and genotypic variations among rhizobia have been extensive. Another reason for the extensive efforts to classify and study this group since the dawn of bacteriology is their ability to interact with legumes in a readily identifiable manner i.e., production of root nodules.

For bacteria from nodules, *Bacillus radicicola* was probably the first name used (Beijerinck, 1888) which was subsequently changed to *Rhizobium* (Frank, 1889). Subsequent studies show that rhizobial host has preferences for nodulation and nitrogen fixation. For example, Nobbe and co-workers (1891, 1895) observed that bacteria isolated from *Pisum sativum* were unable to nodulate plants belonging to the legume tribes Genisteae and Hedysareae. Hiltner and Störmer (1903) therefore, proposed to name root nodule bacterium after the host plant. Three decades later, Fred *et al.*, (1932) proposed the name *Rhizobium* for these bacteria on the basis of their host range for nodulation, with major emphasis on host legume from which the bacterium was recovered (Wilson, 1939; Lim and Burton, 1982; Trinick 1982). However, a major problem faced with this approach was variation in "host range" of both, bacteria and plant with pairs that were more or less faithful to those where all traces of specificity had vanished (Broughton *et al.*, 2000; Perret *et al.*, 2000).

Norris (1965) observed that fast and slow growing rhizobia differed in their symbiotic affinity. Thus, slow growers were largely associated with the tropical legumes and fast growers with temperate legumes, (Allen and Allen, 1981; de Lajudie *et al.*, 1994; George *et al.*, 1994). In addition, there are several reports (Dreyfus and Dommergues, 1981; Scholla and Elkan, 1984; Jenkins *et al.*, 1987; Fulchieri *et al.*, 1999) of the presence of fast and slow growing rhizobia in a single nodule from any one legume. Another major problem with the above approach is that within a total of 19,700 species of legumes that have been identified in 750 genera, large variation in "host range" of both bacteria and plants is known to occur e.g., *Phaseolus* and *Vigna* within the tribe *Phaseolae* form nodules with about half of all the rhizobia (Lewin *et al.*, 1987; Michielis *et al.*, 1998) whereas broad host range *Rhizobium* species NGR 234 nodulates about 50% of the known legumes (Pueppke and Broughton 1999). These observations have been revalidated based on the use of more modern tools (Broughton *et al.*, 2000, Perret *et al.*, 2000). Thus, it is clear that classification of rhizobia on the basis of host range, and biochemical and physiological properties does not reflect the true phylogeny of the group.

Nineties was the beginning of an era of polyphasic taxonomy. The sequence comparisons of 16S rRNA genes and genetic fingerprinting methods based on the use of PCR have been used extensively for characterizing rhizobia (Willems and Collins 1993; Chen *et al.*, 1995; Wang *et al.*, 1999a; Wang *et al.*, 1999b; Willems *et al.*, 2001). In the first edition of Bergey's Manual of Systematic Bacteriology (1984) only two rhizobial genera and four species were described. Since 1984 however,there has been considerable advancement in rhizobial taxonomy. Use of PCR tools and sequencing methods have led to description of new, and reorganization of the existing genera. Rhizobia discovered till 2001 (Sy *et al.*, 2001) belonged to α– proteobacteria. The placement of rhizobia in α– proteobacteria is depicted in Figure 1.

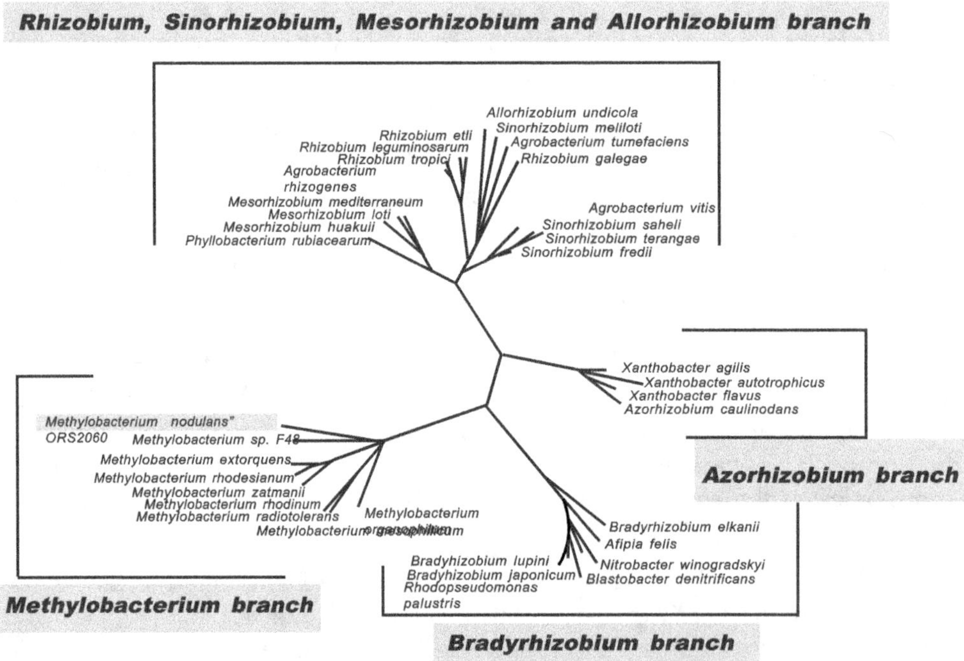

Figure 1 Unrooted phylogenetic tree showing the different rhizobial branches, in the a subdivision of Proteobacteria (Source : Sy *et al.* 2001)

Some examples in which rhizobial classification decisions were made largely on the basis of analysis of 16S rRNA gene include, separation of *Sinorhizobium* from *Rhizobium* (de Lajudie *et al.*, 1994), proposal of the new genus *Allorhizobium* (de Lajudie *et al.*, 1998), suggestion for amalgamating *Agrobacterium* and *Allorhizobium* into genus *Rhizobium* (Young *et al.*, 2001a), proposition for separate genera for *Bradyrhizobium japonicum* and *B. elkanii* (Willems *et al.*, 2001), identification of a species of *Methylobacterium* forming symbiosis with specific species of *Crotolaria* (Sy *et al.*, 2001), and nodulation of the South African legume *Aspalathus carnosa* by *Burkholderia* spp. which is placed within the β subdivision as against the rhizobial majority placed within α Proteobacteria (Moulin *et al.*, 2001).Till 2003, 38 species in seven genera of rhizobia, *Allorhizobium*, *Azorhizobium*, *Bradyrhizobium*, *Mesorhizobium*, *Rhizobium*, *Sinorhizobium* and *Methylobacterium* were known (van Berkum and Eardly, 1998; Sahgal and Johri, 2003). Since then, four more have been added to the list; *Rhizobium loessense* sp. nov. (Wei *et al.*, 2003), Bradyrhizobium betae sp. nov. (Rivas *et al.*, 2004), *Mesorhizobium temperatum* and *M. temperatum* (Gao *et al.*, 2004).

3. 16S rRNA Gene Sequence : A Taxonomic Chronometer Revisited !

Taxonomists have used characters such as DNA base ratios, aminoacid sequence of proteins, DNA-DNA and DNA-RNA hybridizations, constituents of ribosomes, and cell wall composition with surprising consequences for systematics of bacteria. Reviewing earlier work, Trüper and Kramer (1981) had expressed with anticipation; "***Which systematic basis will prevail : morphology, physiology or chemical composition of cellular components?*** and his answer was "***Sequencing conserved genes (or parts of genes)***". This provides insights that elude morphological and physiological methods. Since then improvements in sequencing methods and development of PCR have accelerated the use of sequence data in bacterial taxonomy and has profoundly affected how relationships among the bacteria are portrayed (Olsen *et al*, 1994). The 16S rDNA gene is useful for taxonomy and systematics because it is slowly evolving and the gene product is universally and functionally conserved. However, basing bacterial phylogeny on 16S rRNA gene (or any other gene) sequence variation presupposes that evolution of the genome progresses at a constant rate and that genes are inherited in a strictly hierarchical manner. Practical implications of this hypothesis are that each genome harbours a single copy of 16S rRNA gene or that multiple alleles within a single genome have identical sequences. However, exceptions to this hypothesis have now been described in various taxa viz., *Clostridium* (Rainey *et al.*, 1996), *Escherichia coli* (Carbon *et al.*, 1979), *Haloarcula* (Amman *et al.*, 2000), and *Rhodobacter* (Dreyden and Kaplan, 1990), which contain multiple and often, divergent 16S rRNA genes. For example an actinomycete, *Thermobispora bispora* contains two similar copies of 16S rRNA gene that show a mismatch of 6.4 % at the nucleotide level, (Wang *et al.*, 1997). Since genes under consideration are on the same chromosome within the same cell, their sequence divergence suggests that 5 % mismatch which was previously used to place bacteria in separate genera or species is questionable ?

4. Taxonomic Status of Agrobacterium & Rhizobium

Over the past three decades there has been considerable controversy with respect to taxonomy of *Agrobacterium*, *Allorhizobium* and *Rhizobium* within the family Rhizobiaceae. Genus *Agrobacterium* is differentiated from *Rhizobium* as a group of nitrogen non-fixing species of rhizobia that produce other types of hypertrophies (Jordan, 1984). Both the genera are polyphyletic and constitute a paradoxically diverse group of related members of α-Proteobacteria. This has led recently to the division of the genus *Rhizobium* into *Rhizobium, Mesorhizobium* and *Sinorhizobium* (de Lajudie *et al.*, 1994). Young *et al* (2001a) explored the history of the taxonomic position of *Agrobacterium* and proposed that all species of *Agrobacterium* be amalgamated into the single genus *Rhizobium*, comprising both pathogenic and symbiotic nitrogen-fixing bacteria. However, Farrand *et al.*, 2003 have maintained different opinion on this issue. These researchers argue that agrobacteria are phenotypically different from other members of the family Rhizobiaceae (Holmes and Roberts, 1981; de Lajudie *et al*, 1994 and Young *et al.*, 2001b). In this context, it is of interest to note that Allen and Allen (1950) had presented a list of 18 traits by which fast growing rhizobia could be differentiated from genus, *Agrobacterium*. Furthermore, agrobacteria are differentiated into three biovarieties based on differential biochemical and physiological tests, which is supported by 16S rRNA sequence analysis. Evidence from classical and molecular tools show that biovar 1 and biovar 3 agrobacteria are sufficiently different from members of the genus *Rhizobium*. While biovar 2 agrobacteria cluster more closely to the genus *Rhizobium*, some studies (Jarvis *et al.*, 1989; Soberün-Chávez and Nájera, 1989; Segovia *et al.*, 1991) suggest, that these differ in their capacity to interact with plants and hence, recommended for retention in the genus *Agrobacterium*. However, Young et al., (2003) reporting their work in the same issue of *International Journal of Systematic and Evolutionary Microbiology* argue in support of amalgamation of *Agrobacterium* into *Rhizobium*. According to them Farrand and his team have cited a report of Allen and Allen, a study made at a time when bacterial systematics was relatively unsophisticated and none of the tests described by them are acknowledged today. Moreover, conclusions were based on comparison of *Agrobacterium radiobacter* with only 3-4 rhizobial species recognized today. Furthermore, since clades 1 and 2 described by Tan *et al.*, (2001), Young *et al.* (2001a), Squartini *et al.*, (2002), and Wei *et al.*, (2002) correspond to *Agrobacterium* and *Rhizobium* respectively, any nomenclature change followed thereafter would have to conform to the above phylogenetic classification. The sequence data from additional loci alone, according to Young *et al.*, (2001a) does not guarantee establishment of different clades of *Rhizobium* and *Agrobacterium* described by Farrand and others.

5. Discordant Phylogenies and Lateral Gene Transfer

Several recent studies have highlighted the anomalies regarding taxonomic state of certain genera and that of species within the genera. For example, Young *et al.* (2001) claimed that close relatedness of 16S rRNA sequences of *Agrobacterium* and *Rhizobium* species (< 7% mismatch) warranted regrouping of both the genera into a single genus, *Rhizobium*. Similarly, differences in partial 23S rRNA genes of *Sinorhizobium* and *Rhizobium* are considered insufficient to warrant their separation into different genera (van Berkum *et al.*, 1999). Similarly, placement of *Mesorhizobium loti* was uncertain on the basis of comparative analysis of partial 23S rRNA and 16S rRNA sequences (van Berkum *et al.*, 1999). van Berkum *et al.* (2003) compared evolutionary relationships within a representative set of α-Proteobacteria to confirm the phylogenetic placement of rhizobial genera, *Allorhizobium, Azorhizobium, Bradyrhizobium, Mesorhizobium, Rhizobium,* and *Sinorhizobium* by application of sequence analysis of three loci, 16S rRNA, 23S rRNA and ITS region within their *rrn* operons. Tree topologies generated with 16S rRNA gene sequences were significantly different from the corresponding trees assembled with 23S rRNA gene and internally transcribed (ITS) region sequences. In other words, trees obtained from three different datasets could not be combined into a single tree. The possible reason for this deficiency could be gene conversion in the 16S rRNA. Recombination was also observed between short fragments of the 16S rRNA genes of *B. elkanii* with species of *Mesorhizobium* and also between species of *Sinorhizobium* and *Mesorhizobium*. Thus, detection of genetic transfer and recombination among divergent alleles of the 16S rRNA genes across the genera of α-Proteobacteria contradicts inferences based on genetic isolation of genera by Guant *et al.* (2001). Since phylogenetic placement based on new information of 16S rRNA gene sequence divergence is perhaps questionable, van Berkum and coworkers, propose a conservative approach in which taxonomic decisions ought to be based on analysis of a variety of loci and comparative analytical methods such as NJ trees from Jukes-Cantor distances using MEGA package version 1.01, Parsimony trees using Paup version 4.0b 8a; determining congruence between tree topologies by using Shimodaira-Hasegawa test, significance in difference among likelihood scores and determining possibility of history of recombination among 16S rRNA loci to estimate phylogenetic relationships among genomes.

6. Gene Sequence useful in Predicting relationships in Rhizobia and other Bacteria.

Although many techniques exist (Gürtler and Mayall, 2001), DNA-DNA hybridization still remains a "gold standard for defining bacterial species (Wayne *et al.*, 1987; Stackebrandt *et al.*, 2002); the procedure is however expensive and difficult to reproduce in different laboratories. Recently, an adhoc committee for reevaluation of species definition in bacteria (Stackebrandt *et al.*, 2002) reached a consensus that analysis of at least five genes (house keeping genes) of diverse chromosomal loci and wide distribution could serve as predictors of genome relatedness.

Publicly available genome sequences of 44 genomes, representing 16 genera were used by Zeigler (2003) to identify 34 genes (Table 1). Candidate genes for a "species prediction set" were identified based on four criteria: First, genes must be widely distributed among genomes. Second, each of the 'prediction set' genes must be unique within a given genome. Third, individual gene sequences should be long enough to contain phylogenetically useful information but short enough to be sequenced with small set of primers. Finally, the sequences must predict whole genome relationships with acceptable

Table 1 Candidate genes useful for whole genome sequence identity
(Source : *Zeigler, 2003*)

Sl. No.	*Gene*	*Product*	*Exception*
1.	recN	Recombination/repair protein	BUC, CHA, CHO, MYP
2.	Lig	DNA ligase	BUC, CLO, MYP-PULM
3.	dnaX	DNA polymerase III subunits g, t	
4.	glyA	Serine hydroxmethyltransferase	MYB
5.	cysS	Cysteine tRNA synthetase	
6.	thdF	GTP-binding, thiophene oxidation	
7.	uvrC	Excinuclease ABC, subunit C	CHA
8.	ruvB	Holliday junction helicase subunit A	BUC
9.	metG	Methionine tRNA synthetase	
10.	dnaB	Replicative DNA helicase	
11.	dnaJ	Chaperone with DnaK	MYB
12.	lepA	GTP-binding elongation factor	STR-8232
13.	argS	Arginine tRNA synthetase	
14.	Ffh	GTP-binding export factor	
15.	–	All similar sequences in genome	
16.	serS	Serine tRNA synthetase	
17.	nrdE	NDP reductase 2, a-subunit	BAC-ANTH
18.	ftsZ	Tubulin-like division protein	CHA, CHO
19.	metK	Methionine adenosyltransferase	CHA, CHO, RIC
20.	trpS	Tryptophan tRNA synthetase	
21.	atpA	ATP synthase, F1, a-subunit	CHA, CHO
22.	dxs	Deoxyxylulose-phosphate synthase	MYP, RIC, STA, STR
23.	uvrB	Excision nuclease subunit B	CHA
24.	atpD	ATP synthase, F1, b-subunit	
25.	aspS	Aspartate tRNA synthetase	
26.	Pgi	Glucose phosphate isomerase	NEI, RIC
27.	Tig	Trigger factor	
28.	rho	Transciption termination factor	BAC-ANTH, MYP, STR
29.	proS	Proline tRNA synthetase	
30.	recA	DNA strand exchange and renaturation	BAC-ANTH, BUC
31.	rpoA	RNA polymerase, alpha subunit	
32.	eno	Enolase	RIC
33.	trps	Tryptophan tRNA synthetase	
34.	pgk	Phosphoglycerate kinase	RIC
35.	rrsH	16S rRNA	

Abbreviations : BAC-ANTH, Bacillus anthracis; BUC, Buchnera; CHA, Chlamydia; CHO, Chlamydophila; MYB, Mycobacterium; MYP, Mycoplasma; MYP-PULM, Mycoplasma pulmonis; NEL, Neisseria; RIC, Rickettsia; STA, Staphylococcus; STR, Streptococcus; STR-8232, Streptococcus pyogenes MGAS8232.

precision and accuracy. Based on linear regression procedures, the best single-gene combination was *recN,* and the best two-gene combination was *recN* and *thdF* whereas the best three-gene combination was *recN, thdF* and *rpoA.* In conclusion, a small number of carefully selected gene sequences can indeed equal, or perhaps even surpass, the precision of DNA-DNA hybridization for quantification of genome relatedness. With the availability of this new set of tools and techniques, bacterial systematics may have to look back at some of the species that were established utilizing 16S sequence as the sole criterion.

7. Rhizobia Associated with Annual Legumes

Economically important legumes are commonly inoculated with efficient nitrogen fixing strains of rhizobia with the objective of optimizing crop productivity. However, agricultural soils often contain diverse populations of rhizobia that interfere with the establishment of inoculant strains in nodules of the host plant (Barran and Bromfield, 1997). To alleviate this problem it is important to understand the diversity and dynamics of natural populations of rhizobia and influence of temporal, biotic and environmental variables on their composition.

A few studies (Bromfield *et al.,* 1986; Souza *et al.,* 1994; Hagen and Hamrick, 1996; Handley *et al.,* 1998) have been undertaken to assess the effect of temporal and other environmental variables on field populations of rhizobia. Recently, Bromfield *et al.* (2001) indicated significant temporal shifts in the composition of *Sinorhizobium meliloti* populations associated with *Medicago sativa* and *Melilotus alba* at a single field site located at Ottawa, Canada. These observations support earlier findings (Bromfield *et al.,* 1986; Thurman and Bromfield, 1988; Bromfield *et al.,* 1995) that legume genotype may significantly affect the composition and density of the associated rhizobial populations. No distinct relationship between the size of rhizobial population and level of genetic diversity has however, been observed (Strain *et al.,* 1994).

Selection of specific rhizobia by host plant may be a key factor contributing towards isolation of different biovarieties of the same rhizobial species from different legumes. For example, Wang and co-workers (1999b) identified *Rhizobium etli* from root nodules of *Mimosa affinis* growing in Mexico but these were different from *R. etli* strains nodulating *Phaseolus vulgaris* L. in *nifH* gene organization and host specificity and were hence placed in a new biovariety, *mimosae.* Another implication of host-plant selection of specific rhizobia would be reduced bacterial diversity, if the same legume were cultivated at a particular site over a period of time. In contrast, common bean (*Phaseolus vulgaris*) is promiscuous and effectively nodulates with wide diversity of rhizobia. At least six rhizobia (*Bradyrhizobium* spp., Lange 1961; *R. leguminosarum* bv. *phaseoli*, Jordan 1984; *R tropici*, Martínez-Romero *et al.* 1991; *Rhizobium etli*, Segovia *et.al.* 1993;and *R giardinii* and *R. gallicum*, Amarger *et al.* 1997) nodulate *Phaseolus* (Michelis *et al.* 1998).

While characterizing bean isolates from two different regions of Spain with staircase electrophoresis of LMW RNA, Velázquez *et al.* (2001) observed that rhizobial strains nodulating same legume growing at different eco-regions were genetically variable. At one of the sites, León, strains belonging to bv. *viciae* and *trifolii* of *R. leguminosarum* were present whereas at another site, Andalucia, bacterial population was more heterogenous and comprised of *R. etli*, *R. gallicum*, *R. giardinii*, *R. leguminosarum* bv. *viciae* and bv. *trifolii* and *Sinorhizobium fredii*. Andrade *et al.* (2002) examined the diversity within rhizobial populations along gradients of increasing acidity stress in Brazilian soil which had been cultivated with common bean *(Phaseolus vulgaris)*. Based on RFLP analysis of ITS (intergeneric region between 16S-23S rRNA) region and 16S rRNA gene, Shannon diversity indices were calculated which indicated that rhizobial populations recovered from *Phaseolus vulgaris* nodules growing in unlimed soils had less diversity than those from plots with higher rates of liming. Total rhizobial populations were divided amongst *R. tropici* II B and *R. leguminosarum* bv. *phaseoli*. The relationships between genetic diversity as a number of ITS groups and size of rhizobial populations was species-specific with no change for *R. leguminosarum*, and positive correlation in case of *R. tropici* II B genotype. The predominance of *R. tropici* was expected because this species is a broad host range *Rhizobium* (Martínez- Romero *et al.*, 1991) and best adapted to acidic conditions (Graham *et al.*, 1994; Anyango *et al.*, 1995). Several studies attribute shift in genetic diversity among populations of *Sinorhizobium meliloti*, *R. leguminosarum* and other *Phaseolus* rhizobia to chemical and physical differences between soils (Paffeti *et al.*,1996), history of N - fertilization (Caballero- Mellado *et al.*,1999), and land management (Palmer and Young, 2000). Another study of bean rhizobia from Ecuador and Northern Peru (Bernal and Graham, 2001) shows that strains of *R. etli* from Mesoamerican region show significant phenotypic and phylogenetic separation from those associated with beans in the Andean region of Latin America.

Recently Pandey *et al.* (2004) assessed the genetic diversity of the rhizobial population in nodules of medicinal legumes, *Abrus precatorious, Mucuna pruriens, Melilotus officinalis, Trigonella foenum-graecum* and *Vicia angustifolia* collected from the sub-himalayan tract of Uttaranchal. Nine isolates from five medicinal legumes clustered into six rDNA types based on ARDRA of 16S region. The phylogenetic relationship of these clusters was intertwined within the rhizobial genera, *Bradyrhizobium, Rhizobium* and *Sinorhizobium*. Isolates from *Trigonella* nodules were related to *Bradyrhizobium* as well as *Sinorhizobium*.

In last years, rhizobial isolates from *Amorpha fruticosa, Glycyrrhiza* spp., *Gueldenstaedtia* spp, *Astragalus* and *Lespedeza* spp. located in north-western region of China, with semi-arid climates and soils poor in organic matter, have been characterized (Tan *et al.*, 1999; Wang *et al.*, 1999a; Peng *et al.*, 2000). Novel forms within the genera, *Bradyrhizobium* (Yao *et al.*, 2002), *Mesorhizobium* (Wang *et al.* 1999a), *Rhizobium* (Tan *et al.* 2001; Wei *et al.* 2002; 2003), and *Sinorhizobium* (Wei *et al.*, 2002), have been described. Thus, rhizobia in temperate regions are as diverse as those in tropical areas. Zhang and co-workers (1999), while assessing genetic diversity among rhizobial isolates from field grown *Astragalus sinicus* from 52 ecological different sites of seven southern provinces of China, reported that genetically diverse rhizobial strains could be detected at single site, and strains with the same genetic type could be found in different geographic locations. Moreover, none of the representative strains belonged the *Mesorhizobium huakuii* identified earlier in nodules of *A. sinicus* from China and Japan (Jarvis *et al.*, 1997).

Studies by Rome and co-workers (1996) and Roumaintseva *et al* (1999) show that alfalfa rhizobia fall into two closely related species, *Sinorhizobium meliloti* and *S. medicae* and effectively nodulate *Medicago* (Alfalfa), *Melilotus* (sweet clover) and *Trigonella* (fenugreek). If central Asia is the centre of origin of alfalfa rhizobia as well as their host plants, a wider range of genetic variants would be expected. However, Roumiantseva (2002) identified only 7 chromosomal, 9 pSym A and 3 pSymB types, all belonging to *S. meliloti*, amongst the central Asian isolates, incontrast to 22, 33, and 18 respectively reported in Canadian isolates (Bromfield *et al.*, 1998). The possible reason is recombination between and within the chromosome and megaplasmids.

8. Tree Legumes and Biological N_2 Fixation

Ecological interactions between tree and rhizobacteria are beneficial for three major reasons; trees increase biomass and ameliorate the soil of degraded sites because of improved water and nutrient uptake, prevent soil erosion, increase soil fertility through N_2 fixation, greater organic matter production and recycling of nutrients. Major Indian trees of economic importance are *Acacia, Albizzia lebbeck, Aegle marmelos, Azadiracta indica, Calliandra calothyrus, Gliricidia sepium, Leucaena leucocephala, Prosopis chilensis, P. glandulosa* and *Sesbania grandiflora*, most being leguminous.

There is considerable specificity of rhizobia towards the tree legume host in terms of nodulation and nitrogen fixation. Inoculation with highly effective strains at nursery stage appears to be an effective way to enhance nodulation and nitrogen fixation. Thus, correct estimates of diversity in natural ecosystems as well as nodules are essential.

Rhizobia Associated with Tree Legumes

Several species of *Acacia* are found in Africa, Asia, Australia and Central America and most nodulate; *A. brevispica* from Africa is an exception (Odee and Sprent, 1992). A wide range of strains, fast-to slow-and even very slow-growing have been reported to nodulate *Acacia* spp in Australia (Barnet and Catt, 1991). The fast growing strains are very diverse with a few strains closely related to *Bradyrhizobium* (Barnet *et al.*, 1993). Several other reports suggest that fast growing rhizobia isolated from the root nodules of *Acacia senegal* and *Prosopis chilensis* are phenotypically and genotypically diverse (Zhang *et al.*, 1991; Haukka, and Lindströms, 1994; Haukka *et.al.*, 1996; Nick, 1998). Based on numerical analysis, Zhang *et al.* (1991) placed Sudanese strains in nine different clusters; most strains were high temperature and salt tolerant and largely in agreement with the findings of Barnet (1991) in Australia. Lafay and Burdon (2001) grouped nodule isolates from Australian Acacias into nine genomospecies, eight representing novel forms. These nine genomospecies are spread in the rhizobial genera, *Bradyrhizobium, Mesorhizobium* and *Rhizobium*. In a similar study of Kenyan rhizobia, Odee (1992) reported that fast growing rhizobia were generally more salt tolerant (3% NaCl) than slow-growing (*Bradyrhizobium*) strains. Haukka and Lindström (1994), using pulse field gel electrophoresis, also concluded that the Sudanese and Kenyan strains were very diverse. Later, 16S rRNA gene analysis by the same group (Haukka *et al.*, 1996) revealed that almost all of the Sudanese and Kenyan strains belonged to the genus *Sinorhizobium*; some strains belonged to *S. terangae* and *S. saheli* species, and two strains belonged to the genus *Mesorhizobium*. Nick and co-workers (1999) utilised DNA-DNA hybridization on Sudanese and Kenyan isolates from root nodulates of *Acacia senegal* and *Prosopis chilensis* and grouped them into two clusters which showed low similarity with already described species of other tree legumes.

Moreira *et al.* (1998) assessed the rhizobia nodulating diverse pool of forest legume species in Brazil and observed that although strains studied were taxonomically diverse, they were similar to those described previously. A total of 15 sequences including six novel ones were identified amongst 44 strains from 29 legume tree species belonging to 13 tribes of the three subfamilies of Leguminosae. Sequences from fast growing strains belonged to genera *Rhizobium* spp., *Sinorhizobium*, or *Mesorhizobium*, slow-growing forms to *Bradyrhizobium* whereas, one isolate belonged to *Azorhizobium*. Other studies carried out with trees namely *Acacia*, *Dalbergia sissoo*, *Leucaena leucocephala*, *Mimosa* and *Prosopis* reveal that isolates recovered from them constitute an important source of rhizobial diversity (Dupuy *et al.*, 1994, de Lajudie *et al.*, 1998 and Nick *et al.*, 1999). Amongst 42 rhizobial species described till date, 10 species are reported from tree legumes. These are *Mesorhizobium chacoense* from *Prosopis alba* (Velázquez *et al.*, 2001), *M. plurifarium* from *Acacia* and *Leucaena* (de Lajudie *et al.*, 1998), *Rhizobium huautlense* from *Sesbania herbaceae* (Wang *et al.*, 1998), *Sinorhizobium arboris* and *S. kostiense* from *Acacia senegal* and *Prosopis chilensis* (Nick *et al.*, 1999). *S saheli* from *Sesbania* sp., *S. terangae* from *Acacia* (de Lajudie *et al.*, 1994), *R. tropici* from *Leucaena* (Martínez-Romero *et al.*, 1991), *Sinorhizobium morelense* from nodules of *Leucaena leucocephala* growing in Mexican soils (Wang *et al.*, 2002) and *Ralstonia taiwanensis* from *Mimosa* sps in Taiwan (Chen *et al.*, 2001). In studies carried out by the authors on the diversity of rhizobia of *Dalbergia sissoo* representing various ecozones of northern India, considerable variability was observed (Sahgal, 2002; Sahgal and Johri, 2003). In a screen of 35 isolates, all were able to nodulate the homologous host, *D. sissoo*; 22 of these could nodulate heterologous host *S. aculeata* whereas only three isolates nodulated *L. leucocephala* and *V. mungo* (Sahgal *et al.*, 2004). Based on ARDRA of 16S and IGS, above isolates were grouped into seven rDNA types wherein none was identical to reference strains representing *Azorhizobium*, *Bradyrhizobium*, *Mesorhizobium* and *Rhizobium* (Figure 2). This work was further extended by Samant (2003). Based on combined analysis of IGS and 16S he observed that six isolates from defined clones of *Dalbergia sissoo* CPT5 and CPT6 were genetically different than those from previous study (Sahgal, 2002) and also did not match with any of the authentic strains, representing different rhizobial genera.

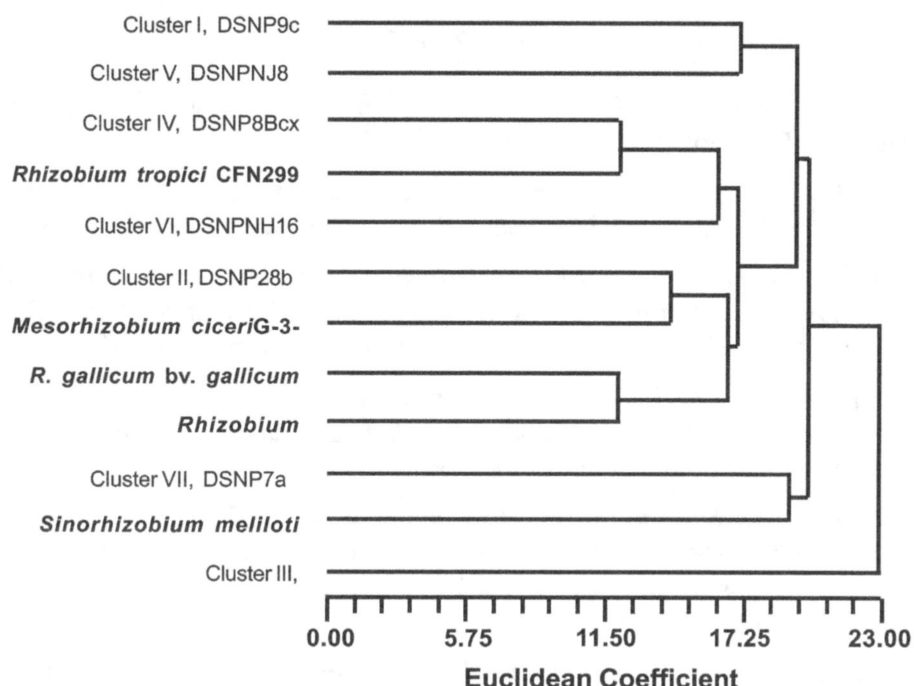

Figure 2 UPGMA cluster showing phylogenetic relationships of Dalbergia sissoo isolates on the basis of ARDRA of 16S & IGS (*Source :* Sahgal, 2002)

9. Future Scenario

Rhizobial diversity scenario is fast changing ever since it was reported that forms other than "*bium*" are involved in nodulation, viz., *Burkholderia phymatum*, *B. tuberum*, methylotropic *Methylobacterium* sp. and *Ralstonia taiwanensis*. This inter-alia changed the fixed perception that all rhizobia belonged to α - proteobacteria; members belonging to β - proteobacteria capable of nodulation are being reported with rapidity. This changed scenario definitely points towards further thrust on survey of wild legumes for 'rhizobia' since so far only 10% of legumes have been evaluated for the nodulating partner. What has indeed come as a surprise is the changing scenario of bacterial taxonomy in general and rhizobia in particular where 16S sequence analysis must be coupled to additional housekeeping genes and multilocus enzyme electrophoresis to substantiate the sequence data. Diversity searches for the new inoculants cannot now merely rely on routine nodulation data since multistrain nodule occupancy is a reality wherein competition between the partners can not be ruled out and chances of successful exploitation, therefore, remote. Considering large availability of wild legumes in the country as also use of age-old traditional varieties by farmers, the field of rhizobial diversity is wide open for exploration, conservation, and exploitation. This path has to be treaded however, with care, keeping in mind that a mix of traditional and modern tools alone shall provide the desired answers to taxonomical interpretations chang exploitative technologies.

References

Allen, E.K. and ***Allen, O.N.*** 1950. Biochemical and symbiotic properties of the rhizobia. *Bacteriological Reviews* 14 : 273-330.

Allen, O. N. and ***Allen, E. K.*** 1981. *The Leguminosae : A Source Book of Characteristics. Uses and Nodulation. University of Wisconsin Press Madison. WI.*

Amann, G., Stetler, K.O., Llobet-Brossa, E., Amann, R. and ***Anton, J.*** 2000. Direct proof for the presence and expression of two 5 % different 16S rRNA genes in individual cells of *Haloarcula marismortui. Extremophiles.* 4 : 373-376.

Amarger, N., Macheret, V., and ***Laguerre, G.*** 1997. *Rhizobium gallicum sp.* nov., and *Rhizobium giardinii sp. nov.,* from *Phaseolus vulgaris* nodules. *International Journal of Bacteriology.* 47: 996-1006.

Andrade, D.S., Murphy, P.J. and ***Giller, K.E.*** 2002. The Diversity of *Phaseolus* nodulating rhizobial population is altered by liming of acid soils planted with *Phaseolus vulgaris* L. in Brazil. *Applied* and *Environmental Microbiology.* 68 : 4025-4034.

Anyango, B., Wilson, K.J., Beynon, J.L. and ***Giller, K.E***. 1995. Diversity of rhizobia nodulating *Phaseolus vulgaris* L. in two Kenyan soils with contrasting pH. *Applied and Environmental Microbiology.* 6: 4016-4021.

Barnet, Y. M. and ***Catt, P.C***. 1991. Distribution and characteristics of root-nodule bacteria isolated from Australian *Acacia spp. Plant and Soil.* 135: 109-120.

Barnet, Y., Catt, P. C., Jenzaniontham, R. and ***Mann, K.*** 1993. Fast growing root-nodule bacteria from Australian Acacia spp. In: *New Horizons in Nitrogen Fixation* (Eds. Palacios, R., Mora, J. and Newton, W.E.) Kluwer Academic Publishers, Dordrecht. p.594.

Barran, L.R. and ***Bromfield, E.S.P.*** 1997. Competition among rhizobia for nodulation of legumes. In: *Biotechnology and the improvement of forage legumes.* (Eds Mc Kersie, B.D. and Brown, D.C.W.) CAB International, Oxford. p. 343-374.

Beijerinck, M.W. 1888. Diebacterien der Papilionaceen Knolchen. *Bot. Ztg.* 46: 726-804.

Bernal, G. and ***Graham, P.H.*** 2001. Diversity in the rhizobia associated with *Phaseolus vulgaris* L. in Ecuador, and comparisons with Mexican bean rhizobia. *Canadian Journal of Microbiology.* 47:526-534.

Bromfield, E.S. P., Butler, G. and ***Barran, L.R.*** 2001. Temporal effect on the composition of a population of *Sinorhizbium meliloti* associated with *Medicago sativa* and *Melilotus alba. Canadian Journal of Microbiology.* 47: 567-573.

Bromfield, E.S.P., Barran, L.R., and **Wheatcroft, R.** 1995. Relative genetic structure of a population of *Rhizobium meliloti* isolated directly from soil and from nodules of alfalfa (*Medicago sativa*) and sweet clover (*Melilotus alba*). *Molecular Ecology.* 4 : 183-188.

Bromfield, E.S.P., Behara, A.M.P., Singh, R.S. and **Barran, L.R.** 1998. Genetic variation in local populations of *Sinorhizobium meliloti. Soil Biology and Biochemistry.* 30: 1707-1716.

Bromfield, E.S.P., Sinha, I.B. and **Wolynetz, M.S.** 1986. Influence of location, host cultivar and inoculation on the composition of naturalized populations of *Rhizobium meliloti* in *Medicago sativa* nodules. *Applied and Environmental Microbiology.* 51: 1077-1084.

Broughton, W.J., Jabbouri, S. and **Perret, X.** 2000. Keys to symbiotic harmony. *Journal of Bacteriology.* 182: 5641-5652.

Caballero-Mellado, J. and **Martínez-Romero, E.** 1999. Soil fertilization limits the genetic diversity of *Rhizobium* in bean nodules. *Symbiosis.* 26: 111-121.

Carbon, C., Philips, J., Fu, Z. Y., Squires, C. and **Squires, C.L.** 1999. The complete nucleotide sequence of ribosomal 16S rRNA from *Escherichia coli.* EMBOJ. 11 : 4175-4185.

Chen, W., Wang, E., Wang, S., Li, Y., Chen, Y. and **Li, Y.** 1995. Characteristics of *Rhizobium tianshanense* sp. nov., a moderately and slowly growing root nodule bacterium isolated from an arid saline environment in Xinjiang, People's Republic of China. *International Journal of Systematic Bacteriology.* 45: 153-159.

Chen, Wen-Ming, Laevens, Severine, Lee, Tsong-Ming, Coenye, T., deVos, P., Mergeay, M. and **Vandamme, P.** 2001. *Ralstonia taiwanensis* sp. nov., isolated from root nodules of Mimosa species and sputum of a cystic fibrosis patient. International Journal of Systematic and Evolutionary Microbiology. 51: 1729-1735.

Danso, S.K.A., Bowen, G.D and **Sanginga, N.** 1992. Biological Nitrogen Fixation in trees in agro-ecosystems. *Plant and Soil* 141 : 177-196.

de Lajudie, P., Willems, A., Nick, G., Moreira, F., Molouba, F., Hoste, B., Torck,U., Neyra, M., Collins, M.D., Lindström, K., Dreyfus, B. and **Gillis, M.** 1998. Characterization of tree rhizobia and description of Mesorhizobium plurifarium sp. nov. *International Journal of Systematic Bacteriology.* 4 : 368-392.

de Lajudie, P., Willems, A., Pet, B., Dewettinck, D. Maestrojuan, G., Neyra, M., Collins, M.D., Dreyfus, B., Kersters, K. and **Gillis, M.** 1994. Polyphasic taxonomy of rhizobia : emendation of the genus *Sinorhizobium* and description of *Sinorhizobium meliloti comb* nov., *Sinorhizobium saheli* sp. nov., and *Sinorhizobium terangae* sp. nov. *International Journal of Systematic Bacteriology.* 44 : 715-733.

Dreyden, S.C. and ***Kaplan, S.*** 1990. Localization and structural analysis of the ribosomal RNA operons of *Rhodobacter sphaeroides*. *Nucleic Acid Research.* 18 : 7267-7277.

Dreyfus, B.L. and ***Dommergues, Y.R.*** 1981. Nodulation of *Acacia species* by fast and slow growing tropical strains of *Rhizobium*. *Applied and Environmental Microbiology.* 41: 97-99.

Dupuy, N., Willems, A., Pet, B., Dewettinck, D., Vandenbruaene, I., Maestrojuan, G., Dreyfus, B., Kersters, K., Collins, M.D. and ***Gillis, M.*** 1994. Phenotypic and genotypic characterization of bradyrhizobia nodulating the leguminous tree *Acacia albida*. *International Journal of Systematic Bacteriology.* 44 :461-473.

Farrand, S.K., van Berkum, P.B. and ***Oger, P.*** 2003. *Agrobacterim* is a definable genus of the family *Rhizobiaceae*. *International Journal of Systematic and Evolutionary Microbiology.* 53 : 1681-1687.

Frank, B. 1889. Ueber die pilzymbiose der Leguminosen. Ber Deut. Bot. Bessell. 7: 332-346.

Fred, E. B., Baldwin, I.L. and ***McCoy, E.*** 1932. Root nodule bacteria and leguminous plants. University of Wisconsin Press, Madison, Wis. p.343.

Fulchieri, M., Oliva, L., Fancelli, S. and ***Bazzicalupo, M.*** 1999. Characterization of *Rhizobium lupinus* from near the Parana river (Argentina) by PCP-RFLP. In : *Nitrogen Fixation : From Molecules to Crop Production.* Proc. 12th Int. Cong. N2 Fixation Brazil Sept. 12-17. p 189.

Gao, J.-L., Turner, S.L., Kan, F.L., Wang, E.T., Tan Z. Y., Qiu, Y.H. Gu, J., Terefework, Z., Young, J.P. W., Lindström, K. and ***Chen, W.X.*** 2004. *Mesorhizobium septentrionale* sp. nov. and *Mesorhizobium temperatum* sp. nov. isolated from *Astragalus adsurgens* growing in the northern regions of China. *International Journal of Systematic and Evolutionary Bacteriology.* Published online 14 May 2004 as http : //dx. doi.org/10.1099/ijs. 0.02840-0.

Gaunt, M.W., Turner, S.L., Rigottier –Gois, L., Loyd-Macgilp, S.A. and ***Young, J.P. W.*** 2001. Phylogenies of atpD and rec A support the small subunit rRNA-based classification of rhizobia. *International Journal of Systematic Bacteriology.* 51 : 2037-2048.

George, M. L. C., Young J.P. W. and ***Borthakur, D.*** 1994. Genetic Characterization of *Rhizobium* sp. strain TAL 1145 that nodulates tree legumes. *Canadian Journal of Microbiology.* 40: 208-215.

Graham, P.H. 1988. *Principles and Applications of Soil Microbiology* (eds. Sylvia, D. M. *et al.*) p 322-345.

Graham, P.H., Draeger, K.J., Ferrey, M.L., Conroy, M.J., Hammer, B.E., Martínez, E., Aarons, S. R. and **Quinto, C.** 1994. Acid pH tolerance in strains of *Rhizobium* and *Bradyrhizobium* and initial studies on the basis for acid tolerance of *Rhizobium tropici* UMR 1899. *Canadian Journal of Microbiology.* 40: 198-207.

Gürtler, V. and **Mayall, B.C.** 2001. Genomic approaches to typing taxonomy and evolution of bacterial isolates. *International Journal of Systematic and Evolutionary Microbiology.* 51: 3-16.

Hagen, M.J. and **Hamrick, J.L.** 1996. A heirarchical analysis of population genetic structure in *Rhizobium leguminosarum bv. trifolii. Molecular Ecology.* 5: 177-186.

Handley, B.A., Hedges, A. J. and **Beringer, J.E.** 1998. Importance of host plants for detecting the population diversity of *Rhizobium leguminosarum* biovar *viciae* in soil. *Soil Biology and Biochemistry.* 30: 241-249.

Haukka, K. and **Lindström, K.** 1994. Pulsed field gel electrophoresis for genotypic comparison of *Rhizobium* bacteria that nodulate leguminous trees. *FEMS Microbiology Letters.* 119: 215-220.

Haukka, K., Kindström, K. and **Young. J.P.W.** 1996. Diversity of partial 16S rRNA sequences among and within strains of African rhizobia isolated from *Acacia and Prosopis. Systematic and Applied Microbiology.* 19: 352-359.

Hiltner, L. and **Störmer, K**. 1903. Neue Untersuchungen über die Wurzelknöllchen der leguminosen and deren Erreger Arbeiten aus der Biologischem Abteilung fur Land-und Forst wirthschaft, Kaiserlichen Gesundhiet samte, Berlin 3: 151-307.

Holmes, B. and **Roberts, P.** 1981. The classification, identification, and nomenclature of agrobacteria. *Journal of Applied Bacteriology.* 50 : 443-467.

Jarvis, B. D. W., Ward, L.J. H. and **Slade, E.A**. 1989. Expression by soil bacteria of nodulation genes from *Rhizobium leguminosarum* biovar *trifollii. Applied and Environmental Microbiology.* 55 : 1426-1434.

Jarvis, D. D. W., van Berkum, P., Chen, W.X., Nour, S. M., Fernandez, M.P., Cleyet-Marel, J.C., and **Gillis, M.** 1997. Transfer of *Rhizobium loti, R. huakuii, R. ciceri, R. mediterraneum* and *R. tianshanense* to *Mesorhizobium* gen nov. *International Journal of Systematic Bacteriology.* 47: 895-898.

Jenkins, M.B., Virginia, R. A. and **Jarrell, W. M.** 1987. Rhizobial ecology of the woody legume mesquite (*Prosopis glandulosa*) in the Sonoran desert. *Applied and Environmental Microbiology* 53: 36-40.

Jordan, D.C. 1984. Genus I. *Rhizobium* Frank 1889. 338[Al] In: *Bergey's Manual of Systematic Bacteriology* Vol. I. (Eds. Kaeig, N.R. and Holt, J.G.) Williams and Wilkins. Baltimore p. 235-242.

Lafay, B. and **Burdon, J.J.** 2001. Small subunit rRNA genotyping of rhizobia nodulating Australian *Acacia spp. Applied and Environmental Microbiology.* 67: 396-402.

Lange, R.T. 1961. Nodule bacteria associated with the indigenous leguminosae of South-western Australia. *Journal of General Microbiology.* 61: 351-359.

Lewin, A., Rosenberg, C., Meyer, H., Wong, C.H., Nelson, L., Manen, J. F., Stanley, J., Dowling, D.N. Dénarié, J. and ***Broughton, W.J.*** 1987. Multiple host specificity loci of the broad host range Rhizobium sp. NGR 234 selected using the widely compatible legume Vigna unguiculata. Plant Molecular Biology. 8: 447-459.

Lim, G. and ***Burton, J.C.*** 1982. Nodulation status of the Leguminosae. *In Nitrogen Fixation II. Rhizobium.* W. J. Broughton (ed.) Oxford University Press, Oxford, United Kingdom p.1-34.

Martínez-Romero, E., Segovia, E., Mercante, F.M., Franco, A.A., Graham, P.H. and ***Pardo, M.A.*** 1991. *Rhizobium tropici*, a novel species nodulating *Phaseolus vulgaris* L. beans and *Leucaena* sp. trees. *International Journal of Systematic Bacteriology.* 41: 417-426.

Michielis, J., Dombrecht, B. Vermeiren, N., Xi, C.-W., Luyten, E. and ***Vanderleyden, J.*** 1998. *Phaseolus vulagris* is a non-selective host for nodulation. *FEMS Microbiology and Ecology.* 26: 193-205.

Moreira, F.M.S., Haukka, K. and ***Young, J.P.W.*** 1998. Biodiversity of rhizobia isolated from a wide range of forest legumes in *Brazil. Molecular Ecology.*7 : 889-895.

Moulin, L., Munive, A., Dreyfus, B. and ***Boivin-Masson, C.*** 2001. Nodulation of legumes by members of the β-subclass of Proteobacteria. *Nature.* 411: 948-950.

Nick, G. 1998. Polyphasic taxonomy of rhizobia isolated from tropical tree legumes. Dissertationes Biocentri iikki Universitatis Helsin giensis. Department of Applied Chemistry and Microbiology, University of Helsinki.

Nick, G., Jussila, M., Hoste, B., Niemi, R.M., Kaijalainen, S., de Lajudie, P., Gillis, M., de Bruijn, F. J. and ***Lindström, K.*** 1999. Rhizobia isolated from root nodules of tropical leguminous trees characterized using DNA-DNA dot-blot hybridisation and rep-PCR genomic fingerprinting. *Systematic and Applied Microbiology.* 22: 287-299.

Nobbe, F., Hiltner, L. and ***Schmid, E.*** 1895. Versucheiiber die Biologie der Knollchenbakterien der Leguminosen, insbesondere iiber die Frage der Arteinheit derselben. *Landwirtschaftlichen Versuchstationer.* Dresden 45, 1-27.

Nobbe, F., Schmid, E., Hiltner, L. and ***Hotter, E.*** 1891. Versuche iiber die stickstoff-Assimilation der Leguminosen, *Landwirtschaftlichen* Dresden 39: 327-359.

Norris, D. O. 1965. Acid production by *Rhizobium*, a unifying concept. *Plant and Soil.* 22 : 143-166.

Odee, D.W. 1993. The Ecology of nitrogen-fixing symbiosis under arid conditions of Kenya. *Ph.D. Thesis.* University of Dundee. 145p.

Odee, D.W. and ***Sprent, J.I.*** 1992. *Acacia brevispica*, a non-nodulated mimosoid legume? *Soil Biology and Biochemistry.* 24: 717-719.

Olsen, G. J., Woese, C.R. and ***Overbeek, R.*** 1994. The winds of (evolutionary) change: breathing new life into microbiology. *Journal of Bacteriology.* 176: 1-6.

Paffeti, D., Scotti, C., Gnocchi, S., Fancelli, S. and ***Bazzicalupo, M.*** 1996. Genetic diversity of an Italian *Rhizobium meliloti* population from different *Medicago sativa* varieties. *Applied and Environmental Microbiology.* 62 : 2279-2285.

Palmer, K.M. and ***Young, J.P.W.*** 2000. Higher diversity of *Rhizobium leguminosarum* biovar *viciae* population in arable soils than in grass soils. *Applied and Environmental Microbiology.* 66: 2245-2450.

Pandey, P., Sahgal, M., Maheshwari, D. K. and ***Johri, B.N.*** 2004. Genetic diversity of rhizobia isolated from medicinal legumes growing in the sub-Himalayan region of Uttaranchal. *Current Science* 85: 202-207.

Peng, G.X., Tan, Z. Y., Wang, E.T., Reinhold Hurek, B., Chen, W. F. and ***Chen, W. X.*** 2002. Identification of isolates from soybean nodules in Xinjiang region as *Sinorhizobium xinjiangense* and genetic differentiation of *S. xinjiangense* from *S. fredii. International Journal of Systematic and Evolutionary Microbiology.* 52: 457-462.

Perret, X., Staehelin, C. and ***Broughton, W. J.*** 2000. Molecular basis of symbiotic promiscuity. *Microbiology and Molecular Biology Reviews* 64: 180-201.

Pueppke, S. G. and ***Broughton, W. J.*** 1999. *Rhizobium* sp. NGR 234 and *R. fredii* USDA 257 share exceptionally broad, nested host ranges. *Molecular Plant Microbe Interactions.* 12: 293-318.

Rainey, F. A., Ward-Rainey, N.L., Janssen, P.H., Hippe, H. and ***Stackebrandt, E.*** 1996. *Clostridium paradoxum* DSM 7308T contains multiple 16S rRNA genes with heterogeneous intervening sequences. *Microbiology.* 142 : 2087-2095.

Rivas, R., Willems, A., Palomo, J. L., Garcia –Benavides, P., Mateos, P.F., Martínez-Molina, E., Gillis, M. and ***Velázquez, E.*** 2004. *Bradyrhizobium betae* sp. nov., isolated from roots of *Beta vulgaris* affected by tumour-like deformations. *International Journal of Systematic and Evolutionary Microbiology.* 54 : 1315-1324.

Rome, S., Fernandez, M.P., Brunel, B., Normand, P. and ***Cleyet-Marel, J.C.*** 1996. *Sinorhizobium medicae* sp. nov., isolated from annual *Medicago* spp. *International Journal of Bacteriology.* 46: 972-980.

Roumiantseva, M. L., Yakutkina, V.V., Dammann-Kalinowski, T., Sharypova, L.A., Keller, M. and ***Simarov, B.V.*** 1999. Comparative analysis of the genome in alfalfa nodule bacteria *Sinorhizobium medicae* and *Sinorhizobium meliloti. Russian Journal of Genetics.* 35: 128-135.

Roumiantseva, M.L., Andronov, E.E., Sharypova, L. A., Dammann-Kalinowski, T., Keler, M., Young, J.P.W. and ***Simarov, B.V.*** 2002. Diversity of *Sinorhizobium melitoti* from the Central Asian Alfalfa Gene Center. *Applied and Environmental Microbiology.* 68.: 4694-4697.

Sahgal, M. 2002. Rhizobial diversity and heterogeneity of the *Dalbergia* Forest Ecosystems. *Ph.D. thesis.* Barkatullah University. Bhopal. 121p.

Sahgal, M. and ***Johri, B. N.*** 2003. The changing face of rhizobial systematics. *Current Science.* 84: 43-48.

Sahgal, M., Sharma, A., Johri, B.N. and ***Prakash, A.*** 2004. Selection of growth promotory rhizobia for *Dalbergia sissoo* from diverse soil ecosystems of India. *Symbiosis.* 36: 83-96.

Samant, M. 2003. Genetic characterization of root nodule isolates from clones of *Dalbergia sissoo. M.Sc. thesis,* G. B. Pant Univ. of Agric. & Tech., Pantnagar.

Scholla, M.H. and ***Elkan, G. H.*** 1984. *Rhizobium fredii* sp. nov., a fast-growing species that effectively nodulates soybeans. *International Journal of Systematic Bacteriology.* 34: 484-486.

Segovia, L., Pinero, D., Palacios, R. and ***Martínez-Romero, E.*** 1991. Genetic structure of a soil population of nonsymbiotic *Rhizobium leguminosarum. Applied and Environmental Microbiology.* 57 : 426-433.

Segovia, L., Young J.P.W. and ***Martínez-Romero, E.*** 1993. Reclassification of American *Rhizobium leguminosarum* biovar *phaseoli* type I strains as *R. etli* sp. nov. *International Journal of Systematic Bacteriology.* 43: 374-377.

Soberón-Chávez, G. and ***Nájera, R.*** 1989. Isolation from soil from soil of *Rhizobium leguminosarum* lacking symbiotic information. *Canadian Journal of Microbiology.* 35 : 465-468.

Souza, V., Eguiarte, L., Avila, G., Cappello, R., Gallardo, C., Montoya, J., and ***Pinerao, D.*** 1994. Genetic structure of *Rhizobium etli* biovar *phaseoli* associated with wild and cultivated bean plants (*Phaseolus vulgaris* and *Phaseolus coccineus*) in Morelos, Mexico. *Applied and Environmental Microbiology.* 60: 2772-2778.

Squartini, A., Struffi, P., Doring, H. and 10 other authors. 2002. *Rhizobium sullae* sp. nov. (formerly '*Rhizobium hedysari*'), the root-nodule microsymbiont of *Hedysarum coronarium* L. *International Journal of Systematic and Evolutionary Microbiology.* 52 : 1267-1276.

Stackebrandt, E., Frederiksen, W., Garrity, G.M., Grimont, A.D., Kämpfer, P., Maiden, M. C.J., Nesme, X., Ressellü-Mora, R., Swings, J., Trüper, H.G. Vanterin, L., Ward. A.C. and *Whitman, W.B.* 2002. Report of the adhoc committee for the re-evaluation of the species definition in bacteriology. *International Journal of Systematic and Evolutionary Microbiology* 52: 1043-1047.

Strain, S. R., Leung, K., Whittam, T.S., de Bruijn, F. J. and *Bottomley, P.J.* 1994. Genetic structure of *Rhizobium leguminosarum* biovar *trifolii* and *viciae* populations found in two oregon soils under different plant communities. *Applied and Environmental Microbiology.* 60: 2772-2778.

Sy, A., Giraud, E., Jourand, P. Garcia, N., Willems, A., de Lajudie, P., Prin, Y., Neyra, M., Gillis, M., Boivin-Masson, C. and *Dreyfus, B.* 2001. Methylotrophic *Methylobacterium* bacteria nodulate and fix nitrogen in symbiosis with legumes. *Journal of Bacteriology.* 183: 214-220.

Tan, Z. Y., Kan, F. L., Peng, G.X., Wang, E.T., Reinhold-Hurek, B. and *Chen, W.X.* 2001. *Rhizobium yanglingense* sp. nov., isolated from arid and semi-arid regions in China. *International Journal of Systematic and Evolutionary Microbiology.* 51: 909-914.

Tan, Z.Y., Wang, E.T., Peng, G.X., Zhu, M.E., Martínez-Romero, E. and *Chen, W.X.* 1999. Characterization of bacteria isolated from wild legumes in the north western region of China. *International Journal of Systematic Bacteriology.* 49: 1457-1469.

Thurman, N.P. and *Bromfield, E.S.P.* 1988. Effect of variation within and between *Medicago* and *Melilotus* species on the composition and dynamics at indigenous populations of *Rhizobium meliloti. Soil Biology and Biochemistry.* 20 : 31-38.

Trinick, M.J. 1982. Biology In: *Nitrogen Fixation* II. *Rhizobium* W.J. Broughton (ed.) Oxford University Press, Oxford, United Kingdom.

Tripathi, A.K., Verma, S.C., and *Ron, E.Z.* 2002. Molecular characterization of a salt-tolerant bacterial community in the rice rhizosphere. *Research in Microbiology.* 153: 579-584.

Trüper, H.G. and *Krämer, J.* 1981. Principles of characterization and identification of prokaryotes. In : *The prokaryotes: A handbook on habitats, isolation, and identification of bacteria* (eds. Starr, M.P. *et al.,*) Springer-Verlag, Berlin, Germany p 176-193.

van Berkum, P. and *Eardly, B. D.* 1998. In: *The Rhizobiaceae : Molecular Biology of Plant Associated Bacteria* (eds Spaink, H.P., Kondorosi, A. and Hooykass, P.J.J.), Kluwer, Dordrecht p 1-24.

van Berkum, P., Ruihua, F., Campbell, T.A. and *Eardly, B.D.* 1990. Some issues of relevance in the taxonomy of rhizobia. In: *Highlights of nitrogen fixation research* (eds : Martínez, E. and Hernández, G.) Plenum Publishing Corp., New York, N.Y. p. 267-270.

Van Berkum, P., Terefework, Z., Paulin, L., Suomalainen, S., Lindström, K. and **Eardly, B.D.** 2003. Discordant phylogenies with the *rrn* loci of rhizobia. *Journal of Bacteriology*.185: 2988-2998.

Velázquez, E., Martínez-Romero, E., Rodríguez-Navarro, D.M., Trujillo, M.E. Daza A., Mateos, P.E., Martínez-Molina, E. and **van Berkum, P.** 2001. Characterization of Rhizobial isolates of *Phaseolus vulgaris* by staircase electrophoresis of low-molecular weight RNA. *Applied and Environmental Microbiology.* 67: 1008-1010.

Wang, E.T., Tan, Z.Y., Willems, A.Y., Fernandez-Lopez, M., Reinhold-Hurek, B. and **Martínez-Romero, E.** 2002. *Sinorhizobium morelense* sp. nov., a *Leucaena leucocephala* associated bacterium that is highly resistant to multiple antibiotics. *International Journal of Systematic and Evolutionary Microbiology.* 52: 1687-1693.

Wang, E.T., Rogel, A., de los Santos, A.G. A.G., Martínez-Romero, J. Cevallos, M.A., and **Martínez-Romero. E.**1999b. *Rhizobium etli* bv *mimosae*, a novel biovar isolated from *Mimosa affinis. International Journal of Systematic Bacteriology.* 49 :1479-1491.

Wang, E.T., van Berkum, P., Sui, X. H., Beyene, D., Chen, W.X. and **Martínez-Romero, E.** (1999a). Diversity of rhizobia associated with *Amorpha fruticosa* isolated from Chinese soils and description of *Mesorhizobium amorphae* sp. nov. *International Journal of Systematic Bacteriology.* 49: 51-65.

Wang, E.T., van Berkum, P., Beyene, D., Sui, X.H., Dorado, O., Chen, W.X. and **Martínez-Romero, E.** 1998. *Rhizobium huautlense* sp. nov., a symbiont of *Sesbania herbaceae* that has a close phylogenetic relationship with *Rhizobium galegae. International Journal of Bacteriology.* 48: 687-699.

Wang, Y., Zhang, Z. and **Ramanan, N.** 1997. The actinomycete *Thermobispora bispora* contains two distinct types of transcriptionally active 16S rRNA gene. *Journal of Bacteriology.* 179 :3270-3276.

Wayne, L.G. Brenner, D.J. Colwell, R.R. and 9 other authors. 1987. International Committee on Systematic Bacteriology. Report of the adhoc committee on reconciliation of approaches to bacterial systematics. *International Journal of Systematic Bacteriology.* 37: 463-464.

Wei, G. H., Tan, Z.Y., Zhu, M.E., Wang, E.T. Han, S. Z. and **Chen, W.X.** 2003. Characterization of rhizobia isolated from legume species within the genera *Astragalus* and *Lespedeza* grown in the Loess Plateau of China and description of *Rhizobium loessense* sp. nov *International Journal of Systematic and Evolutionary Microbiology.* 53: 1575-1583.

Wei, G.H., Wang, E.T., Tan, Z.Y., Zhu, M.E. and **Chen, W.X.** 2002. *Rhizobium indigoferae* sp. nov. and *Sinorhizobium kummerowiae* sp. nov., respectively isolated from *Indigofera spp.* and *Kummerowia stipulacea. International Journal of Systematic and Evolutionary Microbiology.* 52 : 2231-2239.

Willems, A., Coopman, R. and **Gillis, M.** 2001. Phylogenetic and DNA-DNA hybridization analysis of *Bradyrhizobium* species. *International Journal of Systematic and Evolutionary Microbiology*. 51: 111-117.

Willems, M. and **Collins, M.D.** 1993. Phylogenetic analysis of rhizobia and agrobacteria based on 16S rRNA gene sequences. *International Journal of Systematic Bacteriology*. 43: 305-313.

Wilson, J. K. 1939. Leguminous plants and their associated organisms. Cornell Univ. Agric. Exp. Station Memoir 221. Cornell University Press, Ithaea, N.Y.

Yao, Z.Y., Kan, F.L., Wang, E.T., Wei, G.H. and **Chen, W.X.** 2002. Characterization of rhizobia that nodulate legume species of the genus *Lespedeza* and description of *Bradyrhizobium yuanmingense* sp. nov. *International Journal of Systematic and Evolutionary Microbiology*. 52: 2219-2230.

Young, J. M., Kuykendall, L.D., Martínez-Romero, E., Kerr, A. and Sawada, H. 2001a. A revision of *Rhizobium* Frank 1889, with an emended description of the genus, and the inclusion of all species of *Agrobacterium* Conn 1942 and *Allorhizobium undicola* de Lajudie *et al.*, 1998 as new combinations : *Rhizobium radiobacter, R. rhizogenes, R. rubi, R. undicola* and *R. vitis*. *International Journal of Systematic and Evolutionary Microbiology*. 51: 89-103.

Young, J. M., Bull, C.T., de Boer, S.H., Firrao, G., Gardan, L., Saddler, G.S., Stead, D.E. and **Takikawa, Y.** 2001 b. Classification, nomenclature and plant pathogenic bacteria - a clarification. *Phytopathology*. 91: 617-620.

Young, J.M., Kuykendall, L.D., Martínez-Romero, E., Kerr, A. and **Sawada, H.** 2003. Classification and nomenclature of *Agrobacterium* and *Rhizobium* a reply to Farrand et al. (2003). *International Journal of systematic and Evolutionary Microbiology*. 53: 1689-1695.

Zeigler, D.R. 2003. Gene sequences useful for predicting relatedness of whole genomes in bacteria. *International Journal of Systematic and Evolutionary Microbiology*. 53: 1893-1900.

Zhang, X., Harper, P, Karsisto, M. and **Lindström, K.** 1991. Diversity of *Rhizobium* bacteria isolated from the root nodules of leguminous trees. *International Journal of Systematic Bacteriology*. 41: 104-113.

Zhang, X-X., Guo, X. –W., Terefework, Z., Paulin, L., Cao,Y-Z., Hu, F.-R., Lindström, K. and **Li, F. D.** 1999. Genetic diversity among rhizobial isolate from field grown *Astragalus sinicus* of Southern China. *Systematic and Applied Microbiology*. 22: 312-320.

<table><tr><td>3</td></tr></table>

Aggressiveness of *Corynespora cassiicola* Isolates and Reaction of some *Hevea brasiliensis* Clones to their Infection

Annakutty Joseph, Manju M.J.Arunkumar, C.Kuruvilla Jacob and Ramesh.B.Nair

Rubber Research Institute of India, Kottayam - 686 009, Kerala.

ABSTRACT

Corynespora leaf disease caused by *Corynespora cassiicola* is one of the most important diseases of *Hevea brasiliensis*. An attempt was made to study the aggressiveness of sixteen isolates of *Corynespora* and their interactions with sixteen *Hevea* clones in the laboratory by detached leaf inoculation technique. Aggressiveness of the isolates was also studied by estimating their toxin production. The disease intensity in 13 clones planted at RRII, Hevea Breeding Sub Station, Nettana, Karnataka was compared with their disease severity in the laboratory test. The disease severity measured by the lesion size produced by the isolates on different clones varied significantly. RRII 105 was the most susceptible and GT1 the most tolerant clone. The disease intensity in different clones under field condition were comparable to their reaction observed in the laboratory. The degree of virulence of the isolates on rubber clones was also found to vary significantly. A positive correlation was observed between the pathogenicity and the quantity of toxin produced by each isolate.

KEYWORDS : *Corynespora cassiicola*, aggressiveness, *Hevea brasiliensis*, pathogenicity, toxin production

Corynespora leaf fall disease caused by *Corynespora cassiicola* is one of the most important and destructive diseases of *Hevea brasiliensis*. In India, initially it was recognized as a weak pathogen confined to nursery seedlings and bud wood plants (Ramakrishanan and Pillai, 1961) occasionally infecting mature trees (George and Edathil, 1980). From 1996 onwards the disease has been more damaging on mature trees in Southern Karnataka, adjoining areas in Kerala (Rajalakshmy and Kothandaraman, 1996) and in South West India. The disease is a major constraint for rubber cultivation in South and South East Asian countries and has attained an epidemic form affecting rubber plantations in Indonesia, Sri Lanka, Thailand and Malaysia (Kamar and Hidir, 1996; Jacob, 1977). A breakdown of tolerance of some of the cultivated clones has been reported from many countries, which might be due to the development of new and more virulent races of the pathogen. Host selective toxins play an important role in the virulence/

aggressiveness of fungi (Walton, 1996). Breton *et al.* (2000) found that the toxin produced by *C. cassiicola* plays a significant role as disease determinant and in pathogenicity of isolates. He also observed that high toxin concentration could overcome the disease resistance of a clone.

Severity of the disease depends not only on the existence of virulent pathogen but also to favourable climate and susceptibility of host genotypes under cultivation. Susceptibility or resistance of clones varies in relation to virulence of the isolates. The most economic and successful way to control disease is to use disease resistant clones. Use of disease resistant clones also helps in avoiding the hazards of environmental pollution caused by fungicides. Information on host parasite interrelationship may help to develop effective and economic disease management strategies. Hence an attempt has been made to study the aggressiveness of different isolates of *Corynespora* and their interaction with different *H. brasiliensis* clones.

Materials and Methods

Leaves showing typical symptoms of infection by *C. cassiicola* were collected from different rubber growing regions of Karnataka and Kerala states. Leaf bits with typical symptoms were surface sterilized and pathogen was isolated by planting on potato dextrose agar (PDA) medium. Isolates were purified by single spore extraction and incubated at 25-28 ^{0}C on PDA slants. A total of sixteen isolates were used in the present study.

Pathogenicity of the isolates were estimated by detached leaf technique. Sixteen clones (including the recommended/popular clones and some other clones showing resistance/ susceptible reactions to other major leaf diseases) of *H. brasiliensis* were selected for the study. Three leaflets of each clone were inoculated with 16 isolates of *Corynespora*. The isolates were grown on PDA medium for abundant sporulation and conidial suspension of 2.3×10^4 spores/ml were prepared in sterile water. The leaves at light green stage were inoculated with 20 µl spore suspension on the abaxial side of the leaflets placed in moist petriplates. Pathogenicity was estimated by measuring the lesion size after 96 h of inoculation. The data were analyzed and means were compared by Duncan's multiple range test (DMRT).

Aggressiveness of the isolates was also studied by estimating the toxin production by the isolates. The production of toxin was estimated by leaf wilt bioassay using crude culture filtrates. A modified czapek's medium was used for toxin production. The liquid medium (100 µl) was taken in 250 µl conical flask and inoculated with three mycelial plugs from seven-day-old cultures of each isolate. The flasks were incubated without agitation at 25 ^{0}C (photo period 12 h) for 12 days. Cultures were vacuum filtrated through a 0.22 µm millipore membrane and the filtrate stored at 30 ^{0}C until use. Toxin activity was tested using the most susceptible clone (RRII 105) and the tolerant clone (GT 1). Leaflets were excised from 6-8 day old leaves under water and immediately transferred to 20 ml bottles containing 5 ml of toxic fraction (crude culture filtrate). The experiment was replicated three times and the controls were maintained with distilled water. All the bottles were incubated at 25 ^{0}C for 48 h (alternatively at 12 h of darkness and light) and wilting intensity was estimated by calculating   the ratio of fresh weight to dry   weight and expressing the values as per cent of control.

The disease intensity was assessed on 13 clones planted at the Hevea Breeding Sub Station, Nettana, Karnataka. The assessment was carried out during 1998, 1999 and 2000 disease seasons and compared with the results of laboratory screening. Disease intensity was assessed from five trees from each clone by grading 5 leaves selected at random from the top most whorl of each plant. Scoring was done on 0-5 scale based on the intensity of spotting and deformity of the leaves. The percentage disease index was calculated (Horsefall & Heuberger, 1942) and the mean PDI was analyzed statistically

Results and Discussion

Disease severity induced by the isolates on different clones varied with the isolate and clones (Table 1). The mean lesion size varied form 4.91 mm to 14.04 mm (Table 2). In general, RRII 105 showed susceptible reaction to all the isolates. RRII 105 was the most susceptible clone and the mean lesion size significantly more than that developed on all other clones. The mean lesion size was 14.04 in RRII 105, which was followed by RRIM 701. However, the size of lesion on the clones RRIM 701, RRII 208 and PB 28/59 were on par. These clones can be grouped as highly susceptible. GT 1 was the most tolerant clone, with a mean lesion size of 4.91 mm. The lesion size in GT 1 showed highly significant variation from all other clones. The mean lesion size in RRIM 703, PB 235, PB 86, PB 217 and RIM 600 were at par but lower than other clones. These clones can be grouped as moderately tolerant.

The degree of virulence of isolates on rubber clones varied significantly (Table 3). Almost all the isolates showed virulent reaction to RRII 105. Mean lesion size of the isolate ranged from 10.81 mm to 12.87 mm. The isolate 9 produced lesions of maximum size (12.87 mm) and was at par with the isolate 13. The isolate 15 produced the smallest lesions (10.81mm) and was at par with the isolate 5 and 12. .

When the isolate clone interaction was compared, in general significant variation was observed in the size of the lesions produced by different isolates on the same and different clones (Table1). The lesion size ranged from 11.38 to 15.50 on RRII 105 while it ranged from 3.61 to 7.33 on GT 1. When the aggressiveness of the isolates was estimated by their toxin production similar trends were observed. The isolates which produced larger lesions showed more per cent wilting intensity in both the clones. Percentage wilting of the leaves was high for culture filtrates of the isolates 8, 9 and 13 and less for the isolates 5, 10, 12 and 15. A positive correlation was observed between pathogenicity and quantity of toxin produced by each isolate (Fig. 1). In general the percentage wilting was high for the susceptible clone RRII 105 and less for the tolerant clone GT 1 as was observed in the case of pathogenicity.

Table 1 Mean lesion size produced by *C. cassiicola* isolates on clones of *Hevea brasiliensis*

	1	107	86	217	235	5/51	600	701	33	105	208	10	703	260	311	28/59	
	GT	PR	PB	PB	PB	PB	RRIM	RRIM	RRII	RRII	RRII	BD	RRIM	PB	PB	PB	Mean
1.	4.33	11.61	10.83	9.50	10.77	12.67	12.22	13.11	11.50	13.56	11.56	13.33	11.66	12.83	12.35	14.28	11.63
2.	5.22	10.67	12.61	12.56	12.16	13.06	10.11	11.17	11.11	14.72	12.94	12.83	11.22	11.78	12.28	12.06	11.67
3.	5.11	11.50	10.33	11.50	13.94	10.89	11.56	11.11	11.89	12.94	11.11	11.44	11.28	12.11	12.12	12.27	11.32
4.	4.61	14.43	11.78	10.72	11.67	11.11	10.83	10.50	9.67	15.00	12.89	12.33	11.67	12.72	12.11	13.11	11.57
5.	3.89	10.45	10.72	10.00	12.50	12.55	11.12	11.89	12.89	12.78	12.33	11.22	11.72	11.06	11.28	11.05	11.09
6.	5.72	11.39	11.78	11.11	11.22	11.67	12.39	16.83	10.72	15.00	13.72	13.83	12.00	12.35	12.05	13.72	12.22
7.	6.00	12.61	10.99	12.56	12.22	13.06	10.55	12.56	11.22	14.11	13.22	11.61	12.23	13.24	11.96	13.12	11.95
8.	7.33	13.33	11.89	12.61	11.22	11.67	12.11	14.22	12.44	15.16	12.83	13.05	11.55	12.95	12.23	13.56	12.39
9.	5.11	13.89	10.61	10.72	11.17	11.22	12.61	15.39	15.55	14.50	16.78	15.44	12.55	13.39	13.18	13.83	12.87
10.	3.61	10.94	10.33	13.61	10.89	12.00	11.78	14.11	12.11	11.38	13.06	11.61	11.17	10.67	10.72	12.28	11.27
11.	4.72	13.39	14.55	13.94	10.61	11.89	15.05	15.05	11.22	14.92	11.55	12.17	11.11	12.28	11.89	13.12	12.35
12.	3.61	12.33	12.22	11.50	11.45	12.00	10.94	12.89	11.00	12.89	10.06	12.00	10.89	11.83	10.78	11.68	11.13
13.	5.72	14.17	13.50	12.33	11.28	12.17	13.67	14.67	12.61	15.50	13.78	13.44	12.11	12.85	12.00	13.29	12.61
14.	4.5	11.50	11.88	11.78	11.11	11.83	11.50	11.22	10.83	14.61	13.44	11.22	11.17	11.00	11.95	12.06	11.43
15.	4.39	13.06	10.11	10.50	11.22	10.44	11.22	10.06	10.67	12.95	12.33	10.39	10.50	11.22	12.00	11.95	10.81
16.	4.61	12.06	11.83	13.11	11.50	13.11	11.84	13.83	15.33	14.61	12.67	13.39	11.50	12.28	12.00	12.56	12.26
Mean	4.91	12.33	11.63	11.75	11.56	11.96	11.84	13.04	11.92	14.04	12.77	12.46	11.52	12.16	11.93	12.75	

CD (P=0.05)   Means   0.306
Interaction 1.22

Table 2 The mean of lesion size on *H. brasiliensis* clones

Clones	Mean lesion size (mm)
GT 1	4.91 a
RRIM 703	11.52 b
PB 235	11.56 b
PB 86	11.63 bc
PB 217	11.75 bc
RRIM 600	11.84 bcd
RRII 33	11.92 cd
PB 311	11.93 cd
PB 5/51	11.96 cd
PB 260	12.16 de
PR 107	12.33 e
BD 10	12.46 ef
PB 28/59	12.75 fg
RRII 208	12.77 fg
RRIM 701	13.04 h
RRII 105	14.04 i

Means followed by same letters are not significantly different in DMRT

Table 3 The mean of lesion size produced by isolates of *C. cassiicola*

Isolate No.	Mean lesion size (mm)
15	10.81 a
5	11.09 ab
12	11.13 ab
10	11.27 bc
3	11.32 bcd
14	11.43 bcd
4	11.57 cd
1	11.63 d
2	11.67 de
7	11.95 ef
6	12.22 fg
16	12.26 fg
11	12.34 gh
8	12.39 gh
13	12.61 hi
9	12.87 i

Means followed by same letters are not significantly different in DMRT

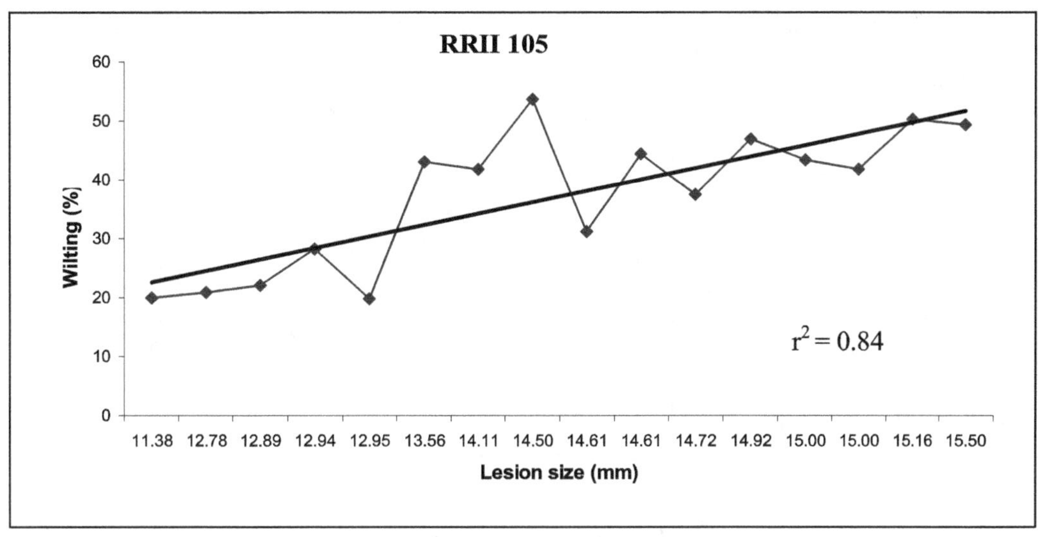

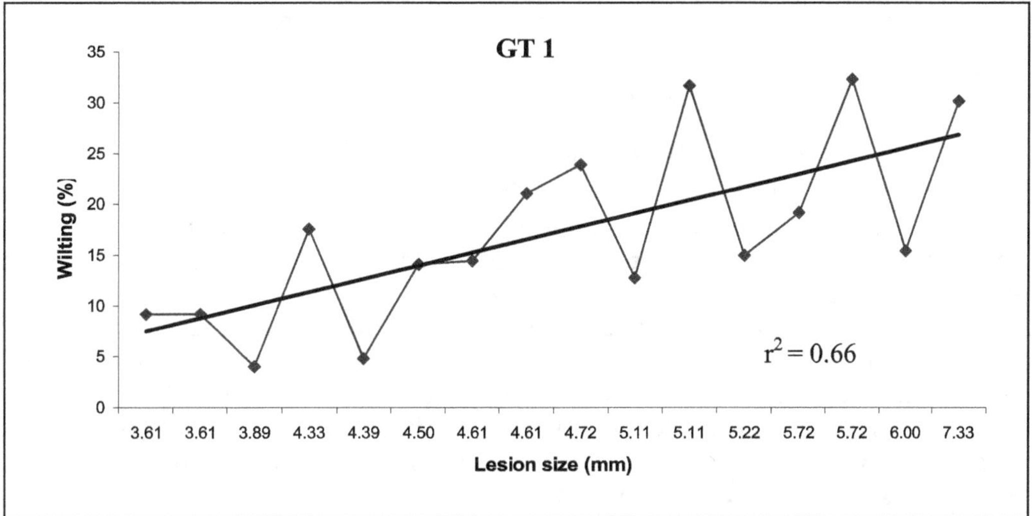

Figure 1 Relationship between pathogenicity and toxin production by *C.cassiicola* isolates

Under field condition also RRII 105 recorded significantly higher disease intensity compared to all other clones (Table 4). It was followed by PB 260 and PB 28/59. GT1 recorded low disease intensity and it significantly differed from all other clones. GT1 was followed by PB 5/51, PB 86 and RRIM 600.

The present study clearly indicated that significant variation in the severity of the disease among different clones. In general, RRII 105 was susceptible to all   isolates and GT 1 was tolerant. Percentage wilting was also high in RRII 105 and less in GT 1. The lower wilting

Table 4 *Corynespora* leaf fall disease intensity in the field during the peak disease pe riod.

Clone	Per cent Disease Intensity			Pooled mean
	1998	*1999*	*2000*	
GT 1	3.00	3.67	2.67	3.11 a
PB 5/51	2.67	8.33	9.00	6.67 b
RRIM 600	3.33	9.00	11.00	7.78 cb
PB 86	6.67	9.33	8.00	8.00 cb
RRII 208	6.00	13.00	11.67	10.22 dc
PB 235	12.67	10.00	12.00	11.56 ed
RRIM 703	13.00	16.00	10.67	13.22 fe
PB 311	10.00	17.33	14.00	13.78 gfe
PB 217	19.00	14.67	12.00	15.22 gf
RRIM 701	18.67	15.00	15.00	16.22 g
PB 28/59	17.67	26.00	17.67	20.44 h
PB 260	26.00	32.00	21.00	26.33 i
RRII 105	64.00	77.33	72.67	71.33 j

CP (P = 0.05) Mean 2.59

observed for GT 1 may be due to its insensitivity to the toxin. Breton *et al.* (2000) observed similar results and reported that the disease resistance could be due to the insensitivity of the clone to the toxin. The susceptibility and resistance of rubber clones depends on the fungal isolates also. In the present study significant variation was observed in the pathogenicity of the isolates on different clones. More pathogenic isolate produced larger lesion and more toxins in almost all the clones. The variation observed in lesion size on the same clone by different isolates can be attributed to the variations in the virulence of the isolates which in turn is related to the quantity of the toxin produced by the isolates. Thus the present findings agree with the findings of Breton *et al.* (2000) that the higher quantity of toxin can overcome the resistance of tolerant clones.

In general, the severity of the disease in different clones in the laboratory test were comparable to the disease severity observed under field conditions. The clones rated as most susceptible (RRII 105) and tolerant (GT 1) in the laboratory screening were the most susceptible and tolerant clones in the field also. The clones RRIM 703, PB 235, PB 86, PB 217 and RRIM 600 were found to be moderately tolerant in the laboratory test, while under field conditions the clones RRIM 600 and PB 86 showed tolerant reaction and PB 217 moderately susceptible. Manju et al. (2001) in a survey carried out in coastal Karnataka and North Malabar regions of Kerala found that GT 1 and RRIM 600 had very low infection while the clones PB 217, PB 235 and PB 260 had moderate infection. Kamar and Hidir (1996) had reported GT 1, PB 217, PB 235 and PB 260 as moderately susceptible clones and RRIM 600 as susceptible clone in the rubber growing regions of

Malaysia. Susceptibility of RRIM 600 and Tjir 1 had also been reported from Sri Lanka (Jayasinghe and Silva, 1996). Sinulingga *et al* (1996)) found *severe disease* incidence in RRIM 600 and GT 1 which were formerly categorized as moderately resistant in Indonesia. Chee (1988 a) found some clones and germplasm materials reported as susceptible under laboratory tests were free from infection under field condition. Tan *et al.* (1992) observed light to moderate infections on some of the clones which were reported to be resistant earlier. The variation in the severity of the disease under field conditions can be due to the variation in the virulence of the pathogen races and the prevailing environmental factors. Early wintering trees usually escape the disease as the environmental conditions are not congenial. Although young leaves (2 days - 4 weeks) are more susceptible to *C. cassiicola* infection, mature leaves also could be infected (Chee 1988 b).

The present study shows variation in the severity of the disease among different clones and that the resistance of a clone is related to the aggressiveness of isolates of *C. cassiicola*. The aggressiveness is determined by the quantity of toxin produced. Hence pathogen isolates capable of producing more toxin should be included in laboratory screening of clones. Since the results of the laboratory and field evaluations were comparable in the present study, particularly with respect to most tolerant and susceptible clones, the detached leaf technique can be used for early evaluation of new *H. brasiliensis* clones for tolerance.

Acknowledgement

The authors express their gratitude to Dr. N. M. Mathew, Director, Rubber Research Institute of India for the encouragement.

References

Breton, F., Sanier, C. and d' Auzac, J. 2000. Role of cassiicolin, a host-selective toxin in pathogenicity of *C. cassiicola*, casual agent of leaf disease of *Hevea*. Journal of Natural Rubber Research, **3**(2): 115-128.

Chee, K.H. 1988 a. Studies on sporulation, pathogenicity and epidemiology of *Corynespora cassiicola* on *Hevea* rubber. Journal of Natural Rubber Research, **3**(1): 21-29.

Chee, K.H. 1988 b. *Corynespora* leaf spot. Planters Bulletin. Rubber Research Institute of Malaysia, **94**: 3-7.

George, M.K. and **Edathil, T.T.** 1980. A report on *Corynespora* leaf spot disease on mature rubber. Paper presented in International Rubber Conference, IRCIND, 1980. Abstract. Session 4 paper 2.

Horsefall, J.G. and **Heuberger, J.W**. 1942. Measuring the magnitude of defoliation desease of tomoto. Phytopathology, **32**:226-232.

Jacob, C.K. 1997. Diseases of potential threat to rubber in India. Planters' Chronicle **2**(10): 451-461.

Jayasinghe, C.K. and ***Silva, W.P.K.*** 1996. Current status of *Corynespora* leaf fall in Sri Lanka. Proceedings of the Workshop on CLF disease of *Hevea* rubber. Indonesian Rubber Research Institute, Medan, Indonesia, pp. 15-19.

Kamar, A.S.S. and ***Hidir, S.M.*** 1996. Current status of *Corynespora* leaf fall in Malaysia. Proceedings of the Workshop on CLF disease of *Hevea* rubber. Indonesian Rubber Research Institute, Medan, Indonesia, pp. 21-28

Manju, M.J., Idicula, S.P. Jacob, C.K., Vinod, K.K., Prem E.E., Suryakumar, M. and ***Kothandraman, R.*** 2001. Incidence and severity of *Corynespora* leaf fall (CLF) disease of rubber in coastal Karnataka and North Malabar region of Kerala. Indian journal of Natural Rubber Research, **14**(2):147-141.

Rajalakshmi, V.K. and ***Kothandaraman, R***. 1996. Current status of *Corynespora* leaf fall in India: The occurrence and management. Proceedings, Workshop on *Corynespora* Leaf fall Disease of *Hevea* Rubber, (Eds. A. Darussamin, S. Pawirosoemardjo, Basuki, R. Azwar, Sadaruddin) 16-17 December 1996, Medan, Indonesia, pp.37-46.

Ramakrishnan, T.S., and ***Pillai, P.N.R***. 1961. Leaf spot of rubber caused by *Corynespora cassiicola* (Berk and Curt) Wei. Rubber Board Bulletin, **5**(1): 32-35.

Sinulingga, W., Suwarto and ***Soepena, H***. 1996. Current status of *Corynespora* leaf fall in Indonesia. Proceedings of the Workshop on CLF disease of *Hevea* rubber. Indonesian Rubber Research Institute, Medan, Indonesia, pp. 29-36

Tan, A.M., Loo, T.P., Vadivel, G., Bachik, M.R. and ***Yoon, K.F.*** 1992. Survey of major leaf diseases of rubber in peninsular Malaysia. Planters Bulletin, Rubber Research Institute of Malaysia, **211**: 51-62.

Walton, J.D. 1996. Host-selective toxins: Agent of compatibility. Plant Cell, **8** : 1723-1733.

<table><tr><td>**4**</td><td># *Some Interesting and Rare Hyphomycetous Fungi from Papikondalu Forest, A.P., India*</td></tr></table>

G. Suresh Kumar, K. Sharath Babu, and I.K. Kunwar

Botany Department, Osmania University, Hyderabad - 500007

ABSTRACT

During our studies on biodiversity of micro-fungi some interesting and rare hyphomycetous taxa were collected from the forests of Papikondalu, A.P., India. The collections were screened and identified as *Phialosporostilbe setosa, Hyphodiscosia jaipurensis, Esdipatilia indica, Spadicoides grovei, Cryptophialoidea secunda, Physalidium elegans, Actinocladium rhodosporum, Virgaria nigra, Speiropsis pedatospora, Bispora betulina, Tetraploa aristata* and *Spegazzinia lobulata*. These fungi are described in the present paper.

KEY WORDS : Biodiversity, taxa, micro-fungi.

We made an attempt to investigate some micro-fungi from Papikondalu forest, Khammam District, A.P., India. The thick forest is in Bhadrachalam and Papikondalu region. This region is undisturbed. Indian hyphomycetes have been extensively studied particularly by C.V. Subramanian(1971), but not much attention was paid by mycologists to investigate the micro-fungi of this region, hence we focussed our study on this region. Our observations on materials from this tropical forest indicate that the knowledge of micro-fungi from these habitats is still in an exploratory phase and many new taxa await to be discovered. In the present study we have described twelve interesting and rare hyphomycetous fungi from Papikondalu forest (Bilgrami *et al.*,1979, 1981, 1991; Sarbhoy *et al.*, 1986, 1996; Jamaluddin *et al.*, 2003). *Spadicoides grovei; Cryptophialoidea secunda; Physalidium elegans* were reported for the first time from India by us (Suresh Kumar *et al.*, 2004).

Phialosporostilbe setosa Bhat and Kendrick, 1993, Mycotaxon, XLI: 57-59 (Figs. 1, 2).

Colonies effuse, hairy, white to grayish brown. Conidiomata synnematal, erect, straight or flexuous, indeterminate, 140-440 X 13-30 μm, composed of 5-20 parallel, thick-walled, septate, brown, conidiophores diverging at their fertile apices, synnemata always incorporating 1 sterile dark brown, smooth, thick-walled, blunt seta, 160-450 X 6-9 μm, septa 15-20 μm apart, seta as long as or slightly longer than the conidiophores and surrounded by them. Conidiogenous cells integrated, terminal, monophialidic, extending percurrently up to 3 times, diverging from the main axis of the synnema, pale to medium brown, cylindrical to cylindric-clavate, 22-45 μm long X 5-7 μm wide below, 3-5 μm at the tip, with an inconspicuous apical collarette. Conidia blastic-phialidic, aggregating in colourless slimy masses and of two kinds: (a) conidia arising from the upper conidiogenous cells are angular, cuneiform, tetrahedral or tetra-radiate, with 3 protuberant corners and a slightly truncate to rounded narrow base, non-septate smooth, thick-walled, colourless, with dense cytoplasm, 9-13 μm long, 6-9 μm at the widest region and on each

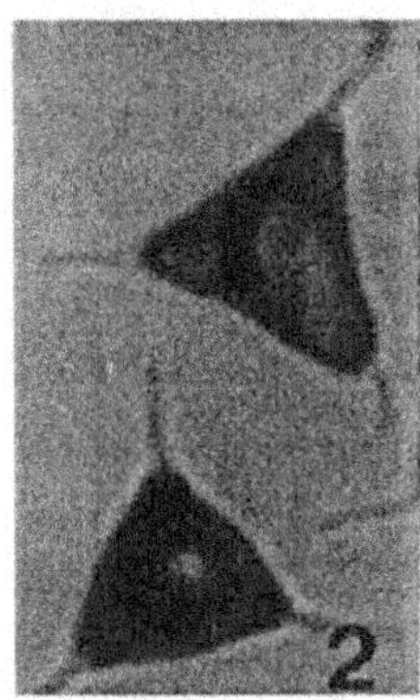

Figure 1 *Phialosporostilbe setosa*-Synnemata, conidiogenous cells, seta and conidia.Bar = 75mm.

Figure 2 Angular conidia with setulae bar = 20mm.

corner having a thin setula, 5-8 μm long; (b) conidia arising from conidiogenous cells in the lower half of the synnema are drop-shaped, with a narrowly pointed base, non-septate, colourless, smooth, 2-4 μm dia.

On decaying bamboo leaves, Papikondalu, Eastern Ghats, Leg. G. Suresh Kumar, K. Sharath Babu, 13. 07. 2004; OUFH 509.

Hyphodiscosia jaipurensis Lodha and Chandra Reddy, 1974, *Trans. Br. mycol. Soc.*,   62: 418-421. (Fig. 3)

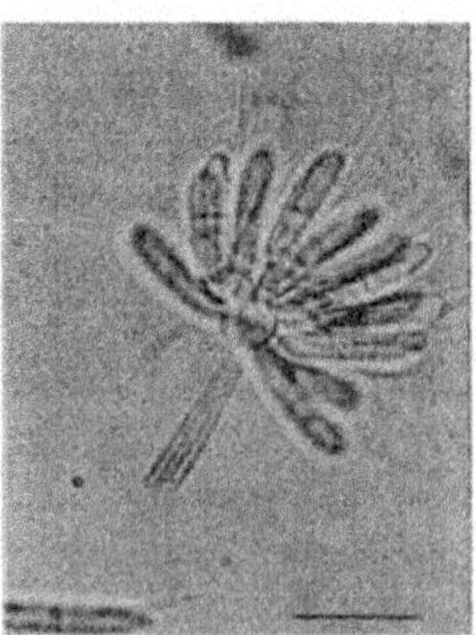

Figure 3 *Hyphodiscosia jaipurenis* - Coniodiophore, conidiogenous cells and conidia with setulae, Bar - 25mm.

Colonies effuse, hairy, pale brown to pinkish. Mycelium mostly immersed, composed of branched, septate, smooth, pale brown, 2-4 μm thick hyphae. Conidiophores arising terminally and laterally from the hyphae, macronematous, mononematous, unbranched, septate, smooth, erect, straight or flexuous, mid dark brown and thick walled below becoming hyaline and thin walled towards apex, up to 95 μm long, 3-5 μm wide at the base and 4-9 μm wide at apex. Conidiogenous cells integrated, terminal, polyblastic, denticulate, denticles broad, cylindrical, clavate, 15-18 μm long and 4-9 μm wide at the widest, sometimes the conidiogenous cells extend upwards and become swollen to form another conidiogenous cell at a higher level in such a way that at two nodes conidiogenous cells (ampullae) are formed. Conidia in colourless masses, solitary, dry, acropleurogenous, simple, cylindrical, rounded at the apex, conicotruncate at the base, hyaline, 18-24 μm long and 4-5 μm wide at the broadest and 1-2 μm wide at base, 0-1 septate, sometimes septum formed supra medially, smooth, guttulate, thin walled with two delicate, flexuous setulae arising from the same side, one from each cell, setula from the upper cell produced sub-apical and the setula of the basal cell from the middle of the cell, 15-18 μm long.

On dead wood, Papikondalu, Eastern Ghats, Leg. G. Suresh Kumar, K. Sharath Babu, 13. 07. 2004; HCIO 45,491; OUFH 488.

Esdipatilia indica C.H. Phadke, 1981, *Tr. Br. mycol. Soc.*, 77(3): 642-643 (Fig. 4)

Figure 4 *Esdipatilia indiaca* - Synnemata with conodial heads, bar = 75mm.

Synnemata arise from septate, branched hyphae, conspicuous, erect, dispersed, cylindrical, pale straw-coloured, 290-675μm long, 45-90μm wide at the base, 25-40μm wide in the middle and 17-28μm wide at the apex just below the mucilaginous conidial head. Individual conidiophores macronematous, erect, filiform, septate, hyaline to sub-hyaline, 2-3μm wide. Conidiogenous cells monoblastic, integrated and terminal, determinate or percurrent. Conidia slimy, formed singly, aggregated into a mucilaginous conidial head which secceeds easily, hyaline to sub-hyaline, 4-7 septate, typically bent, having an enlarged central cell, measuring 51-71μm long (including beak), 9-13μm wide in the middle narrowing to 1-2μm wide at the base and apex, truncate at the base and tapering towards both ends. The terminal portion of the conidium is attenuated to form a multiseptate beak, which forms a wide angle with the broad central cell.

On bark, Papikondalu, Eastern Ghats, Leg. G. Suresh Kumar, K. Sharath Babu, 13. 07. 2004; HCIO 45,603; OUFH 499.

Spadicoides grovei M.B. Ellis, 1963, Mycol. Pap. 93: 6-14. (Fig. 5)

Colonies effuse, dark blackish-brown, hairy. Mycelium partly superficial, partly immersed. Conidiophores macronematous, mononematous, generally unbranched, straight, dark brown, smooth, closely septate, 246-529 X 6-9 μm; Conidiogenous cells polytretic, integrated, terminal and intercalary, determinate, cylindrical. Conidia solitary, dry, acropleurogenous, developing through minute channels in the thick wall of the conidiogenous cells, simple, broadly ellipsoidal, oblong, truncate at the base, dark brown, smooth, 3-septate, with dark brown or black bands at the septa, 20-25 X 12-15 μm.

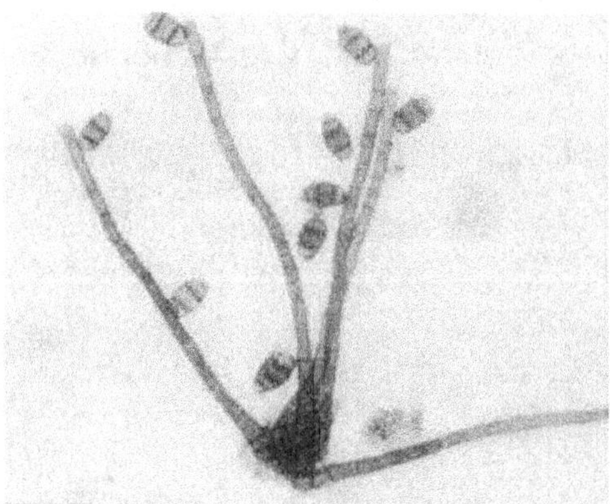

Figure 5 *Spadicoides grovei* - Conidiophores and conidia, Bar = 80mm.

On dead twigs, Papikondalu, Eastern Ghats, Leg. G. Suresh Kumar, K. Sharath Babu, 13. 07. 2004; HCIO 45,554; OUFH 447.

Cryptophialoidea secunda (Kuthubutheen and Sutton) Kuthubutheen and Nawawi, 1987, *Tr. Br. mycol. Soc.*, 89: 581-583. (Fig. 6)

Colonies effuse, pale to olivaceous-brown, slow growing. Mycelium mostly superficial, septate, smooth. Conidiophores macronematous, mononematous, erect, straight or flexuous, subulate, brown, smooth, setiform, solitary, simple, thick-walled, septate with septa arranged more closely in the conidiogenous zone, up to 308 μm long, 6-8 μm wide at the base, tapering to 2-2.5 μm at the apex, conidiogenous zone about halfway up the conidiophore, consisting of a single row (44-61 μm long) of phialides closely arranged on only one side of the conidiophore and not obscured by a shield of sterile cells. Conidiogenous cells monophialidic, phialides semi-endogenous, extrude through minute pores on only one side of the conidiophore, discrete, determinate, lageniform, with collarates, light olivaceous brown, individual phialides 7-9 X 3-4 μm. Conidia hyaline, smooth, 1-septate, falcate, 20-25 μm long, 1-2 μm wide at the broadest part.

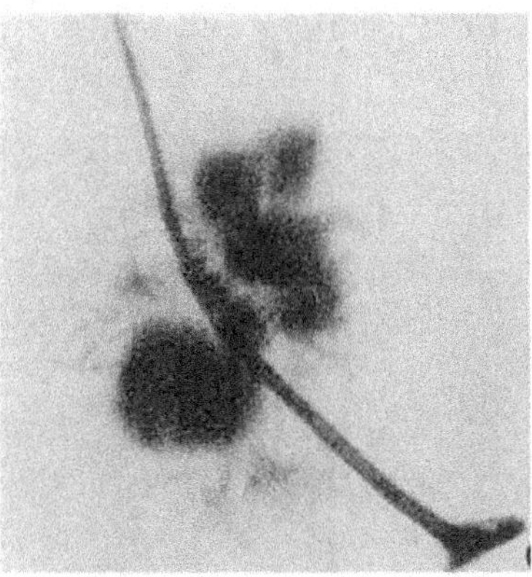

Figure 6 *Crytophialoidea secunda* - Setform conodiophore with conidiogenous cells and conidia, Bar = 60mm.

On wood, Papikon dalu, Eastern Ghats, Leg. G. Suresh Kumar, K. Sharath Babu, 13. 07. 2004; HCIO 45,484; OUFH 483.

Physalidium elegans Mosca, 1965, *Allionia*, 11: 73-79. (Fig. 7)

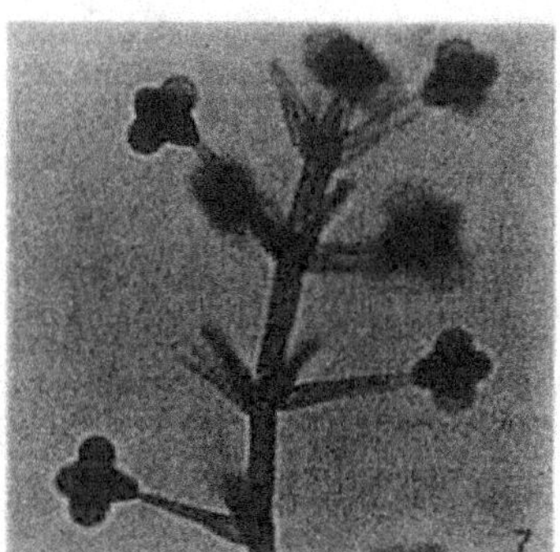

Figure 7 *Physalidium elegans* - Conidiophore with conidiogenous cells and conidia, Bar = 20mm.

Colonies effuse, thin, arachnoid, grayish olive. Conidiophores macronematous, mononematous, up to 542 μm long, 4 μm thick near the base, tapering to 2 μm near the apex, branches short, arranged in verticals, subulate, sometimes swollen at the base, pale brown. Conidiogenous cells integrated, terminal on stipe and branches, determinate or occasionally percurrent, subulate or cylindrical, 11-20 X 2-3 μm. Conidia solitary, smooth, having an obovoid or ellipsoidal central cell, mid to dark brown, 7-10 X 5-6 μm, and two smaller lateral cells, hemispherical or sub-spherical, hyaline or very pale brown, 3-4 μm.

On dead twigs, Papikondalu, Eastern Ghats, Leg. G. Suresh Kumar, K. Sharath Babu, 13. 07. 2004; HCIO 45,415; OUFH 455.

Actinocladium rhodosporum Ehrenberg ex Pers., C. H. Persoon, 1822, *Mycol. eur.*, 1: 32. (Fig. 8)

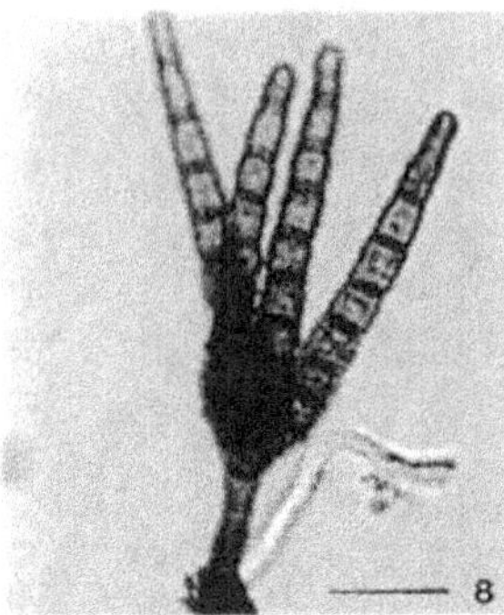

Figure 8 *Actinocladium rhodpsporum*, Bar = 20mm.

Colonies effuse, dark brown to black, hairy. Mycelium partly superficial, partly immersed, composed of branched, septate, smooth, pale to mid brown, 2-4 μm thick hyphae. Conidiophores arise terminally and laterally on the hyphae, single or in small groups, macronematous, mononematous, simple, brown to dark brown, smooth, septate, straight or flexuous, dark brown below, paler above, up to 150μm long, 7-9 μm thick at the base, tapering towards apex, 3-4 μm broad at apex. Conidiogenous cells monoblastic, integrated, terminal, determinate or percurrent, cylindrical, pale brown. Conidia solitary, dry, acrogenous, brown, 1-3 celled, 7-9 μm long with stalk, 3-4 μm wide at the base, surmounted by a group of cells from which 3 or occasionally 2-septate arms radiate upwards, arms up to 70 μm long 5-7 μm thick at base, 3-4 μm thick at apex and the conidium up to 14 μm wide at the broadest.

On dead wood, Papikondalu, Eastern Ghats, Leg. G. Suresh Kumar, K. Sharath Babu, 13. 07. 2004; OUFH 503.

Virgaria nigra (Link) Nees ex S. F. Gray, 1821, Nat. Arr. Br. Pl., 1: 553. (Fig. 9)

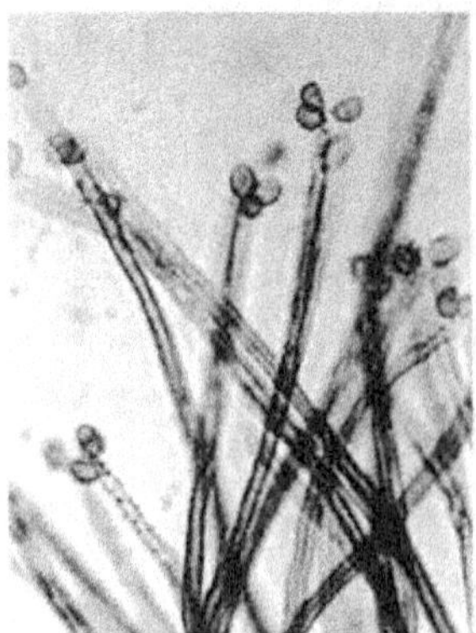

Figure 9 *Virgaria nigra* - Conidiosphore with denticulate conidiogenous cells and reniform conidia, Bar = 20mm.

Colonies effuse, dark olive to brown or almost black, often thick and felt-like. Mycelium partly superficial, partly immersed. Stroma none. Setae and hyphopodia absent. Conidiophores macronematous, mononematous, erect or ascending, extensively branched, straight or flexuous, pale to mid brown, smooth. Conidiophores 60-250 μm long, 2-3 μm thick. Conidiogenous cells polyblastic, integrated, terminal on branches, sympodial, cylindrical, denticulate, denticles cylindrical, numerous. Conidia solitary, dry, acropleurogenous, simple, reniform, often obliquely attenuated at the base, pale to mid brown, smooth, 0-septate, 4-6 X 2-4 μm.

On dead wood and bark, Papikondalu, Eastern Ghats, Leg. . G. Suresh Kumar, K. Sharath Babu, 13. 07. 2004; HCIO 44,776; OUFH 250.

Speiropsis pedatospora Tubaki, 1958, *J. Hattori bot. Lab.*, 20: 171-173. (Fig. 10)

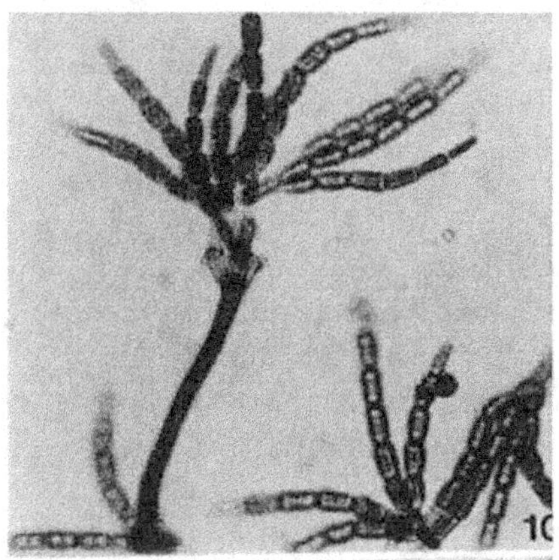

Figure 10 *Speiropsis pedatospora* - Conidiophore with coniodiogenous cells and conidia, Bar = 40mm.

Colonies effuse, brown or blackish brown, hairy or velvety. Mycelium partly superficial, partly immersed. Stroma none. Setae and hyphopodia absent. Conidiophores macronematous, mononematous, with short, pale branches at the apex, stipe straight or curved, mid to dark brown, smooth, septate, conidiophores up to 100 μm long, 4-6 μm thick. Conidiogenous cells discrete or integrated, formed at the ends of the branches, polyblastic, determinate, clavate, cylindrical or ellipsoidal, usually with 2 or 3 terminal protuberances or denticles on which branched chains of conidia are formed. Conidia cylindrical or cuneiform, pale to mid pale brown, 10-14 X 4-7 μm, they remain attached to one another for a long time by narrow connectives forming branched structures up to 90 μm long.

On dead leaves, Papikondalu, Eastern Ghats, Leg. G. Suresh Kumar, K. Sharath Babu, 13. 07. 2004; HCIO 45,500; OUFH 491.

Bispora betulina (Corda) Hughes, 1958, *Can. J. Bot.*, 36: 740. (Fig. 11)

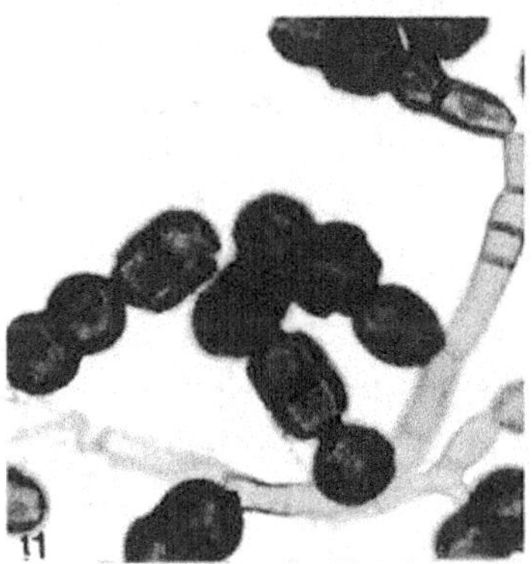

Figure 11 *Bispora betulina* - Conidiophores, conidiogenous cells and catenate conidia, Bar = 15mm.

Colonies punctiform or effuse, usually fuscous or black. Mycelium partly superficial or immersed. Stroma none. Setae and hyphopodia absent. Conidiophores semi-macronematous, mononematous, usually inconspicuous and short on natural substrate, 5-15 X 3-4 μm, much longer in culture, straight or flexuous, unbranched or rarely branched, pale brown or brown, smooth. Conidiogenous cells monoblastic, integrated, terminal, determinate, cylindrical. Conidia catenate, acrogenous, simple, doliiform or cylindrical rounded at the ends, brown or dark brown with very dark band or bands, 2-septate, 9-14 X 4-7 μm.

On dead wood, Papikondalu, Eastern Ghats, Leg. G. Suresh Kumar, K. Sharath Babu, 13. 07. 2004; HCIO 45,425; OUFH 431.

Tetraploa aristata Berkeley and Broome, 1850, *Ann. Mag. Nat. Hist.*, 2, 5: 459. (Fig. 12)

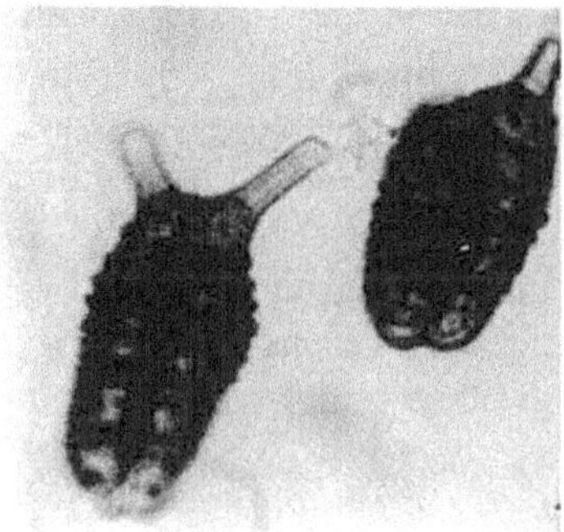

Figure 12 *Tetraploa aristata* - conidia with appendages, Bar = 18mm.

Colonies effuse, brown or dark grayish brown. Mycelium superficial. Conidiophores micronematous, branched and anastomosing to form a network, flexuous, hyaline to pale yellowish-brown, often verruculose. Conidiogenous cells monoblastic or occasionally polyblastic, integrated, intercalary, determinate, cylindrical. Conidia mostly with 4 cells to each column, 25-39 X 14-29 μm, with septate appendages 12-80 μm long, 5-8 μm thick at the base, 2-4 μm at the apex.

On dead wood, Papikondalu, Eastern Ghats, Leg. G. Suresh Kumar, K. Sharath Babu, 13. 07. 2004; HCIO 45,466; OUFH 453.

Spegazzinia lobulata Thrower, 1954, *Aust. J. Bot.*, 2: 362-363. (Fig. 13)

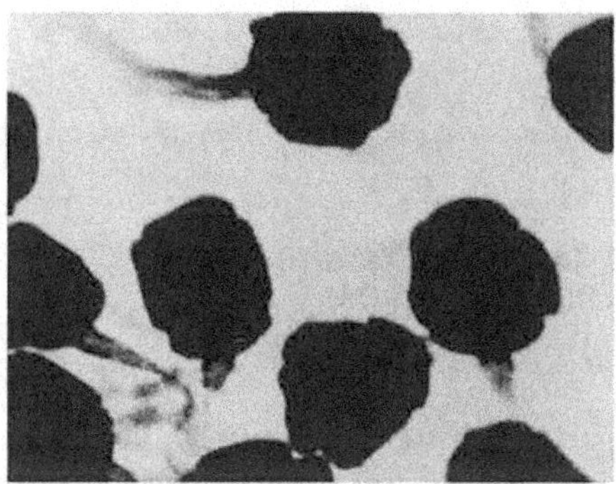

Figure 13 *Spegazzinia lobulata* - lobulate conidia with conidiogenous cells and conidiophores, Bar = 20mm.

Colonies discrete, arbicular or effuse, dark blackish brown to black. Mycelium superficial, hyphae branched and anastomosing to form a close network. Stroma none. Setae and hyphopodia absent. Conidiophores basauxic, macronematous, mononematous, arising usually singly from sub-spherical, ampulliform, cupulate or doliiform conidiophore mother cells, unbranched, straight or flexuous, narrow, sub hyaline to brown, smooth or verrucose, conidiophores up to 140 μm long, 2-3 μm thick at the base, up to 4 μm at the apex. Conidiogenous cells monoblastic, integrated, terminal, narrow, cylindrical. Conidia solitary, dry, acrogenous, flattened in one plane, pale to dark brown, smooth, characteristically lobulate, 16-25 μm wide including lobes.

On bark, Papikondalu, Eastern Ghats, Leg. G. Suresh Kumar, K. Sharath Babu, 13. 07. 2004; HCIO 45,461; OUFH 450.

Acknowlegememt

The authors thank the DBT and MoENF, New Delhi for financial assistance.

References

Bilgrami, K.S., Jamaluddin and Rizvi, M.A. 1979. *The Fungi of India. Part-I*, Today and Tomorrows Printers and Publishers, New Delhi, pp. 467.

Bilgrami, K.S., Jamaluddin and Rizvi, M.A. 1981. *The Fungi of India. Part-II*, Host Index and Addenda. Today and Tomorrows Printers and Publishers, New Delhi, pp. 268.

Bilgrami, K.S., Jamaluddin and Rizvi, M.A. 1991. *The Fungi of India. 2nd ed*, List and References. Today and Tomorrows Printers and Publishers, New Delhi, pp. 798.

Jamaluddin, Goswami, M.G. and Ohja, B.M. 2003. *Fungi of India. (1989-2001).* M/S Scintific Publ. India.

Sarbhoy, A.K., Varshney, J.L. and Agarwal, D.K. 1986. *Fungi of India (1977-81).* Assoc. Publ. Co. New Delhi, pp. 274.

Sarbhoy, A.K., Agarwal, D.K. and Varshney, J.L. 1996. *Fungi of India (1982-92).* CBS Publ. Distri., New Delhi, pp. 350.

Subramanian, C.V., 1971. *Hyphomycetes.* ICAR Publ., New Delhi. pp. 930.

Suresh Kumar, G., Sharath Babu, K., Kunwar, I.K., Manoharachary, C. and Prasad, V. 2004. New Records of fungi from India. *J. Mycol. Pl. Pathol.* (In press).

5 | Diversity and Distribution of VA Mycorrhizal Fungi in Thirteen Wheat Cultivars

Deepak Vyas

Lab of Microbial Technology and plant Pathology Department of Botany, Dr. Hari Singh Gour University, Sagar (M.P.)

ABSTRACT

Thirteen different wheat cultivars selected for the present study are commonly grown in all the four test sites. These included HI-1502, HI-1500, GW-1172, HD-2781, HW-2004, MACS-3208, MPO-1126, A-9-30-1, HI-1501, HD-4694, HD-4672, C-306 and Sujata. Out of 13 wheat cultivars, two cultivars, MACS-3208 and A-9-30-1 at site III and MPO-1126 and A-9-30-1 at site IV were however, not grown during the present study period. Maximum number of VAM species was found at site I (80 species). Out of these *Glomus* was most dominant (61.25%) followed by *Acaulospora* (22.5%), *Scutellospora* (6.25%), *Sclerocystis* (5%), *Gigaspora* (3.75%) and *Entrophospora* (1.25%) respectively. In site II, 73 species of VAM fungi were present with *Glomus* (60.3%), followed by *Acaulospora* (24.7%), *Scutellospora* (5.4%), *Gigaspora* and *Sclerocystis* (4.1%) and *Entrophospora* (1.4%) respectively. In site III, 78 species of VAM fungi occurred with *Glomus* (60.26%), followed by *Acaulospora* (23.07%), *Scutellospora* (6.41%), *Sclerocystis* (5.13%), *Gigaspora* (3.85%) and *Entrophospora* (1.28%) respectively. In site IV, 75 species of VAM fungi were present,out of these Glomus was highest (60%), followed by Acaulospora (22.7%), Scutellospora (6.7%), Sclerocystis (5.3%), Gigaspora (4%) and Entrophospora (1.3%) respectively. These results suggest that selected study sites are rich in VAM frequency and diversity.VAM colonization favours spore population and number of vesicles. VAM colonization was higher in wheat cultivars at site I and IV, while other sites (II and III) showed comparatively less VAM colonization.

Results of the similarity matrix of four different study sites for the occurrence of VAM fungi suggests that the site I and III was closer as they showed highest (96.20%) similarity followed by site I and IV (95.48%), site I & II and site III & IV (92.81%), site II and III (91.39%) and least similarity (90.54%) was found between site II and IV.

KEY WORDS : Mycorrhiza, wheat cultivars

Mycorrhiza, a symbiosis between plants and mutualistic soil fungi, is undoubtedly of extraordinary importance in plant production, plant and soil ecology, and plays a key role in what is generally now-a-days called "sustainable agriculture" (Bethenfalvay and Linderman 1992; Gianinazzi and Schüepp, 1994). The *International Bank of Glomales* (IBG, former *Banque Européenne des Glomales* BEG, 1993) points out that: "The study of plants without their mycorrhizas is the study of artifacts" and "The majority of plants, strictly speaking, do not have roots; they have mycorrhizas." The term "arbuscular mycorrhiza" expresses morphologically the symbiosis of vascular plants with fungi of the order *Glomales* (*Zygomycotina*) which represents the norm of about 90 % of the families of all phyla of land plants (Giovannetti & Sbrana, 1998) including ferns and some mosses (Smith & Read, 1997). These obligate symbiotic fungi persist in soils as thick-walled spores of diameters from 10 to 1000 μm, which are rich in storage lipids and nuclei (Sylvia, 1998). The infection – or better, because of its mutualistic nature, *colonization* - of plant roots arise from spores and other vegetative propagules, which germinate in the rhizosphere. Stimulated by plant exudates, the formation of appressoria, a thickening of hyphae in the epidermis, manifests the recognition between the two symbionts. Subsequently, the fungal hyphae are allowed to penetrate the plant root. Having once ramified, the hyphae form characteristic dichotomously, highly branched arbuscules inside of root cortical cells. The normal *arum*-type is much more common than the paris-type distinguished by extensively coiled intracellular hyphae. The arbuscules emerge continuously but are relatively (< 15 d) short living. They are the only and, unfortunately, merely ephemeral, taxonomically decisive feature of AMF. This interspecific interface consists of the modified plant plasma membrane or periarbuscular membrane and the fungal cell wall. The increased surface area serves both symbiotic partners for an extensive bi-directional metabolic exchange. However, this is proved more for the transfer of P from fungus to plant than for carbohydrates in the opposite direction in which also normal internal hyphae seemed to be involved (Smith & Read, 1997).

AMF have been described as Keystone mutualists in terrestrial ecosystems (O'Neil et al 1991). AMF are responsive to many perturbations that act at ecosystem level, such as global change. (Fitter *et al* 2000, Rilling *et al* 2002) agriculture management practices (Jansa *et al* 2003; Oehl *et al* 2003). AM fungi are known to be distributed throughout the world (Harley, 1969) in environment ranging from tropical rainforest to arctic tundra (Janos, 1980). They are present in the soil, in form of chlamydosopres, zygospores, azygospores and have been recovered from the soils of a variety of habitats e.g. nutrient deficient soils, sand dunes and desserts, industrial wastes, sodic soils, polluted sites, sewage, eroded sites and others like forests, open wood lands, scrub, savanna, heaths, grasslands, anthracite and bituminous coal waste (Chandra, 1992). There are several reports of VAM fungi from diverse habitates also from India (Godse *et al.* 1976; Bagyaraj *et al.*, 1979; Parvathi *et al.*, 1984; Kehri et al. 1987; Kehri and Chandra, 1990; Dwivedi, 2003: Dwivedi et al 2003: Singh et al 2003).

During the past decade it has been established that VAM influence soil fertility and thus the growth and development of plant and therefore, these could be an alternative to rising agricultural and fertilizer costs (Mukerji, 1995). Some plants are highly mycorrhiza dependent, such as onion, citrus and strawberry while others are less dependent such as wheat, oat and potato (Chandra, 1992).

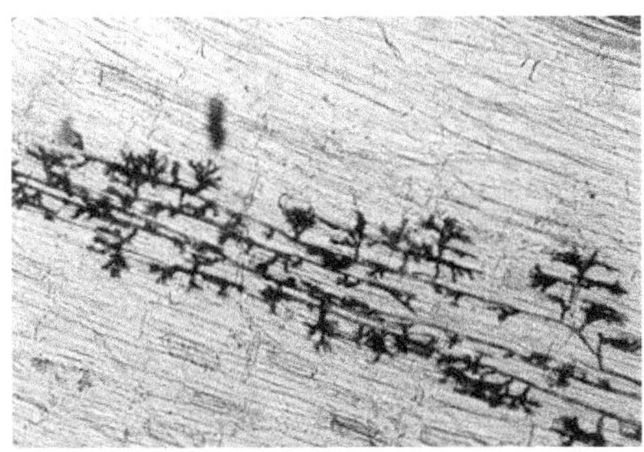

Arum type arbuscule *in the cortex of* Triticum astivum *L. wheat*

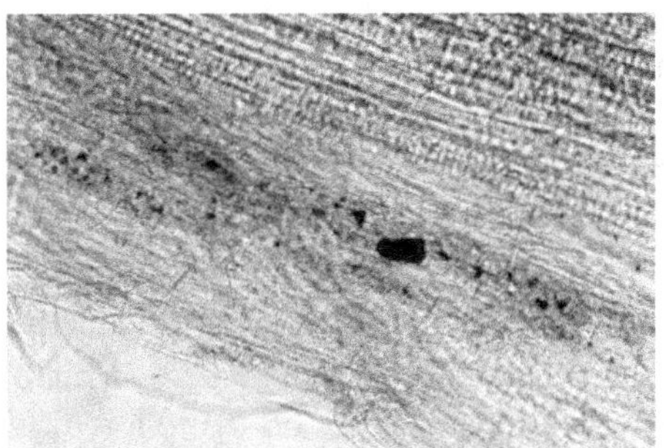

Paris type arbuscule in the cortex of ***Triticum astivum*** L. wheat

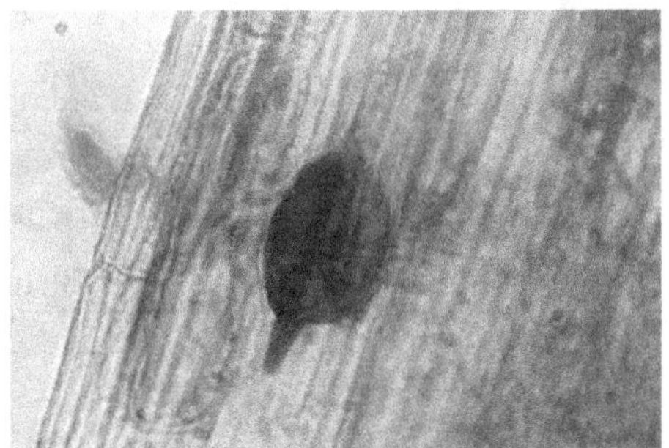

Vesicle in the cortex of ***Triticum astivum*** L. wheat

Root Colonization by VA - Mycorrhiza

Considering the above-mentioned fact it was considered to be most desirable to investigate the VAM association with wheat crop in Sagar region, as it is the principal producer of the wheat in the State of Madhya Pradesh, contributing more than 50% of the total cropped area of the region. Unfortunately, beneficial effect of mycorrhizas, have not been explored earlier in this region. Therefore, the present study was undertaken to find out qualitative and quantitative distribution of VA mycorrhizal fungi, occurring naturally in the agricultural field of wheat at four different study sites of this region.

Material and Methods

The present study represents an attempt to establish the qualitative and quantitative distribution of VAM fungal species in rhizosphere soil of wheat. Four different study sites were taken for the present investigation viz. I-Sagar, II-Khurai, III-Banda and IV-Deori.

Isolation, estimation and identification of VAM Fungi

The rhizospheric soil samples of 13 wheat cultivars were collected from four different study sites up to 10-20 cm. depth. The VAM spores were isolated from the each soil samples by wet sieving and decanting techniques of Gerdeman and Nicolson (1963). VAM spores counted under steriozoom dissecting microscope. Photomicrographs were taken with automatic camera Lieca. Species of VAM fungi were identified with the help of key provided by Schenck and Perez (1987), Mehrotra (1997) and species code of VAM fungi were followed after Perez and Schenck (1990).

Measurement of Mycorrhizal infection

The fine roots were stained by using Phillips and Haymen's (1970). The percent mycorrhizal colonization was calculated by following formula.

$$\% \text{ Root Colonization } = \frac{\text{No. of segments colonized with VAM}}{\text{Total No. of segment observed}} \times 100$$

Relative density and similarity index of VAM fungi

VAM fungal species association and their relative density with each wheat cultivar at different study sites were calculated by following formula:

Relative density

$$= \frac{\text{No. of VAM species associated with the host species}}{\text{Total No. of VAM species associated with all the hosts considered}} \times 100$$

The two sites of presence/absence of VAM species were compared at time and like wise all the sites were compared taking paired data. Similarity index was calculated by using the following formula :

$$S = \frac{2C}{A+B} \times 100$$

where, A = No. of species at site I,
B = No. of species at site II,
C = No. of species common to both the sites

Statistical analysis

Statistical analysis of data for comparison of means and analysis of variance (ANOVA) was followed after Gupta and Kapoor (1997).

Results:

(A) Site - I (Sagar)

1. VAM Spore Population

Table 1 shows the VAM spore population in rhizosphere soils of 13 different wheat cultivars. The maximum spore population was found with C-306 (154 spores) and minimum (65 spores) was found with HD-4672. Others were in between them.

2. *Mycorrhizal colonizaton*

The highest percent root colonization was observed with cultivar C-306 (92%) and lowest was observed with HI-1501 (47%). In cultivar GW-1172 and HD-4672, the root colonization could not be detected due to absence of fine roots in their rhizosphere soils. Others were intermediate between higher and lower range of percent colonization

The average number of vesicles ranged from 19 to 39 cm^{-1} root bit in different wheat cultivar except cultivar GW-1172 and HD-4672 where the average number of vesicle could not be detected due to absence of fine roots. The maximum vesicles were present in cm^{-1} root bit with cultivar A-9-30-1 (39 cm^{-1} root bit) and minimum with HD-4694 (19 cm^{-1} root bit). Others were in between them.

3. *Occurrence of VAM Fungal species*

Table 1 shows, the maximum species occurred with C-306 (51 spp.) and minimum occurred with HD-4672 (29 spp.). Others were in between them.

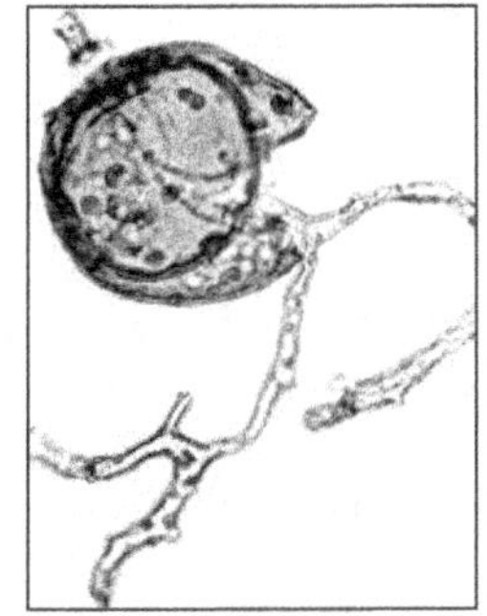

Glomus fecunidisporum
Schenck and Smith

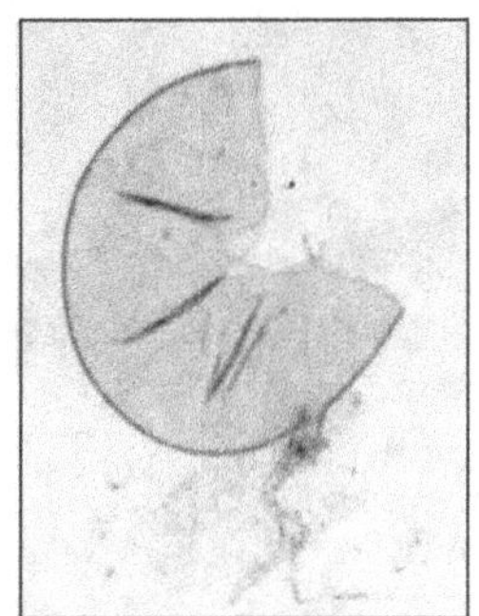

Glomus mosseae
Gerdemann and Trappe

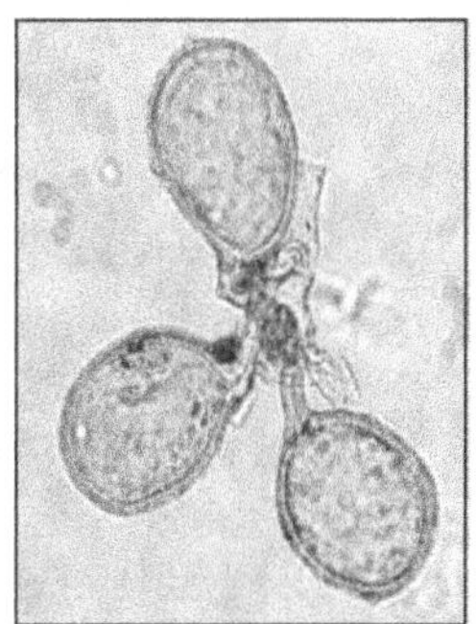

Sclerocystis pachycaulis
Wu and Chen

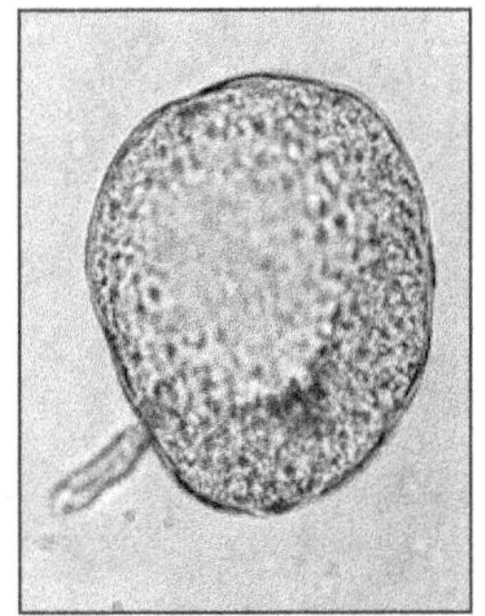

Glomus fecunidisporum
Schenck and Smith

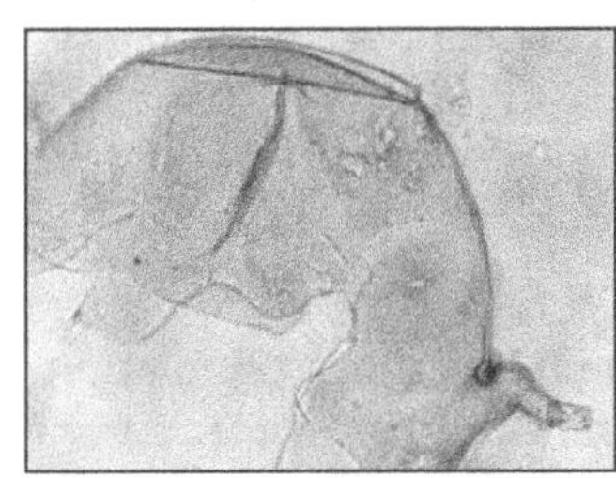

Glomus fecunidisporum
Schenck and Smith

Dominant Mycorrihizal Spores Isolated from Wheat Cultivars

In site I, Glomus was the most dominant VAM fungi in all the cultivars. The maximum species of Glomus occurred (14-33 spp.) followed by Acaullospora (7 to 11 spp.) then Scutellospora (1 to 5 spp.), Sclerocystis (1 to 3 spp.) and Gigaspora (1 to 2 spp.) was found associated with different selected wheat cultivars.

(B) *Site – II (Khurai)*

1. *VAM spore population*

The VAM spore population in rhizosphere soils of 13 wheat cultivars is presented in Table 2. Table shows the maximum spore population with cultivar Sujata (112 spores) and minimum with HD-2781 (54 spores), others were recorded between them.

2. *Mycorrhizal colonization*

The highest percent root colonization was found with cultivar Sujata (72%) and lowest was recorded with HD-4694 (32%), others were found between them.

The average number of vesicles was maximum with cultivar Sujata (32 cm^{-1} root bit) and minimum was recorded with MACS-3208 (18 cm^{-1} root bit), and others were recorded between them.

3. *Occurrence of VAM Fungal species*

The maximum VAM fungi were recorded with the cultivar Sujata (48 spp.) followed by C-306 (46 spp.), then 44 spp. in HI-1502, 33 spp. in HD 4672, 32 spp. in HI 1500, 29 spp. each in GW-1172, HI-1501, MACS–3208 and MPO-1126. 28 spp. in HW 2004, 26 spp. each in HD- 4694 and minimum 23 spp. HD-2781.

In site II also *Glomus* was most dominant fungi. Of *Glomus* (ranges 15-27 species) followed by *Acaulospora* (5 to 14 spp.) then *Scutellspora* (1 to 4 species), *Sclerocystis* (1 to 3 species), *Gigaspora* (1 to 2 species except cultivar HD-2781 and HD-4694) and minimum species was recorded in *Entrophosphora* (one species, which was present in only cultivar HI-1500, GW-1172, MACS-3208, HI-1501, HD-4694 and Sujata) with all the 13 wheat cultivar which is selected for the present study.

(C) *Site – III (Banda)*

1. *VAM spore population*

The maximum spore population was recorded with cultivar C-306 (164 spores) and minimum was recorded with HI-1501 (52 spores), others were recorded between them.

2. *Mycorrhizal colonization*

The highest percent of root colonization was found with cultivar C-306 (91%) followed by Sujata (86%) then HI-1500 (67%), HD-4672 (65%) and lowest with cultivar HD-2781 (37%), others were found between them,

Table 1 Percent of Root Colonization, Average Number of Vessicle, Spore Population, Total Number of VAM Species and Dominant VAM Species Associated with Selected Wheat Cultivars at Site I.

S.No.	Wheat cultivars	Root Colonization (%) ND	Average No. of Vessicle (cm-root bit) ND	Spore Population (50 gm $^{-1}$) Rhizosphere Soil	Total No. of VAM Fungal Species	Dominant VAM Species
1.	HI-1502	86	36	117	41	AGDM, LABS, LCRD, LFSC, LMSS
2.	HI-1500	62	24	98	34	ABRT, ALCN, GABD, LFSC, LHOI
3.	GW-1172	ND	ND	75	30	ADLC, LABS, LHTS, LHOI, LPBS
4.	HD-2781	68	34	103	35	ABRT, LHTS, LMSS
5.	HW-2004	54	21	89	31	ADLC, ALCN, LABS, LHOI
6.	MACS-3208	68	37	108	36	ABRT, LCRD, LFSC, LMSS
7.	MPO-1126	57	26	90	32	ADLT, ASCB, LFSC, LPLD, SPCC
8.	A-9-30-1	76	39	108	35	ADTC, ANCS, LLCT
9.	HI-1501	47	25	86	31	ALCN, ASPN, LHOI, LRDT
10.	HD-4694	72	19	95	32	ARHM, GABD, LHIO, LPBS
11.	HD-4672	ND	ND	65	29	ABRT, LABS, LCRD, LHOI
12.	C-306	92	27	154	51	ABRT, LBTR, LCTC, LPLD
13.	Sujata	88	32	124	49	ADLT, ALCN, LCLR, LHOI

Table 2 Percent of Root Colonization, Average Number of Vessicle, Spore Population, Total Number of VAM Species and Dominant VAM Species Associated with Selected Wheat Cultivars at Site II.

S.No.	Wheat cultivars	Root Colonization (%) ND	Average No. of Vessicle $(cm^{-1}$ root bit)ND	Spore Population $(50\ gm^{-1})$	Total No. of VAM Fungal Rhizosphere Soil	Dominant VAM Species Species
1.	HI-1502	56	31	105	44	ALCN, GABD, LAGR, LCRO
2.	HI-1500	46	28	92	32	LAGR, LFSC, LHOI, LMSS,
3.	GW-1172	43	21	82	29	ALCN, LFSC, LMSS, CCLS,
4.	HD-2781	ND	ND	54	23	LCRD, LFSC, LHOI
5.	HW-2004	37	22	76	28	ALCN, GABD, LCRD
6.	MACS-3208	48	18	87	29	**LFSC, SCCG, CCLS**
7.	MPO-1126	ND	ND	85	29	LAGR, LFSC, LHOI
8.	A-9-30-1	ND	ND	59	26	LCRD, LMSS, LMST
9.	HI-1501	50	19	87	29	ALCN, GABD, CCLS
10.	HD-4694	32	20	62	26	LAGR, LCRD, LFSC, CCLS
11.	HD-4672	52	25	98	33	GABD, LFSC, LHOI, LMSS
12.	C-306	68	30	108	46	ALCN, LCRD, LHOI, LMSS
13.	Sujata	72	32	112	48	GABD, LAGR, LCRD, LHOI

with cultivar HI-1502 (79%), HW 2004 (58%), MPO-1126 (53%), HD-4694 (62%), and Sujata (86%). The percent root colonization could not be detected with the cultivar GW-1172 and HI-1501 due to absence of fine roots in their rhizospheric soils. Cultivar MACS-3208 and A-9-30-1 was not cultivated in the study period.

The average number of vesicle was observed with maximum in cultivar HI-1502 ($39cm^{-1}$ root bit) and minimum with cultivar HD-4672 ($17cm^{-1}$ root bit), others were recorded between them and the number of vessicle with cultivar GW-1172 and HF-1501 could not be detected due to absence of fine root in their rhizospheric soil. The cultivar MACS-3208 and A-9-30-1 was not cultivated (Table 3).

3. *Occurrence of VAM Fungal species*

The maximum number of VAM fungal species occurred with the cultivar C-306 (54 spp.) and minimum with HD-2781 and HI-1501 (25 spp.), others were recorded between them.

Here also, *Glomus* was the most dominant VAM fungi. The maximum species of *Glomus* were found (15 to 31 spp.). This was followed by *Acaulospora* (5 to 12 spp.) then *Scutellospora* (1 to 5 spp.), *Sclerocystis* (1 to 3 spp.), *Gigaspora* (1 to 2 spp.) except cultivar (HD-2781 and HI-1501) and minimum was occured in *Entrophosphora* (1 spp. present in only cultivar HI-1502, HI-1500, MPO-1126 and C-306). The cultivar MACS-3208 and A-9-30-1 was not cultivated.

(D) *Site – IV (Deori)*

1. *VAM spore population*

The maximum VAM spore population was recorded with cultivar C-306 (156 spores) and minimum with HD-2781 (56 spores). Others were found between the range of maximum and minimum

2. *Mycorrhizal colonization*

The percentage of mycorrhizal colonization was recorded maximum with cultivar C-306 (93%), followed by Sujata (88%) then HI-1502 (79%) and minimum with cultivar MACS-3208 (59%), others were recorded between them. The percentage root colonization with cultivars GW-1172, HD-2781, HI-1501 could not be detected.

Maximum vessicles were recorded with cultivar HI-1502 (43 cm^{-1} root bit) followed by HI-1539 (39 cm^{-1} root bit), HD-4672 (35 cm^{-1} root bit), Sujata (34 cm^{-1} rot bit) and minimum with cultivar MACS-3208 (22 cm^{-1} root bit),others were recorded between these values and in the cultivars GW-1172, HD-2781 and HI-1501 there were no vessicles.

3. *Occurrence of VAM Fungal species*

The maximum VAM fungal species occurred with the cultivars C-306 and Sujata (53 spp.) and minimum with cultivar HD-2781 (24 spp.).

Table 3 Percent of Root Colonization, Average Number of Vessicle, Spore Population, Total Number of VAM Species and Dominant VAM Species Associated with Selected Wheat Cultivars at Site III

S.No.	Wheat cultivars	Root Colonization (cm^{-1} root bit)[ND]	Average No. of Vessicle	Spore Population (50 gm $^{-1}$) Rhizosphere Soil	Total No. of VAM Fungal Species	Dominant VAM Species ot bit)[ND]
1.	HI-1502	79	39	109	46	AGDM, LABS, LFSC, LGSP, LHOI
2.	HI-1500	67	29	94	31	AMLL, LABS, G.cardiseptum, LGSP
3.	GW-1172	ND	ND	69	28	AGDM, AMLL, G.cardiseptum, LFSC
4.	HD-2781	37	22	58	25	AGDM, LFSC, LHOI, CCLS
5.	HW-2004	58	23	82	30	ALVS, AMLL, LGSP, LHOI
6.	MACS-3208[NC]	NC	NC	NC	NC	NC
7.	MPO-1126	53	18	72	29	AMLL, LABS, G.cardiseptum, LFSC
8.	A-9-30-1[NC]	NC	NC	NC	NC	NC
9.	HI-1501	ND	ND	52	25	AGDM, LFSC, LHOI, LMCL
10.	HD-4694	62	26	86	31	AGDM, LABS, G.cardiseptum, LGSP
11.	HD-4672	65	17	89	31	ALVS, G.cardiseptum, LFSC, CCLS
12.	C-306	91	34	164	54	AGDM, LABS, LFSC, LGSP
13.	Sujata	87	35	149	50	ALVS, LABS, G.cardiseptum, LFSC,

Like sites I, II and III , Glomus was the most common and dominant species ranges (14 to 34 spp.), followed by Acaulospora (5 to 13 spp.) then Scutellospora (1 to 4 spp.), Sclerocystis (1 to 3 spp.), Gigaspora (1 to 2 spp.) and Entrophospora (one species present with the cultivar HI-1502, HD-2781, HD-4694, HD-4672 and Sujata).

Statistical analysis

Statistical analysis of the experiment is presented in the form of ANOVA (See Table 5). It is clearly evident from the analysis that the site wise variations in the distribution of VAM fungal species are significant at $P < 0.05$ and highly significant for different wheat cultivar at $P < 0.001$.

Relative density of VAM Fungi

Table 6 shows the relative density of VAM fungi at 4 different study sites. In site I, The maximum number of VAM fungal species were found associated with cultivar C-306 (51 spp.) and the density was recorded maximum (10.94) with same cultivar and minimum number of species were occurred with HD-4672 (29 spp.) and density was recorded minimum (6.22) with same cultivar.

In site II, The maximum number of VAM fungal species was found associated with the cultivar Sujata (48 spp.) and relative density was maximum (11.37) recorded with same cultivar whereas, minimum species was recorded with HD-2781 (23 spp.) and relative density was minimum (5.45) with same cultivar.

In Site III, The maximum relative density was found with cultivar C-306 (14.21) and maximum species (54 spp.) was associated with same cultivar however minimum density was recorded with cultivar HD-2781 and HI-1501 (6.57) and species was also recorded minimum (25 spp.) with same cultivar.

In site IV, the maximum relative density (14.20) and maximum species (53 spp.) were recorded with same cultivar C-306 and Sujata; however, the minimum relative density (6.43) and minimum species (24) was recorded with same cultivar HD-2781.

Similarity matrix for the occurrence of VAM Fungi

Table 7 shows the similarity matrix of 4 different sites for the occurrence of VAM fungi. As evident from the result that site I and III was closer as they showed 96.20% similarity followed by site I and IV (95.48%) then site I and II and site III and IV (92.81%), site II and III (91.39%) and least similarity (90.54%) was found between site II and IV.

Discussion

The results of our present study reveals that occurrence of VAM fungi within the root as symbiont and in the vicinity was wide spread irrespective of different habitats as they were

Table 4 Percent of Root Colonization, Average Number of Vessicle, Spore Population, Total Number of VAM Species and Dominant VAM Species Associated with Selected Wheat Cultivars at Site IV

S.No.	Wheat cultivars	Root Colonization (cm^{-1} root bit)[ND]	Average No. of Vessicle	Spore Population (50 gm^{-1}) Rhizosphere Soil	Total No. of VAM Fungal Species	Dominant VAM Species ot bit)[ND]
1.	HI-1502	85	43	117	46	LABS, LCTC, LDMR,LFSC
2.	HI-1500	79	39	102	33	ABRT, LABS, LCRD, LGDM
3.	GW-1172	ND	ND	63	26	ABRT, LCTC, LFSC, LGDM
4.	HD-2781	ND	ND	56	24	AGDM, LCRD, LDMR, LHOI
5.	HW-2004	62	29	82	27	AMLL, LCRD, LFSC, LPST
6.	MACS-3208	59	22	76	27	ABRT, LABS, LDMR, LFSC
7.	MPO-1126[NC]	NC	NC	NC	NC	NC
8.	A-9-30-1[NC]	NC	NC	NC	NC	NC
9.	HI-1501	ND	ND	93	28	AGDM, LCTC, LDMR, LHOI, LPST
10.	HD-4694	65	32	89	27	AMLL, LCRD, LFSC, LHOI
11.	HD-4672	75	35	98	29	ABRT, AMLL, LABS, LFSC
12.	C-306	93	28	156	53	AGDM, LABS, NCTC,LDMR
13.	Sujata	88	34	147	53	ABRT, LCTC, LCRD, LFSC, LPST

Table 5 : Comparison of four different sites for the cultivation of selected wheat cultivars with No. of VAM Species Association

S.No	Study sites	Wheat Cultivars												
		HI 1502	HI 1500	GW 1172	HD 2781	HW 2004	MACS 3208[NC]	MPO 1126[NC]	A-9-30-1[NC]	HI	HD 1501	HD 4694	C-306 4672	Sujata
		No of VAM Fungal Species												
1.	I	41	34	30	35	31	36	32	35	31	32	29	51	49
2.	II	44	32	29	23	28	29	29	26	29	26	33	46	48
3.	III	46	31	28	25	30	X[NC]	29	Z[NC]	25	31	31	54	50
4.	IV	46	33	26	24	27	27	Y[NC]	A[NC]	28	27	29	53	53

ANOVA for occurrence of VAM fungi between wheat cultivors and different sites

Source of Variation	Degree of Freedom	Sum of Square	Mean Square	F-Ratio
Sites	3	108.88	36.29	3.09*
Wheat cultivars	12	3483.6	290.3	24.76**
Residuals (error)	32	375.07	11.72	
Total	47	3967.55		

* Significant at $P<0.05$
**Significant at $P<0.001$

found in all the test sites. In the soils, VAM fungi were found in the form of chlamydospores, zygospores, azygospores and external mycelium. The results suggest that the percent colonization, average number of vessicle, spore population, total VAM fungal species and dominant VAM species differ with different cultivars of wheat in different test sites. The VAM root infection is a dynamic process, which is influenced by edaphic factor such as seasons, VAM strains, soil temperature, soil pH, host cultivar susceptibility to VAM colonization and feeder root condition at tissue of sampling. The quantity and type of VAM propagules also affected the dynamics of root infection, which were also increased by increasing the age of plant (Chandra and Jamaluddin, 1999). However, Saif and Khan (1975) found increased root colonization during the period of maximum vegitative growth and tillering of wheat plants (February-April in W. Pakistan), this being directly influenced by the number of spores in the soil. However, Hayman (1970) found that mycorrhizal colonization in winter wheat was sparse during spring months, but increase during the summer months to a peak at harvest (September). While spore number in the soil increased greatly in July and decreased in September, wheat had become highly mycorrhizal only after tillering and most growth had occurred. Hetrick and Bloom (1983) also found little VAM colonization of fall-planted hard red winter wheat until flowering in May after the cool soils of Kansas had warmed, then a small amount (<1 to 10%) developed just prior to harvest. Similarly, in the Netherlands, Jakobson and Nielsen (1983) found slow and very low levels (<10%) of VAM colonization in winter cereals until mid-April after which there was gradual increase to maximum infection levels approaching 50% at harvest 3 months later. In sharp contrast, infections in spring cereals were early and rapid reaching about 50%, plateau only 15 days after seedlings emergence, a distinctive but as yet little recognized difference in infection patterns for at least autumn vs. spring planted cereals (Young et al., 1985). Detection of few VAM structures in roots of some breeder – plot cultivars contrasted with appreciable (for winter wheat) internal arbuscules/vessicles and external hyphae found with several others. Interestingly, we observed that wheat cultivars C-306 shows maximum colonization more than 90% in three sites i.e. site I, III and IV and 68% at site II. Spore population of VAM fungi was also found higher with same cultivar (C-306) in all the four sites, wheat cultivar Sujata and HI-1502 shows colonization more than 50% in all the four sites, good spore population was also found. Remaining 10 cultivars were found to be colonized more than 50% at one or two sites, minimum colonization of VAM fungi was observed 32% with HD-4694 at site II. This suggests differences in susceptibility of certain cultivars to colonization by particular VAM fungi, as Azcon and Ocampo (1981) reported for thirteen (13) Spanish wheat cultivars tested against a Glomus mosseae isolate in a 10-week experiment. Screening for best VAM x wheat cultivar association should have more promise on nutrient deficient marginal soils. More than the weather, soil-type, extractable nutrients, cropped field or genetic range in wheat cultivars, soil acidity apparently had stronger influence on wheat species of

VAM spores predominated under our field conditions. Here we conclude that, occurrence of VAM fungi varies with host ranges. Though they are ubiquitous yet they shows specificity in association with host plants. According to Mosse (1981); McGraw and Schenck (1980); Barkdoll and Schenck (1987) all the VAM species did not occur in all the areas and not all the species have same effect on their symbiont. Different species and different isolates within species can have different effect of one plant growth. However, biomass production by most of the plant species varied significantly among treatment with different single VAM taxa.

Acknowledgement

Author is thankful to Head, Department of Botany Dr. H.S.G. University Sagar, (M.P). and UGC for the sanction of research award and financial assistance.

References

Azcon, R. and ***Ocampo, J.A.*** 1981. Factors affecting the vesicular-arbuscular infection and mycorrhizal dependency of thirteen wheat cultivars. New Phytol.. 87: 677-685.

Bagyaraj, D.J., Manjunath, A. and ***Patil, R.B***. 1979. Occurrence of vesicular-arbuscular mycorrhizas in some tropical aquatic plants. Trans. Brit. Mycol. Soc. 72 : 164-167.

Barkdoll, A.W. and Schenck, N.C. 1987. Effect of VA mycorrhizal (VAM) fungi selected for aluminium tolerance on bean yield. In: Mycorrhizae in the next decade. (eds.) D. Sylvia, L. Hung and J. Graham. University of Florida. Gainesville : 18.

BEG (March 2002) http://www.bio.ukc.ac.uk/beg/

Bethenfalvay, G.J., and ***Lindermann, R.G.*** 1992. Mycorrhizae in sustainable agriculture. Madison, WI: ASA

Chandra, K.K. and ***Jamaluddin.*** 1999. Distribution of vesicular-arbuscular mycorrhizal fungi in coalmine overburden dumps. Indian Phytopath. 52 : 254-258.

Chandra, S. 1992. VA-Mycorrhiza-Dimensions of its application. Indian Phytopath. 45 : 391-406.

Dwivedi, O.P 2003. Studies on soil microorganisms with special reference to their to vesicular-arbuscular mycorrhizal (vam) fungal association with wheat crop of sagar region , Ph.D Thesis, Dr. hari Singh Gour Viswavidhalaya. Sagar, (M.P)

Dwivedi, O.P, Vyas, D and ***Vyas, K.M.*** 2003. Impact of VAM and bacterial inoculants on wheat genotype C-306 Indian Phytopath. 56:467-469.

Fitter, A.H., Heinemeyer, A. and ***Staddon, P.L.*** 2000. The impact of elevated CO_2 and global climate change on arbuscular mycorrhizas: a mycocentric approach. New Phytol. 147:179-187.

Gerdemann, J.W. and ***Nicolson, T.N. 1963.*** Spores of mycorrhizal Endogon sp. extracted from soil by wet sieving and decanting. Trans. Br. Mycol. Soc. 46 : 235-244.

Gianinazzi, S and ***Schuepp, H.*** 1994. Impact of arbuscular mtcorihizae on sustainable agriculture on natural ecosystems Basel : Birhauser

Godse, D.B., Madhusudan, T., Bagyaraj, D.J. and ***Patil, R.B.*** 1976. Occurrence of vesicular-arbuscular mycorrhizas on crop plants in Karnataka. Trans. Brit. Mycol. Soc. 67 : 169-171.

Gupta, S.C. and ***Kapoor, V.K.*** 1997. Fundamentals of applied statistics (III ed) Sultan Chand and Sons. New Delhi, 6.1-6.113.

Harley, J.L. 1969. The Biology of Mycorrhiza. Leonard Hill. London.

Hayman, D.S. 1970. Endogone spore numbers in soil and vesicular-arbuscular mycorrhiza in wheat as influenced by season and soil treatment. Trans. Br. Mycol. Soc. 54 : 53-63.

Hetrick, B.A.D. and ***Bloom, J.*** 1983. Vesicular-arbuscular mycorrhizal fungi associated with native tall grass prairie and cultivated winter wheat. Can. J. Bot. 61 : 2140-2146.

Jakobsen, I. and ***Nielsen, N.E.*** 1983. Vesicular-arbuscular mycorrhizal infection in cereals and peas at various times and soil depth. New Phytol. 93 : 401-413.

Janos, D.P. 1980. Mycorrihizzae influence tropical succession. Biotropica (Supp.) 12: 56-64.

Jansa, J., Mozafar, A., Kuhn, G., Anken, T., Ruth, and ***Sanders, I.R. 2003. Soil tillage affects the community structure of mycorrihizal fungi in maize roots. Ecol Appl. 13. 1164- 1176.***

Kehri, H.K., Chandra, S. and ***Maheswari,*** S. 1987. Occurrence and intensity of VAM in weeds, ornamentals and cultivated plants at Allahabad and areas adjoining it. Proc. workshop on Mycorrhizae, J.N.U., New Delhi, 273-283.

Kehri, H.K., Chandra, S. and Maheswari, S. 1987. Occurrence and intensity of VAM in weeds, ornamentals and cultivated plants at Allahabad and areas adjoining it. Proc. workshop on Mycorrhizae, J.N.U., New Delhi, 273-283.

Mc-Graw, A.C. and ***Schenck, N.C. 1980.*** Growth stimulation of citrus, ornamental and vegetable crops by select mycorrhizal fungi. Proc. Florida. State Hort. Soc. 93 : 201-205.

Mehrotra, V.S. 1997. Problem associated with morphological taxonomy of AM fungi. Mycorrhiza News. 9: 1-10.

Mosse, B. 1981. Vesicular-arbuscular mycorrhiza Research for Tropical Agriculture Hawaii Institute of Tropical Agriculture and Human Resources, Research Bulletin, 194-196.

O'Neill, E.G., O'Neill, R.V. and ***Norby, R.J.*** 1991. Hierarchy theory as a guide to mycorrhizal research on large-scale problems. Environ. Pollut. 73:271-284.

Oehl, F., Sieverding, E., Ineichen, K., Mader, P., Boller, T. and ***Wiemken, A.*** 2003. Impact of land use intensity on the species diversity of arbuscular mycorrhizal fungi in agrosystems of central Europe. Appl. Environ. Microbiol. 69:2816-2824.

Parvathi, K., Venkateswarlu, K. and ***Rao, A.S.*** 1984. Occurrence of VA-mycorrhizas on different legumes in laterite soil. Curr. Sci., 53: 1254-1255.

Perez, Y. and ***Schenck, N.C.*** 1990. A Unique code for each species of VA- mycorrhizal fungi. Mycologia. 82 : 256-260

Phillips, J.M. and ***Hayman, D.S.*** 1970. Improved Procedures for clearing roots and staining parasitic and vesicular-arbuscular mycorrhizal fungi for rapid assessment of infection. Trans. Br. Mycol. Soc. 55 : 158-161.

Rillig, M.C., Treseder, K.K. and ***Allen, M.F.*** *2002. Global changes and mycorrhizal fungi. In: Mycorrhizal Ecology (Eds Vander Heijden, M.G.A. and I.Sanders) Ecological studies, Vol. 157. Springer Verlag, Berlin, pp.135-160.*

Saif, S.R. and Khan, A.G. *1975. The influence of season and stage of development of plant on endogone mycorrhiza of field grown wheat. Can. J. Microbiol. 21 : 1020-1024.*

Schenck, N.C. and Perez Y. *1987. Manual for identification of VAM fungi.* Synergistic Pub. Gainesville. Fl., U.S.A.

Smith, S and ***Read, D.J.*** 1997. Mycorrhizal Symbiosis. London: Academic Press

Sylvia, D. M. 1998. Mycorihizzal symbioses. In Sylvia, D. M, Hartel, Fuhrmann J., Zuberer, D. (eds) Principles and applications of soil microbiology. Upper Saddle. NJ Prentice Hall. 408-426.

Young, J.L., Davis, E.A. and ***Rose, S.L.*** 1985. Endomycorrhizal fungi in breader wheats and triticale cultivars field grown on fertile soil. Agron. J. 77 : 219-224.

Young, J.L., Davis, E.A. and ***Rose, S.L.*** *1985. Endomycorrhizal fungi in breader wheats and triticale cultivars field grown on fertile soil. Agron. J. 77 : 219-224.*

6

Taxonomy of AM Fungi in Association with Apocynaceae and Asclepiadaceae

S. Swarupa Rani and C. Manoharachary

Applied Mycology and Molecular Plant Pathology Laboratory Department of Botany, Osmania University, Hyderabad - 500 007

ABSTRACT

Arbuscular mycorrhizal (AM) fungal association of ornamental and medicinal plants belonging to the families Apocynaceae and Asclepiadaceae from Hyderabad, Warangal and Rajahmundry (East Godavari dt.) were examined. Spores of 48 AM fungal species belonging to *Acualospora, Archaeospora, Entrophospora, Gigaspora, Glomus and Scutellospora* were isolated. The descriptions of 20 frequently occurring AM fungi is given. They were placed under the recently evolved monophyletic classification. This includes Glomeromycota, Glomeromycetes, Glomerales, Paraglomerales, Diversisporales and Archaeosporales.

KEY WORDS : AM Fungi, Apocynaceae and Asclepiadaceae, Glomeromycota

Members of Apocynaceae and Asclepiadaceae are well known for their medicinal values and ornamental beauties. These plants have not been explored much for Arbuscular Mycorrhizal (AM) fungal association. Information available on the taxonomy of AM fungal association of the above said families is still fragmentary and far from adequate. Thus the present work was aimed at making a significant contribution in understanding AM fungal association, especially the identification up to species level which was done in the light of the present scenario of the changes in glomalean fungi. Taxonomy of Glomalean fungi continues to change as a result of new information derived from all scales (molecular, organismal, morphological, ecological).

Until recently, the order Endogonales (Zygomycotina) consisted of only one family the Endogonaceae (Benjamin, 1979; Morton, 1988) having several genera viz. *Acualospora* Gerdemann and Trappe emend Berch, *Entrophospora* Ames and Schneider, *Gigaspora* Gerdemann and Trappe, *Glomus* Tulasne and Tulasne, *Sclerocystis* Berkeley and Broome and *Scutellospora* Walker and Sanders whose members formed an arbuscular mutualistic symbiosis with many terrestrial plant families (Trappe, 1977). One genus *Endogone* Link Fr. whose members were saprobic (Gerdemann and Trappe, 1974) or formed putative ectomycorrhizal associations (Chu-Chou and Grace, 1979; Fassi et al, 1969). Pirozynski and Dalpe (1989), described a new family Glomaceae containing two genera, *Glomus* and *Sclerocystis*. Morton and Benny (1990) emended the family Glomaceae and erected a new order Glomales and two new families, Acaulosporaceae and Gigasporaceae.

Recently, Schubler et al (2001) based on comprehensive SSU rRNA analysis and on the basis of natural relationship of arbuscular mycorrhizal fungi and the related fungi, recognized a new fungal phylum Glomeromycota with a single class Glomeromycetes (Cavalier–Smith, 1998) circumscribed as for the phylum, containing more than 150 described species, some of which are synonyms (Walker and Trappe, 1993; Walker and Vestberg, 1998).

Information obtained from the present study will offer ample scope for further research to understand biodiversity, taxonomy and mycorrhizal dependency of medicinal as well as ornamental plants, which in turn improve the productivity of plants. Arbuscular Mycorrhizal Fungi (AMF) indeed are one of the gifts from "the nature to nurture, nourish and flourish the plant life".

Materials and Methods

The spores were isolated by wet, sieving and decanting technique (Gerdemann and Nicolson, 1963). Microscopic observations were made with a Leitz Microscope by mounting the spores in polyvinyl lactophenol and sealed with DPX (Dinitro Paraxylene). The characteristic features of each AM fungal species such as wall layers, arrangement of spores in sporocarps and surface ornamentation were photographed with a Nikon SLR camera.

The AM fungal species were identified using the relevant literature, Hall (1984), Morton (1988), Raman and Mohan Kumar (1988), Schenck and Perez (1990), Trappe (1982), Walker (1983, 1986, 1987).

Results

In the present investigation, some ornamental and medicinal plants belonging to the families Apocynaceae and Asclepiadaceae were surveyed for AMF spores and their identification was made up to species level. Although a total of 48 AMF were isolated and identified, in the present paper descriptions of only 20 species of frequently occurring AMF is given along with the photographs (Figs. 1-20). List of frequently isolated AMF is presented in Table 1. They are *Acaulospora* (4 species) *Entrophospora* (1 species), *Gigaspora* (2 species), *Glomus* (11 species), *Scutellospora* (1 species) and *Sclerocystis* (1 species).

Table 1 Twenty frequently occurring arbuscular mycorrhizal fungi (AMF) isolated from rhizosphere soil samples of host plants belonging to the families Apocynaceae (Apo.) and Asclepiadaceae (Ascle.)

AMF	*Apo.*	*Ascle.*
Acaulospora bireticulata	+	+
Acaulospora delicata	+	+
Acaulospora mellea	+	+
Acaulospora scrobiculata	+	+
Entrophosphora shenckii	+	-
Gigaspora gigantea	+	+
Gigaspora margarita	+	+
Glomus australe	+	+
Glomus citricola	+	+
Glomus fasciculatum	+	+
Glomus fistulosum	-	+
Glomus macrocarpum	-	+
Glomus mosseae	+	-
Glomus multicaule	+	+
Glomus sinuosum	+	+
Glomus tenebrosum	+	+
Glomus tortuosum	+	+
Glomus warcupii	+	+
Scutellospora pellucida	+	-
Sclerocystis pakistanica	-	+

Acaulospora bireticulata Rothwell & Trappe Mycotaxon 8 : 471-475. 1979.(Fig. 1)

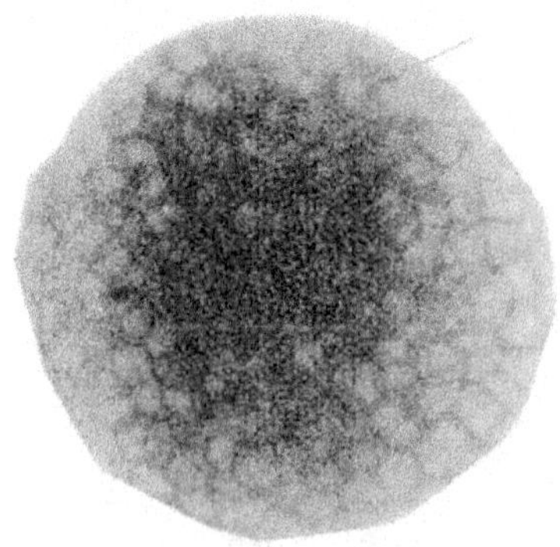

Figure 1

Azygospores formed singly in soil, globose, size ranged from 161-193 × 126-186 μm; colour light brown, spore surface ornamented with polygonal reticulum. The ridges 2 × 1.5-2 μm with sinuos, dark grayish green sides and paler depressed central stratum; Spore walls of 3 layers, each 1 μm thick.

They differ from the original description (Rothwell & Trappe, 1979) in size (150 -155 μm in diam), shape (globose) and in colour of the spores.

Acaulospora delicata Walker, Pfeiffer & Bloss Mycotaxon 25:621-628. 1986 (Fig. 2)

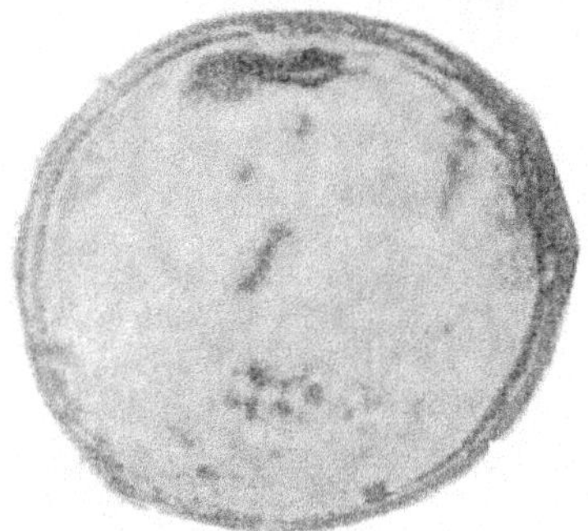

Figure 2

Spores borne singly in the soil, hyaline to pale yellowish, cream, sparkling from the nature of the spore contents. globose, 42-125(-161) x 43-116(-164) μm, spore walls thin, fragile and delicate.

They differ from the original description (Walker, Pfeiffers Bloss, 1986) in size 80 -110 (-140) μm and in shape (sub-globose) of the spores.

Acaulospora mellea Spain & Schenck Mycologia 76:685-699. 1984 (Fig. 3)

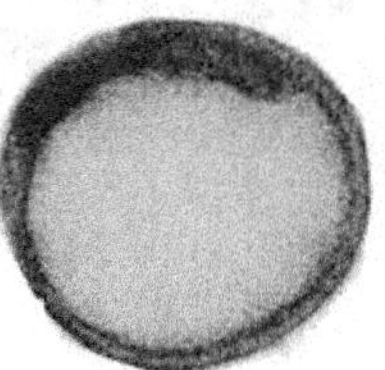

Figure 3

Azygospores formed singly in soil. Honey cloured to yellow-brown, globose to sub-globose 67 - 121 x 62 –116 μm. Spore wall 3 – 4 (16) μm thick, spore contents yellow.

These spores differ from the original description (Schenck, Spain Sieverding and Howeler, 1984) in size (72-) 95-105(- 126) μm diam and in shape (irregular) of the spores.

Acaulospora scrobiculata Trappe Mycotaxon 6:359-388. 1977 (Fig. 4)

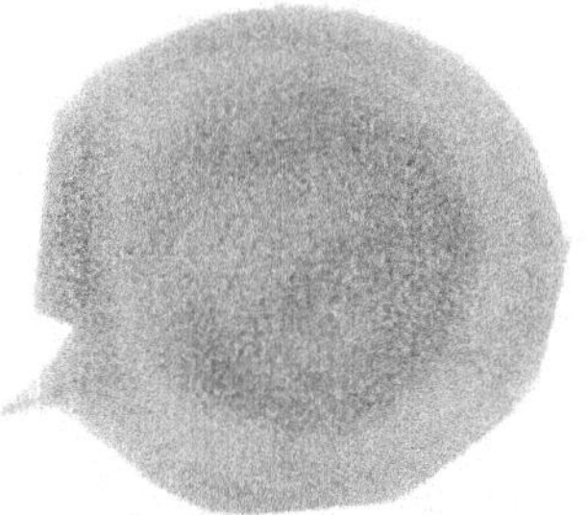

Figure 4

Azygospores forming singly in soil, sessile, globose, 188 X 89 –183 μm, light olive to light brown, spore surface evenly pitted with depressions, separated by ridges. Spore wall 3-9 μm hick, continuous with four layers. Outer wall rigid, pitted, sub-hyaline to yellow with orange tinge. Spore contents small; relatively uniform guttules.

These spores differ from the original description (Trappe 1977) in size (100-240 X 100 – 220 μm) and in colour (sub–hyaline) of the spores.

Entrophospora   schenckii Sieverding & Toro Angewandto Botanik 61:217-223. 1987(Fig. 5)

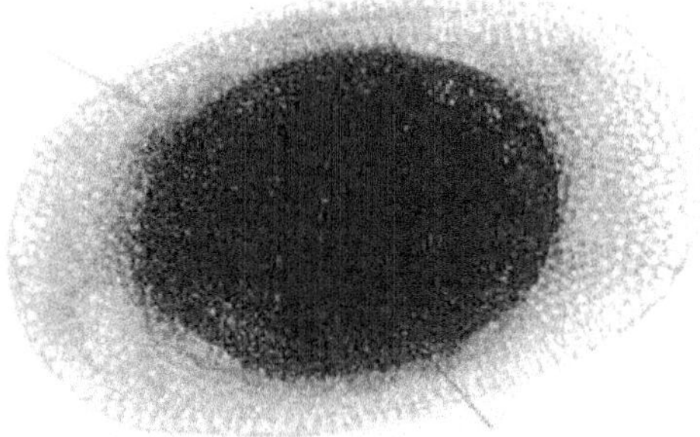

Figure 5

Spore reddish brown to dark brown with orange tinge, globose to sub-globose 243 – 384 X 233 – 291 μm diam; formed singly in soil, soporiferous saccule sub-globose (38-) 5 –116-μm diam; white reticulated, spore wall 4.96 μm thick.

These fungal   spores differ from the original description (Sieverding and Toro, 1987) in size {(37-) 50-60 (-77) μm diam}, in colour (hyaline) and sporiferous saccule diam {37-50- 60(-68) μm}.

Gigaspora gigantea (Nicolson & Gerdemann) Gerdmann & Trappe Mycol. nem. 5:76. 1974 (Fig. 6)

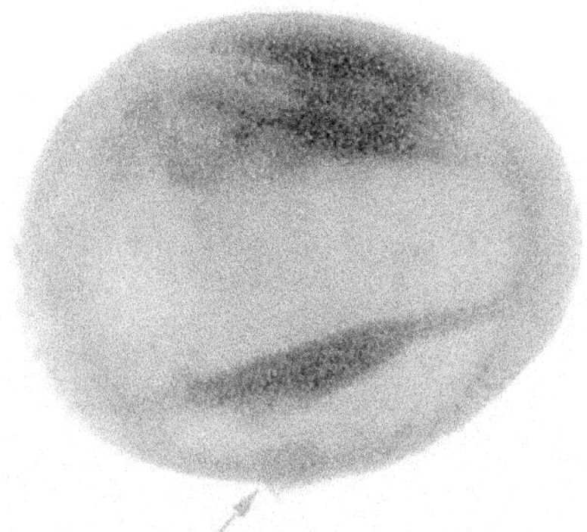

Figure 6

Zygospores formed singly in soil, large dark brown with orange tinge, ellipsoidal, 301 – 525 X 320 – 603 μm, with thin outer wall, composite spore – wall width 3.10 μm; a single bulbous suspensor 20-30 μm diam; with a slender hypha extending from the suspensor to the base of the spore.

These spores differ from the original description (Gerdemann and Trappe, 1974) in size (183–500 X 291-612 μm), shape (spherical) and colour (bright yellow with a greenish tinge) of spores.

Gigaspora margarita Becker & Hall Mycotaxon 4:155-160. 1976 (Fig. 7)

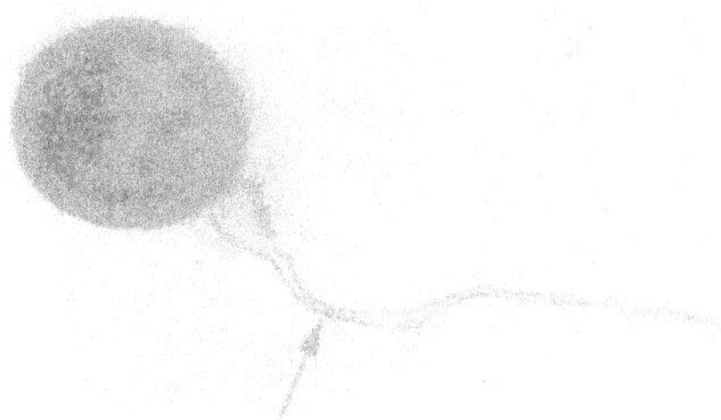

Figure 7

Azygospores formed singly in the soil, globose, 298-420 μm diam. Spore wall smooth and hyaline, 3–5 μm thick; contents of spore's white, composed of many oil droplets which tend to coalesce with age. Spores teminal on the subtending hypha. It is septate below the suspensor like cell. Suspensor like cell 37–49 μm broad, hyaline to light brown, smooth, walls 105 μm thick, thicker at the point of attachment to the spore.

These spores differ from the original description (Becker and Hall, 1976) in size (260-480- μm), and in suspensor like cell breadth (27-58 μm).

Glomus australe (Berkeley) Berch (Fig. 8)

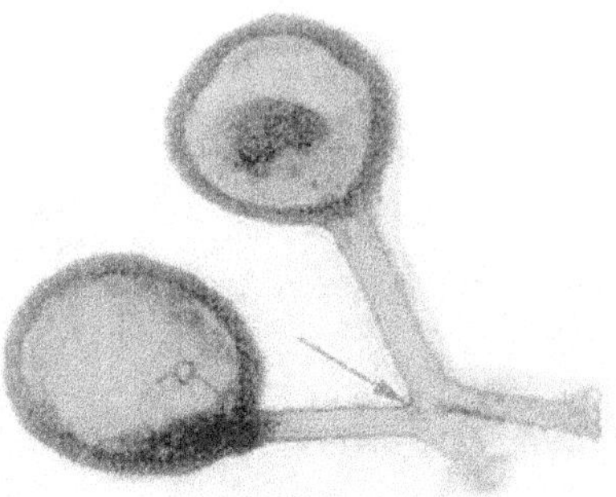

Figure 8

Spores form in loose cluster. Each cluster arises from a central, branched hypha. Spores measure 94–134 μm and have two wall layers. Spore wall double layered; 2.4-4.9 μm thick; pale yellow to light brown. The subtending hypha at the point of attachment to the spore is broad (11-15 μm) and thick walled. The inner wall continuous into the subtending hypha and extends into the central branched hypha of the cluster. The pore in the subtending hypha is open.

These spores slightly differ from the original description (Berch and Fortin, 1983) in spore size (120-180 μm) and in wall thickness (7-I5 μm) of the spores.

Glomus citricola Tang & Zang Acta Botanica Yunnanica 6:295-304. 1984 (Fig. 9)

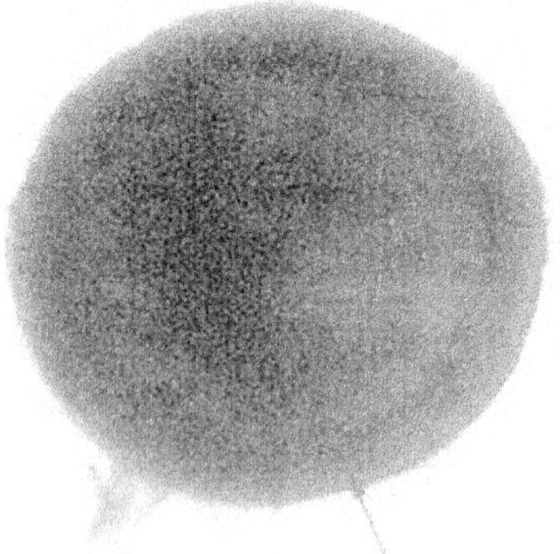

Figure 9

Chlamydospores (large) formed singly in soil. Globose to sub-globose; 69-128 μm width; 72-126 μm length; yellow to brown; the thicker walls are often minutely perforate with thickened inward projections; spore wall 4.96 μm thick; hyphal diam 12-14 μm.

They differ from the original description (Tang and Zang, 1984) in size (35-65 × 60-90 μm) and in spore wall thickness (4-10 μm) of the spores.

Glomus fasciculatum (Thaxter) Gerdemann & Trappe emend, Walker & Koske Mycotaxon 30:253-262. 1987 (Fig. 10)

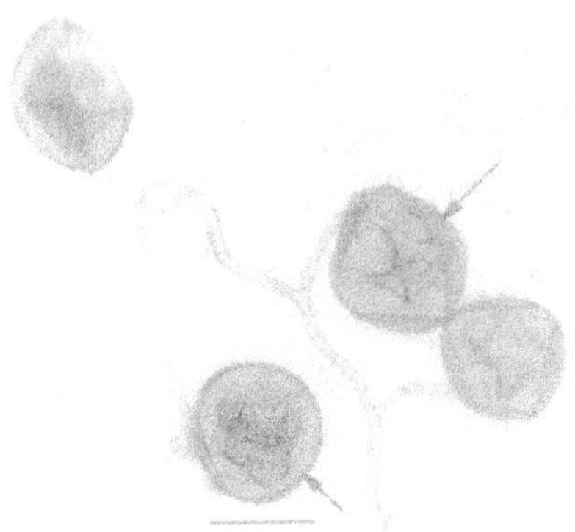

Figure 10

Chlamydospores borne free in soil, in loose sporocarps. Spores globose to irregularly globose, 67–104 μm when globose, 74-163 X 21-128 μm; when sub-globose-obovate; sporewalls highly variable in thickness (3072–12.4 μm) light yellow or yellow brown; hyphal attachments 6-17 μm diam; occluded at maturity.

Glomus fistulosum Skou & Jacobsen Mycotaxon 36:273-282. 1989 (Fig. 11)

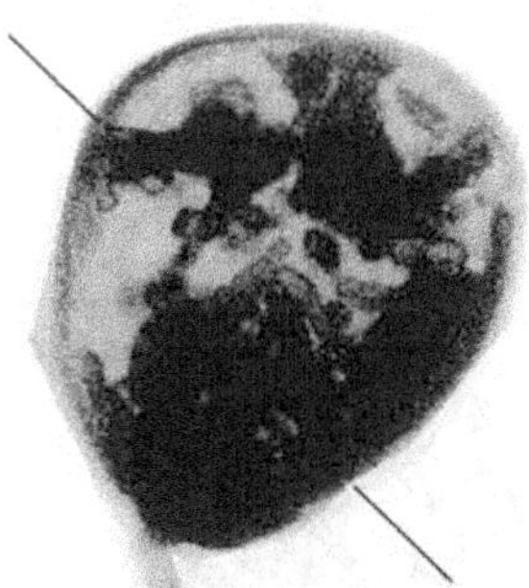

Figure 11

Spores formed singly in soil; pale yellow, broadly ellipsoid, 107-200 X 112-178 μm diam. Spore wall 4.96 μm thick, inner wall with the pronounced fistular laminations; spore contents are colourless and appear as variable-sized globules.

These spores differ slightly from the original description (Skou and Jacobsen, 1989) in size (78–137–200 μm diam) and in spore wall thickness (5.5-8.9–13 μm thick) of the spores.

Glomus macrocarpum Tulasne & Tulasne G.Bot. Ital. Part I. 2:63. 1845 (Fig. 12).

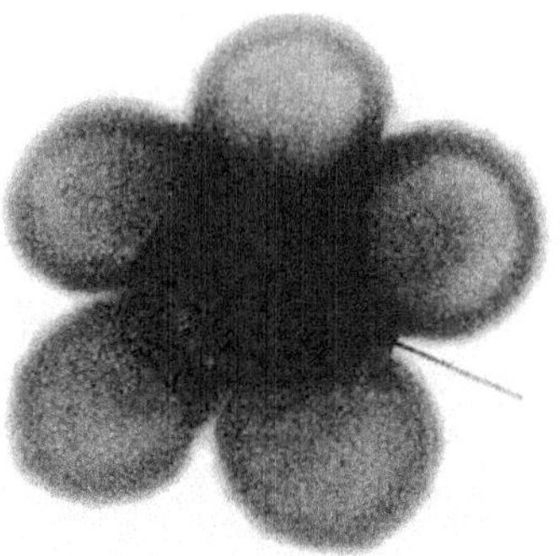

Figure 12

Sporocarps are fragmentary as well as densely compacted. Spores are slightly longer than wide, sub–globose; 81–131×78-141 μm; spore wall composed of two distinct layers; 6.82–12.40 μm thick, yellow brown to brown; Spores taper to the point of attachment of the single persistent hypha. The average diameter of the hypha at this point is 10 μm.

These fungal spores differ from the original description (Tulasne & Tulasne, 1845) in size (90–140 X 110–130 μm), spore wall thickness (7-14 μm thick) and the hyphal diameter (16 μm) of the spores.

Glomus mosseae (Nicolson & Gerdemann) Gerdemann & Trappe Mycol. memo. 5:76. 1974 (Fig. 13)

Chlamydospores light brown, globose to ovoid, 79–186 × 79–82 μm with one funnel shaped base 7-13 μm diam. Spore wall 2 μm thick. (Fig. 13)

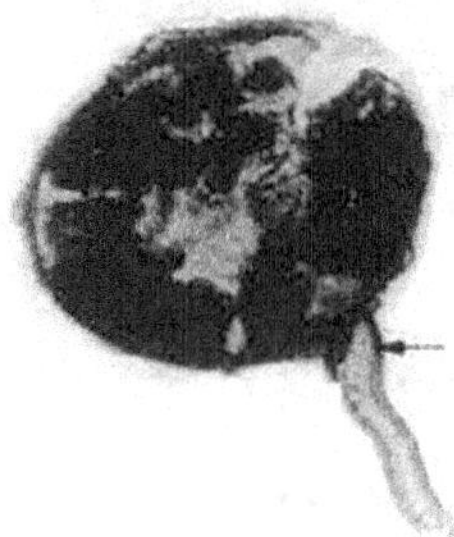

Figure 13

These spores differ from the original description (Gerdemann and Trappe, 1974) in size (105–310 X 110–305 μm) in shape (obovoid, irregular) and funnel shaped base diam (20-30(-50) μm) of the spores.

Glomus multicaule Gerdemann & Bakshi Tr. Brit. Mycol. Soc. 66:340-343. 1976 (Fig. 14)

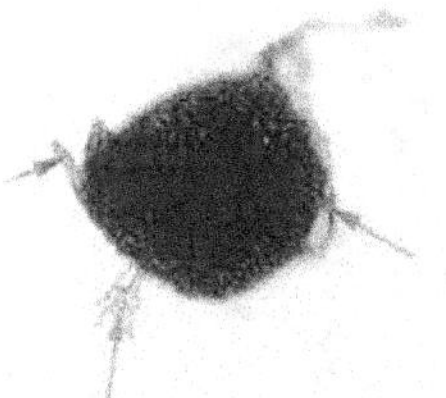

Figure 14

Chlamydospores dark-brown 193–248 X 176–243 μm, broadly ellipsoidal to sub-globose with 3–4 hyphal attachments occurring at opposite ends of spores. Sporewall 19-37 μm, thickest at the points of hyphal attachments.

The spores differ from the original description (Gerdemann and Bakshi, 1976) in size (149–249 X 134–162 μm), shape (triangular) and spore wall thickness (8.6-34 μm) of the spores.

Glomus sinuosum (Gerd & Bakshi) Almeida & Schenck Mycologia 82:703-714. 1990. (Fig. 15)

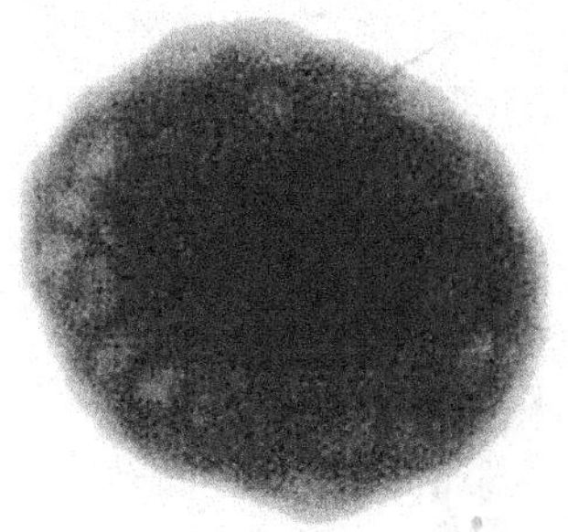

Figure 15 (a)

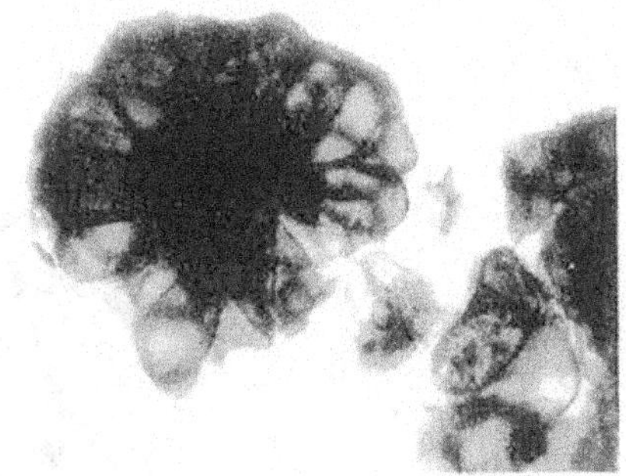

Figure 15 (b)

Sporocarps, dark–brown, 210-360μm diam, globose to sub-globose. Peridium tightly enclosing sporocarps composed of thick walled sinuous hyphae. Chlamydospores 20–90 X 25-65 μm, obovate, radiating out in a single layer from a central plexus of hyphae. Chlamydospore walls brown, 1-2 layered, 48 μm thick; generally thickest near spore-base.

The sporocarps differ from the original description (Gerdemann and Bakshi, 1976) in sporocarp size (248–412 μm diam) and chlamydospore size (45-118 X 30-83 μm) of the spores.

Glomus tenebrosum (Thaxter) Berch Can. J. Bot. 61:2608-2617. 1983 (Fig. 16)

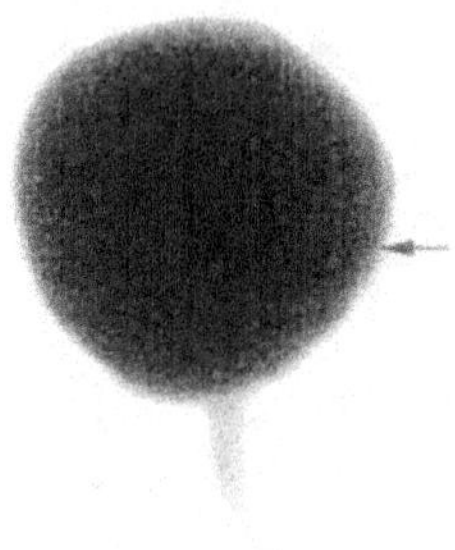

Figure 16

Spores are sub–globose 107–1123 X 94–114 μm yellow to very dark brown. The spore wall is single 0.62–4.96 μm thick. The outer surface of the spore is smooth. The hypha is 8.68–12 μm at the point of attachment to the spore and fades from dark brown or yellow near the spore within approximately 100 μm of the attachment. Subtending hypha thin walled and branched.

The spores differ from the original description (Berch and Fortin, 1983) in size (200-) 240 (-270)X(205-) 230 (-270 μm) wall thickness (13–26 μm) and in hyphal diameter (16–42 μm) of the spores.

Glomuss tortuosum Schenck & Smith Mycologia 74:77-92. 1982 (Fig. 17).

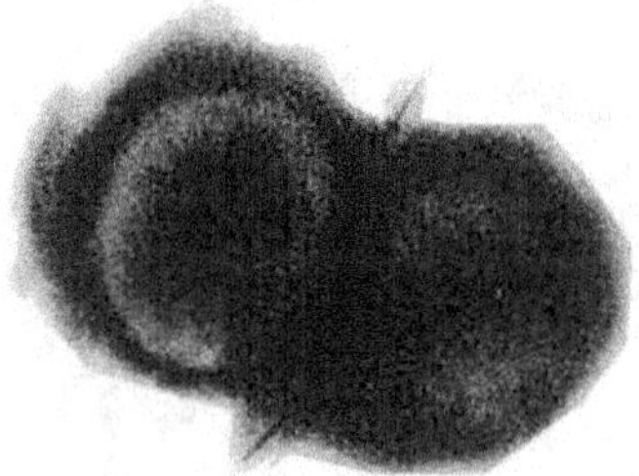

Figure 17

Chlamydospores borne singly in the soil, but occasionally adhering in pairs. Yellow to dull - grey brown with a mantle of sinuous hyphae closely appressed to the spore and flattened. Chlamydospores largely globose, 94 X 94 μm diam. Spores with a single laminate, thin wall, 1-4.98 μm diam.

The spores differ from the original description (Schenck and Smith, 1982) in size (120–210μm diam) and in spore wall diameter (1–2 μm) of the spores.

Glomus warcupii McGee Trans. Brit. Mycol. Soc. 87:123-129. 1986. (Fig. 18).

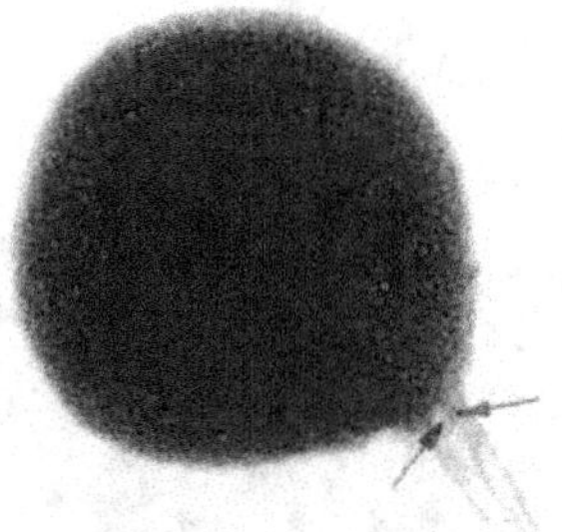

Figure 18

Spores terminal on hyphae, 119-277 X 114-277 μm diam, globose to sub-globose with an ever–present outer hyaline mucilaginous layer. Mature spores dark brown to reddish brown. The outer and inner spore wall laminated 10–24 μm   thick. Contents of spores hyaline and cut off by a brown laminated septum upto 5 μm thick, which concave on the both sides. Subtending hypha cylindrical to swollen. 14-17 μm diam with a hyaline outer wall and a brown to pale yellow inner wall.

The spores differ from the original description (McGee 1986) in size (200-320μm) in wall thickness (16–68(-90) μm) and in subtending hyphal diam (20-50 μm) of the spores.

Scutellospora pellucida _(Nicol. & SchencK)_ Walker & Sanders Mycotaxon 27:219-235. 1986 (Fig. 19)

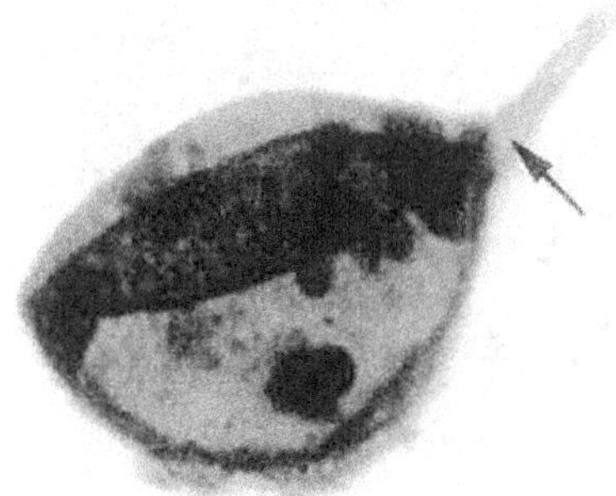

Figure 19

Azygospores formed singly in the soil, borne terminally on a bulbous suspensor like cell, glistening with oil droplets, hyaline to white, llipsoid, 99 X 72 μm. Spore wall smooth, composed of two separable walls; walls < 0.5 μm thick. Suspensor like cell hyaline, subtending hyphae hyaline 7 μm wide, hyphal walls 0.5-l.0 μm thick; germination shield oval.

They differ from the original description (Koske and Walker, 1986) in size (58-183 (-250) × 58-241(–410) μm) and shape (globose) of the spores.

Sclerocystis pakistanica Iqbal & Bushra Trans. Mycol. Soc. Japan 21:57-63. 1980 (Fig. 20 a and b).

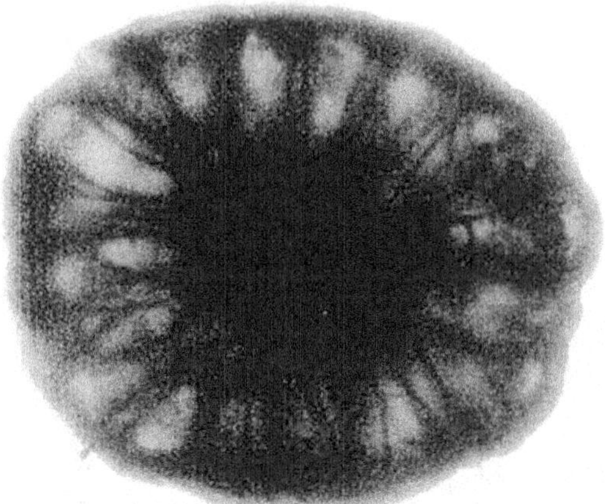

Figure 20 (a)

Figure 20 (b)

Sporocarps   pale   brown to dark–brown 438-488(-550) μm in diam, sub-globose; Peridium tightly enclosing sporocarp, composed of thick walled inner woven hyphae. Chlamydospores 35–150 X 15–38 μm, clavate, elongated radiating out in a single layer from a central plexus of hyphae. Chlamydospore wall pale-brown to dark brown 1μm thick.

These sporocarps differ from the original description (Iqbal and Perveen, 1980) in sporocarp size (520–590(-700) μm in diam) and in xhlamydospore size (65-205 X 33-65 μm) of the spores.

The spores have been deposited in Osmania University Fungal Herbarium (OUFH), Botany Department, Osmania University, Hyderabad.

Discussion

In the present research, though slight differences are there in spore morphology, ornamentation, wall characteristics etc., majority of the *Acaulospora*, *Gigaspora* and *Scutellospora* species identified and isolated from the rhizosphere soil samples of the plants belonging to the families Apocynaceae and Asclepiadaceae fitted well into the known descriptions.

Almeida and Schenck (1990) transferred most of the species of *Sclerocystis* (except *Sclerocystis coremioides*) to *Glomus* based on spore ontogeny. However Wu (1993 a) argued that distribution of shared characters is justified to group all species with highly organized sporocarp into *Sclerocystis*. Redecker *et al.* (2000) transferred *Sclerocystis coremioides*, the only species retained by Almeida and Schenck (1990) in the genus *Sclerocystis* to *Glomus* based on molecular studies.

Explanation of Figs.

Figs. 1 – 10. Spores of frequently occurring AM fungi associated with Apocynaceae and Asclepiadaceae. Scale bar= 50 μm; except fig.5=100 μm 1. Light brown, globose spore showing complex spore ornamentation (polygonal reticulum)(arrow). 2. Hyaline spore of *Acaulospora delicata* with thin spore walls (arrow). 3. Intact, honey coloured spore of *Acaulospora mellea*. 4. Slightly broken spore of *Acaulospora scrobiculata* with minute evenly pitted surface with a rigid pitted outer spore layer (arrow). 5. Spore of *Entrophospora schenckii* with sub-globose, dark-brown, reticulated spore in white reticulated sporiferous saccule (arrows). 6. The largest spore of *Gigaspora gigantea* borne on a single bulbous suspensor with slender hyphae (arrow). 7. Globose, pearl like spore of *Gigaspora margarita* borne terminally on a broad, hyaline suspensor like cell with long, septate subtending hyphae (arrow). 8. A pair of clustered spores of *Glomus australe* formed from a central branched hypha (arrow). 9. Globose chlamydospore *of Glomus citricola* with thicker walls, minutely perforated with thickened inward projections (arrow). 10. Loose sporocarp of *Glomus fasciculatum* with spherical to irregularly globose chlamydospores (arrows).

Figs. 11 – 20 Scale bar = 50 μm; except figs.12,14,19=100 μm

11. Pale-yellow spore of *Glomus fistulosum* Skou & Jakobsen with characterstic fistules across the spore (arrowed) 12. Densely compacted spores of *Glomus macrocarpum* Tulasne & Tulasne. 13. Globose, brown coloured spore of *Glomus mosseae* (Nicol & Gerd.) Gerd. & Trappe with one funnel- shaped base (arrowed). 14. Dark-brown spore of *Glomus multicaule*

Gerdemann & Bakshi with 3-4 hyphal attachments occurring on opposite ends of spores (arrowed). 15 (a) Dark brown sporocarp of *Glomus sinuosum* (Gerd & Bakshi) Almeida & Schenck with peridium tightly enclosing sporocarp composed of thick-walled sinuous hyphae. (b) Obovate Chlamydospores radiating out in a single layer from a central plexus of hyphae (arrowed) 16. Dark-brown, sub-globose Chlamydospore of Glomus *tenebrosum*(Thaxter) Berch with smooth spore surface and thin walled branched hyphae. 17. A pair of adhered spores of *Glomus tortuosum* Schenck & Smith with a mantle of hyphae closely appressed to the spore and flattened (arrowed). 18. Mature dark-brown spore of *Glomus warcupii* McGee with its contents cut off by a brown laminated septum, which concave on the both sides (arrowed). 19. Intact spore of *Scutellospora pellucida* (Nicol. & Schenck) Walker & Sanders with a oval shaped germination shield (arrowed). 20 (a) Dark-brown, sub-globose sporocarp of *Sclerocystis pakistanica* Iqbal & Bushra with peridium tightly enclosing sporocarp (arrowed). (b) Pale-brown cylindro-clavate to elangated Chlamydospores of *Sclerocystis pakistanica* Iqbal & Bushra radiating out in a single layer from a central plexus of hyphae.

References

Almeida, R.T. and Schenck, N. C. 1990. A revision of the genus *Sclerocystis* (Glomaceae, Glomales). *Mycologia,* 82: 703-714.

Becker, W.N. and I.H.Hall. 1976. *Gigaspora* margarita, a new species in the Endogonaceae. *Mycotaxon* 4: 155-160.

Benjamin, R.K. 1979. Zygomycetes and their spores. In: **The Whole Fungus. Vol II, (B. Kendriek ed.),** pp 573 – 622 . National Museums of Canada, Ottawa, Canada.

Berch, S.M. and J.A. Fortin. 1983. Leptotypification of *Glomus macrocarpum* and proposal of new combinations: *Glomus australe, Glomus versiforme* and *Glomus tenebrosum* (Endogonaceae). *Can. J. Bot.* 61: 2608-2617.

Cavalier – Smith ,T. 1998. A revised six – kingdom system of life. *Biol. Rev.* 73: 203-266.

Chu–Chou, M. and Grace, L.G. 1979. *Endogone flammicerona* as a mycorrhiza symbiont of Douglas fir in New Zealand. *N.Z.J. For. Sci.* 9: 344-347.

Fassi , B., Fontana, A. and Trappe, J.M. 1969. Ectomycorrhizae formed by *Endogone lactiflua* with species of *Pinus* and *Pseudotsuga*. *Mycologia* 61:412-414.

Gerdemann, J.W. and Trappe, J.M. 1974. The Endogonaceae in the Pacific North West, *Mycol. Mem.* 5:1974.

Gerdemann J.W and Bakshi, B.K. 1976. Endogonaceae of India; two new species. *Trans, Br. Mycol. Soc.* 66: 340-343.

Gerdemann, J.W., Nicolson, T.H. 1963. Spores of mycorrhizal *Endogone* Species extracted from soil by wet-sieving and decanting. *Tr. Br. Mycol. Soc.* 46:235-244.

Hall, I.R. 1984. Taxonomy of Vesicular Arbuscular mycorrhizal fungi. In : **Vesicular arbuscular mycorrhiza** (C.L. Powell and D.J. Bagyaraj, eds), pp57-94. LRC Press. Boca Raton, Florida.

Iqbal, S.H. and B. Perveen. 1980. Some species of Sclerocystis (Endogonaceae) from Pakistan. *Trans. Mycol. Soc. Japan.* 21 : 57 – 63.

Koske, R.E. and C. Walker. 1986. Species of *Scutellospora* (Endogonaceae) with smooth-walled spores from maritime sand dunes: two new species and a redescription of the spores of *Scutellospora pellucida* and *Scutellospora calospora*. *Mycotaxon.* 27: 219-235.

McGee, P.A. 1986. Further sporocarpic species of *Glomus* (Endogonaceae) from South Australia. *Trans. Br. Mycol. Soc.* 87: 123 - 129.

Morton, J.B., and Benny, G.L. 1990. Revised classification of arbuscular mycorrhizal fungi (Zygomycetes) : A new order Glomales, two new suborders, Glominae and Gigasporinae, and two new families, Acaulosporaceae and Gigasporaceae, with an emendation of Glomaceae. *Mycotaxon.* 37: 473 – 491.

Pirozynski, K.A. and Dalpe, Y. 1989. Geological history of the Glomaceae with particular reference to mycorrhizal symbiosis. *Symbiosis.* 7: 1 – 36.

Raman, N. and Mohan Kumar, V. 1988. Techniques in mycorrhizal research. CAS in Botany, Univ. of Madras. 279 P.

Rothwell, F.M. and J.M. Trappe. 1979. *Acaulospora birecticulata* sp. nov. *Mycotaxon.* 8 : 471 – 475.

Schenck, N.C. and Smith, G.S. 1982. Additional new and unreported species of mycorrhizal fungi (Endogonaceae) from Florida. *Mycologia.* 74: 77 – 92.

Schenck, N.C. and Perez, Y. 1990. Manual for the Identification of VA Mycorrhizal Fungi. Synergistic Publications, Gainsville, Florida, USA . pp 283.

Schubler, A., Schwarzott, D., Walker, C. 2001. A new fungal phylum Glomeromycota : Phylogeney and evolution. *Mycol. Res.* 105. 1413 – 1421.

Sieverding, E., and S. Toro, T. 1987. *Acaulospora denticulata* sp.nov. and *Acaulospora rehmii* sp. nov. (Endogonaceae) with ornamental spore walls. *Angewandto botanik* 61 : 217 – 223.

Schenck, N.C., Spain, J.L., Sieverding E. and Howeler, R.H. 1984. Several new and unreported vesicular – arbuscular mycorrhizal fungi (Endogonaceae) from Colombia. *Mycologia* 76: 685 – 699.

Skou, J.P. and Jakobsen, I. 1989. Two new species from arable land. *Mycotaxon* 36 : 273 – 282.

Tang, Z. and M. Zang. 1984. Additions to the Key of Endogonaceae and a new species of mycorrhizal fungus *Glomus citricolus*. *Acta Botanica Yunnanica* 6: 295 – 304.

Trappe, J.M. 1977. Three new Endogonaceae : *Glomus constrictus, Sclerocystis clavispora* and *Acaulospora scrobiculata*. *Mycotaxon* 6 : 359 – 388.

Trappe, J.M. 1982. Synoptic Key to the genera and species of zygomycetous vesicular arbuscular mycorrhizal fungi. *Phytopathology* 72 : 1102 – 1108.

Walker, C. 1983. Taxonomic concepts in the endogonaceae spore wall characterstics in species descriptions. *Mycotaxon* 18 : 443 – 455.

Walker, C. 1986. Taxonomic concepts into Endogonaceae II. A fifth morphological wall type in Endogonaceous spores. *Mycotaxon* 25: 95-99.

Walker, C. 1987. Current concepts in the taxonomy of Endogonaceae. In : Mycorrhizae in the next decade. Practical Applications and Research priorities. *Proc. of 7th NACOM C.D.M.* Sylvia, L.L.Hung and J.H.Graham, eds.). Florida. 300 – 302.

Walker, C. and E. Koske. 1987. Taxonomic concepts in the Endogonaceae: IV. *Glomus fasciculatum* redescribed. *Mycotaxon* 60 : 253 – 262.

Walker, C., M. Pfeiffer and H.E. Bloss. 1986. *Acaulospora delicata* sp. nov an endomycorrhizal fungus from Arizona. *Mycotaxon* 25 : 621 – 628.

Walker, C. and Trappe, J.M. 1993. Names and epithets in the Glomales and Endogonales. *Mycol. Res,* 97 (3) : 339 – 344.

Walker, C. and Vestberg, M. 1998. Synonymy amongst the Arbuscular mycorrhizal fungi. *Glomales claroideum, Glomus maculosum, Glomus multisubtensum* and *Glomus fistulosum*. *Ann. Bot.* 82 : 601 – 624.

Wu, C. (1993a) Glomales of Taiwan: II. A comparative study of spore ontogeny in *Sclerocystis* (Glomaceae, Glomales). *Mycotaxon* 47: 25-39.

7

Seasonal Variations of Nematode Population in the Rhizosphere of Guava and Papaya

Mansoor A. Siddiqui

Plant Pathology/Nematology Section, Department of Botany, Aligarh Muslim University, Aligarh - 202 002, India

ABSTRACT

Two economical plants *viz.*, guava, (*Psidium guajava*), an important small fruit tree and papaya, (*Carica papaya*), an important rapidly growing and heavily yielding tree were selected for the study. These two plants showed different population behaviour of plant parasitic nematodes viz., *Hoplolaimus indicus, Helicotylenchus indicus, Rotylenchulus reniformis, Tylenchorhynchus brassicae, Tylenchus filiformis, Meloidogyne incognita* and *Hemicriconemoides mangiferae* at 10, 20 and 40 cm depths during various seasons of the year. Plant parasitic nematodes attain an annual peak in the month of August and decline in the month of June. These nematode population fluctuations were observed directly affected by soil moisture, temperature, plant growth and indirectly affected by soil pH. The nematode population was higher at higher soil moisture content. The nematode population was drastically low in dry seasons. The nematode population was high in upper 10 cm layer followed by middle 20 cm and lower 40 cm layers.

KEY WORDS : Seasonal fluctuation, soil pH, temperature, moisture, guava, mango, nematode population.

The environmental factors especially soil temperature, soil moisture, pH value, aeration and osmotic potential play direct or indirect roles in regulating phytophagous nematode population (Wallace, 1965; Norton, 1978; Ferris, 1981; Egunjobi, 1981; Van Gundy, 1985). The population density of phytoparasitic nematodes is correlated with the amount of damage to a plant. Knowledge about the mean density of nematode species inhabting the rhizosphere of plants is of paramount importance in embarking nematode management programmes (Huang *et al.*, 1984; Ferris, 1985; Almeida *et al.*, 1987; Gaur *et al.*, 1988, 1993; Singh and Gaur, 1996). In perennial plants, it is necessary to study the nematode population during different seasons. Because changes in environmental conditions are important factors besides the host itself (Kornobis, 1981a, b; Siddiqui, 1980; Sabova *et al.*, 1987; Das and Mukhopadhyaya, 1990; Mukhopadhyaya and Sarkar, 1990).

Intensive location specific interdisciplinary research on the ecology of plant parasitic nematode pest species is necessary to understand the environmental relations, survival strategies and methods of increasing local abiotic and biotic environmental resistance to nematodes. Exhaustive studies are still needed to more knowledge about the effect of seasonal conditions on the changes in nematode population. The present investigation is, therefore, an attempt in this direction wherein the population of soil inhabiting nematodes around the roots of two fruit trees viz., guava, an important small fruit and papaya, an important rapidly growing and heavily yielding tree, were selected for the study. The soil pH, temperature and moisture during the period of investigation have been recorded to determine their correlation, if any, on the population densities of nematodes from January 2001 to December 2001.

Materials and Methods

The groves of guava, (*Psidium guajava* L.), at Aligarh Muslim University Farm and papaya (*Carica papaya* L.), at Grand Trunk road near Sarsaul District Aligarh were selected. Both the groves were heavily infested with the lance nematode, *Hoplolaimus indicus* Sher, the spiral nematode, *Helicotylenchus indicus* Siddiqi, the reniform nematode, *Rotylenchulus reniformis* Linford & Oliveira, the stunt nematode, *Tylenchorhynchus brassicae* Siddiqi, the filiform nematode, *Tylenchus filiformis* Butschli, the root-knot nematode, *Meloidogyne incognita* (Kofoid & White) Chitwood and the ring nematode, *Hemicriconemoides mangiferae* Siddiqi. The soil samples were taken at monthly intervals in polythene bags from the roots of five guava and papaya plants. The soil samples were taken at 10, 20 and 40 cm depths and 25 cm distance from the stem with the help of a soil sampler. These depths were described as upper, middle and lower layers. The soil samples were processed by Cobb's sieving and decanting technique alongwith modified Baermann funnel method (Southey, 1986). Each value was an average of five replicates. Nematode population, soil pH, soil temperature (°C) and soil moisture (%) were recorded at the time of each soil sampling. Data was statistically analysed for Critical Difference (C.D.) at $P = 0.05$ and $P = 0.01$ (Panse and Sukhatme, 1978).

Results

The results presented in Tables (1-4) and Figures (1, 2) show that the seasonal environmental factors such as soil temperature and soil moisture have a direct effect on the activity and rate of development and spread of infection of nematode pests whereas pH value has indirect effect on nematode multiplication throughout the year.

In guava, the average soil population of *Hoplolaimus indicus* (average of 10, 20 and 40 cm depths) was 7, 17, 76, 70, 57, 17, 80, 97, 62, 47, 22 and 23 during January, February, March, April, May, June, July, August, September, October, November and December respectively. The corresponding figures for *Helocotylenchus indicus* were 5, 14, 26, 19, 13, 3, 35, 38, 87, 40, 25 and 39; for *Rotylenchulus reniformis* 10, 27, 77,

33, 6, 7, 40, 252, 185, 113, 73 and 24; for *Tylenchorhynchus brassicae* 26, 30, 60, 25, 17, 7, 32, 80, 68, 44, 31 and 27; for *Tylenchus filiformis* 13, 23, 71, 32, 17, 5, 27, 67, 32, 22, 18 and 10; for *Hemicriconemoides mangiferae* 8, 18, 57, 51, 14, 4, 159, 24, 25, 16, 5 and 17. The average total population of plant parasitic nematodes (Tylenchids) was 69, 132, 366, 230, 123, 43, 369, 778, 458, 282, 174 and 141 (Table 1, Fig. 1).

The average soil moisture (%) around the roots of guava was 21.13, 12.64, 10.89, 9.00, 5.30, 2.43, 7.33, 17.93, 11.00, 7.20, 2.97 and 6.73 during January, February, March, April, May, June, July, August, September, October, November and December respectively. The corresponding figures for soil temperature was 17, 19, 21, 29, 33, 41, 39, 33, 35, 32, 28 and 22 °C. The pH ranged between 5.20 to 8.50 (Table 3, Fig. 1).

In papaya, the average soil population of *Hoplolaimus indicus* was 9, 9, 23, 21, 11, 3, 22, 37, 19, 14, 12 and 8 during January, February, March, April, May, June, July, August, September, October, November and December respectively. The corresponding figures for *Helicotylenchus indicus* were 9, 15, 22, 17, 10, 5, 30, 111, 29, 13, 7 and 13; for *Rotylenchulus reniformis* 55, 102, 168, 151, 99, 92, 182, 193, 138, 103, 97 and 62; for *Tylenchorhynchus brassicae* 21, 48, 68, 28, 20, 8, 48, 32, 28, 13, 13 and 27; for *Tylenchus filiformis* 13, 22, 37, 25, 8, 0, 28, 48, 32, 28, 13 and 32; for *Meloidogyne incognita* 69, 124, 99, 37, 50, 27, 22, 181, 95, 43, 11 and 13. The average total soil population of plant parasitic nematodes (Tylenchids) was 176, 321, 417, 273, 198, 136, 331, 568, 274, 213, 162 and 152 (Table 2, Fig. 2).

For papaya, the average soil moisture (%) was 18.43, 16.83, 14.03, 12.26, 12.73, 7.83, 16.90, 19.37, 14.10, 11.43, 10.13 and 15.63 during January, February, March, April, May, June, July, August, September, October, November and December respectively. The corresponding figures for soil temperature were 16, 18, 20, 26, 30, 40, 38, 32, 34, 31, 27 and 21 °C. The soil pH ranged between 5.2 to 8.5 (Table 4, Fig. 2).

The nematode population was maximum in upper 10 cm layer of soil. Nematode multiplied on guava and papaya. Population of plant parasitic nematodes attained an annual peak in August and declined in June both in guava as well as papaya. *M. incognita*, *Hel. indicus* and *R. reniformis* highly multiplied on guava and papaya. It shows that these guava and papaya are good hosts for these nematodes. Whereas, *Hoplolaimus indicus*, *Tylenchorhynchus brassicae* multiplied moderately and *Hemicriconemoides mangiferae* multiplied poorly on guava and papaya (Tables 1, 2 & Figs.1, 2).

In August when the soil moisture was high the nematode population was maximum. In dry weather nematode population decreased and again increased when the guava and papaya groves were irrigated. In the groves of guava the nematode population increased when the soil pH were 5.20 (March), 6.00 (April), 5.50 (August), 5.80 (September), 6.10 (October), 6.20 (November) and 6.30 (December). The nematode population declined when the soil pH was 7.90 (January), 8.20 (February), 8.30 (May) and 8.50 in December (Table 3, Fig. 1). Similar trend was observed in case of papaya (Table 4, Fig. 2).

Table 1 Distribution of plant parasitic nematodes associated with guava (*Psidium guajava* L.) in three soil (10, 20, 40 cm) depths.

PPN	Depth (cm)	Average nematode population per 200 g soil												C.D.	
		Jan	Feb	Mar	Apr	May	Jun	Jul 0.01	Aug	Sep	Oct	Nov	Dec	$p = 0.05$	$p =$
Hop	10	12	28	107	100	90	50	110	130	95	80	50	40	13.38	17.79
	20	6	17	87	80	60	–	80	90	60	50	15	20	13.72	18.25
	40	4	7	33	30	20	–	50	70	30	10	–	10	13.90	18.48
	Av.	7	17	76	70	57	17	80	97	62	47	22	23	14.30	19.01
Hel	10	10	20	41	39	30	10	73	80	210	80	47	80	8.00	10.64
	20	5	17	27	10	10	–	23	25	33	30	23	33	9.74	12.95
	40	–	6	10	8	–	–	8	10	17	10	5	5	9.54	12.68
	Av.	5	14	26	19	13	3	35	38	87	40	25	39	9.86	13.11
Rot	10	17	50	110	60	12	10	62	474	373	170	110	47	19.48	25.90
	20	7	20	90	30	3	5	45	191	138	130	80	15	21.26	28.27
	40	5	10	30	10	3	5	12	92	43	40	30	10	19.63	26.11
	Av.	10	27	77	33	6	7	40	252	185	113	73	24	19.82	26.36
Trh	10	46	48	97	43	30	20	58	140	130	87	53	50	22.00	29.26
	20	25	28	53	23	20	–	27	57	45	33	30	20	20.88	27.77
	40	7	13	30	10	–	–	12	43	28	13	10	10	20.74	27.58
	Av.	26	30	60	25	17	7	32	80	68	44	31	27	23.00	30.59
Tyl	10	26	33	102	55	30	10	61	113	51	45	43	20	13.34	17.74
	20	8	24	90	30	10	3	17	63	32	13	8	5	11.32	15.05
	40	4	11	20	10	10	3	3	25	12	7	2	5	9.32	12.39
	Av.	13	23	71	32	17	5	27	67	32	22	18	10	11.98	15.93

Table 1 Contd...

PPN	Depth (cm)	Average nematode population per 200 g soil												C.D.	
		Jan	Feb	Mar	Apr	May	Jun	Jul $p=0.01$	Aug	Sep	Oct	Nov	Dec	$p=0.05$	
Hem	10	16	43	122	112	22	7	255	393	58	29	13	28	18.00	23.94
	20	4	10	37	30	10	3	130	230	13	10	3	12	19.04	25.32
	40	4	–	13	10	10	3	92	107	5	8	–	12	15.48	20.58
	Av.	8	18	57	51	14	4	159	243	25	16	5	17	15.94	21.20
Total	10	127	222	579	409	214	107	619	1330	917	491	316	265	21.36	28.40
	20	55	116	384	203	113	11	322	656	321	266	159	105	20.66	27.47
	40	24	57	136	78	43	11	167	347	135	88	47	52	19.98	26.57
	Av.	69	132	366	230	123	43	369	778	458	282	174	141	22.68	30.16
Sap	10	250	530	455	640	450	250	700	1000	585	560	510	180	–	–
	20	150	250	280	290	340	160	600	600	390	200	185	160	–	–
	40	80	140	230	140	160	80	510	540	210	190	110	120	–	–
	Av.	160	307	322	357	317	163	603	713	395	317	268	153	–	–

Data represents an average of five replicates.

PPN = Plant parasitic nematodes, Hop = *Hoplolaimus indicus*, Hel = *Helicotylenchus indicus*, Rot = *Rotylenchulus reniformis*, Trh = *Tylenchorhynchus brassicae*, Tyl = *Tylenchus filiformis*, Hem = *Hemicriconemoides mangiferae*, Total = Total plant parasitics nematodes, Sap = Total saprozoic, Av = Average of 10, 20 and 40 cm depths.

C.D. = Critical difference

Table 2 Distribution of plant parasitic nematodes associated with papaya (*Carica papaya* L.) in three soil (10, 20, 40 cm) depths.

PPN	Depth (cm)	Average nematode population per 200 g soil												C.D.	
		Jan	Feb	Mar	Apr	May	Jun	Jul 0.01	Aug	Sep	Oct	Nov	Dec	p = 0.05	p =
Hop	10	12	28	107	100	90	50	110	130	95	80	50	40	13.38	17.79
Hop	10	13	14	39	26	13	10	35	50	30	23	20	13	10.60	14.09
	20	13	12	17	23	10	–	20	47	20	12	10	8	10.78	14.33
	40	–	–	12	14	10	–	10	13	8	7	5	4	10.92	14.52
	Av.	9	9	23	21	11	3	22	37	19	14	12	8	11.06	14.70
Hel	10	10	22	36	25	15	10	40	160	47	22	13	17	11.16	14.84
	20	10	15	18	15	10	5	30	147	32	10	7	17	11.22	14.92
	40	7	7	11	10	5	–	20	27	7	7	–	5	11.40	15.16
	Av.	9	15	22	17	10	5	30	111	29	13	7	13	11.86	15.77
Rot	10	90	127	231	210	133	127	241	233	187	123	123	88	9.70	12.90
	20	47	100	173	133	112	80	177	227	158	103	105	80	9.86	13.11
	40	27	80	100	110	51	70	127	120	70	82	62	17	9.58	12.74
	Av.	55	102	168	151	99	92	182	193	138	103	97	62	9.84	13.08
Trh	10	24	63	88	50	30	15	73	50	33	30	30	55	15.99	21.25
	20	23	57	80	25	20	10	43	27	27	8	10	20	16.86	22.42
	40	17	23	37	10	10	–	27	18	23	–	–	5	17.38	23.11
	Av.	21	48	68	28	20	8	48	32	28	13	13	27	17.10	22.74
Tyl	10	26	45	60	50	20	–	60	73	50	33	30	43	15.04	20.00
	20	11	13	33	20	5	–	20	43	27	27	8	33	13.34	20.40
	40	2	8	19	5	–	–	5	27	18	23	–	20	15.88	21.12
	Av.	13	22	37	25	8	–	28	48	32	28	13	32	15.72	20.90

Table 2 Contd...

PPN	Depth (cm)	Average nematode population per 200 g soil												C.D.	
		Jan	Feb	Mar	Apr	May	Jun	Jul 0.01	Aug	Sep	Oct	Nov	Dec	p = 0.05	p =
Mel	10	80	137	133	57	100	40	33	273	125	80	43	20	10.68	14.20
	20	80	123	127	35	40	30	23	153	110	40	15	10	10.92	14.52
	40	47	117	37	–	10	10	10	117	50	10	5	10	12.94	17.21
	Av.	69	124	99	37	50	27	22	181	95	43	11	13	12.36	16.44
Total	10	243	408	587	418	311	202	482	739	372	311	259	236	19.34	25.72
	20	184	320	448	251	197	125	313	644	274	200	145	168	19.76	26.28
	40	100	135	216	149	86	80	199	322	176	129	82	61	20.42	27.15
	Av.	176	321	417	273	198	136	331	568	274	213	162	152	19.98	26.57
Sap	10	335	335	410	511	335	155	847	949	735	650	670	327	–	–
	20	230	245	215	330	295	67	612	633	450	590	590	249	–	–
	40	195	325	130	217	310	93	210	422	225	220	333	135	–	–
	Av.	253	302	252	353	313	105	556	668	470	487	531	237	–	–

Data represents an average of five replicates.

PPN = Plant parasitic nematodes, Hop = *Hoplolaimus indicus*, Hel = *Helicotylenchus indicus*, Rot = *Rotylenchulus reniformis*,

Trh = *Tylenchorhynchus brassicae*, Tyl = *Tylenchus filiformis*, Mel = *Meloidogyne incognita*, Total = Total plant parasitics nematodes, Sap = Total saprozoic, Av = Average of 10, 20 and 40 cm depths.

C.D. = Critical difference

Table 3 Data of soil pH, temperature and moisture from January 2001 to December 2001 around the roots of guava, *Psidium guajava* L.

Period	pH	Temperature (°C)	Moisture (%) at different depths (cm)			
			10	20	40	Av.
January	7.90	17	16.10	21.80	26.05	21.13
February	8.20	19	14.41	11.12	12.40	12.64
March	5.20	21	12.27	10.00	10.41	10.89
April	6.00	29	12.00	8.00	7.01	9.00
May	8.30	33	3.70	5.90	6.30	5.30
June	8.50	41	2.40	2.90	2.00	2.43
July	5.70	39	8.50	5.70	7.80	7.33
August	5.50	33	18.00	19.10	16.70	17.93
September	5.80	35	10.00	12.00	11.00	11.00
October	6.10	32	5.60	7.00	9.00	7.20
November	6.20	28	3.00	3.50	2.40	2.97
December	6.30	22	8.90	4.30	7.00	6.73

Av = Average soil moisture of 10, 20 and 40 cm depths.

pH = Hydrogen ion potential.

Table 4 Data of soil pH, temperature and moisture from January 2001 to December 2001 around the roots of papaya, *Carica papaya* L.

Period	pH	Temperature(°C)	Moisture (%) at different depths (cm)			
			10	20	40	Av.
January	7.7	16	19.00	18.60	17.70	18.43
February	7.9	18	16.00	17.30	17.20	16.83
March	5.3	20	15.80	12.00	14.30	14.03
April	5.8	26	12.80	11.00	13.00	12.26
May	8.3	30	12.70	13.50	12.00	12.73
June	8.5	40	8.90	8.00	6.60	7.83
July	5.7	38	17.00	15.00	18.70	16.90
August	5.2	32	20.70	19.40	18.00	19.37
September	5.6	34	14.50	13.00	14.80	14.10
October	6.2	31	11.00	12.30	11.00	11.43
November	5.7	27	12.60	9.70	8.10	10.13
December	6.4	21	14.30	17.00	15.60	15.63

Av = Average soil moisture of 10, 20 and 40 cm depths.

pH = Hydrogen ion potential.

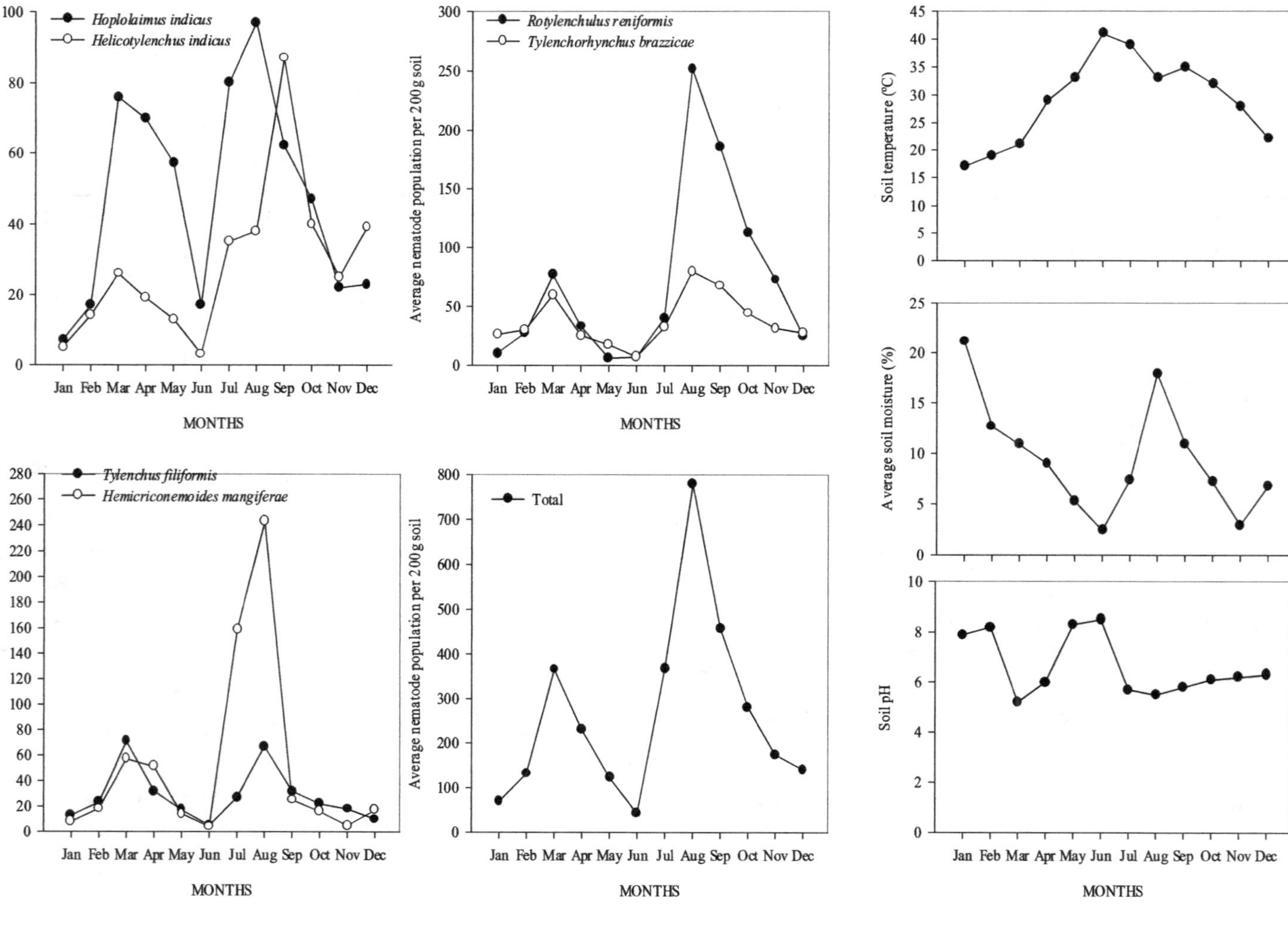

Figure 1 Influence of soil pH, moisture and temperature on nematode population densities in different seasons on guava (*Psidium guajava* L.)

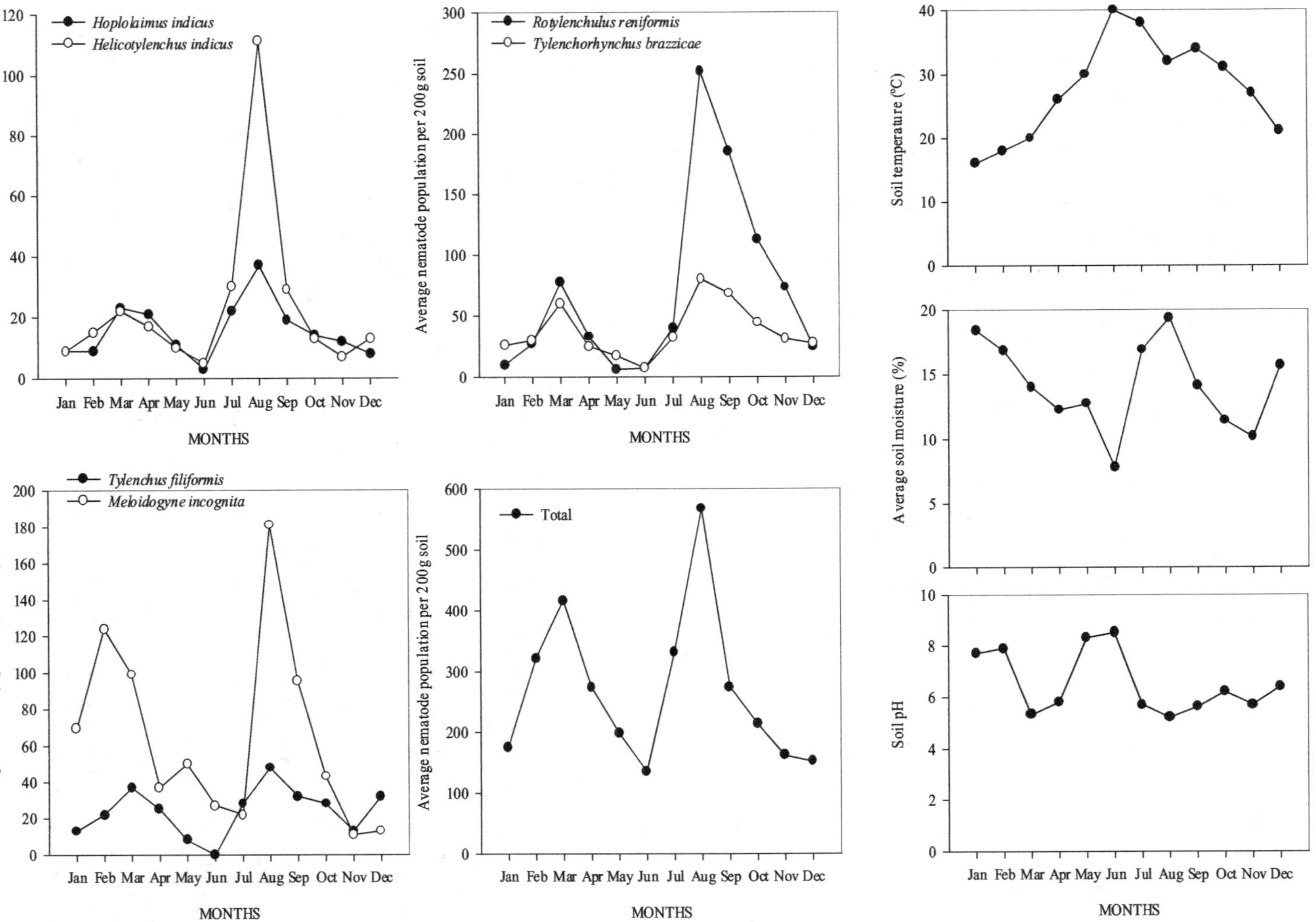

Figure 2 Influence of soil pH, moisture and temperature on nematode population densities in different seasons on papaya (*Carica papaya*)

Discussion

The drought condition may be responsible for the decrease in nematode population. The dry conditions in April, May and June were due to high soil temperature (29 °C – 41 °C). Desiccation of soil and other substrates can be lethal to nematodes as they require thin film of water around them for their normal activity. They can slow down the rate of multiplication. In dry soil conditions decline in nematode population may be due to desiccation or migration of nematodes deep into the soil. Physiological stress on plants due to soil moisture deficit affects the nitrogen content of plants (Grandison, 1973; Grandison and Wallace, 1974; Gaur, 1995). The cuticular sheaths help the nematodes to reduce the rate of water loss when the soil begins to dry. During dehydration and rehydration, nematodes consume considerable amounts of their energy reserves and can reduce their population densities (Gaur, 1994). Less soil moisture severely affect the nematode population when the moisture contents dries the condition of soil solution increases and the nematodes are subjected to increase stress (Dropkin, 1980). In December and January low soil temperature combined with soil moisture can be responsible for nematode population decline. Besides this growth of host plant may be related to the changes in nematode population (Oostenbrink, 1966).

The nematode population was highest in upper 10 cm followed by 20 and 40 cm depth. It can be due to the presence of more feeder roots in upper 10 cm. Spatial distribution of plant parasitic nematodes is one of the most characteristic properties (Taylor, 1971).

Pratylenchus penetrans multiplied at pH 5.5 to 5.8 and upto 6.0 but decline at 6.6 (Kimpinski and Willis, 1981; Morgan and Mac Lean, 1968). The low pH 4.0 was associated with high levels of potassium, manganese and phenols in the soybean plant and the suberized outer layer developed in soybean roots (Burns, 1971). The pH below 5.0 and above 8.0 may be inhibitory to nematodes (Wallace, 1965).

Thus from the above results it can be concluded that seasonal fluctuation in nematode population inhabiting the rhizosphere of guava and papaya might be due to the direct effect of soil temperature, moisture or their combination on the activity and rate of development of nematode pests. The pH effect indirectly on nematode population (Lowenberg *et al.*, 1960; Gnanapragasam, 1987; Urek, 1998). Our results are in conformity with Dao (1970), Saeed and Ghaffar (1986).

The infection of nematodes on plant can be managed to adopt life cycle when the soil moisture present is less. This will help avoidance of infestations on the plant. Thus the ecological factors can be essential while applying components such as chemicals on limited scale and biocontrol agents. These research findings indicate that plant parasitic nematodes associated with guava and papaya shows different population fluctuation behaviour during various seasons of the year. These studies may go a long way in Nematode Pest Management Programme.

References

Almeida, V.F. De, Campos, V.P. and **Delima, R.D.** 1987. [Population fluctuations of *Meloidogyne exigua* in coffee rhizosphere]. Fluctuacao populacional de *Meloidogyne exigua* na rizosfera do cafeiro. *Nematol. Brasileria* **11**: 159-175.

Burns, N.C. 1971. Soil pH effects on nematode populations associated with soyabean. *J. Nematol.* **3**, 238.

Dao, D.F. 1970. Climatic influence on the distribution pattern of plant parasitic and soil inhabiting nematode. *Meded. Landb. Hoggesch*, Wageningen **70**, 21-181.

Das, T.K. and **Mukhopadhyaya, M.C.** 1990. Preliminary observations on the population of some plant parasitic, predatory and free living nematodes in cultivated soil and grassland. *Indian J. Nematol.* **20**: 108-111.

Dropkin, V.H. 1980. Introduction to Plant Nematology. John Wiley and Sons (Publishers), New York, Toronto, pp. 293.

Egunjobi, O.A.I. 1981. The ecology of *Meloidogyne* spp. in Nigeria. In *Proc. 3rd Res. Plan. Conf. on Root-Knot Nematodes, Meloidogyne* spp., Regions IV and V, 16-20 Nov., 1981 (IPM). *Int. Inst. Tropic Agric.* Ibadan, Nigeria.

Ferris, H. 1981. Dynamic action threshold for diseases induced by nematodes. *Ann. Rev. Phytopathol.* **19**: 427-436.

Ferris, H. 1985. Density dependent nematode seasonal multiplication rates and over winter survivorship: a critical point model. *J. Nematol.* **17**: 93-100.

Gaur, H.S. 1994. Ecology of plant parasitic nematodes. In: *Nematode pest management in crops.* (Eds.) Bhatti, D.S. and Walia, R.K. CBS Publ. & Distr, pp. 31-65.

Gaur, H.S. 1995. Some ecological considerations in integrated nematode management. In: *Nematode pest management: an appraisal of eco friendly approaches.* (Eds.) Swarup, G., Dasgupta, D.R. and Gill, J.S., Nematological Society of India, I.A.R.I., New Delhi, 28-36.

Gaur, H.S., Singh, J. and **Gupta, H.C.** 1993. Distribution of reniform nematode, *Rotylenchulus reniformis* in an eggplant *Solanum melongena* field. *Ann. Pl. Prot. Sci.* **1**: 99-106.

Gaur, H.S., Sigh, U.C. and **Gupta, H.C.** 1988. Spatial distribution and precision of sampling of *Pratylenchus zeae* in maize field. *Indian J. Nematol.* **18**: 150-152.

Gnanapragasam, N.C. 1987. Effect of soil acidity on the population levels of *Pratylenchus loosi. Int. Nematol. Network Newsl.* **4**: 23-24.

Grandison, G.S. 1973. Soil moisture and nematode parasitism of plants. Proc. Soil and Plant Water Symp., Palmerston N., New Zealand Information Services No. **96**: 115-119.

Grandison, G.S. and ***Wallace, H.R.*** 1974. The distribution and abundance of *Pratylenchus thornei* in fields of strawberry clover (*Trifolium fragiferum*). *Nematologica* **20**: 283-290.

Huang, S.P., Desouza, S.P. and ***Campos, V.P.*** 1984. Seasonal variation of a *Meloidogyne exigua* population in coffee plantation. *J. Nematol.* **16**: 115-117.

Kimpinski, J. and ***Willis, C.B.*** 1981. Influence of soil temperature and pH on *Pratylenchus penetrans* and *P. crenatus* in alfalfa and timothy. *J. Nematol.* **13**: 333-338.

Kornobis, S. 1981a. [Vertical distribution of nematodes in soil]. Pionowe rozmieszczenie nicieni w glebie. [Materially Konf. Nematol., Skierniewice, maj. 1978]. Zeszyty Problemowe Postepow Nauk Rolniczych. No. **249** 7-17.

Kornobis, S. 1981b. [Hortizontal distribution of nematodes in soil and comments on methods of investigation]. Poziome rozmieszczemie nicieni w glebie. [Materially Konf. Nematol., Skierniewica, maj. 1978]. Zeszyty Problemove Postepow Nauk Rolniczych (1981) No. **249**: 19-26.

Lowenberg, J.R., Sullivan, T. and ***Schuster, M.L.*** 1960. The effect of pH and minerals on the hatching and survival of *Meloidogyne incognita* larvae. *Phytophathology* **50**: 215-217.

Morgan, G.T. and ***MacLean, A.A.*** 1968. Influence of soil pH on an introduced population of *Pratylenchus penetrans*. *Nematologiea* **14**: 311-312.

Mukhopadhyaya, M.C. and ***Sarkar, P.K.*** 1990. Population behaviour and distribution pattern of *Tylenchorhynchus zeae* and *Hoplolaimus indicus* in cultivated soil receiving industrial pollutants. *Indian J. Nematol.* **20**: 152-160.

Norton, D.C. 1978. Population fluctuations of *Xiphinema americanum* in Iowa. *Phytopathology* **53**: 66-68.

Oostenbrink, M. 1966. Major characteristics of the relation between nematodes and plants. Meded Landb Hogesch, Wageningen **66**, 46pp.

Panse, V.G. and ***Sukhatme, P.V.*** 1978. Statistical methods for agricultural workers (Revised by Sukhatme, P.V. & Amble, V.N.), I.C.A.R., New Delhi, p. 347.

Sabova, M., Liskova, M. and ***Valocka B.*** 1987. [Seasonal dynamics of parasitic nematodes in cereal monocultures and in crop rotation to climatic and soil conditions]. Sezonna dynamika parazitckych nematod V monokulturach obilnin av osevynch postpoch vo vztalm ku klimatickym a polnym podmienkam. Biologia B (Zoologia), Czechoslovakia 42: 121-126.

Saeed, M. and ***Ghaffar, A.*** 1986. Seasonal population fluctuations of *Hemicriconemoides mangiferae* and other nematodes associated with sapodilla (*Achras zapota*) in Karachi. *Pak. J. Nematol.* **4**: 67-74.

Siddiqui, M.A. 1980. Population changes of plant parasitic nematodes associated with eight fruit trees. M.Sc. Disser. Dept. of Botany, A.M.U., Aligarh, pp 14, 1980.

Singh, J. and ***Gaur, H.S.*** 1996. Seasonal variation in the horizontal spatial pattern of the root-knot nematode *Meloidogyne incognita* and application of the negative binomial and Taylor's power law models in developing sampling schemes. *Indian J. Nematol.* **26**: 226-236.

Southey, J.F. 1986. *Laboratory methods for work with plant and soil nematodes*, Ministry of Agriculture, Fish Food, HMSO, London, pp 202.

Taylor, L.R. 1971. Aggregation as a species characteristics. In *Statistical ecology*. (Eds.) Patil, G.P., Piclon, E.C. and Water, W.E. Pennsylvania University Press **1**: 357-377.

Urek, G. 1998. The influence of soil (pH) reaction on nematode population in soil (Vpliv talne reakeije (pH) na populacijo ogorne V tteh, Research Reports Biotechnical Faculty University of Ljubljana Agriculture Issue No. **71**: 39-46.

Van Gundy, S.D. 1985. Ecology of *Meloidogyne* spp. emphasis on environmental factors affecting survival and pathogenecity. In *An advanced treatise on Meloidogyne*. Vol. I. Biology and Control (Eds.) Sasser, J.N. and Carter, C.C.). North Carolina, Raleigh, USA, 177-182.

Wallace, H.R. 1965. The soil environment. In: *Plant Nematology*, (Ed.) Southey, J.F. Tech. Bull. No. 7, Ministry of Agriculture, Fish Food, HMSO, London.

<table><tr><td>8</td><td>

Studies on the Thermophilic Fungi from Different Substrates of Dharward, Karnataka

</td></tr></table>

Ch. Ramesh and Anil Kumar

*Department of Botany, Karnatak University, Dharwad 580 003, Karnataka

ABSTRACT

Thermophilic fungi are known to play a significant role in breaking down complex molecules of organic matter in nature, under elevated temperature conditions. These are distributed widely in nature but they are abundant in some places only. Thermophilic fungi were isolated from various substrata such as composting plant materials, mushroom composts, manure soils, herbivores dung, hay stacks etc. The present paper deals with the isolation of 22 species belonging to 8 genera on different substrates collected from different regions of Dharwad, Karnataka state. It also deals with occurrence and role of thermophilic fungi .

KEY WORDS : Thermophilic fungi, dung, mushroom compost, hay stack.

Thermophilic microorganisms form a diverse group of organisms sharing chiefly a com mon ability to grow at higher temperatures. Among the eukaryotes some fungi possess exceptional ability to grow at the temperatures up to 60 °C. Cooney and Emerson defined thermophilic fungi as those, which have a maximum temperature for growth at or above 50 °C and a minimum temperature for a growth at or above 20 °C. A majority of thermophilic fungi were isolated and described from composting plant materials where in microbial thermogenesis heats up the compact mass of material and favors the development of thermophilic microflora in the warm and humid environment. Cooney and Emerson(1964) have stressed the need for a thorough search of these microorganisms in the tropical regions of the world. India being a tropical country the need for detailed studies on the occurrence and role of thermophilic fungi.

Materials and methods

Collection of Samples

With a view to isolate thermophilic fungi different samples from each of the following substrates were collected and a composite sample in each case was used for estimation of thermophilic fungi.

Compost *:* Samples of compost prepared out of cow dung and farm residues were collected from different places of Karnataka viz Dharwad, Haliyal, UAS Campus, Dharwad, Navalur, Sirsi, Kumata, Ankola, Ganeshgudi, Alnavar etc. collection of the sample from the compost pit was done by taking about 100 g of the sample from each compost pit at various depths.

Buffalo manure *:* Samples of buffalo manure were collected from the buffalo sheds of places in and around Dharwad. Fresh dung mixed with paddy husk was heaped over the ground and when the temperature inside the heap was 55- 60 °C the collection was done as in case of compost.

Bagasse *:* Sugarcane bagasse samples were obtained from some farms at Dharwad and Belgaum. From the two month old heaped bagasse samples were obtained as in case of compost.

Hay *:* Samples of mouldy paddy hay which was baled and stored were collected from different regions of Dharwad and from other areas.

Jowar grain from Hagar *:* Jowar grain samples from the under ground storage pit Hagar were obtained from some private farms at Dharwad, Alnavar, Dandeli. etc. The grains were stored in the underground structure called Hagar, which consisted of a circular pit dug with a depth of 8 ft and a diameter of 6 ft. The pit was filled with jowar grains and the mouth was plastered with mud. The temperature inside the pit was nearly 40 °C at the time of collection of the sample.

Isolation of Thermophilic Fungi

In all 102 samples collected from 8 different sources as detailed above were used for the isolation of thermophilic fungi using four different techniques as follows.

Dilution Plate Technique

Ten grams of the compost sample was transferred to a flask containing 100 ml sterile water. The contents were shaken on a wrist action mechanical shaker for 5 min. then the contents were serially diluted to obtain 10^{-4}, 10^{-5} and 10^{-6} dilutions. One ml of each dilution was transferred to sterile petriplates. Triplicates were kept for each dilution. Martin rose Bengal agar, Emerson's modified yeast starch or YpSs agar and Dextrose peptone yeast extract agar were used for the isolation of fungi.

Warcup's Soil Plate Method

A small quantity of the samples was placed in to a sterile petridish and 20 ml of sterile Czapek's Dox's medium was added after cooling to 40 °C. The contents were thoroughly mixed and the plates were incubated at 50 °C under high humidity. Further isolation was done as detailed above.

Cooney and Emerson's Method

A 10 cm petridish lined with thick damp toweling was sterilized at 15 lb pressure for 15 minutes. About 10 g of the sample was placed in the dish and was incubated at 50 °C. High humidity was provided by placing the dish over a small stand kept in the plate containing water and covering the whole with a bell jar, inner side of which was lined with a damp blotting paper. Observations on the growth of thermophilic fungi were made frequently and isolations were made whenever visible growth was observed.

Waksman's Direct Inoculation Method

Twenty ml of modified Emerson's Y_pS_S medium was poured into a sterile petridish and was allowed to solidify. Small quantities of the sample were sprinkled over the medium in the dishes, such plates were incubated in inverted position at 50 °C under high humidity. Observations on the fungal growth, if any were made every day and the fungi were isolated as detailed above.

Results

The distribution of thermophilic fungi isolated from 102 samples collected from eight different substrata is presented in Table. No 1. All the substrata showed the presence of thermophilic fungi and many thermophilic fungi occurred in more than one source. The fungi *Aspergillus fumigatus*, *Mucor pusillus* and *Humicola lanuginosa* were present in all the sources. *Thermoascus aurantiacus* was not commonly found in many sources and was isolated only from the soil. Compost, manure and cow manure were rich in thermophilic fungi.

The four techniques used for isolation of thermophilic fungi were compared for their efficiency to obtain different species of fungi and the result is shown in Table No. 2. It is seen from the table that the dilution plate technique using antibiotic YpSs medium and Cooney Emerson's method are most suited to obtain different types of fungi from these sources.

The population of thermophilic fungi as estimated by dilution plate technique in substrata are presented in Table No 3. The population was highest in the buffalo manure, followed by cow manure, compost, bagasse, litter hay, grains and Manure.

The population in the soil samples was the least containing 17 to 34 fungi /gram of dry soil, while the buffalo manure contained as high as 1.17 million fungi / g of dry sample. Although YpSs medium gave higher counts of fungi in most of the sources, there was no difference in the populations obtained on the three media.

The distribution of thermophilic fungi belonging to eight different genera in the eight substrata was studied and the percentage of their population in each of the eight sources is presented in Table No 4. The species of the genus *Humicola* were abundant in most of the sources, being highest in the cow manure. *Mucor* spp. were also widely distributed

Table 1 Distribution of Thermophilic fungi in eight(8) substrata

S.No.	Name	1	2	3	4	5	6	7	8
1	*Aspergillus sp*	-	-	+	-	-	-	+	+
2	*Humicola lanuginosa*	+	+	-	+	+	+	+	-
3	*Mucor rufescens*	+	+	-	-	+		-	+
4	*Mucor pusillus*		+		+	-	+	-	
5	*Thermoascus aurantiascus*	+	-	+	-	+	-	-	+
6	*Humicola insolans*	+	-	+	-	-	+	+	-
7	*Humicola griseus var thermoidea*		+	+	-	-	-	-	+
8	*Humicola sp*	+			+				
9	*Humicola sp*	-	+	+	-	+	-	-	
10	*Humicola sp*	+			+		+		+
11	*Humicola sp*		+			+			
12	*Humicola sp*	+			+		+		+
13	*Aspergillus glaucus*	+	+		+		+	+	+
14	*A.fumigatus*	+		+	+		+	+	-
15	*Rhizomucor sp*		+	-	+	-	-		+
16	*Rhizomucor pusillus*		+	+		+		+	+
17	*Malbranchea sp*	+		+		+		+	
18	*M. pulchella*		+		+		+		+
19	*Chaetomium sp*	+		+		+		+	
20	*C. thermophile*	+	+		+	+	+	-	-
21	*C.thermophile var. coprophile*	+		+		+		+	
22	*Torula sp*	+		+		+		+	+
23	Unidentified 1		+		+		+		
24	Unidentified 2			+		+		+	
25	Unidentified 3		+		+			+	
26	Unidentified 4	+		+		+		+	

1. Compost
2. Buffalo manure
3. Bagasse
4. Manure
5. Poultry litter
6. Cow manure.
7. Hay
8. Jowar grain

Table 2 A comparison of the total number of Thermophilic fungal species obtained by using four different isolation techniques.

S.no	Source Warcup's	Dilution plate technique Cooney						Waksman method	method and Emerson method
		MBR	YpSs	DPYA	MBR	YpSs	DPYA	method	and Emerson method
1	Compost	7	19	5	4	5	5	4	8
2	Buffalo manure	8	11	6	5	4	3	5	6
3	Bagasse	3	4	3	1	2	2	2	6
4	Manure	1	2	1	1	1	1	2	1
5	Poultry litter	-	2	1	1	1	2	3	4
6	Cow manure	5	6	4	2	3	1	5	6
7	Hay	1	2	3	1	2	1	2	3
8	Jowar grains	1	2	1	2	1	2	3	4

MRB-Martin's Rose Bengal agar.

YpSs- Yeast Starch agar.

DPYA- Dextrose Peptone Yeast extract agar.

Table 3 The population of Thermophilic fungi in eight substrates as estimated by the dilution plate technique, on three media. (Incubation temperature 50°C).

S.no	Source	Fungal population in 10^{-3}/g of dry sample		
		MRB	YpSs	DPYA
1	Compost	571-5000	791-4000	597-8000
2	Buffalo manure	1071	117	969-2000
3	Bagasse	6.6280	9.8470	7.3840
4	Manure	0.0204	0.0340	0.0170
5	Poultry litter	3.9900	4.4600	3-1690
6	Cow manure	704-2000	856-9000	586-8000
7	Hay	1.0980	1.0980	0.8241
8	Jowar grains	0.0352	0.0660	0.0616

MRB-Martin's rose Bengal agar.

YpSs- Yeast starch agar.

DPYA- Dextrose peptone yeast extract agar

being highest in the grains. The genera *Torula, Malbranchea* found smaller portion of the total population. The thermotolarent and Thermophilic *Aspergillus* sp were found in all the sources. *Humicola lanuginosa* showed rapid growth at 50 °C which was the highest optimum temperature range among all the fungi. *Thermoascus aurantiacus* had the optima between 47 °C to 50 °C. *Humicola insolens* showed rapid growth at 45 °C. *H.grisea* and *Aspergillus fumigatus* had their optima between 40-43 °C *Malbranchea pulchella* showed best growth at 50 °C. Most rapid growth of *Mucor pusillus* was seen between 50-55 °C. In general the optimum temperature for the growth of thermophilic fungi ranged from 50-55 °C.

Table 4 Percentage populations of eight genera of fungi in eight substrata as estimated by the dilution plate technique on YpSs medium.

S.no.	Source	% Population							
		I	*II*	*III*	*IV*	*V*	*VI*	*VII*	*VIII*
1	Compost	37.8	15.5	7.8	4.4	6.9	15.6	5.6	4.4
2	Buffalo manure	43.5	7.8	7.0	0.0	11.3	17.4	6.9	6.1
3	Bagasse	21.2	42.3	17.3	11.5	7.7	0.0	0.0	0.0
4	Manure	30.0	40.0	10.0	10.0	0.0	10.0	0.0	0.0
5	Poultry litter	23.7	32.0	30.0	5.3	0.0	0.0	0.0	9.0
6	Cow manure	45.5	19.5	16.0	0.0	5.0	6.0	4.0	2.0
7	Hay	12.5	50.0	28.0	9.5	0.0	0.0	0.0	0.0
8	Jowar grains	6.0	46.7	40.6	6.7	0.0	0.0	0.0	0.0

I-	Humicola. sp	V-	Thermoascus sp.
II-	Aspergillus. Sp	VI-	Rhizomucor sp.
III-	Mucor sp.	VII-	Chaetomium sp.
IV-	Torula sp.	VIII-	Malbranchea sp.

Discussion

Abundance of thermophilic fungi in the buffalo manure and compost obtained in the present investigation further supports the earlier findings of many workers (Waksman et al 1939, Blom and Emerson 1962, Fergus 1964, Cooney and Emerson 1964). The thermophillic fungi obtained from compost in this study remarkably agree with those of Klopotek (1962) except that the presence of *Humicola insolans* was doubted by Cooney and Emerson (1964).

In their comprehensive book on thermophilic fungi Cooney and Emerson (1964) make no mention of the bagasse as a source of thermophilic fungi. The bagasse baled or heaped, develops enormous heat inside and forms congenial environment for the development of thermophilic microorganisms (Ramabhadram, 1967). In the present study four obligately thermophilic fungi were isolated from bagasse. The occurrence of two obligately Thermophilic fungi viz; *Humicola lanuginosa* and *Mucor pusillus* forms the first record.

The occurrence of *Humicola lanuginosa and Thermoascus aurantiacus* in the manure is also reported by Apinis (1960,1963a, b) but *Humicola grisea* has not been reported earlier from the manure/soil. The meager thermophilic fungal population in the soil observed in the present study is in conformity with the reports of the Waksman et al (1939) and Apinis (1960). Some reports have recently appeared on the occurrence of thermophilic fungi in birds nest (Apinis and Pugh, 1967). Cooney and Emerson (1964) have reported on the occurrence of *Mucor pusillus* in the living fowl. In the present study *Malbranchea pulchella* and *H.lanuginosa* were found to occur in poultry litter.

Various herbivores dung like horse dung, elephant dung, have been used to isolate thermophilic fungi. Buffalo manure/cow manure is also well known for self heating to become fire fang (compost appearing as lime coated due to the growth of Actinomycetes and fungi). Cooney and Emerson (1964) have reported the occurrence of *Chaetomium thermophile var coprophile* and *Malbranchea pulchella* in goat dung and *H.insolens* and *H.lanuginosa* in the buffalo manure and cow manure. The present analysis of cow manure adds few obligatory thermophilic fungi.viz; *Torula, Mucor* to their list. Mouldy hay yielded few obligately thermophilic fungi. Gregory and Lacey (1963 a & b) isolated *H.stellata* from mouldy hay.

Crisan (1964) has given an account of the possible methods for collecting and isolating thermophilic fungi. He found that Waksman's direct inoculation techniques using Emerson's YpSs medium containing 30 units of streptomycin/ml gave excellent results. The present study also showed the suitability of using antibiotic YpSs medium for isolation of thermophilic fungi. Crisan (1964) also found that initial incubation at higher temperature is advantageous. Accordingly, an initial incubation temperature of 50 °C was preferred in the present study. The optimum and maximum temperatures of growth for all the thermophilic fungi were found to be slightly higher than those given by Cooney and Emerson (1964).

Acknowledgement

Authors are very thankful to University Grants Commission, New Delhi for their financial support.

References

Apinis, A.E 1960. Thermophilic microorganisms in some green land communities. *Abs. .Mycology,* 1:87280,1967.

Apinis, A.E 1963 a, Occurrence of thermophilous micro-fungi in certain alluvial soils near Nottingham. *Nova Hedwigia* 5:57-58.

Apinis, A.E 1963 b, Thermophilous fungi of coastal grasslands. In soil organisms proceeding, of the colloquium on soil fauna. Soil Microflora and their Relationships, Editors J.Doeksen and J. Vander Drift, North Holland, Amsterdam.

Apinis, A.E and Pugh,G.J.F 1967, Thermophilous fungi of birds nests. Mycopath. *Mycol. Appl.* 33:1-9.

Blom, B.D and Emerson, R. 1962. Studies on thermophily in fungi with particular reference to a new Thermophilic Penicillium. *Am. J. Bot.*49:665.

Cooney,D.G and Emerson.R.1964. *Thermophilic fungi,* W.H. freeman, Sanfrancisco.

Crisan,E.V. 1964. Isolation and culture of *Thermophilic fungi.* Contrib. BoyceThompson Inst.22: 291-302.

Fergus,C.L 1964. Thermophilic and thermotolerent molds and actinomycetes of mushroom compost during peak heating. *Mycologia* 61:267-284.

Gregory,P.H and Lacey,M.E. 1963a. Mycological examination of dust from mouldy hay associated with farmer's lung disease. *J. Gen. Microbiol.* 30:75-88.

Gregory,P.H and Lacey,M.E. 1963b. Liberation of spores from mouldy hay. *Trans. Br. Mycol. Soc.* 46:73-80.

Klopotek,A.1962. Uber das vorkommen and verhalten von schimmelpilzen heider kompostierung stadtischer *Abfallstoefe.Antonie von leauvenhock,* 28:141-160.

Ramabhadram,R 1967. Studies on Thermophilic microorganisms in Bagasse, compost and horse dung.Ph.D thesis,Annamalai University Madras state.

Waksman,S.A; umbreit W.W and Cordon, T.C 1939 Thermophilic actinomycetes and fungi in soil and composts. *Soil Sci.*47:37-54.

Studies on Some Aquatic Fungi, *Gonapodya*

R.V. Gandhe and Anagha Kurne

Postgraduate Research Centre, Department of Botany, Modern College,
Pune 411 005, Maharashtra, India.

e-mail : rvgandhe@hotmail.com

ABSTRACT

Two species of *Gonapodya* namely, *Gonapodya prolifera* (Cornu) Fischer and *Gonapodya polymorpha* Thaxter were isolated from closed water bodies on submerged decaying *Acacia* pods during summer and rainy seasons. Sexual and asexual stages were isolated during summer season at the high temperature and low oxygen content. However in rainy season only asexual stages were observed. Both the species growing on submerged *Acacia* pods, form a new host record.

KEY WORDS : *Gonapodya*, New host, Zoosporic fungi

While working on fresh watermoulds from closed water bodies such as ponds, lakes and ditches several interesting Zoosporic Fungi were isolated in different seasons. They were maintained in the laboratory on different sterile baits. However, the species of *Blastocladia* and *Gonapodya* did not show progressive growth on sterile *Opium* and Hemp baits but rapidly developed on decaying *Acacia* pods. These species were very frequently isolated during summer from Lakaki pond, Pune, Maharashtra, India on a new host substrate i.e., *Acacia* pods. They form typical colourless distinct pustules on submerged decaying *Acacia* pods. These decaying *Acacia* pods also developed similar pustules in laboratory conditions in dark. The luxuriant growth of *Gonapodya* species on the pods was observed within 2-3 weeks associated with some microorganisms. The occurrence of *Gonapodya* species was detected by the characteristic brownish colour.

Another important saprophytic fungus, *Blastocladia* was also present along with the *Gonapodya* sp. Many workers have recorded these two genera occurring at same time on decaying plant parts. The species described in this paper are *G. prolifera* (Cornu) Fischer and *G. polymorpha* Thaxter.

Material and Methods

Along with the regular water sample collection, decaying submerged *Acacia* pods were collected from the Lakaki pond to observe exogenous and endogenous growth of Zoosporic fungi. They were inoculated in sterile distilled water and maintained at laboratory conditions. pH and temperature were recorded at the collection site using portable pH meter and the thermometer. Samples were baited with autoclaved *Acacia* pods (Butler, 1907) collected from the same spot. Cultures were observed every day for their progress and maintained.

Identification of different species of Zoosporic fungi were made with the help of monographs and relevant literature (Coker, 1923; Johnson, 1956; John, 1958; Sparrow, 1960,1968,1973; Seymour, 1970; Dick, 1973;Usha Kiran and Dayal, 1988; Khulbe, 2001)

Physicochemical parameters were studied by applying standard methods (APHA, AWWA, and WEE, 1992).

Observations

Gonapodya polymorpha Thaxter

Thallus eucarpic, polycentric, branched, characteristic brownish in appearance, mycelium composed of hyphae which are divided into long segments by pseudoseptae, sometimes slender hyphae without septation, septations without constrictions, (Fig. 1); sporangia born racemosely or in clusters, 3 - 4 sporangia growing at the rounded tip of hypha; (Fig. 2), shorter, stouter, distal portion of sporangia not elongated, tapering gradually to a blunt apex with an apical perforation (Fig. 3), mostly 48-64 × 32-48 µm, hyphal growth beyond the zoosporangial apical perforation (Fig. 4); zoospores variable in shape, release one by one from the apical pore of the sporangia; female gametangia spherical to ovoid but mostly spherical (Fig. 5, 6, 7), born in racemes or catenulate, terminal or intercalary; characteristic yellowish brown in appearance, 16-47 µm in diameter, mostly 28 - 38 µm, female gametes discharged by forming beak like papillae from gametangial wall (fig.8,9), length ranges from 9- 53 µm, diameter ranges from 9-13 µm, female gametes 9 µm in diameter, 8 to 18 in number; male gametangia ovoid with short distal portion tapering to blunt apex, 38 µm in diameter; male gametes are with refractive granules, 3 –5 µm in diameter.

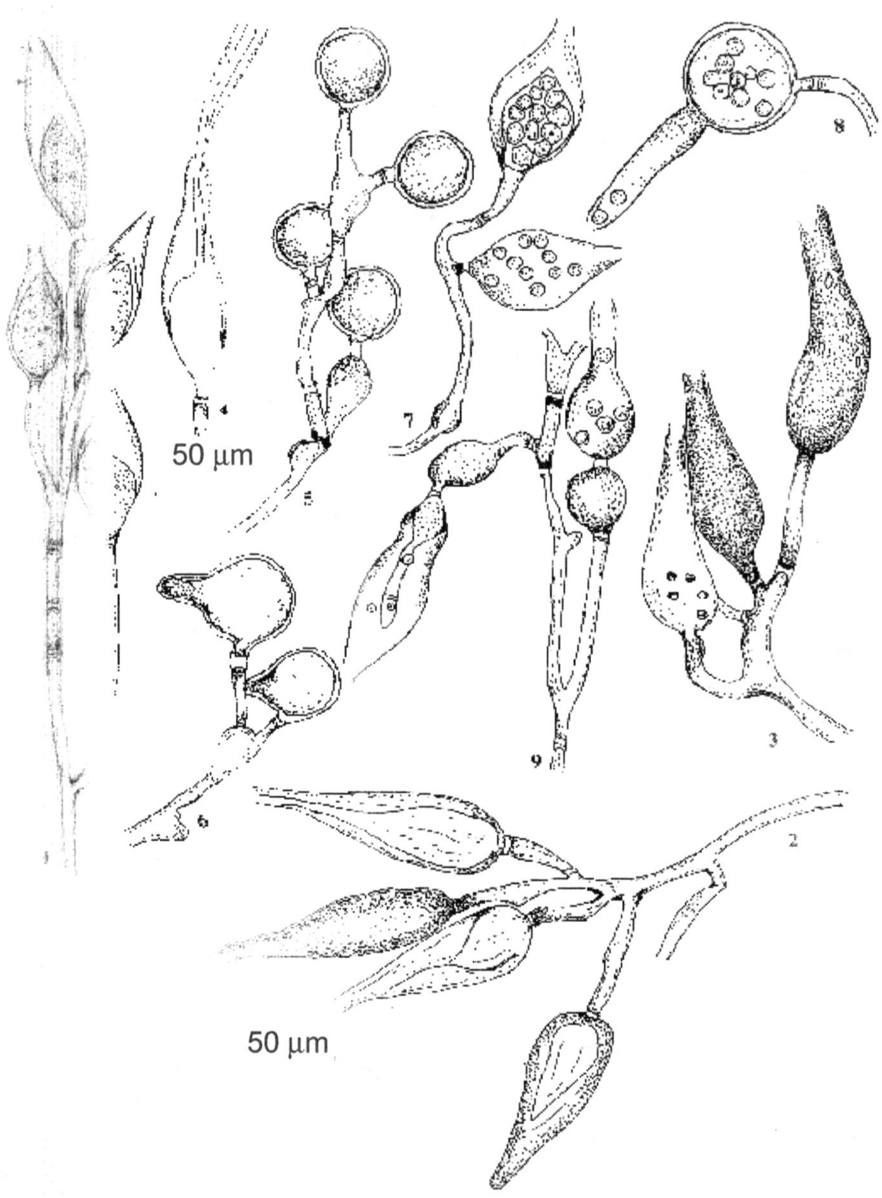

Figure 1 *Gonapodya polymorpha Thaxter* **1.** Hyphae divided into long segments by pseudoseptae. **2.** Sporangia in cluster **3.** Zoosporangia with blunt apex and zoospores **4.** Hyphal growth beyond the zoosporangial apical perforation **5 and 6.** Spherical female gametangia **7.** Female and Male gametangia on one branch **8 and 9.** Female gametangia with beak like papillae

Gonapodya prolifera (Cornu) Fischer

Thallus eucarpic, polycentric, branched, characteristic brownish in appearance. Mycelium composed of hyphae, divided into clavate segments by pseudoseptae, constrictions are clear in this specie as compared to that of *G.polymorpha* (Fig. 10); sporangia terminal, solitary or born in clusters, 3-4 sporangia growing at rounded tip of hyphae (Fig. 11), distal portion elongated, tapering gradually to a narrow apex with an apical perforation (Fig. 12), one to four times proliferous, (Fig. 13), some terminal sporangia are dark brown in colour, 50 to 122 µm in length mostly 63 to 82 µm, 16 to 22 µm in diameter; below each sporangium pseudoseptum is present; zoospores variable in shape, release one by one from the apical pore of the sporangium; female gametangia terminal, elongate to ovoid, with long tips (Fig. 14), 53 µm- 79 µm in length, diameter of female gametangia at base is 19 µm to 22 µm while at apex it is 6 to 9 µm 2-3 times proliferous (Fig. 14); 10-15 in number, spherical, 9 µm in diameter; male gametangia terminal, elongate to ovoid (fig.15 and 16), 50- 69 µm in length and 19 to 25 µm in diameter, male gametes are with refractive granules, 3–6 µm in diameter.

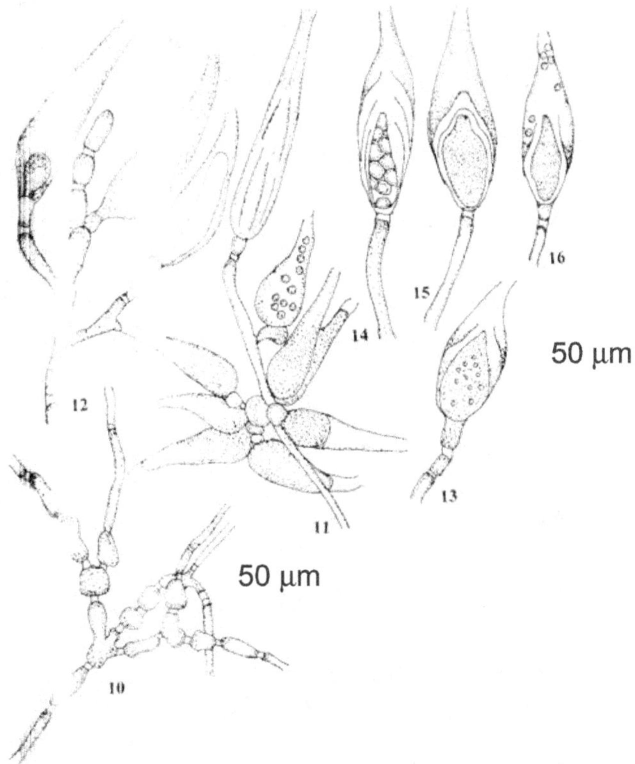

Figure 2 *Gonapodya polymorpha Thaxter* **10.** Hyphae with constrictions **11.** Cluster of sporangia, note terminal, solitory, tapering sporangium **12.**Zoosporangia **13.** Proliferating zoosporangium with zoospores **14.** Proliferating female gametangium **15 & 16** Male gametangia

Discussion

Von Minden (1916) and A. Lund (1934) pointed out that *Gonapodya, Blastocladia* may grow under conditions of low Oxygen content and generally foul environment. Emerson and Cantino (1948, 1949 a, b) conducted laboratory studies on *Blastocladia* and *Pythiogeton* and showed that such fungi occupy these habitats because they are capable of carrying on a fermentative type of metabolism. We have isolated both asexual and sexual stages under low oxygen content 0.28 to 2.76 ppm in April to May 2002. However, during April to June 2003 asexual stage was frequently isolated at range of high oxygen content 5.6–8.1 ppm and in the temperature range of 26 °C to 31 °C from the same site. Therefore our observations differ to the observations made by earlier workers. We have pointed luxuriant Gonapodial growth in both the summer and rainy seasons. During summer season we observed sexual as well as asexual stages. In rainy season when the temperature falls down and humidity increases in surroundings only asexual stages were isolated with luxuriant growth of female gametangia. However these female gametangia failed to develop oospores. Rachel John (1953) has recorded typical pale gray colour hyphae of the *G.polymorpha* but we have observed brownish yellow appearance. The morphology of the species is more or less similar to the published descriptions of Thaxter (1895), Rachel John (1953), Benjamin and Johns (1954), Sparrow (1960), Usha Kiran and Dayal (1992) and Fuller and Clay (1993). We have isolated both the species on the same host. Although it was very difficult to study them separately, the main difference in both the species was the divisions of the segments, which was distinct in *G.prolifera* and almost indistinct in *G.polymorpha*

According to Karling (1977) *Gonapodya spp.* occur naturally on submerged fruits under fairly foul environmental conditions .We have also isolated these two species on decaying submerged *Acacia* pods under fairly foul environment. Our observations are similar to that of Karling.

Acknowledgement

The authors are grateful to the authorities of Modern College, Pune for providing laboratory facilities during this work.

References

APHA, AWWA and **WEE** 1992. *Standard methods for the examination of water and wastewater.* Amer. Health Assoc. Publ. New York. 18th Edition.

Butler, E.J. 1907. *An account of the genus Pythium and some Chytridiaceae.* Mem. Dept. Agri. India. Bot. Ser. 1:pp.1-160.

Coker, W.C. 1923. *The saprolegniaceae with notes on other watermoulds.* The university of North Carolina Press, Chapel Hill, North Carolina, pp. 22-76.

Dick, M.W. 1973. *Saprolegniales. In* : The Fungi. An Advanced Treatise. (Eds. Ainsworth, G.C., F.K. Sparrow and A.S. Sussman,) Academic Press, New York and London. Vol. IVB, pp. 113-144.

Fuller, M.S. and **Clay, R.P.** 1993. Observations of Gonapodya in pure culture: growth, development and cell wall characterization. *Mycologia,* 85 (1): pp. 38-45.

Johns, R.M. and **Benjamin, R.K.** 1954. Sexual reproduction in Gonapodya. *Mycologia,* 46 : pp. 201-08

Johnson, T.W. 1956. *The genus Achlya morphology and taxonomy.* Univ. Michigan Press, Ann. Arbor, Michigan pp.180

John, Rachel 1953. Studies in the Indian aquatic fungi. III. Sexual reproduction in *Gonapodya polymorpha* Thaxter. *Proc. Ind. Acad. Sci.*, B, Vol. 50, 38, pp.253-259.

Khulbe, R.D. 2001. *A manual of aquatic fungi (Chytridiomycetes & Oomycetes).* Daya publishing house, Delhi, pp. 204

Sparrow, F.K. 1960. *Aquatic phycomycetes* (2nd edition). University of Michigan press. Ann. Arbor.1187 pp.

Sparrow, F.K. Jr. 1968. *Ecology of fresh water fungi, In* 'The Fungi, An Advanced Treatise' (Eds. G.C.Ainsworth and A.S. Sussman) Academic Press, New York. Vol. III: pp.95-105.

Seymour, R.L. 1970. The genus Saprolegnia. *Nova Hedwigia* 19: pp. 1-124.

Sparrow, F.K. Jr. 1973. Mastigomycotina (Zoosporic Fungi), In the Fungi. (Eds. G.C.Ainsworth, F.K. Sparrow and A.S. Sussman.) Academic Press, New York and London. Vol. VI B. pp. 61-73

Thaxter, R. 1895. New or peculiar aquatic fungi. II. Gonapodya Fischer and Myrioblepharis, nov. gen. *Bot. Gaz.,* 20 : pp.477-485.

Usha Kiran and **Dayal, R.** 1992. A new host record of Blastocladia and Gonapodya. *Proc. Nat. Acad. Sci., India* 62 (B) III. pp. 503-506.

Von Minden (1916), A. Lund (1934), Emerson and Cantino (1948) and (1949 a, b), Karling (1977) *In* Sparrow, F.K. Jr. (1968). *Ecology of fresh water fungi, In* 'The Fungi, An Advanced Treatise' (Eds. G.C.Ainsworth and A.S. Sussman) Academic Press, New York. Vol. III: pp.95-105.

10 | *Phoma Sacc. Distribution and Secondary Metabolite Production*

M.K.Rai, S.Sonar, Prajakta Deshmukh, A.K.Pandey*
G.J.Kövics and R.C.Rajak***

Deptt. of Biotechnology, Amravati University, Amravati - 444 602, Maharashtra, India

*Deptt. of Biological Science, Rani Durgawati University, Jabalpur, M.P., India

**Debrecen University, Faculty of Agriculture, Deptt. of Plant Protection H-4015 Debrecen, P.O. Box 36, Hungary,

ABSTRACT

Phoma Sacc. is an ubiquitious fungus, which has been reported from plants, soil, human beings, animals and air. Some species of *Phoma* secret secondary metabolites. The pigment producing species of *Phoma*, viz., *P. sorghina, P. herbarum, P. exigua* var. *exigua, P. macrostoma* var. *macrostoma* and *P. glomerata* have great potential for the production of dyes. Some species produce pharmaceutically active metabolites, like-squalestatin-1 (S1) and squalestatin-2 (S2). The secondary metabolites secreted by species of *Phoma* are antitumour, antimicrobial and anti HIV. Equisetin and phomasetin obtained from species of *Phoma* are useful against AIDS. The *Phoma* spp. can be exploited for the production of dyes, mycopesticides and agrophytochemicals. The secondary metabolite production by species of *Phoma* and their utilization as dyes, antibiotics and as biocontrol agents is discussed.

Keywords : *Phoma,* Secondary metabolites, antibiotics, dyes.

There are many fungi which produce a variety of metabolites. These metabolites may influence the functioning of the fungus and are thus important indirectly for its growth and survival. There is diversity in chemical structure and function of these secondary metabolites. Secondary metabolites can be defined as the compounds produced by the higher plants or fungi that are not essential to their basic metabolism. The compounds are generally produced following active growth, and may have an unusual chemical structure. Some metabolites are found in many related fungi, while others are only found in one or a few species. The restricted distribution implies a lack of general function of secondary metabolites in fungi (Mann,1986; Vining, 1990). In fact, six of the twenty most commonly prescribed medications are of fungal origin. These metabolites have been subjected to combinatorial chemistry following growth in selective media. Some metabolites are toxic to humans and other to animals. Yet others can modify the growth and metabolism of plants. Interestingly, the most important secondary metabolites seem to be synthesised from one or a combination of three biosynthetic pathways: polyketides arising from Acetyl Coenzyme A, mevalonate pathway also arises from Acetyl Coenzyme A, and from amino acids.

Phoma species are reported from plants (Tandon and Bilgrami, 1960; Bilgrami, 1963; Agarwal and Sahni, 1964; Dutta and Ghosh, 1965; Chandra and Tandon, 1965, 1966; Hasija, 1966; Pavgi and Singh, 1966; Shreemali, 1972, 1973, 1978; Jamaluddin *et al.*, 1975; Bhowmik and Singh, 1976; Raut, 1977; Kanaujia, 1978; Maiti *et al.*, 1978; Kamal and Singh, 1979; Rai and Misra, 1981; Agrawal and Misra, 1981; Rao and Thirumalachar, 1981; Wadje and Deshpande, 1981), as saprophytes from soil (Mathur and Thirumalachar, 1959; Dorenbosch, 1970).   There are few cases of *Phoma* infection in human beings (Perazzi, 1925; Pollacci, 1935; Agostini and Tredici, 1937; Shukla *et al.*, 1984; Rai, 1989).

Boerema and his co-workers gave a significant contribution in the field of *Phoma* (Boerema, 1964, 1976, 1993, 1997, Brewer and Boerema, 1965; Boerema *et al.*, 1965, 1968, 1971, 1973, 1977, 1981, 1994, 1996, 1997, 1999; Boerema and Howeler, 1967; Boerema and Dorenbosch,1970, 1973; Boerema and Bollen, 1975; Boerema and van Kesteren, 1981; Boerema and Loerakker, 1981,1985; Gruyter and Nooderloos, 1992; Gruyter *et al.*, 1993, 1998; Nooderloos *et al.*, 1993; Boerema and Gruyter, 1998, 1999).

The cultural and morphological studies of Indian species of *Phoma* were made and on the basis of these studies 20 broad morphological groups were formed (Rai, 1985, 1986a, 1986b, 1987, 1989, 2000; Rai and Rajak, 1982a, 1982b, 1983a, 1983b, 1985, 1986a, 1986b, 1993; Rajak and Rai, 1982a, 1982b, 1983a, 1983b, 1984a, 1984b, 1985). These groups include *P. medicaginis* Malbr. and Roum. var. *pinodella* (Jones) Boerema, *P. medicaginis* var. *medicaginis* Malbr. and Roum., *P. pomorum* Thum., *P. herbarum* Westend., *P. exigua* var. *exigua* Desm., *P. tropica* Schneider and Boerema, *P. glomerata* (Corda) Wollenweber Hochapfel, *P. sorghina* (Sacc.) Boerema, *P. multirostrata* (Mathur *et al.*) Dorenbosch and Boerema, *P. capitulum* Pawar *et al.*, *P. betae* Frank, *P. jolyana* Pirozynski and Morgan-jones, *P. fimeti* Brun., *P. chrysanthemicola* Hollos., *P. complanata* (Tode ex Fr.) Desm., *P. destructiva* Plowr., *P. eupyrena* Sacc., and *P. arachidicola* Marasas *et al.*

The present chapter discusses the secondary metabolite production by different species of *Phoma* in general and toxins, dyes and antibiotic production in particular..

Role in Biotechnology

Various species of the genus *Phoma* produce several phytotoxic secondary metabolites, some of them are as follows :

***Phoma lingam* (Tode ex. Fr.) Desm.**

The fungus produces two important phytotoxic compounds i.e. Sirodesmin PL and Deacetylsirodesmin PL (Ferezou *et al.*, 1977, Fig. 1). Phomamide (cyclo-O-(yy-dimetheylallyl)-L-serine a phytotoxic intermediate compound has been extracted from this fungus during Sirodesmin synthesis (Vining and Wright 1977; Curtis *et. al.* 1977; Ferezou *et al* 1980a, b., Stoessl, 1981).

Pedras and Biesenthal (2000) presented secondary metabolite profiles of 26 isolates of the blackleg fungus *Leptosphaeria maculans* (Desm.) Ces. Et de Not., sexual stage of *Phoma lingam*. These secondary metabolites are used for differentiation of *Phoma* blackleg isolates. Based on HPLC analyses, 26 isolates can be placed in three main groups as below :

   (i)   isolates producing phomamide and sirodesmins

   (ii)   isolates producing indolyldioxopiperazines

   (iii)   isolates producing polyketides

Phytotoxin or metabolite	Reference
Phomalide	Pedras *et al.*, 1993
Phomapyrone A	Pedras *et al.*, 1994
Phomamide and Sirodesmins	Pedras, 1998
L-valyl-L-tryptophan anhydride	Pedras *et al.*, 1998
Phomalairdenone	Pedras *et al.*, 1999

Phoma herbarum

Brefeldin A, a component with pronounced phytotoxic property has been isolated from several fungi including this fungus (Mortimer *et al* 1979). Structure and synthetic pathways of the compound have been discussed in many publications (Sigg, 1964; Suzuki *et.al.* 1970; Weber *et al*, 1971; Mabuni *et.al.* 1979). Pandey *et. al.* (2002) reported severe phytotoxicity of cell free culture filtrate obtained from *P. herbarum* FGCC# 3 and 4 against *L. camara*. They recorded severe chlorosis, curling and finally complete collapse of leaves within 48 hours of treatment. Toxic metabolites are produced by FGCC# 3 are thermostable and thermotolerant. Compound isolated with benzene solvent showed maximum phytotoxicity. Recently, Vikrant (2002) reported pronounced phytotoxicity of CFCF of *P. herbarum* FGCC#75 against *Parthenium. hysterophorus*.

Phoma exigua

Phomenon is a highly phytotoxic eremophilane compound isolated from this fungus. Rothweiler and Tomm(1966, 1970) has purified and characterized the *Phomin* and Dehydrophomin from the culture filtrate of *P. exigua* var *exigua* (Desm.) Boerema.

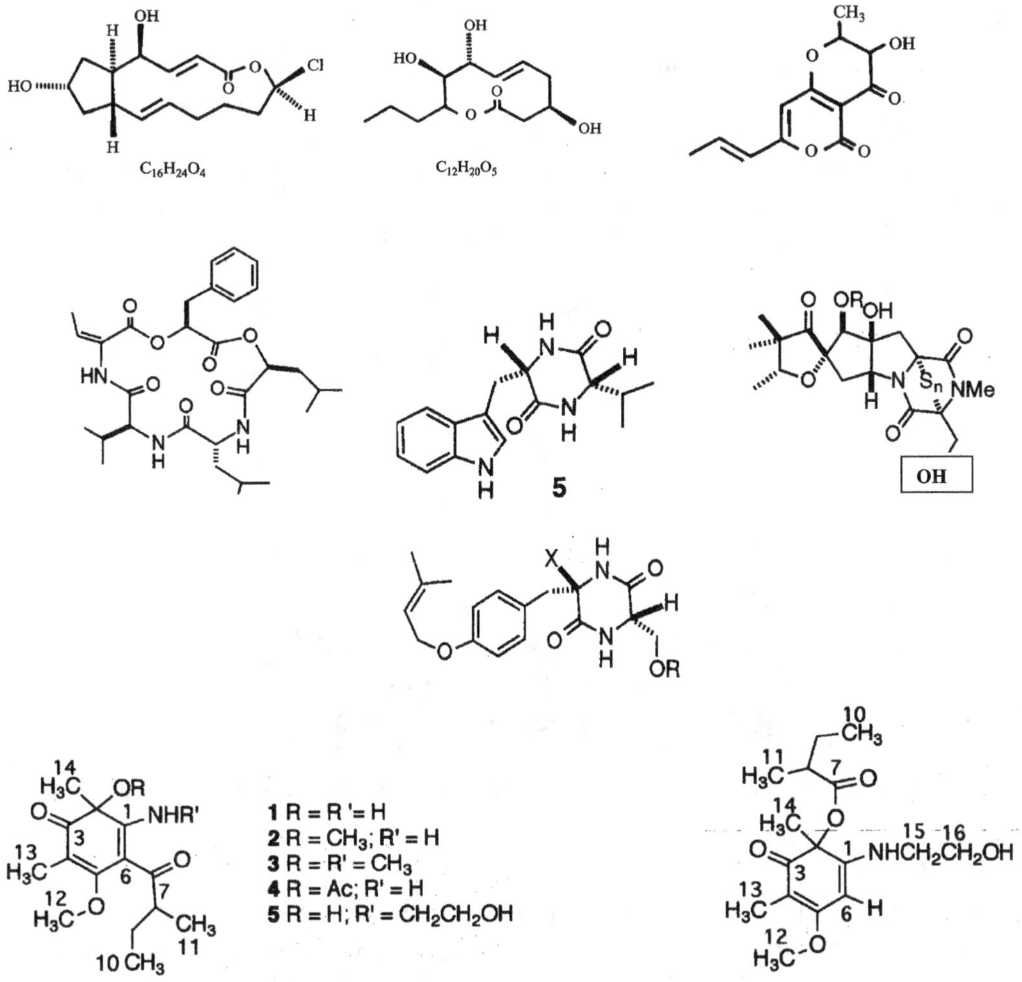

Source: Noordeloos et al. (1993), Pedras et al. (1995, 1999)

Figure 1 Pigments and crystals from *P. medicaginis* 1-3 (Breveldin A, Pinodellalide A, Radicinin) ; 4-7 Metabolites from *P. lingam* (Phomalide, sirodesmins, phomalairdenone); 8-9 Compounds from *P. lingam* and *P. wasabi.*

Phytotoxicity of Phomin has been demonstrated. Other phytotoxic Cytochalasins viz. Deoxaphomin, Proxiphomin, Protophomin and Cytocnalasin A, B, have also been extracted from this pathogen (Binder andTomm 1973a,b; Scottt *et al* 1975; Rothweiler and Tomm 1966). Cytochalasin A,B,F,T, UV, Assochalasin, Deoxaphomin and 7-0-acetylcytochalasin B have also been extracted from *Phoma exigua* var *heteromorpha* (Capasso *et al* 1991a, b; Evidenente *et al* 1992). Vurro *et al* (1997) have extensively reviewed the technological and biological aspects of these compounds.

Phoma sorghina

The fungus incites leaf infection in Pokeweed (*Phytolacca americana* L.). Several phytotoxins viz., diphenyl ether, epoxydon, desoxyepoxydon, phyllostine, 6-methyl salicyalate ether have been produced by this pathogen. Among these epoxydons have shown very high broad spectrum toxicity to both mono and dicot weed species (Venkatasubbaian et al 1992; Abbas and Duke, 1997). Tenuazonic acid in the form of Mg and Ca chelates was also obtained from this species (Steyn and Rabie, 1976).

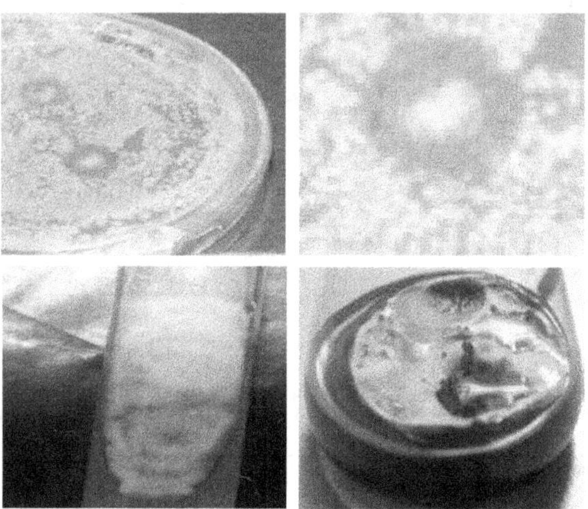

Figure 2   In vitro production of anthraquinone pigments by P. sorghina

Phoma maconaldii

Zinnol a broad spectrum phytotoxin has been extracted from this pathogen responsible for leaf and stem blight and withering of cut seedlings of many plants (Sugawara and Strobel 1986). The toxin binds to specific binding sites and associated with stimulation of Ca^{++} uptake by the affected cells at only 0.1-1.0 mm levels. Calcium regulated cell processes may be disrupted by this toxin. It's relatively simple structure could be the basis for herbicide discovery studies (Abbas and Duke. 1997).

Phoma trachciphilia

Culture filtrate of this pathogen is very rich in secondary metabolites of high phytotoxic properties (Yoder, 1980; Gentile et. al. 1992;Nachmias et al 1977, 1979).

Phoma foveata

Pachybasin, a phytotoxin has been extracted from this pathogen by Strange (1997).

It is evident from the above discussion that *Phoma* spp. could be the novel agents for many weed problems. They have both mycoherbicidal and biorational properties. Looking to the number of known species, species associated with weeds are very less. This might be because of ignorance of weed pathogens. Mycologists as well as plant pathologists in the past have given attention to economically important plant diseases. Therefore, urgent attention towards herbicidal potentiality of *Phoma* is needed.

Antibiotic production

There are some important antibiotics produced by some species of *Phoma* (Pearce 1997; Singh *et al.*, 1997; Baxter *et al.*, 1998). Pharmaceutically active metabolites squalestatin-1 (S1) and squalestatin-2 (S2) are produced by a *Phoma* sp. (IMI 332962) (Baxter *et al.*, 1998). Singh *et al.* (1997) reported production of some antitumour agents by *Phoma* sp (MF 6118). These include Fusidienol-A, which is the second member of the fusidienol family of inhibitors to possess a novel tricyclic oxygen-containing heterocycle with a 7/6/6 ring system. Equisetin and phomasetin obtained from species of *Phoma* are useful against AIDS (Singh *et al.*, 1998). Dombrowski *et al.* (1999) isolated equisetin derivatived from *Phoma* sp. (MF6070). This showed HIV virus integrase inhibitor. Due to its potential, the fungus can be used against AIDS. Some marine micro-organisms including *Phoma* are reported to possess antibiotic activities in them (Sponga *et al.*, 1999). The marine species of *Phoma* showed antimycotic activities against *Enterococcus faecium*, *E. coli*, and *Candida albicans*.

Table 1 Antibiotic production by sopecies of *Phoma*

Phoma sp.	Active chemical/product	Reference
Phoma exigua var. exigua	Antibiotic E (antibacterial and antifungal) Cytochalasin B	Boerema and Howeler (1967)
Phoma pigmentivora	LL-D253alpha	McIntyre *et al.* (1984)
Phoma lingam	Phomenoic acid and phomenolactone (antifungal and antibacterial)	Topgi *et al.* (1987)
Phoma sp.	Equisetin (anti HIV)	Dombrowski *et al.*, (1999)
Phoma sp. Marine	Phomactin A	Sugano *et al.* (1991)
P. exigua var. heteromorpha	Cytochalasin F	Capasso *et al.* (1991)
Phoma sp.	Squalestatin (antiinfective agents)	Dawson *et al.*, 1992 Sidebottom *et al.* (1992); Pearce (1997)
Phoma sp.	Squalestatin 1, 2 (S1, S2)	Baxter *et al.* (1992)
Phoma sp.	Antiviral agent (against HIV) Equisetin	Hazuda *et al.* (1999)
Phoma sp.	Antiviral agent (against HIV) Equisetin and Phomasetin	Singh *et al.* (1998)
Phoma sp.	Antibiotic AB 5362-A Antibiotic AB 5362-B Antibiotic AB 5362-C	–

Table 1 *Contd...*

Phoma sp.	Active chemical/product	Reference
Phoma sp.	Antitumour (Fusidienol A)	Singh *et al.* (1997)
Phoma sp.	Antibiotic, antitumour, RKS-1778	Kakeya *et al.* (1997)
Phoma sp.	Antitumour (Phomacins)	Alvi *et al.* (1997)
Phoma sp.	Pesticide (Octahydronaphthol derivative MK8383, Patent)	Wakui *et al.*, 1999
Phoma sp.	Antibiotic activities	Sponga *et al.* (1999)
Phoma sp. (NRRL 25697)	Phomadecalins A-D and Phomapentenone A	Yongsheng *et al.*, (2002)
Phoma sp	YM-202204, antifungal	Nagai *et al.* (2002)
Phoma sp.	FOM-8108, inhibitors of neutral sphingomyelinase	Yamaguchi *et al.* (2002)

Active biocontrol agents

Weeds are serious problems to not only the agricultural and forestry fields, but also responsible for several major problems to human and animal health World over including India. Synthetic chemical herbicides have been the mainstay for weed control practices since the end of World War II and no doubt are responsible for much of the unparalleled increased crop productivity that has occurred during this period. The high costs involved in developing and registering chemical herbicides and recent trends in environmental awareness have prompted researchers to investigate alternative systems of weed control. Ideally, such a system would control target weeds at or near the same levels as that achieved with chemical herbicides while at the same time not posing a threat to either the environment or non-target organisms (Pandey *et. al.*, 2001).

The Science and Technology of weed control by using plant pathogens more especially fungal pathogens as an effective alternative to chemical herbicides have gained significant attention and momentum in 1970s. Century old concept in weed control and plant disease epidemiology were successfully put to text and few economically important weeds were controlled by fungal pathogens used under classical and mycoherbicidal strategies. With the advancement in the knowledge biorationals and integrated management strategies have also came into foray. Biological, technological and economical perspectives of various strategies have been extensively reviewed in several publications (Auld, 1990; Abbas and Duke, 1995, 1997; Boyette and Abbas, 1995; Charudattan, 1991, 1996; Hasija *et.al* 1994; Hoagland, 1990ab, 1999, 2001; Pandey, 1999, 2000; Pandey *et al.* 1996ab, 1997, 2001; Saxena and Pandey 2000; Saxena *et al* 2001).

Phoma is a cosmopolitan fungus, species of this occurs in wide variety of substrates ranging from various parts of plants, soil, water, air and even human and animals. Several species of the genus live parasitically or saprophytically on plants and often associated with distinct disease symptoms viz., leaf spots, lesions on stem, fruits or even on roots (tubers), damping off, die-back, seed and fruit rots and seedling blight etc. it is surprising that despite of excellent phytopathogenic potential shown by various species or varieties of the genus, their mycoherbicidal potential has been ignored significantly. There are only few attempts have been made to evaluate them as mycoherbicides. Heiny (1990, 1994) isolated a highly host specific strain of *Phoma proboscis* from diseased parts of field bindweed (*Convolvulus arvensis*). Heiny and Templeton (1991) have reported very high mycoherbicidal efficiency when the agent was applied to the seedling of the weed with a temperature range from 16-28 ^{0}C and more than 9 hrs dew period. Heiny (1994) have extensively evaluated the compatibility of synthetic herbicides for integration with the mycoherbicidal agent.

Rajak *et al.* (1990) isolated a strain of *P. herbarum* from diseased leaves of *Parthenium* (*Parthenium hysterophorus* L.) collected from Central India. The fungus when applied @ 2.3 x 10^2 spores/ml caused more than 90% inhibition in seed germination, seedling mortality and leaf damage followed by reduction in height of the weed. Pandey and Pandey (2000) recovered three strains of *P. herbarum* (LC#32,37,39) from diseased leaves and stem of *Lantana camara*. All the three strains incites severe infection in the weed specially at seedling stage (Pandey 2000). Recently Pandey (2002) also isolated a strain of *P. herbarum* FGCC# 70 from disease seedlings of an invasive weed, *Hyptis suaveolens* (L) poit. She recorded very high mycoherbicidal potential when seedlings were treated @ 2.0x10^5 spores/ml at 28 ^{0}C + 1 ^{0}C, 24 hrs dew period. Several other species viz., *Phoma campaulata* on *Cassia fistula* (Rajak and Rai. 1982), *P. exigua* on *Sesamum indicum* (Singh and Agarwal. 1973), *P. eupyrena* on *Achyrenthus aspera* (Khanna and Chandra.1977), *P. glomerata* on *Crotalaria juncia* (Pathak and Chauhan. 1976), and *Parthenium hysterophorus* (Padmbai Luke.1976), *P. lantanae* on *Lantana camara* (Singh and Agarwal. 1974), *P. palmarum* on *Calotropis procera* (Khanna and Chandra. 1977; Kamal and Singh.1979), *P. tridacis* on *Tridax procumbens* (Wehmeyer. 1964), *P. herbarum* var. *ipomoeae* (Kamal and Singh.1979), and *P. euphorbiae* on *Euphorbia hirta* (Rangaswami *et al.*, 1970) have been reported from various parts of India.

Excellent pesticidal activity against some phytopathogenic fungi was reported in octahydronaphthol derivative MK8383 of a *Phoma* sp. (Wakui *et al.*, 1999). Club root of cruciferous plants can be controlled by *Phoma glomerata* and its product epoxydon (Arie *et al.*, 1998, 1999). Raymond *et al.* (2000) reported that *P. glomerata* has ability to colonize and finally inhibit the growth of powdery mildew on oak and thus can be used as a biocontrol agent. In fact, powdery mildews are common plant pathogens, which can be easily identified by formation of profuse white powder-like conidia and mycelia. For control of powdery mildews, only one fungus *Ampelomyces quisqualis* Ces. is known. Therefore, enormous opportunities and possibilities are there to utilize this mycoparasitic fungus as a potential bio-control agent. Pandey (2000) evaluated the mycoherbicidal potential of indigenous fungal pathogens against *Lantana camara* L., which is an obnoxious weed of India. He isolated about 40 fungi from different parts of Madhya Pradesh.

Out of these isolates, the maximum efficacy was shown by *Alternaria alternata* followed by *Phoma multirostrata*. The later is thermophillic and originally reported from India by Mathur and Thirumalachar (1959). It is widely distributed in subtropical, tropical regions and warm green houses.

Production of Dyes

Earlier all colorants were obtained from natural sources: plants, lichens, insects or shellfish (Hobson *et al.*, 1997). Anthraquinones were predominants among these as exemplified by red dyes kermes from the insect *Kermocococcus ibicish* (Thompson, 1971). The natural dyes are better than the man-made dyes in fastness and brilliance (Taylor, 1986). Today, synthetic anthraquinone dyes are the second most important group of organic colorants listed in the colour index and tend to predominate in the violet, blue and green hue sectors.

The chemical synthesis of anthraquinones require the use of strong acids at high temperature and heavy metal catalysts, as a consequence of which environmentally hazardous effluents and byproducts are produced. With increasing awareness of the environment degradation by industry the disposal of industrial effluent is becoming more costly and strictly regulated. A return to dyes extracted from molluscs, insects and plants grown in their environment is not the intention. These methods are not commercially practical and can in many ways be regarded as environmentally unfriendly and totally impractical. For example 150 000 insects reared on 0.16 hectares of cactus plants are required to produce 1 Kg of cochineal dye. In order to obtain the dye, the insects are put in bags and dried slowly by long exposure to the sun. However, a microbial source cultivated in bioreactors is by no means out of question.

There are many fungi which produce anthraquinones as a secondary metabolite. Fungal anthraquinones as polyketide-derived secondary metabolite occur widely in many genera of fungi. Compared with the commercially available hydroxyanthraquinones most possess an adiitional methyl substitution in position three, e.g. emodin and this allows a study of the effect of such a group on the dyeing properties of dyestuffs derived from them. A fungal anthraquinone cynodontin (1,4,5,8-tetra hydroxy-3-methylanthraquinone) was produced in sufficient purity to allow it to be transformed using a simple chemical step to a dye product and this was compared with a comercially available close analogue (Hobson *et al.*, 1997).

Phoma exigua Desm. produces pigments. Bick and Rhee (1966) have reported that *P. exigua* var. *foveata* contains many anthraquinone pigments, such as, pachybasin, chrysophanol, emodin and phomarin. In acid condition, this complex of pigments becomes yellow and in alkaline conditions red. This character is based on ammonia test described by Logan and Khan (1969). On malt-agar, *P. exigua* var. *foveata* gives pinkish colour after exposure to ammonia. This is due to reaction of diffusible anthraquinones and their reaction with ammonia. In old cultures, anthraquinone pigments crystallize out as yellow-green crystals. Tichelaar (1974) found that the fungicide thiophanate-methyl accelerates and increases the crystallization process of the pigments. Both *P. exigua var exigua* and *P. exigua* var. *inoxydabilis* produce cytochalasin B, which is also known as "phomine" (Bousquet and Barbier, 1972; Scott *et al.*, 1975)

Some isolates of *P. exigua* var. *foveata* produce antibiotic substance E similar to isolates of the ubiquitous *P. exigua* var. *exigua* (Boerema and Howeler, 1967). It is a colourless substance and can easily be demonstrated in cultures by sodium hydroxide test. On application of a drop of sodium hydroxide at the margin of colonies on malt agar oxidation takes place and pigment alpha converts into pigment beta. Pigment alpha is red-purple at pH <10.5 and blue-green at pH >12.5. Pigment beta is yellow at ph <3.5 and red at pH >5.5. Anthraquinone pigments are also produced by some other *Phoma* species.

Some other species of ***Phoma*** also produce anthraquinones and phomalgin A (Pedras *et al.*, 1995). The potential of these pigments can be utilized commercially.

Conclusion and Future Outlook

The application of microbial secondary metabolites in various fields of biotechnology has attracted the researchers. The reasons for the production of secondary metabolites in species of *Phoma* are yet to be investigated. It is assumed that these active metabolites are mediators of interactions between organisms. However, their role in fitness of the producer and evolutionary implications of these effects remain unknown. Because the synthesis of secondary metabolites imposes substantial metabolic costs, tight control of the synthesis improves the fitness of the producer. Knowledge of conditions inducing the synthesis of a particular metabolite helps developing hypotheses about its biological function.

In one hand, the metabolites of some species of *Phoma* are responsible for heavy economic loss and generate health problems as contaminants of plant and plant-based products. New insights into the effect of environmental conditions on the synthesis of undesirable metabolites may help limit the exposure of humans and farm animals to these toxins. On the otherhand, there is increasing demand of metabolites of some other species of *Phoma* as these are exploited in medicine and industry. However, the identification of these metabolites is not easy. Modern MS and NMR instrumentation can greatly assist the analyst in providing high quality data in a more rapid and time-efficient manner. Screening of different species of *Phoma* for the production of different antibiotics and dyes would be an excellent idea, which may open up new vistas in the field of secondary metabolite production. Finally, preparation of a database of different species of *Phoma* will facilitate the selection of proper species for the production of beneficial metabolites.

Acknowledgements

MKR is grateful to Dr G H Boerema, former Head, Plant Protection Institute, Wageningen, The Netherlands; and to Professor M. Soledade C. Pedras, Department of Chemistry, University of Saskatchewan, Saskatchewan, Canada for supply of literature. Thanks are also due to Dr Gruyter, Director, Plant Protection Institute, Wageningen, The Netherlands.

References

Abbas, H.K. and **Duke, S.O.** 1995. Phytotoxins from plant pathogens as potential herbicides. J. Toxicol Toxin Review 14: 523-543.

Abbas, H.K. and **Duke, S.O.** 1997. Plant pathogens and their phytotoxins as herbicides.pp1-20 In: Toxins in Plant Diseases Development and Evolving Biotechnology (eds. R. K. Upadhyay and K.G. Mukherjee) Oxford and IBH Publishing CO. Pvt. Ltd. New Delhi.

Agostini, A. and **Tredici, V.** 1937 Sopra una nuova specie di micete commensale (Phoma hominis Agostini et Tredici) isolato da forme cliniche del derma. Atti Ist Bot. Univ. Lab. Crittog Pavia, Ser 4a, 9:179-189.

Agrawal, S. C. and **Misra, J. K.** 1981. Phoma lucknowensis sp. Nov. From Indian alkaline soils. Curr. Sci. 50(13): 600.

Agarwal, G. P. and **Sahni, V. P.** 1964. Fungi causing plant diseases at Jabalpur (M.P.). Mycopath. et Mycol. Appl. 22(2): 245-247.

Arie, T., Kobayashi, Y., Kono, Y., Okada, G. and **Yamaguchi, I.** 1999. Control of clubroot of cruciferous by Phoma glomerata and its product epoxydon. Pestic. Sci. 55: 602-604.

Arie, T., Kobayashi, Y., Okada, G., Kono Y. and **Yamaguchi, I.** 1998. Control of soilborne clubroot disease of cruciferous plants by epoxydon from Phoma glomerata. Plant Pathology 47: 743-748.

Baxter, A., Fitzgerald, B.J, Hutson, J.L, McCarthy, A.D, Motteram, J.M, Ross, B.C, Sapra, M, Snowden, M.A. and **Watson, N.S.** 1992. Squalestatin 1, a potent inhibitor of squalene synthase, which lowers serum cholesterol in vivo. J. Biol. Chem. 267 (17): 11705-11708

Auld, B. A. 1990. Mycoherbicides: One alternative to chemical control of Weeds, In: Alternatives to the Chemical Control of Weeds. pp. 71-73(C. Basset and L.G.White House and J. A Zabkiewicz, eds) Proc. In. Conf. Roturcia, New Zealand, Ministry of Forestry, No.155: 71-73.

Baxter, C.J., Magan, N., Lane, B. and **Wildman, H.G** 1998. Influence of water activity and temperature on in vitro growth of surface cultures of a Phoma sp. And production of the pharmaceutical metabolites, squalestatin S1 and S2. Applied Microbiol. Biotechnologl. 49(3):328-332.

Bhowmik, T. P. and **Singh, A.** 1976. Occurrence of Phoma exigua Desm. On sunflower. Curr Sci 45: 776-777.

Bick, I.R.C. and **Rhee, C.** 1966. Anthraquinone pigments from Phoma foveata Foister. Biochemical Journal 98:112-116.

Bilgrami, K. S. 1963. Association of a new species of Phoma with Pleuospora herbarum (Pers.) Rahb. Curr sci 32: 174-175.

Binder, M. and **Tamm, C.** 1973a. Deoxyaphomin des erste(13)-Cytochalasan, ein moglicher biogenetischer vorlaufer der 24-Oxa-(14) Cytochalasane. Helv.Chem. Acta. 56: 966-724.

Binder, M. and **Tamm, C.** 1973b. Proxiphomin und protophomin, zweineue Cytochalasane. Helv. Chem. Acta. 56: 2387-2396.

Boerema, G.H. 1964. Phoma herbarum Westend., type-species of the form-genus Phoma Sacc. Persoonia 3: 9-16.

Boerema, G.H. 1976. The Phoma species studied in culture by Dr R.W.G.Dennis. Trans.Br.mycol.Soc. 67: 289-319.

Boerema, G.H. 1993. Contributions towards a monograph of Phoma (Coelomycetes) – II. Section Peyronellaea. Persoonia 15: 197-221.

Boerema, G.H. 1997 Contributions towards a monograph of Phoma (Coelomycetes) – V. Subdivision of the genus in sections. Mycotaxon 64: 321-333.

Boerema, G.H. and **Bollen, G.J.** 1975 . Conidiogenesis and conidial septation as differentiating criteria between Phoma and Ascochyta. Persoonia 8: 111-144.

Boerema, G.H., **Cruger, G., Gerlach, M.** and **Nirenberg, H.** 1981 Phoma exigua var. diversispora and related fungi on Phaseolus beans. Z.für Pflkr.Pflschutz (J.Plant Dis. Prot.) 88: 597-607.

Boerema, G.H. and **Dorenbosch, M..M..J** 1970. On Phoma macrostomum Mont., an ubiquitous species on woody plants. Persoonia 6: 49-58.

Boerema, G.H. and. **Dorenbosch, M..M.J.** 1973 .The Phoma and Ascochyta species described by Wollenweber and Hochapfel in their study on fruit-rotting.Studies in Mycology 3: 1-50.

Boerema, G.H., Dorenbosch, M.M.J. and **van Kesteren, H.A.** 1977. Remarks on species of Phoma referred to Peyronellaea V. Kew Bull. 31: 533-544."1976"

Boerema, G. H. and **Howeler, L. H.** 1967. Phoma Exigua Desm And Its Varieties. Persoonia 5: 15-28.

Boerema, G.H. and **Gruyter, J. de** 1998. Contributions towards a monograph of Phoma (Coelomycetes) – VII. Section Sclerophomella: Taxa with thick-walled pseudoparenchymatous pycnidia. Persoonia 17: 81-95.

Boerema, G.H. and **Gruyter, J. de** 1999. Contributions towards a monograph of Phoma (Coelomycetes) III – Supplement. Additional species of section Plenodomus. Persoonia 17: 273-280.

Boerema, G.H., Gruyter, J. de. and **Graaf, P. van de.** 1999. Contributions towards a monograph of Phoma (Coelomycetes) IV – Supplement. An addition to section Heterospora: Phoma schneiderae spec. nov., synanamorph Stagonosporopsis lupini (Boerema and R. Schneid.) comb. nov. Persoonia 17: 281-285.

Boerema, G.H., Gruyter, J. de and. ***van Kesteren, H.A*** 1994. Contributions towards a monograph of Phoma (Coelomycetes) – III. 1. Section Plenodomus: Taxa often with a Leptosphaeria teleomorph. Persoonia 15: 431-487.

Boerema, G.H., Gruyter, J. de and ***Noordeloos, M.E.*** 1997. Contributions towards a monograph of Phoma (Coelomycetes) – IV. Section Heterospora: Taxa with large sized conidial dimorphs, in vivo sometimes as Stagonosporopsis synanamorphs. Persoonia 16: 335-371.

Boerema, G.H. and ***Höweler, L.H.*** 1967. Phoma exigua Desm. and its varieties. Persoonia 5:15-28.

Boerema, G.H. and. ***van Kesteren***, H.A. 1964. The nomenclature of two fungi parasiting Brassica. Persoonia 3: 17-28.

Boerema, G.H. and ***van Kesteren, H.A.*** 1981. Nomenclatural notes on some species of Phoma sect. Plenodomus. Persoonia 11: 317-331.

Boerema, G.H. and ***Loerakker, W.M.*** 1981. Phoma piskorzii (Petrak) comb. nov., the anamorph of Leptosphaeria acuta (Fuckel) P.Karst. Persoonia 11: 311- 315.

Boerema, G.H. and ***Loerakker, W.M.*** 1985. Notes on Phoma 2. Trans.Br.mycol.Soc. 84: 289-302.

Boerema, G.H., Loerakker, W.M. and ***Maria E.C. Hamers*** 1996. Contributions towards a monograph of Phoma (Coelomycetes) – III. 2. Misapplications of type species name and the generic synonyms of section Plenodomus (Excluded species). Persoonia 16: 141-190.

Bousquet, J.F. and ***Barbier, M.*** 1972. Sur l'activite' phytotoxique de trois souches de Phoma exigua et al. presence de la cytochalasine B (ou phomine) dans leur milieu de culture. Phytopathologische Zeitschrift 75;365-367.

Boyette, C.D. and. ***Abbas, H..K*** 1995. Weed control with mycoherbicide and phytotoxins: A nontraditional applications of allelopathy.pp.280-299 In Allelopathy organisms,processes and applications (K. Inderjit, M.M.Dakhini and F.A. Einhelling eds), ACS Symp Ser No. 582, Washington DC.

Brewer, J.G. and ***Boerema, G.H.*** 1965. Electron microscope observations on the development of pycnidiospores in Phoma and Ascochyta spp. Proc.K.ned.Akad. (Wet. C) 68: 86-97.

Capasso, R., Evidente, A., and ***Vurro, M.*** 1991a. Cytochalasins from Phoma exigua var. heteromorpha. Phytochemistry 30(12):3945-3950.

Capasso, R., Evidente, A., Randazzo, G. Vurro, M. and ***Bottalico, A.*** 1991b. Analysis of cytochalasins in culture of Ascochyta spp.and in infected plants by High Performance Liquid and Thin Layer Chromatography. Phytochemical Analysis 2:87-92.

Chandra, S. and ***Tandon, R. N.*** 1965. Two new leaf spot fungi. Curr Sci 34: 365-366.

Chandra, S. and ***Tandon, R. N***. 1966. Three new leaf infecting fungi from Allahabad. Mycopath et Mycol appl 29: 273-276.

Charudattan, R. 1991. The Mycoherbicide approach with Plant Pathogens: 25-57, In: Microbial Control of Weeds. (TeBeest DO. ed.) Chapman and Hall Inc. USA.

Charudattan, R. 1996. Recent development in the field of microbial herbicides.pp.63-84 In Proc. Agri.Bio.Symp. Biopesticides for crop protection (K. H. Park ed), Seoul National University Sewon, South Korea.

Curtis, P.J., Greatbanks, D, Hesp, B., Forbes Cameron, A. and ***Freer, A.A.*** 1977. Sirodesmins A, B, C, and G antiviral epipolythiopiperazine-2,5-diones of fungal origin: X- ray analysis of sirodesmin A diacetate. J. Chem. Soc. Perkin Trans.1: 180-189.

Dawson, M.J., Farthing, J.E., Marshall, P.S., Middleton, R.F., O'Neill, M.J., Shuttleworth, A., Stylli, C., Tait, R.M., Taylor, P.M., Wildman H.G., Buss, A.D., Langley, D. and ***Hayes, M.V.*** 1992. The squalestatins, novel inhibitors of squalene synthase produced by a species of Phoma I. Taxonomy, fermentation, isolation, physico-chemical properties and biological activity. J. Antibiot 45: 639-647.

Dennis, R.W.G. 1946. Notes on some British fungi ascribed to Phoma and related genera. Trans.Br.mycol.Soc. 29: 11-42.

Dombrowski, A.W., Hastings, J.C., Polishook, J.D., Singh, S.B. 1999. New equisetin derivatives are HIV virus integrase-inhibitors. Biotechnological abstract.

Dorenbosch, M.M.J. 1970. Key to nine ubiquitous soil-borne Phoma-like fungi. Persoonia 6: 1-14.

Dutta, B. G. and ***Ghosh, G. R*** . 1965. Soil fungi from Orissa (India)-IV. Soil fungi of paddy fields. Mycopath et Mycol appl 25: 316-322.

Evidente, A., Lanzetta, R., Capasso, R., Vurro, M. and. ***Bottalico, A.*** 1992. Cytochalasins U and V, two new cytochalasanes form Phoma exigua var. heteromorpha. Teterahedron 48: 6317-6324.

Ferezou, J.P., Riche, C., Quesneau-Thierry, A., Pascard-Billy, C., Barbier, M., Bousquetand, J.F., Boudart, G. 1977. Structures de deux toines ioslees des cultures du champignon Phoma lingam Tode: la sirodesmine PL *et al* desacetylsirodesmine PL. Nouv. J. Chim.1: 327-324.

Ferezou, J.P., Quesneau-Thierry, A., Barbier, M., Kollmann, A., Bousquetand, J.F. 1980a. Structure and synthesis of phomamide, a new piperazine-2, 5-dione related to the sirodesmins, isolated from the culture medium of Phoma lingam Tode. J. Chem. Soc. Perkin Trans. 1:113-115.

Ferezou, J.P, Quesneau-Thierry, A, Servy, C, Zissmann, E and **Barbier, M.** 1980b. Sirodesmin PL biosynthesis in Phoma lingam Tode. J. Chem. Soc. Perkins Trans. 1: 1739-1746.

Gentile, A., Tribulato, E., Continella, G., Vardi, A. 1992. Differential responses of Citrus calli and protoplasts to culture filtrate and toxin of Phoma tracheiphila. Theor. Appl. Genet. 83: 759-764.

Gruyter, J. de and **Noordeloos, M.E.** 1992. Contributions towards a monograph of Phoma (Coelomycetes) – I. 1. Section Phoma: Taxa with very small conidia in vitro. Persoonia 15: 71-92.

Gruyter, J. de., Noordeloos, M.E. and **Boerema, G.H.** 1993. Contributions towards a monograph of Phoma (Coelomycetes) – I. 2. Section Phoma: Additional taxa with very small conidia and taxa with conidia up to 7 μm long. Persoonia 15: 369-400.

Gruyter, J. de., Noordeloos, M.E. and **Boerema, G.H.** 1998. Contributions towards a monograph of Phoma (Coelomycetes) – I. 3. Section Phoma: Taxa with conidia longer than 7 μm. Persoonia 16: 471-490.

Hasija, S. K. 1966. Additions to the fungi of Jabalpur, M.P.-V. Mycopath et Mycol appl 28: 33-41.

Hasija, S.K., Rajak, R.C. and **Pandey, A. K.** 1994. Microbes in the management of obnoxious Weeds. In: Vistas in Seed Biology (Singh T. and Trivedi, P. C. eds). Print Well, Jaipur: 82-104

Hazuda, D., Blau, C.U., Felock, P., Hastings, J., Pramanik, B., Wolfe, A., Bushman, F., Farnet, C., Goetz, M., Williams, M., Silverman, K., Lingham, R., Singh, S. 1999. Isolation and characterization of novel human immunodeficiency virus integrase inhibitors from fungal metabolites. Antiviral Chem Chemother 10 (2): 63-70.

Heiny, D.K. 1990. Phoma probocis sp. nov. pathogenic on Convoulus arvensis Mycotaxon 36: 457-471.

Heiny, D.K. and **Templeton, G. E.** 1991. Effects of Spore concentration, temperature and dew period on disease of field bindwind caused by Phoma proboscis. Phytopathology 81:905-909.

Heiny, D. K. 1994. Field survival of Phoma proboscis and synergism with herbicides for control of field bindweed. Plant Disease 78:1156-1164.

Hoagland, R.E. 2001. The genus Streptomyces: A rich source of novel phytotoxins. pp139-169 In Ecology of Desert Environments (Ishwari Parkash ed) Scientific Publishers, Jodhpur India.

Hoagland, R.E. 1990a. Microbes and Microbial Products as herbicides: An overview.pp2-52. Amer. Chem. Symp. Ser. No. 439, Washington, D.C., American Chemical Society.

Hoagland, R. E. ed 1990b. Microbes and Microbial Products as Herbicides: Amer. Chem. Symp. Ser. No. 439, Washington, D.C. , American Chemical Society.

Hoagland, R.E. 1999. Plant pathogens and microbial products as agents for Biological weed control.pp.214-255. In Advances in Microbial Biotechnology (J.P. Tiwari, T.N. Lakhanpal, Jagjit Singh, Rajni Gupta and B.P.Chamda eds) APH publishing corporation, New Delhi – 110002, India.

Hobson, D.K., Edwards, R.L., and **Wales, D.S.** 1997. Cynodontin: a secondary metabolite and dyestuff intermediate. J. Chem. Tech. Biotechnol. 70:343-348.

Jamaluddin, M., Tandon, P. and **Tandon, R. N.** 1975. A fruit rot of Aonla (Phyllanthus emblica L.) caused by Phoma sp. Proc Nat Acad Sci India 45 (B): 75-76.

Johnston, P.R. 1981. Phoma in New Zealand grasses and pasture legumes. N.Z.Jl Bot. 19: 173-186.

Kakeya, H., Morishita, M., Onozawa, C, Usami, R, Horikoshi, K, Kimura, K, Yoshihama, M, Osada, H. 1997. RKS-1778, its preparation, a cell cycle inhibitor and an antitumour agent produced by Phoma sp. J. Nat. Prod. 60(7): 669-672.

Kamal and **Singh S.** 1979. Fungi of Gorakhpur IX. Indian Phytopath.29: 188.

Kanaujia R. S. 1979. Notes on a new fungal disease of Alocacia indica. Indian Phytopath 32: 463-464.

Kaur, Surinder 1999. Microbial herbicides: Factors in development In: Advances in Microbial Biotechnology (Tiwari J.P., T.N. Lakhanpal, J. Singh, R. Gupta, and B.P. Chamola eds) APH Publishing Cooperation, New Delhi: 258-277.

Khanna, K.K. and **Chandra, S.** 1977. Some new leaf spot disease II. Proc. Nat. Acad. Sci. India 47(B): 251-253.

Kövics G.J. and **Gruyter, J. de.** 1995. Comparative studies of esterase isozyme patterns of some Phoma species occurre on soybean. DATE Tudományos Közleményei (Bulletin of Debrecen Agricultural University) 31: 191-207.

Kövics, G.J., de Gruyter, J. and **van der Aa, H.A.** 1999. Phoma sojicola comb. nov. and other hyaline-spored coelomycetes pathogenic on soybean. Mycological Research 103 (8) 1065-1070.

Logan, C. and **Khan, A. A.** 1969. Comparative studies of Phoma sp. associated with potato gangrene in Northern Iyreland. Trans Br mycol Soc 52: 9-17.

Mathur, P. N. and **Thirumalachar, M. J.** 1959. Studies on some Indian soil fungi.I. Some new or noteworthy Sphaeropsidales. Sydowia 13: 143-147.

Mabuni, C. T., Garlaschelli, L., Ellison, R.A. and **Hutchinson, C.R.** 1979. Biosynthesis of the C15 macrolide antibiotics.1. Biochemical origin of the four oxygen atoms in brefeldin A. J. Am. Chem. Soc. 101: 707-714.

Maiti, S., Dinesh, C., Khatua and **Sen, Chitreshwar.** 1978. A new leaf-spot disease of betel vine. Indian J Mycol Pl Pathol 8(2): 217.

Mann, J. 1986 Secondary Metabolism.OUP, Oxford.

Morgan-Jones, G. and **Burch., K.B.** 1988. Studies on the genus Phoma. X. Concerning Phoma eupyrena, an ubiquitous, soilborne species, Mycotaxon 31: 427-434.

Nachmias, A., Barash, I., Solel, Z. and. **Strobel, G.A.** 1977. Purification and characterization of a phytotoxin produced by Phoma tracheiphila the causal agent of mal secco disease of citrus. Physiol. Plant Pathol. 10: 147-158.

Nachmias, A., Barash, I. Buchner, V., Solel, Z. and **Strobel, G.A.** 1979. A phytotoxic glycopeptide from lemon leaves infected with Phoma tracheiphila. Physiol. Plant Pathol. 14: 135-140.

Nagai, K., Kamigiri K., Matsumoto, H., Kawano, Y., Yamaoka, M., Shimoi, H., Watanabe, M., Suzuki, K. 2002. YM-202204, a new antifungal antibiotic produced by marine fungus Phoma sp., J Antibiot (Tokyo). 55(12):1036-1041

Noordeloos, M.E., de Gruyter, J., van Eijk, G.W. and **Roeijmans, H.J.** 1993. Production of dendritic crystals in pure cultures of Phoma and Ascochyta and its value as a taxonomic character relative to morphology, pathology and cultural characteristics. Mycol. Res. 97: 1343-1350.

Padmbai, Luke. 1976. Fungi in the root region of Parthenium hysterophorus L. Curr. Sci. 45: 631-632.

Pandey, A. K., Luka, R. S. , Hasija, S.K. and **Rajak, R.C.** 1991. Pathogenicity of some fungi to Parthenium and obnoxious weed in Madhya Pardesh, J. Bio Control 5: 113-115.

Pandey, A.K., Mishra, Jyoti, Rajak, R.C. and **Hasija, S.K.** 1996. Potential of indigenous strains of Sclerotium rolfsii Sacc. for the management of Parthenium hysterophorus L: A serious threat to Biodiversity in India:104-138.In: Herbal Medicines, Biodiversity and Conservation Strategies (R.C.Rajak and M.K.Rai eds,) International Book Distributors, DehraDun: 104-138.

Pandey, A..K., Gaythri, S., Rajak, R.C. and **Hasija, S.K.** 1996. Possibilities, problems and prospects of microbial management of Parthenium hysterophorus L. in India:253-269. In Perspectives in Biological Science (Eds. Rai, V., M.L. Naik and C. Manoharachary). Dazale Offset Printers, Raipur (M.P.): 253-269.

Pandey, A. K., Rajak, R.C. and. Hasija, S. K. 1997. Potential of indigenous Colletotrichum species for the management of Weeds in India. In Achievements and Prospects in Mycology and Plant Pathology(S. S.Chahal, I.B. Parashar, H.S.Randhawa, and S. Arya eds.). International Book Distributors, DehraDun : 59-85.

Pandey, A.K. 1999. Herbicidal potential of microorganism: Present status and Future prospects. In: Microbial Biotechnology for sustainable developments and productivity (Rajak R.C. ed). , Scientific Publications Jodhpur Rajsthan: 87-105.

Pandey, A.K. 2000. Microorganism associated with Weeds: Opportunities and challenges for their exploitation as herbicides. Int. J. Mendel 17(1-2): 59-62.

Pandey, Santosh and Pandey, A.K. 2000. Mycoherbicidal potential of some fungi against Lantana camara L.: A preliminary observation. Journal of Tropical Forestry 16: 28-32.

Pandey, A.K., Rajak, R.C. and Hasija, S.K. 2001. Biotechnological development of ecofreiendly mycoherbicides. In Innovative approaches in Microbiology (Maheshwari, D.K. and R.C. Dube eds), Bisen singh Mahendra pal singh, Dehra Dun, India: 1-21.

Pandey, A. K., Chandla Poonam and Rajak, R.C. 2002. Herbicidal potential of secondary metabolites of some fungi against Lantana camara L. J. Mycol. Pl. Pathol. 32(1): 100-102.

Pandey, A. 2002. Screening and Characterization of Mycoherbicides for Management of *Hyptis suaveolens*(L). Poit. Ph. D Thesis. R.D. University, Jabalpur

Pathak, P.D. and Chauhan, R.K.S. 1976. Two new leaf spot disease of Crotalaria juncea L caused by Pleospora infectoria fuckel and Phoma glomerata corda. Curr sci. 45: 206.

Pandey, S.K. 2000. Mycoherbicidal potential of some fungal pathogens for the management of Lantana camara L., Ph.D. Thesis (unpublished) submitted to R.D. University, Jabalpur, M.P., India.

Pavgi, M. S. and Singh, U. P. 1966. Parasitic fungi from north India. VII. Mycopath Mycol Appl 30: 261-270.

Pearce, C. 1997. Biologically active fungal metabolites. Adv. Appl. Microbiol. 44: 1-80.

Pedras, M.S.C. 1996. The chemistry of cyclohexenediones produced by the blackleg fungus. Can. J. Chem. 74:1597-1601.

Pedras, M. and Soledade, C. 1996. The chemistry of cyclohexenediones produced by the blackleg fungus. Can. J. Chem. 74:1597-1601.

Pedras, M. and Soledade, C. 1998. Towards an understanding and control of plant fungal diseases in Brassicaceae. In, Recent Res. Dev.Agric. Food Chem. 2:513-532.

Pedras, M., Soledade, C., and **Biesenthal, C.J.** 1998. Production of the host-selective toxin phomalide by isolates of Leptoshaeria maculans and its correlation with sirodesmin PL production. Can. J. Microbiol. 44:547-553.

Pedras, M., Soledade C., Taylor, Janet L. and **Victor M. Morales.** 1995. Phomaligin A and other yellow pigments in Phoma lingam and P. wasabiae. Phytochemistry 38 (5):1215-1222.

Pedras, M., Soledade C., Erosa-Lopez, G.C., Quail, J.W. and **Taylor, J.L.** 1999. Phomalairdenone:A new host-selective phytotoxin from a virulent type of the blackleg fungus Phoma lingam. Bioorg. Med. Chem. Lett. 9:3291-3294.

Pedras, M., Soledade, C., and **Biesenthal, C.J.** 2000. HPLC analyses of cultures of Phoma spp.:Differentiation among groups and species through secondary metabolite profiles, Can. J. Microbiol. 46:685-691.

Pedras, M., Soledade, C., Morales, V.M., and. **Taylor, J.L.** 1994. Phomapyrones: three metabolites from the blackleg fungus, Phytochemistry 36:1315-1318.

Pedras, M. Soledade C., Taylor, J.L. and **Nakashima,T.T.** 1973. A novel chemical signal from the blackleg fungus:Beyond phytotoxinsand phytoalexins. J. Org. Chem., 58:4778-4780.

Perazzi, P. 1925. I miceti dimoranti nella regione genitale della donna. Atti Accad Fisiocrit Siena 385-387.

Pollacci, G. 1935. Sopra una nuova specie de micete commensale (Phoma hominis Agostini et Tredici), isolato da forme cliniche del dema. Atti Ist Bot Univ Lab crittog Pavia, Ser 4, 153-155

Rai, M K. 1981. Studies on some Indian Sphaeropsidales with special reference to Phoma and related fungi. Ph.D. Thesis, University of Jabalpur, Jabalpur, Madhya Pradesh, India.

Rai, M. K. 1989. Phoma sorghina infection in human beings. Mycopathologia 105, 167-170.

Rai, M.K. 2003. Diversity and Biotechnological Applications of Indian species of Phoma:79-204. In Frontiers of Fungal Diversity in India(G.P.Rao, C.Manoharachari, D.J.Bhat, R.C.Rajak and T.N.Lakhanpal eds), International Book Distributing Co.,Lucknow, India.

Rai, J. N. and **Misra, J. K.** 1981. A new species of Phoma from Indian alkaline soil. Curr Sci 50: 377.

Rai, M. K. and **Rajak, R. C.** 1984. Phoma jatropae, a synonym of Phoma medicaginis var. medicaginis. J Econ Tax Bot 5(2): 395-396.

Rai, M. K. and **Rajak, R. C.** 1986a. Phoma multirostrata on three new hosts. J Econ Tax Bot 8(1): 229-230.

Rai, M. K. and **Rajak, R. C.** 1993. Distinguishing characteristics of selected Phoma species. Mycotaxon XL VIII: 389-414.

Rajak, R. C. 1981. New Sphaeropsidales from India. Indian J Mycol Pl Pathol 11: 73-79.

Rajak, R.C., Farkya, S., Hasija, S.K. and **Pandey, A. K.** 1990. Fungi associated with Congress Weed (Parthenium hysterophorus L.) Proc, Nat Acad Sci, India 60: 165-168.

Rajak, R .C. and **Rai, M. K.** 1982a. Species of Phoma from legumes. Indian Phytopath 35(4): 609-611.

Rajak, R. C. and **Rai, M. K.** 1983a. Effect of different factors on the morphology and cultural characters of 18 species and 5 varieties of Phoma .I. Effect of different media. Bibl. Mycol 91: 301-317.

Rajak, R. C. and **Rai, M. K.** 1983b. Cholesterol content in twenty species of Phoma. Nat Acad Sci Letters 6 (4): 113-114.

Rajak, R. C. and **Rai, M. K.** 1984a. Effect of different factors on the morphology and cultural characters of Phoma. II. Effect of different hydrogen-ion-concentrations. Nova Hedwigia 40: 299-311.

Rajak, R. C. and **Rai, M. K.** 1985. A key to the identification of species of Phoma in pure culture. J Econ Tax Bot 7(3): 588-590.

Rangaswami, G., Seshadri, V. S. and **Lucky K A Channamma**. 1970. Fungi of south India, University of Agriculture sciences, Bangalore pp 193.

Rao, S. and **Thirumalachar**, U. 1981. Phoma exigua infecting brinjal leaves. Indian Phytopath. 34:37.

Raut, J. G. 1977. Phoma leaf blight of sunflower. Indian Phytopath 30: 269-270.

Rothweiler, W.C. and **Tomm, C.** 1966. Isolation and Structure of Phomin. Experentia 22: 750-752.

Rothweiler , W.C. and **Tomm, C.** 1970. Isoleirung und struktur der antibiotica phomin und S. dehydrophomin. Helv. Chim. Acta. 53: 696-724.

Saxena, Sanjai and **Pandey, A.K.** 2000. Preliminary evaluation of fungal metabolites as natural herbicides for the management of Lantana camara. Indian Phytopath. 53(4): 490-491.

Saxena, Sanjai and **Pandey, A. K.** 2001. Microbial metabolites as ecofriendly agrochemical for the next millennium. Appl. Microbiol. Biotechnol. 55: 395-403.

Scott, P.M., Harwig, J, Chen, Y.K. and **Kennedy, B.P.C.** 1975. Cytochalasins A and B from strains of Phoma exigua var. exigua and formation of cytochalasin B in potato gangrene. Journal of general Microbiology 87: 177-180.

Shreemali, J. L. 1972. Two new pathogenic fungi causing diseases on indian medicinal plants. Ind J Mycol Pl Pathol 2: 84-85.

Shreemali, J.L. 1973. Sphaeropsidean fungi obtained from Indian phylloclades. Indian Phytopath 26: 56-62.

Shreemali, J. L. 1978. Two new species of Phoma from India. Indian J Mycol Pl Pathol 8: 220-221.

Shukla, N.P., Rajak, R.K., Agarwal, G.P. and ***Gupta, D.K.*** 1984 Phoma minutispora as a human pathogen, Mykosen 27, 255-258.

Sidebottom, P.J, Highcock, R.M, Lane, S.J., Procopion, P.A. and ***Watson, N.S.*** 1992. The squalestatins, novel inhibitors of squalene synthase produced by a species of Phoma II. Structural elucidation. J. Antibiot. 45: 648-658.

Sigg, H.P. 1964. Die Konstitution von Brefeldin A. Helv. Chim. Aca 47: 1401-1415.

Singh, S.B., Ball, R.G., Zink, D.L., Monaghan, R.L., Polishook, J.D., Sanchez, M., Pelaez, F., Silverman, K.C., and ***Lingham, R.B.*** 1997. Fusidienol - A: a novel Ras farnesyl-protein –transferase inhibitor from Phoma sp. J. Org. Chem. 62(21):7485-7488.

Singh, S.B., Zink, D.L., Goetz, M.A., Dombrowski, A.W., Polishook, J.D., Hazuda, D.J.S. 1998. Equisetin and a novel opposite stereochemical homolog phomasetin, two fungal metabolites as inhibitors of HIV-1 integrase, Biotechnology abs No. 98—06361.

Singh, S.M. and ***Agarwal, G.P.*** 1973. On some foliicolous Sphaeropsidales. Proc. Nat. Acad. Sci. India 43(B): 73-80.

Singh, S.M. and ***Agarwal, G.P.*** 1974. Some Sphaeropsidales new to India. Indian Phytopath 27: 244-246.

Sponga, F., Cavaletti, L., Lazzarini, A., Borghi, A., Ciciliato, I., Losi, D., and ***Marinelli, F.*** 1999. Biodiversity and potentials of marine-derived microorganisms. J. Biotechnol 70 (1-3):65-69.

Stoessl, A. 1981. Structure and Biogenetic relations: Fungal non-host specific pp. 109-220. In Toxins in Plant diseases (ed Durbin R.O.), Academic Press Inc.

Steyn, P.S. and. ***Rabie, C.J.*** 1976. Characterization of Mangnesium and Calcium tenuazonate from Phoma sorghina. Phytochemistry 15:1978-1979

Strange, R.N. 1997. Phytotoxin associated with Ascochyta species pp. 167-181. In Toxin in Plant Disease development and evolving Biotechnology (R.K Upadhyay and K.G. Mukherjee ed.) Oxford and IBH Publishing Co. Pvt. Ltd. New Delhi.

Sugano, M., Sato, A., Lijima, Y., Oshima, T., Furuya, K. Kuwano, H., Hata, T., Hanzawa, H. 1991. Phomactin A: a novel PAF antagonist from a marine fungus Phoma sp. J. Am. Chem. Soc. 113(14): 5463-5464.

Sugawara, F. and **Strobel, G.** 1986. Zinniol, a phytotoxin produced by Phoma macdonaldii. Plant Sci. 43: 19-23.

Sutton, B.C. 1977. Coelomycetes VI. Nomenclature of generic names proposed for Coelomycetes. Mycological Papers 141: 1-253.

Sutton, B.C. 1980. The Coelomycetes. CAB International Mycological Institute, Kew. 696 pp.

Suzuki, Y., Tanaka, H., Aoki, H. and **Tamura, T.** 1970. Ascotoxin (decumbin), a metabolite of Ascochyta imperfecta Peck. Agric Biol. Chem 34: 395-413.

Taber, R.A., Pettit, R.E. and **Philley, G.L.** 1984. Peanut web blotch: I. Cultural characteristics and identity of causal fungus. Peanut Sci. 11: 109-114.

Tandon , R. N. and **Bilgrami, K. S.** 1960. A new host for Phyllosticta cycadina Pass Sci and Cult 26: 277-278.

Taylor, G.W. 1986. Natural dyes in textile applications : Rev. Prog. Coloration 16; 53-61.

Thompson, R.H. 1971. Naturally occurring quinones, 2nd edn., Academic press, London, New York.

Tichelaar, G.M. 1974. The use of thiophanate-methyl for distinguishing between the two Phoma varieties causing gangrene of potatoes. Netherlands J of Plant Pathology 80: 169-70.

Topgi, R.S., Devys, M., Bousquet, J. F., Kollmann, A. and **Barbier, M.** 1987. Phomenoic acid and phomenolactone, antifungal substances from Phoma lingam (Tode) Desm.: kinetics of their biosynthesis, with an optimization of the isolation procedures. Appl. Environ. Microbiol. 53 (5): 966–968

Venkatasubbaiah, P., Van Dyke, C.G. and **Chilton, W.S.** 1992. Phytotoxic metabolites of Phoma sorghina a new foliar blight pathogen pokeweed. Mycologia 84: 19-23.

Vikrant, P. 2002. Management of Parthenium employing secondary metabolites of Phoma herbarum FGCC# 75. Dissertation work, R.D. University, Jabalpur.

Vining, L.C. 1990. Functions of secondary metabolites. Annual review of Microbiology 44: 395 – 427.

Vining, L.C. and **Wright, J.L.C.** 1977. Biosynthesis of Oligopeptides. Biosynthesis 5: 240-305.

Vurro, M., Bottalico, A., Capasso, R. and **Evidente, A.** 1997. Cytochalsin from phytopathogenic Asochyta and Phoma spp. In: Toxin in Plant Disease development and evolving Biotechnology(R.K Upadhyay and K.G. Mukherjee ed.) Oxford and IBH Publishing Co. Pvt Ltd. New Delhi: 127-147.

Wadje, S. S. and ***Deshpande, K. S.*** 1981. A new leaf-spot of cotton caused by Phoma exigua. Indian Phytopath 33: 126-127.

Wakui, F., Mikoshiba, H., and ***Tachibana, K.*** 1999. Light-stable agrochemical microbicide preparation, biotechnological abs No. 99-06952.

Weber, H.P., Hauser, D. and ***Sigg, H.P.*** 1971. Structure of Brefeldin A. Helv. Chim. Acta 54: 2763-2766.

Wehmeyer, L.E. 1964. Some fungi imperfecti form Punjab and Kashmir. Mycologia 55: 313-335.

Yamaguchi , Y., Masuma, R., Uchida, R.,. Arai, M., Tomoda, *H.* and ***Ômura, S.*** 2002. Phoma sp. FOM-8108, a producer of gentisylquinones, isolated from sea sand. Mycoscience, 43(2): 127 – 133

Yoder, O.C. 1980. Toxins in Pathogenesis. Ann. Rev.Phytopathol.18:103-129.

 Black Mildews on *Goniothalamus* Species in Agasthyavanam National Park, Kerala, India

V.B. HOSAGOUDAR, G. BAGYANARAYANA* and H. BIJU

Microbial Taxonomic Unit, Microbiology Division, Tropical Botanic Garden and Research
Institute, Palode 695 562, Thiruvananthapuram, Kerala
˙Department of Botany, Osmania University, Hyderabad 500 007, Andhra Pradesh

ABSTRACT

Three black mildew fungi, namely, *Amazonia goniothalami*, *Irenopsis goniothalami* and *Trichasterina goniothalami* infecting endemic host plants, *Goniothalamus rhynchantherus* and *G. wightii* in Agasthyavanam National park have been described and illustrated in detail.

KEY WORDS : Black mildews – *Gomiothalamus* spp – Kerala.

The genus *Goniothalamus* belongs to the family Annonaceae comprises about 115 species and of which 10 are in India. *Goniothalamus cardiopetalus*, *G. rhynchantherus*, *G. thwaitesii*, *G. wightii* and *G. wynaadensis* occur in the Western Ghats region of Peninsular India (Mitra, 1995). *Goniothalamus rhynchantherus* and *G. wightii* are endemic to southern Western Ghats and were found infected with 'black mildews' in the present study area.

Agastyamala, named after the sage Agastya, is located towards the southern part of the Thiruvananthapuram district in Kerala state. The hilltop is at an altitude of 1868 m, descends steeply towards the western side. This region harbours tropical evergreen forest and is known for the endemic flora of the flowering plants.

'Black mildews' are the ectophytic fungi which cause dense, black velvet colonies on the surface of the leaves, and on soft and tender parts of plants. Taxonomically, these fungi include the members of Meliolaceae, Asterinaceae, Englerulaceae, Hyphomycetes, etc. However, in the present study, the black mildews on *Goniothalamus* belong to the family Asterinaceae (the genus *Trichasterina*) and Meliolaceae (the genera *Amazonia* and *Irenopsis*).

Key to the genera of black mildews on Goniothalamus species

1. Fruiting body thyriothecia   ...   *Trichasterina*
1. Fruiting body perithecia   2
2. Fruiting body flattened-globose   ...   *Amazonia*
2. Fruiting body globose with perithecial setae ...   *Irenopsis*

Descriptions of the Species

1. *Amazonia goniothalami* Hosag., Rajkumar, Biju & Abraham, Mycotaxon 72: 431, 2001. (Fig. 1)

Colonies predominantly hypophyllous, subdense to dense, up to 5 mm in diameter, confluent. Hyphae straight, branching alternate to opposite at acute angles, loosely to closely reticulate, cells 9-16 × 6-8 μm. Appressoria alternate, antrorse to closely antrorse, straight, 18-26 μm long; stalk cells cylindrical to cuneate, 6-8 μm long; head cells ovate, oblong to cylindrical, entire, 12-15 × 8-12 μm. Phialides not seen. Perithecia flattened-globose with radiating cells, up to 160 μm in diameter; ascospores oblong to ellipsoidal, 4-septate, constricted at the septa, 44-48 × 20-23 μm.

Specimens examined : On leaves of *Goniothalamus wightii* Hook.f & Thoms (Annonaceae), Agasthyamalai, Kerala, India, Dec.9, 1999, G. Rajkumar HCIO 43359, TBGT 257.

This is the only species of the genus *Amazonia* on the members of the family Annonaceae (Hansford, 1961; Hosagoudar, 1996; Hosagoudar *et al.* 1997; Hu *et al.* 1997, 1999; Mibey & Hawksworth, 1997). This species is known only from the type collection

2. *Irenopsis goniothalami* Hosag. & Abraham, Kavaka 24: 15, 1996. (Fig. 2)

Colonies hypophyllous, thin, spreading, up to 10 mm in diameter, rarely confluent. Hyphae straight to crooked, branching irregular at acute angles, loosely reticulate, cells 32-34 x 6-8 μm. Appressoria alternate, mostly straight, rarely curved, antrorse to subantrorse, rarely recurved, 19-20 μm long; stalk cells cylindrical to cuneate 6-10 μm long; head cells ovate, globose, entire, angular to lobate, 11-15 × 11-16 μm. Phialides mixed with appressoria, alternate to unilateral, ampulliform, 17-20 × 4-7 μm. Perithecia scattered, mostly immature, up to 160 μm in diameter. Perithecial setae 0-6, simple, straight to curved, mostly prostrate, erect, 2-6-septate, rarely nodulose in the middle, bulbous at the base, rarely obtuse to broadly rounded at the apex, up to 150 μm long, ascospores obovoidal, 4-septate, constricted at the septa 32-35 × 14-16 μm.

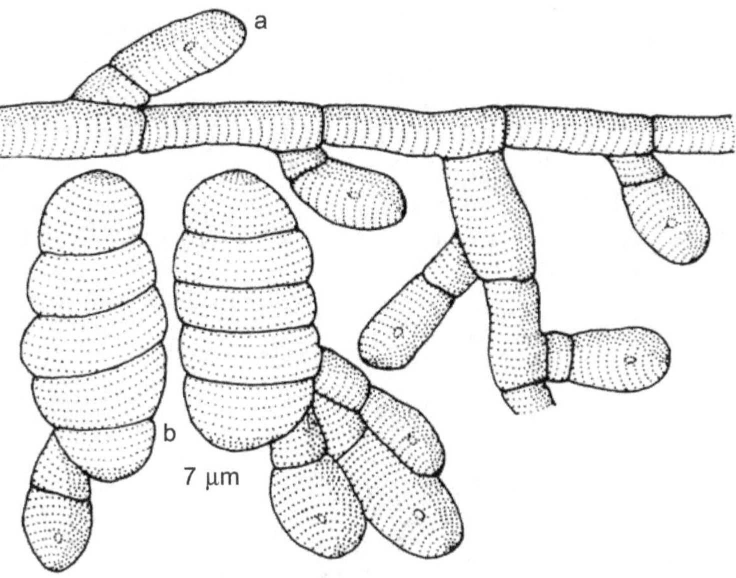

Fig. 1- *Amazonia goniothalami* Hosag.. *et al.*
a- Appressorium, b- Ascospores

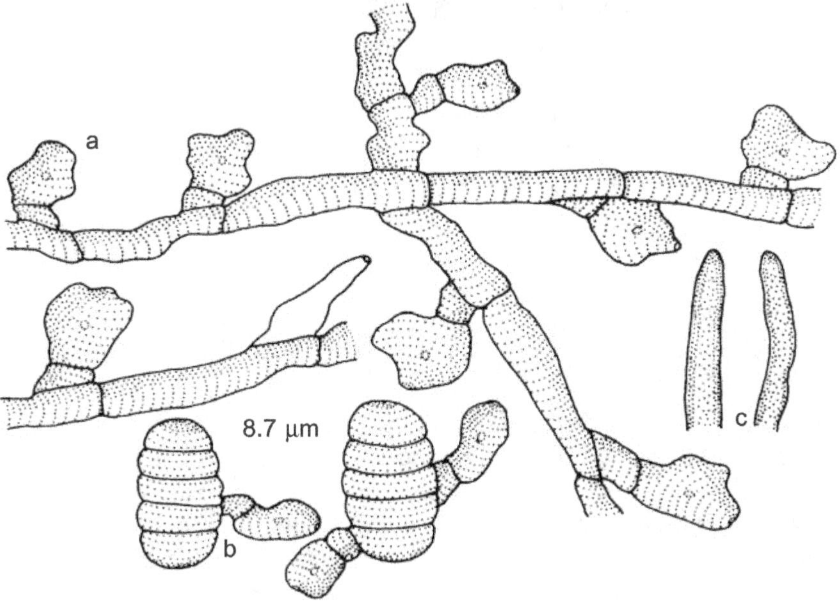

Fig. 2- *Irenopsis goniothalami* Hosag. & Abraham
a- Appressorium, b- Ascospores, c- Perithecial setae

Specimes examined : On leaves of *Goniothalamus rhynchantherus* Dunn (Annonaceae), in the evergreen forest of Chemunji, Bonaccaud, Thiruvananthapuram, Kerala, India, March 11, 1997, V.B. Hosagoudar TBGT 200, ILLS.

This is the first report of the genus *Irenopsis* on the members of the family Annonaceae (Hansford, 1961; Hosagoudar *et al.* 1977). This species is known only from the type collection.

3. *Trichasterina goniothalami* Hosag. & Goos, Mycotaxon 59: 165, 1996. (Fig. 3)

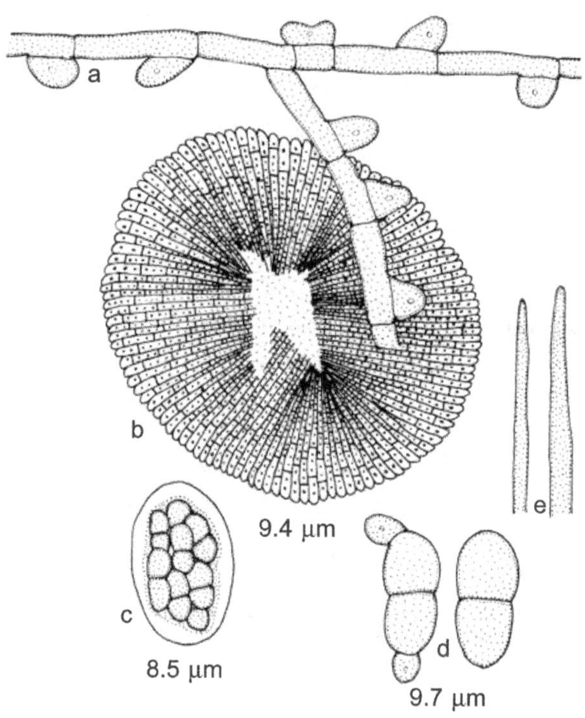

Fig. 3- *Trichasterina goniothalami* Hosag. & Goos
a- Appressoriate mycelium, b- Thyriothecium, c- Ascus, d- Ascospores, e- Mycelial setae

Colonies amphigenous, dense, velvety, up to 10 mm in diameter, rarely confluent. Hyphae straight to substraight, branching irregular at acute angles, loosely to closely reticulate, cells 18-31 x 5-7 µm. Appressoria unilateral, alternate and about 5% opposite, seated on a broad base, unicellular, ovate, cylindrical, mammiform, rarely truncate, 6-13 x 6-10 µm. Mycelial and thyriothecial setae numerous, simple, straight, erect, obtuse at the tip, up to 110 µm long. Thyriothecia scattered, orbicular, carbonaceous black and opaque, up to 110 µm in diameter; margin crenate, dehisce stellately at the centre; asci many, globose to slightly ovate, eight spored, 31-40 x 27-31 µm; ascospores brown, conglobate, 1-septate, deeply constricted at the septum, 31-34 x 12-16 µm, wall smooth.

Specimens examined : On leaves of *Goniothalamus wighti* Hook.f. & Thoms. (Annonaceae), Chemunji, March 11, 1997, V.B. Hosagoudar HCIO 44076, TBGT 510; April 12, 1999, G. Rajkumar HCIO 43368, TBGT 258; Feb. 8, 1998, G. Rajkumar HCIO 43367, TBGT 259; April 23, 1998, G. Rajkumar HCIO 43365, TBGT 260; July 16, 1998, G. Rajkumar HCIO 43366, TBGT 261.

This species differs from *Trichasterina polyalthiae* Hansf. in having alternate appressoria, smaller setae and larger ascospores (Hansford, 1955). This species is known only from the present locality

Acknowledgements

Thanks are due to the Director, Tropical Botanic Garden and Research Institute, Palode for the facilities.

References

Hansford, C.G. 1955. Tropical fungi – V. New species and revisions. *Sydowia* 9: 1-88.

Hansford, C.G. 1961. The Meliolaceae. A Monograph. *Sydowia.* Beih. 2: 1-806.

Hosagoudar, V. B.1996. *Meliolales of India.* Botanical Survey of India, Calcutta, pp. 363.

Hosagoudar, V. B., Abraham, T. K. and ***Pushpangadan, P.*** 1997. *The Meliolineae - A Supplement.* Tropical Botanic Garden and Research Institute, Palode, Thiruvananthapuram, Kerala, India, pp. 201.

Hu.Y., Ouyang, Y. Song Bin and ***Jiang. G.*** 1996. *Flora Fungorum Sinicorum.* Vol. 4. Meliolales (1) Science Press Beijing, pp. 270, plate IV.

Hu.Y., Song Bin, Ouyang, Y. and ***Jiang, G.*** 1999. *Flora Fungorum Sinicorum.* Vol. 11. Meliolales (2) Science Press Beijing, pp. 252.

Mibey, R.K. and ***Hawksworth, D.L.*** 1997. Meliolaceae and Asterinaceae of the Shimba Hills, Kenya. *Mycol. Pap.* 174: 1-108.

12 | The Status of *Chaetomium* from the Soils of Jalgaon

Gauri Rane[1] and R.V. Gandhe[2]

1. Postgraduate Research Centre, Deptt. of Botany, M.J. College, Jalgaon-425 002.
2. Postgraduate Research Centre, Deptt. of Botany, Modern College, Pune-411 005.

ABSTRACT

The present paper deals with the present status of the genus *Chaetomium* from the soils of Jalgaon district (MS) India. A total of eight species were isolated from the different established sites and four species are described as new. These new species were described on the basis of morphology of perithecia, terminal hairs and ascospores with asci. The species described showed seasonal variation in their occurrence and distribution.

KEYWORDS : *Chaetomium*, new species, variations, soil fungi, taxonomy.

Chaetomium Kunze ex Fries

= *Chaetomiotricha* Peyr.1914

Mycelium thread like, thick and thin walled, septate, hyaline to yellowish brown, uninucleate to multinucleate, perithecium translucent to opaque or dark coloured, covered with a dense mass of hairs, superficial, globose to subglobose or barrel shaped, thin walled, wall becoming brittle with age, distinctly cellular, ostiolate. Terminal hairs light yellowish brown, thick or thin walled, septate or aseptate or indistinctly septate, smooth to finely or coarsely roughened. Simple or branched, straight to arcuate, undulate or coiled, may end in a rounded or pointed hyaline tip. Lateral hairs hyaline to yellowish brown, septate or aseptate, smooth or finely or coarsely roughened, simple or rarely branched, straight to undulate, rarely coiled, may end in a rounded or pointed hyaline tip.

Ascomata mostly superficial attached to the substrate by rhizoids, subglobose to vasiform, with lateral and terminal septate hairs that may be variously branched or contorted. Ascoma wall pseudoparenchymatous, asci cylindrical to clavate, eight spored or rarely four spored, uninucleate, paraphyses hyaline, greatly reduced, evanescent, ascus deliquesces before or after the maturity of ascospores. Ascospores one celled, olive brown to chocolate brown, globose to subglobose, fusiform to broadly lemon shaped, sometimes tri or quadrangular in shape, rounded or umbonate or faintly apiculate at one or both ends. Few species produce aleuorospores and chlamydospores but seldom any conidia.

Type species : *Chaetomium globosum* Kunze

Since the establishment of the genus *Chaetomium* by Gustav Kunze (1817), several species have been added. Ames (1968) and Seth (1970) published monographic and illustrative detailed account of the different species of the family Chaetomiaceae and genus *Chaetomium* respectively.

The genus has lot of economic importance in various industries like textiles, seeds, wood, paper and pulp because of its property of decomposition of cellulose and destruction of economically valuable material. Bilgrami et al (1991) reported 97 species from India and Bhide et al (1987) reported 19 species from Maharashtra many of them have been isolated from soils.

Total of 8 species were isolated in different seasons from different soil types of Jalgaon district and are described here.

Study Area

Jalgaon district is one of the largest district in Maharashtra and belongs to Deccan upland. It has 3 chief hill ranges, Satpuda at north, Hatti in South West and Ajanta and Satmala which covers the southern part of the district. The main river of the district is Tapi which flows from western highlands of Madhya Pradesh and ends in Surat at Gujarat. Waghur, Girna, Purna, Bhogavati,and Bori are the tributaries of Tapi. The climate of the Jalgaon district is generally dry, except in monsoon. Being far away from the coast, there are extremes of temperature in the district. The temperature ranges from 10°C–30°C in winter and 30°C – 47°C during summer. The average rainfall is 740 mm. The soil of the district is derived from basalt. The soil mainly falls into 5 categories namely, medium black soils, deep black soils, forest soils, loamy soils and sandy soils. Agriculture is the prominent occupation in Jalgaon district. The land falls in to two categories, Jirayat (Dry cropland) and Bagayat (Wet cropland). The reserved forest of Jalgaon district forms about 17% of the total area of the district and exhibits different types of soils. The forests are dry deciduous type.

Area of Soil Sample Collection

The localities selected for soil sample collection were mainly on the basis of soil type and convenience. Four different soil types as agricultural soils, river soils, uncultivated soils and forest soils were considered. Agricultural soils from Mamurabad and Shirsoli, river soils of Girna and Wagur, uncultivated soils from Paldhi and Bhusawal and forest soils of Pal and Manudevi.

Materials and Methods

Soil samples were collected from selected localities and brought to the laboratory. Physicochemical factors of soil like pH, temperature, texture, water holding capacity and density of soil particles were also recorded. The dilution plate method (Waksman, 1916) was used for isolation and purification of soil fungi using suitable culture media like Czapek dox Agar, Potato Dextrose Agar and Lactose Yeast Agar media. Pure cultures were often maintained on Potato Dextrose Agar medium. The different topographical characters of the colonies were at regular time intervals and recorded for growth studies. The semi permanent slides of the isolated fungi were prepared using 1% cotton blue and lactophenol. Identification of species was made by referring relevant literature and monographs. Camera lucida drawings of the species were made.

*Key to the Species of **Chaetomium*** :

I -   Terminal hairs coiled 1

II -   Terminal hairs straight...2

1.A - Terminal hairs loosely coiled, perithecia dark brown, subglobose, ascospores brown, ovoid, 9.5 –10.5 µm × 3.5-4.0 µm........................ *C. cuniculorum*

B -   Terminal hairs roughened, 2-3 coiled perithecia dark brown, ascospores olive brown, fusiform, 9-10 µm × 6-6.5 µm...................*C. subspirilliferum*

C -   Terminal hairs 5-7 coiled, perithecia dark brown, ascospores, light brown, oval, 6-6.5 µm × 3-4 µm......................................*C. haetomium*

2.A - Perithecia yellowish to pale brown, terminal hairs slightly undulate, ascospores pale brown, elliptical to slightly fusiform, 6-7 µm × 3-4 µm.........*C. osmaniae*

B -   Perithecia dark brown, terminal hairs slightly undulate, ascospores brown, variable in morphology, 12-13.5 µm × 5-6.5 µm.......... *C. haetomium*

C -   Perithecia dark brown to black, terminal hairs straight, ascospores dark brown, ellipsoid, 12-13.5 µm × 5-6.5 µm.........................*C. fusisporale*

D -   Perithecia black, subglobose, terminal hairs facing towards the body of perithecium, scospores brown, elliptical, 8-10 µm × 5-6.5 µm....................
....*C. haetomium*

E -   Perithecia dark brown black, terminal hairs sinuous, ascospores pale brown, 10-12 µm × 6-7 µm...*C.*

Chaetomium cuniculorum Fuckel Symb. Myc. 89.1869

=Chaetomium cristatum Ames 639,1949

Colonies were isolated on Czapek Dox Agar growing slowly about 0.5 cm in five days in the form of black dots, reverse colourless.

Mycelium pale yellowish, septate, branched, perithecia dark brown, few, mostly subglobose, globose, 325-375 µm × 275-310 µm in size, pseudoparenchymatous, profusely covered with hairs. Terminal hairs brown, unbranched, septate, irregularly sinuous, 3.3 µm in diameter, straight at the base but undulated at the apices; walls of the hairs dark, finely roughened. Lateral hairs straight, unbranched, septate, 3.3 µm in diameter, smooth, brown. Ascospores brown, ovoid, 9.5-10.5 µm × 3.5-4.0 µm, apiculate at one end. Rhizoids and asci not seen.

This species was isolated only once from cultivated soil of Shirsoli during summer. The soil temperature was 38^0c and pH was 7.4. This is one of the isolated species of *Chaetomium* showing restricted occurrence.

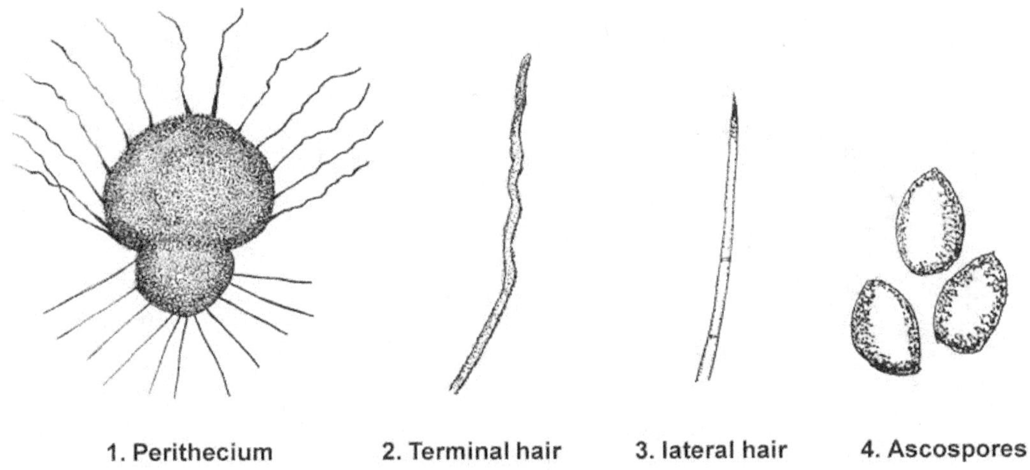

1. Perithecium   2. Terminal hair   3. lateral hair   4. Ascospores

Figure 1-4 *Chaetomium cuniculorum*

Remarks : The species differs from the species described by Seth (1970) in having larger perithecia and only one kind of terminal hairs.

It is not earlier reported from India so this is new addition to Indian soil fungi.

C. haetomium subspirilliferum Sergejeva

174, 1960, Figs. 5 – 8

Colonies were isolated on Czapek Dox Agar growing slowly about 0.4 cm in ten days, reverse colourless.

Mycelium hyaline, branched, septate. Perithecia black, gregarious, subglobose, 350-375 μm × 175-250 μm in size, pseudoparenchymatous, covered with hairs. Terminal hairs unbranched, rarely septate, straight below 2-3 loosely coiled up wards, brown, brown, 3.5 μm wide, finely verrucose. Lateral hairs straight, septate, unbranched, 3.0 μm wide, smooth. Ascospores mostly fusiform, olive brown, 9-10 μm × 6-6.5 μm, apiculate at both ends. Rhizoids and asci are not seen.

This species was isolated only once from cultivated soil of Mamurabad during summer. The soil temperature was 38^0c and pH was 6.8 that is slightly acidic. This is also one of the isolated species of *Chaetomium* showing restricted occurrence.

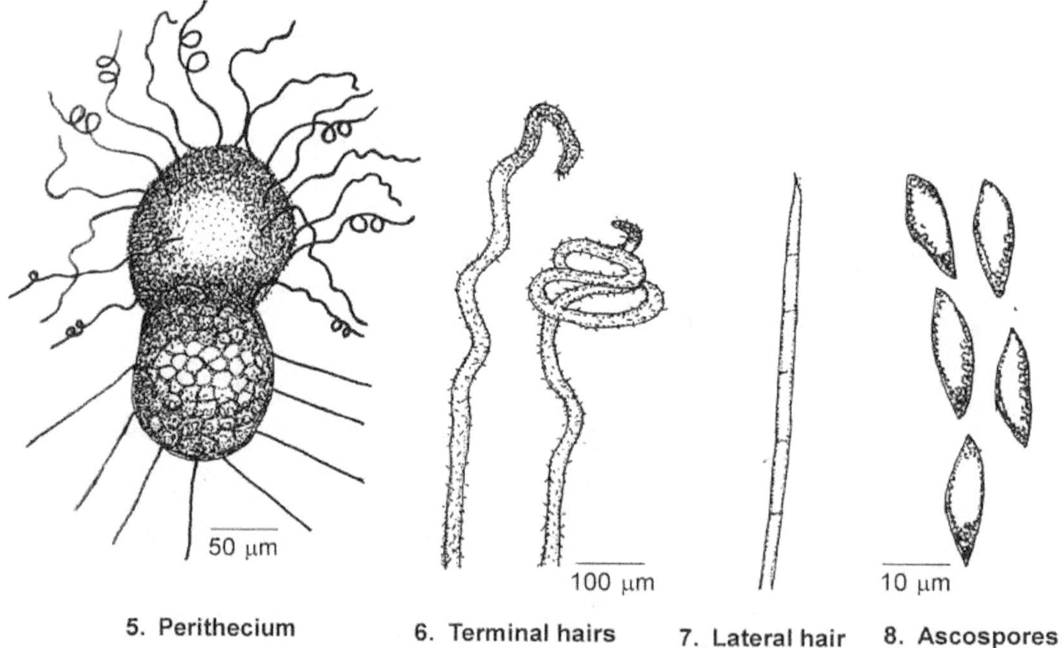

Figure 5-8 *Chaetomium subspirilliferum*

Remarks : The species described here is accommodated under *C. subspirilliferum* though the isolated species showed very large perithecia than the established species. The species isolated from Jalgaon soil also developed smaller spores.

This species is recorded for the first time from Maharashtra and India.

Chaetomium (Figs. 9 – 12)

Colonies were isolated on Czapek Dox Agar growing slowly about 0.3 cm in ten days, reverse colourless.

Mycelium pale yellow branched, septate. Perithecia mostly globose, dark brown,

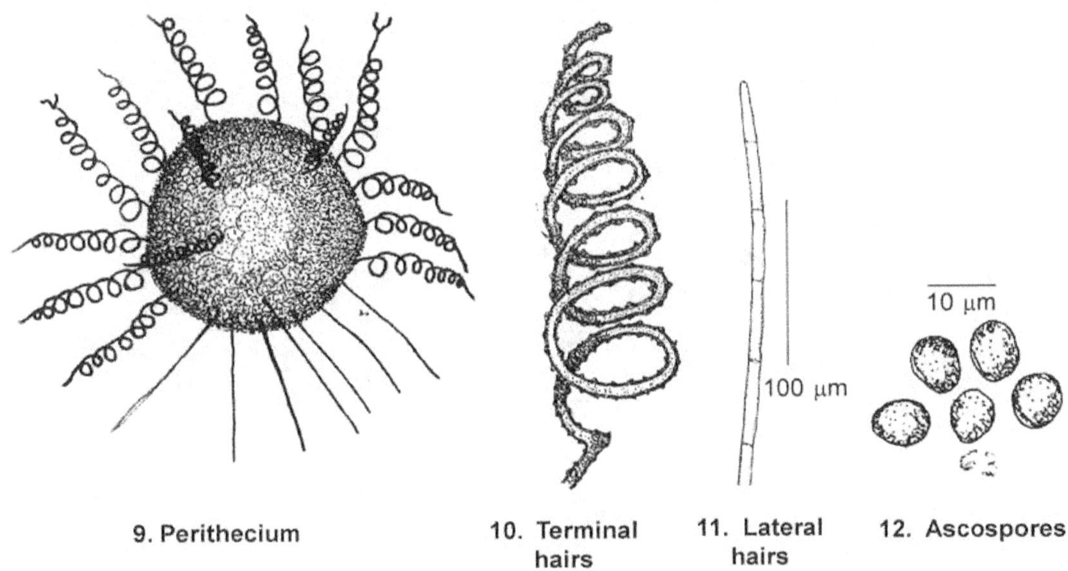

9. Perithecium   10. Terminal   11. Lateral   12. Ascospores
                            hairs   hairs

Figure 9-12 *Chaetomium*

250-280 μm × 250 μm, pseudoparenchymatous, covered with much coiled and straight hairs, gregarious. Terminal hairs much coiled, 5-7 coiled, brown, 3.3 μm wide, thick walled, unbranched and slightly roughened. Sometimes these terminal hairs bifurcated at the tips. Lateral hairs straight, brown, 2.5 μm thick, straight, smooth walled, unbranched, septate. Ascospores oval, light brown, 6-6.5 μm × 3-4 μm, non apiculate with rounded ends.

This species is isolated only from cultivated soil of Shirsoli during summer. The temperature of soil was 38^0c and the pH was 7.4. This is the isolated species of *Chaetomium* showing restricted occurrence.

Remarks : The isolated species showed some morphological similarities to the described species *C.perlucidum*. However, the perithecia are remarkably larger in this species. Terminal hairs are also much coiled and differ from the *C. perlucidum*. Morphology of ascospores also differs than other species so far described. All these morphological differences put apart this species than the described one. Therefore this species is proposed as a new species.

Chaetomium osmaniae Rama Rao and Rama Reddy, Mycopath. Mycol. Appl. 31,1967. (Figs. 13-16)

=*Chaetomium elongatum* Rama Rao and Ram Reddy, Mycopath. Mycol. Appl.24, 113-118, 1964.

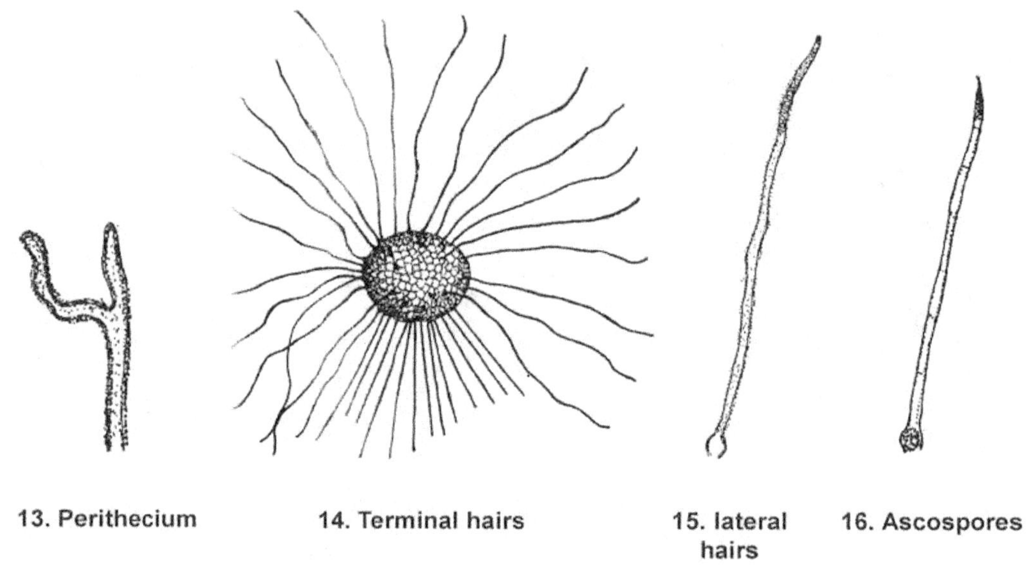

Figure 13-16 *Chaetomium osmaniae*

Colonies were isolated on Lactose Yeast Extract Agar growing slowly about 1cm in ten days, pale to olive brown; reverse of colony is black with colourless periphery.

Mycelium hyaline, branched, septate, develops perithecia after 5-6 days, Perithecia yellowish to pale brown, ovoid, 100-110 µm, pseudoparenchymatous, covered with long and short hairs. Terminal hairs long, slightly undulate, brown, 3.3 µm, finely roughened, unbranched, base swollen end in to finely roughened hyaline tip. Lateral hairs slightly undulate, pale brown, 3.0 µm, and unbranched, septate, base prominently and characteristically swollen. Ascospores pale brown, elliptical to slightly fusiform, 6.5-7 µm × 3.5-4 µm and apiculate at one or both ends.

This species was the only species of *Chaetomium*, which was isolated from six localities. From Mamurabad, Shirsoli and Girna river soils during rainy season and summer, Bhusaval soil during winter and Paldhi soil during summer. The temperature range of this species was very wide and it was in between 20^0c-38^0c as well as the pH range was also very wide from 6.8-7.4.

Remarks : The isolated species showed smaller spores than the species described by Rama Rao and Ram Reddy (1967).

It is earlier recorded from soil Hyderabad, Andhra Pradesh. This species is recorded for the first time from Maharashtra.

Chaetomium (Figs. 17 – 21).

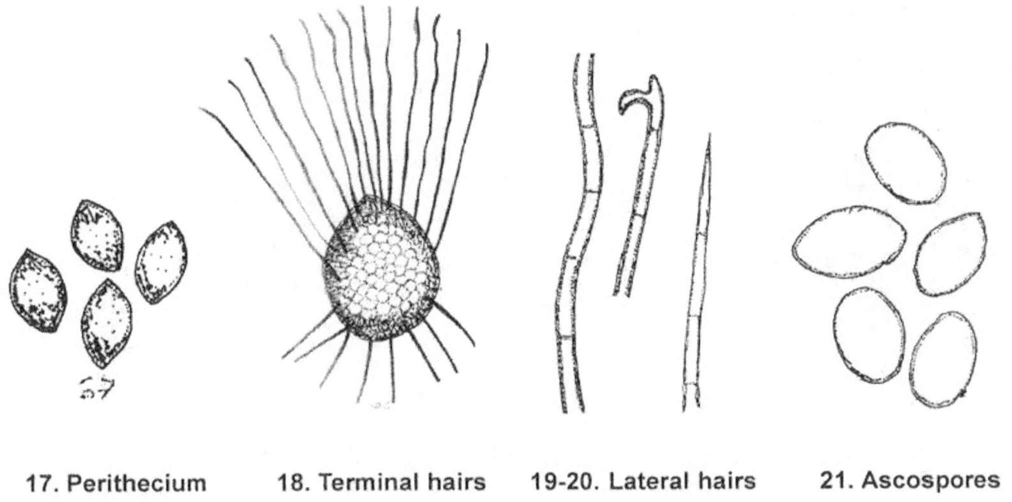

17. Perithecium   18. Terminal hairs   19-20. Lateral hairs   21. Ascospores

Figure 17-21 *Chaetomium sp. nov.*

Colonies were isolated on Lactose Yeast Extract Agar slow growing about 1cm in diameter in ten days, appears olive green grayish with blue reverse.

Mycelium septate branched, hyphae with pale brown mat, which develops perithecia. Perithecia dark brown, mostly globose to subglobose, gregarious, 100-125 µm × 75-100 µm in size, pseudoparenchymatous, profusely covered with hairs, Terminal hairs long, slightly undulate, brown, 3.0 µm wide, thick walled, smooth, septate, rarely short branched at the tip. Lateral hairs short, straight, 2.5 µm, unbranched, thin walled, smooth. Ascospores brown, thin walled, variable in morphology, mostly ellipsoidal with rounded ends, slightly apiculate, 12 –13.5 µm × 5-6.5 µm in diameter.

This species was isolated only from Mamurabad soil during rainy season. This was isolated at slightly low temperature of 28^0c and at pH 6.8 which was slightly acidic. This is also one of the isolated species of *Chaetomium* showing restricted occurrence.

Remarks : The isolated species broadly showed more or less resemblance to the described species *C. perlucidum* but differs in some characters. The present species has smaller perithecia, slightly undulate terminal hairs, and ellipsoidal slightly appeculate ascospores. Therefore, this species is proposed as a new species.

Chaetomium fusisporale Rai and Mukerji, Can. J. Bot. 40: 1379 -1384, 1962. Figs. 22 – 25

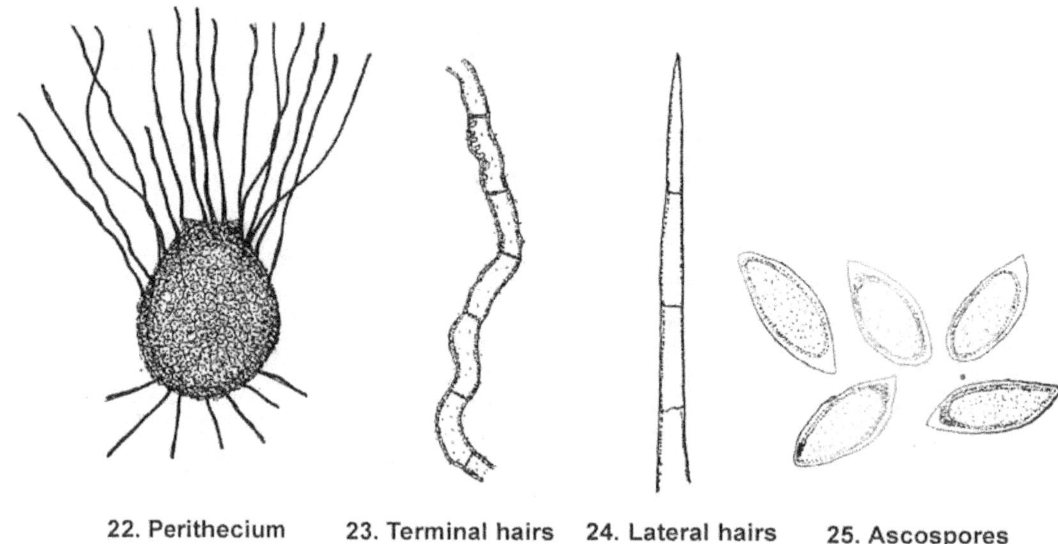

Figure 22-25 *Chaetomium fusisporale*

Colonies were isolated on Lactose Yeast Extract Agar and Lactose Yeast Extract Agar, growing slowly about 1.5 cm in diameter in ten days forming dark brown dots, reverse colourless.

Mycelium branched, septate, pale yellowish, perithecia black, subglobose to ovate, gregarious, 120-140 µm × 100-120 µm in size, covered profusely with long, brown, undulate, hairs, pseudoparenchymatous, distinctly ostiolate. Terminal hairs long, undulate, dark brown, unbranched, 3.5 µm, septate, smooth to very minutely roughened, basal cell slightly swollen ends in rounded tip. Lateral hairs dark brown, short, straight, 3.0 µm wide, septate, smooth to irregularly roughened, ends in blunt rounded tip. Ascospores mostly fusiform, olive brown, 12-13.5 µm × 6-6.5 µm, slightly pointed and prominently apiculate at one end, which is hyaline.

This species was isolated from four soils that is Pal and Manudevi forest soils during winter and Waghur River and Paldhi soil during summer. This species was also showing tolerance to the wide range of temperature from 18^0c-38^0c and pH range was 6.9-7.4.

Remarks : The species described here is similar to the established species by Rai and Mukerji (1962) in several characters such as perithecia, terminal hairs but the ascospores in the isolated species were larger as well as clearly apiculate.

It is earlier recorded from soils of Lucknow and Kanpur Uttar Pradesh. This species is recorded for the first time from Maharashtra.

Chaetomium (Figs. 26 – 28)

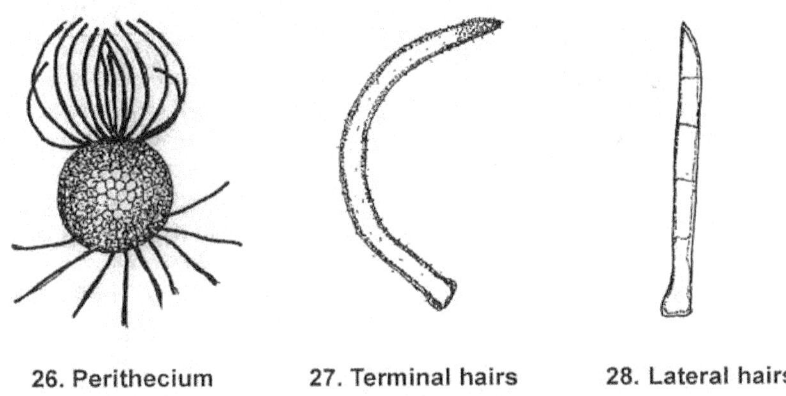

26. Perithecium   27. Terminal hairs   28. Lateral hairs

Figure 26-28   *C. sp. nov.*

Colonies were isolated on Lactose Yeast Extract Agar, growing slowly appeared in form of irregular black dots on mycelial mat after seven days, reverse colourless.

Mycelium hyaline, septate, branched, perithecia dark brown, superficial, mostly globose, 100 μm in diameter pseudoparenchymatous; terminal hairs brown, arcuate from base, basal cell slightly swollen, 3.5 μm wide, unbranched, finely and sparsely, roughened. lateral hairs short, straight with slightly swollen base, brown, and 2.5-3.0 μm wide, unbranched, smooth, septate. Ascospores ellipsoidal with apiculate at one end, 10.5 μm × 8.0 μm, brown.

This species was isolated only from Manudevi forest soil during summer. The temperature was 28^0c and pH was 7.1 and slightly alkaline. This is another isolated species of *Chaetomium* showing restricted occurrence.

Remarks : The isolated species is more or less resembles to the *C.gracile* and *C.venezuelens* as described by Seth (1970) in habit but totally differs in morphology of perithecia. The perithecia are often elongated and covered with dense hairs in *C.gracile* and *C.venezuelens*. However, the present species shows few terminal and lateral hairs with often globose perithecia. The morphology of ascospores is similar to that of *C.gracile* but widely differs from *C.venezuelens*. The present species also shows some specific characters very different than so far described species. Therefore this species is proposed as a new species.

Chaetomium sp.nov. (Figs. 29 – 33)

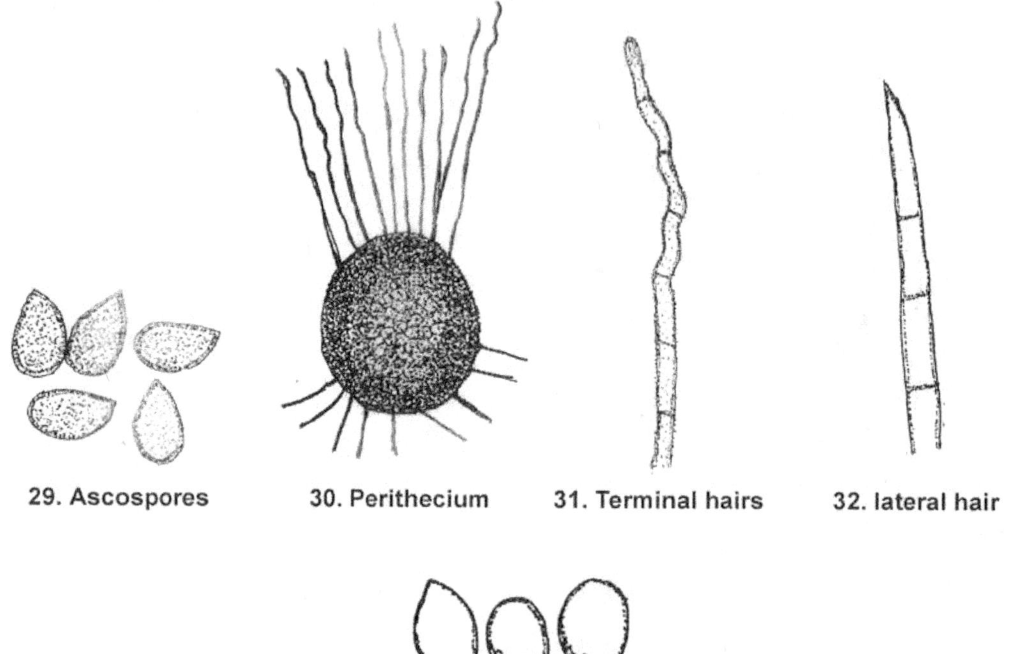

Figure 29-33   *C. sp. nov.*

Colonies were isolated on Lactose Yeast Extract Agar, dark brown dotted structures growing slowly about 0.5 cm in ten days, reverse colourless.

Mycelium colourless septate, branched, Perithecia dark brown, globose, 150-180 μm × 130-15 μm in size, pseudoparenchymatous, gregarious, numerous hairs giving hazy appearance. Terminal hairs brown, straight near base loosely coiled upwards, 3.3 μm wide, finely verrucose, unbranched, septate. Lateral hairs light brown, almost straight, 3.3 μm wide, septate, unbranched, smooth. Ascospores light brown, oval with rounded or apiculate at both the ends, 10.0 μm × 6.5 μm in size, thick walled, smooth.

This species was isolated only from Shirsoli soil during summer. The temperature was 38^0c and pH was 7.4. This is one of the isolated species of *Chaetomium* showing restricted occurrence.

Remarks : The species is totally dissimilar with all the species so far described by Seth (1970) in case of morphology of perithecia, terminal and lateral hairs, and ascospores therefore it is proposed as a new species.

References

Ames, L.M. 1963. A monograph of Chaetomiaceae. The United States Army Research and Development Ser.2.

Bhide, V.P., Pande, A., Sathe, A.V., and Rao V. 1987. Fungi of Maharashtra (Supplement I). MACS Research Institute Publication, pp 146.

Bilgrami, K.S., Jamaludin and Razvi, M.A. 1991. Fungi of India. List and refrences (2nd revised edition). Today and Tomorrow's Printers and Publishers, New Delhi. pp 798.

Kunze, G. and Schmidt, J.K. 1817. *Chaetomium* Mycologische Hefte nebste inen allgemein- botanischen Anzeiger, 1 – 2, 15 Leipzig.

Rai, J.N. and Tiwari, J.P. 1962; New species of *Chaetomium* from Indian soils. Can. J. of Bot., 40: 857 – 860

Rama Rao and Rama Reddy. 1967. Studies on Soil Fungi II, Additions to the Ascomycetes from soils of Hydrabad, India. Mycopath. mycol. appl., 24: 113-118

Seth, H.K. 1970. A monograph of genus *Chaetomium*. Nova Hedwigia, 16 : 495-599

Waksman S.A. 1916. Soil Fungi and their activities. Soil Science 2: 103 – 115.

13

Some Interesting Conidial Fungi from Andhra Pradesh, India

Vasant Rao and B. Bhadraiah

Deptt. of Botany, Osmania University, Hyderabad - 500 007, A.P. India.

ABSTRACT

A new genus belonging to conidial fungi is being described in honour of Prof. C. Manoharachary, a well known mycologist, on the occasion of his superannuation as *Caeamchariomyces* with *C. indica* as its type species. Further, another new species of *Annellodochium* as *A. manohara* and yet another new genus *Parapsuedofusidium* with *P. indicum* as its type species are also being proposed as they are new to science.

KEYWORDS : *Caeamchariomyces indica, Annellodochium manohara, Parapseudofusidium indicum.*

During the survey of microfungi of Andhra Pradesh, Vasant Rao et al., (2004) have surveyed the fungi around some aquatic bodies in Andhra Pradesh and concluded that this region is rich in mycoflora. These fungi though collected very frequently were not described and published anywhere, hence are being presented in this communication on the occasion of superannuation of Prof C. Manoharachary, our esteemed colleague, an excellent teacher, recipient of several awards and a well known mycologist of the country along with other new taxa of fungi

1. *Caeamchariomyces* gen. et sp. nov.

Caeamchariomyces Vasant Rao & Bhadraiah **gen. nov.**

Deuteromycotina, hyphomycetes, anamorph.

Coloniae effusae, synnematae, conidiomata pedicellata, globosa, capitata, albo-grisea et rubra, pilosae ad capitis. Mycelium partim immersum et partim superficiale ex hyphis septata, rami laevia, compositum stroma pseudoparenchymatica, et synnemata, pedicellata compacta, hyalina, vel subhyalina, recta, simplicae ad basim, ramosa ad apicem. Conidiophora macronemata, simplicea, vel ramosa, compacta, septata. Cellae conidiogenae, polyblastic, terminale, sympodici, discretei, cylindricae, hyalina vel minuta, protuberentiata. Conidia holoblastica, acropleurogenae, filiformia, rami profusa, aseptata, acuto ad apicem, truncata ad basim, digitao in formia, laevia hyalina.

Colonies effuse, synnematous, short, stalked, hairy, globose, whitish grey or pinkish. Mycelium partly immersed and partly superficial to form synnematous conidiomata, hyphae hyaline to subhyaline, septate, branched, smooth, erect. Conidiophores macro-nematous, and synnematous with simple or branched head, closely adpressed, septate. Conidiogenous cells polyblastic, terminal, sympodic, discrete, cylindrical, hyaline with minute protuberences. Conidia holoblastic, acropleurogenous, filiform always properly branched several times, digitate, aseptate, apex acute, base truncate, smooth, hyaline.

Species typica : *Caeamchariomyces indica*

Caeamchariomyces indica Vasant Rao & Bhadraiah **sp. nov. (Plate 1a, b)**
Coloniae effusae, albo-greisae. Conidiomata synnematica, pedicellata ad compacta, orinda capita, pedicellae 30-100 μm longa, 30-50 μm crassa, compacta 250-350 μm crassa. Mycelium partim immersum, partim superficiale, formentibus, stroma compositum. Cellae pseudoparenchymatica, exhyphis, 2-3 μm crassa, hyalina vel subhyalina. Synnema albogrisea, conidiomata 30-50 μm crassa ad apicem. Conidiophora macronemata, synnemata, simplicea vel rami 30-100 μm longa, 2-3.5 mm crassa lata 1-3 septata, laevia hyalina. Cellae conidiogenae polyblastici, terminale, sympodici ad minuta, protuberentia, 1-cellae, rarier septati, hyalina, laevia. Conidia holoblastica, acropleurogenae, filiformia, rami profusa, formentibus, digitato, dendroid, configurata, 2-3 μm at basim, 5-25 μm longa, inclusio rami dendoid configurata, laevia hyalina. Conidio secessio schiozolytico.

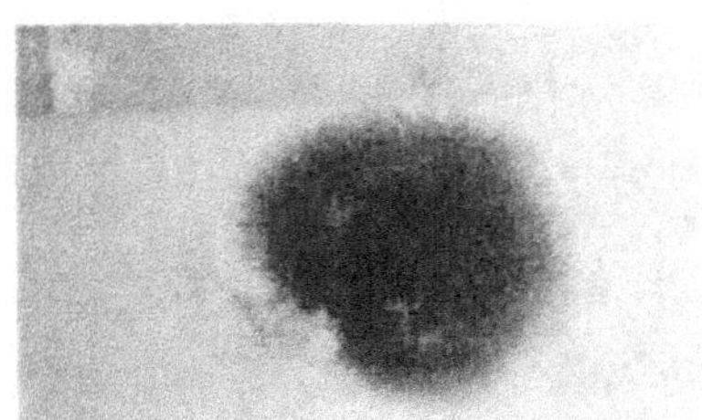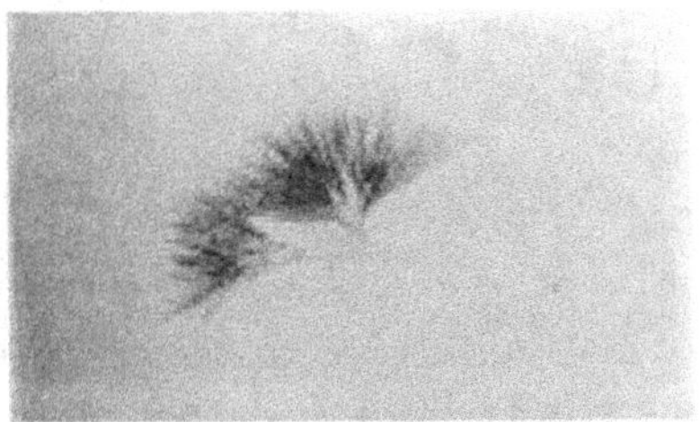

Plate 1. (a) *Caeamchariomyces indica* gen. et sp. nov. a. Whole colony, 100x :
(b) Conidiophores with conidia , 450x.

Typus emortio folio, Chintoor regio, A.P., Indiae. NCMH.leg VR & PSV.No. 1119. Oct, 1994.

Colonies effuse, cottony, pinkish or whitish grey. conidiomata synnematous, short, stalked with compact round head, stalk 30-100 μm long, 30-50 μm broad, head 250-350 μm broad. Mycelium partly immersed and partly superficial, hyphae delicate, thin walled, 2-3 μm broad, septate forming small submerged stroma made up of thin walled or thick walled, hyaline to subhyaline pseudoparenchymatous cells producing synnemata.

Synnemata whitish grey, stout, pin-head like, conidiomata 30-50 µm broad at base, 30-100 µm long, 200-350 µm broad at the head region. Conidiophores macronematous, simples, rarely branched 30-100 µm long, 2-3.5 µm broad, 1-3 septate, smooth, hyaline, conidiogenous cells terminal, polyblastic, sympodic with minute protuberences 1-celled rarely septate, hyaline, smooth. Conidia holoblastic, acropleurogenous, thread like, branched profusely to form digitate structure, tree like, 2-3 µm at base, 5.25 µm long including the branched configuration, smooth, hyaline, thin walled. Conidia secession schizolytic but usually the whole digitate structure gets separated.

Caeamchariomyces is close to *Linodochium* Hohnel and some aquatic Ingoldean Hyphomycetes like *Articulospora* Ingold and others. *Linodochium* produces sporodochia, macronematous conidiophores with polyblastic, septate, simple, filiform, scolecosporous and hyaline conidia. *Articulospora* produces hyaline and branched conidia. Its dendroid conidia are reminescent of *Speriopsis* Tubaki, *Weisnerimyces* Koord. and *Ramaraomyces* Rao *et al.* (1989) which are septate or connected with small separating cells in between the transverse septa of the conidia. There is no known hyphomycete to combine these characters. Therefore, it is being proposed in honour of Prof. C. Manoharachary, a leading mycologist of our time, colleague and a well known teacher of mycology and plant pathology taking his initials of the name in latinised form C = Cae, am = M = Manohara, Chario = Chary, Myces = fungus. The specific epithet is after the country India – indica.

2. *Annellodochium* Deighton, F.C. Mycol. Pap. No. 118; 28-30, 1969.

Deighton (1969) created *Annellodochium* with *A. ramulisporum* for the sporodochial anamorphs probably hyperparasiting *Diatrype* on dead branches of *Chlorophora regia* in Sierra Leone, Africa. *Annellodochium* is characterised by sporodochial conidiomata, conidiophores macronematous, compact conidiogenous cells, percurrent, monoblastic. Conidia terminal, bifurcate, solitary or in short branched chains, pale dilute olivaceous brown, smooth. *Annellodochium* was originally described from Africa. These collections form the first new generic record to India and eventually with a new species because of the conidial and conidiophore characters. Hence, it is being described as *Annellodochium manohara* meaning the beautiful. Specific epithet is being described after the Prof. C. Manoharachary.

Annellodochium manohara Vasant Rao & Bhadraiah **sp. nov.** (Fig. 1, Plate 2).

Coloniae sporodochia, pulvinata, erumpenta, irregulariter, albigio, 300-700 µm lata, 150-300 µm alta, stromatica. Mycelium ad compositum ex hyphis, septati, 3-5mm crassis vel radiatum formentibus conidiophora. Conidiophora macronemata semimacronemato, simplicea, compacta vel radiata, septate, cylindricea 10-20 µm longa. 3-5 µm lata. Cellae conidiogenae monoblastici. Conidia holoblastica, terminale, inconspicua percurrenta, 2-3 cellulae, beifurcata, cellae globosa, constricta, laevia, inflatilus, hyalina vel subhyaline albida, 10-25 µm lata vel 10-15 µm longa.

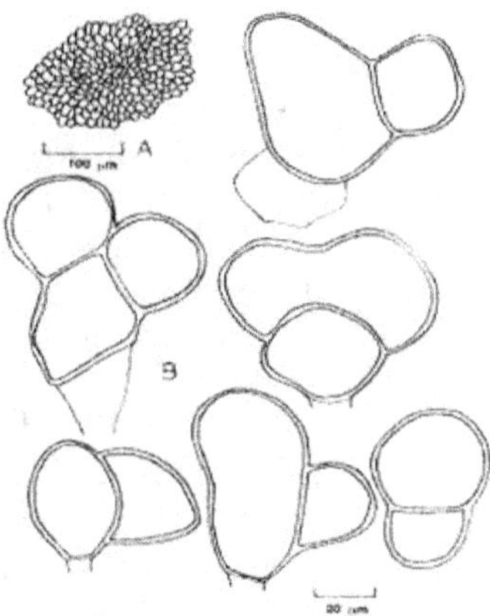

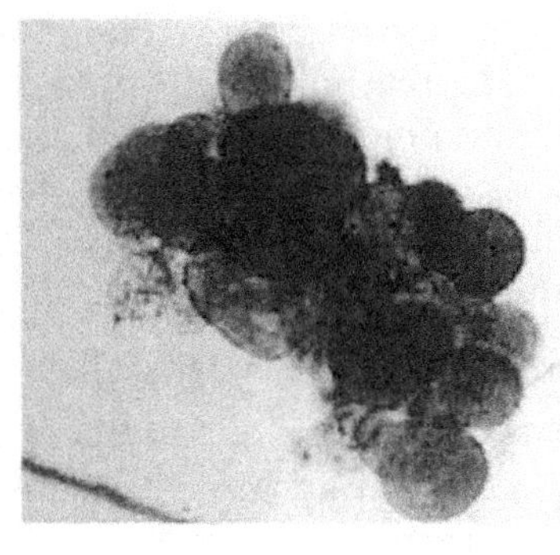

Figure 1 *Annellodochium manohara*
A. Sporodochium and B. Conidiophores
with conidiogenous cells & conidia

Plate 2 *Annellodochium manohara*
sp. nov. conidiophores and conidia, 450x

Habitat in folia emortius Chintoor regio, A.P., ad positus NCMH. leg VR, & PSV (typus). No. 993. Dec, 1993.

Colonies sporodochial, raised, pulvinate, erumpent, irregular, whitish grey, 300-700 μm broad. 150-300 μm long, consisting of inner stroma. Mycelium composed of densely packed septate hyphae, 3-5 μm wide with radiating tips which become conidiophores.

Conidiophores macronematous or semimacronematous, simple, compact or radiating 10-20 μm long 3-5 μm broad septate, smooth, cylindrical. Conidiogenous cells terminal, monoblastic, inconspicuously percurrent. Conidia holoblastic, terminal, 2-3 celled, bifurcate, thin walled, cells globular, septate, deeply constricted, smooth, cells inflated and rounded off. 10-25 μm broad and 10-15 μm long. Conidial secession rhexolytic. No germination was observed. Material examined on unidentified leaves. Chitoor forest, Dt. Khammam, A.P. NCMH. Coll. PSV. No. 993 Dec 1993 and canconum forest, Goa 2004.

3. *Parapseudofusidium gen.* et sp. nov.

Parapseudofusidium Vasant Rao & Bhadraiah gen. nov.

Deuteromycotina, Hyphomycetes, anamorphae.

Colonia globuslata, albinosa, conidiomata synnematica. Mycelium partim immersum, partim submersum, ex hyphis sepatati, rami. Conidiophoris macronematous, synnematica, ramosae, hyalina vel subhyalina, 1- cellae, Cellae conidiogenae polyblastici, terminate et integrata. Conidia holoblastica, acrogena in simplicia, catenua or rami, cylindricia, vel fusiformia laevia hyalina vel subhyalina, conidio secessio schizolytico.

Species typica : *Parapseudofusidium indicum*

Colonies cottony, globular, whitish synnematous. Mycelium partly immersed and partly superficial, septate branched hyphae, no clamp connections. Conidiophores macronematous, synnematous, branched, hyaline to subhyaline, septate, short, 1-celled. Conidiogenous cells polyblastic terminal observed or integrated. Conidia holoblastic, acrogenous in simple or branched chains, cylindrical to fusiform, smooth, hyaline to subhyaline. Conidial secession schizolytic. No germination and teleomorph was observed.

Parapseudofusidium indicum Vasant Rao and Bhadraiah **sp. nov.** (Fig.2, Plate 3).

Coloniae villosae usque pilosae, globulatae. Conidiomata synnematica conspicua, erecta capitata 150-500 µm lata ad apicem, 100-150 µm ad pedicellto. Mycelium partim immersum, partim submersum ex hyphis 2-3.5 mm crassis. Conidiophora macronemata, synnemata, adhaerentibus, parallelis, septatis, ramosis 1- cellae, elongatis, cylindricia. Cellae conidiogenae indestintis, terminale vel integratis polyblastici 2.5-3.5 mm lata.

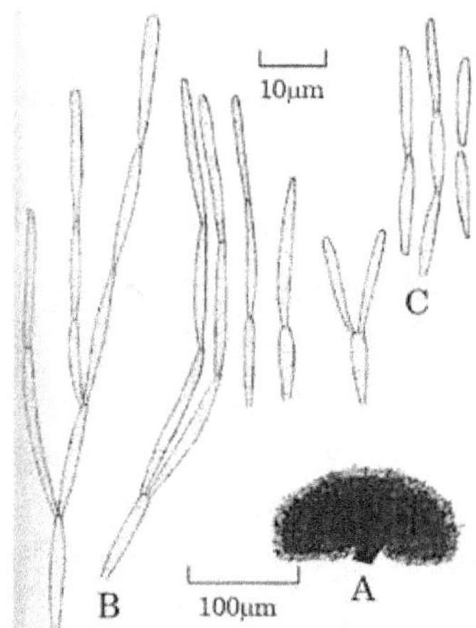

Figure 2 *Parapseudofusidium indicum*; A. Colony, B. Conidiophores with conidiogenous cells and C. Conidia

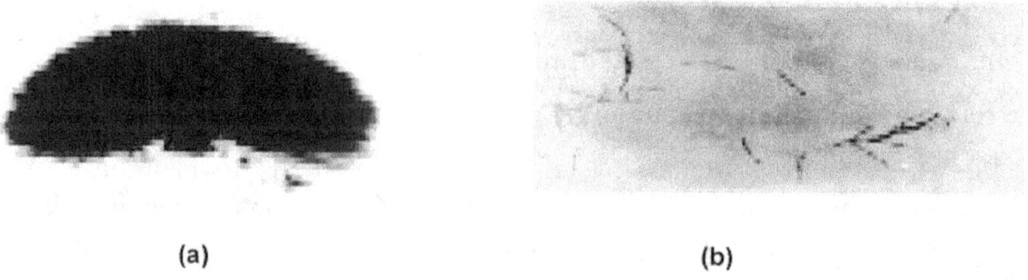

(a)   (b)

Plate 3 *Parapseudofusidium indicum*. gen et. sp. nov. (a) Whole colony, 100x (b) Conidiophores and Conidia, 450x

Conidia holoblastici, cylindrici, fusiformiae, vel filiformae in catenua simplicia usque rami 2-3.5 µm lata, 15-25 µm longa, laevia, hyalina vel subhyalina, conidio secessio schizolytico. Holotypus emortio folio, Chitoor regio, A.P., Indiae, NCMH. leg PSV. Num. 998. Oct, 1993.

Colonies, cottony globular, whitish, conidiomata synnematous with short, stout stalk and expanding globular head composed of radiating conidial mass. 150-500 µm broad in head region and 100-150 µm in the stalk region. Mycelium partly immersed and partly superficial composed of septate, hyaline to subhyaline, branched hyphae, 2-3.5 um um broad, thread like, No clamp connections any where in the conidiomata. Conidiophores macronematous, synnematous, stalk of the synnema made up of several adpressed hyphal threads branched, septate, smooth 2.5-3.5 um broad and mostly 1-celled elongated, cylindrical. Conidiogenous cells indistinct, termina mostly polyblastic and (not phialidic as in *Fusidium*) and *Pseudofusidiums* denticles indistinguishable.

Conidia holoblastic, cylindrical or fusiform, thread like 1-celled in chains catenate or branched 2.3.5 µm broad and 15-25 µm long, smooth, hyaline to subhyaline. Conidial secession schizolytic at septa. No germination and teleomorph observed.

Parapseudofusidium indicum Vasant Rao & Bhadraiah **sp. nov** (Fig. 2, Plate 3).

Parapseudofusidium is being proposed as a new genus tentatively because it produces a stalked synnemata with expanding crown of compact conidiophores and conidial mass having a basal stipe like structure. It seems several allied taxa like *Fusidium* Link, *Acremonium* Link, *Polyscytalum* Riess, *Hormiactis* Preuss, *Septonema* Corda, *Heteroconium* Petrak and others which are combination of moniliaceous and dematiaceous forms were described from soil and decomposing leaf litter. The phialidic nature is absent hence, it is not worthwhile to consider as, *Acremonium* and *Paecilomyces* and others. Its moniliaceous conidiomata distinguish it from *Polyscytalum*, *Septonema*, and *Heteroconium*. Its conidial chains however, remind *Hormiactis* which is ill defined and poorly described. Until further studies the author is hesitant to place this taxon in *Hormiactis*, *Polyscytalum* and *Parapleurotheciopsis* group. Therefore it is thought befitting to create a separate new genus viz. *Parapseudofusidium* with *P. indicum* as its type species. It is pertinent to compare this taxon with *Pseudofusidium* Deighton (1969) which is mononematous and its chain of conidia as *P. clusiae* is superfluous and different. Therefore, proposition of *Parapseudofusidium* with *P. indicum* is justified

Acknowledgments

The author expresses his gratitude to Prof. D.J. Bhat of Goa University; Prof Y.S. Rao and Prof. (Mrs.) Alka Pande of MACS, Pune, Prof. S. M. Reddy, K.U. Warangal and Dr. (Mrs.) I. K. Kunwar of OU for going through the manuscript and suggestions. Further,

Prof. (Mrs.) J. Uma Bala, Prof. M. Prabhakar, Prof. K. Janardhan Reddy and other colleagues and authorities of Osmania University for providing facilities and encouragement and UGC, New Delhi for financial support and travel.

References

Carmichael, J.W., Kendrick, W.B., Conners, I.L., and Sigler Lyrine. 1980. *Genera* of *Hyphomycetes*, The University of Alberta, Alberta Press, Edmonton, pp 1-386.

Deighton, F.C. 1969. Some Hyper parasitic Hypomycetes and a note on *Cercosporella uredinophila* Sacc. Microfungi IV, Mycol. Pap. 118: 28 – 30.

Ellis, M.B. 1971. *Dematiceous Hyphomycetes*, IMI, CAB, Surrey, UK.

Ellis, M.B. 1976. *More Dematiceous Hyphomycetes*, IMI, CAB, Surrey, UK.

Matsushima, T. 1975. *Icones microfungorum Matsushima lectorum*, Kobe, Japan.

Subramaniam, C.V. 1971. *Hyphomycetes*, ICAR, New Delhi

Vasant Rao, Manohara Chary, C., Kumar G.S. and Subodh, V. K. 2004. *Fungi around aquatic bodies in Andhra Pradesh*, B. S. Publ. Hyderabad .

Rao, K., Manohara Chary, C. and Goos R.D. 1989. Forest litter Hyphomycetes from A.P. India. *Mycological Res.* 81(5) : 790-793.

Occurrence of Lichen Flora of Nasik City and Effect Of Sulphur Dioxide Pollutant on *Pyxine petricola var. petricola* Nyl.

P.P. Sethy

Department of Botany, Bhonsala Military College, Rambhoomi, Nashik-05
e-mail : bhonsala_nsk@sancharnet.in

ABSTRACT

About 60 specimens of foliose and crustose lichens were collected from different localities of Nashik city of Maharashtra State and examined critically in the laboratory. Nine species were identified chemotaxonomically. *Pyxine petricola* var. *petricola* Nyl., the only species belonging to the family Physciaceae was found to grow on the barks of varieties of angiospermic plants and also on rocks.

Effect of Sulphur dioxide pollutant on the thallus of this species was studied on different samples. Deformation and blackening of some of the lichen thalli was quite noticeable. Degradation of chlorophyll pigments has been studied in various samples by using spectrophotometer. The result analysis shows that the percentage of degradation of chlorophyll pigment was maximum in heavy traffic areas and comparatively very negligible in the open areas. Phytoecological aspects of lichens were also studied.

KEYWORDS : Lichens, chemotaxonomy, Sulphur dioxide, chlorophyll degradation, spectrophotometer.

Nashik is one of the pilgrim city of Maharashtra covering a total area of 259.13 sq. km and the total population of 10,76,967 (2001 census). The city is located between 19.35' and 20.52' North latitude and 73.16' and 74.56 East longitude. The annual average rainfall is from 560 mm to 700 mm. The temperature rises more in the month of April and May. The highest temperature is recorded till now is 43 degree centigrade. The air is humid during rainy season up to December.

All metropolitan and big cities of India show considerable amount of environmental degradation (Das et al. 1986). Nashik city is not exception to this. Since last ten years, the population of city has increased almost two folds and is increasing in a faster rate. There is a direct relationship between increase in population numbers and environmental impact. Nashik being a holy place, Lakhs of devotees visit it every year. Due to heavy vehicular traffic in the city and emission of house hold gases, pollute the air into maximum level along with gases from industries, brick manufacturing units etc. The green belt cover has been considerably reduced because of various developmental projects like new settlement areas, widening of roads in the city for Kumbha mela and establishment of new industrial units. As a result harmful gases level of the city increased considerably.

In recent years, Lichens are used for monitoring purposes (Dubey, 1999; Ponmurugon et al, 2002; Richardson, 1992). The lichen thallus is a perennial, stable and long-lived structure on the bark of a tree or on barks. Air borne pollutants are absorbed very effectively over the entire surface of the thallus because there are no protective structures such as waxy cuticle or stomatas. These elements are bound in the hyphal walls and store harmlessly or are taken up by the metabolically active algae, which die off or are damaged. Lichens are very sensitive to air pollution and very few lichens grow in a city area and almost absent in a heavily polluted area.

The present study has been undertaken to know the distribution of lichen species in different parts of Nashik City. Some aspects of Phyto-ecological association and climatic factors are also studied which are responsible for the occurrence of the species. Effect of Sulphur dioxide pollutant on the lichen thalli is also studied.

Materials and Methods

During the course of the present study about 60 specimens were collected from different parts of the city. Collections usually were done during the months of August to January. Specimens were kept in blotting paper having high absorption capacity. All the specimens were examined and accession numbers were given to them.

The external morphological details were studied by using binocular microscope. The anatomical studies of the thallus and ascocarps were studied under research compound microscope after taking fine hand sections and mounting in lactophenol on a slide (Aswasthi, 2000). Colour tests and Thin Layer Chromatography were carried out to determine the presence or absence of chemical substances in the lichen thallus.

Pyxine petricola var. *petricola* Nyl, was the only species found to occur in different localities of Nashik City. Therefore, this species is selected for biomonitoring studies. Various samples were prepared and chlorophyll estimation studies were carried out. Marginal lobes of the thallus (about 15 mg) were placed in 3 ml of dimethyl sulphoxide for 45 minutes at 60 degree centigrade in the dark. The extracts were filtered and the absorbance of the extract was then measured in a spectrophotometer at 415 and 435 nm. The amount of degradation of chlorophyll a to phaeophytin is calculated from the optical density ratio 435/415 (Ronen and Galen, 1984).

Results and Discussion

Different localities were surveyed for occurrence of different species of lichens. It has been observed that very few lichen species were growing on limited Angiospermic plants and also on rocks. All the lichen specimens were identified taxonomically. A few species were confirmed with the expert in the Department of Mycology and Plant Pathology, Agharkar Research Institute, Pune. Nine species of lichens were found growing in the city.

Names of lichens	*Nature of substratum*
1. *Arthothelium saxicolam* Massal	On rocks (saxicolous)
2. *Bulbothrix meizospora* (Nyl.) Hale	On rocks (saxicolous)
3. *Diploschistes scruposus* (Schreb.) Norm.	On rocks (saxicolous)
4. *Graphina acharii* (Fee) Mull. & Arg.	On barks (corticolous)
5. *Leptogium* sp.	On barks (corticolous)
6. *Parmotrema praesorediosum* (Nyl.) Hale	On rocks (saxicolous)
7. *Parmotrema tinctorum* (Nyl.) Hale	On rocks & barks
8. *Pertusaria* sp.	On barks
9. *Pyxine petricola* var. *petricola* Ny.	On barks & rocks

The only lichen species, *Pyxine petricola* var. *petricola* Nyl. is well distributed in different parts of Nasik city. This species grows luxuriantly on the barks of *Mangifera indica, Ficus benghalensis, Plumeria alba, Rolstonia regia* and occasionally on other angiospermic plants. Hence this species is selected for studying the effect of Sulphur dioxide on the thallus.

Pyxine petricola var. *petricola* Nyl., a foliose lichen belongs to the lichen family Physciaceae. The thallus has distinct lobes and whitish gray in colour. Since there is no protective structures such as waxy cuticle or stomata, the soluble sulphur dioxide accumulats on the hyphal wall and takes part in the metabolic pathway. The degradation of chlorophyll l-a into phaeophytin-a is measured with a Spectrophotometer (Brown and Hooker 1977).

Estimation of chlorophyll pigments were carried out in different samples which were collected from both polluted and open areas of the city in different seasons. Sulphur dioxide gas is effectively absorbed by the thallus during humid atmosphere.

The following data and results were recorded in five samples in the month of September 2003. The degradation of Chlorophyll is expressed by the OD 435/ OD 415 values.

Sr. No	Area	Wavelength in nm	OD	Chlorophyll Degradation value
01	Pandav Leni	435	.462	
		415	.259	1.78
02	Panchavati	435	.465	
		415	.257	1.80
03	CIDCO	435	.464	
		415	.256	1.81
04	Nasik Road	435	.573	
		415	312	1.83
05	C.B.S.	435	.581	
		415	.307	1.89

Like wise estimation of chlorophyll pigments of different samples are carried out in the month of October, December, March and July for two years. It has been observed that the degree of chlorophyll degradation is maximum in C.B.S. area where vehicular pollution is maximum. The least degradation shows in Pandav Leni area where vehicular pollution is minimum and also is an open area.

Following observations were made in the present work : -

1. Only nine species of lichens are growing in Nasik City. The limited number of species is due to absence of suitable old angiospermic plants. The distribution of each species is also in very limited areas. Many old trees have been cut down for widening of roads and other developmental works in the city.

2. Since last five years, the average temperature of the city is increasing. Last summer experienced 43 degree centigrade temperature which was the highest in this decade. High temperature and less relative humidity in the air prevent lichen communities to colonize and stabilize on the substratum.

3. There is constant increase in the number of vehicles in the city. Nashik being a pilgrim centre more devotees visit the city. New housing complexes, motor vehicles and bricks manufacturing units caused more pollution in the air. The polluted air does not allow juvenile and delicate lichen thalli to grow and establish on the substratum because they are very sensitive to air pollution.

4. It seems that *Pyxine petricola* var. *petricola* Nyl. is a pollution tolerant species and grows in adverse climatic condition.

5. Blackening of lichen thalli and deformation of lobes was observed in some areas. It may be due to accumulation of various pollutants.

6. Biological monitoring of the quality of air of an urban area has lot of significance. Use of lichens as bioindicator has the advantage of relatively low cost. The degree of error can compensate by the density of sampling points. An integrated approach is probably the most appropriate solution for monitoring air pollution.

Acknowledgements

I would like to express sincere gratitude to Dr. (Miss) U.V. Mkhija, Scientist, Department of Mycology and Plant Pathology, Agharkar Research Institute, Pune for her help and constant encouragement. I am also thankful to the Joint Director, University Grants Commission, Western Regional Office, Ganeshkhind, Pune for the financial assistance provided for the Minor Research Project.

References

Awasthi, D.D. 2000. Lichenology in Indian subcontinent. *A hand book of lichens.* pp. 1-124.

Brown, D.H. and **Hooker, T.N.** 1977. The significance of acidic lichen substances in the estimation chlorophyll and phaeophytin in Lichens. *New Phytologist.* 78:617-624.

Das et.al.1986. Studies on the plant responses to air pollution I. Occurrence of lichens in relation to traffic load of Calcutta City. *Indian Biologist.* 26-29.

Dubey, A.N. 1999. Accumulation of lead by lichens growing in and around Faizabad, Uttar Pradesh. *Indian. J. Environment Biology.* 20:223-225.

Ponmurugon, P. et.al. 2002. Effect of environmental factors on lichen distribution. Ecology. *Environment and conservation.* 8:147-150.

Richardson, D.H. 1992. *Pollution monitoring with lichens.* A Naturalists Handbooks. 19:1-75.

Ronen, R and **Galun, M.** 1984. Pigment extraction from lichens with dimethyl Sulphoxide (DMSO) and estimation of chlorophyll degradation. *Environmental and Experimental Botany.* 24:234-245.

15 | *Two New Hyphomycetes from Forests of Karnataka, India*

B. Satyanarayana Reddy[1] and Vasant Rao[2]

[1]Dept. of Botany. Govt Degree College, Ramannapet.
[2]Dept. of Botany, Osmania University, Hyderabad.

ABSTRACT

Two interesting Hyphomycetes namely *Stanjehughesia manohara*, and *Alysidiopsis indica* occurring on dead bark and branches respectively, are described.

KEY WORDS : Hyphomycetes, lignicolous fungi, new species.

During the survey (1987) of Hyphomycetous fungi, two interesting taxa were collected from the forests of Karnataka, India. The collections were screened and identified as the species belonging to *Stanjehughesia* Subramanian and *Alysidiopsis* Sutton. The present fungal isolates are different from the earlier reported species in conidial morphology, size, shape and colouration etc, (Ellis, 1971,1976; Sutton, 1973). Hence they are being described as new taxa. The present paper deals with the description of these taxa.

Stanjehughesia manohara Reddy, B.S. & V. Rao. sp. nov.(Fig. 1)

Colonies effuse, greyish, black, lignicolous. *Mycelium* mostly immersed, *Stroma* often present, pseudo parenchymatous, small, black, 20-30 μm x 10-25 μm. *Conidiophores* little differentiated, not more than a single conidiogenous cell or 1-septate, smooth walled, pale brown to dark brown, 8-20 μm long, 8-12 μm broad *Conidiogenous cells* solitary, short, simple, monoblastic, discrete, terminal, determinate, obclavo-cylindrical. *Conidia* holoblastic, solitary, simple, dry, acrogenous, straight or slightly curved, truncate at the base, 10-20 euseptate, dark brown at the base, apical region sub-hyaline to pale brown, aseptate with curved or circinate apex , 200-230 μm long, 5-10 μm wide at the base and 2 μm at the apex.

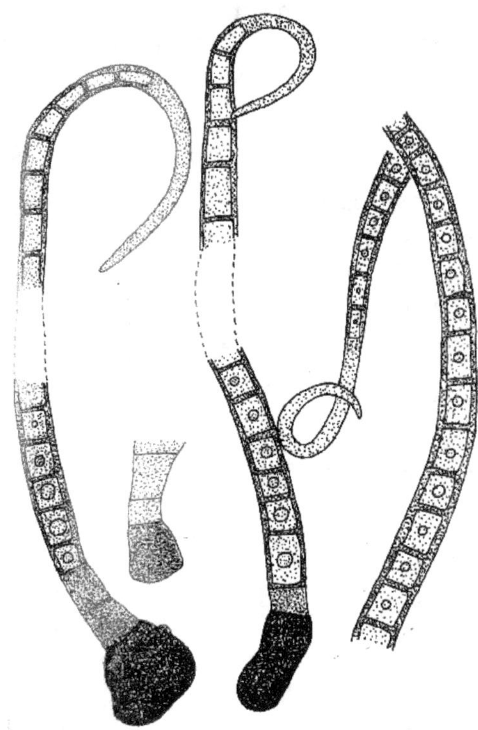

Figure 1. *Stanjehughesia manohara*

Coloniae effusae, griaese, nigrae, *Mycelium* plerumque in substrato immersum, *Stromata* saepe presentia, pseudoparenchymata, atrae, 20-30 μm x 10-25 μm. *Conidiophora* leviter conspicua, ex singularia cellula conidiogena composita vel 1-septata, pallide brunnea vel atro brunnea, 8-20 μm longa, 8-12 μm crassa. *Cellulae conidiogenae* solitaria, brevia, simplica, monoblasticae, discretae, terminale, determinata, obclavo-cylindricae. *Conidia* holoblastica, solitaria, simplica, sicca, acrogena recta vel curvata, basi truncate, 10-20 euseptata, basi atrobrunnea, regione apicali sub-hyalina vel pallide brunnea, apice aseptata, curvata vel circinata, 200-230 μm longa, basi 5-10 μm,apice 2 μm lata,

Material collected on unidentified bark, Gajnur & Bhagavathi, Karnataka, V.M.R.L No 1186,(Type) Nov 1987.coll. B.S.N.R.H.C.I.O (Co-Type).

The hyphomycetes with solitary, more or less obclaviform phragmo conidia emerging from monoblastic conidiogenous cells integrated in conidiophores with different degrees of differentiation have been included in *Sporidesmium* Link. This large and heterogeneous genus was reassessed by Subramanian (1992) and by Hernandez- Gutjerrz and Suttan (1997), who split *Sporidesmium* Link. in to different genera. Those species producing acrogenous, solitary, dry and euseptate conidia originating from simple conidiogenous cells were placed in the new genus *Stanjehughesia* by Subramanian (1992).

The present taxon shows some similarity with *Janetia curviopsis* Goh and Hyde (1996) in its circinocaudate conidial apex but differs in the other aspects like polyblastic, denticulate conidiogenous cells, conidial size, number of septa ete.

Repetophragma subulata (Cooke& Ellis) Subramanian (1992) shows similar conidial structure but differs in having distinct conidiophores, conidial size etc.

Stanjehughesia hermiscoides (Corda) Subramanian (1992) resembles with *S. manohara* sp. nov. in conidiophore size, and external morphology of conidia but differs in circinate nature of the conidial apex and conidial septation.

In the light of its long, obclavo-cylindrical conidia with circinate, hyaline aseptate apex, it is described as *S. manohara* sp. nov.

It gives us great pleasure to name this species after Prof. C.Manoharachary, a well known mycologist.

Alysidiopsis indica Reddy B.S & V. Rao. sp. nov. (Fig. 2).

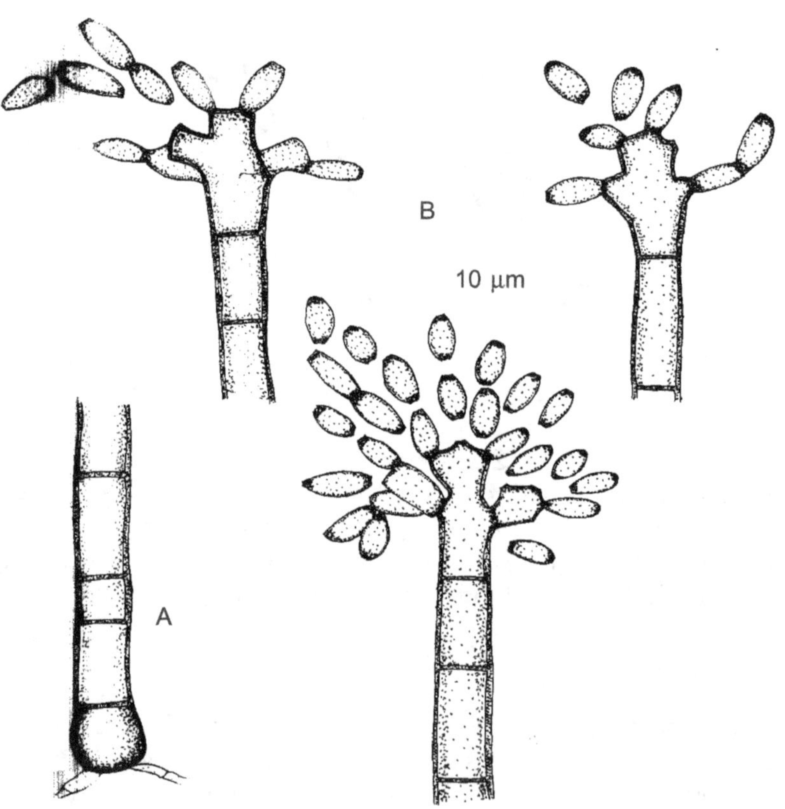

A : Conidiophore
B : Conidiophore with conidiogenous cells and conidia.

Figure 2 *Alsydiopsis indica*

Colonies effuse, brown, hairy. *Mycelium* mostly immersed, composed of branched, septate, hyaline to pale brown, smooth, 3-6μm thick hyphae. *Conidiophores* solitary, macro nematous, mononematous, erect, straight or flexuous, 5-10 septate, pale brown, thick walled, 100-150μm long, 3-7μm wide, irregularly branched near the apex. *Conidiogenous cells* monoblastic or polyblastic, integrated and terminal on branches, denticulate, denticles short. *Conidia* holoblastic, formed in simple or branched chains, ellipsoidal, lemoniform, doliiform or oblong, 0-septate, pale brown, smooth truncate and dark at the ends, 4-5μm long, 2-3μm broad.

Colonies effusae, brunneae, pilosae *Mycelium* plerumque immersum ex hyphis ramosis, septatis, hyalinis vel pallide brunneis, laevibus, 3-6μm crasis compositum. *Conidiophora* singular, macronemata, mononemta, erecta, recta vel flexuosa, 5-10 septata, pallide brunnea, crassa tunicata, 100-150μm longa, 3-7μm crassa, ramosis ad apicem, *Cellae conidiogenae* monoblasticae vel polyblasticae, in conidiophoris incorporatae, terminales in ramulisque, denticulatae, denticulis breve. *Conidia* holoblastica, in catenulas simplices vel ramosae, ellipsoidea, lemoniformia, doliformia, vel oblonga, 0-septata, pallide brunnea, laevia vel atra ad apicem, 4-5μm longa, 2-3μm crassa.

Material collected on unidentified branches, S.K. border, Karnataka, V.M.R.L. No. 678 (Type) Nov. 1987. Coll. B.S.N.R., H.C.I.O (Co-Type).

The present taxon shows some similarity with *Alysidiopsis pipsissewae* Sutton, but differs in conidiophore and conidial morphology like smaller conidiophores, shape of conidiogenous cells, conidial shape, size and 0-septate conidia

The first auther thanks V. Krishna Murthy, Principal, Govt. Degree College, Ramannapet, Nalgonda. Dist. A.P. for providing facilities. He is also grateful to Prof. S. M. Reddy and Dr. S. J. Chary for their constant encouragements.

References

Ellis, M.B. 1971. *Dematiaceous Hyphomycetes.* Commonwealth Mycological Institute, Kew, Surrey, England.

Ellis, M.B.1976. *More Dematiaceous Hyphomycetes.* Commonwealth Mycological Institute, Kew, Surrey, England.

Goh, T. K. and **Hyde, K.D**. 1996. *Janetia curviopsis,* a new species, and an emended description of the genus. *Mycologia* 88 6:1014-1021

Hernandez-Gutierrez, A. and **Sutton, B.C.** 1997. *Imimyces* and *Linkosia,* two new genera segregated from *Sporidesmium sensu lato,* and redescription of *Polydesmus. Mycol. Res.* 101 (2): 201-209.

Subramanian, C.V. 1992. A reassessment of *Sporidesmium* (Hyphomycetes) and some related taxa. *Proc. Ind. Nat. Sci. Acad. B, Biol. Sci.* 58: 179-189.

Sutton. B.C. 1973. Hyphomycetes from Manitoba and Saskatchewan, Canada. *Mycol. Pap* 132: 5-8.

16 | Survey of AM Fungi Associated with Pearl Millet

Shinde B.P. and *Nair L.N.

Department of Botany, Fergusson College, Pune 411 004 (MS),

* 38/18. Vinayak Aparts, Prabhat Road, Lane no. 10, Pune 411 004.

ABSTRACT

The vesicular arbuscular mycorrhizal (AM) fungal propagules were isolated from the rhizosphere and non-rhizosphere soil of pearl millet at an interval of 30 days from the time of sowing to harvesting from four different localities of Nasik district of Maharashtra. The results obtained showed maximum (360) number of propagules per 25 g of soil during September at Thengode and minimum (144) during July at Satana. The number of propagules increased gradually in all the four localities as the plant grew. The maximum number of propagules and cent percent root colonization was reported at the time of active growth stage of the plants. The number of propagules was less in non-rhizosphere soil as compared to the rhizosphere soil. Twelve species of four genera were reported from rhizosphere soil of pearl millet. The genus *Glomus* was dominant with six species. Three species of *Sclerocysti*s, two of *Acaulospora* and one of *Gigaspora* were reported from these four localities.

KEY WORDS : *VAM fungi, Pearl millet, Glomus, Acaulospora, Sclerocystis, Gigaspora.*

Pearl millet (*Penninsetum typhoides,* Linn.) belongs to the family Graminae. It is cultivated in Maharashtra as one of the most important cereal crops. Arbuscular mycorrhizal (AM) fungi are reported to inhabit about 3,00,000 hosts in the world flora (Kendrick and Berch, 1985). The AM fungi form a network of hyphae that covers the surrounding soil, beyond the root zones, which increases the surface area of absorption resulting in increased yield.

The quantitative examination of soil samples from rhizosphere of pearl millet and other hosts for the distribution of major genera of AM fungi was carried out by Singh and Pandya (1995). The present investigation was undertaken for the survey of AM fungi associated with pearl millet.

Materials and Methods

The root and soil samples were collected at an interval of 30 days from the rhizosphere of pearl millet from four different localities of Nasik district (Maharashtra) with non-rhizosphere samples as control. The soil samples were analysed for the presence of AM fungal propagules by Wet Sieving and Decanting method (Gerdemann and Nicolson, 1963). The assessment of mycorrhizal infection was carried out by using Phillips

Table 1 VAM propagules and percentage root infection.

Month	Satana			Mulane			Munjwad			Thengode		
	Propagules /25 g soil		% root infection	Propagules / 25 g soil		% root infection	Propagules /25 g soil		% root infection	Propagules /25 g soil		% root infection
	R.	NR		R.	NR		R.	NR		R.	NR	
July	144	45	40	152	46	40	204	45	40	272	47	60
August	180	46	70	288	48	80	252	45	70	348	49	90
September	184	46	90	352	48	100	312	45	90	360	50	100

R- Rhizosphere, NR- Non-rhizosphere

and Hayman's (1970) method. The percentage of root infection was evaluated by the method of Giovanetti and Mosse (1980). Identification of spores was carried out using the manual of Schenck and Perez (1990) and standard set of slide collections (Hall and Abbott, 1981).

Results and Discussion

Soil type in all the four localities was loamy clay. The AM propagules were isolated from the rhizosphere soil of pearl millet after every 30 days upto harvesting from Satana, Mulane, Munjwad and Thengode for three months for three consecutive years. The results obtained (Table 1) showed maximum (360) number of propagules per 25g soil in September from Thengode and minimum (144) from Satana during July. The number of propagules increased gradually at all the four localities as the growth of plant increased. It was found that during its active stage of growth there was maximum number of propagules (360) in soil.

Similar kinds of results were obtained (Table -1) for percentage root colonization. It was minimum (40%- 60%) in July at all the four localities. It was cent percent during September at Mulane and Thengode. The non –rhizosphere soil had only 45-50 spores per 25g soil. Similar types of results are obtained in other cereals (Shinde and Nair, 1995, 2000a,b.).

Twelve species of four genera of AM fungi were reported from the rhizosphere soil of pearl millet from all the four localities (Table -2). The genus *Glomus* was dominant with six species. It was followed by three species of *Sclerocystis,* two of *Acaulospora* and one of *Gigaspora.* As compared to Satana, Mulane and Munjwad the soil samples from Thengode were rich in VAM propagules. From above results it can be concluded that along with the active growth stage of the plant environmental factors also affect the number of propagules and percentage root infection.

Table 2 Distribution of VAM species.

Sr. No.	Name of species	Satana	Mulane	Munjwad	Thengode
1	*Glomus aggregatum*	+	+	-	+
2	*Glomus constrictum*	-	+	+	+
3	*Glomus fasciculatum*	+	+	+	+
4	*Glomus fistulosum*	+	-	+	+
5	*Glomus manihot*	-	+	-	+
6	*Glomus mosseae*	+	+	+	+
7	*Sclerocystis dussii*	-	-	+	+
8	*Sclerocystis coremiodis*	-	-	-	+
9	*Sclerocystis Taiwanensis*	+	-	+	+
10	*Acaulospora laevis*	-	+	+	+
11	*Acaulospora thomii*	+	+	+	+
12	*Gigaspora margarita*	-	-	-	+

+ Present, - Absent

References

Gerdemann, J.W. and **Nicolson T.H.** 1963. Spores of mycorrhizal Endogone extracted from soil by wet sieving and decanting. *Trans. Br. myco. Soc.* 46: 235-244.

Givannetti, M., and **Mosse, B.** 1980. An evaluation of techniques for measuring vesicular- arbuscular mycorrhizal infection in roots. *New Phyto.* 71: 287-295.

Hall, I. R. and **L.K. Abbott** 1981. *Photographic slide collection illustrating features of the Endogonaceae*, University of Western Australia.

Kendrick, B. and **S. Berch** 1985. Mycorrhizae, Applications in Agriculture and Forestry. In: *Comprehensive Biotechnology*. Vol. 4 (Edt. C. W. Robinson, Pergamon), Oxford pp. 109-150.

Phillips, J.M. and **D.S. Hayman** 1970. Improved procedures for clearing and staining parasitic and vesicular- arbuscular mycorrhizal fungi for assessment of infection. *Trans. Br. myco. Soc.* 55: 158-161.

Schenck, N. C. and **Y. Perez** 1990. *Manual for the identification of VA – Mycorrhizal fungi*, University of Florida, Gainesville, FL, pp. 1-286.

Shinde, B.P. 1995. *Studies on VAM fungi and Rhizobium of Phaseolus spp. from Baglan area (Nasik Dist.).* Ph.D. thesis, University of Pune.

Shinde B. P. and **Nair L. N.** 2000a: Effect of continuous cropping on VAM association of maize. *In: Ecology of Fungi* (Eds. D.J. Bhat and S.Raghu kumar) Published by Goa University, Goa, India, 140-142.

Shinde, B. P. and **Nair, L. N.** 2000b. Vesicular Arbuscular Mycorrhizal fungi from Baglan area (Nasik district) of Maharashtra. In: *Vistas in Mycology and Plant Pathology* (Eds. L. V. Gangavane, et al. Commonwealth Publishers, New Delhi, 227-239.

Singh R. and **Pandya, R. K.** 1995. The occurrence of vesicular arbuscular mycorrhiza in pearl millet and other hosts. In: *Mycorrhiza: Biofertilizers for the future* (Eds. A. Adholeya and S. Singh) TERI, New Delhi, pp 56-58.

 # Marine Fungi from Sundarbans-(India)–III

N. S. Pawar[1] and B. D. Borse[2]

1. S.S. V. P. S's Arts, Comm. and Science College, Shindkheda-425 406. (Maharashtra), India

2. S. S. V. P. S's Dr. P. R. Ghogarey Science College, Dhule - 424 001. (Maharashtra), India

ABSTRACT

Ten species of higher marine fungi collected on submerged parts of mangroves and salt marsh plant from the coast of Sundarbans (the Bay of Bengal, Indian Ocean) are reported. They are *Bathyascus tropicalis, Ceriosporopsis caduca Eutypella naqsii, Swampomyces triseptatus, Savoryella paucispora, Thalassogena sphaerica, Tirispora mandoviana* from Ascomycota, *Nia vibrissa* from Basidiomycota and *Botryophialophora marina, Phoma suaedae* from mitosporic fungi. *Botryophialophora* and *Thalassogena* form new generic records to the fungi of India.

KEY WORDS : Ascomycetes, mangroves, marine fungi, mitosporic fungi, Sundarbans.

The information on marine fungi of the Sundarbans coast is scanty. However, East coast was explored for the marine fungi (Raghukumar, 1973; Ravikumar and Vittal, 1987, 1996; Sarma and Vittal 2000, 2001; Borse, 2002; Borse and Borse 2001; Borse and Pawar 2001; Borse, B. D *et al.* 2001, 2002, Borse K. N. *et al.* 2001, 2002.). *Bathyascus tropicalis Kohlmeyer, Ceriosporopsis caduca Jones* and Zainal, *Eutypella naqsii* Hyde, *Savoryella paucispora* (Cribb and Cribb) Koch, Swampomyces triseptatus Hyde and Nakagiri, *Thalassogena sphaerica* Kohlmeyer and Volkmann-Kohlmeyer, *Tirispora mandoviana* Sarma and Hyde and *Botryophialophora marina* Linder, and *Phoma suaedae* Japp are described and illustrated.

Materials and Methods

In the course of survey of marine fungi along the Sundarbans (W. B.), samples of intertidal wood, dead and decaying substrates of mangroves were collected from the mangroves like Avicennia marina Vierh. Excoecaria agallocha L. Rhizophora mucronata Lamk. Sonneratia apetala Buch.- Ham. and salt marsh plant Suaeda maritima Dumort along the coast of Sundarbans. The intertidal mangrove wood, driftwood, pneumatophores and roots along the mangrove creeks, as well as totally submerged wood or water logged were collected at low tide by cutting or sawing off pieces of the substrates.

Collected samples were placed in polythene bags and sealed well to prevent loss of moisture, on returning to the laboratory the specimens were examined for the sporulating structures (ascomata, pycnidia, conidiophores, etc.).

Incubated material was periodically examined for the presence of fungi. The identification of the fungi were made with the help of Hyde and Sarma (2000), Hyde et al. (2000), Kohlmeyer and Kohlmeyer (1979) and Kohlmeyer and Volkmann-Kohlmeyer (1991). The reports of marine fungi were confirmed with the help of Bilgrami et al. (1991) and Sarbhoy et al. (1996)

Results and Discussions

Ascomycota

Bathyascus tropicalis Kohlmeyer : Fig.1.

Ascomata 140-220 m high, 120-160 m diameter, solitary, ellipsoidal, superficial, ostiolate, papillate, coriaceous, thin- walled, light brown. Necks 190-200 m, cylindrical, eccentric, ostiolar canal without periphyses. Hamethecium absent. Asci 90-100 × 6-9 m, eight-spored, ellipsoidal, unitunicate, thin-walled without apical apparatuses, deliquescing before ascospores maturity. Ascospores 80-110 × 6.5-10 μm, elongate, fusiform, rounded at the upper end, tapering toward the base, straight, slightly curved, one-septate below the center, not or slightly constricted at the septum, without appendages, hyaline.

Habitat : on intertidal mangrove wood, Frasarganj, 8 Nov. 2000, MFFS No.1.

Other specimens examined : on intertidal wood of Avicennia marina, Pakhirala, 27 May 2000, MFFS NO.1A, on intertidal wood of Rhizophora mucronata Pakhirala, 20 May 2001. MFFS 1B.

Distribution along the Indian coast : West Coast : Maharashtra (Borse, 1985, 1988; Ramesh and Borse, 1989). Lakshadwip Islands (Raghukumar, 1973).

Notes : The description of the present collection completely agrees with that of the type description of B. tropicalis given by Kohlmeyer (1980). Hence, it is assigned to that species. It is for the first time reported from the East coast of India.

Ceriosporopsis caduca Jones and Zainal: Fig 2.

Ascomata 200-300 μm high, 250-325 μm diameter, globose to subglobose, superficial, ostiolate, papillate, coriaceous, thin-walled, brown to reddish brown. Necks 200-430 μm long, 40-46 μm diameter, cylindrical, ostiolar canal without periphyses, pale brown. Paraphyses absent. Asci not observed. Ascospores 23.5-28.5 × 11.5-15.5 μm, ellipsoidal, one-septate slightly or not constricted at the septum, hyaline, slender, appendaged. Polar appendages are long up to 180 μm, subcylindrical and tapering.

Habitat : on intertidal mangrove wood, Frsarganj, 8 Nov. 2000. MFFS 2.

Distribution along the Indian coast : West Coast : Maharashtra (Borse, 2002)

It is for the first time collected from the East coast of India.

Eutypella naqsii Hyde : Fig 3.

Stroma thick, forming a blackened crust on the host surface, with ectostromatal cushions occurring around each ascoma composing host cells and intracellular blackened fungal hyphae. Ascomata forming beneath, raised, blackened, crust-like stroma, on the surface, with 3-5 ascomata per stroma; in vertical section 650-840 µm in diameter, 520-650 µm high, globose or subglobose, with periphysate necks, collectively erumpent through pustulate disc. Peridium up to 50 µm wide, Hamathecium 8 µm wide at the base, hypha like, sparse, septate, hyaline and tapering distally. Asci not observed. Ascospores 6.4-8 × 1.2-1.6 µm, allantoid, straight, or mostly curved, hyaline to pale yellow, unicellular.

Habitat : on stem of Avicennia marina, Jambu Dwip, 23 May 2001, MFFS NO. 3.

Hyde (1995) described the Eutypella naqsii from Australia, North Queensland, (Pacific Ocean) on intertidal branch of Avicennia sp. The morphology of the present collection agrees well with the type description given by Hyde (1995). However, material was apparently old and no asci were observed. The present collection was collected on intertidal wood of Avicennia marina and seems to be host specific. It is found to be rare in occurrence. The fungus makes a new record for the fungi of India.

Savoryella paucispora (Cribb and Cribb) Koch: Fig 4.

Ascomata 90-125 µm diameter, solitary or gregarious, flask shaped, immersed or partly immersed, ostiolate, Papillate, 52-140 µm long, 40-60 µm in diameter, simple or racemose, periphysate paraphyses not observed. Asci 60-100 × 12-26 µm, two -spored cylindrical, clavate, with a short foot, thin-walled at maturity, persistent. Ascospores 45-54 × 14-16 µm, fusicoid, ellipsoidal, three-septate, slightly constricted at the septa, central cells brown, apical cells hyaline, inconspicuous verrucose.

Habitat : on intertidal mangrove wood, Canning, 25 May 2000, MFFS NO. 6

Distribution along the Indian Coast : Andaman and Nicobar Islands (Chinnaraj, 1993). West coast: Maharashtra (Borse, 2000, 2002); Goa (Prasannarai and Sridhar, 2001); Karnataka (Prasannarai and Sridhar 2001). Lakshadweep Islands (Chinnaraj, 1992)

The description of the present collection is completely agrees with the type description of S. paucispora given by Koch (1986). It has been for the first time reported from the East coast of India.

Swampomyses triseptatus Hyde and Nakagiri: Fig 5.

Ascomata 230-340 µm, high, 160-220 µm in diameter, subglobose or pyriform, immersed with long axis horizontal, oblique or vertical to the host surface, brown-black, coriaceous, ostiolate, contents apricot coloured in mass, accuring mostly signally, developing under a thin darkened superficial pseudostroma, up to 20 µm thick, and composed of host cells with darkened fungal hyphae. Necks 70 µm in diameter, dark with periphyses. Peridium up to 25 µm thick, composed of six - eight layers of brown flattened angular cells, these cells less flattened near the ostiole. Hamathecium 1.2 - 2.8 µm wide, branched appearing to anastomous, numerous, filamentous, hyaline in a gel and fusing with the perisphyses in the neck. Asci 115-56 × 8-10 µm, eight-spored, cylindrical, thin-walled, short pedunculate, apically thickened but with a pore and non-amyloid. Ascospores 16-24 × 8-12 µm, overlapping, unseriate, hyaline yellowish, ellipsoidal, three-spetate, weakly constricted at the septa, without sheath or appendages.

Habitat : on intertidal mangrove wood, Ganga Sagar, 9 Nov. 2000. MFFS NO. 7.

The description and measurements of Ascomata, Asci and Ascospores are completely agrees with the description of S. triseptatus given by Hyde and Nakagiri (1992). The fungus is rare in occurrence. It makes an addition to the fungi of India.

Thalassogena Kohlmeyer and Volkmann- Kohlmeyer:

The genus Thalassogena was erected by Kohlmeyer and Volkmann-Kohlmeyer (1987) with T. sphaerica as its type species. The genus belongs to the Halosphaeriaceae and is monotypic. The genus is characterized by having Ascomata sub globose, periphysate, papillate, cream colored, single. Peridium thin, forming a *textura angularis*. Catenophyses developing from the thin-walled pseudoparenchyma. Asci eight-spored, clavate, pedunculate with an apical pore, non-amyloid, thin-walled, unitunicate, persistent, maturing successively on a ascogenous tissue at the bottom of the locule, Ascospores subglobose, one-celled, hyaline, without sheaths or appendages.

Type Species : *Thalossogena sphaerica* Kohlm. and Volkmn.- Kohlm.

Thalossogena sphaerica Kohlm. and Volkm.- Kohlm: Fig 6.

Ascomata 150-260 µm in diameter, subglobose, immersed to superficial, ostiolate, periphysate, papillate, coriaceous, cream-colored, single, surrounded by brown hyphae. Necks 230–570 µm long, 50-85 µm wide, ostiolar canal filled with periphyses. Catenophyses 3–11 µm diam., developing from the pseudoparenchyma that originally fills the central cavity. Asci 72-89 × 22-27 µm, eight-spored, clavate, pedunculate, flattened at the apex and with a pore, thin-walled, unitunicate, ± persistent, maturing successively on an ascogenous tissue at the bottom of the locule. Ascospores : 12.5-16.5 µm in diameter, subglobose, rarely ellipsoidal, one-celled, hyaline, without sheath or appendages, with one large lipid globule surrounded by numerous small non-lipoid droplets.

Habitat : on intertidal wood of Excoecaria agallocha, Sajanakhali, 4 Nov. 2000, MFFS No. 4.

This fungus was described by Kohlmeyer and Volkmann Kohlmeyer (1987) from Belize (Atlantic Ocean) on wood submerged at 3m. depth. Hyde (1988) reported on prop roots of *Rhizophora apiculata* and exposed subterranean roots of *Rhizophora* sp. from intertidal region of Kampung, Kapok and Danua mangroves Brunei (Pacific Ocean). In the present work, it was collected on intertidal wood of Excoecaria agallocha and Avicennia marina. The present fungus is being recorded for the first time from India.

Tirispora mandoviana Sarma and Hyde: Fig 7.

Ascomata 260-240 μm high, 250-430 μm in diam., globose to subglobose, superficial or semi-immersed, hyaline to pale brown, solitary, ostiolate, papillate. Necks 220-466 μm long, 60-90 μm wide, periphysate. Catenophyses not seen. Asci 80-115 × 19-31 μm, 8–spored, clavate, pedicellate, unitunicate, persistent, thin-walled, with an indistinct apical pore. Ascospores : 15-22 × 8-12 μm, ellipsoidal, bicelled with large oil globules, hyaline, thick-walled, with a single appendages at one pole, 2-4.5 μm wide and 1.5-3.5 μm high, apical, cap like, appraised, arranged randomly in relation to the apex of the ascospores.

Habitat : on intertidal wood of Avicennia marina Sunasda 22 May, 2001, MFFS NO.5.

Distribution along the Indian Coast : West coast: Goa (Sarma and Hyde, 2000a).

The description of the present collection is completely agrees with that of the Tirispora mandoviana given by Sarma and Hyde (2000a). It is being for the first time collected from the East coast of India.

Basidiomycetes

Nia vibrissa Moore and Meyers: Fig 8.

Hyphae 2-3 μm in diameter, septate, ramose, hyaline, Basidiocarps 1-3 μm in diameter, subglobose, superficial, anchored in the substrate with an inconspicuous cylindrical pedicel, whitish, yellowish, pinkish, and finally orange coloured, soft, thin-walled, villous or smooth. Containing a homogenous gleba; opening by irregular rapture of the peridium; solitary or gregarious. Peridium 10-15 μm thick, composed of hyphae with their thick walls united, merging toward the center into a 20 μm thick layer of separate tangential hyphae and bearing on the outside long hairs, external hyphae up to 275 μm long, 4-6 μm in diameter, thick-walled, straight or curved, sometime slightly curled apically and uncinate. Basidia not observed. Basidiospores 9-14 μm long, 7-10 mm in diameter, ovoid or ellipsoidal, one-celled, hyaline, appendaged; at the apex provided with a single, slender, flexible, attenuate, hyaline appendage, 25-40 μm long, four (3-5) similar subterminal radiating appendages around the base, 25-34 μm long, at the point of attachment to the basidium with a short cylindrical projection.

Habitat : on the intertidal wood of Excoecaria agallocha, Ganga Sagar, 7 Nov. 2000. MFFS NO. 8.

Other Specimen Examined : on intertidal mangrove wood, Pakhirala, 4 Nov. 2000, MFFS, 8A.

Distribution along the Indian coast : West Coast: Maharashtra (Borse, 1985, 1988; Ramesh and Borse, 1989), Goa (Nandan et al., 1993), Gujarat (Patil and Borse, 2001); Karnataka, (Prasannarai and Sridhar, 2001); Kerala (Prasannarai and Sridhar, 2001).

The morphology of the present collection agrees well with the description of Nia vibrissa given by Kohlmeyer and Kohlmeyer (1979). The fungus has been for the first time collected from the East coast of India.

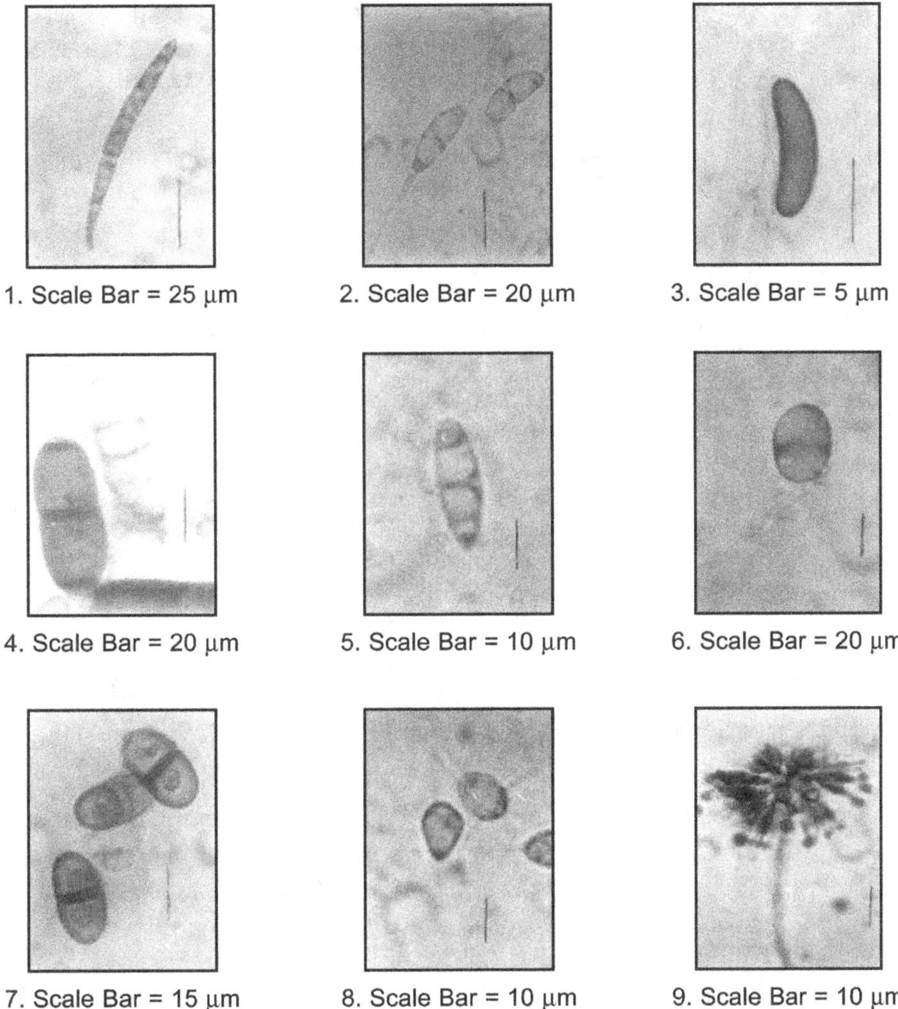

1. Scale Bar = 25 μm   2. Scale Bar = 20 μm   3. Scale Bar = 5 μm

4. Scale Bar = 20 μm   5. Scale Bar = 10 μm   6. Scale Bar = 20 μm

7. Scale Bar = 15 μm   8. Scale Bar = 10 μm   9. Scale Bar = 10 μm

Figsure 1 – 7 *Ascospores* **8.** *Basidiospore* **9.** *Conidiophores and Conidia.*

1. *Bathyascus tropicalis,* 2. *Ceriosporopsis caduca,* 3. *Eutypella naqsii*

4. *Savoryella paucispora* 5. *Swampomyces triseptatus,* 6. *Thalassogena sphaerica*

7. *Tirispora mandoviana* 8. *Nia vibrissa* 9. *Botriophialophora marina*

Mitosporic Fungi

Botryophialophora Linder. The genus *Botryophialophora* was erected by Linder with *B. marina* as its type species. The genus is monotypic. The genus is characterized by having, Hyphae septate, simple or branched, hyaline, or light colored. Conidiophores subglobose or irregular, lateral or apical on the main hyphae, bearing clusters of conidiogenous cells. Conidiogenous cells phialidic, flask-shaped, apex narrow, trumpet-shaped, simple, hyaline, on conidiophores or rarely produced singly on a hypha. Conidia (enteroblastic- phialidic) globose or subglobose, one-celled, hyaline.

Botryophialophora marina Linder: Fig 9.

Hyphae 4-10 µm in diameter, septate, simple or branched, hyaline, or light colored. Conidiophores subglobose or irregularly polygonal, lateral or apical on the main hyphae, bearing clusters of up to 20 or more conidiogenous cells. Conidiogenous cells 3.5-8.5 µm long, 1.5-3.3 µm in diameter, phialidic, flask-shaped, apex narrow, trumpet-shaped, simple, hyaline, on conidiophores or rarely produced singly on a hypha. Conidia (enteroblastic-phialidic) 2-3.3 µm in diameter, globose or subglobose, one–celled, hyaline, containing a large vacuole.

Habitat : on intertidal wood of Sonneratia apetala, Jharkhali, 7 Nov. 2000.

MFFS 9.

The description of the present collection is agrees with the description of Botryophialophora marina given by Kohlmeyer and Kohlmeyer (1979). The genus makes new records for the fungi of India.

Phoma suaedae Jaap. Pycnidia 150 µm high, 150 - 230 µm in diameter, subglobose to ellipsoidal, immersed, becoming erumpent, ostiolate, short papillate or epapillate, coriaceous, light to dark brown, darkest around the ostiole, solitary or gregarious. Peridium composed of an outer layer with thick-walled, dark cells and an inner layer with hyaline cells, bearing conidiogenous cells. Papillae about 15 µm in diameter, ostiolar canal about 10 µm in diameter. Conidiogenous cells 5-8 µm long, phialidic, flask shaped, hyaline originating along the inner wall of the peridium, similar to the inner cells of the peridium. Conidia 6-8 × 3-4.5 µm, ellipsoidal to subglobose, one-celled, often biguttulate, hyaline to yellowish, originating in basipetal succession.

Habitats :   on dead stems of Suaeda maritima, Canning, 25 May 2000. MFFS, NO. 10.

The description of the present collection is completely agrees well with the description of Phoma Suaedae given by Kohlmeyer and Kohlmeyer (1979), therefore, it is referred to that species. The fungus is rare in occurrence. It has been first time reported from the Indian marine waters.

Acknowledgments

We are grateful to the Hon'ble Chairman Shri. P. M. Sisode S. S. V. P. S's Arts, Commerce and Science College, for the laboratory facilities and encouragement. Our sincere thanks are due to Dr. B. V. Kamble, Principal, S. S. V. P. S's Arts, Commerce and Science College, for encouragement and Laboratory facilities. The junior author (NSP) is also grateful to the authorities of the University Grants Commission (WRO), Pune, for the financial help during the present work.

References

Bilgarmi, K. S., Jamaluddin, S. and **Rizwi, M. A.** 1991. *Fungi of India: List And references*. Today and Tomorrow's Printers and Publishers, New Delhi, pp. 798.

Borse, B. D. 1985. Marine fungi from India.-II. *Curr. Sci.* 54:881- 882.

Borse, B. D. 1988. Frequency of occurrence of marine fungi from Maharashtra coast, India. Indian J.Marine Sci. 17:165-167.

Borse, B. D. 2000. Marine fungi from India. X. Savoryella Jones and Eaton (Ascomycotina). In: *Ecology of Fungi.* (eds. D. J. Bhat and S. Raghukumar) Goa Univ. Press, Goa, pp. 163-165.

Borse, B. D. 2002. Marine fungi from India- XI. A Check list. *J. Ind. Bot. Soc.* 81: 83-92.

Borse, B. D. and **Borse, K. N.** 2001. New reports of Marine Ascomycetes from Orissa, India. *Geobios.* 28 : 62 – 64.

Borse, B. D. and **Pawar, N. S.** 2001. Corbosphgerella and Dryosphera two new generic records of marine Ascomycetes from West Bengal. *Geobios.* 28 : 117-120.

Borse, B. D., Pawar, N. S. and **Borse, K. N.** 2001. Marine fungi from Sundarbans (India)- I: Sea foam spores. *J. Ind. Bot. Soc.* 80: 275-278.

Borse, B. D., Pawar, N. S. and **Borse, K. N.** 2002. Marine fungi from Sundarbans (India)- II: The genus Arenariomyces Hohnk. *J. App. Basic. Mycol.* 1: 8-10.

Borse K. N., Pawar, N. S. and **Borse, B. D.** 2001. Marine fungi from Orissa (India). Arenicolous group. *BRI's JAST.* 4: 17-22.

Borse, K. N., Pawar, N. S. and. **Borse, B. D.** 2002. *Marine fungi from Orissa- II. The genus* **Corollospora** Werdermann. *Geobios* 29:258-263.

Chinnaraj, S. 1992. Higher marine fungi of Lakshadweep Islands and a note on Quintaria lignatilis. *Cryptogamae Mycol.* 13: 312 – 319.

Chinnaraj, S. 1993. Higher marine fungi from mangroves of Andaman and Nicobar Islands. *Sydowia* 45 : 109 – 115.

Hyde, K. D. 1988. Observations on the vertical distribution of Marine Fungi on Rhizophora sp.at Kampong Danau managrove. *Asian Marine Biology.* 5:77–81.

Hyde, K. D. 1995. *Eutypella naqsii* sp. nov. from intertidal Avicennia. *Mycol. Res.* 99: 1462 – 1464.

Hyde, K. D. and ***Nakagiri, A.*** 1992. Intertidal fungi from Australia. The genus *Swampomyces* including *S. triseptatus sp. nov. Sydowia.* 44 : 122 – 130

Hyde, K. D., and ***Sarma, V. V.*** 2000. Pictorial Key to Higher marine fungi. In: *Marine Mycology - A Practical Approach* (eds. K. D. Hyde and S. B. Pointing). Fungal Diversity Research Series. I. Fungal Diversity Press, Hong Kong. pp. 205 – 270.

Hyde, K. D., Sarma, V. V. and ***Jones, E.B.G.*** 2000. Morphology and Taxonomy of Higher marine fungi. In: *Marine Mycology.- A Practical Approach* (eds. K. D. Hyde and S. B. Pointing). Fungal Diversity Research Series. I Fungal Diversity Press, Hong Kong, pp. 172 – 204.

Jones, E. B. G. and ***A., Zainal*** 1988. *Ceriosporopsis caduca* sp. nov. a marine lignicolous Ascomycotina from submerged wood. *Mycotaxon* 32: 237- 244.

Koch, J. 1986. Some lignicolous fungi from Thailand including two new species. *Nord. J. Bot.* 6 : 497 – 499.

Kohlmeyer, J. 1980. Tropical and Sub-tropical filamentous fungi of the Western Atlantic ocean. *Bot. Mar.* 23: 529-544.

Kohlmeyer, J. and ***Kohlmeyer, E.*** 1979. *Marine mycology-The Higher Fungi.* p - 698. Academic Press, New York, pp 698.

Kohlmeyer, J and ***Volkmann – Kohlmeyer, B.*** 1987. *Thalassogena* new ascomycetes genus in the Halosphoerials. *Systema Ascomycetum.* 6 : 223 – 228.

Kohlmeyer, J. and ***Volkmann – Kohlmeyer, B.*** 1991. Illustrated Key to the filamentous marine fungi. *Bot. Mar.* 34: 1 – 61.

Nandan, S. N., Shinde, D. N. and ***Borse, B. D.*** 1993. Marine fungi from Goa coast (India). *Biol. Ind.* 4 : 29 – 34.

Patil, K. B. and ***Borse, B. D.*** 2001. Studies in higher marine fungi from Gujarat coast (India). *Geobios.* 28 : 41 – 44.

Prasannarai, K. and ***Sridhar, K. R.*** 2001. Diversity and Abundance of Higher marine fungi on woody substrates along the West coast of India. *Curr. Sci.* 81: 304-311.

Raghukumar, S. 1973. Marine lignicolous fungi from India. *Kavaka.* 1: 73 – 85.

Ramesh, Ch. and **Borse, B. D.** 1989. Marine fungi from Maharashtra coast (India). *Acta Botanica Indica.* 17 : 143 – 146.

Ravikumar, D. R. and **Vittal, B.P.R.** 1987. Studies on mangrove fungi of India -I. note on some Ascomycetes. *Kavaka.* 15: 99 – 103.

Ravikumar, D. R. and **Vittal B.P.R.** 1996. Fungal diversity on decomposing of mangrove plant Rhizophora in Pichavrum estuary, East Coast of India *Ind. J, of marine Sci.* 25 : 142 – 144.

Sarbhoy, A. K., Vershey, **J. L.** and **D. K. Agrawal** 1996. *Fungi of India.* 1982-1992. S. K. Jain Publishers and Distributors, New Delhi.

Sarma, V. V. and **Hyde, K. D.** 2000 a . Tirispora mandoviana sp. nov. from Chorao mangroves, Goa, the West coast of India. *Australasian Mycologist* 19: 52-56.

Sarma, V. V. and **Hyde, K. D.** 2000 b. Synoptic Key to higher marine fungi. In: *Marine Mycology – A Practical Approach* (eds. K. D. Hyde and S. B. Pointing). Fungal Diversity Research Series. I, Fungal Diversity Press, Hong Kong. Pp. 271 – 307.

Sarma, V. V. and **Vittal, B.P.R.** 2000. Biodiversity of mangrove fungi on different substrata of Rhizophora apiculata and Avicennia sp. from Godavari and Krishna deltas, east Coast of India. In : *Aquatic Mycology across the millenium* (Eds.. K. D. Hyde, W. H. Ho K. D. and S. B. Pointing). Fungal Diversity. 5: 23 – 41

Sarma V. V., and Vittal , B. P. R. 2001. Biodiversity of mangrove fungi on selected plants in the Godavari and Krishna Deltas, East Coast of India. *Fungal diversity.* 6 : 115 – 130.

18 Growth Pattern and Antimicrobial Spectrum of an Actinomycete Strain Isolated from Laterite Soil

K.J.P. Narayana, V.V.K.V. Rao and M. Vijayalakshmi*

Dept. of Botany and Microbiology, Acharya Nagarjuna University,
Nagarjuna Nagar - 522510, A.P.

* Corresponding author: vijaya_muvva@yahoo.com

ABSTRACT

While enumerating the actinomycete population from different soils, a strain of actinomycete from laterite soil was found to be prevalent. Growth pattern and antimicrobial spectrum of the isolate were studied in yeast extract malt extract dextrose and starch casein media. The impact of different carbon and nitrogen sources on growth pattern and antimicrobial spectrum of the isolate was also investigated. The strain exhibited antimicrobial activity against bacteria like *Bacillus cereus*, *Escherichia coli*, *Pseudomonas aeruginosa*, *Staphylococcus aureus* and *Candida albicans*.

KEY WORDS : Actinomycete isolate, growth pattern, antimicrobial spectrum.

Actinomycetes are ubiquitous gram positive bacteria that occur in nature and man made environment. They comprise a major group of antibiotic producing microorganisms distributed in nature. The number and importance of actinomycetes in the soil microflora depend on factors such as soil type, depth, water content and agricultural treatment (Lacey, 1971). The present paper deals with the growth pattern and antimicrobial spectrum of an actinomycete strain isolated from larterite soil

Materials and Methods

A laterite soil sample collected from nearby area of Nagarjuna Nagar was air dried, powdered and sieved. The factors such as moisture, pH, organic matter and total nitrogen of the soil sample were determined (Moore and Chapman, 1986). Actinomycetes from soil samples were isolated following soil dilution plate method (Waksman, 1959) using asparagine glycerol salts agar medium (Pridham and Lyons, 1980).

A strain of actinomycete found to be abundant in the plates was isolated, purified and maintained on yeast extract malt extract dextrose agar medium. Growth pattern and antimicrobial spectrum of the isolate was studied on media such as yeast extract malt extract dextrose broth (YMD) containing yeast extract 0.4 %, malt extract 1 %, dextrose

1 % and calcium carbonate 0.2 % with pH 7.5 and starch casein broth (SCS) containing starch 0.1 %, sodium caseinate 0.2 %, di potassium hydrogen ortho phosphate 0.02 %, magnesium sulphate 0.02 % and ferrous sulphate 0.005% with pH 7.5 (Wu and Chen, 1995). The cultures were incubated at 30°C for one week. The biomass of the strain and its antimicrobial spectrum were determined at 24 h intervals for one week.

The impact of different carbon and nitrogen sources on the growth pattern and antimicrobial spectrum of the isolate was studied in the basal medium consisting of, dipotassium hydrogen ortho phosphate 0.5%, magnesium sulphate 0.1%, calcium carbonate 0.02 % and ferrous sulphate 0.005% with pH 7.0. Different carbon sources (1%) were added separately to the basal medium with sodium nitrate (0.2%) as nitrogen source. The effect of different nitrogen sources was studied by adding a nitrogen source (1%) to the basal medium containing glucose (1%) as the carbon source. The growth pattern and antimicrobial spectrum of the isolate were determined at 24 h interval for one week.

In order to screen the isolate for antimicrobial metabolites, the culture filtrates were extracted with dichloro methane. The solvent extracts were concentrated and preserved for further study. Microorganisms employed for antimicrobial activity include bacteria like *Bacillus cereus* (MTCC 430), *Escherichia coli* (MTCC 40), *Pseudomonas aeruginosa* (MTCC 424), *Staphylococcus aureus* (MTCC 96) and *Candida albicans* (MTCC 183). Filter paper discs soaked in dichloro methane extract were placed on nutrient agar medium seeded with the test organism. The seeded plates were incubated at 30°C and the area of inhibition zone was recorded after 18 h of incubation (Alexander and Strete, 2001).

Results and Discussion

The characteristics of the soil from which actinomycete population enumerated were as follows : pH (5.7), moisture (7%), organic carbon (0.87%) and total nitrogen (0.22%). The actinomycete count from laterite soil was found to be 1.8×10^5 g^{-1} dry soil .

The isolate grew well on asparagine glycerol salts agar medium. The isolate showed similar growth pattern in two culture media viz., YMD and SCS. But the biomass produced in YMD was greater than that of SCS medium. The log phase of the isolate lasted for 72 h. The stationary phase in YMD continued up to 120 h and in SCS medium, it extended up to 144 h.

Data on the antimicrobial spectrum of the isolate are presented in Table 1. The solvent extract of one day old culture grown in two media did not exhibit any antimicrobial activity. The extracts obtained from 48 h old culture were less inhibitory to *Bacillus cereus, Escherichia coli, Pseudomonas aeruginosa, Staphylococcus aureus* and *Candida albicans*. A gradual increase in the antimicrobial activity was found with increasing age of the culture up to five days. Growth of all test organisms was inhibited to a great extent by the extracts obtained from 5 day old culture. *Bacillus cereus* and *Pseudomonas aeruginosa* are highly sensitive to the products of the isolate. In YMD medium, better antibiotic production was noticed than SCS medium (Figure 1).

Table 1 Antimicrobial spectrum of the actinomycete isolate incubated for one week.

Incubation Period (h)	Area of inhibition zone (mm²)									
	Candida albicans		Bacillus cereus		Escherichia coli		Pseudomonas aeruginosa		Staphylococcus aureus	
	YMD	SCS	YMD	SCS	YMD	SCS	YMD	SCS	YMD	SCS
24	–	–	–	–	–	–	–	–	–	–
48	50.2	28.2	50.2	78.5	28.2	50.2	78.5	50.2	38.4	–
72	78.5	50.2	78.5	113.1	78.5	78.5	113.1	113.1	78.5	28.2
96	78.5	78.5	201.1	314.2	78.5	78.5	113.1	153.9	78.5	78.5
120	78.5	78.5	314.2	314.2	78.5	78.5	201.1	153.9	78.5	78.5
144	28.2	28.2	314.2	113.1	50.2	50.2	50.2	78.5	28.2	28.2
168	–	–	28.2	28.2	–	–	–	50.2	–	–

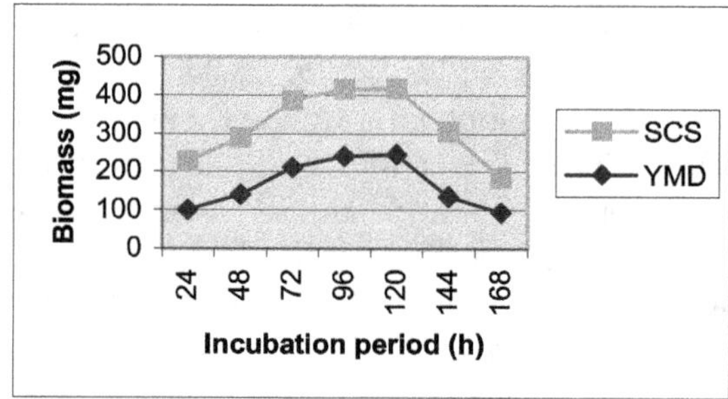

Figure 1. Growth pattern of the strain of actinomycetes on starch casein salts agar (SCS), yeast extract malt extract dextrose agar (YMD) medium.

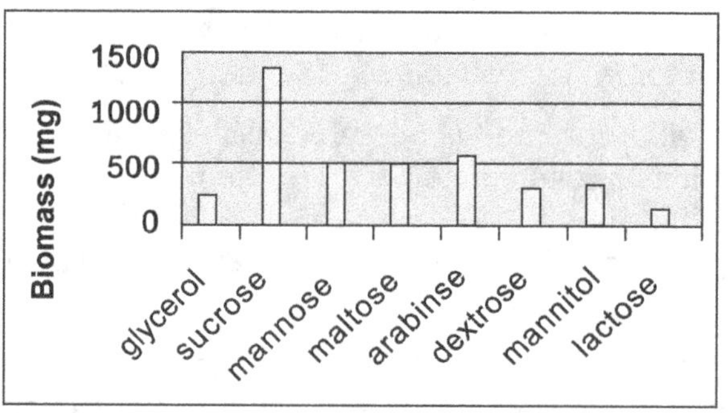

Figure 2. Effect of different carbon sources (1% w/v) on growth pattern of actinomycete isolate.

The strain exhibited good growth in carbon sources such as sucrose followed by mannose, arabinose and maltose, but lactose could not support the growth of the isolate (Fig. 2). The strain showed good growth in inorganic nitrogen sources such as sodium nitrate, ammonium sulfate and ammonium nitrate and in organic nitrogen sources like glutamine. Moderate growth was observed in ammonium oxalate, asparagine and tyrosine. The strain did not show any growth in urea (Fig. 3).

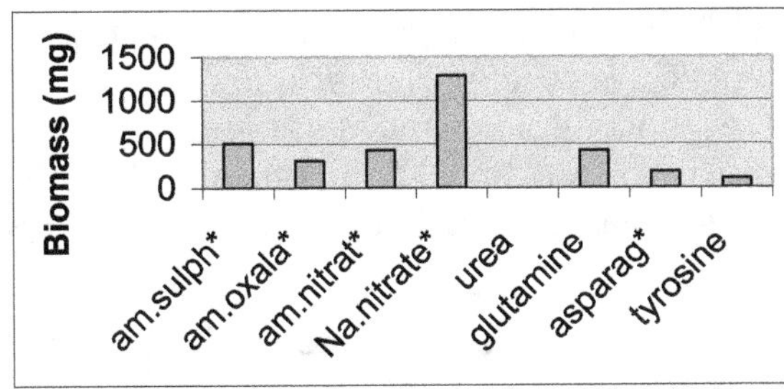

Figure 3. Effect of nitrogen source (0.2 % w/v) on growth pattern of actinomycete isolate.
 • am.sulph-ammonium sulphate, am.oxala-ammonium oxalate, am.nitrat-ammonium nitrate, Na.nitrate sodium nitrate, asparag asparagine

In antibiotic fermentations, maltose, glycerol, and mannitol are better carbon sources than dextrose, arabinose, mannose, and lactose (Table 2). The inorganic nitrogen sources like ammonium sulfate, ammonium nitrate and sodium nitrate and the organic nitrogen sources such as glutamine and asparagine proved to be good for antibiotic production by the isolate (Table 3).

Table 2. Effect of carbon source (1% w/v) on antimicrobial spectrum of actinomycete isolate

Area of inhibition zone (mm^2)					
Carbon source	Candida albicans	Bacillus cereus	Stphaphylococcus aureus	Escherichia coli	Pseudomonas aeruginosa
Glycerol	50.2	452.5	78.5	–	201.1
Sucrose	28.2	113.1	28.2	–	113.1
Mannose	28.2	113.1	–	–	28.2
Maltose	78.5	531.1	78.5	–	153.9
Arabinose	28.2	254.5	28.2	–	50.2
Dextrose	50.2	201.1	–	–	78.5
Mannitol	28.2	314.2	28.2	–	28.2
Lactose	–	113.1	–	–	28.2

Table 3. Effect of nitrogen source (0.2 % w/v) on antimicrobial spectrum of actinomycete isolate

Area of inhibition zone (mm^2)					
Nitrogen source	Candida albicans	Bacillus cereus	Stphaphylococcus aureus	Escherichia coli	Pseudomonas aeruginosa
Ammonium sulphate	50.2	452.5	78.5	–	113.1
Ammonium oxalate	–	153.9	–	–	28.2
Ammonium nitrate	113.1	314.2	28.2	–	–
Sodium nitrate	28.2	201.1	28.2	–	113.1
Urea	–	–	–	–	–
Glutamine	28.2	201.1	–	–	78.5
Asparagine	28.2	153.9	–	–	28.2
Tyrosine	–	113.1	–	–	–

Rangaswamy *et al.*, (1967) reported the prevalence of actinomycetes in laterite soils of South India. Most of the actinomycetes present in different soils had the antagonistic properties and should act as biocontrol agents. Many strains of actinomycetes are known to produce either antibacterial or antifungal antibiotics. In the present investigation the strain of the actinomycetes isolated from laterite soil exhibited antifungal as well as antibacterial activity.

Sahay and Srivastava (1977) reported that the media containing glycerol, glucose, sucrose and maltose as carbon sources supported good growth of several strains of actinomycetes. The isolates showed better growth in the media containing inorganic nitrogen sources as compared to those with organic nitrogen sources. Among several carbon sources tested by them, strains of actinomycetes growing on sucrose showed better antimicrobial spectrum than those in glycerol and glucose. The actinomycete isolate of the present study exhibited good growth with sucrose. But antimicrobial metabolites were high when maltose was used as carbon source. Inorganic nitrogen sources supported the growth as well as the production of antimicrobial compounds by this isolate when compared with organic nitrogen sources.

References

Alexander, S.K. and **Strete, D**. 2001. *Microbiology, photographic atlas for the laboratory*, Addison Wesely Longman. Inc., New York, pp. 103-104.

Lacey, J. 1973. Actinomycetes from soil, composts and fodders, In: *Actinomycetes, characteristics and practical importance* (Eds Sykes G and Skinner FA), Academic press, London, pp. 235-237.

Moore, P.D. and **Chapman, S.B**. 1986. *Methods in Plant Ecology*, Black Well scientific publication, pp 291-315.

Pridham, T.G. and **Lyons, A.J.** 1980. *Actinomycete Taxonomy*, SIM special publication number 6, pp 169- 170.

Rangaswamy, G. Oblisamy, G. and **Swaminathan, R**. 1967. *Antagonistic actinomycetes in the soils of South India.* Univ of Agric Sci. Bangalore and U.S. Dept of Agric.

Sahay, B.N. and **Srivastava, K.P**. 1977. Study of growth, antibiotic and pigment production of six strains of *Streptomyces* on media with varying carbon and nitrogen sources. *Ind J Microbiol.* 17: 141-143.

Waksman, S.A. 1959. *The actinomycetes. Vol.* I *Nature, occurrence and activities,* Baltimore, The Williams & Wilkins Co.

Wu, R.Y. and **Chen, M.H.** 1995. Identification of the *Streptomyces* strain KS 3-5. *Bot–Bull-Acad-Sin.* 36:201-230.

Unit – II

Plant Pathology

19 | Diseases of Tea in Southern India and their Management

B. Chandra Mouli

Rallis Research Centre, 21 & 22, Phase II, P.B. No. 5813, Peenya Industrial Area, Bangalore - 560 058, Karnataka, India.
e-mail : bhogu.mouli@rallis.co.in

ABSTRACT:

Tea diseases by and large are of fungal origin, comprising about 209 genera and 385 species. Mainly root diseases, stem diseases, rust, blight, sooty mold and bird's eye spot affect the tea plants. Blister blight is the most important disease in tea cultivation. With proper precautions and fungicides the diseases could be controlled.

KEY WORDS : Tea, root diseases, stem diseases, rust, blight, sooty mold

The earliest planting of tea in Southern India dates back to 1859 in the Nilgiris and subsequently tea estates were opened up in other areas of Tamil Nadu, Kerala and Karnataka (Muthiah, 1993). Presently tea is being cultivated in an area of about 88,068 hectares in Tamil Nadu, Kerala and Karnataka (Table 1).

Table 1 Tea : Area and Production (2002) in S. India

State	*Area (Ha)*	*Production × 000 kg*
Tamil Nadu	76007	128963
Kerala	36944	59679
Karnataka	2129	5768
Total	115080	194410

Source : Planters' Chronicle Feb 2004

In Southern India, tea is grown between latitudes from 8^0S and 13^0N, at altitudes ranging from 300 to 2300 meters above MSL (Mean Sea Level). Tea growing regions receive a mean annual rainfall of 900 to 8000 mm. The drought period generally ranges from 120 to 150 days during December to April.

Tea belongs to the family Theaceae and all the cultivated plants belong to two distinct species, namely, the small leaved *Camellia sinensis* (L.) O. Kuntze and large leaved *Camellia assamica* (Masters) Wight. The Cambod type, a sub species of the latter is termed *Camellia assamica* spp. *lasiocalyx* (Planchon ex Wight) Wight. Commercially cultivated tea belong to the above taxa or hybrids between them and closely allied species.

Tender shoots comprising two/three leaves and a bud form the harvest. The crop is plucked at five to 14-day intervals depending on the location, growth and the season. The life span of the tea bush is long, extending beyond 100 years. During this long period it is prone to several diseases. All parts of the tea bush are the targets of attack by microbes.

Tea Diseases : Nature and Occurrence

Tea diseases, by and large, are of fungal origin. A total of 209 genera comprising 385 species of fungi, representing all major groups, were reported to occur on tea (Agnihothrudu, 1964); subsequently a few more were added (Rattan and Pawsey, 1981; Chandra Mouli and Ravi Kumar, 1988). These fungi consisted of both pathogens and saprophytes, and the disease causing fungi comprised 74 genera (Chandra Mouli, 1994). Apart from fungi, the existence of 17 genera of nematodes, five bacteria and one each of a virus and parasitic green alga were also reported.

Tea Diseases in Southern India

Several diseases affect tea plants, right from the time they are put out in the nurseries, and during the long life span of the tea bushes in the field. Some of these diseases could inflict serious damage causing capital losses of the bushes leading to crop losses. The major diseases afflicting tea bushes in S. India and their management are discussed hereunder.

Nursery Disaeases

Nursery constitutes an important stage in the life of the tea plant, as well put out nursery plants when transplanted in the field survive better and contribute to good growth and productivity.

Bitten Off

Tea plants prefer a soil, which has pH in the range of 4.5-5.5. When the soil pH exceeds this value, tea plants do not grow well; they become weak and the leaves are crowded as a cluster at the apex. Pre-mature defoliation occurs and the plants die due to extensive decaying of roots.

If the soils are found to have a high pH, treatment either with one or two per cent aluminium sulphate would rectify the situation.

Stalk Rot

The vegetatively propagated stem cuttings in the nursery are often infected by fungal pathogens, namely, *Pestalotia theae* Sawada and *Colletotrichum camelliae* Massee, which results in the dropping of the mother leaf and subsequent rotting of the cuttings (Venkata Ram, 1960a). Application of either 0.35 per cent of copper oxychloride, 0.3 percent or 0.3 per cent mancozeb at 7-10 day intervals effectively controls the disease.

Root Rot

As the roots emerge from the cutting and start growing, they are occasionally invaded by species of *Cylindrocladium, Fusarium* and *Pythium*. The affected roots degenerate rapidly.

To check these pathogens, treatment either with captan or mancozeb at 0.3 per cent is necessary. It is advantageous to use the above two fungicides, alternately, for better control.

Blister Blight

In nurseries, where large number of seedlings, biclonal seedlings and susceptible clonal materials are raised, blister blight disease caused by *Exobasidium vexans* Massee assumes importance. Under favourable wet weather conditions, during the monsoon, the pathogen infects the emerging and tender foliage. When infection is severe the affected leaves drop.

Regular spraying of copper oxychloride at 0.2 per cent during the disease season will be helpful in preventing the development of the disease. The spray interval and the number of sprays largely depend on the severity of the problem.

Diseases in the Main Field

The diseases of tea that occur in the main field will be discussed under three categories, namely, the root diseases, stem diseases and leaf diseases.

Root Diseases

The root diseases in tea, in almost all cases, are legacy from the jungle, as tea is planted after clearing the jungles. Some of the jungle trees are known to harbour fungal parasites on their roots. At the time of felling if the affected roots of these jungle trees are not properly removed and if allowed to remain, they directly transmit disease to the tea plants grown therein.

Root diseases of tea are insidious and losses caused by them are staggering. These diseases are slow spreading and could be identified only at the terminal stages of infection. Although exact records of crop losses are not available, one of the estates in S. India admitted that they have uprooted one hectare of tea affected by root diseases

(Venkata Ram, 1964a). Chandra Mouli (1988) reviewed the situation of root diseases in S. India. From 1991 to 1996 surveys were carried out by the UPASI Tea Research Institute on the existence of primary root diseases in the tea growing areas in S. India. The details of the survey are presented in Table 2.

Table 2 Occurrence of Primary Root Diseases of Tea in S. India

State	Total area surveyed (Ha)	Primary root diseases (Ha)			
		Red	Brown	Black	Xylaria
Tamil Nadu	20,184	2.2	1.6	0.4	0.4
Kerala	26,146	17.7	3.8	6.6	0.4
Karnataka	1,606	21.6	0.1	0.4	–
Total	47,936	41.5	5.5	7.4	0.8

Source : Project Report, National Tea Research Foundation 1996.

Primary Root Diseases

The critical survey revealed the occurrence of four major primary root diseases, namely, the red root disease, the brown root disease, the black root disease and the xylaria root disease. Root splitting disease caused by *Armillaria mellea* Vahl. common in earlier years, was non-existent.

Red Root Disease

This disease is incited by the basidiomycetous fungus *Poria hypolateritia* (Berk) Cooke is extremely common in S. India as reported by the survey. Of the total root disease affected area, this disease constituted 75 per cent of the area.

The fungal mycelium is initially white, then becomes pink to pale red and in later stages it is deep red and almost black in color. The mycelium is generally present as a sheet on the root surface and the characteristic red colour always appears when the root is rubbed vigorously in water and held to bright light. The diseased tea roots are en-crusted with soil and they become soft and spongy in advanced stages, which crumble under gentle pressure. The disease mainly spreads by root contact and to a limited extent by free mycelial spread in the soil. It takes almost five years for the tea bush to succumb to the disease and thus the red root disease is a slow killer. The basidiocarp of *P. hypolateritia* is seldom observed and when it is seen it is pinkish, leathery, soft and flattish, and contains numerous tubules, which produce basidia.

Black Root Disease

The causal pathogen *Rosellinia arcuata* Petch, an ascomycete, occurs generally in soils, which are rich in organic litter, in cool or cold habitat. Though it is major root disease problem at elevations above 1500 MSL, it is also noticed at mid and low elevations. Of the total root disease affected area, about 14 per cent is affected by this disease.

The fungal mycelium is wooly, whitish grey or black in appearance. It spreads rapidly on the surface of the soil on organic debris. When the mycelium comes in contact with tea roots, it penetrates the bark and produces typical star shaped radiating mycelial structures on the surface of wood, below the bark. Very rarely, black and spherical perithecia are seen at the collar region of the plant. The disease also affects certain green manure and shade tree species, which are commonly grown in tea fields (Chandra Mouli and Parthiban, 1991).

Brown Root Disease

This disease, caused by the basidiomycetous fungus, *Fomes noxius* Corner (*Fomes lamaoensis* (Murr) Sacc. & Trott.) is generally common in mid and low elevations. This pathogen mainly spreads by root contact, hence the disease occurs in tea fields in distinct concentric patches (Venkata Ram, 1975). In S. India, of the total root disease affected area, brown root disease is known to occur in about 10 per cent of the area. Fungal mycelium is fawny brown in colour.

The surface of the infected root is heavily encrusted with soil, sand and stone particles, which are difficult to remove, even by abrasion. The diseased roots become spongy and disintegrate in advanced stages of infection. When such roots are split longitudinally typical brown honey-comb like reticulations are seen in the wood.

Xylaria Root Disease

Unlike aforementioned primary root diseases, this disease is fairly of recent occurrence (Venkata Ram and Joseph, 1974a). It has been reported in about two per cent of the total area affected by the root diseases in S. India. The pathogen, *Xylaria* sp., which incites this disease, produces typical black ribbon-like compact strands of fungal hyphae. The growth of the fungus is superficial on the surface of the bark. From the diseased roots, characteristic, finger-like erect, dark and corky fungal stroma emerge, which contain perithecia with asci producing ascospores.

Control of Primary Root Diseases

Before the introduction of soil fumigation, the method practiced in earlier years to contain the spread of the primary root diseases was by isolating the disease affected patch from the surrounding healthy tea, by taking a four feet deep and 18" wide trench all around.

As a phytosanitary method this was quite effective in preventing the spread of the disease to some extent. However, mere trenching did not serve the purpose, as in many instances, the disease jumped the trenches when the latter were not kept clean after digging. These unkept trenches got filled with debris and led to the merger of the diseased and healthy areas (Venkata Ram, 1972), thus defeating the very purpose of the operation.

Soil fumigation has been the most effective method in the management of the primary root diseases. The diseased patch earmarked for soil fumigation should be cleared by uprooting all the dead bushes along with the diseased roots. Soil fumigation could be carried out by injecting at 30 cm depth either with 8.0 ml of metham sodium or 16 ml of carbon disulphide per 30-cm^2 areas. Methyl bromide can also be used at four grams per the same unit, but as the compound is highly volatile and toxic, it has to be released under polythene tarp, erected over the affected area and the polythene tarp can be removed seven days after treatment.

Replanting the fumigated areas may be taken up three months after the treatment. Soil fumigation can be best done either in April/May or October/November when the soil is not dry and the holes can be punched with ease.

In the recent years, certain species of *Trichoderma* were found to be antagonistic to the root disease pathogens. (Baby and Chandra Mouli, 1996). Experiments are on the anvil to evolve suitable integrated management strategies in the management of the primary root diseases.

Secondary Root Diseases

Secondary root diseases are caused by microorganisms, which invade the root systems of tea plants when the latter become debilitated. Debilitation can be brought about by adverse environmental factors. Some of the important secondary root diseases encountered in tea in S. India are charcoal stump rot, diplodia root disease and violet root rot.

Charcoal Stump Rot

This disease, caused by ascomycetous fungus *Ustulina zonata* (Lev.) Sacc., has been observed to affect the roots of tea plants in pockets, injured by lightning. In most instances a circular patch of 50 to 500 bushes is invaded by the fungus, approximately two months following the lightning strike (Chandra Mouli and Venkata Ram, 1976 ; Venkata Ram and Joseph 1974b). The symptoms of this disease are easily recognized by the presence of irregular black parallel lines in the wood and fan shaped radiating mycelial structures on the surface of the wood beneath the bark. The disease derives its name from the typical charcoal-like encrustations at the plant collar, which comprise brittle stroma with perithecia embedded. The tiny pores noticeable on the surface of the stroma are the ostioles of the perithecia.

Diplodia Root Disease

It is caused by deuteromycetous fungus, *Botryodiplodia theobromae* Pat., attacking tea roots that are depleted of starch reserves owing to exhaustion (Venkata Ram, 1960b), generally due to pruning after heavy crop production. It has also been observed that fungal infection takes place when the soil temperatures are relatively high. The disease can be recognized by the black pycnidia, which are encircled by a white mycelium.

Violet Root Rot

When tea plants suffer from water logging, their roots are invaded by *Sphaerostilbe repens* Berk & Br. and the infected roots become violet in colour. The fungus is easily identified by thick ribbon shaped flat mycelial strands which are seen profusely both under the bark and on the surface of the wood. The diseased roots also emit rancid smell due to rotting. This disease is generally seen in tea fields that get inundated during monsoon and in water logged soils.

Management of Secondary Root Diseases

Secondary root diseases do not warrant specific remedial measures. As certain pre-disposing factors cause these problems, alleviation of such factors would remedy the situation. In the case of charcoal stump rot, the dead and moribund tea bushes affected by lightning should be uprooted and disposed off. The area, thus cleared, may be put either under thorn less *Mimosa* (*Mimosa invisa* Mart.) or Guatemala grass (*Tripsacum laxum* Nash) for about 18 months and then replanted with tea. Avoiding pruning immediately after taking the heavy rush crop would prevent invasion by the fungus causing diplodia root disease. Draining excess of water by making trenches is necessary for protection of plants from violet root rot.

Stem Diseases

The fungal pathogens that incite stem diseases in tea are generally traumatic parasites. They usually gain entry either through injuries, wounds are debilitated bark. The following are the common stem diseases in S. India.

Collar Canker

With the introduction of vegetative propagation in tea in the sixties, several tea estates adopted planting of clonal cuttings, and with this the collar canker disease, which was hitherto of minor importance, came to prominence. The causal pathogen, *Phomopsis theae* Petch infects the bark either at the collar or occasionally on the primary and rarely even secondary branches, killing the bark. When the infection causes girdling of entire collar, the plant dies. The healthy tissue at the periphery of the invaded zone forms a thick callus. The disease is severe on susceptible cultivars (Clones) following severe drought. *P.theae* has been found to invade at relatively high temperatures (32-35^{0}C) and at this temperature wound/injury on the plant is not a pre-requisite for infection (Venkata Ram, 1973). The low moisture regimes in the soil brings about severe water stress in plants, which in turn make the plants susceptible to *Phomopsis* infection. Hence, any cultural operation, which brings about moisture stress in plants, would pre-dispose the plants to *Phomopsis* invasion. Shanmuganathan and Bopearatcly (1972) reported a much higher incidence of the disease in clonal tea plants that were brought into bearing by pegging. Preventive measures by avoiding operations viz. planting susceptible clonal material, planting in gravely soils, injuring plant collar by weeding implements and placing chemical fertilizers too close to collar of the plants. Adopting measures such as using proper shade to mitigate drought and planting at correct nursery level would also help in protecting the plants from *Phomopsis* infection.

Wood Rot

In the mid fifties and till early sixties, one of the important factors for decline in the crop was attributed to severe wood rot of the primary frames of the bushes by two species of *Hypoxylon, H. serpens* (Pers.) Kickx and *H. nummularium* Bull. Ex Fr. the pathogen occurs only on mature tea bushes and produces typical fructifications, black rough encrustations as in the case of *H. serpens* and smooth, coin-like as if painted with tar, in case of *H. nummularium*. The causal pathogen does not produce any externally visible mycelium but the infected wood dries up and crumbles in to a powder on gentle pressure. Fungal invasion takes place on bark affected by sun scorch. Chemical control of the disease is beset with difficulties of finding a suitable fungicide that would penetrate several layers of plant tissue to eradicate the fungus *in situ,* hence the only recourse is drastic surgery aimed at the removal of all the infected tissues. This operation, commonly termed as rejuvenation pruning, is curative and ensures rejuvenation of the frame with the concomitant increase in the crop (Venkata Ram, 1970).

Die-Back

In S. India, tea fields recovering from pruning, located at and above 1500 m above MSL are prone to die-back disease. The base of the shoot, emerging from the pruning cut, develops canker and the entire shoot becomes weak and easily detached from the frame. When the base of such affected branch is examined, thread-like exudations (Spore mass) of the fungus *Leptothyrium theae* Petch are readily seen. A thorough spraying of the bush frames, immediately after pruning with copper fungicide (0.2%) would protect plants from infection by *L.theae*.

Thorny Stem Blight

Thorny stem blight disease affects stems and branches of the tea plant and in several instances, has been noticed to kill single branches. In situations wherein the causal pathogen *Tunstalla aculeata* (Petch) Agni. penetrates deep down the main stem or collar, the affected bush dies. The fungus is noticed during its fruiting period as black thorns on dead wood. These thorny projections constitute the apices of the stroma of the fungus. The area surrounding the thorns often raises and cracks. Agnihothrudu (1961) carried out critical taxonomic studies on this fungus and assigned it to a new genus *Tunstallia*; which was earlier known as *Aglaospora*. The life cycle of the pathogen has been worked out by Chandra Mouli and Parthiban (1992).

Pink Disease

This disease has been reported from S. India by Subba Rao (1936). The causal pathogen *Corticium salmonicolor* B & Br., a basidiomycete, has been frequently reported on tea in earlier years and is rather rare now. Pink disease is identified by the occurrence of a rosy pink crust or even as a thin film of white mycelium on the affected stems.

Branch Canker

Macrophoma theicola Petch is known to invade the bark of the stem of tea bushes, generally, following damage caused by hail. After invasion, the bark collapses to varying extent and a callus is formed at the periphery of the infecting zone resulting in typical longitudinal cankers on the stem. The disease is seen in tea areas affected by hail.

Red Rust

This disease is caused by a parasitic green alga, *Cephaleuros parasiticus* Karst. It is recognized by the presence of the algal sporangia on the stems and leaves of the affected plants. The sporangia are bright orange in color, and when they occur in profusion a rust-like deposit is visible to the naked eye, hence the disease derives its name as red rust. The branches affected by red rust become weak and leafless.

This disease is secondary and the pre-disposing factors are drought, water logging, improper soil pH, lack of available soil potash, lack of shade and poor drainage (Satyanarayana and Barua, 1983). These factors should be properly ascertained and alleviated to minimize the severity of the disease. These measures can be effectively supplemented by application of proprietary copper fungicides.

Thread Blight

In southern India thread blight is observed only on unhealthy and moribund tea bushes. The fungus, *Marasmius pulcher* (B & Br.) Petch is the causal organism of the epiphytic thread blight.

Leaf Diseases

Fungal pathogens, capable of infecting tender succulent leaves and shoots which form the harvestable crop are economically very important as failure to control them leads to direct crop loss. The others, which cause secondary leaf diseases, are weak parasites, which gain entry when the host is under a state of debilitation.

Blister Blight

This is the most important leaf disease in tea cultivation. The existence of this disease in North East India is known as early as in 1855, but unknown to S. India until 1946 (Subba Rao, 1946). After introduction in S. India, this disease spread to all tea growing regions within a short span of time and the S. Indian Tea industry lost 18 million kilograms of tea during the first five years of the epidemic (Venkata Ram, 1964b), before the chemical control measures were fully implemented. The fungus *Exobasidium vexans* Massee is a basidiomycete with a short life span of 11-28 days. Reitsma and Van Emden (1949) classified the visible symptoms of the disease right from oil spot lesion to heavily sporulating matured blisters into seven stages. It is an obligate pathogen, which completes its life cycle on tender tea leaves and has no other alternate hosts. The basidiospores of the pathogen when lodge on the leaf surface directly penetrate causing lesions. High relative humidity, of 80% and above, moderate temperatures ($15\text{-}25^0$C) and protracted leaf wetness of 13 hours or more favour disease incidence (Venkata Ram, 1979). Being a fungus with a short life cycle, it is a multiple cycle disease and several generations of the pathogen are completed in a single disease season. The pathogen is unable to infect the mature leaves. The disease spreads through air-borne basidiospores.

Foliar spraying of fungicides is the most effective and economic mode of blister blight management. A combination of copper oxychloride and nickel chloride offers good protective and curative control of the disease (Venkata Ram and Chandra Mouli, 1983). Several organic fungicides were evaluated (Chandra Mouli, 1979 ; Chandra Mouli and Premkumar, 1986) and those with systemic action offered good disease control even on extended rounds. Among the triazole containing systemic fungicides, hexaconazole (Contaf 5 E) in combination with copper oxychloride was found to be the best and economical

(Chandra Mouli and Premkumar, 1995, 1997). Bitertanol (Chandra Mouli, 1991) and propiconazole (Premkumar, 1997) also provided good disease control.

Brown Blight

The fungus *Colletotrichum camelliae* Massee is observed to be associated with necrotic blotches on mature tea leaves, to which the name brown blight is given. However, this fungus does not infect healthy leaf tissue; and is noticed to colonise the necrotic regions on the leaves caused by sun scorch or old blister blight lesions.

Grey Blight

This is similar to brown blight, but the causal fungus is *Pestalotia theae* Sawada. The affected patches on mature tea leaves are either dull grey or ash coloured in appearance. The fungal fructifications (acervuli) occur along irregular concentric rings approximately 0.1 cm apart. The entry of this fungus is associated either with sun scorch injury or scorch caused by chemical sprays. The typical grey patches are seen either at the tip, margin or the center of the leaf.

Sooty Mould

Several sooty mould fungi, the commoner being *Capnodium theae* Boed. and *Meliola camelliae* (Catt.) Sacc. occur on the surface of the tea leaves and grow as a black film; hence the name sooty mould. The fungal film being superficial can be easily rubbed off the surface. The sooty mould fungi establish and grow on the honey dew deposit of scale insects and mealy bugs.

Spraying a suitable insecticide to reduce the infestation of scale insects and mealy bugs would ward off the occurrence of the sooty moulds.

Bird's Eye Spot

The leaf spot which resembles a bird's eye is caused by *Cercosporella theae* Petch ; but there is some sort of confusion with another fungus, namely, *Cercospora theae* Breda de Haan, which also causes a leaf spot. The spots occur on young leaves, three to five millimeters in diameter, grayish to brown with the reddish or dark brown ring encircling it.

Though this disease is of less economic significance, in recent years it has been found to cause concern in tea nurseries located at high elevations. Any copper fungicides at 0.25%, captan (0.2%) or mancozeb (0.2%) would reduce the severity of the disease.

References

Agnihothrudu, V. 1961. Notes on fungi from North-East India. VII : *Tunstallia* gen. nov. causing the thorny stem blight of tea (*Camellia sinensis* (L.) O. Kuntze). *Phytopath. Z.* 40(3): 277-282

Agnihothrudu, V. 1964. A world-list of fungi reported on tea (*Camellia* spp) *J. Madras Univ.* B 34 : 155-217

Baby, U.I. and **Chandra Mouli, B.** 1996. Biological antagonism of *Trichoderma* and *Gliocladium* spp against certain primary root pathogens of tea. *J. Plantation Crops.* (Suppl) 24: 249-255

Chandra Mouli, B. 1979. Efficacy of protectant, eradicant and systemic fungicides in tea blister blight control in South India *Proc. Placrosym II* : 205-211

Chandra Mouli, B. 1988. Root diseases of tea and their control. *J. Coffee Res.* (Suppl.) **18** : 29-37

Chandra Mouli, B. 1991. Bitertanol for blister blight control. *Newsletter.* UPASI Tea Res. Instt. **1** : 5-6

Chandra Mouli, B. 1994. Common names for the disease of tea. Phytopathology Newsletter. American Phytopathological Society. 28(11): 177.

Chandra Mouli, B. and **Venkata Ram, C.S.** 1976. Invasion of *Ustulina zonata* on the roots of tea plants injured by lightning. *Indian Phytopath.* 29 (3) : 339-340

Chandra Mouli, B. and **Premkumar, R.** 1986. Control of tea blister blight (*Exobasidium vexans Massee*) with organic fungicides. *J. Plantation Crops* (Suppl.) 16 : 249-258

Chandra Mouli, B. and **Ravi Kumar, P.** 1988. Fungi occurring on tea and *Grevillea robusta* in the plantations of Southern India. *Indian Phytopath.* 41(3): 503

Chandra Mouli, B. and **Parthiban, M.** 1991. Black root disease of tea. *UPASI* tea *Sci.Dep. Bull.* 44: 14-16.

Chandra Mouli, B. and **Parthiban, M.** 1992. Thorny stem blight disease of tea. *Newsletter UPASI Tea Res. Instt.* **2** : 6-7

Chandra Mouli, B. and **Premkumar, R.** 1995. Hexaconazole, a novel fungicide in tea blister blight management. UPASI Tea Sci. Dep. Bull. 48 : 59-66

Chandra Mouli, B. and **Premkumar, R.** 1997. Comparative evaluation of fungicide spray schedules against blister blight (*Exobasidium vexans Massee*) disease of tea.

Pestology 21(2) : 19-21

Muthiah, S. 1993. *A Planting Century* 1893-1993. Affiliated East-West Press Pvt. Ltd., New Delhi, 495pp

Premkumar, R. 1997. Tea blister blight control – Recommendations on the usage of propiconazole. *Handbook of Tea Culture* Sec. 13, *UPASI Sci. Dep. Tea Res. Instt.* 1pp

Rattan, P. S. and **Pawsey, R. G.** 1981. Death of tea in Malawi caused by *Pseudophaeolus baudonii.* Trop. Pest. Mgmt. 27 : 225-229

Reitsma, J. and **Van Emden, J. H.** 1949. De Bladpokkenziekte van de Thee. *Bergcult.* 18 : 218-231

Satyanarayana, G and **Barua, G. C. S.** 1983. Leaf and stem diseases of tea in North East India with reference to recent advances in control measures. *J. Plantation Crops* (Suppl.) 11 : 27-31

Shanmuganathan, N. and **Bopearatchy, R.N.** 1972. Studies on collar and branch canker disease of young tea (*Phomopsis theae* Petch). 4.. Effects of some cultural treatments on disease incidence and severity. *Tea Quart.* 43 : 36-44

Subba Rao, M.K. 1936. *Pink Diseaese.* Bull. No.10. Tea Scientific Department, UPASI, pp 25

Subba Rao, **M.K.** 1946. *Blister blight of Tea in South India.* United Planters' Association of Southern India, Paper No.4, pp 14

Venkata Ram, C.S. 1960a. Fungi causing stalk-rot and failure of tea cuttings in vegetative propogation nursery beds and their control. *Annu. Admin. Rep. UPASI Sci Dep. Tea Sect.* 1959-1960 : 59-62

Venkata Ram, C.S. 1960b. A reassessment of Diplodia root disease of tea plants. *UPASI Tea Sci. Dep. Bull.* 19 : 18-28

Venkata Ram, C.S. 1964a. Disease control in tea-recent developments. *Proc. Symp. Pests and Diseases in Tea*, UPASI, Coonoor, 8-13 pp

Venkata Ram, C.S. 1964b. Plant protection problems in tea – blister blight control in Southern India. *Plant Prot. Bull.* 16 : 1-12

Venkata Ram, C.S. 1970. Diseases as limiting factor in tea crop production. Wood-rot and zinc deficiency in Southern India. *UPASI Tea Sci. Dep. Bull.* 28 : 6-11

Venkata Ram, C.S. 1972. Control of red and brown root diseases in tea. *Planters' Chron.* 67 : 163-165

Venkata Ram, C.S. 1973. Influence of genetic factors and cultural practices on *Phomopsis* canker disease in tea. *UPASI Tea Sci Dep. Bull.* 30 : 5-17

Venkata Ram, C.S. 1975. Brown root disease of tea. *Planters' Chron.* 70 : 217-218

Venkata Ram, C.S. 1979. Factors influencing germination, penetration and lesion development on tea leaves following deposition by *Exobasidium vexans* spores. *Proc. Placrosym II* : 193-204

Venkata Ram, C.S. and ***Joseph, C.P.D.*** 1974a. Development in the occurrence and control of root diseases in tea. 1. A new root disease incited by a species of Xylaria. ***UPASI Tea Sci. Dep. Bull.*** 31 : 6-10

Venkata Ram, C.S. and ***Joseph, C.P.D.*** 1974b. Development in occurrence and control of root diseases in tea.2. Soil fumigation against *Armillaria mellea* and status of *Ustulina zonata* as a root pathogen. UPASI Tea Sci. Dep. Bull. 31 : 11-18.

Venkata Ram, C. S. and ***Chandra Mouli, B.*** 1983. Interaction of dosage, spray interval and fungicide action in blister blight disease control in tea. *Crop. Prot.* 2 : 27-36.

20

Soil and Tuber-Borne diseases of Potato and their Management in India

A. K. Somani[1], S. M. Paul Khurana[2], Ashok V. Gadewar[2] and R. K. Arora[3]

[1] Central Potato Research Station, Gwalior (Madhya Pradesh) 474 006
[2] Central Potato Research Institute, Shimla (Himachal Pradesh) 171 001
[3] Central Potato Research Station, Jalandhar (Punjab) 144 003.

ABSTRACT

Some of the important soil and tuber-born diseases of potato in India are soft rots, common scab, brown rot, black scurf, charcoal rot, dry rot, powdery scab wart, and Sclerotium rot. Majority of the potato diseases spread through seed potatos. These diseases cannot be eliminated completely but managed through integrated approach *viz.*, growing resistant varieties, use of disease free seed, seed tuber treatment through chemical, biocontrol, domestic quarantine, crop rotation and grren manuring, fertilizer and micro-nutrient amendments, soil application of fungicidies, soil solarization, adjustment of planting and harvesting dates etc.

KEY WORDS : Potato, diseases, soil, tuber-borne, bio-control, resistant varities.

Soil and tuber-borne diseases of potato are distributed widely and occur year after year once the field becomes infested with the pathogen(s). They may cause rots and turn the tubers unfit for consumption or may disfigure, reduce shelf life and diminish market value. The affected tubers are unfit for seed purpose. Present review deals with the work done in India over the past four decades and other references available are - Arora, 2003; Arora *et al.* 2003; Arora and Khurana, 2004; Khurana, 2000.

Potato tuber contains about 80 per cent water and high carbohydrates that provide the pathogens a suitable medium for their growth. Moreover, tuber becomes vulnerable to attack by pathogens when it is present in the vicinity in soil. The tuber-borne pathogens survive in cold stores till next season on tuber surface as well as in the adhering soil. The pathogens also survive in soil in field as saprophytes in the form of resting spores/ structures, in infected plant residues or in rhizosphere of other crops which makes it difficult to eliminate them.

Tuber rots-soft, dry, charcoal, *Sclerotium* and mouldy have been reported to be influenced by relative humidity, storage temperature and physical damage besides dry matter and reducing sugars in tuber (Somani and Chauhan, 2000). Some of the important soil and tuber borne diseases in India are, soft rots, common scab, brown rot, black scurf, charcoal rot, dry rot, powdery scab, wart, and *Sclerotium* rot. These are discussed here.

Common Scab (*Streptomyces scabies*)

Common scab, though earlier restricted to hilly areas, occur throughout India in moderate to severe form. The disease has little or no effect on yield but scabby lesions on tuber surface give an ugly look, reduce the market value, and require deep peeling for purpose of food. Peeling reduces nutritive value as vitamins and minerals present in outer region of the tubers are lost. Infected tubers become unfit for use as seed. However, adverse effect of such tuber on human health has not yet been reported.

Symptoms

The disease appears on potato tubers in the form of small, round brown lesions around the lenticel but no symptoms are found on the above ground parts of plant. However, sometimes, roots and stolons are affected. The infection causes elongation and division of invaded cells resulting in development of corky tissues (Paharia and Pushkarnath, 1963). In India, five types of symptoms have been observed : (i) mere brownish roughening or abrasion of tuber skin, (ii) proliferated lenticels with hard, corky deposition around them resulting in formation of star-shaped lesions, (iii) Irregular concentric series of wrinkled layers or cork arranged around the central brown spot, (iv) Raised, rough and corky pustules, (v) Depressed (3-4 mm deep) pits surrounded by hard corky tissue.

Pathogen and Epidemiology

Besides *Streptomyces scabies*, several other species of *Streptomyces* viz., *S. collinus* (Dey, et al. 1981), *S. griseus*, *S. longisporoflavus*, *S. cinereus* (Dey and Singh, 1983), *S. volanceoruber*, *S. alborgriseolus* and *S. catenulae* (CPRI, 1983) *S. griseoviridis*, *S. californicus* and *S. albus* (Masoodi and Singh, 1985) have been found associated with common scab. On culture medium colonies of *S. scabies* are white to yellowish and appear translucent to opaque. Conidial chains may be straight to flexuous. Diseased tubers and sick soil serve as a source of inoculum (Paharia and Pushkarnath, 1963). *Streptomyces* spp. survives in soil for more than 5 years (Singh *et al.* 2002). Common scab infection increase at high soil temperature and low soil moisture (Vashisth and Chaubey, 1975). Sahai (1989) reported that *Streptomyces* spp. predominate in sandy loam soils under aerobic conditions. High soil moisture that reduce oxygen from the soil affect them adversely (Singh and Singh, 1981).

Brown Rot (*Ralstonia solanacearum*)

Brown rot was first recorded in 1891 from Poona district and other parts of Maharashtra Subsequently it was recorded from hills of Uttar Pradesh, Tamil Nadu, Himachal Pradesh and plains of Maharashtra, Uttar Pradesh, Bihar, Karnataka, and several other parts of the country.

Symptoms

The disease affects both the foliage and tubers. Initially, foliage of the infected plants droop down showing water hunger and subsequently the plants wilt permanently and die. The infected tubers may show no external symptoms but when cut, show circular browning in vascular region. Some times slimy mass (may be mixed with soil) containing bacteria ooze from the eyes of infected tubers.

Pathogen and Epidemiology

Erwin F. Smith first established *Bacillus solanacearum* as the pathogen responsible for the disease. The bacterium later was christened as *Pseudomonas solanacearum* and presently named *Ralstonia solanacearum*. Infected seed tubers are the most effective source of inoculum carrier. Affected plants decay in soil releasing masses of the bacterium that form small pellets glued in soil. The pathogen also survives through infecting more than 115 other plants. Secondary spread takes place by root to root contacts, irrigation water and agriculture implements. Cut tubers used as seed may also spread the disease. High temperature (28-30^{0}C) and high soil moisture favours wilt. Hingorani *et al.* (1956) observed that 50 to 100 per cent water holding capacity of the soil enhances expression of wilt symptoms.

Soft Rot (*Erwinia carotovora* and other bacteria)

Soft rot of potato is an important disease that can be observed at all the stages of potato crop cultivation and production. Potato as seed in field rot in soil. Infected potato rot during transport, storage as well as when the tubers are still attached to the mother plant (Somani and Shekhawat, 1990). Several pathogens, fungi and bacteria, cause soft rot but bacterial soft rot caused by *Erwinia* spp is common and is important. Extensive surveys of seed and table potatoes carried from 1982 to 1987 at Jalandhar (Punjab) revealed prevalence of soft rot in 78.4 per cent of the samples. Incidence ranged 0.2 per cent in freshly harvested crops in farmers fields to 3. 8 per cent in cold stored tubers.

Symptoms

Initially, a small area of tuber around lenticel or stolon end becomes water soaked and under high humidity it becomes soft and pulpy. It also emits foul smell. Under low humidity progress of infected area get restricted to sunken type of lesion. Often the skin of the rotted tuber remains intact, tuber gets swollen due to gas formation and burst. Bursting sound is common during sorting or grading, Infected seed tubers fail to germinate as they rot and that gives gap in crop stand (Somani and Shekhawat, 1990). Bacteria (*Erwinia*) from infected tuber may cause black leg of shoots. The affected stem becomes black at ground level. Black leg has been reported from Shimla, Kumaon and Ootacamund hills in India (Singh, 1968; CPRI 1949-56; 1971-73; 1981). Tissues in the affected part shrivel and rot, consequently the affected plants become yellow, wilt and die. Seed piece decay developing from soft rot affected tubers have also been observed both in hills and plains (Dutt, *et al.* 1972; Thomas *et al.* 1978; Jalali *et al.* 1982; Chaudhary, 1983).

Pathogens and Epidemiology

Earlier only *Erwinia carotovora* (Jones) Holland *var. carotovora* and *var atroseptica* were considered as the cause of soft rot. Later on other pectolytic bacteria viz. *Bacillus polymyxa, B. subtilis, B. mesentericus, B. megatherium, Pseudomonas marginalis, P. viridiflava, Clostridium spp., Micrococcus spp.* and *Flavobacterium spp.* were found associated with tuber decay (Singh and Jain, 1953; CPRI 1949-56; 1971-73; 1978; 1979; 1982; 1983). Evidence gathered during 1978 to 1985 in India clearly showed that different types of bacteria may cause soft rot of potato tubers at different temperatures (Hirgorani and Addy, 1953, CPRI, 1982; 1983; Ghanekar *et al.* 1984-a). Shekhawat and Perombelon (1985) observed that tuber decay in temperate climate is caused by *Erwinia* and to some extent by *Clostridium spp.* while *B. polymyxa* and *P. marginalis* predominated in hot climates.

Excessive moisture in field predisposes the tubers to soft rot (CPRI 1949-56). This may be due to proliferation of lenticels which provide an easy entry to bacteria. This could also be due to the anaerobic conditions or low supply of oxygen (Tripathi, 1979). High temperature during harvesting (Singh and Verma, 1979) or exposure of tubers to fluctuating extreme temperatures (Hingorani and Addy, 1953) predisposes the tubers to soft rot. Storage temperature of 4^0C and relative humidity upto 35 per cent prevents the bacteria from causing soft rot (Hingorani and Addy, 1953). However, about 1.0 per cent tuber rot has been recorded in cold stores (2-4 ^{0}C and 90% RH) (Somani, unpublished)

Potato tubers harvested from a wet heavy soil may develop severe rot while tubers from sandy soil develop negligible soft rot (Dutt, 1979). Poor ventilation in storage and transit may also result in heavy soft rot (Hingorani and Addy, 1953). Tubers which are immature, large, damaged by hail, exposed to sun, infected with late blight or dry rot are prone to development of soft rot (Hingorani and Addy, 1953; CPRI, 1979; Singh and Verma, 1979; Somani, 1993). Injured potato seed tubers, crop grown under higher nitrogen fertilization or with nitrogen in the form of ammonium chloride is more prone to soft rot in country stores (Somani, 1990; Somani and Shekhawat, 1988).

Other Potato tuber rots

Tuber rots, somewhat similar to bacterial soft rot have been reported caused by several fungal pathogens. The pathogens involved are *Pythium aphanidermatum* (Thakur and Chenulu,1970), *P. ultimum* (Prasad, 1979), *P. butleri* (Somani, 1986), *Verticillium alboatrum* (Bhargava and Hari krishore, 1953), *Rhizopus arhizus* (Thakur and Chenulu, 1970), *Aspergillus fumigatus* and *Geotrichum candidum* (Khanna and Chandra, 1980), *Pestalotia versicolor* (Gaur and Chenulu, 1981), *Rosellinia sp.*(Phadtare, *et al.*1983) and *Alternaria alternata* (Somani, 2004).

Black Scurf (*Rhizoctonia solani*)

Kuhn first reported the disease in 1858 from Germany while from India by Shaw in 1912. Presently it is found throughout the potato growing areas in the country (Dutt,

1979). The basidial stage of the pathogen has also been reported from Himachal Pradesh, Bihar, Punjab and Uttar Pradesh (CPRI 1971 to 73; Dutt, 1979, Singh *et al.* 1984). The disease does not affect the yield of potato (Dutt, 1979) but considerably reduces market value (Somani, 1986). Extensive surveys (1980-83) of seed and ware potatoes in Punjab, revealed that the black scurf was prevalent in 96.5 per cent samples with an average incidence of 39.4 per cent in freshly harvested potato. In cold stored potato prevalence was 94.8 per cent with an average disease incidence of 45.4 per cent (Verma, *et al.* 1990).

Symptoms

Black scurf is a tuber disease but sometimes other plant parts viz., stolon, stem or even foliage are also affected (CPRI, 1976; Prasad and Agarwal, 1983). Stem canker and wilt have also been reported from West Bengal, Punjab and Bihar (Dar and Dutta, 1980). Several chocolate to black coloured, pin head to pea size, sclerotia (resting bodies of the pathogen) are formed on the surface of mature tubers. In some cases pathogen may form thin black mycelial crust, it may some time, cover large portion of tuber. These structures are superficial and can easily be removed without damaging the tuber tissue. The pathogen often girdles the stem at basal level or just below the soil surface. In such cases, aerial tubers may form due to blockage of phloem tissue restricting the translocation of photosynthates. Under high humid conditions, basidial stage may be seen in the form of white mycelial growth at base of the stems..

Pathogen and Epidemiology

Black scurf is caused by *Rhizoctonia solani* (*Thanatephorus cucumeris*). Basidiospores are produced when night temperature is below 22^0C and RH 88 per cent (Singh *et al.* 1984). Variations in cultural characters and pathogenic behaviour in *R. solani* have been reported and grouped (Singh, 1964; Dutt et al., 1974; Sheoraj et al. 1974). Aversion reaction has been used to group *R. solani* isolates (Thirumalachar, 1953; Dutt *et al.* 1974). Sclerotial patterns have been found to be useful in grouping the isolates (Sheoraj *et al.* 1974). Pattern of sclerotial formation is a consistent character in classifying cultural races. In all, five types and four sub-types have been established. Suresh and Mall (1977) studied 25 isolates and found that they were of AG3 and AG4 types. Optimum temperature is 25^0C for mycelial growth, 23^0C for sclerotial germination, 21 to 25^0C for basidiospore germination and 18^0C for stem lesion formation. High summer temperatures in sub-tropical plains are not suitable for sclerotia production and their survival (Singh, 1964). Therefore, *R. solani* has to over summer either as saprophytic mycelium or by infecting the crops grown during summer season. Since survival of inoculum in soil in the plains is less likely, the infected seed tubers serve as the main source of inoculum (Small, 1945; Somani, 1986). However, in hills, the pathogen survives in soil throughout the year and serves the source of inoculum.

Charcoal Rot (*Macrophomina phaseolina*)

Charcoal rot is a disease of warm and dry climate. In India the disease occurs in Bihar, West Bengal, Eastern Uttar Pradesh, Madhya Pradesh, Maharashtra, Karnataka, Orissa and Punjab (Thirumalachar, 1955; Dutt, 1979; Dutt and Sheoraj, 1970; Upadhya, 1976; Sharma, 1980; Somani, 1996). In Madhya Pradesh, it is a major problem in late harvested potatoes however the disease can be brought under control just by adjusting dates of planting and harvesting.

Symptoms

Three types of symptoms have been reported viz. charcoal tuber rot, dry tuber rot and stem blight (Paharia, 1960) Charcoal tuber rot is the most common and important disease in India. Potato tubers get infected in soil and symptoms may develop immediately or after harvesting. The diseased tubers show black areas around proliferated lenticels and eyes. Initially, a black spot (2-3 mm) develop around a lenticel with a whitish speck at the center. Black spots develop further into a blotch which finally covers the whole tuber. The tuber skin remains intact but the flesh beneath the infected lenticels gets blackened up to a depth of 2-5 mm. In case of the dry tuber rot, black sunken areas develop on the tuber with cavities filled with fungus mycelium and sclerotia underneath the skin. Later on secondary pathogens attack such tubers and turn these into foul smelling wet rot (Pushkarnath, 1976).The third type is the stem blight which is not common in India. In this case stems at or below the ground level are infected and becomes hollow. Black sclerotia appear on the lesions and pycinidia also be formed (Thirumalachar, 1955).

Pathogen and Epidemiology

Charcoal rot is caused by *Macrophomina phaseolina* (Tassi) Goid. It has two sub-species viz., *sesamica* and *typica* which differ in culture characters (Thirualachar, 1953). Perfect stage is *Botryodiplodia solani tuberosi* Thirum. which is rare (Thirumalachar and O'Brien, 1977). *M. phaseolina* has been reported from 284 host plants (Young, 1949). Both infected tubers and soil serve as primary source of inoculum however, soil is the main source. Temperature below 28^0C checks the disease. Atmospheric temperature 30^0C or above predisposes the tubers to charcoal rot. Storage temperature below 10^0C does not allow charcoal rot development even if the tubers are infected. But such tubers start rotting when taken out of cold store (Thirumalachar, 1953). Disease development is faster in sandy to sandy loam soil as compared to clay loam and clay soils (Paharia, 1960) since light soil become dry and warm quickly. Exposure of harvested tubers to sun increases the development of charcoal rot.

Dry Rot (*Fusarium spp.*)

Dry rot is mainly a storage disease of potato (Boyd, 1972). It has been observed in freshly harvested and heaped tubers at Gwalior in Madhya Pradesh (Somani, 1993) and also develop after storage (Dutt,1979). It is main problem in country store or other traditional methods but not in cold stores.

Symptoms

The fungus prefers wound and proliferated lenticels for infection. Initially, the infected portion is slightly depressed which increases in size and the skin develops wrinkles to form irregular concentric circles. Tissue beneath the infected region becomes mealy and brown in colour. Cavities are formed which are lined with fluffy white to brown and sometimes blue mycelium. The fungus also forms fluffy cushions on the affected tissue. Ultimately, the whole tuber is dried and becomes light in weight. Symptoms reported on potato plants include wilt, stem rot and seedling damping off. Sometimes wilting of plants is accomplished with rotting of stem base. Singh (1986) reported brown lesions on tubers and stolons due to *Fusarium* spp.

Pathogen and Epidemiology

Smith and Swingle (1904) first reported dry rot caused by *Fusarium oxysporum*. Later on, more than 20 species of *Fusarium* have been reported to cause dry rots (O'Brien and Rich, 1976). In India, following 13 species of *Fusarium* have been reported to cause dry rot on potato tubers and they are *F. coeruleum* (Ajrekar and Kamat, 1923),   *F. radicicola* (Butler and Bisby, 1931), *F. trichothecioides* (Uppal, 1935), *F. solani* (Mitter, 1935), *F. eumartii* (Dey, 1947), *F. oxysporum* (Agarwal, 1949), *F. moniliforme var. subglutiaeus* (Singh, 1958; CPRI, 1978), *F. equiseti* (Rai, 1979-a), *F. avenaceum* (Rai, 1981-a), *F. redolens* (Rai, 1981-b), *F. acuminatum* (Rai and Singh, 1981). *F. semitectum* (Singh, 1986) and *F. pallidoroseum* (Somani, 1993).

Besides *Fusarium* spp, other fungi reported to cause dry rot are, *Apergillus* spp (Thirumalachar and Ramamurthi, 1952), *Cylindrocarpum tonkinensis* (Nayak, 1964), (Thakur and Chenulu, 1970), *A. flavus* (Somani, 1993), *Oospora pustulans* (CPRI, 1971-73; 1980). *Colletorichum coccoides* (Rai, 1975), *Alternaria alternata* (Sharma, 1978; Khanna and Chandra, 1980), *Trichothecium roseum* (CPRI, 1980), *Phacidiopycnis tuberivora* (CPRI, 1981), *Phoma* sp., *Penicillium* sp. (Singh, 1986), *Verticillium* sp. and *Acremonium* sp. (Somani, 1993).

Infected tubers and soil are the primary source of inoculum. *F. oxysporum* and *F. solani* have good saprophytic ability to survive in soil. Generally tubers get contaminated at harvest (Mall, 1977). Storage temperature between 20-28^{0}C is congenial for dry rot development (Mann and Joshi, 1925; Mitra, 1934; Agarwal, 1949). High humidity favours the infection (Moore, 1945). Susceptibility to dry rot increases with increase in tuber size (Singh, 1986). Tubers exposed to herbicide paraquat that is used for killing haulms may also favour development of dry rot by *F. oxysporum* at harvest or during storage in heaps (Somani and Chauhan, 1994). Dry rot incidence was significantly and positively correlated with phenolics and reducing sugar in tubers, while it was negatively correlated with tuber dry matter (Somani, 1993).

Powdery Scab (*Spongospora subterranea*)

The disease affects all the underground parts of potato plant but observed severe on tubers. The disease is confined to high hills, first reported from Mahabaleshwar and Nilgiris. Since then it has been reported from most of the hills. It does not affect tuber yield but reduces market value of the produce and make it unfit for seed purpose.(Dutt and Pushkarnath, 1960).

Symptoms

Disease appears on the surface as small pimple like circular light brown spots (in monsoon season) which increase in size and rupture, exposing a cavity containing brown powdery mass of spore balls. The cavity is surrounded by ruptured periderm. Development of warty protruberances on tubers have also been reported (Bhattacharyya and Sheoraj, 1978).

Pathogen and Epidemiology

The disease is caused by *Spongospora subterranea* (Wallr.) Lagerheim. The spores of the pathogen are adhered to each other forming spongy spore balls. The fungus over winters through spores in soil on infected seed tubers. Low soil temperature (10-15^{0}C) along with high soil moisture (0.1-0.2 atmosphere) are optimum for disease development (Bhattacharyya and Sheoraj, 1981). Acidic soils, clay and loam soils are more suitable for disease development. The disease has not been reported from any other crop.

Wart (*Synchytrium endobioticum*)

Wart is a dreaded disease of potato. In India, it was reported from Darjeeling district in West Bengal in 1953 on 'Furore' variety imported from Denmark (Ganguly and Paul, 1953). It remained restricted to Darjeeling district due to timely imposition of quarantine legislation (Phadtare, *et al.*, 1990). Regular, periodical surveys are conducted by the Directorate of Plant Protection, Quarantine and Storage (Govt. of India), Department of Agriculture (West Bengal) and Central Potato Research Institute, Shimla to monitor the disease. In summer crop, the disease incidence ranged between 2 to 75per cent at different locations during 1958 and 1962 (Mehta, et al., 1962). Wart is generally not observed in winter crop however, Phadtare and Kulshreshtha (1983) recorded 5-10 per cent incidence in winter crop as well. The disease has also been recorded in Sikkim and Nepal.

Symptoms

Cauliflower like warty growth develops on tubers, stolons, stem bases and even on leaves near soil. Initially, it appear as small white granular swellings on the eyes. Its size may increase in field as well as during storage. Initially, the warts are white or pink but later become black. Wart in sun exposed tubers may become light green in colour.

Pathogen and Epidemiology

The disease is caused by *Synchytrium endobioticum* (Schilb) Perc. The fungus do not develop any mycelium. It forms thin-walled summer sporangia which may germinate immediately or form thick walled resting spores and over winter. Resting spores remain in soil and on tubers. Resting spores can survive in soil even up to 30 years (Anonymous, 1969) and can withstand harsh conditions like momentary dip in boiling water or exposure to freezing temperature or passage through the intestines of cattle. The disease is worst in wet seasons since 100 per cent soil moisture is an ideal condition for the germination of sporangia. Resting spores may germinate at 10 to 30^0C with optimum being 19-20^0C. Wart infection may take place between pH 3.9 to pH 8.5. Soils rich in humus that retain moisture for a longer period are suitable for disease.

Sclerotium Rot (*Sclerotium rolfsii*)

Sclerotium rot is a disease of warmer climate. Tubers rot both during transit and storage. In India, Shaw and Ajrekar (1915) first reported the disease from Bengal and Bombay.

Symptoms

Small slightly sunken yellow spots with brownish margin appear on potato tubers. Tuber skin generally remain firm but may get invaded by secondary organisms and the tubers may become soft. The internal tissues decay and cavities are formed. A fine, filmy and white mycelium grows over the spots. The hyphae grow radially forming fan-shaped structures which are closely oppressed to the surface of the tuber. Mustard seed like sclerotia are formed in abundance on mycelium over the tubers (Dutt, 1979). Stem, roots and all other underground parts of the plant can be attacked by the fungus. Wilting of plants may also occur.

Pathogen and Epidemiology

Sclerotium rolfsii is the cause of this disease. Mycelium of the pathogen is silky white. The resting structure, sclerotia are like mustard seed and globose in shape. The mycelium and sclerotia remain in soil and may be carried to uninfected fields through infected seed tubers. High moisture and high soil temperature (30 to 35^0C) favours the disease (Abeygunawardena and Wood, 1957). Sclerotia can remain viable even at freezing temperature. The pathogen thrives well in acidic sandy or loam soils. Ammonical nitrogen checks growth of *Sclerotium rolfsii*.

Some other little known potato tuber diseases in India are: Pink rot (*Phytophthora erythroseptica*) from Darjeeling and Shimla hills (Phadtare, 1978; Rai, 1979-b); *Oospora* tuber rot (*Oospora pustulans*) from Shimla hills, Shillong and Ooty (Rai, 1974; CPRI, 1981; 1982); Silver scruf (*Spondylocladium astrovirens*) from Nilgiri and Darjeeling hills (Srivastava, 1963; 1965; Pushkarnath, *et al.*, 1966); *Colletotrichum* tuber rot (*Colletotrichum atramentarium*) from different parts of Himachal Pradesh and Eastern plains (CPRI, 1949-56; 1978; Rai, 1975); *Gilmaniella* rot (*Gilmaniella humicola*) from Bihar (Sahai, 1967), Phomopsis rot (*Phomopsis tuberivora*) from Shimla hills (Rai, 1983).

Management

Most of the important potato tuber diseases are both seed and soil-borne. The infected tubers used as seed spread the diseases in new uninfected fields. Similarly, the produce of a crop raised from healthy seed in infested soil gets infected and the cycle continues (Somani, 1986-b). These diseases can not be eliminated completely but managed through integrated approach.

Management through Seed

1. ***Growing resistant varieties :*** Disease control through use of resistant varieties is the best but presently, a very few varieties are available which are resistant to range of tuber borne diseases. Multiple disease resistance in a single variety is desirable (Nagaich, 1983) but it is difficult to achieve. Some diseases are location/ region specific and require special attention. However, resistance against some diseases is available. Varieties Kufri Jyoti, Kufri Bahar, Kufri Sherpa, and Kufri Kanchan are immune to wart and are recommended for Darjeeling district (Gaur et al., 1999). Kufri Dewa, Kufri Bahar and Kufri Sherpa show resistance to black scurf (CPRI, 1989-90; 1998-2000; Gupta et al., 1983). Kufri Dewa shows resistance against soft rot also.

2. ***Use of disease free seed :*** Planting of healthy seed free from tuber-borne diseases in disease free soil is most effective strategy to manage these diseases. Diseased tubers if any should be sorted out and discarded before planting.

3. ***Seed tuber treatment with chemicals :*** Earlier organo-mercurial compound-Methoxy Ethyl mercuric chloride with 6 per cent mercury (Emisan-6, Aretan-6) @ 0.5 per cent for 30 min dip was found effective against common scab and black scurf and was in practice. It is also effective against dry rot, charcoal rot (Sahai, 1969), and powdery scab (Bhattacharyya and Sheoraj, 1982; 1986). Twenty minute dip of potato tubers in 0.25 % solution was found non-toxic. The same solution can be used for 20 dips. The treatment was effective against common scab and black scurf (Somani, 1990).

   Somani (1990), however, observed injurious effect of this treatment during 1982 at Jalandhar (Punjab) and recorded blind eye tubers and gaps in emergence and low yields in seed treated with MEMC. Organomercurials are hazardous and has been banned in most of the developed countries. Therefore, efforts were made to find safe chemicals (Dutt and Thaplyal, 1978; Dutt et al., 1980). Acetic acid 1.0 per cent plus zinc sulphate 0.05 per cent for 15 min dip was found to be as effective as MEMC in controlling black scurf disease (Somani, 1986-a). But a chemical treatment effective against several diseases was more practical and thus desirable. Efficacy of boric acid against black scurf and common scab was studied during 1984 to 1987. Boric acid at 1.0 per cent was effective against black scurf only but at 3.0 per cent it proved highly effective against both black scurf and common scab and was at par with MEMC. Tuber treatment before cold storage was better over treatment carried after cold storage. (Somani, 1998). The treatment has since then recommended for the control of tuber-borne diseases (AICPIP, 1987). Boric acid is also effective against soft rot (Somani and Shekhawat, 1985) and dry rot (Singh *et al.*, 1989). Boric acid 3.0 per cent spray treatment has been reported effective as the dip treatment against black scurf (Khanna and Sharma, 1996). Since the spray treatment require less chemical (45g/q) and the treatment is cost effective and has now been recommended by the Institute (Singh, *et al.*, 2002).

4. ***Bio-control :*** Use of bio-control agents is considered a good alternative to chemicals for disease control. Several fungal isolates were screened against *R. solani, S. rolfsii, F. oxysporum, F. solani* and *M. phaseolina.* One isolate each of *Trichoderma viride, T. koningii, T. harnatum, T. harzianum* and two isolates of *Gliocladium virens* could control all the pathogens *in vitro* from 72 to 100 per cent (Arora, 1999). The antagonists controlled the pathogens through hyper parasitism, lysis or disintegration of host cells. A bio-preparation of *Trichoderma viride* effective in reducing black scurf incidence by 62.5 per cent and disease index by 77.2 per cent has been developed (CPRI, 1999-2000). Field trials conducted at Jalandhar and Gwalior during 1999-2001 confirmed the effectiveness of the bio-preparation against black scurf (Arora and Somani, 2001).

The antagonistic activity of *T.koningii* has been utilized in USSR against dry rot (Boyd, 1972). Roy et al. (1996) isolated a mycoparasite (*F. equiseti*) from sclerotia of *R. solani* which over grew the sclerotia quickly and made them ineffective. It was better than 3.0 per cent boric acid. *Bacillus subtilis* was found effective in controlling *R. solani* (Ischen, 1987). Application of *T. harzianum, P. fluorescens* and *T. viride* in *R. solani* sick plots was found to be effective in reducing black scurf (Hazarika, *et al.*, 2000). Bacterial antagonists like *P. fluorescens, Bacillus* spp. and *B. polymyxa* have been found effective against *R. solanacearum* causing bacterial wilt (CPRI, 1983; 1988). Thirumlachar and O'Brien(1977) reported *B. subtilis* effective against charcoal rot.

5. ***Domestic Quarantine :*** The Government of India imposed a legal ban on movement of potatoes from the state of West Bengal to other parts of the country in 1959 under DIP Act 1914, II Section A (Mehta, 1963). The Government of West Bengal imposed a ban on the movement of potatoes from most severly wart infested areas viz., Pulbazar PS and Jorebunglow to other areas in Darjeeling vide notification No. 5481.Inpt.dt. 20.05.80. Due to the imposition of domestic quarantine and simultaneous introduction of wart immune varieties, further spread of wart from this region has been restricted.

6. ***Tuber maturity and tuber injury :*** Immature tubers are more prone to bruising and thus to soft rot development. Tubers injured during harvesting may also develop dry rot and mouldy rots. Therefore, to prevent the tubers from contracting infection the tuber skin need to be firm and can be achieved by harvesting crop at full maturity and removing or killing (using herbicide) foliage 7-10 before harvest. Further, after harvesting tubers be heaped for 10-12 days in cool ventilated and shady place for suberization. Large size tubers have more chances of getting tuber injury and contracting dry rot and are also more prone to charcoal rot (Paharia,1960).

Management through Soil

Major pathogens causing tuber borne diseases also survive in soil. Hence control through seed alone is not sufficient. To check the soil-borne inoculum, following methods are effective. Most of them are just manipulation of normal cultural practices.

7. ***Greening of tubers :*** Green tubers are less prone to storage diseases (Thirumalachar and Narsimhan, 1970). Sukumaran *et al.* (1975) found that tubers exposed to diffused sunlight turn green and have more glycoalkaloids viz., solanin and charonine. Upadhya *et al.* (1976) observed that greening of tubers delay development of charcoal rot. Sharma and Chauhan (1978) found low incidence of dry rot, Selerotium rot and charcoal rot in green tubers. Therefore, this method can be useful for developing seed tubers.

8. ***Crop rotation and green mannuring :*** Common scab can effectively be controlled by growing wheat, oat, green gram and sorghum (CPRI, 1964; 1965; 1983). The disease can also be controlled through paddy-wheat rotations and potato-onion-maize or clover – Pearl millet – potato rotations. *Streptomyces* spp. population was minimized by potato-hot weather cultivation– paddy–potato; potato-calocassia-green manure–potato and potato–onion-green manure-potato (Singh *et al.,* 2000). Vashistha and Chaubey (1975) recommended 3-7 years' break in potato cultivation to reduce common scab disease. Green mannuring before potato also reduces common scab.

   Charcoal rot and black scurf can be managed by sunhemp and maize rotations (Bhattacharyya and Malhotra, 1979). Sunhemp was very effective in reducing black scurf however, application of farm yard manure increased the disease. Sikka *et al.,* 1971; Singh and Jeswani, 1983). Soft rot was minimum in green manure– potato – wheat rotation (CPRI, 1981).

   Brown rot could be reduced in crop roatations: potato–wheat–potato, potato–carrot/beet root/ onion–potato, potato–sorghum–potato, potato–finger millet–knoll khol /onion/cabbage/carrot–potato, potato–maize–potato and potato–finger millet–potato in Karnataka (Shekhawat *et al.,* 1988) and other rotations (Shekhawat *et al.*, 2000). Powdery scab disease could be managed by rotation with wheat for 2 years or pea – wheat for 3 years (Bhattacharyya and Sheoraj, 1982).

9. ***Addition of straw, saw dust and oil cake :*** Application of saw dust @ 25 q/ha (Singh *et al.,* 1972) or @ 30 q/ha (Dutt, 1979) could control black scurf disease. Mustard oil cake or ground nut oil cake @ 25 q/ha were effective against black scurf (Singh *et al.,* 1972). Singh and Jeswani (1983) found mustard straw and gram straw effective in reducing black scurf. Ground nut or cotton seed cakes controlled *Fusarium* wilt (Singh, *et al.,* 1988).

10. ***Fertilizer and micro-nutrient amendments :*** Application of high doses of nitrogen (200-300 kg/ha) to potato crop increased the incidence of soft rot in country store (Somani and Shekhawat, 1988). Source of nitrogen applied also affected soft rot incidence. Ammonium chloride induced highest disease incidence followed by ammonium sulphate, calcium ammonium nitrate and least by liquid fertilizer nitrogen (Somani and Shekhawat, 1988). However, foliar application of micro-nutrients did not affect keeping quality of tubers (Harikishore *et al.,* 1964).

Common scab is suppressed in acidic soils and the pathogens does not survive below pH5.0. Therefore, lime and fertilizers which increases the pH also increase common scab and therefore, should be avoided. Ammonium sulphate lowers the pH and reduces common scab (Vashisth and Chaubey, 1975; Pushkarnath, 1976). Trehan and Grewal (1980) reviewed the effect of fertilizers and micro-nutrients on common scab. High nitrogen doses increased the incidence while phosphorus reduced the scab incidence. Black scurf disease was reduced with the application of zinc, manganese, molybdenum and boron (Singh et al., 1972). *Sclerotium* wilt was controlled by using nitrogen 160 kg per hectare (CPRI, 1971-73).

11. ***Soil application of fungicides :*** Soil application of Penta-chloro-nitro benzene (PCNB) @ 30 kg/ha controlled common scab when the disease free tubers were planted in sick soil (CPRI, 1975). PCNB @ 50 kg/ha was the best treatment against black scurf however, the lower doses proved less effective (Singh, *et al.,* 1978). PCNB application in soil is however, not economical the chemical has been banned and is no longer available in the market.

12. ***Hot or cold wheather cultivation :*** In the subtropical plains leaving the fields fallow during hot summer season and ploughing reduced soil-borne pathogens (Sikka *et al.,* 1971) and control black scurf (Dutt, 1979). One to two deep ploughings of infested soil during May-June in plains and plateaus and in October-November in hills also reduced bacterial wilt (Kumar, 1970; Shekhawat *et al.,* 1979; CPRI, 1985).

13. ***Soil solarization :*** It is improvement over hot weather cultivation. In this case, the fields are lightly irrigated and after 48 hr covered with transparent polythene mulching during hot summer for a period of 4 to 6 weeks. Soil solarization can successfully be carried out during 3rd week of May to 2nd week of June in north-western plains when the soil temperature between 45 to 61^0C could be achieved in the solarized plots. The treatment has been found to control black scurf, sclerotium wilt and russet scab. In addition to the control of the pathogens, it also drastically reduce weeds population and brings abut a significant increase in potato tuber yield (CPRI, 1997-98; Arora & Sharma, 2003; Arora, *et al.,* 1997). Seed treatment with T. viride or boric acid along with soil solarization improved further control of black scurf (Arora, 2000).

14. ***Irrigation :*** Effective control of common scab and charcoal rot can be achieved through light irrigation given frequently from beginning of tuberization. Flooding the fields was also found to control *Sclerotium wilt* (IARI, 1953-54). Frequent irrigation create anaerobic conditions in soil, reduces oxygen level and thus control common scab (Lapwood and Wellings, 1971; Lapwood *et al.,* 1971; CPRI, 1980; 1981; Singh and Singh, 1981). The irrigation also reduces the soil temperature and control charcoal rot (Thirumalachar, 1955, Dutt and Sheoraj, 1970). However, frequent irrigation may predisposes the developing tubers to soft rot (Singh and Verma, 1979).

15. ***Adjustment of planting and harvesting dates :*** In north-eastern plains potato crop harvested in February- March exhibited a high incidence of soft rot compared to that harvested in January (CPRI, 1969). In northern plains, late harvested (March) tubers exhibited highest incidence of soft rot in store as compared to those harvested

in January. A decline of 5 to 10 per cent in emergence was recorded from produce of late harvested crop (Somani, *et al.*, 1988). Black scurf is reduced on harvesting the crop just before or immediately after full maturity. Delay in harvesting after haulm cutting increases incidence of black scurf (CPRI, 1963; 1965). Charcoal rot is a disease of high soil temperature. Hence, the crop should be harvested before the soil temperature crosses 28°C (Thirumalachar, 1955). *Fusarium wilt* is also favoured by high temperature, the disease is more in early planted crop when the soil temperatures are high. It can be reduced by delay in planting of the crop (Singh, 1986).

Conclusions

Majority of potato diseases (viral, fungal and bacterial) spread through seed potato. The spread is not only local but also transnational as the potato seed tuber provides congenial niche for pathogens' survival. Once the pathogen/ disease is introduced in to the disease free area through seed potatoes, the pathogen may multiply under congenial conditions astronomically. Sound research based extensive knowledge of pathogens' biology will enable us to manage them. All the world intellectuals are becoming averse to the use of chemicals therefore integrated eco-friendly measures like boric acid, bio-agents, solarization etc would find greater applications in coming days. Rapid developments are also taking place in the field of technologies for detection of the pathogen so that risky pathogens are kept away and/or excluded right at the entry point. The method of preventing further spread by quarantine restrictions has worked well in India for the diseases like potato wart and golden nematodes. The pathogens carried through the potato seed also have the ability to perpetuate in field either in soil or on other host and/or non host plants. Therefore, management of these pathogens requires, either one or multiple or integrated, practices like crop rotation, solarization etc. that keep pathogen population in the field to a negligible level so that the damages are not economically high.

References

Abeygunawardana D.V.W. and **Wood, R.K.S.** 1957. Factors affecting the germination and mycelial growth of *Sclerotium rolfsii* Sacc. *Trans. Br. mycol. Soc.* 40 : 221-231.

Agarwal, S.C. 1949. Storage diseases of potato I. The role of *Fusarium oxysporum* in the rot of potatoes. *J. Indian Bot. Soc.* 28 : 50-62.

AICPIP. 1987. *Sixteenth Progress Report.* All India Co-ordinated Potato Improvement Project (CPRI, Shimla), 69.

Ajrekar, S.L. and **Kamat, M.N.** 1923. The relationship of the species of Fusarium causing wilt and dry rot of potatoes in western India. *Agric. J. India* 18 : 515-520.

Anonymous. 1969. Wart disease of potatoes. *Minist. Agric. Fish. Fd. Advisory Leaflet* 274 : 1-5.

Arora, R.K. 1999. Evaluation of bio-agents for control of soil and tuber borne diseases of potato. *Indian Phytopath* 52 : 310 (Abstr.).

Arora, R.K. 2000. Combined effect of soil solarization and seed treatment on control of potato black scurf. In : *Potato, Global Research & Development* – Vol. II (Eds. S.M. Paul Khurana, et al.). Indian Potato Assoc., Shimla. 390-392.

Arora, R.K. 2003. Management of soil and tuber borne diseases through eco-friendly strategies. In: "*Integrated Plant Disease Management through eco-friendly strategies*" (Eds. R.S.Singh et al.) Indian Phytopathological Society, New Delhi 44-51.

Arora, R.K. and ***Jyotsana Sharma*** 2003. Management of russet scab of potato through soil solarization. *J. Indian Potato Assoc.* 30: 139-140

Arora, R.K. and ***Paul Khurana S.M.*** 2004. Major Fungal and Bacterial Diseases of Potato and Their Management. In: *Disease Management of Fruits and Vegetables Vol. I. Fruit and Vegetable Diseases* (Ed. K.G.Mukerji), Kluwer Academic Publishers. The Netherlands. 189- 231.

Arora, R.K. and ***Somani A.K.*** 2001. A bio-formulation for control of soil and tuber borne diseases of potato. *J. Indian Potato Assoc.* 28 : 88-89.

Arora, R.K., Singh, P.H. and ***Singh, B.P.*** 2003. Fungal Diseases and Their Management. In : *Potato – Production and Utilization in Sub-Tropics* (Eds. S.M.Paul Khurana et al.) Mehta Publishers, New Delhi. 193-209.

Arora, R.K., Trehan, S.P., Sharma, J. and ***Khanna, R.N.*** 1997. Soil solarization for improving potato health and production. *Indian Phytopathological Society Golden Jubilee Conference. Integrated Plant Disease Management for Sustainable Agriculture.* Vol. I. 1190-1191.

Bhargava, K.S. and ***Hari Kishore*** 1953. Occurrence of *Verticillium* in potato tubers in India. *Nature, Lond.* 171 : 800-801.

Bhattacharyya, S.K. and ***Sheoraj*** 1982. Studies on of potatoes III. Chemical and cultural control. *J. Indian Potato Assoc.* 9 : 111-115.

Bhattacharyya, S.K. and ***Sheoraj***1986. Field screening of different chemicals against powdery scab of potatoes. *Pesticides* 2 : 23-26.

Bhattacharyya, S.K. and ***Mahlotra V.P.*** 1979. Control of charcoal rot of potato through crop rotations. *J. Indian Potato Assoc.* 6 : 1999-204.

Boyd, A.E.W. 1972. Potato storage diseases. *Rev. Plant Pathol.* 51 : 297-321.

Butler, E.J. and ***Bisby G.R.*** 1931. *The Fungi of India* (Revised by R.S. Vasudeva in 1960) ICAR, New Delhi. Mono 1, 237.

Chaudhary, R.G. 1983. Causes of potato seed decay in low hills of Arunachal Pradesh, *J. Indian Potato Assoc.* 10: 111-115.

CPRI, 1949-56. Progress Rept of Scientific work, Central Potato Research Institute, Shimla, 28-38.

CPRI, 1963. Ann. Sci Rept, Central Potato Research Institute, Shimla 65-68.

CPRI. 1964. Ann. Sci. Rept., Central Potato Research Institute, Shimla 90-95.

CPRI, 1965. Ann. Sci. Rept., Central Potato Research Institute, Shimla. 71-77.

CPRI, 1969, Ann. Sci. Rept., Central Potato Research Institute, Shimla 45.

CPRI, 1971 to 1973. Progress Rept, Central Potato Research Institute, Shimla 112-122.

CPRI, 1975. Ann. Sci. Rept., Central Potato Research Institute, Shimla. 90-115.

CPRI, 1976. Ann Sci Rept, Central Potato Research Institute, Shimla. 62-65.

CPRI, 1978. Ann Sci Rept, Central Potato Research Institute, Shimla. 60-93.

CPRI, 1979. Ann Sci Reprt, Central Potato Research Institute, Shimla. 80-82.

CPRI, 1980. Ann Sci Rept, Central Potato Research Institute, Shimla. 79 -110.

CPRI, 1981. Ann Sci Rept, Central Potato Research Institute, Shimla. 61-74.

CPRI, 1982. Ann Sci Rept, Central Potato Research Institute, Shimla. 100-110.

CPRI. 1983. Ann Sci Rept, Central Potato Research Institute, Shimla. 93-101.

CPRI. 1988. Ann. Sci. Rept., Central Potato Research Institute, Shimla. 117-121.

CPRI. 1989-90. Ann. Sci. Rept., Central Potato Research Institute, Shimla. 74-76.

CPRI. 1992-93. Ann. Sci. Rept., Central Potato Research Institute, Shimla. 41.

CPRI. 1997-98. Ann. Sci. Rept., Central Potato Research Institute, Shimla. 66-67.

CPRI, 1999-2000. Ann. Sci. Rept., Central Potato Research Institute, Shimla 98-135.

Dar, B.A. and *Dutta, A.K.* 1980. Investigation on black scurf and stem necrosis of potato. *Indian J. Mycol. Plant Path.* 10 : 101(Abstr.).

Dey, N.R. 1947. Cultivation and storage of potatoes in Bihar with special reference to diseases prevalent in stores and fields. *Allahabad Fmr.* 21.177-204.

Dey, S.K. and *Singh, N.* 1983. Studies on variability of seven isolates of Streptomyces spp. causing common scab of potato in Punjab (India). *Phytopath Mediterranea* 22 : 59-66.

Dey, S.K., Singh, N. and *Singh H.* 1981. *Streptomyces collinus* Lin. – a new species causing common scab of potato. *Curr. Sci.* 50 : 830.

Dutt, B.L. 1979. *Bacterial and fungal diseases of potato.* Indian Council of Agricultural Research, New Delhi. 196.

Dutt, B.L. and **Pushkarnath**. 1960. Resistance of potato varieties to powdery scab. *Indian Potato J.* 2 : 78-82.

Dutt, B.L. and **Thaplyal P.** 1978. Laboratory evaluation of efficiency of chemical treatment of seed potatoes for control of black scurf. *J. Indian Potato Assoc.* 5 : 167-169.

Dutt, B.L. and **Sheo Raj** 1970. Potato charcoal rot and its control in Punjab. *Indian Farming.* 20(2) : 31-32.

Dutt, B.L., B.P. Singh, Jeswani, M.D., Kumar, A., Teotia, S.S., Ranjeet Singh and **Lakshman Prasad** 1980. Potato (Solanum tuberosum L.) Black scurf *Rhizoctonia solani. Fungicide and Nematicide Test Results.* V.35.

Dutt, B.L., Mukhtar Singh and **Verma, S.M**. 1972. Potato seed decay (*Fusarium* spp. Aspergillus spp. and *Erwinia* spp.) Fungicidal and Nematicidal test results of 1971. *Amer. Phytopath Soc.* 27 : 149.

Dutt, B.L., Sheo Raj and **Hari Kishore** 1974. Relationship among isolates of *Rhizoctonia solani* Kuhn. *Mycopathol. Mycol. Appl.* 52 : 77-84.

Ganguly, A. and **Paul, D.N.** 1953. Wart disease of potato in India. *Sci. Cult.* 18 : 605-606.

Gaur, A. and **Chenulu V.V.** 1981. Two unrecorded storage dis orders of *Citrus reticulata* L. and *Solanum tuberosum* L. *Curr. Sci.* 50 : 322-323.

Gaur, P.C., Naik P.S., Kaushik, S.K. and. **Chakrabarti S.K**. 1999. Indian Potato varieties *Tech. Bull.* No.51, Central Potato Research Institute, Shimla. pp 38.

Ghanekar, A.S., Padwal, S.R.– Desai and G.B. Nandkarni 1984 (a) Etiology of soft rot in stored potato. *J. Food Sci. & Tech. India* 21(3) : 127-131.

Gupta, R.B.L., Bhatnagar, G.C. and **Bhatt B.N.** 1983. Reaction of potato varieties/ hybrids to diseases in Rajasthan. *In : Potato in Developing countries* (B.B. Nagaich *et al.*, Eds.). Indian Potato Assoc., Shimla pp 399-402.

Harikishore, Anand S.K. and **Ram Phal** 1964. Storage behaviour of potatoes as affected by irrigation and micro-nutrient application. *Indian Potato J.* 6 : 87-89.

Hazarika, D.K., Phookan, A.K., Das K.K., Dubey, L.N. and **Das B.C**. 2000. Biological management of black scurf of potato. *In : Potato, Global Research & Development* – Vol. .I (Eds. S.M. Paul Khurana, *et al.*). *Indian Potato Assoc.* Shimla. pp.401-404.

Hingorani, M.K. and **Addy S.K.** 1953. Factors influencing bacterial soft rot of potatoes. *Indian Phytopath.* 6 : 110-115.

Hingorani, M.K., Mehta, P.P. and *Singh, N.J.* 1956. Bacterial brown rot of potatoes in India. *Indian Phytopath.* 9 : 67-71.

IARI. 1953-54. Sci. Rept., Indian Agricultural Research Institute, New Delhi, 87-99.

Ischen, J.S.M. 1987. Control of *Rhizoctonia solani* in potato by *Bacillus subtilis*. *Trans. Mycol. Soc. Japan* 284 : 483-493.

Jalali, Indu, Pandita M.L. and *Hooda R.S.* 1982. Effect of fungicidal treatment, cutting and seed size on seed piece decay, growth and yield of potato. *J. Indian Potato Assoc.* 9 : 116-120.

Khanna, K.K. and *Chandra, S.* 1980. Studies on storage diseases of fruits and vegetables V-some new storage diseases of potato. *Indian J. Mycol. Pl. Pathol.* 10 : 81-82.

Khanna, R.N. and *Jyotsana Sharma*. 1996. Effect of boric acid treatment on seed and soil borne *Rhizoctonia solani* inocula and rhizosphere microflora. *J. Indian Potato Assoc.* 23 : 01-07.

Paul Khurana, S.M. 2000. Diseases and pests of potato – a manual. Central Potato Research Institute, Shimla, 66.

Kumar, S.M. 1970. Note on effect of some cultural practices on the incidence of bacterial wilt of potato. *Indian J. Agric. Sci.* 40 : 854-855.

Lapwood, D.H. and *Wellings W.R.* 1970. The control of common scab of potatoes by irrigation. *Ann. appl. Biol.* 66 : 397-405.

Lapwood, D.H., Wellings W. R. and *Wellings H.W.* 1971. Irrigation as a practical mean to control potato common scab (*Streptomyces scabies*). Pl. Path. 20 : 157-163.

Mall, S. 1977. Saprophytic colonization by *Rhizoctonia solani* and *Fusarium* spp. causing black scurf and wilts of potato. *Indian Phytopath.* 30 : 559-560.

Mann, H.H. and *Joshi W.V.* 1925. Further investigations on potato cultivation in western India. VI Storage experiments with potatoes. *Dept. of Agric. Bombay*, Bull. 121, 17-25.

Masoodi. S.D. and *Singh D.V.* 1985. Interaction of some additional species of *Streptomyces* causing common scab of potato in Uttar Pradesh. *Indian Phytopath.* 38 : 604.

Mehta, P.R. 1963. Biological basis of domestic quarantine against the wart disease. *Indian Potato J.* 5 : 76-79.

Mehta, P.R., Sundaram N.V. and ***Renjhen P.L.***. 1962. Survey of wart disease of potato in Darjeeling district. *Indian Potato J.* 4 : 107-109.

Mitra, A. 1934. A new wound parasite of potato tubers. *Nature* (Lond.) C33, 3350. 67.

Mitter, J.H. 1935. India : New Plant diseases recorded in 1934. *United Provinces Int. Bull. Pl. Prot.* 9 : 177.

Moore, F.J. 1945. A comparison of *Fusarium avenaceum* and *Fusarium coeruleum* as causes of wastage in stored potato tubers. *Ann. appl. Biol.* 32 : 304-309.

Nagaich, B.B. 1983. Disease resistance in potato in India. *Indian Phytopath.* 36 : 1-10.

Nayak, M.L. 1964. A new organism causing dry rot of potatoes. *Sci. Cult.* 30 : 143-144.

O. Brien, M.J. and ***Rich A.E.*** 1976. Potato Diseases. *U.S. Dept. Agric Handb.*, 474.

Paharia, K.D. 1960. Charcoal rot of potato in India. *Indian Potato J.* 2 : 1-11.

Paharia, K.D. and ***Pushkarnath,*** 1963. Occurrence of common scab on potatoes in Bihar. Indian Potato J. 5 : 104-105.

Phadtare, S.G. 1978. Pink rot of potato – a new report from India. *J. Indian Potato Assoc.* 5 : 174-175.

Phadtare, S.G. and ***Kulshreshtha U.V.*** 1983. Occurrence of wart disease in the winter crop in Darjeeling district. *J. Indian Potato Assoc.* 10 : 68-69.

Phadtare, S.G., Singh P.H. and ***Shekhawat G.S.*** 1990. Wart disease of potato in Darjeeling hills. *Tech. Bull.* No.19. Central Potato Research Institute, Shimla 47.

Phadtare, S.G., Singh, P.H. and ***Subba, P.B.*** 1983. Black rot disease of potato in Darjeeling. *J. Indian Potato Assoc.* 10: 186-188.

Prasad, B. 1979. *Pythium ultimum* inciting blight and tuber rot in potatoes in India. *J. Indian Potato Assoc.* 6 : 186-188.

Prasad, B. and ***Agarwal H.O.*** 1983. New foliar diseases due to *Rhizoctonia* and *Chaetomium* spp. *J. Indian Potato Assoc.* 10 : 116-120.

Pushkarnath. 1976. Potato in sub-tropics. *Orient Longman, New Delhi*, 289.

Pushkarnath, Sahai, D. and ***Srivastava, S.N.S.*** 1966. The rot and deterioration of potatoes in storage. In S.P. Raychaudhary *et al*(ed.) *Plant Disease Problems Proc. Int. Symp. Pl. Pathol.*, New Delhi, pp.324-328.

Rai, R.P. 1974. Occurrence of *Oospora* sp. inciting tuber rot, root rot and collar rot of the potato in India. *Z. Pflanzenkrank Pflschutz*, 81 : 463-467.

Rai, R.P. 1975. *Colletotrichum coccoides* infecting potatoes in India. *J. Indian Potato Assoc.* 2 : 7-9.

Rai, R.P. 1979-a. *Fusarium equiseti*(Cord.) Sacc. Causing dry rot of potato tubers – a new report. *Curr. Sci.* 48 : 1043-1045.

Rai, R.P. 1979-b. Pink rot of potato in Shimla hills. *J. Indian Potato Assoc.* 6 : 36-40.

Rai, R.P. 1981-a. *Fusarium avenaceum* (Fr.) Sacc. Inciting basal stem rot and tuber rot of potato in India. *J. Indian Potato Assoc.* 8 : 41-44.

Rai, R.P. 1981-b. *Fusarium redolens* Wollenw – a new pathogen of potato, *Curr. Sci.* 50 : 379-380.

Rai, R.P. 1983. *Phacidiopycnis tuberivora* – a pathogen of potato new to India. *Indian Phytopath.* 36 : 582-583.

Rai, R.P. and **Singh, B.P.**1981. A new disease of potato incited by *Fusarium acuminatum* Ell. and *Ev. Curr. Sci.* 50 : 1037-1038.

Roy, S., Singh, B.P. and **Shekhawat. G.S.** 1996. Control of black scurf disease of potato. Effect of boric acid on R. solani sclerotia viability and association of antagonist fungus. 48th *IPS Annual Meeting, PAU, Ludhiana.* 13-16 Feb., 1996.

Sahai, D. 1967. A new disease of potato tubers caused by *Gilmaniella humicola* Barron. *Curr. Sci.* 36 : 645-646.

Sahai, D. 1969. Evaluation of certain fungicides against *Macrophomina phaseoli* from charcoal rot infected potato. *Sci. Cult.* 54 : 686-687.

Sahai, D. 1989. The common scab disease of potato and its management in Eastern plains(Bihar). *Indian Phytopath.* 42 : 311.

Sharma, B.L. 1978. Phytopathological studies on storage diseases of potatoes. *Ph.D. Thesis. Jiwaji Univ. Gwalior*, 197 pp.

Sharma, B.L. 1980. Post-harvest diseases of potato at Gwalior (MP) *J. Indian Potato Assoc.* 7 : 38-40.

Sharma, B.L., Chauhan, R.K.S. 1978. Effect of greening of potato tubers on susceptibility to post-harvest diseases. *J. Indian Potato Assoc.* 5 : 38-39.

Sharma, Jyotsana. 1996. Susceptibility of potato cultivars to tuber rot caused by *Sclerotium rolfsii Sacc. J. Indian Potato Assoc.* 23 : 162-163.

Shaw, F.J.F. 1912. Morphology and parasitism of *Rhizoctonia* in India. *Mem. Deptt. of Agric. India* 4 : 115-153.

Shaw, F.J.F. and **Ajrekar, S.L.** 1915. The genus *Rhizoctonia* in India. *Mem. Dep. Agric. India Bot. Ser.* 7(4) : 177-194.

Sheoraj, Dutt B.L. and **Sahai, D**. 1974. Cultural variations in potato isolates of *Rhizoctonia solani* Kuhn. *Acta Phytopath. Acad. Sci. Hung.* 9 : 71-79.

Shekhawat, G.S., Kishore, V., Singh, D.S., **Khanna, R.N., Singh, R.** and **Bahal, V.K**. 1979. Survival of *Pseudomonas solanacearum* under diverse agroclimates in India. *Indian J. Agric. Sci.* 49 : 735-738.

Shekhawat, G.S. and. **Perombelon, M.C.M.** 1985. Pathogenicity of soft rot bacteria. *Ann. Sci. Rept. Scottish crop Research Institute Invergowre, Dundee, Scotland* p.100.

Shekhawat, G.S., Gadewar, A.V., Bahal, V.K. and **Verma, R.K**. 1988. Cultural practices for managing bacterial wilt of potato. *In : Bacterial diseases of potato :* report of the planning conference on bacterial diseases of potato, 1987. *International Potato Centre, Lima, Peru.* pp. 65-84.

Shekhawat, G.S., Gadewar, A.V. and **Chakrabarti, S.K**. 2000 . Potato Bacterial Wilt in India. Technical Bulletin (Revised) No. 38, *Central Potato Research Institute Shimla,* 56pp

Sikka, L.C., Srivastava, S.N.S., Singh, A.K. and **Bhardwaj, V.P.** 1971. Integrated approach to the control of *Rhizoctonia solani* on potato. *Indian Phytopath.* 24 : 54-57.

Singh, Babu and **Jain, S.S.** 1953. Diseases of potato in storage and their control. *Agriculture Anim. Husb.* Uttar Pradesh 3 : 64-67.

Singh, B.P. 1986. Studies on Fusarium wilt and dry rot of potatoes (*S. tuberosum* L.). Ph.D. Thesis, Dept. of Botany, Aligarh Muslim Univ., Aligarh. pp.273

Singh, B.P. and **Jeswani, M.D.** 1983. Managing black scurf by tuber treatment and soil application of organic wastes. *J. Indian Potato Assoc.* 10 : 50-55.

Singh, B.P., Nagaich, B.B. and **Saxena, S.K.** 1987. Fungi associated with dry rot of potatoes, their frequency and distribution. *Indian J. Plant Pathol.* 5 : 142-145.

Singh, B.P., Arora, R.K. and **Paul Khurana, S.M**. 2002. Soil and Tuber Borne Diseases of Potato. Central Potato Research Institute, Shimla. Technical Bulletin No. 41 (Revised) 74.

Singh, B.P., Saxena, S.K. and **Nagaich, B.B.**1989. Chemical control of dry rot of potatoes caused by *Fusarium oxysporum* and wilt of potato. *J. Indian Potato Assoc.* 14 : 141-143

Singh, K., Mishra, S.R. and **Chaubey, R.D.** 1984. Occurance of *Thanatephorus cucumaris* (Fr.) Donk on potato in India. *Indian J. Plant Pathol.* 2 : 132-134.

Singh, M. 1964. Studies on four isolates of *Rhizoctonia solani* Kuhn the causal agent of black scurf disease of potato in Punjab. I Symptoms, strains disease and pathogenicity. *Indian Phytopath.* 17 : 23-26.

Singh, Mukhtar and **Verma, S.C.** 1979. Post harvest technology and utilization of potato in Hari Kishore(ed.) *Post Harvest Technology and Utilization of potato.* International Potato Centre. Region VI New Delhi.1-114.

Singh, N. and **Singh, H.** 1981. Irrigation : A cultural practice to control common scab of potato. *J. Indian Potato Assoc.* 8 : 35-36.

Singh, R.K. 1958. *Fusarium monitiforme* sheld. causing rot of vegetables. *Indian Phytopath.* 11 : 189-191.

Singh, R.P., Shekhawat, G.S. and **Khanna, R.N.** 2000. Potato based cropping sequences and their effect on common scab. *In : Potato Global Research and Development*, Vol.I(eds. S.M. Paul Khurana *et al.*). *Indian Potato Assoc.* 463-465.

Singh, R.S., Chaubey, H.S. and **Singh, N.** 1972. Studies on the control of black scurf disease of potato. *Indian Phytopath.* 25 : 343-349.

Singh, R.S., Dubey, G.S. and **Chaudhary, R.G.** 1978. Control of black scurf of potato through chemicals and organic amendments in Tarai(U.P.). In : B.B. Nagaich *et al.*(ed.). *Potato in Developing countries. Indian Potato Assoc.* CPRI, Shimla. 338-347.

Small, T. 1945. Black scurf and stem-cauker of potato (*Corticum solani*). Field studies on the use of clean and contaminated seed potatoes and on the contamination of crop tubers. *Ann. appl. Biol.* 32 : 206-209.

Smith, E.F. and **Swingle, D.B.** 1904. The dry rot of potatoes due to *Fusarium oxysporum. U.S. Dept. Agric. Bur. Plant Indus. Bull.* 55.

Somani, A.K. 1986-a Non-hazardous chemical control of black-scurf of potato. *Indian J. Agric. Sci.* 56 : 366-369.

Somani, A.K. 1986-b. *Pythium butleri* causing tuber rot of potato. *Indian Phytopath.* 39 :529-531.

Somani, A.K. 1988-a. Control of black scurf (*Rhizoctonia solani*) and common scab (*Streptomyces scabies*) of potato(Solanum tuberosum) with boric acid. *Indian J. Agric. Sci.* 58 : 693-698.

Somani, A.K. 1988-b. Potato seed and soil health. *Seeds and Farms.* 14 (9) : 28-32.

Somani, A.K. 1990. Tuber injury due to organomercurial compound and standardization of tuber treatment. *J. Indian Potato Assoc.* 17 : 181-191.

Somani, A.K. 1993. Epidemiology and physico-chemical studies on tuber rots of potato. Ph.D. Thesis, 1993, School of Studies in Botany, Jiwaji University, Gwalior, Madhya Pradesh. 213 p.

Somani, A.K. 2004. Potato tuber rots and associated incitants. *J. Indian Potato Assoc.* 31(3-4).

Somani, A.K. and *Shekhawat, G.S.* 1985. Evaluation of chemicals to prevent soft rot of potato. *J. Indian Potato Assoc.* 12 : 214-216.

Somani, A.K. and *Shekhawat, G.S.* 1988. Influence of levels and sources of nitrogen on the development of bacterial soft rot in potato tubers. *Indian Phytopath* 41 : 238-240.

Somani, A.K. and *Shekhawat, G.S.* 1990. Bacterial soft rot of potato in India. *Tech. Bull.* No.21. Central Potato Research Institute, Shimla. 25.

Somani, A.K. and *Chauhan, R.K.S.* 1994. Herbicide paraquat spray as a predisposing factor for dry rot development in freshly harvested potatoes. *In : Potato Present and Future.* Proceedings of the National Symposium held at Modipuram March 1-3, 1993 (eds. G.S. Shekhawat *et al.*) Indian Potato Association, Shimla, 219-220.

Somani, A.K. and *Chauhan, R.K.S.* 1996. Potato tuber rots in Gwalior, Madhya Pradesh. *J. Indian Potato Assoc.* 23 : 144-148.

Somani, A.K. and *Chauhan, R.K.S.* 2000. Correlation studies on potato tuber rots with storage conditions and tuber compostion. *In : Potato Global Research & Development Vol.I* (Eds. S.M. Paul Khurana et al.*)* pp.466-471.

Somani, A.K., Sharma, V.C. and *Shekhawat, G.S.* 1988. Influence of date of harvesting of potato on the development of soft rot. *Indian Phytopath.* 41 : 517-520.

Srivastava, S.N.S. 1963. Out breaks and new records. *Pl. Prot. Bull. FAO* 11 : 69.

Srivastava, S.N.S. 1965. The occurrence of silver scurf of potato in India. *Sci. Cult.* 31 : 537.

Sukumaran, N.P., Kaul, H.N., Uppal, D.S. and *Grewal, S.S.* 1975. Effect of post harvest greening of potato on their chlorophyll and glycoalkaloid contents and keeping quality during storage at room temperature. *J. Indian Potato Assoc.* 2 : 34-37.

Suresh, K. and *Mall, S.* 1982. Anastomosis groups of potato isolates of Rhizoctonia solani in India. *Mycologia* 74 : 337-338.

Thakur, D.P. and **Chenulu, V.V.** 1970. Survey of Delhi Market for fungal diseases of fruits and vegetables. *In : Plant Disease Problems* (Eds. Raychaudhury, S.P. et al). *Indian Phytopathological Society*. New Delhi, 308-315.

Thirumalachar, M.J. 1953. R. solani infection of potato tuber in India. *Phytopathology* 43 : 645-647.

Thirumalachar, M.J. 1955. Incidence of charcoal rot of potato in Bihar(India) in relation to cultural conditions. *Phytopathology* 45 : 91-93.

Thirumalachar, M.J. and **Ramamurthi G.S**. 1952. Dry rot and sprout injury in potato tubers by species of *Aspergillus. Proc. Indian Sci. Confr.* IV 39 : 38.

Thirumalachar, M.J. and **Narsimhan, M.J.** 1970. Greening of seed potato tubers and its effect on resistance to fungal diseases during storage. 1st *Intn. Symp. Pl. Pathol*. Indian Phytopathol. Society, New Delhi. 879-881.

Thirumalachar, M.J. and **O'Brien.** 1977. Suppression of charcoal rot in potato with a bacterial antagonist. *Plant Dis. Rept.* 61 : 543-546.

Thomas, P., Srirangarajan, A.N., Padwal – Desai S.R., Ghanekar, A.S., Shrisat, S.G., Pendharkar M.B., Nair, P.M. and **Nandkarni, G.B.** 1978. Feasibility of radiation processing for post-harvest storage of potatoes under tropical conditions. *In : Food Preservation by Irradiation*, vol.1, IAEA, Vienna. And BARC, Bombay 1978, 71-82

Trehan, S.P. and **Grewal, J.S.** 1980. Cultural practices for the control of potato common scab (*Streptomyces scabies*) - A review. *Agric. Reviews.* 1 : 103-113.

Tripathi, R.K. 1979. Mechanism of resistance in potato tubers to bacterial soft rot (*Erwinia carotovora var. carotovora*). *In : Symp. Post Harvest Tech. Util. Potato*, Shimla & New Delhi. Aug-Sept. 1979.

Upadhya, M.D. 1976. Breeding potato varieties resistant to charcoal rot. Final Tech. Rept of a PL-480. Project. Central Potato Research Institute, Shimla 31.

Uppal, B.N. 1935. Annu. Rept. Agric. Bombay Presidency 1933-34, 174.

Vashisth, K.S. and **Chaubey, I.P.** 1975. Problem of common scab disease of potato in Punjab. *The Punjab Grower* 10 : 35-38.

Vashistha, K.S., Sharma, V.C. and **Somani, A.K.** 1990. Sources of resistance in andigena against common scab disease of potato. *Asian Potato J.* 1 : 20-21.

<table><tr><td>**21**</td><td>

*Integrated Management of Sorghum Diseases in India**

</td></tr></table>

T.G. Nageshwar Rao

National Research Centre for Sorghum, Rajendranagar, Hyderabad-500 030

tgnrao@rediffmail.com

ABSTRACT

Sorghum diseases can be broadly divided into four categories based on plant part infected viz., seed and seedling diseases, panicle diseases, stem and root diseases and foliar diseases. The genotypes found resistant to a particular disease at one location are susceptible at other locations. The sorghum diseases can be managed in an integrated manners by cultural practices viz., adjustment of date of sowing, crop rotation, destruction of wild sorghum, harvesting at physicalogical maturity, deep ploughing, mechanical uprooting of downy mildew infected and viral infected plants as and when they appear are observed and burning them, balanced fertilization, maintaining optimum plant population, moisture management, avoidance of collateral host; chemical management; genetic improvement by host plant resistance and integrated disease management.

KEYWORDS : Sorghum diseases, management, cultural practices, chemical management, genetic improvement.

Sorghum (*Sorghum bicolor* L. Moench) is grown as food, feed and fodder crop in India over an area of 10.2 million ha with a production of 9 million tones and a productivity of 833 kg/ha. The crop is grown mostly in Maharashtra, Karnataka, Andhra Pradesh and Madhya Pradesh states. The crop is also grown in other states like Uttar Pradesh., Rajasthan, Gujarat and Tamil Nadu to a lesser extent mostly for fodder purposes. Sorghum is also used to produce several industrial products like sugar, starch, syrup, alcohol, besides confectionary items like biscuits and cakes. The crop is affected by number of diseases, but only a dozen of them are very serious With the change in climate, change in cropping pattern and with the introduction of high yielding varieties and hybrids of sorghum for cultivation, the disease scenario has also changed. The diseases which were hitherto considered to be of minor in importance have assumed as major diseases. With the introduction of high yielding dwarf photosensitive hybrids, the grain molds have assumed serious now as the crop at flowering is caught in late rains Downy mildew of sorghum caused by *Peronosclerospora sorghi*, (Weston and Uppal) Shaw is potentially destructive disease in Karnataka, Andhra Pradesh and Tamil nadu states. During rabi season, charcoal rot incited by *Macrophomina phaseolina* a soil borne disease is most important one as rabi crop is raised with the residual soil moisture received in kharif season.

The crop at maturity will be subjected to soil moisture depletion coupled with high temperatures which leads to the development of charcoal rot. Among fodder sorghum, foliar diseases viz., Leaf blight, zonate leaf spot, sooty stripe, rust, anthracnose, rough leaf spot are very important. The minor diseases on panicle are smuts which occur occasionally. The smuts being seed borne diseases can be managed very well by seed treatment. The diseases of sorghum, their causal organisms and occurrence in different parts of the country are presented in Table 1. Sorghum diseases can broadly be divided into four categories based on the plant part infected viz., seed and seedling diseases, panicle diseases, stem and root diseases and foliar diseases. The diseases which are found to have economic significance are presented here.

Table 1 Occurrence and distribution of Diseases of Sorghum in different regions of India

Part affected	Disease	Casual Organism	Occurrence
	Major		
Panicle Diseases	Grain Molds	*Fusarium moniliforme Curvularia lunata* J.Sheld Lisea Fujikuroi Sawada, (Wakker) Boedijn, and Alternaria alternata (Fr). Keissler, *Phoma sorghina* (Sacc.) Boerema, Dorenbosch, & Van Kesteren,	Parts of Andhra Pradesh, Karnataka, Maharashtra Tamil Nadu and Gujarat
	Smuts		
	Grain smut	*Sporisorium sorghi* Link in Willd	Sporadic in different regions
	Loose smut	*Sphacelotheca cruenta* (Kuhn) K.Vanky	Parts of Maharashtra and Karnataka
	Head smut	*Sporisorium reliana* (kuhn) G.P.Clinton	Parts of Maharashtra and Karnataka Andhra Pradesh and Tamil Nadu
	Long smut	*Tolyposporium ehrenbergii*	Sporadic in Rajasthan, Tamil Nadu, Karnataka, Andhra Pradesh
Foliar Diseases	Downy mildew	*Perenosclerospora sorghi* Weston & Uppal	Parts of Karnataka, Tamil Nadu, Maharashtra, Andhra Pradesh, Sporadic in Madhya Pradesh and Uttar Pradesh
	Leaf blight	*Exerohilum turcicum* (Pass.) K.J.Leonard and E.G.Suggs	Parts of Madhya Pradesh, Rajasthan, Tamil Nadu, Karnataka, and A.P
	Rust	*Puccinia purpurea* Cooke	All sorghum growing regions in India
	Sooty stripe	*Ramulispora sorghi* (Ellis & Everh.) Olive & Lefebvre in Olive et al.	Parts of Maharashtra, Andhra Pradesh, Rajasthan and Karnataka
	Anthracnose	*Colletotrichum graminicola* (Ces.) G.W.Wils.	Parts of Karnataka, Madhya Pradesh, Uttar Pradesh, Andhra Pradesh, Tamil Nadu, and Rajasthan

Table 1 Contd....

Part affected	Disease	Casual Organism	Occurrence
	Major		
	Rough leaf spot	*Ascochyta sorghina* Sacc.	Parts of Andhra Pradesh and Madhya Pradesh
	Gray leaf spot	*Curvularia sorghi* Ellis & Everh.	Parts of Karnataka, Madhya Pradesh, Uttar Pradesh, Andhra Pradesh, Tamil Nadu, and Rajasthan
	Tar spot	*Phyllachora saccari* P.Henn.	Parts of Karnataka and Tamil Nadu
	Sorghum red stripe	Virus (SCMV)	Parts of Karnataka, Madhya Pradesh, Maharashtra and Andhra Pradesh
	Rough leaf spot	*Ascochyta sorghina* Sacc.	Parts of Andhra Pradesh and Madhya Pradesh
	Sorghum red stripe	Virus (SCMV)	Parts of Karnataka, Madhya Pradesh, Maharashtra and Andhra Pradesh
	Anthracnose	*Colletotrichum graminicola* (Ces.) G.W.Wils.	Parts of Karnataka, M.P., UP, Andhra Pradesh, Tamil Nadu, and Rajasthan
	Rough leaf spot	*Ascochyta sorghina* Sacc.	Parts of Andhra Pradesh and Madhya Pradesh
	Gray leaf spot	*Curvularia sorghi* Ellis & Everh.	Parts of Karnataka, Madhya Pradesh, Uttar Pradesh, Andhra Pradesh, Tamil Nadu, and Rajasthan
	Tar spot	*Phyllachora saccari* P.Henn.	Parts of Karnataka and Tamil Nadu
	Sorghum red stripe	Virus (SCMV)	Parts of Karnataka, Madhya Pradesh, Maharashtra and Andhra Pradesh
Stem and Root Diseases	Fusarium root & stalk rot	*F.moniliforme* J.Sheld.	Maharashtra, Karnataka and Andhra Pradesh
	Charcoal rot	*Macrophomina phaseolina* (Tassi) Goidanich	Maharashtra, Karnataka and Andhra Pradesh
	Wilt	*Acremonium strictnum* W.Gams *Cephalosporium acremonium* Auct. Non Corda	Parts of Nothern India
	Crazy top	*Scleropthora macerospora* (Sacc.) Thirumalachar *et al.* *Scleropthora macerospora* Sacc.	Maharashtra and Karnataka

Disease Survey and Surveillance

Disease survey and surveillance is the most important activity that monitors the incidence and severity of various diseases in different locations so that this information will be utilized to advocate proper control measures and also for the researchers to concentrate more on their efforts to take up research programmes in that direction. In India sorghum disease surveys are being taken up in Andhra Pradesh, Maharashtra, Karnataka, Rajasthan, Gujarat, Tamilnadu and Uttaranchal states regularly and the disease situation is being constantly monitored. This also reveals the information on variability of the pathogenic populations in different agro climatic zones of India.

Variabilility in Pathogens

The genotypes found resistant to a particular disease at one location are susceptible at other locations. This is due to the existence of variability in the pathogens. Pathogenic variability has been demonstrated very well in sorghum anthracnose caused by *Colletotrichum graminicola*, downy mildew (*Peronosclerospora sorghi* and leaf blight (*Exserohilum turcicum*). The isolates of *C. graminicola* exhibit wide variability in morphology, pathogenicity (Mathur *et al* 2001; Latha *et al.*, 2002, Thakur *et al.*, 1998) Ali and Warren (1992). Rao *et al.*, 1998 grouped all available isolates into six groups based on pathotype. Cultural , morphological, and pathogenic variability among isolates of grain mold pathogens viz., *Fusarium moniliforme, Curvularria lunata and Phoma sorghina* has been demonstrated. (Bhaskar, 2004)

Seed and Seedling Diseases

The death of sorghum in the first week of planting is a problem in some of the sorghum growing areas. Seed may be attacked by one or more seed borne or soil borne pathogens either prior to germination or emergence. This usually occurs when conditions are not optimum for plant development, such as in poorly-drained, cold, wet soils, or in very dry, crusted soils. Ironically, these conditions are generally favorable both for pathogen activity and for disease development. Sorghum seedlings are very delicate during the emergence period and are slow to establish a permanent root system. As a result, sorghum depends upon its primary, temporary root system for a period longer than many other crops. Under less than optimal growing conditions, this primary root system is extremely vulnerable to soil borne pathogens. This is when damping-off in a field is most visible and damaging. The deteriorating tissue may appear anywhere from whitish-gray in color to pink to dark brown. In some instances, especially after emergence in cooler soils, plants may show a purplish coloration of the leaves. This problem is generally nutritional in nature and is caused by a temporary phosphorus deficiency, rather than attack by a soil pathogen. Both seed and seedling diseases can greatly reduce stand densities in the field. This failure to get good stands is frequently the reason for excessively high planting rates in subsequent seasons. High planting rates may also cause additional stress and stress-related diseases at later stages of growth. Seed and seedling diseases can be managed by using disease free high quality seed, use adequate seed rate with proper spacing, plant seed when soil is warm, where moisture is not a limiting factor, treat the seed with seed dressers, avoid planting in low areas, or in areas with poor drainage.

Stalk Rots

Stalk rots are common in rabi sorghums and generally follow root rot or certain combinations of specific environmental and plant developmental sequences which predispose plants to be attacked by stalk rotting organisms. Since the rabi sorghum is grown purely on residual soil moisture conditions the crop at maturity stage is subjected to severe moisture stress conditions. Stalk rots can greatly reduce yield either directly through plant lodging, or indirectly through reduced head and grain fill. Charcoal rot is the most frequently encountered stalk rot disease of sorghum in India. Charcoal rot is incited by a soil borne fungus, *Macrophomina phaseolina*, (Tassi) Goid present wherever sorghum is grown. The disease is more severe especially in hybrids compared to straight varieties and in shallow soils than in deep soils. The same organism causes a similar disease in corn and soybean. The disease usually occurs during seed development as the crop is subjected to severe soil moisture stress coupled with high temperatures. As a result, the disease is more widespread in some years and localities than in others. Since water stress is important for disease development, the first plants affected by charcoal rot will be those growing in soils that do not retain water well. In addition, pre bloom rainfall, tillage method, cropping sequence, plant spacing and soil fertility also influence patterns of charcoal rot in the field. Under optimal conditions, severe damage in a field can occur in a few days. If water stress conditions are interrupted, however, the disease process may be retarded or stopped entirely. Splitting the stalks of suspect plants lengthwise easily identifies charcoal rot. If the disease is present, the stalk's interior will be shredded, and very small, dark, fungal bodies will be visible. These fungal structures give the split stalk a charcoal dusted appearance and hence the disease is thus named as charcoal rot. The most practical way to control charcoal rot is to grow grain sorghum varieties which are tolerant to charcoal rot and/or predisposing stress factors. No resistant varieties are available.

Panicle Diseases

Grain molds, ergot and smuts are the panicle diseases of which grain mold is a major disease followed by ergot. Smuts occur very rarely and are easily manageable being seed borne diseases with seed dressing fungicides.

Grain Mold

In India, grain mold is a major disease of sorghum grown in Kharif season especially in the Southern India. The disease develops when the crop at flowering to maturity stages coincides with high rainfall received during late stage of the crop. These late rains develop high humid conditions which is congenial for the increase in the disease severity. Grain mold has become a major and widespread disease problem because it not only reduces yield but also reduces the quality of grain by way of discoloration and production of several mycotoxins like fumanosins which makes it unfit for human/ pultry consumption.

Grain mold is a complex disease as it is caused by number of fungal species. The infection results in moldy growth on the grains frequently referred as 'blackening'. Several fungi are implicated in grain molding and some of these produce harmful mycotoxins (Bhat et. al., 2000). Furthermore, molded grains are sold in the markets at a discounted price up to 40 % compared to healthy grains. More than 40 fungal genera are associated with molds, but only few of them will infect flower tissues in the erly stages of grain development (Bandyopadhyay,2000) The principal grain mold fungi in India include *Fusarium moniliforme* J.Sheld Lisea Fujikuroi Sawada, *Curvularia lunata* (Wakker) Boedijn, and *Alternaria alternata* (Fr). Keissler, *Phoma sorghina* (Sacc.) Boerema, Dorenbosch, & Van Kesteren, and Aspergillus spp. (*A.niger* Van Tieghem, *A.flavus* Link). Among these *Fusarium moniliforme, Curvularia lunata and Phoma sorghina are* the most common fungi which occur very frequently in the sorghum growing areas. The research work carried on grain molds for the last twenty years is reviewed by Chandrashekar et. al., 2000. Management of Grain molds is very difficult in the filed level as the crop is grown mostly in the poor and marginal soils with meager inputs. Chemical sprays to the crop at maturity stage is practically not feasible as continuous rains will not allow for spraying of the chemicals. Growing mold resistant cultivars is one of the best and most reliable and practically feasible solutions over different environments, and is effective and economical to farmers. This method requires availability of germplasm, resistance screening techniques, and strong breeding programme to transfer disease resistance into high yielding cultivars. However in case of mold, deriving disease resistance is very difficult proposition as it involves all facultative parasites and a complex of organisms are involved.

In the last 25 years there has been a great deal of research directed towards determining resistance mechanisms. A pigmented test, where condensed tannins are present, is the most important trait conferring grain mold resistance. Red pericarp, where flavan-4-ols are located, also confers resistance to grain mold, but not as strongly as pigmented testa. Pigmented testa and red peicarp, when combined provide additive effects on resistance (Bandyopadhyay et. al., 2000). However not all sorghum lines with red pericarp are resistant to grain mold. The association of flavan-4-ols and grain mold resistance has been demonstrated in Asia (Jambunathan et al., 1990). Direct relationship between grain hardness and mold resistance has been shown (Jambunatham et. al., 1991). Seed characters such as hard seed, low percent floaters, long glumes (Indira and Rao, 1997) and surface wax on the seed (Somani and Indira, 1997) are utilized as criteria in selecting for resistance to grain mold.

Late infection by mold fungi is often restricted to the pericarp with little internal colonization of endosperm, such grain superficially appear moldy (discolored) and fetch low price in the market. If the pericarp is removed, such grains appear almost as clean as non-moldy grains. Removal of pericarp is possible by several means including mechanical dehulling and using seed pearler. This approach is yet to be tested both for technical and socio-economical aspects and it may lead to reduction of mycotoxin contamination. Commercialization of grain processing if found feasible can open opportunities for use of high level of mold resistance found in red sorghum with colored pericarp but without testa layer that contains anti nutritional condensed tannins.

Ergot (Sugary disease)

Sorghum ergot has been reported from many sorghum growing countries (Tarr, 1962). The disease severity has been increasing in India with the introduction of short duration high yielding hybrids with male sterile lines. The disease is severe due to non synchronization of male and female flowers in the hybrid seed production plots. In some cases seed setting is completely destroyed due to severity of infection in male sterile lines. In 1975, extensive areas in seed production were severely damaged by this disease in Maharashtra (Satara district), Madhya Pradesh, Gujrat and Andhra Pradesh states. The severity of the disease depends mostly upon weather conditions at flowering time. Rapid spread of the disease takes place when the crop at anthesis coincides with warm temperature (19 °C-21°C) coupled with high degree of relative humidity and cloudy weather. The dry weather conditions prevail after the honey dew stage, favors sclerotial development The disease is caused by *Sphacelia sorghi* Mac Rac (Imperfect stage) *Claviceps sorghi* Kulkarni *et al* (Perfect stage).

Foliar Diseases Caused by Fungi

Numerous fungi can attack the foliage of grain as well as forage sorghums at various stages of crop growth. Severe damage caused by these fungi is most likely to occur under periods of extended humidity. Fortunately, grain sorghum can sustain considerable damage to its foliage without incurring significant yield losses. However, yield loss can occur if damage occurs to the upper leaves of plants at the time the grain is filling. The congenial conditions for the development of fall the foliar diseases include warm temperature coupled with high relative humidity. During most years, disease development occurs from the whorl stage through maturity. Important foliar diseases that affect sorghum include downy mildew, anthracnose, leaf blight, rust, zonate leaf spot, sooty stripe, rust, rough leaf spot. The foliar diseases not only reduce the yields by way of interfering in photosynthetic efficiency, but also reduce the quality of fodder considerably

Downy Mildew

Downy mildew was first reported in India by Butler, 1907. Under high humid conditions, the disease develops and becomes a limiting factor in cultivation of susceptible yet promising variety. The most conspicuous symptom is the appearance of vivid green and white stripes on the leaves. Whole leaves may become chlorotic at advance stage and plants usually fail to head, even if heads are exerted they are small compact or club shaped and have little or no seed set. In severe cases losses may be upto 100 % or in epidemic form the disease causes more than 20 % yield losses (Anahosur and Patil, 1983). The disease is caused by *Peronosclerospora sorghi* Weston & Uppal = *Sclerospora sorghi* Weston & Uppal.

Anthracnose

Sorghum anthracnose is economically important in the areas where sorghum is grown under warm and humid environmental conditions. The disease can limit grain and fodder sorghum production in most of the regions of the world. A reduction in grain yield due to this disease is estimated to exceed 50 % or more on susceptible cultivars in severe epidemics. The pathogen attacks mainly on the foliage, but also on grains, peduncle and stems. The foliar phase is most destructive on forage sorghum and causes substantial losses to fodder quality in north- west plain zones. The pathogen *Colletotrichum graminicola (Ces.) Wils (P.Hem)* causes small circular, red lesions with a distinct margin may appear on susceptible leaves in midsummer and cause defoliation. In severe cases, plants may die before maturation. These lesions can enlarge slightly in humid weather. Tan spots with red-purple margins may develop on upper portions of stalks. The inside of these stalks will be brick red interspersed with the normal white colour of healthy tissue. Diseased stalks lodge at the point of infection, and yields may be reduced. The pathogen survives as sclerotia in the sorghum stalk residues (Casela and Fredericksen, 1993)

Root Diseases

Striga and its management

One of the most common and severe root disease is caused by two phanerogamic parasite *Striga asiatica* Kuntze and *S. densiflora* Benth. Striga occurs in India, Burma and several parts of Africa. These are the root parasites and their presence becomes apparent only when they appear above ground level, by which time the host plants are well grown. The leafy shoots emerge around the plant, one or more shoots to each plant. Since the parasite forms well developed leaves and synthesize carbohydrates, it is classified as a partial parasite, but when there is heavy incidence, the host plants are stunted and the grain yield is reduced. The parasite blossoms and fruits within 3-4 weeks of emergence, producing several hundred thousands of minute seeds per plant. The seeds are dispersed by wind, finally settling on the soil, where they germinate through the influence of root exudates from the host plant without which stimulant they remain dormant for many years. On germination they infect the host roots, establish complete contact as a parasite and develop to produce shoots. Striga can be managed by growing resistant cultivars, pulling out striga shoots before they set seeds and destroying them by burning, application of weedicide like 2,4-D.

Viral Diseases

Among viral diseases, chlorotic stripe stunt disease caused by maize stripe virus is important which reduces grain yield, plant height, internodal length considerably (Revuru & Garud 1998, Narayana and Muniyappa, 1995 a&b)

MANAGEMENT OPTIONS

Cultural Management

Cultural management of diseases like adjusting dates of sowing, crop rotation, balanced fertilization, maintaining optimum plant population, destroying collateral hosts, deep summer ploughings, optimum moisture management, harvesting at physiological maturity of the crop is essential to reduce the initial inoculum of the pathogen present on plant debris or on seed.

Adjustment of date of sowing to avoid wet weather for the development of serious epidemic of grain mold pathogens and early sowings in rabi sorghum will reduce the incidence of charcoal rot as the crop at terminal stage can escape moisture stress and high temperatures. Early sowing in Maharashtra and Gujrat avoided the ergot disease (Sangitrao *et. al.*, 1979). Early planting escapes the disease (Sundaram *et. al.*, 1973).

Crop rotation with a species other than sorghum for at least two to three years and burial of crop residues with mold board plough helps in decomposing the inoculum from debris in the absence of sorghum (Frederiksen, 1986).

Destruction of wild sorghum on which *C. graminicola* can survive during the off season also offer reasonable control of the disease and also elimination of other susceptible plants such as Johnson grass.

Harvesting at physiological maturity and artificial drying of seeds to reduce the grain moisture content to <10% reduces grain mold, enhances seed germination, and improves grain quality by escaping the late infection of grain mold pathogen (Indira, et. al., 1997). Sorghum grains harvested under adverse wet conditions needs to be dried immediately otherwise all saprophytic fungi will grow on the produce and deteriorates the grain quality considerably. Drying can be done under sun or in the community dryers if available.

Deep ploughing especially in summer months helps to expose the resting structures of the pathogen like sclerotia and oospore and ultimately its death (Singh, 1978).

Mechanical uprooting of downy mildew infected plants before leaf shredding and viral infected plants as and when they are observed and burning them helps in reducing disease (Ranagaswami, 1988).

Balanced Fertilization is to be applied. Nitrogenous fertilizers have profound influence on the incidence of several sorghum diseases: The least incidence of molds and maximum germination was recorded at 80 kgs of nitrogen application /ha (Anahosur and Reddy,1976). Indira and Rana (1980) observed significant differences on the incidence of head molds. The incidence was lowest with 120 kg of nitrogen/ ha. Potassium on the other hand contributes for resistance to diseases.

Maintaining optimum plant population by adopting proper spacing is essential as excessive plant populations; deplete soil moisture which enhances the diseases like charcoal rot.

Tillage and mulching systems done to reduce soil moisture loss will be free from infection of charcoal rot pathogen Soil mulching with wheat straw, 3 hoeings to conserve soil moisture.

Moisture management is the most important aspect in the management of sorghum diseases like charcoal rot especially at critical periods of crop growth,. The disease could be managed well provided such irrigation facilities are available.

Avoidance of collateral host plants: *Pennnisetum typhoides, P. americanum, Zea mays, Panicum maximum, Ischaemum pilosum* are collateral hosts of this disease in India. Of these *I. pilosum* is very common in Vidarbha region of Maharashtra state and seems to be the chief source of the pathogen. Avoidance of such collateral host plants on the field bunds is useful to check disease spread. Removal of collateral hosts in and around fields, rouging of infected plants is cultural method to control ergot (Sundaram. 1968).

Chemical Management

Chemical management can be done as prophylactic as well as curative. The former one involves seed dressing with fungicides before sowing which protect the seeds from seed borne diseases and to enhance seedling vigour and viability, while the latter involves spraying with fungicides in the standing crop, which reduces the incidence of the diseases. This can be done depending on the severity of the disease and also the economic gain which we get by managing the diseases: In India chemical management in an established crop is not a common practice in sorghum production as the crop is mostly grown mostly on marginal lands with minimum inputs. Combination of Captan 0.2% +Aureofungin (200ppm) is effective in the management of grain molds. First spray is to be given immediately after grain maturity and second one at 15 days after gives effective control. Two sprays of Tilt 25% EC, a triazole compound (Propiconazole) @ 0.25 ml also reduces grain molds as well as ergot and increases yields considerably., Downy mildew can be managed by seed dressing with metalaxuyl followed by sprays with fungicides.i.e ridomil 25 as seed dressing or apron 35 1g a.i./kg at 7 days interval beginning from germination. Foliar diseases can be managed well by fungicides. However, seed dressing with thiram can reduce the seed borne inoculum of *C. graminicola.* (Sharma 1980; Anahosur 1992). Spray of 0.3% Captan or Thiram / dithane M-45 on 30 days old crop is essential for the management of foliar diseases. The seeds are to be treated with 30 % salt solution to remove ergot sclerotia (Bhagwat and Pedalonkar, 1973). Systemic fungicides like Thiophanate (Topsin) have shown promise to control ergot, but it was considered not feasible on practical scale except in seed production plots of high commercial value (Gangadharan *et al.*, 1976; Anahosur, 1979). Two sprays of Benlate 0.1 % at 50 % flowering stage and later after a fortnight is more effective.

Biological Management

Trichoderma viride and *Trichoderma harzianum* have been identified as potential biocontrol agents for the management of soil borne diseases like charcoal rots, The efficacy of these bio agents has been amply proved in invitro conditions but its utility in the field conditions is in the infancy stages. Seed treatment with biocontrol agents viz., *Trichoderma viride, T. harzianum, Bacillus subtilis, Pseudumonas flourescence*, was found to be effective for the management of grain mold (AICSIP, Annual report 1998). Molded grains are treated with 25% crude garlic extract to reduce mold contamination(Singh and Navi 2000)

Genetic Improvement by Host plant resistance

Efforts are being made to identify potential sources of resistance to grain molds, charcoal rot and foliar diseases and to incorporate these sources of resistance to the high yielding background. Monogenic and polygenic resistance to anthracnose pathogen have been identified and is used individually or in combination to provide adequate disease control Green house and field screening techniques to identify resistance to anthracnose disease are available (Pande *et al.*, 1993 and 1994). Using these techniques, source of resistance to anthracnose have been identified and successfully deployed in the several national and international sorghum improvement programmes (Thakur and Pande, 1995). The sources of resistance to downy mildew have been identified in several genotypes and some have shown stable resistance (Shivanna and Anahosur, 1990; Anahosur *et al.*, 1990).. A selection from Nandyal in Andhra Pradesh DMRS-1 IS 2702) was developed and found to posses possible stable resistance to downy mildew (Anahosur *et al.*, 1984). From International Sorghum Downy mildew Nursery (ISDMN) the stable resistant genotypes are IS 3443, 2702, 22230, 22231 and 18757 these were completely free from infection. For the typical rabi tracts in Maharashtra, Karnataka and Andhra Pradesh, hybrids CSH 8R, CSV 8R, and CSH 13 R are suitable. Two hybrids SPH 221 and CSH 12R are identified especially suitable for Maharashtra and Karnataka. A new promising hybrid CSH 13 (K & R) with good grain and fodder yields established its superiority over M 35-1, Swati and CSV 8R tolerent to drought, lodging and was released as CSH 13R for All India cultivation. Similarly SPV 913 was released in Andhra Pradesh as NTJ -2. GRS 1 - a new rabi sorghum variety, developed as an alternative to M35-1, has resistance to charcoal rot (*Macrophomina phaseolina*) and is superior in grain and fodder yield. Genotypes that exhibited < 30% infection under artificial inoculation such as CS 3541, IS 3443, IS 3547, IS 14332 and CSH 5, 6 and 9 have field tolerance (Anahosur. 1990). Genotypes Co 25, TNS 23, TNS 28 showed multiple resistances including sugary disease (Laxmanan *et al.*, 1990). Anahosur and Laxmanan (1981) identified multiple resistance in IS 8283, Sb 1085, CSV 4, IS 7528 and IS 8185 to grain mold, downy mildew, rust, anthracnose, zonate leaf spot, ergot and charcoal rot.

Integrated Disease Management

The Integrated Disease Management of sorghum involves combination of host plant resistance in conjunction with cultural practices, chemical control such as use of fungicides for disease management program, particularly in developing countries. Sources of resistance to anthracnose have been identified. Efforts have been made to incorporate resistance into plants with good agronomic background and to strengthen the levels of resistance. Resistance to foliar and grain phase of the disease along with yield should be given equal consideration in the development and release of new cultivars. Resistant forage sorghum cultivars may be developed for leaf anthracnose endemic areas especially for resource poor farmers. Disease avoidance through timely planting as decided by crop history, disease epidemiology and rainfall patterns, balanced fertilizer application, clean cultivation, and proper crop combination should be popularized among the farmers to reduce losses. Readily available and inexpensive fungicides may be used along with partially resistant high yielding cultivars as and when necessary based on economic threshold. A combination of fungicides and insecticides seed dressing should be advocated to the farmers along with the importance of planting healthy seeds to minimize seed and seedling anthracnose. Early sowing, growing resistant cultivars, synchrony of flowering, Fungicidal control, coupled with eradication of collateral hosts, will provide effective control of this disease

Conclusions and Future Strategies

Wide range of diversity in the pathogen populations makes it difficult to breed for durable and stable disease resistance in sorghum for deploying resistant cultivars. Conventional breeding takes longer time for transferring resistance in to agronomically sound background. Hence there is a need to utilize biotechnological methods for deriving disease resistance. There is a need to monitor pathogenic populations constantly by disease surveys and surveillance in different agro climatic zones of the country so that we can deploy the resistant cultivars based on the existence of the pathogenic race or pathotypes in a given locality. Identification of multiple disease resistance lines plays a significant role in identifying genes controlling resistance for various diseases. Greater emphasis should also be given to understand host pathogen system at micro level and mechanism of resistance which can be used as selection criteria for breeding for resistance to diseases which ultimately utilized in the development of effective integrated disease management strategies.

References

Ahmed, R. and **Majumder, S.K.** 1992. Host factor in the production of conidia and oospores of *Peronosclerospora sorghi*. *Indian Phytopath*. 45(3) :369-370.

Ali, MEK and **Warren, H.L.** 1987. Physiological races of *Colletrotrichum graminicola* on *sorghum*. *Plant Dis*. 71: 402-04

Ali, MEK., Warren, H.L and **Latin, R.X.** 1987. Relationship between anthracnose leaf blight and losses in grain yield of sorghum . *Plant Dis*. 71:803-806.

All India coordinated Sorghum improvement Project, Annual Report, National Research Centre for Sorghum, Hyderabad

Anahosur, K.H. and **Laxman, M.** 1991 Estimation of loss in grain yield in sorghum genotypes due to downy mildew. *Indian Phytopath*. 44:520-522.

Anwar, M.N., Majumder, S.K. and ***Shetty, H.S.*** 1995. Changes in phenolic acids in sorghum and maize leaves infected with *Peronosclerospora sorghi, Indian Phytopath.* 48(1):21-26

Bandyopadhyay, R., Butler, D.R., Chandrashekar, A., Reddy, R.K. and ***Navi, S.S.*** 2000. Biology, epidemiology and management of sorghum grain mold. Pages 34-71 In Technical and Institutional options for the management of sorghum grain mold management: proceedings of an international consultation, 18-19 May 2000, ICRISAT, Patancheru India (eds Chandrashekar,A., Bandyopadhyay,R and Hall,AJ) Patancheru, Andhra Pradesh.

Bandyopadhyay, R., Mughogho, L.K and ***Satyanarayana, M.V.*** 1991. Occurrence of airborne spores of fungi causing grain mould over a sorghum crop. *Mycol.Res.* 95(11):1315-1320

Bhaskar Rao, I. 2004. *Variability of major grain mould fungi of sorghum, Sorghum bicolor (L) Moench.* M.Sc.(Ag) thesis submitted to Acharya NG Ranga Agricultural University, Hyderabad. pp 107.

Bhat, R.V., Shetty, H.P.K. and ***Vasanthi, S.*** 2000. Human and animal health significance of mycotoxin in sorghum with special reference to fumanosins. Pages 107-115 In: *Technical and Institutional options for the management of sorghum grain mold management:* proceedings of an international consultation, 18-19 May 2000, ICRISAT, Patancheru India (eds Chandrashekar,A., Bandyopadhyay,R and Hall,AJ) Patancheru,AP.

Cardwell, K.F., Hepperly, P.R. and ***Frederiksen, R.A.*** 1989. Pathotypes of *Colletotrichum graminicola* and seed transmission of sorghum anthracnose 73: 255-257

Casela, C.R. and ***Frederiksen, R.A.*** 1993. Survival of *Colletotrichum graminicoala* sclerotia in sorghum stalk residues. *Plant Dis.* 77:825-827

Chandrashekar, A., Bandypopadhyay, R. and ***Hall, A.J***. (eds).2000. Technical and institutional options for sorghum grain mold management. Proceedings of an international consultation, 18-19 May 2000, ICRISAT, Patancheru, India. 299 pp

Duncan, R.R. and ***Milliano, WAJ de*** Plant Disease control in sorghum and pearl millet.

Fredericksenj, R.A. 1986. *Compedium of sorghum Diseases.* American Phytopathological Society. St. Paul. Minnesota.

Frederiksen, R.A. and ***Renfro, B.L***. 1977. Global status of maize downy mildew. *Annual Review of Phytopathology* 15: 249-275

Frederiksen, R.A. Thomas, M.D., Bandopadhyay, R. and *Mughogho, L.K.* 1993. Variable pathogens of Sorghum. *In: Disease Management through Biotechnology: Improving sorghum and millet crops in Asia and Africa.* 15-19 Nov. 1993, ICRISAT, Patancheru, Andhra Pradesh

Frederiksen, R.A. 2000 Diseases and Disease management in sorghum. In: *Sorghum Origin, History, Technology and Production.* Edited By C.Wayne Smith. ISBN 0-471-24237=3. John Wiley and Sons. Inc. 497-533

Garud, T.B., Aglave, B.N. and *Ambekar. S.S.* 1994. Integrated approach to tackle the grain mold problem in Maharashtra. *International Sorghum and millets news letter 35:101-102*

Garud, T.B. and *Nagargoje, S.R.*. 1997. Integrated approach to tackle charcoal rot of sorghum in Maharashtra. *International Sorghum and Millets News letter* 38: 80-81.

Garud, T.B., Mali, V.R., Kohire, O.D. and *Choudhari, S.D.* 1990. Approaches to control sorghum red stripe virus. *Indian Phytopath.* 43(1):10-12

Garud, T.B., Kohire, O.D. and *Choudhari, S.D.* 1990. Pokkah boeng or twisted top a new disease to rabi sorghum in Maharashtra. *J. Maharashtra Agric. Univ.* 15(1):127-128

Jambunathan, R. and *Kherdekar, M.S.* 1991. Flavan-4-ol concentration in leaf tissues of grain mold susceptible and resistant sorghum plants at different stages of development. *J. Agric. Food Chem.* 39(6):1163-1165

Jambunathan, R., Kherdekar, M.S. and *Vaidya, P.* 1991. Ergosterol concentration in mold susceptible and mold resistant sorghum at different stages of grain development and its relationship to flavan-4-ols. *J. Agric. Food Chem.* 1990. 39(10):1866-1870

Jambunathan, R., Kherdekar, M.S. and *Bandyppadhyay, R.* 1990. Flavan-4-ols concentration in mold susceptible and mold resistant sorghum at different stages of grain development. *J. Agric. Food Chem.* 1990. 38: 545-548

Jambunathan, R., Kherdekar, M.S. and *Stenhouse, J.W.* 1992. Sorghum grain hardness and its relationship to mold susceptibility and mold resistance. *J. Agric. Food Chem.* 40(8):1403-1408

Karunakar, R.I., Narayana, Y.D., Pande, S., Mughogho, L.K. and *Singh, S.D.* 1994 Evaluation of early flowering sorghum germplasm accessions for downy mildew resistance in the greenhouse. *International Sorghum and millets news letter* 35: 102-103

King, S.B. 1971. Sorghum diseases and their control. *In Sorghum in Seventies* Eds NGP Rao and Leland R.House. Oxford and IBH Publishing Co. Delhi., Bombay and Calcutta. Pp411-435

Kokate, S.B., More, W.D. and ***Perane, R.R.*** 1989. Storage study on efficacy of different fungicides in controlling mycoflora of mouldy seeds of Jowar (Sorghum bicolor L Moench in CSH-9. . *Indian J. Mycol. Pl.Pathol.* 19(1):116-117

Latha, J., Mathur, K., Mukherjee, P.K., Chakrabarti, A., Rao, V.P. and ***Thakur, R.P.*** 2002. Morphological, pathogenic and genetic variability amongst sorghum isolates of *Colletotriachum graminicola* from India. *Indian Phytopath.* 55(1) : 19-25

Laxmanan, P. and ***Mohan, S.*** 1989. Effect of fungicides against sorghum downy mildew in Tamilnadu. *Indian J. Mycol. Pl.Pathol.* 19(1): 111-113.

Mahaling, D.M and ***Anahosur, K.H.*** 1998. Invitro evaluation of fungicides against grain molds and stalk rots of sorghum. *Journal of Mycology and Plant Pathology* 28(2):174-176

Mathur, K., Thakur, R.P. and ***Rao, V.P.*** 2000. Pathogenic variability and vegetative compatibility among isolates of *Colletotrichum graminicola* and *C.gloeosporiodes* causing foliar and grain anthracnose in sorghum. *Indian Phytopath.* 53(3) : 407-414

Mathur, K., Thakur, R.P. and ***Rao. V.P.*** 2001. Characterization of *Colletotrichum graminicola* populations from a sorghum hybrid CSH-9 for morphological and pathogenic variability. *Indian Phytopath.* 54 (3): 299-303

Narayana, Y.D. and ***Muniyappa, V.*** 1995 a. Effect of sorghum stripe virus on plant growth and grain yield of sorghum. *Indian Journal of Virology* 11(20:53-58

Narayana, Y.D. and ***Muniyappa, V.*** 1995 b. Survey, symptomatology and detection of an isolate of maize stripe virus on sorghum in Karnataka. *Indian Phytopath.* 48 : 1-6

Padule, D.N., Mahajan, P.D., Perane, R.R. and ***Patil, R.B.*** 1999. Efficacy of fungicides for increasing storability of grain molds infected seed of sorghum hybrid CSH-14. *Seed Research.* 27(1)95-99

Rao, V.P., Thakur, R.P. and ***Mathur, K.*** 1998. Morphological and pathogenic diversity among grain sorghum isolates of *Colletotrichum graminicola* in India. *Indian Phytopath.* 51()2): 164-174

Raut, J.G. and ***Wangikar. P.D.*** Control of seed borne fungi of sorghum. Effect of selected fungicides on fungi and seed germination. *Pesticides* 23(10):29-32

Revuru, S.S. and ***Garud, T.B.*** 1998. Effect of chlorotic stripe stunt disease on plant growth and grain yield of different sorghum cultivars. *Journal of Maharashtra Agricultural Universities* 23(3):253-255.

Singh, D.P. and ***Agarwal, V.K.*** 1988. Control of seed borne infection of grain mold pathogens of sorghum by fungicidal seed treatment. *Indian Journal of Plant Pathology* 6(2):128-132.

Singh, S.D. and **Navi, S.S.** 2000. Garlic as biocontrol agent for sorghum ergot. *J. Mycol. Pl. Pathol.* 30(3):350-354

Singh, S.D, Rao, K.E.P., Navi, S.S. and **Satyanarayana, M.V.** 1993. Identification o resistance to grain mold in white grain sorghum. *Sorghum Newsletter* 34: 24

Somani, R.B., Randrangi, RB, Wankhade, S.G. and **Patil, D.B.** 1992 Amino acid spectra of healthy and mouldy grains of sorghum hybrid SPH 388. *Indian Phytopath.* 249

Sundaram, N.V. 1977. *Diseases of sorghum.* Presented at International workshop at ICRISAT, Hyderabad, 19.

Tarr, SAJ. 1972. *The Principles of Plant Pathology.* The Scientific and Medical Division, Mc Millan Publishers Ltd 632pp

Thakur, R.P., Mathur, K., Rao, V.P., Chandra, S., Shivramkrishnan, S., Kannan, S., Hiremath, R.V. Tailor, H.C., Kushwaha, U.S., Dwivedi, R.R. and **Indira, S.** Pathogenic and genetic characterization of six Indian populations of *Colletotrichum sublineolum,* the causal agent of sorghum anthracnose. *Indian Phytopath.* 51(4): 338-348

Thakur, R.P. Status of International sorghum anthracnose and pearl millet downy mildew virulence nurseries. *In : Disease analysis through genetics and biotechnology. Inter disciplinary bridges to improved sorghum and millet crops.* Eds John F. Leslie and Richard A Frederiksen. Iowa State University press 359 pp

Thomas, M.D. and **Frederiksen, R.A.** 1995. Dynamics of oval and falcate conidium production of *Colletotrichum graminicola* from sorghum. *Mycologia* 87(1), 1995 pp 87-89

Williams, R.J. and **Rao, K.N.** A review on sorghum grain molds 1981. *Tropical Pest Management* 27(2):200-211

Zummo, N. 1978. Fusarium disease complex of sorghum in west Africa in sorghum diseases A worlds Review. Proc. 11-15 Dec pp 297-299

22

Seed Mycoflora of Pigeon Pea and their Management

Krishnappa M.,* Chakravarthy C.N. and Thippeswamy B.
Department of P. G. Studies and Research in Applied Botany, Kuvempu University,
Jnana Sahyadri, Shankaraghatta - 577 451, Shimoga (District), Karnataka, INDIA
*Corresponding author

ABSTRACT

Pigeon pea (*Cajanus cajan* L. Millsp.) is a major pulse crop in Karnataka. Crop is affected by number of fungal, bacterial and viral diseases. Among them fungal disease causes sever yield loss. During kharif-2000 and 2001 field survey was carried out in Karnataka and fifty-two seed samples were collected and screened by standard blotter method for the mycoflora. 21 different fugal species belongs 14 genera were recorded, the major fungus are, *Alernaria alternata, Fusarium oxysporum, Fusarium udum, Macrophomina phaseolina, Phoma cajani, Trichothecium roseum, Penicillium* sp., and *Aspergillus* sp., out of fifty-two samples, which are (Ten samples) showing higher incidence of pathogens were selected and subjected to different seed health testing methods. Among them Standard blotter method was found to be best for seed-borne fungal pathogen detection in pigeon pea. Location of pathogen and control of the seed-borne fungi are discussed in the main paper.

KEY WORDS : Pigeon pea, Seed health testing methods, Mycoflora, Location, Seed treatment

Pigeon pea is a tropical and sub tropical crop, it is commonly known as red gram. In Karnataka, area covered under pigeon pea cultivation, production and yield is 42,1,810 ha. 98,473 tones and 246 kg/ha respectively (Anon, 1999). Pigeon pea is known to be affected by number of fugal disease, which are play an important role in reducing the yield and quality of the seeds. Major fungal disease are Alternarial blight, Collar rot. Fusarium wilt (Butler, 1906), Colletotrichum leaf spot, rust, Sclerotium stem rot and wet root rot (Nene, 1985). Present study was under taken to collation of seed samples in major pigeon pea growing areas in Karnataka and subjected to different seed health testing methods for their mycoflora, find out the location of the pathogen and control the seed borne fungi by using different fungicides through seed treatment.

Materials & Methods

Collection of Seed Samples

Fifty-two pigeon pea seed samples were collected from different agro-climatic regions of Karnataka. From the farmer field and commercial market. The collected seeds were dried under sunlight to reduce the seed moisture and were stored in cloth bags at room temperature. The seeds were subjected for seed health test as per the recommendation of international seed health testing association (ISTA, 1976)

Standard Blotter Method (SBM)

Blotter paper discs were dipped in sterilized distilled water and placed in sterilized petriplate. Excess of water removed from the blotter by inverting the plate. Ten seeds were placed equidistantly on moist sterilized blotter discs using sterilized forceps in a laminar airflow hood. The seeds were incubated 26 ± 2^0C for one week in alternating cycles of 12 h NUV light and 12 h darkness. The seeds were examined on 7th day for associated mycoflora (ISTA, 1976) using stereo binocular microscope. Ten samples showing higher incidence of seed-borne fungi were selected for different seed health testing methods.

2, 4-D Method (Neergaard, 1968)

This method is similar to the standard blotter method here 0.2% solution of sodium salt of 2, 4-Dichlorophenoxy acetic acid is used instead of distilled water. Then the seeds were incubated as in case of standard blotter method.

Deep-Freeze Blotter Method (Limonard, 1968)

The procedure is similar to standard blotter method except that after 24-h. of incubation under 12-h. NUV and 12-h. darkness at 26 ± 2^0C, then petriplates were transferred to deep-freeze at -20^0C in complete darkness for 24-h. and again incubated for 5 days under 12-h. alternative cycle of NUV and darkness at 26 ± 2^0C.

Water Agar Method (ISTA, 1976)

In this method known quantity of agar (12.5 gm) powder was dissolved in 1 litre of boiling distilled water to prepare water agar media. This media is then autoclaved at about 15 lbs, at 21^0C for 20 min. After few minute approximately 10-15 ml. of media was poured in to a previously sterilized petriplates in the laminar airflow hood. After solidification, seeds were placed equidistantly using sterilized forceps. These seeds are previously surface sterilized with 1% sodium hypochlorite. Incubation procedure in similar to standard blotter method.

Screening of Seeds for Associated Mycoflora

The incubated seeds were screened on seventh day using stereo binocular microscope. The fungi on the seeds were identified by observing characteristic features such as the form, length and arrangement of conidiophores, the form, size, septation, colour, chain formation, etc., of conidia and their arrangement on the conidiophores, appearance of spore masses, characteristic of mycelium, density of colony etc., The associated mycoflora were recorded and identified with the help of standard manuals (Barnett, 1960 and Booth, 1977).

Location of Pathogenic Fungi by Component Plating Method

This method was adopted to find out the location of the pathogen in different components of the seed as described by Moden *et al.,* (1975). One hundred seeds having heavy load of dominant pathogenic fungi like Fusarium species were selected randomly. The seeds were soaked separately in sterilized distilled water for 24 hours. The soaked seeds were then dissected into three parts viz, pericarp (Seed coat), endosperm/cotyledons and embryo with the help of sterilized forceps and needles under aseptic condition. The components were placed on moist blotter discs. The plates were incubated for seven days under NUV light and darkness cycles of 12/12 hour at 26 ± 2^0C. On the 7th day these components were observed under stereo-binoculars microscope for the occurrence of mycoflora.

Seed Treatment

Ten samples showing higher incidence were selected from the seed health testing for the seed treatment. The test fungicides were treated individually. According to the recommendation (3 gm/kg seeds). Uniform seed dressing was achieved by adding required amount of fungicides into suitable airtight container by constant rotation for 30 minute. The treated seeds were plated on three layer wet blotters in petriplates and then incubated at 26 ± 2^0C under 12/12 hour alternate cycle of NUV light and darkness. Untreated seeds served as control.

The fungicides are used

1. Indofil M-45 (Mancozeb 75% W.P.)
2. Vitavax (Carboxin 75% W.P.)
3. Bavistin (Carbendazim 50% W.P.)
4. Captan (Captaf 50% W.P.)

Results

In analysis of seed-borne pathogen of different pigeon pea samples showed the occurrence of 21 fungal species belongs to 14 genera. The major fungus are, *Alernaria alternata, Fusarium oxysporum, Fusarium udum, Macrophomina phaseolina, Phoma cajani, Trichothecium roseum, Cladosporium herbarum, Penicillium* sp., and *Aspergillus* niger, *A. flavus, Rhizopus nigaricans,* and *Curvularia sp.,* Many worker have reported on the seed-borne fungi of pigeon pea, Nakkeeran *et al.,* (1997), Charya *et al.,* (1983), Rankole *et al.,* (1996), Lokesh *et al.,*(1993) and Alankapande *et al.,*(1992)

Ten samples showed higher incidence of pathogen in SBM was selected for different seed health testing and record the mycoflora (Table 1). Among all the method SBM is best method for the seed-borne pathogen detection in pigeon pea.

In the location of pathogen studies, Fusarium species was located in all the components i e., seed coat, cotyledons, and embryonic axis. Embryonic axis shows a lower per cent of infection compared to the other components, here *Fusarium oxyspoum* percentage was more compare to *F. udum* (Table 2). All the test fungicides significantly reduced the total percentage of fungi but their effect varied against the individual fungal incidence.

Discussion

In the present investigation pigeon pea seeds were subjected to different seed health testing. Among them SBM was found to be best method. The major fungi were recorded are, *Alernaria alternata, Fusarium oxysporum, Fusarium udum, Macrophomina phaseolina, Phoma cajani, Trichothecium roseum, Cladosporium herbarum, Penicillium* sp., and *Aspergillus* niger, *A. flavus, Rhizopus nigaricans,* and *Curvularia sp.,* most of the seed samples associated with both pathogenic and saprophytic fungi, are adversely effect on seed germination, seed rotting and seed discolouration was initiated by the heavy colonization of fungus. Fusarium species infected seeds are prompted to work on its location in different components of seeds. These studies reveled that, higher percent of Fusarium species harbored on the cotyledons.

Many scientist worked on the efficacy of some fungicides on seed-borne fungi of pigeon pea, Herjinder singh *et al.,* (1996), Sumith *et al.,* (1995), Solanke *et al.,*(1990), and Lokesh *et al.,*(1988) have reported the control of mycoflora up to 85 % using different fungicides. Present study also revealed that in case of Indofil M-45 treatment, the fungi such as *Alternaria alternata, Macrophomina phaseolina, Fusarium oxysporum* and *Trichothecium rosium* were inhibited and germination was not adversely affected at the tried concentration. Bavistin effectively controlled the sever systemic fungi such as *Fusarium oxysporum* and *Fusarium udum.* In case of Vitavax treatment, controlled the *A. alternata, Curvularia sp., Aspergillus sp., Phoma cajani, Macrophomina phaseolina* and *Trichothecium rosium.* Captan effectively controlled the *Fusarium oxysporum, Fusarium udum* and *Aspergillus sp.,* but it has not affected the *Cladosporium herbarum,* and *Rhizopus nigaricans.* Indofil M-45 and Vitavax are promising over other chemicals in the total percent inhibition of fungi.

Table 1 Seed Mycoflora of Pigeon pea in different seed health testing methods.

Place of collection	Methods	A.ni	A.flav	Alt.at	Cla.sp	Cu.Sp	F.ud	F.ox	M.Ph	Ph.sp	Rh.sp
Bangalore*	SBM	128	29	42	0	10	32	40	51	38	51
	WA	256	84	12	4	0	0	0	0	0	184
	2,4-D	164	24	0	0	0	128	0	0	0	268
	DF	320	208	4	0	0	28	0	0	16	372
Byadgi*	SBM	184	124	76	60	0	44	28	40	0	32
	WA	276	80	0	60	0	0	0	8	0	16
	2,4-D	300	116	88	188	0	0	0	0	0	4
	DF	88	56	8	208	0	40	0	0	12	4
Halebidu*	SBM	0	32	0	156	56	68	32	0	128	240
	WA	28	48	28	16	0	0	152	0	56	300
	2,4-D	20	8	16	28	0	164	0	8	64	328
	DF	24	20	96	176	0	180	24	16	20	108
H.N.pura*	SBM	116	20	16	0	16	16	80	16	0	48
	WA	80	24	0	8	0	0	0	0	0	132
	2,4-D	8	0	0	0	0	52	0	0	76	176
	DF	36	32	0	68	16	52	0	0	56	68
Madugiri*	SBM	148	24	140	148	0	84	20	100	100	28
	WA	84	24	24	56	0	0	232	0	16	44
	2,4-D	276	0	140	288	0	0	100	32	92	4
	DF	44	108	152	244	0	204	8	0	24	0
Nagamangala*	SBM	0	0	300	304	0	24	40	104	80	156
	WA	16	0	124	92	0	0	20	0	0	8
	2,4-D	100	44	284	268	0	72	0	0	104	112
	DF	28	16	344	316	0	92	4	0	44	12

Table 1 Contd....

Place of collection	Methods	A.ni	A.flav	Alt.at	Cla.sp	Cu.Sp	F.ud	F.ox	M.Ph	Ph.sp	Rh.sp
Shiggaon*	SBM	52	8	64	76	32	20	36	52	78	24
	WA	192	56	0	0	0	0	0	0	0	124
	2,4-D	216	128	72	156	0	12	0	20	28	88
	DF	136	36	72	72	0	0	16	0	28	44
Sharavanavelagola*	SBM	0	0	76	216	40	20	88	32	0	96
	WA	24	0	240	72	0	0	252	0	16	12
	2,4-D	28	0	44	68	0	0	156	0	24	104
	DF	0	8	136	148	0	260	40	0	76	0
Shikaripura*	SBM	60	52	0	104	0	32	24	0	220	212
	WA	64	0	0	12	0	0	0	0	56	36
	2,4-D	228	128	0	28	6	0	8	0	72	56
	DF	140	140	36	88	0	48	0	0	8	48
Sira*	SBM	180	0	172	140	0	84	8	96	104	64
	WA	76	12	84	0	0	20	0	56	0	0
	2,4-D	12	0	16	0	0	172	0	0	20	16
	DF	32	16	52	300	0	236	0	0	24	0

Data based on 400 seeds per sample

* Local variety

A.ni	=	*Aspergillus niger*
A.flav	=	*Aspergillus flavus*
Alt. at	=	*Alternaria alternata*
M ph	=	*Macrophomia phaseolina*
F.ud	=	*Fusarium udum*
F.ox	=	*Fusarium oxysporum*
Cla.sp	=	*Cladosporium herbarium*
Cu.sp	=	*Curvularia* species
Rh.sp	=	*Rhizopus nigricans*
Ph.sp	=	*Phoma cajani*

Table 2 Location of the Fusarium Species (F. *oxysporum*, F.*udum*) in different seed components

Sl. No.	Place of Collection (Local Variety)	Seed Components					
		Seed coat		Cotyledons		Embryonic axis (Plumule & Radical)	
		F.ox	F.ud	F.ox	F.ud	F.ox	F.ud
1.	Bangalore	20	10	08	04	06	04
2.	Byadgi	22	16	16	12	12	18
3.	Halebidu	18	28	10	16	10	12
4.	Holenarsipur	26	06	16	10	04	-
5.	Madhugiri	10	22	08	20	06	12
6.	Nagamangala	18	12	22	12	04	-
7.	Shiggaon	14	06	24	12	12	08
8.	Shravanabelagola	24	08	20	10	10	02
9.	Shikaripura	10	12	08	14	04	02
10.	Sira	08	22	04	14	02	04

F.ox = *Fusarium oxysporum*, F.ud= *Fusarium udum*
Data based on 100 seeds for each sample. (Each sample in 4 replicates)

References

Alanka Pande., Varma. K. V. R. V and **Pande. A** .1992. Seed borne fungi of pigeon pea: their pathogenicity and fungicidal control. *Biovigyanam* .18: 33-38.

Anonymous.1999. Agricultural crop statistical data recording book. Department of statistics, Bangalore.

Barnett, H.L. 1960. Illustrated Genera of imperfect fungi. *Burgess publishing company, Second edition, West Virginia.*

Booth, C. 1977. Fusarium laboratory guide to the identification of the major species. *Common wealth mycological institute, Kew, Surrey, England.* PP, 1-58.

Butler, E.J. 1906. The wilt disease of pigeon pea and pepper. *Agric. J. India.* 1 : 25-36.

Charya, M. A. S. and **Reddy, S. M.** 1983. Mycoflora of pigeon pea. *IPN.* 5: 56-58.

Harijinder Singh., Sharma, Y. R. and *Singh, H.* 1996. Efficacy of fungicides for the control of phytophthora stem blight of pigeon pea. *Plant dis. Res.* 11: 59-62.

ISTA. 1976. International rules for seed testing Annexes. *Seed Sci. Technol.* 4:51-77.

Limonard, T. 1968. Ecological aspects of seed health testing. Reprint from proc, *Int. seed test. Assoc.* 33: 167 p.

Lokesh, M. S. and *Hiremath , R. V.* 1993. Effect of relative humidity on seed mycoflora and nutritive value of red gram *(Cajanus cajan* (L.) Millsp.). *Mysore J. Agric. Sci.* 27: 268-271.

Moden, S., Singh, D., Mathar, S. B. and *Neergaard, P.* 1975. Detection and location of seed-borne inoculum of *Ascochyta rabiei* and its transmission in Chickpea *(Cicer arietinum). Seed Sci. Technol.* 3:667-681.

Nakkeran. and *Devi, P. R.* 1997. Seed borne mycoflora of pigeon pea and their management. *Plant dis. Res.* 12: 197-200.

*Neergaard, P. 1*969. *Plenodomus lingam* black leg of crucifers occurrence in Danish seed lots for export, and control by germisan-hot water treatment. *Friesia.* 9: 167-169.

Nene, Y.L. 1985. A proposed list of common names for diseases of pigeon pea. *IPN.* 4: 31-32

Rankole, S.A., Esseighe, D. A. and *Enikuemchin, O. A.* 1996. Mycoflora and Aflotoxin production in pigeon pea stored in Jute sacks and Iron bins. *Mycopathol.* 132: 155-160.

Salanke, R. B., Choulwar., S. B. and *Kore, S. S.* !990. Efficacy of some fungicides on seed mycoflora seed germination and seedling vigour of pigeon pea. *Res. Bull. Punjab Uni. Sci.* 41: 109-111.

Sumitha, R. and *Gaikwad, S.J.* 1995. Checking Fusarium wilt of pigeon pea by biological means, *Journal of Soils and crops,* 5: 163-165.

23 | *Evaluation of Sorghum against Grain Mold and its Impact on Seed Viability*

Bharati N. Bhat, R. Sudhakar and Pooranchand

Acharya NG Ranga Agricultural University, Regional Agricultural Research Station, Palem - 509 215, Mahaboobnagar dist. A.P.

ABSTRACT

Grain mold is one of the serious problems of the *kharif* season sorghum (*Sorghum bicolar*) in India. Severe quantitative as well as qualitative losses are recorded when the crop is exposed to rains at grain formation and developmental stages. During *kharif* 2003, several advanced yield trial material and grain mold resistant lines were evaluated for resistance to grain mold with respect to intensity of disease, percent grain affected, germination percentage, fungi associated and 100- grain weight of individual entry at RARS, Palem, Mahaboobnagar district, A.P. Good rainfall received during the months of September and October provided congenial conditions for the development of the mold at research station. Grain mold incidence was severe in advanced varietal and hybrid trial that ranged from 4.0-5.0 field grade and 3.8 - 5.0 threshed grade. The percent grain affected was least in G-03-03 and it recorded highest germination (72%). *Fusarium moniliforme* infection ranged from 22-63 % and *Curvularia lunata* infection from 25-42% in standard blotter test. Regarding grain mold resistant nursery, entries GMRP 9, GMRP 102 and GMRP 13 recorded moderate level of resistance (3.1- 3.3 field grade). Also GMRP 9 exhibited good germination (68%), less percent grain affected (25%) and only 13% *Fusarium* infection. Among the breeding material (segregating progenies) nearly twelve lines were found to be promising for further utilization in resistance breeding programme.

KEYWORDS : Sorghum, grain mold, resistance, germination, *Fusarium, Curvularia*

Sorghum (*Sorghum bicolar*) is affected by many plant diseases which reduce grain and fodder yield, also stover quality (Bandyopadhyay *et al.,* 2001). Among them, grain mold of sorghum caused by a complex of fungi, mainly species of *Alternaria, Curvularia, Drechslera, Fusarium,* and *Phoma* are particularly wide spread (Navi *et al.,* 1999). Grain mold fungi have repeatedly been associated with reduction in seed mass, grain density, germination and seedling vigour. Apart from reducing grain quality and market value, species of *Furarium* produce mycotoxins, which affect human and animal health after consumption of moldy grains (Bhat *et al.,* 2000).

Chemical control appears to provide some protection against grain mold, but it is neither practical nor economical. Employment of host-plant resistance is the most preferred method of control and economic strategy. Efforts to produce sorghum genotypes with tolerance to grain mold by conventional breeding methods have yielded partially successful results. Major research efforts have been focused on development of resistant cultivars, using diverse genetic base. In this article, an attempt was made to evaluate the sorghum breeding material and selected germplasm lines against grain mold disease and to elucidate the effect of major pathogenic on fungi on seed germination.

Materials and Methods

Seeds of yield trial material viz. Advanced varietal/hybrid entries, Intial varietal/hybrid entries and Grain mold resistant nursery received from National Research Centre from Sorghum (NRCS), Rajendranagar, Hyderabad, Andhra Pradesh were grown in the regular plot at the Regional Agricultural Research Station (ANGRAU), Palem, Mahaboobnagar district, Andhra Pradesh during the rainy season 2003. For comparison, two resistant checks IS 14332 and IS 14338 and released checks CSV 15 / CSH 9 were planted randomly along with test entries. The sorghum genotypes were planted in a single row of 4 metre (45 cm X 15 cm spacing) with three replications. Standard agronomic practices for fertilizers (80 kg N; 40 P kg /ha), and pesticides (spraying Endosulphan @ 2 ml/lit at 7, 14 and 21 days of seeding emergence and application of carbofuran granules in leaf whorl of 25 day old plants) were followed but no fungicide was sprayed.

Data Recording : All entries including selected crosses of sorghum breeding material of RARS, Palem were evaluated for grain mold, following visual appraisal method which is the quickest and earliest (Bandyopadhay and Mughogho, 1988).

Grain mold severity score : Sorghum lines were evaluated for grain mold, 14 days after physiological maturity stage on 1-5 scale (1 = no mold, 2 = 1-10%, 3 = 11-25%, 4 = 26-50% and 5 = >50% grain molded on a panicle). Genotypes with score 1-2 were considered resistant, 2. 1-3 as moderately resistant, 3. 1-4 as moderately susceptible and 4. 1-5 as highly susceptible.

Threshed grain mold severity score : threshed grain (10 gm) from each panicle, spread in a Petri dish was scored for visual rating on 1-5 scale using a magnifying lens under proper lighting. The threshed grain samples were subjected to standard blotter test to determine the frequency of different pathogenic mold fungi on the grains and also germination percentage.

100-grain weight : Manual counting of 100 grains from each sample was done and weighed in electronic balance.

Results and Discussion

At Palem, the total precipitation received during kharif 2003 was 602.9 mm spread over 41 rainy days. Maximum and minimum temperature ranged between 27-35 °C and 17-24 °C respectively. Good rainfall received during the months of Sept-October provided congenial conditions for the development of molds. Incidence of grain mold was heavy.

In blotter test, two predominant fungi viz., *Fusarium moniliforme* and *Curvularia lunata* were observed in more frequency. Seed viability reduced with the increase in frequency of these fungi.

Evaluation of Advanced Hybrids / Varieties for Resistance to Diseases (AV/HT)

The results of reaction to advanced hybrids/ varieties for resistance to diseases is depicted in Table 1.Two resistant checks IS 14332 and IS 14338 were used here. The grain mold incidence was severe, that ranged from 4- 5 field grade (FG) and 3.8 –5.0 threshed grade (TG). None of the entries performed well. The percent grain affected was least in G-03-03 (43%) and it recorded highest germination 71.7 %. *Fusarium* infection ranged from 22 to 63 % and *Curvularia* infection from 25-42 % in test entries.

Evaluation of Initial Varieties for Resistance to Diseases (IVT)

Initial Varietal Trial I: The grain mold incidence ranged from 4.8-5.0 (FG & TG). None of the entries found to be resistant. The percent grain affected ranged from 54-71% and germination 35-63 %. *Fusarium* infection ranged from 25-35% in test entries.

Initial Varietal Trial II: As indicated in Table 2, grain mold incidence ranged from 3.8– 5.0 (FG) and 4.2 – 5.0 (TG). SPV 1644 was found to be promising (3.8 TG). Also SPV 1644 recorded least percent grain affected and good germination.

Evaluation of Initial Hybrids for Resistance to Diseases (IHT)

Initial Hybrid Trial I: Test entries (Table 3) recorded heavy incidence of grain mold (4.2- 5.0 FG and 3.8- 5.0 TG). SPH 1455 was found to be promising (3.8 TG) with least percent grain affected (48 %) and good germination (60 %). *Fusarium* infection ranged from 30-52 %.

Initial Hybrid Trial II: Grain mold incidence was severe (4.8- 5 FG and TG). All test entries were found to be susceptible and recorded less germination (20-38%) and more percent grain infection (69-89%).

Evaluation of Grain Mold Resistant Nursery for Resistance to Diseases (GMRN)

Screening of GMRN with 26 entries as mentioned in Table 4, enabled to identify few genotypes with moderate level of resistance (3.1-3.3 FG). They were GMRP 9, GMRP 102, followed by GMRP 13. Also GMRP 9 recorded less percent grain affected (25 %) and good germination (68 %). Also *Fusarium* infection was less in GMRP 9 (13%).

Screening Breeding Material for Grain mold

Selected lines from breeding material were evaluated for resistance to mold. About 12 crosses showed moderate resistance to mold, which were depicted in Table 5.

These will be advanced further and evaluated for grain mold.

Table 1 *Advanced Varietal/ Trial (AV/HT)* : Kharif 2003 data at R.A.R.S., Palem
Grain molds (Grade- resistant 1 to highly susceptible 5)

Sl. No.	Treatment	Field grade	Threshed grade	Percent grain affected	Germination Percentage	Fungi associated (%)		100 seed weight (g)
						F. moniliforme	*C. lunata*	
1.	G-01-03	5	5	78	30	50	33	1.54
2.	G-02-03	5	5	79	28	48	33	1.6
3.	G-03-03	4	3.8	43	72	25	27	1.7
4.	G-04-03	4.2	4.6	48	35	32	32	1.6
5.	G-05-03	5	5	72	28	35	30	1.65
6.	G-06-03	5	5	77	30	38	30	1.55
7.	G-07-03	5	5	72	28	32	28	1.57
8.	G-08-03	5	5	80	27	47	42	1.59
9.	G-09-03	5	5	75	30	43	33	1.66
10.	G-10-03	5	5	89	18	63	30	1.58
11.	G-11-03	5	5	83	23	60	28	1.66
12.	G-12-03	5	5	87	18	55	28	1.69
13.	G-13-03	5	5	79	28	47	18	1.58
14.	G-14-03	5	5	80	30	52	23	1.68

Table 1 *Contd...*

Sl. No.	Treatment	Field grade	Threshed grade	Percent grain affected	Germination Percentage	Fungi associated (%)		100 seed weight (g)
						F. moniliforme	*C. lunata*	
15.	G-15-03	5	5	80	33	43	22	1.64
16.	G-16-03	5	5	85	23	60	23	1.61
17.	G-17-03	5	5	89	18	60	27	1.67
18.	G-18-03	5	5	86	20	48	28	1.59
19.	G-19-03	5	5	88	20	65	25	1.68
20.	G-20-03	5	5	85	32	63	28	1.69
21.	G-21-03	4	5	70	52	37	28	1.72
22.	G-22-03	5	5	79	35	32	30	1.59
23.	G-23-03	5	5	80	30	37	23	1.68
24.	G-24-03	5	5	73	33	55	27	1.72
25.	IS 14332	2	2	14	87	8	3	1.77
26.	IS 14338	1.4	1.7	8	83	3	1.7	1.81
	CD (5%)	0.3	0.2	5.3	4.2	4.8	4.7	0.08
	CV	3.35	2.03	4.2	7.25	7.1	9.6	2.79

Table 2 *Initial Varietal Trial II (IVT II):* Kharif 2003 data at R.A.R.S., Palem
Grain molds (Grade- resistant 1 to highly susceptible 5)

Sl. No.	Treatment	Field grade	Threshed grade	Percent grain affected	Germination Percentage	Fungi associated F. moniliforme	(%) C. lunata	100 grain weight (g)
1.	SPV 1638	5	4.9	58	55	35	37	1.60
2.	SPV 1639	4.8	4.8	55	57	28	32	1.61
3.	SPV 1640	5	5	63	50	32	37	1.68
4.	SPV 1641	5	4.7	54	62	27	32	1.81
5.	SPV 1642	4.5	4.6	52	52	32	28	1.69
6.	SPV 1643	5	5	65	53	35	33	1.61
7.	SPV 1644	3.8	4.2	48	62	27	33	1.67
8.	SPV 1645	5	5	66	50	32	35	1.66
9.	SPV 1646	4.7	5	53	55	30	38	1.60
10.	SPV 1647	4.5	5	57	57	32	30	1.56
11.	SPV 1648	5	5	70	38	38	30	1.53
12.	SPV 1649	5	5	70	33	37	33	1.61
13.	SPV 462	4.5	4.8	55	43	32	37	1.72
14.	CSV 15	4.7	4.8	55	43	35	30	1.70
	CD (5%)	0.5	0.4	3.4	3.4	3.4	5.1	0.1
	CV	6.74	5.34	3.47	4.17	6.28	9.56	2.68

Table 3 *Initial Hybrid Trial I (IHT I):* Kharif 2003 data at R.A.R.S., Palem
Grain molds (Grade- resistant 1 to highly susceptible 5)

Sl. No.	Treatment	Field grade	Threshed grade	Percent grain affected	Germination Percentage	Fungi associated (%)	
						F. moniliforme	C. lunata
1.	SPH 1441	4.8	5	78	35	42	40
2.	SPH 1442	5	5	81	30	42	37
3.	SPH 1443	5	5	88	30	52	35
4.	SPH 1444	4.5	5	72	40	42	33
5.	SPH 1445	5	5	78	32	40	42
6.	SPH 1446	4.8	5	73	33	35	38
7.	SPH 1447	5	5	85	28	52	35
8.	SPH 1448	4.8	5	86	32	50	32
9.	SPH 1449	5	4.8	76	35	42	23
10.	SPH 1450	4.7	5	86	30	45	30
11.	SPH 1451	5	5	76	42	37	40
12.	SPH 1452	5	4.2	69	35	37	37
13.	SPH 1453	4.8	4.8	78	35	35	48
14.	SPH 1454	5	5	90	22	35	42
15.	SPH 1455	4.2	3.8	48	60	30	35
16.	SPH 1456	4.5	4.7	58	42	38	28
17.	SPH 1457	5	5	73	33	42	32
18.	SPH 1469	5	5	74	28	42	30
19	SPH 1471	4.8	5	82	30	50	40
20.	CSH 9	5	5	81	28	50	38
21.	CSH 14	5	5	86	30	48	35
22.	CSH 16	4.8	5	76	33	37	23
23.	CSH 17	5	5	75	33	42	32
24.	IS 14332	2.1	2.2	10	88	3	3
	CD (5%)	0.4	0.3	5.1	4.9	4.1	4.2
	CV	4.66	3.54	3.53	8.14	6.35	7.37

Table 4 *Grain Mold Resistant Nursery (GMRN) :* Kharif 2003 data at R.A.R.S., Palem
Grain molds (Grade- resistant 1 to highly susceptible 5)

Sl. No.	Treatment	Field grade	Threshed grade	Percent grain affected	Germination Percentage	Fungi associated (%)	
						F. moniliforme	*C. lunata*
1.	GMRP 9	3.2	3.1	25	68	13	13
2.	GMRP 13	3.5	3.6	39	62	17	18
3.	GMRP 34	3.6	4	46	57	23	27
4.	GMRP 84	4	3.9	43	42	22	20
5.	GMRP 88	3.8	3.9	44	48	22	30
6.	GMRP 90	4.2	4	47	55	18	27
7.	GMRP 91	3.8	3.8	38	53	27	32
8.	GMRP 92	4	4	45	55	18	23
9.	GMRP 99	4	3.9	48	62	22	18
10.	GMRP 100	3.8	3.8	44	43	22	17
11.	GMRP 101	4	4	46	53	25	24
12.	GMRP 102	3.3	3.3	33	55	17	20
13.	PVK 834	5	5	55	47	32	27
14.	PVK 937	4	4	51	43	32	22
15.	PVK 841	5	5	56	57	28	28
16.	AKGMR 57	4.7	4.6	54	53	28	27
17.	AKGMR 58	4.5	4.5	55	48	35	27
18.	AKGMR 59	5	5	65	25	37	25
19.	AKGMR 60	5	5	65	30	33	17
20	AKGMR 61	5	5	56	58	32	23
21	KBN-528	4.7	4.7	52	65	27	28
22	KBN-589	5	5	58	63	32	27
23	Kekri local	4.7	4.7	55	68	23	32
24	IS 14332	1.3	1.3	8	88	3	1.7
25	IS 14338	1.0	1.0	4	87	1.7	1.7
	CD (5%)	0.5	0.4	3.6	4.7	4.3	4.3
	CV	7.96	6.73	5.21	5.91	9.32	9.51

Table 5 Screening Breeding material for Grain mold (moderately resistant)

SI.No.	Cross	Grain mold (FG)
1	CSV 15 x 30034	2.4
2	PSV 54 x 94017	2.0
3	PSV 51 x 94017	2.8
4	PSV 54 x 46582	2.8
5	PSV 56 x ICSV 111	2.5
6	P-2 x ICSV 111	2.6
7	PSV 49 x ICSV 233	2.2
8	M 35-1 x 46582	2.6
9	P-2 x 46582	2.5
10	7411 x 46582	2.4
11	P-2 x 46581	2.4
12	P-2 x 46592	2.9

From the foregoing discussion, it can be concluded that we have to identify resistant varieties/hybrids for grain mold, which is a continuous process. It is difficult to obtain high yielding, bold, and white sorghum with complete protection against grain mold. Hence, it is the need of the hour to develop resistant varieties/hybrids to grain mold through resistance breeding. Future focus should be on conventional breeding, exploiting grain hardness in improving mold resistance and also through biotechnological approaches.

References

Bandyopadhyay, R. and **Mughogho, L.K.** 1988. Evaluation of field screening techniques for resistance to sorghum grain molds. *Pl. dis.* 72: 500-503.

Bandyopadhyay R, Pande S., Blummel N, Thomas D and **Rama Devi K.** 2001. Effect of plant disease on yield and nutritive value of sorghum and groundnut crop residues. *In proceedings; 10th Animal Nutrition conference, NDRI, Karnal,* India. p. 28.

Bhat R.V., Shetty H.P.K. and **Vasanthi S.** 2000. Human and animal health significance of mycotoxins in sorghum with special reference to fumonisins. Pages 107-115 in Technical and Institutional options for Sorghum Grain mold Management. *Proceedings of an International Consultation,* 18-19 May 2000, ICRISAT, Patancheru, India.

Navi S.S., Bandyopadhyay, R, Hall A.J. and **Bramel-Cox P.J.** 1999. A pictorial guide for the identification of mold fungi on sorghum grain. *Information bulletin No.59.* Patancheru, 502324 A.P. India, ICRISAT, pp 118.

24 | Incidence of Diseases on Tomato Fruits in Marathwada and their Biocontrol Measures

C.S. Swami and D.S.Mukadam

Department of Botany, Dr. Babasaheb Ambedkar Marathwada University, Aurangabad - 431004 (M.S)

ABSTRACT

Tomato (*Lycopersicon esculentum* Mill.) is an important vegetable crop grown all over the country . The fruits of tomato are affected by many fungal pathogens. In the present study, Different fungal pathogens were isolated from the fruits collected from different places of Marathwada. Studies were also made on the in vitro control of these pathogens using *Trichoderma viride* and some plant extracts. Totally 13 fungi were found pathogenic to tomato fruits. The degree of incidence was found to be variable with tomato varieties, stages of the tomato fruits and the localities.

KEYWORDS : Tomato, fungal pathogens, biocontrol, *Trichoderma viridi*

Tomato (*Lycopersicon esculentum* Mill.). is reported   to be affected by about twenty diseases of microbial origin. Among them, the fungal pathogens have been found to infect and damage tomato fruits both in field at different developmental stages as well as in the market during storage. This may result in the qualitative and quantitative loss of tomato fruits. Therefore, the present studies were undertaken with special reference to the crop grown in Marathwada region.

Material And Methods

Diseased tomato fruit samples shoppwing different symptoms werep collected from field and market places of various localitipes in Marathwada.

Tomato fruits of different varieties at green (young) and red (ripe) stages with different symptoms as described here were collected and studied for the association of fungi. For this infected tissues of the fruits were surface sterilized with 0.1% mercuric chloride, washed thrice with sterile distilled water and plated on agar. Growth and sporulation of the fungi were seen on seventh day.

The fungal pathogens were identified on the basis of sporulation characteristics. Pure cultures were prepared & maintained on potato dextrose agar medium.

Antagonistic potential of *Trichoderma viride* against tomato fungi was studied by dual culture method used by Sudhamoy Mandal et. al., (1999). While fungi toxicity of different plant extracts was studied by poisoned food technique described by Nene and Thapliyal (1993).

Results And Discussion

A. I - Various Symptoms on Fruits by different Fungi (Plate I)

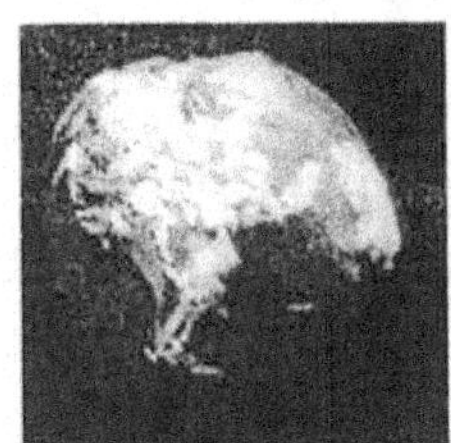

Geotrichum candidum

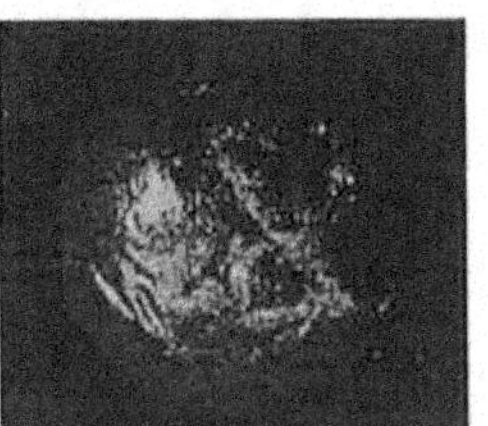

Phytophthora infestans

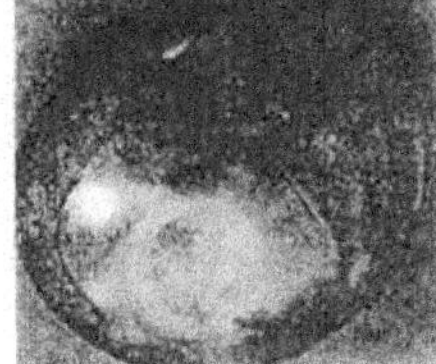

Colletotrichum dematium

Phytophthora parasitica

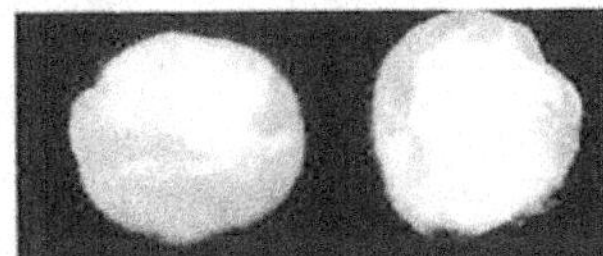

Fusarium oxysporum

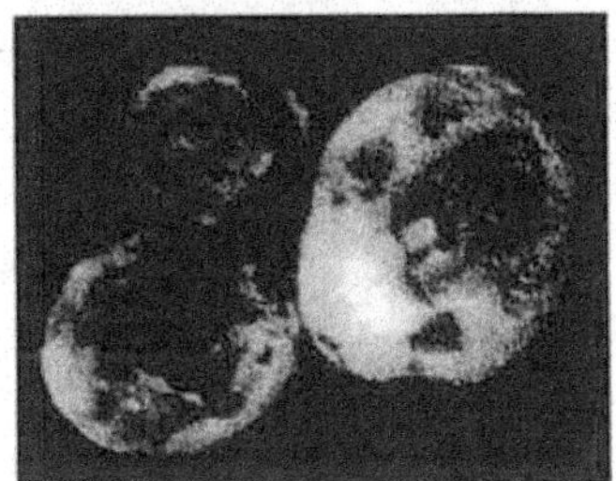

Phoma destructiva

Alternaria solani

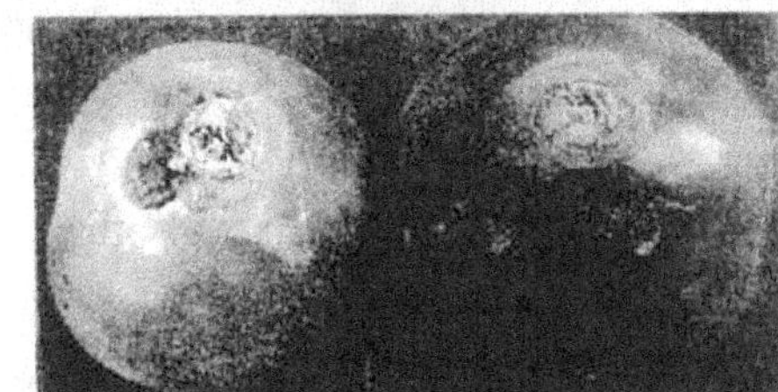

Rhizoctonia solani

Cladosporium fulvum

Symptoms on Fruits by different Fungi

Totally thirteen fungi appeared on tomato fruits. They were Alternaria solani, *Geotrichum candidum*, *Fusarium roseum*, *F. oxysporum*, *Phoma destructiva*, *Rhizoctonia solani*, *Phytophthora* SP, *Cladosporium fulvum*, *Curvularia lunata*, *Aspergillus niger*, *A. flavus*, *Penicillium expansum* and *Rhizopus stolonifer*. The symptoms produced by them are described here.

1. Alternaria solani : It causes early blight of tomato. Fruit spot at stem-end, brown to black, up to 1 inch in diameter, usually having concentric rings. Spot may occur leathery and may be covered with a velvety mass of black spores.

Geotrichum condidum : It causes sour rot of tomato. On green fruit, lesions appear dull, greasy and water soaked. Decay often begins at edge of stem scar and gets uniformly spreaded. As fruit ripens, shows a white fungal growth in the scar or at the point when skin is broken. In ripe fruit, tissues become soft; water soaked and filled with a white, cheesy or scum like fungus, often followed by bacterial soft rot.

Fusarium roseum and Fusarium oxysporum : On ripe fruit, spot pale, turns brown and soon include entire fruit surface. In dry weather tissue remains firm and with plenty moisture, lesion becomes water soaked and covered with whitish / pinkish fungal growth. On green fruit, decay area is slightly sunken but firm. A tuft of white mold present on surface.

Phoma destructive : Spots begin usually at stem scar, may also be associated with cracks. Spots on green fruits brown to black, unto ¼ inch in diameter. As fruit ripen, lesion diameter can increase up to 1 inch. Spots on ripe fruits leathery, somewhat sunken and dark.

Rhozoctonia solani: It causes soil rot of tomato. Brown rot with alternating light and dark coloured zonate bands and sharply defined margins in areas where fruit contacts soil. The affected fruit later on becomes mushy.

Phytophthora infestans : It causes late blight of tomato. Greenish brown greasy spots may enlarge until entire fruit is involved. The fruit tissue remains firm at first. In moist weather, a white fungal growth appears on affected surface.

Phytophthora parasitica : It causes buckeye rot of tomato. Greyish green or brown water soaked spot appears where fruit touches the soil. In warm weather, it enlarges rapidly and may cover half or more of the fruit and exhibit pale brown concentric rings.

Cladosporium fulvum (Fulvia fulva) : A conspicuous black, leathery stem ends rot. Rot margins may be irregular. The fruit may have blackened radial furrows and may be lop-sided, with fruit remaining green or yellow on the retarded side.

Curvularia lunata : Black or brown spots on fruit, fruit tissue firm at first. In moisture condition, fungal growth develops on fruit.

Aspergillus niger : Lesions large, with water soaked portions. Rot develops affecting entire fruit. There is appearance of the fungal growth on fruit.

Aspergillus flavus : Water soaked large spots, affecting half or entire fruit and show fungal growth on it.

Rhizopus stolonifer : It causes soft rot of tomato. Lesions large, affected areas puffy, water soaking. Rot develops rapidly and affects entire fruit. Fruit splits open and collapses into a wrinkled soft mass over run by grey black fungus. Fruits have fermented odour.

Penicillium expansum : Large water soaked spots, affecting half or entire fruit, fungal growth develops on fruit.

II Incidence of fungi on tomato fruits of different varieties at green stage .

Incidence of fungi was studied on fruits at green stage of seven varieties of tomato, viz., Avinash-2, Naveen-2000, Pusa Ruby, Rashmi, Rohini, Vaishali and Local. The results are given in Table 1.

It is clear from the results that, fruits of all the seven varieties at green stage showed presence of fungi ranging from one to many. *Alternaria solani* was present on fruits of all varieties. *Geotrichum candidum* was associated only with Local variety. Incidence of *Fusarium oxysporum* was observed on Naveen - 2000, Rashmi, Rohini and Local. Rhizoctonia solani was found to be associated with fruits of varieties Avinash-2 and Rohini. The fruits of Rashmi and Local were associated with *Phytophthora* sp. It was interesting to note that at this stage none of the varieties showed presence of *Aspergillus niger* and *Rhizopus stolonifer.*

III Incidence of fungi on tomato fruits of different varieties at reddish stage

Incidence of fungi on tomatoes of the seven varieties at reddish stage was studied and results are given in Table 1.

It is evident from the results that, *Alternaria solani* was present on red fruits of all the varieties. While, *Geotrichum candidum* was associated with the fruits of var. Naveen-2000, Rashmi and Local. All the varieties except Avinash-2 and Vaishali were associated with *Fusarium oxysporum.* Similarly, *Rhizoctonia solani* showed its association with the fruits of var. Avinash-2, Rohini and Local. Whereas, *Phytophthora* sp. was found to be present on fruits of variety Avinash-2, Rashmi and Local.

IV Incidence of fungi on tomato fruits of variety Vaishali at different localities in Marathwada

Incidence of fungi was studied on tomato fruits of variety Vaishali at 4 localities of Marathwada region, viz., Aurangabad, Latur, Beed and Osmanabad. The results are summarised in Table 2.

It is clear from the Table 2 that, fruits of var. Vaishali showed presence of *Alternaria solani, Aspergillus niger* and *Rhizopus stolonifer* irrespective of the locality. *Geotrichum candidum, Phoma destructiva, Cladosporium fulvum, Curvularia lunata, Aspergillus flavus* and *Penicillium expansum* were not found to be associated with tomato fruits of var.

Table: 1: Incidence of fungi on green and red stages of fruits in different varieties of tomato

Fungi	Tomato varieties												
	Avinash-2		Naveen-2000		Pusa Ruby		Rashmi		Rohini		Vaishali		Local	
	G	R	G	R	G	R	G	R	G	R	G	R	G	R
Alternaria solani	+	+	+	+	+	+	+	+	+	+	+	+	+	+
Geotrichum candidum	–	–	–	+	–	–	–	+	–	–	–	–	+	+
Fusarium oxysporum	–	–	+	+	–	+	+	+	+	+	–	–	+	+
Rhizoctonia solani	+	+	–	–	–	–	–	–	+	+	–	–	–	+
Phytophthora sp.	–	+	–	–	–	–	+	+	–	–	–	–	+	+
Aspergillus niger	–	+	+	–	–	–	–	–	+	–	–	–	–	+
Rhizopus stolonifer	–	–	–	+	–	–	–	+	–	–	–	–	–	+

+ Fungus present   – Fungus absent

G= Green stage of fruits.   R = Red stage of fruits

Table 2 Incidence of fungi on tomato fruits of var. Vaishali from different Localities of Marathwada

Fungi	Localities			
	Aurangabad	Latur	Beed	Osmanabad
Alternaria	+	+	+	+
Geotrichum	--	--	--	--
Fusarium roseum	--	+	--	--
Fuarium oxysporum	+	+	--	+
Phoma destructiva	--	--	--	--
Rhizoctinia solani	--	+	--	--
Phytophthora sp.	--	--	+	--
Cladosporium fulvum	--	--	--	--
Curvularia lunata	--	--	--	--
Aspergillus niger	+	+	+	+
Aspergillus	--	--	--	--
Penicillium expansum	--	--	--	--
Rhizopus stolonifer	+	+	+	+

+ Fungus present   -- Fungus absent

Vaishali at any locality. The incidence of Fusarium roseum and Rhizoctonia solani was recorded only at Latur. Fusarium oxysporum was recorded at all the localities except Beed. Phytophthora sp. was recorded on fruits of var. Vaishali only at Beed.

B) Biocontrol studies:

(a) Use of *Trichoderma viride*

(i) Antagonistic nature of Trichoderma viride in the centre of plate against the tomato fungi inoculated

In order to study the antagonistic nature of *Trichoderma viride* against tomato fungi, the fungal spore suspensions were spread separately in petriplates containing solidified Rose Bengal agar medium. At the centre of these plates, an agar disc (5mm) containing mycelial growth of *Trichoderma viride* was placed. The plates were incubated for 7 days and observed for inhibition of the growth of tomato fungi and results are given in Table 3.

It is clear from the Table 4 that, *Trichoderma viride* caused maximum growth inhibition of *Alternaria solani* and *Rhizopiis stolonifer*, followed by *Phytophthora* sp., *Fusarium oxysporum* and *Curvularia lunata*. In case of *Geotrichum candidum*, *Aspergillus niger* and A. *flavus* there was less inhibition. *Phoma deslructiva*, *Rhizoctonia solani*, *Cladosporium fulvum* and *Penicillium expansum* showed somewhat uniform inhibition by *Trichoderma viride*.

(b) Use of plant extracts

(i) Effect of leaf extracts on growth of tomato fungi

Table 3 Antagonistic nature of Trichoderm viride in the cnetre of plate aganist tomato fungi inoculated.

Fungi unoculated in plate	Control (Growth without *Trichoderma viride* ((mm)	Zone of inhibition due to *Trichodera viride* (mm)	Percent inhibition
Alternaria Solani	68	46	67.6
Geotrichum	90	12	13.3
Fusarium roseum	78	38	48.7
Fusarium oxysporum	76	32	42.1
Phoma destructiva	60	18	30.0
Rhizoctonia solani	62	22	35.4
Phytophthoru sp.	!0	42	52.5
Cladosprorium fulvum	78	24	30.7
Curvularia lunata	68	27	40.3
Aspergillus niger	72	16	22.2
Aspergillus flavus	70	12	17.1
Penicillium expansum	68	22	32.5
Rhizopus stolonifer	85	52	61.1

Table 4 Effect of leaf extracts on growth of Tomato Fungi

Plant extract	Diameter of fungal growth (mm)					
	Aso	Clu	Gca	Rso	Fox	Phy
Polyalthia	33	41	52	42	28	23
	(49.2)	(41.4)	(34.1)	(50.5)	(58.2)	(62.9)
Annon squamosa	47	48	I	45	44	39
	(27.6)	31.4)		(47.0)	(34.3)	(37.1)
Callistemon rigidus	30	35	29	58	40	26
	(53.8)	(50.0)	(63.2)	(31.7)	(40.3)	(58.0)
Psidium guajava	37	45	30	66	45	30
	(43.0)	(35.7)	(62.0)	(22.3)	(32.8)	(51.6)
Datura stramonium	33	41	31	60	58	I
	(49.2)	(41.4)	(60.7)	(29.4)	(13.4)	
Solanum melongena	65	70	I	85	45	55
	(00.0)	(00.0)		(00.0)	(32.8)	(11.2)
Adathoda vasica	35	36	50	40	33	32
	(46.1)	(48.5)	(36.7)	(52.9)	(50.7)	(48.3)
Tagetes erecta	29	41	35	69	28	45
	(55.3)	(41.3)	(55.7)	(18.8)	(58.2)	(27.4)
Ocimum sanctum	34	35	I	35	53	35
	(47.6)	(50.0)		(58.8)	(20.9)	(43.5)
Azadirachta indica	31	29	42	42	38	34
	(52.3)	(58.5)	(46.8)	(50.5)	(43.2)	(45.1)
Control	65	70	79	85	67	62

(Figures in parentheses indicate percent inhibition of mycelial growth over control)

I - Induction of mycelial growth over control

  Aso - *Alternzaria solani* **Clu** - *Curvularia lunata*
  Gca - *Geotrichum candidum* **Rso** - *Rhizoctonia solani*
  Fox - *Fusarium oxysporum* **Phy** - *Phytophthora sp.*

Ten fresh plant leaf extracts at 10% (aqueous) concentration were tested against six tomato fungi for inhibition of growth in solid medium. They were added in the solid medium to get 1:1 final concentration. The fungi were grownon this medium. The medium without plant extract served as control. The results are summarised in Table 4.

The data given in table 5 reveals that 10% aqueous fresh leaf extracts of *Polyalthia longifolia* for *Phytophthora* sp. and *Fusarium oxysporum, Annona squamosa* for *Rhozoctonia solani, Callistemon rigidus* for *Geotrichum candidum, Phytophthora* sp. and *Alternaria solani, Psidium guajava* for *Geotrichum candidum* and *Phytophthora* sp. were inhibitory for the mycelial growth. Similarly, *Datura stramonium* for *Geotrichum candidum, Alternaria solani, Solanum melongena* for *Fusarium oxysporum* and *Phytophthora* sp., *Adathoda vasica* for *R. solani, F. oxysporum* and *C. lunata, Tagetes erecta* for *F. oxysporum, G. candidum* and *A. solani, Ocimum sanctum* for *R. solani* and *C. lunata* and *Azadirachta indica* for *C. lunata, A. solani* and *R. solani* were found to inhibit the mycelial growth. However, *Annona squamosa, Solanum melongena* and *Ocimum sanctum* stimulated mycelial growth of *Geotrichum candidum* and *Datura stramonium* stimulated *Phytophthora* sp.

II) *Effect of gymnospermic plants on growth of tomato fungi*

Fresh gymnospermic leaf extracts at 10% (aqueous) concentration were tested for their effect on mycelial growth of six tomato fungi in solid medium and results are given in Table 5.

Table 5 Effect of gymnospermic plants on growth of tomato fungi

Plant extract	Diameter of fungal growth (mm)					
	Aso	Clu	Gca	Rso	Fox	Phy
Thuja occidentails	I	50	I	63	I	47
	(25.3)		(21.2)		(22.9)	
Cupressus sp.	39	54	77	57	70	42
	(35.0)	(19.4)	(00.0)	(28.7)	(00.0)	(31.1)
Araucaria afaucarianna	49	53	77	68	51	49
	(18.3)	(20.9)	(00.0)	(15.0)	(27.1)	(19.6)
Taxoclium distichum	41	49	I	70	41	39
	(31.60	(26.8)		(12.5)	(41.4)	(36.0)
Control	60	67		80	70	61

Figures in parentheses indicate the percent of mycelial growth over control.

I - Induction of mycelial growth over control

Aso - *Alternaria solani*   Clu - *Curvularia lunata*

Gca - *Geotrichum candidum*   Rso - *Rhizoctonia solani*

Fox - *Fusarium oxysporum*   Phy - *Phytophthora* sp.

Data from table 5 reveals that, gymnospermic leaf extracts were less inhibitory than that of angiospermic leaf extracts. *Thuja occidentalis* for *Curvularia lunala*, *Rhizoctonia solani* and *Phytophthora* sp., *Cupressus* sp. for all fungi except *Geotrichum candidum* and *Fusarium oxysporum*, *Araucaria* sp. and *Taxodium distichum* for all fungi except *G. candidum* were inhibitory for mycelial growth. *Thuja occidentalis* for *A. solani*, *G. candidum* and *F. oxysporum* and *Taxodium distichum* for *G. candidum* proved stimulatory for mycelial growth.

Studies were undertaken to understand the qualitative and quantitative pathogenic and non-pathogenic fungi on tomato fruits during their developmental stages (field). The findings are mainly on isolation of fungi from tomato fruits of different varieties at different development stages and use of plant extracts and Trichoderma viride for the control of fruit pathogens.

Tomato fruits of seven varieties, when screened for the association of fungal pathogens, yielded thirteen fungal species. Among them, fungi associated with tomato fruits in field were mainly Alternaria solani, Geotrichum candidum, Fusarium roseum, F. oxsyporum, *Phoma destrutiva*, *Rhizoctonia solani*, *Phytophthora* sp. and *Cladosporium fulvum*. It is also clear from the earlier literature that, on field tomatoes show presence of, *Phoma destructiva* was reported by Kogan (1977), *Curvularia lunata* and *C. trifolii* by Kore and Bhide (1977), *Cercospora canescens* by Garcia (1978), *Phytophthora nicotianae* var. *nicotianae* by Skadow (1978), *Fulvia fulva* by Kuznetsov (1979), *Fusarium roseum* var. *avenaceum* by Marras *et al.*, (1981), *Cladosporium tennissimum* by Narain and Rout (1981), *Atlernaria tenuiissima*, *Pseudocercospora fulgena* and *P. diffusa* by Belyaeva *et al.*, (1982), *Alternaria alternata* and *A. solani* by Marcinkowswa (1982), *Fusarium semitectum* by Neduraman and Vidhyasekaran (1982), *Sclerotium rolfsi* by Prasad and Thakur (1984), *Phytophthora parasitica* by Hoy *et al.*, (1984), *Rhizoctonia solani* by Mathey and Oriolani (1987) and Rahiman (1988), *Alternaria alternata* by Kore and Gole (1990), *Phytophthora infestans* and *Alternaria solani* by Fontem (1993), *Fusarium oxysporum* f.sp. *lycopersici*, *Phylophthora nicotianae* var. *parasitica*, *P. infestans*, *Cladosporium fulvum Alternaria solani* and *Colletotrichum coccodes* by Gullino (1995), *Botrytis cinerea*, *Glomerella phomoides*, *Fulvia fulva* by Mizubuti and Brommonschenkel (1996), *Alternaria solani* and *Penicillium notatum* by Ramgiry *et al.*, (1997) and *Alternaria alternata* by Bhatt *et al.*, (2002).

The associated mycoflora was found to be variable as reported by different workers (Garcia, 1978; Belayeva *et al.*, 1982; Prasad and Thakur, 1984; Jamaluddin and Tondon, 1976; Chary *et al.*, 1981; Sonoda *et al.*, 1982; Okoli and Erinle, 1989; Sharma, 1994). The variation in mycoflora may be due to different in environmental conditions at different, localities and different varieties of tomato. Some fungi which were not isolated from the tomatoes of this region may be their non-occurrence in the area. The present report on the fungi suggests favourable conditions of the region for the mycoflora isolated.

The variation in the composition of mycoflora in different varieties of tomato reflects about the variable degree of resistance or susceptibility for the establishment of particular group of fungi in the varieties as varieties yielding maximum number of fungal species indicate their susceptibility. On the contrary, the varieties with poor incidence of fungi represent their resistant capacity. In the present study variety local showed high incidence of many fungi, indicating its susceptibility while, variety Pusa Ruby proved to be much resistant as it has low incidence of fungi.

The variety Vaishali is grown all over the Marathwada region, therefore, incidence of fungi on this was studied at 4 different localities of Marathwada and was noted that, fruits of this variety were susceptible to *Alternaria solani, Fusarium oxysporum, Aspergillus niger* and *Rhizopus stolonifer*, irrespective of the locality. The performance of the variety in relation to fungal pathogens was found to be uniform at all the localities. Many workers have studied on resistance of different varieties of tomato to fungal diseases. Waraitch *et al.*, (1976) recorded fruits of cvs T0-1 and Kalyanpur to be resistant and HS-102, Pioneer - II, Gemed, Sweet-72, Punjab Chhuhara, EC 55054, EC 55055, Marzano P-4 and Angulata as moderate resistant. Chauhan (1987) reported cvs. HS 101, HS 102, Gambde and A142 resistant to *Cladosporium fulvum* fruit rot, while resistant to *Phoma* fruit rot was shown by HS 101, HS 102 and Gambde Kaur *et al.*, (1988) reported varieties Ottowa 30, Ottowa 31, PI 112835, PI 260404 and PI 270407 to be resistant to late blight. Thareja and Srivastava (1989) found the varieties EC 129171, EC 128964 and EC 129166, EC 72901, EC 122962-1 and Hybrd-10 to be resistant to buckey rot.

References

Belyaeva, V, B., Levkina, L. M. and Kh. Kasteiyanos 1982. Infection of tomato by Alternaria and Cercospora sp. In Cuba. Mikologiya i. Fitopatologia 16(2): 139-143.

Bhatt, J.C., Gahlain, A. and Pant, S .K. 2000.Record of Alternaria alternata on tomato, Capsicum and Spinach in Kumaon hills. *Indian Phytopath.* 53(4): 495.

Chary, S.J., Kumar, B.P. and Reddy, S. M. 1981. Hitherto unrecorded post harvest diseases of tomato. *Indian Phytopath.*33(4): 624-625.

Chauhan, M.S. 1987. Reaction of tomato varieties to leaf spots and fruit rots in Haryana. Indian *J. Mycol. Pl. Pathol.* 17(3): 355.

Fontem, D. A. 1993. Survey of tomato diseases of Camroon.*Tropicultura* 11 (3): 87-90.

Garcia, J.L., Diaz, H. and Gonzalez, L. A. 1978. A new disease of tomato (Lycopersicon esculentum L.) Karsten fruits. Ciencias de la Agricultura No. 2: 47-52.

Gullino, M.L., Minuto, G. and Garibaldi, A. 1995. Fungal diseases of tomato grown in the green house: Development of the problems and possible solutions. Informatore Fitopatologico 45(9); 30-39.

Hoy, M. W. and Ogawa, J.M. 1984. Toxicity of surfacatant Naccotol to four decay causing fungi of fresh market tomatoes. *Plant Disease* 68(8): 699-703.

Jamaluddin and Tondon, M. P. 1976. Some New market diseases of vegetables and fruits. *Indian Phytopath.* 29.

Kogan, E.D. 1977. The pathogen of black rot of tomato fruits in Moldavia. Lzv. AN Mold. SSR, Biol. I Khim N No.2: 81-82. From Referativnyl Zhuranal 9:55.1248. (3): 385-386.

Kore, S.S. and Bhide, V.P. 1977. *Curvularia trifolii* and C. *lunata var.aeria* causal agents of leaf spot and fruit rot of tomato. *Indian J.Mycol.Pl.Patho.*6(2):194-195.

Kore, S.S. and Goel, M. D. 1990. Investigation on fruit rot of tomato causd by Alternaia alternata. *Indian Phytopath*. 43

Kuznstov, N.N. 1979. Brown spot of tomato in Sakhalin.Zaschita Rastenii No.3:30.

Marcinkowska, J. 1982. Fungi of the genus Alternaria occurring on tomato. *Acta Agronomica*.34(2):261-276.

Marras, F.P., Corda and Fiori, M. 1981. *Fusarium roseum* var.*avenaceum* (Sacc).Synd.and Hans,casual agent of soft rot of greenhouse tomatoes. *Studi Sessaresi* III 27:233-225.

Mathey, M.G. de and Oriolani, E.J.A. 1987. Fruit rot of tomato caused by *Rhizoctonia solani* Kuhn.in the Republic of *Argentina*.IDIA No.441-444:48-49.

Mizubuti, E.G.S. and Brommonschenkel, S.H. 1996. Diseases caused by fungi in tomato. Informe *Agropecuario* 18(4):7-14.

Narain, A. and Rout, G.B. 1989. Atomato rot caused by *Cladosporium tenuissimum.Indian Phytopath*.34(2):237-238.

Nene Y.L. and Thapliyal, P.N. 1993. Fungicides in plant disease control. Oxford and IPH publishing CO. Pvt.Ltd.,New Delhi., PP.531.

Okoli, C.A.N. and Erinle, D.D. 1989. Factors rsposible for market losses of tomato fruits in the Zaria area of Nigeria. *J.Hort.Sci*.64(1):69-71.

Rahiman, H. 1988. Anastomosis group of Rhizoctonia solani causing soil rot of tomato fruit in Majandaran. *Iranian J. Pl.Patho*.24(1-2):9-11.

Ramgiry, S.R., Tomar, I.S. and Tawar, M. L. 1997. Studies on mycoflora associate with pre and post harvest tomato fruits. *Crop Research* 13(1):231-233

Sharma R.L.1994. Prevalence of post harvest diseases of tomato in Himachal Pradesh. *Pl. Dis.Res*.9(2):195-197.

Skadow, K. 1978. Phytophthora nicotianae. Breda de Hans.var. nicotianae on green house tomatoes. *Archive fur phytopathologie und pflanzenschutz* 14(5):291-299.

Sonoda, R.M., Hayslip, N.C. and Stoffella, P.J. 1982. Tomato fruit rot infection cycle in a fresh market packing operation. *Proceedings of the Florida state Horticultural Society*. 94:281-282.

Sudhamoy Mandal, Srivastava, K.D., Rashmi Aggarwal and Singh, D.V. 1999. Mycoparasitic action of some fungi on spot bloch pathogen *(Drechslera sorokiniane)* On wheat. *Indian Phytopath*. 52(1) : 39-43.

Thareja, Rameshkumar and Srivastava, M.P. 1989. Varietal resistance of tomato against buckeye rot. *Indian Phytopath*.42(2):297.

Waraitch, K.S., Munshi, G.D. and Nandapuri, K.S. 1976. Screening of tomato cultivars for resistance against Rhizoctonia fruit rot. *Phytopathologia Mediterranea* 15(1): 68-69.

25

Screening of Efficient *Bradyrhizobium Japonicum* Strains for the Improvement of Soybean Production

S.K. Mahna[1], M.K. Meghavansi[1], D. Harwani[1], K. Prasad[1] and D. Wernner[2]

[1] Department of Botany, Maharshi Dayanand Saraswati University, Ajmeer-305009, Rajasthan (India) [2]FG Zellbiologie Und Angewandte Botanik, Fachbereich Biologie, Philipps-Universitat, marburg (Germany) Email : Incodevinindia@hotmail.com

ABSTRACT

Fourteen bradyrhizobial strains were isolated from five different soybean cultivars (JS 335, JS 71-05, NRC 2, NRC 7 and NRC 12) grown in four different agro-climatic regions of India viz., Kota, Bundi, Indore and Jabalpur. Efficiency of fourteen strains, after their authentication in growth pouch experiment, in terms of frequency of nodules, plant biomass, shoot nitrogen and shoot phosphorus was determined in sterile soil (mist house experiment) collected from the soybean fields of Agricultural Research Station (ARS), Ummedganj, Kota (Rajasthan) as well as in unsterile soil (field trial; conducted at ARS, Ummedganj, Kota). In the field trial, data pertaining to number of pods and seed weight were also taken into account for determination of efficiency of bradyrhizobial strains. Results showed that nodular frequency, plant biomass, shoot N and shoot P were significantly increased in soybean plants raised in sterile soil after bacterization while both increase as well as decrease in the studied parameters of plants was recorded in unsterile soil due to inoculation. Accordingly, all the fourteen bradyrhizobial strains were ranked in terms of their efficiency in sterile as well as in unsterile soil. In the present study, foreign bradyrhizobial strains have been observed to be more effective than indigenous strains for the studied parameters. Cultivar JS 335 is identified as better cultivar than other four cultivars as it showed relatively better interaction with its five bradyrhizobial strains. Looking to the overall performance of fourteen bradyrhizobial strains, tentatively five strains have been identified as most effective and competitive strains which can be used as bioinoculants for the improvement of soybean production in India.

KEY WORDS : *Bradyrhizobium japonicum*, soybean cultivars, efficiency, mist house experiment, field trial.

Soybean (*Glycine max* L. Merrill) is an agronomic crop of global importance. Soybean seeds are the main source of dietary vegetable oil which is nutritionally excellent oil and inexpensive source of protein (Sharma and Kothari, 1994). India ranks fifth in the world after United States, Brazil, China and Argentina regarding soybean area and production. In India, it is grown mainly in Madhya Pradesh, Maharashtra and in small pockets in other states which are Uttar Pradesh, Andhra Pradesh, Punjab, Tamil Nadu, Uttaranchal, Gujarat, Karnataka and Chattisgarh (Mahna, 2003). Soybean forms root nodules as part of a symbiotic association with *Bradyrhizobium* or *Rhizobium* (*Sinorhizobium*) bacteria (Hirsch and LaRue, 1997). The symbioses between *Rhizobium* or *Bradyrhizobium* and legumes are a cheaper and usually more effective agronomic practice for ensuring an adequate supply of N for legume based crop than the application of fertilizer-N (Zaharan, 1999). Production of soybean can be improved by application of symbiotically fixed nitrogen using effective *Bradyrhizobium japonicum* strains in pre-seeding bacterization. The improvement of soybean production in India shall have positive impact on the rural economy and socio-economic status of the Indian farmers. In the present investigation, experiments were conducted in sterile soil (mist house experiment) and unsterile soil (field trial) for the selection of efficient and competitive *Bradyrhizobium japonicum* strains belonging to five soybean cultivars to be used as bioinoculants for improvement of soybean production in India.

Materials and Methods

Nodule collection and isolation of bacterial strains : Soybean plants belonging to five different cultivars namely JS 335, JS 71-05, NRC 2, NRC 7 and NRC 12 were excavated from various field sites in Rajasthan (Kota and Bundi) and Madhya Pradesh (Jabalpur and Indore) during the 50% flowering stage of the crop and transported in plastic bags to the laboratory, where bacterial strains were isolated from the root nodules. In the process, nodules were separated from the roots and washed in sterilized distilled water for several times and following the serial dilution agar plate technique as described by Somasegaran and Hoben (1994), isolation of root nodule bacteria was done using the 20E medium (Werner *et al.*, 1975). Pure cultures were obtained with one or more further sub culturing steps.

Plant assay test (authentication of bacterial isolates): Nodulation ability of the representative isolates of each soybean cultivar was verified by their inoculation onto the seedlings of the parent cultivar. Surface sterilized seeds (4% NaOCl for 10 minutes) of five soybean cultivars were grown on plates of water agar (1%). Six day old seedlings were dipped into the mid log phase broth cultures of different bradyrhizobial strains. The bacterized seedlings were then planted in the growth pouches (procured from Mega International, USA) and were maintained in a growth chamber under controlled environmental conditions (photoperiod 14 hrs, light intensity 12000 lux, temperature 28 $\pm$ 1^{0}C and humidity 70-80%). Sterile N free Jensen medium (Somasegaran and Hoben, 1994) was given to the growing plants, twice each week. After four weeks, nodules were observed on the plant roots. Uninoculated plants served as control and they were observed unnodulated. Authenticated bradyrhizobial strains were nomenclatured so as to indicate their region of origin and name of cultivar (Table 1) and they are being maintained in glycerol vials at -40^{0}C using 20 E medium.

Mist house experiment : Air dried and sieved autoclaved soil (Vertisol; pH 8.0; EC

Table 1: Nomenclature of *Bradyrhizobium japonicum* strains with their respective cultivars and region of origin

S. No.	Bradyrhizobial strains	Parent cultivars	Region of origin
1	KJ 335-1	JS 335	Kota (Rajasthan)
2	KJ 335-7	JS 335	Kota (Rajasthan)
3	IJ 335-1	JS 335	Indore (Madhya Pradesh)
4	JJ 335-3	JS 335	Jabalpur (Madhya Pradesh)
5	BJ 335-1	JS 335	Bundi (Rajasthan)
6	KJ 71-3	JS 71-05	Kota (Rajasthan)
7	IJ 71-2	JS 71-05	Indore (Madhya Pradesh)
8	JJ 71-3	JS 71-05	Jabalpur (Madhya Pradesh)
9	IN 2-1	NRC 2	Indore (Madhya Pradesh)
10	JN 2-5	NRC 2	Jabalpur (Madhya Pradesh)
11	IN 7-2	NRC 7	Indore (Madhya Pradesh)
12	IN 7-3	NRC 7	Indore (Madhya Pradesh)
13	JN 12-1	NRC 12	Jabalpur (Madhya Pradesh)
14	JN 12-5	NRC 12	Jabalpur (Madhya Pradesh)

0.41 mmhos; organic matter 0.89%; available N 264.7 kg/ha; available P_2O_5 46 kg/ha and K_2O 200 kg/ha) collected during the month of June 2003 from the soybean cultivated fields of Agricultural Research Station Ummedganj, Kota (Rajasthan) was used to fill in earthen pots (8 kg/pot).

Seeds of five soybean cultivars procured from Soybean Breeding Research Centre, JNKVV, Jabalpur (Madhya Pradesh) were surface sterilized with 4% NaOCl for 10 minutes and then were inoculated with mid log phase cultures of fourteen authenticated *Bradyrhizobium japonicum* (BJ) strains using slurry method (Vincent, 1970). Amongst fourteen BJ strains, each strain was tested in its parent cultivar and uninoculated seeds of the same served as control. Five seeds per pot were sown for each strain and after emergence plants were thinned to two plants per pot. Plants were maintained in the mist house (light intensity 15000 lux – 19000 lux, temperature 27^0C - 35^0C and humidity 70% - 80%) and irrigated with water as needed.

After 45 days, plants were uprooted and data pertaining to number of fix +ve nodules (pink nodules when cut into half) and plant biomass were collected. Shoot N and shoot P of dried plants were estimated using Kjeldhal procedure and vando-molybdate method respectively (Jackson, 1973).

Field Trial : Experiment was conducted during Kharif 2003 at Agricultural Research Station (ARS), Ummedganj, Kota, Rajasthan (India).

Inoculant preparation and seed treatment : A mixture of phosphorus free sterilized charcoal (pH 7.00) and sand (3:1) was used as carrier for inoculant preparation. Sterilized carrier was inoculated with mid log phase cultures of bradyrhizobia (30 ml broth culture/ 100 g of carrier) aseptically. Bradyrhizobial multiplication and survival in the carrier was determined by performing CFU (colony forming units) count. Carrier inoculant having 10^{10} - 10^{11} bradyrhizobial cells g^{-1} was applied to soybean seeds of different cultivars using jaggery solution as a sticker material before the sowing. As the amount of bradyrhizobial culture to be added in the carrier depends upon its moisture holding capacity, therefore in the present work also, a specific amount of bradyrhizobial culture was standardized for mixing with used carrier.

Experimental design : Experiment was set up in a field plot of 25 X 15 meter in a complete randomized block design (RBD) in two replicates. For each strain 800 inoculated dry seeds were sown in four rows. Length of each row was kept five meter, with 50 cm distance between adjacent rows.

Plant harvesting was done at two stages of growth. Data on nodular frequency and plant biomass were recorded after 45 days at the 50% flowering stage and data on number of pods and grain yield were recorded after 90 days at maturation stage. Nitrogen and phosphorus content of 45 days old shoot were estimated in three replicates.

Statistical analysis : Analysis of variance (ANOVA) was performed on the data pertaining to nodular frequency, plant biomass and number of pods per plant, and least significant difference (P = 0.05) was calculated. Accordingly significant difference between the data values has been marked in the Tables 2 and 3.

Table 2 Influence of *Bradyrhizobium japonicum* strains on nodular frequency, plant biomass, shoot N and shoot P of soybean grown in earthern pots under mist house conditions

Bradyrhizobial strains	No. of fix +ve nodules[1]	Plant biomass (g)[1]	Shoot N (%)[2]	Shoot P (%)[2]
Control	0.00	6.06[b] ± 0.10	1.38	0.181
KJ 335-1	32.40[b] ± 0.81	12.56[a] ± 0.33	2.49	0.230
KJ 335-7	25.80[c] ± 0.73	10.52[a] ± 1.29	2.12	0.192
IJ 335-1	42.00[a] ± 1.80	11.54[a] ± 0.95	2.55	0.226
JJ 335-3	28.40[c] ± 1.83	10.14[a] ± 0.73	2.16	0.194
BJ 335-1	42.40[a] ± 0.74	13.01[a] ± 1.46	2.69	0.232
Control	0.00	5.58[b] ± 0.25	1.09	0.178
KJ 71-3	32.40[a] ± 1.54	9.50[a] ± 0.44	2.17	0.213
IJ 71-2	29.40[b] ± 1.03	8.69[a] ± 0.54	2.10	0.182
JJ 71-3	22.80[c] ± 1.16	8.47[a] ± 0.30	1.85	0.193
Control	0.00	5.18[b] ± 0.67	1.21	0.163
IN 2-1	25.00[a] ± 1.30	11.50[a] ± 0.36	2.06	0.186
JN 2-5	25.60[a] ± 1.11	12.24[a] ± 0.82	2.09	0.193
Control	0.00	5.59[b] ± 0.33	1.17	0.171
IN 7-2	29.00[a] ± 1.14	10.17[a] ± 0.91	2.09	0.211
IN 7-3	25.80[b] ± 1.28	11.38[a] ± 1.44	2.05	0.234
Control	0.00	5.06[c] ± 0.20	0.89	0.167
JN 12-1	26.80[a] ± 1.07	10.25[b] ± 0.44	1.93	0.186
JN 12-5	28.60[a] ± 1.44	12.47[a] ± 0.68	2.10	0.19

1. Data recorded after 45 days of sowing and are average means of 5 plants.
2. N and P contents of 45 days old plant shoots (values are means of 3 replicates).
± Standard error; Values without common letters differ significantly at LSD P = 0.05.

Table 3 Influence of *Bradyrhizobium japonicum* strains on nodular frequency, plant biomass, shoot N, shoot P and productivity of soybean grown under field conditions

Bradyrhizobial strains	No. of fix +ve nodules[1]	Nodules fresh weight (g)[1]	Plant biomass (g)[1]	Shoot N (%)[2]	Shoot P (%)[2]	No. of pods/ plant[3]	Seed yield / plant (g)[3]
Control	$21.40^b \pm 0.35$	$0.130^b \pm 0.018$	$22.78^b \pm 0.34$	2.51	0.296	$28.00^b \pm 0.43$	5.06
KJ 335-1	$22.80^{ab} \pm 0.43$	$0.157^a \pm 0.013$	$25.89^a \pm 0.47$	2.66	0.312	$30.80^a \pm 0.88$	5.75
KJ 335-7	$15.00^d \pm 0.55$	$0.100^b \pm 0.015$	$22.49^b \pm 0.52$	2.62	0.284	$24.20^c \pm 0.82$	4.83
IJ 335-1	$23.60^{ab} \pm 0.50$	$0.171^a \pm 0.015$	$23.41^b \pm 0.48$	2.56	0.312	$29.60^{ab} \pm 0.47$	5.91
JJ 335-3	$18.80^c \pm 00.47$	$0.130^b \pm 0.021$	$21.12^{bc} \pm 0.44$	2.42	0.281	$25.40^b \pm 0.83$	5.00
BJ 335-1	$24.80^a \pm 0.69$	$0.176^a \pm 0.031$	$26.74^a \pm 0.43$	3.32	0.317	$31.00^a \pm 0.61$	6.16
Control	$21.40^a \pm 0.50$	$0.096^b \pm 0.011$	$20.72^a \pm 0.51$	2.18	0.271	$24.40^a \pm 0.54$	5.16
KJ 71-3	$22.20^a \pm 0.62$	$0.103^b \pm 0.017$	$19.95^a \pm 0.89$	2.30	0.283	$24.20^a \pm 0.71$	5.00
IJ 71-2	$20.00^a \pm 0.63$	$0.124^a \pm 0.021$	$18.04^b \pm 0.49$	2.36	0.316	$17.00^b \pm 0.65$	4.50
JJ 71-3	$18.40^a \pm 0.54$	$0.095^b \pm 0.022$	$16.15^b \pm 0.65$	2.16	0.249	$18.80^b \pm 0.56$	4.75
Control	$12.00^b \pm 0.39$	$0.069^b \pm 0.031$	$17.77^b \pm 0.75$	2.14	0.258	$19.00^b \pm 0.70$	4.02
IN 2-1	$14.80^a \pm 0.45$	$0.098^a \pm 0.009$	$22.72^a \pm 0.70$	2.68	0.292	$24.20^a \pm 0.71$	4.58
JN 2-5	$14.60^a \pm 0.51$	$0.076^b \pm 0.010$	$20.03^a \pm 0.75$	2.48	0.356	$23.20^a \pm 0.62$	4.83
Control	$18.40^a \pm 0.57$	$0.120^a \pm 0.014$	$16.12^b \pm 0.55$	2.38	0.268	$20.80^b \pm 0.75$	4.33
IN 7-2	$18.20^a \pm 0.54$	$0.119^a \pm 0.016$	$18.00^b \pm 0.61$	2.52	0.279	$24.40^a \pm 0.68$	4.16
IN 7-3	$14.00^a \pm 0.48$	$0.084^b \pm 0.022$	$22.16^a \pm 0.59$	2.28	0.298	$23.00^a \pm 0.62$	4.66
Control	$14.40^b \pm 0.42$	$0.080^b \pm 0.019$	$18.93^b \pm 0.70$	2.60	0.277	$21.00^c \pm 0.61$	4.27
JN 12-1	$20.40^a \pm 0.54$	$0.127^a \pm 0.008$	$19.25^b \pm 0.51$	2.60	0.299	$25.60^b \pm 0.43$	4.55
JN 12-5	$18.60^a \pm 0.63$	$0.159^a \pm 0.011$	$26.82^a \pm 0.58$	2.94	0.366	$30.20^a \pm 0.61$	5.03

1. Data recorded after 45 days of sowing and are average means of 25 plants.
2. N and P contents of 45 days old plant shoots (values are means of 3 replicates).
3. Data recorded after 90 days of sowing and are average means of 10 plants.
±   Standard error; Values without common letters differ significantly at LSD P = 0.05.

Results and Discussion

Isolation and authentication of bradyrhizobial strains: To begin with, a good number of isolates were obtained from root nodules of soybean plants grown in Madhya Pradesh and Rajasthan. All the isolates were subjected to authentication test before performing any experiment. In the present investigation, fourteen strains isolated from five different cultivars were used for inoculation of seeds after they were authenticated in growth pouches. In the initial stage, plant assay test for few bradyrhizobial strains was performed in growth pouches as well as in Leonard jars but results clearly revealed that use of growth pouches is more convenient, cheaper and less time consuming than other conventional methods (Mahna *et al.*, 2003).

Microscopic observation of the authenticated BJ strains revealed that all the strains were Gram-ve, long, motile rods and formed small, circular, glistening and whitish colonies on 20 E medium.

Mist house experiment : In the mist house experiment successful nodulation was observed on roots of BJ inoculated soybean plants and no nodulation was recorded on the uninoculated plants. In general all the strains infected the upper most root system as majority of the nodules were clustered asymmetrically on the upper most part of primary/ lateral roots. Considerable variability in the frequency of fix +ve nodules was observed in the plants of five cultivars raised after treatment with fourteen BJ strains. Amongst tested strains, highest frequency of nodules (Av. 42.40) was recorded in the plants of JS 335 cultivar inoculated with BJ335-1 strain while lowest frequency of nodules (Av. 22.80) was observed in the plants of JS 71-05 cultivar inoculated with JJ71-3 strain.

Similarly, variable influence of seed inoculation with fourteen BJ strains was also observed on plant biomass. A significant per cent increase in the plant biomass over control ranging 67.33% - 114.69 % in JS 335 cultivar; 51.79 % - 70.25 % in JS 71-05 cultivar; 122.00 % - 136.29 % in NRC 2 cultivar; 81.93 % - 103.58 % in NRC 7 cultivar and 102.57 % - 146.44 % in NRC 12 cultivar was recorded. Likewise, noticeable enhancement in the nitrogen and phosphorus contents of shoots of BJ inoculated plants over control was assessed. Four BJ strains namely BJ 335-1, KJ 71-3, JN2-5 and JN 12-5 caused maximum accumulation of both shoot N and shoot P in the plants of parent cultivars while highest values of shoot N after IN7-2 inoculation and shoot P after IN 7-3 treatment were estimated in the plants of NRC 7 cultivar.

Field trial : Under the field condition, variable performance of soybean plants of five cultivars in terms of vegetative, reproductive and biochemical characters was observed after bactization with fourteen BJ strains. Comparative data pertaining to frequency of fix +ve nodules formed on the roots of inoculated and uninoculated plants revealed that high frequency of nodules was formed on roots of eight BJ strains inoculated plants in the following order: BJ 335-1, IJ 335-1, KJ 335-1, KJ 71-3, JN 12-1, JN 12-5, IN2-1 and JN 2-5.Contrary to above, low frequency of nodules was formed in the plants raised after inoculation with BJ strains in the following order: IJ 71-2, JJ 335-3, JJ 71-3, IN7-2, KJ 335-7 and IN7-3. Significant increase over control in nodule fresh weight was registered after inoculation of BJ 335-1, IJ 335-1, JN 12-5, KJ 335-1, IJ 71-2, JN 12-1 and IN2-1 BJ strains. Non-significant increase/decrease over control in nodule fresh weight was noticed in the plants grown after inoculation with remaining seven BJ strains.

Evaluation of data pertaining to plant biomass exhibited that bacterization of seeds of soybean with KJ 335-1, BJ 335-1, IN 2-1, JN 2-5, IN 7-3 and JN 12-5 BJ strains, significantly enhanced the plant biomass. On the other hand, treatment of seeds with IJ 335-1, IN 7-2 and JN 12-1 BJ strains could also increase the plant biomass but the enhancement was statistically non significant. Contrary to above, significant decrease in plant biomass after seed inoculation with IJ 71-2 and JJ 71-3 strains and non significant decrease in plant biomass after seed inoculation with KJ 335-7, JJ 335-3 and KJ 71-3 strains were recorded.

Alterations in shoot N and shoot P of different soybean cultivars after seed bacterization were observed. Majority of the BJ strains showed potentiality to enhance the shoot N (1.99 % - 32.27 % increased shoot N) except bacterization with JN12-1 (unchanged shoot N value) and JJ 71-3, JJ 335-3 and IN 7-3 (0.92 % - 9.52 % decreased shoot N) when compared with their respective control plants. An increase over control in shoot P ranging 4.10 % - 37.98 % was observed after the seeds inoculation with majority of BJ strains except in three strains (KJ335-7, JJ 335-3 and JJ71-3) where a decrease (4.05 % - 8.76 %) was assessed in shoot P.

As per the data computed in Table 3 on the number of pods formed on the plants of five soybean cultivars, significant increase in the pod number was noted in the plants of different cultivars grown after inoculation with KJ 335-1, BJ 335-1, IN 2-1, JN2-5, IN7-2, IN 7-3 and JN12-5 strains while non significant increase/decrease in pod number was observed over control in the plants inoculated with remaining BJ strains. Taking into account the data on seed yield, fourteen BJ strains exhibited variable impact on the five soybean cultivars. KJ 335-1, IJ335-1, BJ 335-1, IN 2-1, JN 2-5, IN7-3, JN12-1 and JN 12-5 BJ strains enhanced the seed yield (6.56 % - 21.74 %) over the control while a decrease in seed yield (1.19 % - 12.79 %) was recorded in the plants raised from the seeds bacterized with KJ 335-7, JJ 335-3, KJ 71-3, IJ 71-2, JJ 71-3 and IN7-2 strains.

Present results showed that number of fix +ve nodules, plant biomass, shoot N and shoot P were significantly increased in the soybean plants raised after seed inoculation with fourteen BJ strains in sterilized soil (mist house experiment). However variations in the increased value of fix +ve nodules and plant biomass as per the LSD test ($P < 0.05$) and increase in per cent shoot N and per cent shoot P were observed. Taking into account data of all the four studied parameters in sterilized soil, fourteen strains were grouped into three categories i.e., highly efficient strains (Indigenous; KJ 335-1 and KJ 71-3, Non-Indigenous; BJ 335-1, IJ 335-1, IN7-3 and JN 12-5), moderately efficient strains (Non-Indigenous; IJ 71-2, IN 2-1, JN 2-5, IN 7-2 and JN 12-1) and efficient strains (Indigenous; KJ 335-7, Non-Indigenous; JJ 71-3). On the other hand, both increase as well decrease in the number of fix +ve nodules, plant biomass, shoot N, shoot P and yield of soybean plants after bacterization with fourteen BJ strains were observed in the unsterile soil (field trial). Accordingly, bradyrhizobial strains were ranked as highly efficient strains (Non-Indigenous; BJ 335-1, IN 2-1, JN 2-5 and JN 12-5), moderately efficient strains (Indigenous; KJ 335-1, Non-Indigenous; IJ 335-1 and JN 12-1) and poor strains (Indigenous; KJ 335-7 and KJ 71-3, Non-Indigenous; JJ 335-3, IJ 71-2, JJ 71-3, IN 7-2 and IN 7-3). Some what similar to present finding, Zdor and Pueppke (1988) have pointed out more nodules in autoclaved soil than unsterile soil in their experiment pertaining to interaction of soybean and *Bradyrhizobium japonicum* 123 and 138 serogroups and formation of high number of nodules in sterile soil is attributed to lack of antagonists in the rhizosphere. In addition to this, Lukiwati and Simanungkalit (2002) have also reported that dry matter, N and P uptake of soybean plants were significantly increased by bradyrhizobial inoculation in sterilized soil experiment.

After assessing the ranking of bradyrhizobial strains in mist house experiment and in field trial, the present study suggests that foreign bradyrhizobial strains are more effective than indigenous strains for the studied parameters. Parallel to our results, Okereke *et al.* (2001) have also described more effectiveness of foreign bradyrhizobial strains than native strains in enhancing nodulation, dry matter and seed yield of soybean cultivars in Nigeria.

While analyzing the recorded values of studied parameters in the plants of five soybean cultivars bacterized with their respective strains both in sterile soil and unsterile soil, cultivar JS 335 is identified as better cultivar than the other four cultivars as it showed relatively better performance after interacting with its five BJ strains. Okereke *et al.* (2001) have considered cultivar as a main factor and bradyrhizobial strains as subfactor and concluded that inoculum response was cultivar and site specific. Likewise, cultivar Jackson was identified as the most tolerant for N_2 fixation of eight cultivars evaluated (Sall and Sinclair, 1991). Contrary to above, Lukiwati and Simanungkalit (2002) have reported non significant variations in interaction between varieties of soybean and bradyrhizobial inoculation.

Taking into account the ranking of fourteen bradyrhizobial strains in mist house experiment as well as in field trial, a few important conclusions were derived. Three strains namely JN 12-1, BJ 335-1 and JN 12-5 showed consistency in their performance in both sterile and unsterile soil thereby indicating their better competitiveness in facing the antagonistic effects of soil microflora. Efficiency values of remaining strains except two namely IN 2-1 and JN 2-5 as recorded in the mist house experiment were variably reduced to different levels in the field experiment thus clearly indicating the antagonistic effect of microflora in rhizospheric soil.

In view of the results obtained in the present study, tentatively five bradyrhizobial strains (BJ 335-1, IN 2-1, JN 2-5, JN 12-1 and JN 12-5) have been identified as most efficient and competitive strains. A field trial is under progress utilizing selected strains as bioinoculants for confirming their efficiency and their future use for the improvement of soybean production in India.

Acknowledgements

This research work was supported by the financial aid of Euro Commission, Brussels (Contract No. ICA 4-CT-2001-10057). Thanks are also due to Soybean Breeding Research Centre, JNKVV, Jabalpur (Madhya Pradesh) for providing seeds of different soybean cultivars and to Zonal Director, ARS Ummedganj, Kota (Rajasthan) for providing land for the field trial..

References

Hirsch, A.M. and *LaRue, T.A.* 1997. Is the legume nodule a modified root or stem or an organ *sui generis* ? *Crit Rev plant Sci* **16**: 361-392.

Jackson, M. L. 1973. *Soil Chemical Analysis.* Prentice Hall of India. New Delhi.

Lukiwati, D. R. and *Simanungkalit, R. D. M.* 2002. Dry matter yield, N and P uptake of soybean with *Glomus manihotis* and *Bradyrhizobium japonicum.* 17th WCSS, 14-21 August, Thailand pp. 1190 (1-8).

Mahna, S.K. 2003. Production, regional distribution of cultivars and agricultural aspects of soybean in India. In: *Agriculture, Forestry, Ecology and the Environment.* (Eds. Werner, D and Newton, W.E.) Kluwer Academic Publishers. Printed in the Netherlands.

Mahna, S.K., Prasad, K., Vedi, S. and *Meghvansi, M.K.* 2003. Isolations, authentication and efficiency of indigenous Bradyrhizobial strains of Rajasthan. *In: Biosciences: Advances, Impact and Relevance.* (Ed. Singh, V. P.) pp.33-37. Department of Plant Science, M.J.P. Rohilkhand University , Bareilly.

Okereke, G. U., Onochie, C., Onunkwo, A. and ***Onyeagba, E.*** 2001. Effectiveness of foreign bradyrhizobia strains in enhancing nodulation, dry matter and seed yield of soybean (*Glycine max* L.) cultivars in Nigeria. *Biology and Fertility of Soil* ***33 (1)***: 3-9.

Sall, K. and ***Sinclair, T. R.*** 1991. Soybean genotypic differences in sensitivity of symbiotic nitrogen fixation to soil dehydration. *Plant and Soil* ***133***: 31-37.

Sharma, V.K. and ***Kothari, S.L.*** 1994. Somatic embryogenesis in soybean. *Journal of Indian botanical Society* ***73***: 241-243.

Somasegaran, P. and ***Hoben, H.J.*** 1994. *Handbook for Rhizobia : Methods in legume-Rhizobium Technology.* Springer, New York.

Vincent, J.M. 1970. *A manual for the practical study of root nodule bacteria.* IBP handbook 15. Blackwell Scientific Publications, Oxford, England.

Werner, D., Wilcockson, J. and ***Zimmermann, E.*** 1975. Absorption and selection of rhizobia by ion exchange papers. *Arch. Microbiol.* ***105***: 27-32.

Zaharan, H.H. 1999. *Rhizobium*-legume symbiosis and nitrogen fixation under severe conditions and in an arid climate. *Microbiology and Molecular Biology Reviews* ***63 (4):*** 968-989.

Zdor, R. E. and ***Pueppke, S. G.*** 1988. Early infection and competition for nodulation of soybean by *Bradyrhizobium japonicum* 123 and 138. *Applied and Environmental Microbiology* ***54 (8)***:1996-2002.

26 | *Studies on Pathogenicity of Fungi of Tomato*

Ashok Chavan, Shubhada Bharaswadkar and Vasant Rao Sonawane
Botany Research Centre Vasant Rao Naik Mahavidyalaya, Aurangabad

ABSTRACT

Tomato fruits and seeds were artificially inoculated to know the pathogenicity of fungi. Tested fungi showed several abnormalities in the seedlings whereas intensity of rotting at different stages also varied with different fungi.

KEYWORDS : Pathogenicity, tomato, fungi, rotting.

Tomato (*Lycopersicon esculentum* Mill.) is one of the most important solanaceous crops. It is grown in both *kharif* and *rabi* seasons. It has become well known popular vegetable grown throughout India. The fruits of tomato specially are most important due to their specific nutritive values. It is clear from the literature that the crop suffers with variety of diseases, which cause heavy losses in yield. The fungal pathogens have been found to affect and damage the tomato fruits in the field as well as in the market during storage and transport. Such infected fruits and seeds are poor in quality, both for propagation and consumption.

It was found that tomato fruit during their developmental stages in the field showed variation in composition of associated mycoflora. To detect the role of fungi in fruit rotting, experiments were carried out in order to study pathogenicity of fruits and seeds of tomato and few common fungal pathogens were selected for the study.

Materials and Methods

Healthy fruits and seeds of tomato var. Vaishali were selected for pathogenicity test. Pathogenicity was tested using Koch's postulates. In order to study the effect of spore suspension on tomato, freshly harvested tomato fruit at different developing stages (Green, Yellowish red, red) were collected and surface sterilized with mercuric chloride, washed with distilled water for three times and fruits were injected by pin prick method. The inoculated fruits were placed in pre sterilized polythene bag.

To study effect of fungi on seeds and its germinability, seeds were sterilized with 0.1% mercuric chloride and washed with distilled water the seeds were rolled on actively growing fungal culture and and incubated at room temperature for 7 days.

Results and Discussion

Effect of fungi on seeds and its germinability is summarised in Table 1. It appears that maximum seed rotting takes place due to *Fusarium oxysporum* followed by *Curvularia lunata, Aspergillus flavus, A. ustus* and *Drechslera tetramera*. Rate of seed germination is also varied. Rotten seeds did not germinate but other germinated seeds developed seedlings showed several abnormalities. Length of the shoot is reduced due to *Trichoderma viride, Curvularia lunata, Penicillium purpurogenum, Cladosporium oxysporum, Aspergillus ustus, Fusarium oxysporum*. Root lengths also reduced due to *Trichoderma viride, Fusarium oxysporum, Curvularia lunata, Drechslera tetramera, Cladosporium oxysporum*. After 7 days of incubation period root rotting takes place due to *Aspergillus flavus, A. ustus* and *Drechslera tetramera*. Seed colour was changed to brown due to *Alternaria alternata* and *A. solani* and *Curvularia lunata*.

Effect of fungi was tested against the intensity of fruit rotting. It becomes clear from the results given in Table 2, that *Aspergillus flavus* and *Trichoderma viride* started rotting on third day. On fifth day *Aspergillus* species and *Trichoderma viride* showed moderate rotting. As the duration increases fungi showed severe rotting to the fruits.

Mehta *et al.*, (1976) reported that injury to the fruit by pin prick with addition of spore suspension in tomato proved to be best for the fact disease development causing fruit rot of tomato by *Alternaria alternata*. Whereas Swami (2002) noted maximum rotting of tomato fruits after ten days by *Fusarium roseum, F. oxysporum, Aspergiullus flavus, A. niger* and *Rhizopus stolonifer.*

In order to study the effect of spore suspension of test fungi on tomato, tomato fruits at different developing stages were inoculated artificially with different fungi and the results are summarised in Table 3.

It was observed from the results that *Cladosporium oxysporum* and *Penicillium purpurogenum* did not show any symptoms on green stage whereas *Fusarium oxysporum* showed maximum rotting. Moderate rotting was reported with *Alternaria alternata, Curvularia lunata, Drechslera tetramera* at yellow stage of tomato. Whereas severe rotting was reported at red and ripe stage by *Aspergillus flavus, A. ustus, Fusarium oxysporum* and *Penicillium purpurogenum*. This clearly indicates that fungi produces hydrolytic enzymes due to which maceration of fruit might have resulted in the rotting. Tani *et al.*, (1978) reported half ripe to mature tomato fruits were susceptible to *Alternaria alternata* while Arya *et al.*, (1988) reported fruit rotting in inoculated tomato fruits with *Aspergillus* sp., *Fusarium* sp. and *Penicillium* sp. Gallo *et al.*, (1988) also reported rotting due to *Alternaria alternata, Rhizoctonia solani, Collectotrichum coccodes, Botrytis cinera.*

Table 1 Effect of fungal infestation on seeds / seedlings of tomato

Seed mycoflora	% seed rot	Seed Colour	Shoot Length abnormalities	Shoot Root abnormalities	Length	Root
Alternaria alternata	30	Brown	2.2	Stunted	4.5	Normal
Alternaria solani	28	Brown	2.0	Blight	4.7	Normal
Aspergillus flavus	39	—	1.7	Stunted	3.9	Rot
Aspergillus ustus	38	—	1.2	Tip rot	4.0	Rot
Cladosporium oxysporum	27	Grey	1.1	Stunted	3.0	Shortening
Curvularia lunata	40	Brown	1.0	Stunted	2.9	Shortening
Drechslera tetramera	38	Black	2.1	Tip rot	3.5	Rot
Fusarium oxysporum	63	—	1.2	Tip rot	2.6	Shortening
Penicillium purpurogenum	21	Yellow	1.1	Tip rot	3.2	Rot
Trichoderma viride	30	Green	1.0	Blight	2.5	Shortening
Control	00	—	4.2	Normal	6.2	Normal

Table 2 Effect of incubation period on fruit rotting due to fungi

Spore suspension of fungi	Intensity of rotting of tomatoes after days					
	01	03	05	07	09	11
Alternaria alternata	—	—	+	+++	++++	++++
Alternaria solani	—	—	+	+++	++++	++++
Aspergillus flavus	—	+	++	++++	++++	++++
Aspergillus ustus	—	—	++	+++	++++	++++
Cladosporium oxysporum	—	—	+	++	+++	++++
Curvularia lunata	—	—	+	+++	++++	++++
Drechslera tetramera	—	—	+	++	+++	++++
Fusarium oxysporum	—	—	++	+++	++++	++++
Penicillium purpurogenum	—	—	+	+++	+++	++++
Trichoderma viride	—	+	++	+++	++++	++++

— = No Rotting   +++ = Severe Rotting

+ = Minor Rotting   ++++ = Total Rot

++ = Moderate Rotting

Table 3 Pathogenicity of fungi on different stages of tomato fruit

Fungi	Stages		
	Green and young	Yellow	Red
Alternaria alternata	+	++	+++
Alternaria solani	+	+	+++
Aspergillus flavus	+	+++	++++
Aspergillus ustus	+	+++	++++
Cladosporium oxysporum	—	+	++
Curvularia lunata	+	++	+++
Drechslera tetramera	+	++	++
Fusarium oxysporum	+++	+++	++++
Penicillium purpurogenum	—	+++	++++
Trichoderma viride	++	++	+++

— = No Rotting   +++ = Severe Rotting

+ = Minor Rotting   ++++ = Total Rot

++ = Moderate Rotting

References

Arya, C.M., Rivera, G. and **Campos, D.** 1988. Identification and Pathogenicity of fungi associated with tomatoes from the field and Market in Costa Rica *Fitopathologia* **23**(1) : 1 – 4.

Gallo Llobet, L. Barroso Espinosa, J.J. Hernandez, J. and **Manisto Perez P.** 1988. Diseases of high incidence on tomato in the canary Islands II. *Centro de investigacion Y. Technologia Agrarias,* 60 – 64.

Mehta, P., K.M. Vyas and **S.B. Sakena** 1976. Pathological studies on Fruit rot of tomato caused by *Alternaria alternata* and *Alternaria solani. Indian Phytopath, Uni. Saugar, Saugar, M.P., India* **28**(2): 247 – 252.

Swami, C.S. 2002. Studies on Pathogenic and non pathogenic fungi of tomato fruits. Ph.D. thesis, Dr. B.A.M.U., Aurangabad (M.S.).

Tani T., Yamamota H., Marimoto T., Kawada K. and **Kitagava H.** 1978. Fruit rot of tomato caused by *Alternaria alternata* during various storage conditions. *Tech Bull. Agric. Kagawa Univ.* **29**(2): 277 – 282.

Unit – III

Molecular Biology

27

Phylogenetic Relationships of Colletotrichum Species Infecting Rubber (Hevea brasiliensis) based on Nuclear Ribosomal DNA Spacer Sequences

C. Bindu Roy[1, 2], M. Ravindran[1], C. Kuruvilla Jacob[2], M. A. Nazeer[1] and T. Saha[1*]

[1]Genome Analysis Laboratory; [2]Plant Pathology Division, Rubber Research Institute of India, Kottayam 686 009, Kerala.

* Corresponding author (e-mail : saha@rubberboard.org.in)

ABSTRACT

Colletotrichum leaf fall disease in rubber caused by the *Colletotrichum spp.* produces three different disease symptoms, namely, raised spots, anthracnose and circular papery lesions. Involvement of two different species of the pathogen namely *C. gloeosporioides* and *C. acutatum* could be established through rDNA-ITS-RFLP analysis of the fungal isolates from rubber. Ribosomal DNA spacer sequences from both the *Colletotrichum* species infecting rubber were compared to get an idea about the nucleotide divergence existing among them in the spacer regions including 5.8 S gene. Aligned sequence data of the spacer regions revealed the existence of more nucleotide divergence including base substitutions and indels in the ITS 1 compared to the ITS 2 region. Phylogeny of the rubber pathogen with the closely related fungal isolates from different hosts, based on the rDNA spacer sequences, clearly revealed their uniqueness. The bootstrapped consensus tree derived through neighbour-joining method comprised of two major branches having the fungal isolates belonging to two different species indicating clear species delineation of *Colletotrichum*. The fungal isolates belonging to *C. gloeosporioides* and *C. acutatum* infecting *Hevea* appeared to be closely related to the pathogens infecting fragaria and cyclamen respectively.

KEYWORDS : *Hevea brasiliensis, Colletotrichum gloeosporioides, Colletotrichum acutatum,* ITS-RFLP, rDNA spacers, phylogenetic relationship.

The fungal genus *Colletotrichum* Corda includes several pimportant plant pathogens that infect a wide range of both tropical and temperate crop plants (Jeffries *et al.*, 1990). These fungi cause diseases commonly known as anthracnose of grasses, legumes, vegetables, small fruits and perennial tree crops (Waller, 1992). Colletotrichum leaf disease is one of the major causes of declining yields of para rubber (*Hevea brasiliensis*) in the South East Asia. The Colletotrichum leaf disease in rubber was reported first in the year 1906 from Ceylon and *Colletotrichum heveae* (Synonym: *Colletotrichum gloeosporioides*) was identified as the causative agent of this disease (Petch, 1906). Later the disease was reported from many parts of the world. Two different species of *Colletotrichum* have been reported on rubber causing three different disease symptoms: *C. gloeosporioides* (teleomorph: *Glomerella cingulata*) for anthracnose and papery lesions on mature rubber leaves, and *C. acutatum* for raised spot symptoms (Saha *et al.*, 2002). From India as well as Sri Lanka it has been reported that the main pathogen of Colletotrichum leaf disease of rubber is *C. acutatum* (Jayasinghe *et al.*, 1997; Kumar *et al.*, 2002; Saha *et al.*, 2002).

In several *Colletotrichum* species, conventional taxonomic characters such as conidial shape, size, appressorium morphology as well as pathogenicity vary widely (Sreenivasaprasad *et al.*, 1996). Morphologically indistinguishable *Colletotrichum* isolates have been assigned different species names based on their host origin. The existence of more than one species on a single host, like rubber, causing disease complexes, further complicates the situation. Sutton (1992) suggested that relationships within the genus *Colletotrichum* were unlikely to be resolved using morphology alone as morphological plasticity and overlapping phenotypes make traditional taxonomic criteria unreliable for accurate delineation of *Colletotrichum* species. Molecular techniques exploiting variations in the ribosomal DNA (rDNA) are being used extensively for systematic and phylogenetic studies in fungal pathogens (Sreenivasaprasad *et al.*, 1994; Johnston and Jones, 1997; Martin and Garcia-Figueres, 1999; Green *et al.*, 2004). Different regions of the rDNA diverged at different rates allowing the regions to be exploited at different taxonomic levels (Bruns *et al.*, 1991). The non-coding rDNA regions, especially the internal transcribed spacer regions (ITS 1 and ITS 2), are generally more variable than the rRNA genes and also be useful for understanding phylogenetic relationships at a sub-generic level. Use of rDNA restriction digest analysis and sequence data as molecular methods for species delineation in *Colletotrichum* have been reported (Sreenivasaprasad *et al.*, 1994; Sreenivasaprasad *et al.*, 1996; Saha *et al.*, 2002). The ITS sequence data offers potentially enough characters for phylogenetic reconstruction and one of the advantages of ITS is that it is flanked by regions that are highly conserved among genera and species making PCR analysis and sequencing straight forward.

In rubber, two species of *Colletotrichum* were identified through molecular analysis including RAPD markers and restriction fragment length polymorphism (RFLP) of ribosomal DNA spacers (Saha *et al.*, 2002). In the present study, we have revalidated the use of rDNA-ITS RFLPs for species delineation of *Colletotrichum* pathogens infecting *Hevea* and also determined the nucleotide sequences of ribosomal ITS regions including the 5.8 S ribosomal gene from both the species, *C. gloeosporioides* and *C. acutatum*, to evaluate divergence of the nucleotide sequences in this region between these two species. The sequence data were also used to infer the phylogeny of the *Colletotrichum* pathogens from rubber with the same species from other crops.

Materials and Methods

Colletotrichum Isolates and Genomic DNA Extraction

The isolates from three typical symptoms, raised spots, anthracnose and papery lesions on leaflets of different clones of *H. brasiliensis* were collected from rubber plantation of the Rubber Research Institute of India (RRII), Kottayam and other locations in Kerala (Table 1). Twenty-five isolates (14 from raised spots, 6 from anthracnose and 5 from papery lesions) were used in the present study. Isolations were made by plating surface-sterilized pieces of diseased tissue on oat meal agar (OMA). All isolates were purified by single-spore culture, maintained on OMA and stored at 5°C in the dark. Fungal isolates were identified based on culture characteristics and microscopic morphology. Three mycelial plugs (5 mm diameter) were removed from the advancing margins of 5-day-old culture of each isolate, transferred to oat meal broth and incubated at 25°C on an orbital shaker (100 rpm) for 4 days for the extraction of genomic DNA. About 300-500 mg of mycelium was harvested by vacuum filtering through Whatman No.1 filter paper, washed three times with sterile distilled water and were ground to powder in liquid nitrogen. Extraction and purification of the total genomic DNA were carried out following a modified CTAB (hexadecyl trimethyl ammonium bromide, Sigma Co., St. Louis, USA) method optimised for *Corynespora cassiicola* (Saha *et. al.*, 2000).

PCR Amplification of rDNA-ITS Regions

Amplification of the internal transcribed spacer regions between the small (18S) and large (28S) nuclear rDNA was carried out using primers ITS 1 (5'-TCCGTAGGTGAACCTGCGG-3') and ITS 4 (5'-TCCTCCGCTTATTGATATGC-3') (White *et al.*, 1990) synthesised at Sigma Genosys, UK. The amplified fragment includes the 5.8S rDNA gene and the internal transcribed spacers (ITS1 and ITS2). PCR amplification was performed in a total volume of 50 ml containing 100 ng of template DNA with 0.5 mM of each primer, 0.2 mM of each dNTP, 2 units of Taq DNA polymerase (Promega, USA), 5 ml of 10X DNA polymerase buffer (100 mM Tris-HCl (pH 9.0), 500 mM KCl, 20 mM $MgCl_2$). Amplifications were performed in a thermal cycler (GeneAmp PCR System 9600, Perkin Elmer Cetus, USA) with an initial denaturation step at 94°C for 3 min, followed by 45 cycles of 30 sec at 94°C, 1 min at 55°C and 2 min at 72°C with a final extension at 72°C for 7 min. Amplified products were separated on a 1% agarose gel in 1X TAE buffer. The gels were stained with ethidium bromide and viewed on a UV transilluminator.

Restriction Analysis of the Amplified rDNA-ITS Regions

PCR amplified rDNA products were purified and digested with the restriction enzymes *Acc* I, *Alu* I, *Sau* 3A I, *EcoR* I (Roche), *Taq* I and *Xho* I (Promega) using the buffers and conditions recommended by the suppliers. The digested fragments were separated on a 4% agarose gel, stained and visualized as described above.

Table 1 Source of *Colletotrichum* isolates

Isolate's code	Host genotype	Symptom	Location
CG1	BD 10	Raised spots	RRII
CG2	BD 10	Raised spots	RRII
CG3	GT 1	Raised spots	RRII
CG4	GT 1	Raised spots	RRII
CG5	RRII 208	Raised spots	RRII
CG6[1]	RRII 105	Raised spots	RRII
CG7	PB 280	Raised spots	RRII
CG8	PB 310	Raised spots	RRII
CG9	PB 217	Raised spots	RRII
CG10	PB 217	Raised spots	RRII
CG11	RRII 105	Raised spots	RRII
CG12	RRII 105	Raised spots	Cheruvally Estate
CG13	RRII 105	Raised spots	Cheruvally Estate
CG14	RRII 105	Raised spots	Manickal Estate
CG15	Seedling	Anthracnose	RRII
CG16	Seedling	Anthracnose	RRII
CG17	Seedling	Anthracnose	RRII
CG18[2]	Seedling	Anthracnose	RRII
CG19	PR 255	Papery lesions	RRII
CG20[3]	RRII 300	Papery lesions	RRII
CG21	RRII 300	Papery lesions	RRII
CG22	PB 235	Papery lesions	RRII
CG23	Seedling	Anthracnose	Kaliyar Estate
CG24	PB 260	Anthracnose	Manickal Estate
CG25	RRII 105	Papery lesions	Manickal Estate

[1] Culture preserved as IMI 383015

[2] Culture preserved as IMI 383016

[3] Culture preserved as IMI 383017

Cloning and sequencing of rDNA-ITS regions

The amplified rDNA from both the species of *Colletotrichum*, based on their restriction profiles, were purified using GFX column and ligated into pGEM-T vector. The ligated products were subsequently used for transformation of *E. coli* cells (JM 109). Nucleotide sequences of the cloned PCR products from each of the species were determined by sequencing reaction using a BigDye Terminator Cycle Sequencing kit and the products were run with an Applied Biosystems ABI 3700 Sequencer at the Microsynth GmbH, Switzerland. Sequencing reactions were primed on both strands using either the T7 or SP6 promoter sequences of the pGEM-T vector.

Sequence Analysis

Nucleotide sequences obtained in this study were compared with the rDNA-ITS sequences of several *Colletotrichum* isolates belonging to two different species *C. gloeosporioides* and *C. acutatum* infecting different host plants retrieved from the NCBI GenBank Nucleotide Sequence Database (Table 2) (http://www.ncbi.nlm.nih.gov). Initially, the sequences from the rubber pathogens were subjected to homology search with the nucleotide sequences in the GenBank through BLASTN programme available at the web server http://www.ncbi.nlm.nih.gov. The sequences belonging to *C. gloeosporioides* and *C. acutatum* from other crops that made significant alignments with the rubber pathogens were selected and retrieved for further analysis. For each *Colletotrichum* species, sequences were selected from ten different hosts ignoring the sequences from other *Colletotrichum* species although they showed significant homology with the rubber pathogen. For sequence comparison, *Colletotrichum* ITS sequences including 5.8 S rDNA regions were analysed after trimming the original sequences for only ITS and 5.8 S rDNA regions. Sequence alignment was done using the CLUSTALW (Thompson *et. al.*, 1994) alignment programme available on the web-server http://www.ebi.ac.uk/clustalw. Clustal alignments were further used for inferring phylogenetic relationships among the *Colletotrichum* isolates. The neighbour-joining analyses were performed using Treecon 1.15 (Van de Peer and De Wachter, 1994). Distances were calculated under the Kimura two-parameter with insertions separately accounted for. All bootstrap analyses were performed with 1000 replications. The bootstrap replicate trees were re-rooted using the sequences from *C. gloeosporioides* infecting lentil (AF451905) as the outgroup before the bootstrap consensus was computed.

Table 2 Ribosomal DNA spacer sequences (ITS 1-5.8S-ITS2) of the *Colletotrichum* spp. retrieved from the database for the study.

Colletotrichum gloeosporioides		*Colletotrichum acutatum*	
GenBank accession number	**Host plant**	**GenBank accession number**	**Host plant**
AF 488777*	Rubber	AF 488778*	Rubber
AF 090855	Olive	AF 207794	Almond
AF 207792	Almond	AF 411765	Rhododendron
AF 411769	Rhododendron	AJ 301911	Vaccinium
AF451905	Lentil	AJ 301915	Primula
AJ 301907	Hypericum	AJ 301921	Capsicum
AJ 301909	Mango	AJ 301922	Anemone
AJ 301919	Fragaria	AJ 301924	Coffee
AJ 301977	Palm	AJ 301950	Fragaria
AJ 301988	Citrus	AJ 301964	Lupinus
AJ 311884	Yam	AJ 301982	Cyclamen

* Sequence data generated in this study and registered with the GenBank.

Results and Discussion

Ribosomal DNA-ITS RFLPs

The entire ITS region including the 5.8 S ribosomal RNA gene (Fig. 1a), were amplified using the universal primers ITS 1 and ITS 4 (White *et. al.*, 1990). The amplified product was approximately 0.6 kb (Fig. 1b) and there was no detectable length polymorphism among the *Colletotrichum* isolates from rubber on agarose gel as described earlier (Saha *et. al.*, 2002). Amplified rDNA-ITS fragments were subjected to digestion with four different restriction endonucleases (*Alu* I, *Sau* 3A I, *EcoR* I and *Taq* I) to identify rDNA-ITS groups through fragment length polymorphisms, which yielded two specific restriction profiles (Fig. 2a-c). Restriction patterns of the isolates causing raised spots were different from the other two populations originating from anthracnose and papery lesions indicating the involvement of two different species as reported earlier (Saha *et. al.*, 2002).

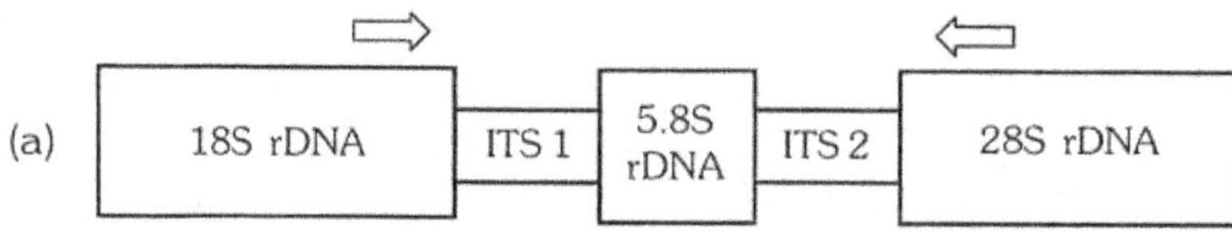

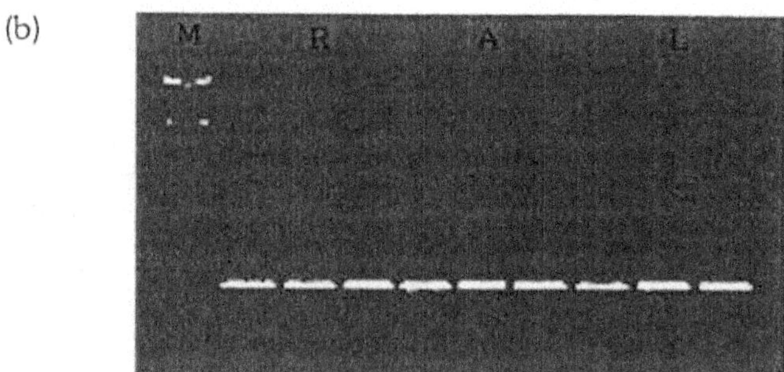

Figure 1 (a) Schematic diagram of repeat unit of nuclear ribosomal DNA showing the internal transcribed spacer regions ITS 1 and ITS 2. Arrowhead indicates the primers (ITS 1/ITS 4) used for PCR amplification of the spacer regions along with 5.8 S rDNA.

(b) Representative gel photograph showing the amplification of ITS region including 5.8 S rDNA (~0.6 kb) of *Colletotrichum* isolates causing raised spots (R), anthracnose (A) and papery lesion (L) symptoms on *Hevea*. Ribosomal DNA-ITS length polymorphisms could not be detected among the isolates. Lane M: molecular weight marker (l-DNA marker/*Eco* RI + *Hind* III).

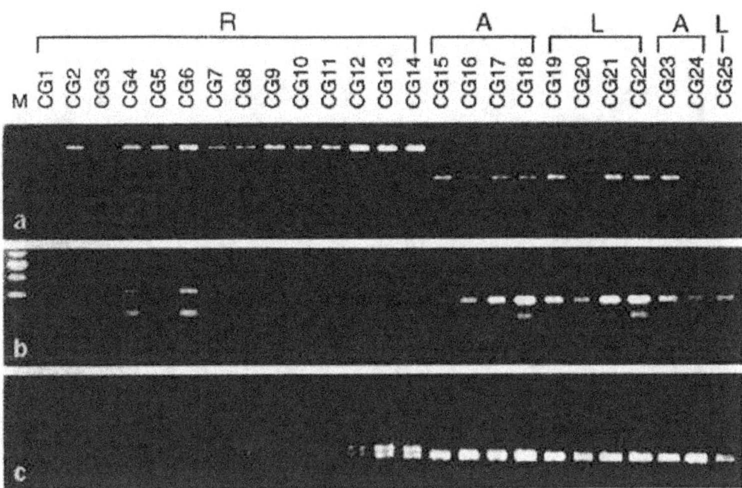

Figure 2   Representative gel photograph of the amplified rDNA-ITS fragments subjected to digestion with three different restriction endonucleases: *Alu* I (a), *Sau* 3A I (b) and *Taq* I (c) to identify rDNA-ITS groups through fragment length polymorphisms. Restriction patterns clearly showed that the isolates causing raised spots (R) were different from the other isolates causing anthracnose (A) and papery lesions (L). The isolates (CG1 to CG25) are indicated in the corresponding lanes. Lane M: molecular weight marker: 100 bp ladder (Promega).

Inter-specific variability in rDNA-ITS sequences of *Colletotrichum* species from rubber

Two rDNA spacer sequences derived from *C. gloeosporioides* and *C. acutatum* infecting *Hevea* were aligned to identify the degree of nucleotide variations in the ITS regions including 5.8 S rDNA (Fig. 3). These two species showed substantial nucleotide divergence in both the spacer regions with a lesser degree of variation in the 5.8 S rDNA as expected. The length of the spacer regions (ITS 1 and ITS 2) including 5.8 S was found to be 481 and 490 bp in *C. gloeosporioides* and *C. acutatum* respectively. ITS 1 of *C. acutatum* contained 185 bp whereas *C. gloeosporioides* had 171 bp. The number of base pairs in ITS 2 (152bp) and 5.8 S (158 bp) regions were found to be the same in both the species. However, sequence alignment of rDNA spacers from both the species resulted in 494 bases/characters as several gaps were introduced indicating the possible insertions/deletions (indels) (Fig. 3). A total 47 base substitutions (transitions and transversions) along with 7 indels ranging from 1 to 5 bases were noticed in the ITS regions including 5.8 S. Out of 47 base substitutions, 29 and 15 base substitutions were noticed in ITS 1 and ITS 2 regions respectively, whereas only 3 base substitutions were detected in the 5.8 S gene, which is highly conserved than the spacer regions. One of the general features noticed in the aligned sequences was the presence of more number of C«T transitions compared to any other base substitution. All the seven indels were noticed in the spacer regions of which, 5 were concentrated in the ITS 1 region with a wide range of variation in the size of the indels.

```
C.acutatum        CTGAGTTACCGCTCTATAACCCTTTGTGAACATACCTAACCGTTGCTTCGGCGGGCAGGGGAAGCCTCTCGCGGGCCTCCCCTCCCGGCGCCGGCCCCACCACGGGGACGGGGCGCCC
C.gloeosporioides CTGAGTTTACGCTCTACAACCCTTTGTGAACATACCTATAACTGTTGCTTCGGCGGGTAGGG--TCTCCGCGA--CCCTCCCGGCCTC-CCGCCTCCGGGCGGGTCGGCGCCC
                  ******  ******  ********************** ** ***************** ****    ****     **********  *  **  *  *  ** *  *******

C.acutatum        GCCGGAGGA-AACCAAACTCTATTTACACGACGTCTCTTCTGAGTGGCACAAGCAAATAATTAAAACCTTTAACAACGGATCTCTTGGTTCTGGCATCGATGAAGAACGCAGCGAAATGC
C.gloeosporioides GCCGGAGGATAACCAAACTCTGATTTAACGACGTTTCTTCTGAGTGGTACAAGCAAATAATCAAAACTTTTAACAACGGATCTCTTGGTTCTGGCATCGATGAAGAACGCAGCGAAATGC
                  *********  **********   **  *******  ***********  *************  ****  **  *************************************************

C.acutatum        GATAAGTAATGTGAATTGCAGAATTCAGTGAATCATCGAATCTTTGAACGCACATTGCGCTCGCCAGCATTCTGGCGAGCATGCCTGTTCGAGCGTCATTTCAACCCTCAAGCACCGCTT
C.gloeosporioides GATAAGTAATGTGAATTGCAGAATTCAGTGAATCATCGAATCTTTGAACGCACATTGCGCCCGCCAGCATTCTGGCGGGCATGCCTGTTCGAGCGTCATTTCAACCCTCAAGCTCTGCTT
                  ************************************************************ ***********************  *  **  ************************* *  ***

C.acutatum        GGTTTTGGGGCCCCACGGCACACGTGGGCCCTTAAAGGTAGTGGCGGACCCTCCCGGAGCCTCCTCTGCGTAGTAACTA-ACGTCTCGCACTGGGATTCGGAGGGACTCTTGCCGTAAAA
C.gloeosporioides GGTGTTGGGGCCCTACAGCTGATGTAGGCCCTCAAAGGTAGTGGCGGACCCTCCCGGAGCCTCCTTTGCGTAGTAACTTTACGTCTCGCACTGGGATCCGGAGGGACTCTTGCCGTAAAA
                  *** ********* ** **  * ** ****** ********************************** ***********   *************** * ***********************

C.acutatum        CCCCCCAATTTTTT
C.gloeosporioides CCCCC-AATTTTCC
                  *****  ****
```

Figure 3   Alignment of the rDNA-ITS sequences of *Colletotrichum gloeosporioides* and *C. acutatum* causing Colletotrichum leaf disease in rubber. Sequences were aligned by the CLUSTALW program. Shaded region of the sequence indicates 5.8 S gene. Asterix symbol (*) indicates the complete match whereas, dash (-) for the insertions/deletions.

Intra-specific variability based on rDNA-ITS sequences

Entire ribosomal spacer sequences of *C. acutatum* and *C. gloeosporioides* infecting different crop species were retrieved from the GenBank database and aligned with the respective *Colletotrichum* isolate from rubber to get an insight of the nucleotide variations existing within the species.

C. gloeosporioides

Spacer regions including 5.8 S rDNA of *C. gloeosporioides* infecting rhododendron, fragaria, almond, citrus, yam, mango, hypericum, lentil, olive and palm, retrieved from the database were aligned with the sequence belonging to *C. gloeosporioides* infecting rubber (Fig. 4). The multiple alignments of the sequences showed the nucleotide variability existing among the isolates from different hosts. Base substitutions along with small indels (1-2 bases) were identified in these sequences. Eight nucleotide substitutions and two indels were noticed in the ITS 1 region. Of the eight substitutions, the most prevalent one was the C«T transition. Two indels noticed in this region was due to the additional bases present in the isolate infecting yam and lentil. In 5.8 S rDNA sequences, a single base polymorphism was noticed in the form of an indel among the isolates, which was due to an additional base present in the isolates infecting olive and lentil. Compared to the ITS 1 region, less number of polymorphic bases was noticed in the ITS 2 region, which was characterised by the presence of only four substitutions and one indel.

C. acutatum

The ten rDNA spacer sequences that produced significant alignment with the rubber isolates belonging to *C. acutatum* were from lupinus, almond, anemone, fragaria, capsicum, rhododendron, cyclamen, primula, vaccinium and coffee. In the aligned sequences of all isolates of the pathogen from different hosts, only two substitutions and one indel were found to exist in the ITS 1 region, whereas, ITS 2 appeared to be relatively more variable than the ITS 1 containing 7 nucleotide substitutions (Fig. 5) in their spacer regions. Only a single base substitution was noticed in the 5.8 S rDNA region and that was due to C«T transition present in rubber pathogen, otherwise this region appeared to be highly conserved while considering the same gene sequences from the other ten *C. acutatum* isolates. Another unique substitution was noticed in the ITS 2 region of the rubber pathogen only. From the alignment result, it is evident that the isolates belonging to *C. acutatum* infecting different hosts are closely related compared to the *C. gloeosporioides* isolates.

```
AJ301907 (Hypericum)    CTGAGTTTACGCTCTACAACCCTTTGTGAACATACC-TATAACTGTTGCTTCGGCGGGCAGGGTCTCCGTGACCCTCCCGGCCTCCCGCCCCC-GGGCGGGTCGGCGCCCGCCGGAGGA
AJ301977 (Palm)         CTGAGTTTACGCTCTACAACCCTTTGTGAACATACC-TATAACTGTTGCTTCGGCGGGTAGGGTCCCCGTGACCCTCCCGGCCTCCCGCCCCC-GGGCGGGTCGGCGCCCGCCGGAGGA
AF451905 (Lentil)       CTGAGTTTACGCTCTACAACCCTTTGTGAACATACC-TACAACTGTTGCTTCGGCGGGTAGGGTCCCCGTGACCCTCCCGGCCTCCCGCCCCCCCGGGCGGGTCGGCGCCCGCCGGAGGA
AF411769 (Rhododendron) CTGAGTTTACGCTCTACAACCCTTTGTGAACATACC-TATAACTGTTGCTTCGGCGGGTAGGGTCTCCGTGACCCTCCCGGCCTCCCGCCCCC-GGGCGGGTCGGCGCCCGCCGGAGGA
AF207792 (Almond)       CTGAGTTTACGCTCTATAACCCTTTGTGAACATACC-TATAACTGTTGCTTCGGCGGGTAGGGTCTCCGTGACCCTCCCGGCCTCCCGCCCCC-GGGCGGGTCGGCGCCCGCCGGAGGA
AJ301988 (Citrus)       CTGAGTTTACGCTCTACAACCCTTTGTGAACATACC-TACAACTGTTGCTTCGGCGGGTAGGGTCTCCGCGACCCTCCCGGCCTCCCGCCTCC-GGGCGGGTCGGCGCCCGCCGGAGGA
AF090855 (Olive)        CTGAGTTTACGCTCTACAACCCTTTGTGAACATACC-TACAACTGTTGCTTCGGCGGGTAGGGTCTCCGCGACCCTCCCGGCCTCCCGCCTCC-GGGCGGGTCGGCGCCCGCCGGAGGA
AJ311884 (Yam)          CTGAGTTTACGCTCTACAACCCTTTGTGAACATACCCTATAACTGTTGCTTCGGCGGGTAGGGTCTCCGCGACCCTCCCGGCCTCCCGCCTCC-GGGCGGGTCGGCGCCCGCCGGAGGA
AF488777 (Rubber)       CTGAGTTTACGCTCTACAACCCTTTGTGAACATACC-TATAACTGTTGCTTCGGCGGGTAGGGTCTCCGCGACCCTCCCGGCCTCCCGCCTCC-GGGCGGGTCGGCGCCCGCCGGAGGA
AJ301919 (Fragaria)     CTGAGTTTACGCTCTATAACCCTTTGTGAACATACC-TATAACTGTTGCTTCGGCGGGTAGGGTCTCCGCGACCCTCCCGGCCTCCCGCCTCC-GGGCGGGTCGGCGCCCGCCGGAGGA
AJ301909 (Mango)        CTGAGTTTACGCTCTACAACCCTTTGTGAACATACC-TATAACTGTTGCTTCGGCGGGTAGGGTCTCCGCGACCCTCCCGGCCTCCCGCCCCC-GGGCGGGTCGGCGCCCGCCGGAGGA
                        ****************** *****************  ** ****************** ***** ***** ****************** **  ** **********************

AJ301907 (Hypericum)    TAACCAAACTCTGATTTAACGACGTTTCTTCTGAGTGGTACAAGCAAATAATCAAAACTTTTAACAAC-GGATCTCTTGGTTCTGGCATCGATGAAGAACGCAGCGAAATGCGATAAGTA
AJ301977 (Palm)         TAACCAAACTCTGATTTAACGACGTTTCTTCTGAGTGGTACAAGCAAATAATCAAAACTTTTAACAAC-GGATCTCTTGGTTCTGGCATCGATGAAGAACGCAGCGAAATGCGATAAGTA
AF451905 (Lentil)       TAACCAAACTCTGATTTAACGACGTTTCTTCTGAGTGGTACAAGCAAATAATCAAAACTTTTAACAACCGGATCTCTTGGTTCTGGCATCGATGAAGAACGCAGCGAAATGCGATAAGTA
AF411769 (Rhododendron) TAACCAAACTCTGATTTAACGACGTTTCTTCTGAGTGGTACAAGCAAATAATCAAAACTTTTAACAAC-GGATCTCTTGGTTCTGGCATCGATGAAGAACGCAGCGAAATGCGATAAGTA
AF207792 (Almond)       TAACCAAACTCTGATTTAACGACGTTTCTTCTGAGTGGTACAAGCAAATAATCAAAACTTTTAACAAC-GGATCTCTTGGTTCTGGCATCGATGAAGAACGCAGCGAAATGCGATAAGTA
AJ301988 (Citrus)       TAACCAAACTCTGATTTAACGACGTTTCTTCTGAGTGGTACAAGCAAATAATCAAAACTTTTAACAAC-GGATCTCTTGGTTCTGGCATCGATGAAGAACGCAGCGAAATGCGATAAGTA
AF090855 (Olive)        TAACCAAACTCTGATTTAACGACGTTTCTTCTGAGTGGTACAAGCAAATAATCAAAACTTTTAACAACCGGATCTCTTGGTTCTGGCATCGATGAAGAACGCAGCGAAATGCGATAAGTA
AJ311884 (Yam)          TAACCAAACTCTGATTTAACGACGTTTCTTCTGAGTGGTACAAGCAAATAATCAAAACTTTTAACAAC-GGATCTCTTGGTTCTGGCATCGATGAAGAACGCAGCGAAATGCGATAAGTA
AF488777 (Rubber)       TAACCAAACTCTGATTTAACGACGTTTCTTCTGAGTGGTACAAGCAAATAATCAAAACTTTTAACAAC-GGATCTCTTGGTTCTGGCATCGATGAAGAACGCAGCGAAATGCGATAAGTA
AJ301919 (Fragaria)     TAACCAAACTCTGATTTAACGACGTTTCTTCTGAGTGGTACAAGCAAATAATCAAAACTTTTAACAAC-GGATCTCTTGGTTCTGGCATCGATGAAGAACGCAGCGAAATGCGATAAGTA
AJ301909 (Mango)        TAACCAAACTCTGATGCAACGACGTTTCTTCTGAGTGGTACAAGCAAATAATCAAAACTTTTAACAAC-GGATCTCTTGGTTCTGGCATCGATGAAGAACGCAGCGAAATGCGATAAGTA
                        *************** *****************************************************  **************************************************

AJ301907 (Hypericum)    ATGTGAATTGCAGAATTCAGTGAATCATCGAATCTTTGAACGCACATTGCGCCCGCCAGCATTCTGGCGGGCATGCCTGTTCGAGCGTCATTTCAACCCTCAAGCTCTGCTTGGTGTTGG
AJ301977 (Palm)         ATGTGAATTGCAGAATTCAGTGAATCATCGAATCTTTGAACGCACATTGCGCCCGCCAGCATTCTGGCGGGCATGCCTGTTCGAGCGTCATTTCAACCCTCAAGCTCTGCTTGGTGTTGG
AF451905 (Lentil)       ATGTGAATTGCAGAATTCAGTGAATCATCGAATCTTTGAACGCACATTGCGCCCGCCAGCATTCTGGCGGGCATGCCTGTTCGAGCGTCATTTCAACCCTCAAGCTCTGCTTGGTGTTGG
AF411769 (Rhododendron) ATGTGAATTGCAGAATTCAGTGAATCATCGAATCTTTGAACGCACATTGCGCCCGCCAGCATTCTGGCGGGCATGCCTGTTCGAGCGTCATTTCAACCCTCAAGCTCTGCTTGGTGTTGG
AF207792 (Almond)       ATGTGAATTGCAGAATTCAGTGAATCATCGAATCTTTGAACGCACATTGCGCCCGCCAGCATTCTGGCGGGCATGCCTGTTCGAGCGTCATTTCAACCCTCAAGCTCTGCTTGGTGTTGG
AJ301988 (Citrus)       ATGTGAATTGCAGAATTCAGTGAATCATCGAATCTTTGAACGCACATTGCGCCCGCCAGCATTCTGGCGGGCATGCCTGTTCGAGCGTCATTTCAACCCTCAAGCTCTGCTTGGTGTTGG
AF090855 (Olive)        ATGTGAATTGCAGAATTCAGTGAATCATCGAATCTTTGAACGCACATTGCGCCCGCCAGCATTCTGGCGGGCATGCCTGTTCGAGCGTCATTTCAACCCTCAAGCTCTGCTTGGTGTTGG
AJ311884 (Yam)          ATGTGAATTGCAGAATTCAGTGAATCATCGAATCTTTGAACGCACATTGCGCCCGCCAGCATTCTGGCGGGCATGCCTGTTCGAGCGTCATTTCAACCCTCAAGCTCTGCTTGGTGTTGG
AF488777 (Rubber)       ATGTGAATTGCAGAATTCAGTGAATCATCGAATCTTTGAACGCACATTGCGCCCGCCAGCATTCTGGCGGGCATGCCTGTTCGAGCGTCATTTCAACCCTCAAGCTCTGCTTGGTGTTGG
AJ301919 (Fragaria)     ATGTGAATTGCAGAATTCAGTGAATCATCGAATCTTTGAACGCACATTGCGCCCGCCAGCATTCTGGCGGGCATGCCTGTTCGAGCGTCATTTCAACCCTCAAGCTCTGCTTGGTGTTGG
AJ301909 (Mango)        ATGTGAATTGCAGAATTCAGTGAATCATCGAATCTTTGAACGCACATTGCGCCCGCCAGCATTCTGGCGGGCATGCCTGTTCGAGCGTCATTTCAACCCTCAAGCTCTGCTTGGTGTTGG
                        ************************************************************************************************************************

AJ301907 (Hypericum)    GGCCCTACGGCTGACGTAGGCCCTCAAAGGTAGTGGCGGACCCTCCCGGAGCCTCCTTTGCGTAGTAACTTTACGTCTCGCACTGGGATCCGGAGGGACTCTTGCCGTAAAACCCCC-AA
AJ301977 (Palm)         GGCCCTACGGCTGACGTAGGCCCTCAAAGGTAGTGGCGGACCCTCCCGGAGCCTCCTTTGCGTAGTAACTTTACGTCTCGCACTGGGATCCGGAGGGACTCTTGCCGTAAAACCCCC-AA
AF451905 (Lentil)       GGCCCTACAGCTGATGTAGGCCCTCAAAGGTAGTGGCGGACCCTCCCGGAGCCTCCTTTGCGTAGTAACTTTACGTCTCGCACTGGGATCCGGAGGGACTCTTGCCGTAAAACCCCCCAA
AF411769 (Rhododendron) GGCCCTACAGCTGATGTAGGCCCTCAAAGGTAGTGGCGGACCCTCCCGGAGCCTCCTTTGCGTAGTAACTTTACGTCTCGCACTGGGATCCGGAGGGACTCTTGCCGTAAAACCCCC-AA
AF207792 (Almond)       GGCCCTACAGCTGATGTAGGCCCTCAAAGGTAGTGGCGGACCCTCCCGGAGCCTCCTTTGCGTAGTAACTTTACGTCTCGCACTGGGATCCGGAGGGACTCTTGCCGTAAAACCCCC-AA
AJ301988 (Citrus)       GGCCCTACAGCCGATGTAGGCCCTCAAAGGTAGTGGCGGACCCTCCCGGAGCCTCCTTTGCGTAGTAACTTTACGTCTCGCACTGGGATCCGGAGGGACTCTTGCCGTAAAACCCCCCAA
AF090855 (Olive)        GGCCCTACAGCCGATGTAGGCCCTCAAAGGTAGTGGCGGACCCTCCCGGAGCCTCCTTTGCGTAGTAACTTTACGTCTCGCACTGGGATCCGGAGGGACTCTTGCCGTAAAACCCCCCAA
AJ311884 (Yam)          GGCCCTACAGCCGATGTAGGCCCTCAAAGGTAGTGGCGGACCCTCTCGGAGCCTCCTTTGCGTAGTAACTTTACGTCTCGCACTGGGATCCGGAGGGACTCTTGCCGTAAAACCCCC-AA
AF488777 (Rubber)       GGCCCTACAGCTGATGTAGGCCCTCAAAGGTAGTGGCGGACCCTCCCGGAGCCTCCTTTGCGTAGTAACTTTACGTCTCGCACTGGGATCCGGAGGGACTCTTGCCGTAAAACCCCC-AA
AJ301919 (Fragaria)     GGCCCTACAGCTGATGTAGGCCCTCAAAGGTAGTGGCGGACCCTCCCGGAGCCTCCTTTGCGTAGTAACTTTACGTCTCGCACTGGGATCCGGAGGGACTCTTGCCGTAAAACCCCCCAA
AJ301909 (Mango)        GGCCCTACAGCTGATGTAGGCCCTCAAAGGTAGTGGCGGACCCTCCCGGAGCCTCCTTTGCGTAGTAACTTTACGTCTCGCACTGGGATCCGGAGGGACTCTTGCCGTAAAACCCCCCAA
                        ********  **  ******************************* *********************************************************************** **

AJ301907 (Hypericum)    TTTTCC
AJ301977 (Palm)         TTTTCC
AF451905 (Lentil)       TTTTCC
AF411769 (Rhododendron) TTTTCC
AF207792 (Almond)       TTTTCC
AJ301988 (Citrus)       TTTTCC
AF090855 (Olive)        TTTTCC
AJ311884 (Yam)          TTTTCC
AF488777 (Rubber)       TTTTCC
AJ301919 (Fragaria)     TTTTCC
AJ301909 (Mango)        TTTTCC
                        ******
```

Figure 4   Clustal alignment of the rDNA-ITS sequence of the rubber pathogen, *Colletotrichum gloeosporioides*, with closely related isolates of the same species infecting different crops. Shaded region of the sequence indicates 5.8 S gene. Asterix symbol (*) indicates the complete match of the aligned bases whereas; dash (-) for the insertions/deletions in the sequences.

```
AF301982  (Cyclamen)       CTGAGTTACCGCTCTATAACCCTTTGTGAACATACCTAACCGTTGCTTCGGCGGGCAGGGGAAGCCTCTCGCGGGCCTCCCCTCCCGGCGCCGGCCCC-ACCACGGGGACGGGGCGCCCG
AJ301921  (Capsicum)       CTGAGTTACCGCTCTATAACCCTTTGTGAACATACCTAACCGTTGCTTCGGCGGGCAGGGGAAGCCTCTCGCGGGCCTCCCCTCCCGGCGCCGGCCCCCACCACGGGGACGGGGCGCCCG
AF488778  (Rubber)         CTGAGTTACCGCTCTATAACCCTTTGTGAACATACCTAACCGTTGCTTCGGCGGGCAGGGGAAGCCTCTCGCGGGCCTCCCCTCCCGGCGCCGGCCCC-ACCACGGGGACGGGGCGCCCG
AF411765  (Rhododendron)   CTGAGTTACCGCTCTATAACCCTTTGTGAACATACCTAACCGTTGCTTCGGCGGGCAGGGGAAGCCTCTCGCGGGCCTCCCCTCCCGGCGCCGGCCCCCACCACGGGGACGGGGCGCCCG
AF207794  (Almond)         CTGAGTTACCGCTCTATAACCCTTTGTGAACATACCTAACCGTTGCTTCGGCGGGCAGGGGAAGCCTCTCGCGGGCCTCCCCTCCCGGCGCCGGCCCCCACCACGGGGACGGGGCGCCCG
AJ301964  (Lupinus)        CTGAGTTACCGCTCTATAACCCTTTGTGAACATACCTAACCGTTGCTTCGGCGGGCAGGGGAAGCCTCTCGCGGGCCTCCCCTCCCGGCGCCGGCCCCCACCACGGGGACGGGGCGCCCG
AJ301950  (Fragaria)       CTGAGTTACCGCTCTATAACCCTTTGTGAACATACCTAACCGTTGCTTCGGCGGGCAGGGGAAGCCTCTCGCGGGCCTCCCCTCCCGGCGCCGGCCCCCACCACGGGGACGGGGCGCCCG
AJ301922  (Anemone)        CTGAGTTACCGCTCTATAACCCTTTGTGAACATACCTAACCGTTGCTTCGGCGGGCAGGGGAAGCCTCTCGCGGGCCTCCCCTCCCGGCGCCGGCCCCCACCACGGGGACGGGGCGCCCG
AJ301911  (Vaccinium)      CTGAGTTACCGCTCTATAACCCTTTGTGAACGTACCTAACCGTTGCTTCGGCGGGCAGGGGAAGCCTCTCGCGGGCCTCCCCTCCCGGCGCCGGCCCCCACCACGGGGACGGGGCGCCCG
AJ301915  (Primula)        CTGAGTTACCGCTCTATAACCCTTTGTGAACGTACCTAACCGTTGCTTCGGCGGGCAGGGGAAGCCTCTCGCGGGCCTCCCCTCCCGGCGCCGGCCCCCACCACGGGGACGGGGCGCCCG
AJ301924  (Coffea)         CTGAGTTACCGCTCTACAACCCTTTGTGAACATACCTAACCGTTGCTTCGGCGGGCAGGGGAAGCCTCTCGCGGGCCTCCCCTCCCGGCGCCGGCCCC-ACCACGGGGACGGGGCGCCCG
                           ********************************  ******************************************************************  ********************

AF301982  (Cyclamen)       CCGGAGGAAACCAAACTCTATTTACACGACGTCTCTTCTGAGTGGCACAAGCAAATAATTAAAACTTTTAACAACGGATCTCTTGGTTCTGGCATCGATGAAGAACGCAGCGAAATGCGA
AJ301921  (Capsicum)       CCGGAGGAAACCAAACTCTATTTACACGACGTCTCTTCTGAGTGGCACAAGCAAATAATTAAAACTTTTAACAACGGATCTCTTGGTTCTGGCATCGATGAAGAACGCAGCGAAATGCGA
AF488778  (Rubber)         CCGGAGGAAACCAAACTCTATTTACACGACGTCTCTTCTGAGTGGCACAAGCAAATAATTAAAACCTTTAACAACGGATCTCTTGGTTCTGGCATCGATGAAGAACGCAGCGAAATGCGA
AF411765  (Rhododendron)   CCGGAGGAAACCAAACTCTATTTACACGACGTCTCTTCTGAGTGGCACAAGCAAATAATTAAAACTTTTAACAACGGATCTCTTGGTTCTGGCATCGATGAAGAACGCAGCGAAATGCGA
AF207794  (Almond)         CCGGAGGAAACCAAACTCTATTTACACGACGTCTCTTCTGAGTGGCACAAGCAAATAATTAAAACTTTTAACAACGGATCTCTTGGTTCTGGCATCGATGAAGAACGCAGCGAAATGCGA
AJ301964  (Lupinus)        CCGGAGGAAACCAAACTCTATTTACACGACGTCTCTTCTGAGTGGCACAAGCAAATAATTAAAACTTTTAACAACGGATCTCTTGGTTCTGGCATCGATGAAGAACGCAGCGAAATGCGA
AJ301950  (Fragaria)       CCGGAGGAAACCAAACTCTATTTACACGACGTCTCTTCTGAGTGGCACAAGCAAATAATTAAAACTTTTAACAACGGATCTCTTGGTTCTGGCATCGATGAAGAACGCAGCGAAATGCGA
AJ301922  (Anemone)        CCGGAGGAAACCAAACTCTATTTACACGACGTCTCTTCTGAGTGGCACAAGCAAATAATTAAAACTTTTAACAACGGATCTCTTGGTTCTGGCATCGATGAAGAACGCAGCGAAATGCGA
AJ301911  (Vaccinium)      CCGGAGGAAACCAAACTCTATTTACACGACGTCTCTTCTGAGTGGCACAAGCAAATAATTAAAACTTTTAACAACGGATCTCTTGGTTCTGGCATCGATGAAGAACGCAGCGAAATGCGA
AJ301915  (Primula)        CCGGAGGAAACCAAACTCTATTTACACGACGTCTCTTCTGAGTGGCACAAGCAAATAATTAAAACTTTTAACAACGGATCTCTTGGTTCTGGCATCGATGAAGAACGCAGCGAAATGCGA
AJ301924  (Coffea)         CCGGAGGAAACCAAACTCTATTTACACGACGTCTCTTCTGAGTGGCACAAGCAAATAATTAAAACTTTTAACAACGGATCTCTTGGTTCTGGCATCGATGAAGAACGCAGCGAAATGCGA
                           *************************************************************** ********************************************************

AF301982  (Cyclamen)       TAAGTAATGTGAATTGCAGAATTCAGTGAATCATCGAATCTTTGAACGCACATTGCGCTCGCCAGCATTCTGGCGAGCATGCCTGTTCGAGCGTCATTTCAACCCTCAAGCACCGCTTGG
AJ301921  (Capsicum)       TAAGTAATGTGAATTGCAGAATTCAGTGAATCATCGAATCTTTGAACGCACATTGCGCTCGCCAGCATTCTGGCGAGCATGCCTGTTCGAGCGTCATTTCAACCCTCAAGCACCGCTTGG
AF488778  (Rubber)         TAAGTAATGTGAATTGCAGAATTCAGTGAATCATCGAATCTTTGAACGCACATTGCGCTCGCCAGCATTCTGGCGAGCATGCCTGTTCGAGCGTCATTTCAACCCTCAAGCACCGCTTGG
AF411765  (Rhododendron)   TAAGTAATGTGAATTGCAGAATTCAGTGAATCATCGAATCTTTGAACGCACATTGCGCTCGCCAGCATTCTGGCGAGCATGCCTGTTCGAGCGTCATTTCAACCCTCAAGCACCGCTTGG
AF207794  (Almond)         TAAGTAATGTGAATTGCAGAATTCAGTGAATCATCGAATCTTTGAACGCACATTGCGCTCGCCAGCATTCTGGCGAGCATGCCTGTTCGAGCGTCATTTCAACCCTCAAGCACCGCTTGG
AJ301964  (Lupinus)        TAAGTAATGTGAATTGCAGAATTCAGTGAATCATCGAATCTTTGAACGCACATTGCGCTCGCCAGCATTCTGGCGAGCATGCCTGTTCGAGCGTCATTTCAACCCTCAAGCACCGCTTGG
AJ301950  (Fragaria)       TAAGTAATGTGAATTGCAGAATTCAGTGAATCATCGAATCTTTGAACGCACATTGCGCTCGCCAGCATTCTGGCGAGCATGCCTGTTCGAGCGTCATTTCAACCCTCAAGCACCGCTTGG
AJ301922  (Anemone)        TAAGTAATGTGAATTGCAGAATTCAGTGAATCATCGAATCTTTGAACGCACATTGCGCTCGCCAGCATTCTGGCGAGCATGCCTGTTCGAGCGTCATTTCAACCCTCAAGCACCGCTTGG
AJ301911  (Vaccinium)      TAAGTAATGTGAATTGCAGAATTCAGTGAATCATCGAATCTTTGAACGCACATTGCGCTCGCCAGCATTCTGGCGAGCATGCCTGTTCGAGCGTCATTTCAACCCTCAAGCACCGCTTGG
AJ301915  (Primula)        TAAGTAATGTGAATTGCAGAATTCAGTGAATCATCGAATCTTTGAACGCACATTGCGCTCGCCAGCATTCTGGCGAGCATGCCTGTTCGAGCGTCATTTCAACCCTCAAGCACCGCTTGG
AJ301924  (Coffea)         TAAGTAATGTGAATTGCAGAATTCAGTGAATCATCGAATCTTTGAACGCACATTGCGCTCGCCAGCATTCTGGCGAGCATGCCTGTTCGAGCGTCATTTCAACCCTCAAGCACCGCTTGG
                           ************************************************************************************************************************

AF301982  (Cyclamen)       TTTTGGGGCCCCACGGCACACGTGGGCCCTTAAAGGTAGTGGCGGACCCTCCCGGAGCCTCCTTTGCGTAGTAACTAACGTCTCGCACTGGGATCCGGAGGGACTCTTGCCGTAAAACCC
AJ301921  (Capsicum)       TTTTGGGGCCCCACGGCACACGTGGGCCCTTAAAGGTAGTGGCGGACCCTCCCGGAGCCTCCTCTGCGTAGTAACTAACGTCTCGCACTGGGATCCGGAGGGACTCTTGCCGTAAAACCC
AF488778  (Rubber)         TTTTGGGGCCCCACGGCACACGTGGGCCCTTAAAGGTAGTGGCGGACCCTCCCGGAGCCTCCTCTGCGTAGTAACTAACGTCTCGCACTGGGATTCGGAGGGACTCTTGCCGTAAAACCC
AF411765  (Rhododendron)   TTTTGGGGCCCCACGGCACACGTGGGCCCTTAAAGGTAGTGGCGGACCCTCCCGGAGCCTCCTTTGCGTAGTAACTAACGTCTCGCACTGGGATTCGGAGGGACTCTTGCCGTAAAACCC
AF207794  (Almond)         TTTTGGGGCCCCACGGCACACGTGGGCCCTTAAAGGTAGTGGCGGACCCTCCCGGAGCCTCCTTTGCGTAGTAACTAACGTCTCGCACTGGGATTCGGAGGGACTCTTGCCGTAAAACCC
AJ301964  (Lupinus)        TTTTGGGGCCCCACGGCACACGTGGGCCCTTAAAGGTAGTGGCGGACCCTCCCGGAGCCTCCTTTGCGTAGTAACTAACGTCTCGCACTGGGATTCGGAGGGACTCTTGCCGTAAAACCC
AJ301950  (Fragaria)       TTTTGGGGCCCCACGGCACACGTGGGCCCTTAAAGGTAGTGGCGGACCCTCCCGGAGCCTCCTTTGCGTAGTAACTAACGTCTCGCACTGGGATTCGGAGGGACTCTTGCCGTAAAACCC
AJ301922  (Anemone)        TTTTGGGGCCCCACGGCACACGTGGGCCCTTAAAGGTAGTGGCGGACCCTCCCGGAGCCTCCTTTGCGTAGTAACTAACGTCTCGCACTGGGATTCGGAGGGACTCTTGCCGTAAAACCC
AJ301911  (Vaccinium)      TTTTGGGGCCCCACGGCCGACGTGGGCCCTTAAAGGTAGTGGCGGACCCTCCCGGAGCCTCCTTTGCGTAGTAACTAACGTCTCGCACTGGGATCCGGAGGGACTCTTGCCGTTAAAACCC
AJ301915  (Primula)        TTTTGGGGCCCCACGGCCGACGTGGGCCCTTAAAGGTAGTGGCGGACCCTCCCGGAGCCTCCTTTGCGTAGTAACTAACGTCTCGCACTGGGATCCGGAGGGACTCTTGCCGTTAAAACCC
AJ301924  (Coffea)        .TTTTGGGGCCCCACGGCAGACGTGGGCCCTTAAAGGTAGTGGCGGACCCTCCCGGAGCCTCCTTTGCGTAGTAACTAACGTCTCGCACTGGGATCCGGAGGGACTCTTGCCGTAAAACCC

AF301982  (Cyclamen)       CCCAATTTTTT
AJ301921  (Capsicum)       CCAAATTTTTT
AF488778  (Rubber)         CCCAATTTTTT
AF411765  (Rhododendron)   CCAAATTTTTT
AF207794  (Almond)         CCAAATTTTTT
AJ301964  (Lupinus)        CCAAATTTTTT
AJ301950  (Fragaria)       CCAAATTTTTT
AJ301922  (Anemone)        CCAAATTTTTT
AJ301911  (Vaccinium)      CCAAATTCTTT
AJ301915  (Primula)        CCAAATTCTTT
AJ301924  (Coffee)         CCAAATTCTTT
                           **  **** ***
```

Figure 5 Clustal alignment of the rDNA-ITS sequences of the fungal isolates belonging to *Colletotrichum acutatum* infecting different crops including rubber. Shaded region of the sequence indicates 5.8 S gene. Asterix symbol (*) indicates the complete match of the aligned bases whereas; dash (-) for the insertions/deletions in the sequences.

Phylogenetic Analysis

ITS sequences along with the 5.8 S rDNA obtained from all the 22 *Colletotrichum* isolates belonging to *C. gloeosporioides* and *C. acutatum* including the rubber pathogen were aligned and the data was used for phylogenetic analysis. This sequence analyses represented a reliable insight into the phylogeny of *Colletotrichum* isolates infecting different hosts. The phylogram clearly showed that all the pathogens were grouped into two delineating the two *Colletotrichum* species viz. *C. gloeosporioides* and *C. acutatum* (Fig. 6). In *C. gloeosporioides*, each isolate appeared as separate branch showing the uniqueness of the pathogen. Phylogenetically, *C. gloeosporioides* isolate infecting rubber was closer to the isolate infecting fragaria than that of citrus, olive and yam. The relationship of the isolates belonging to *C. acutatum* revealed that those infecting rhododendron, almond, lupinus, anemone, fragaria and capsicum were phylogenetically similar and grouped together. The isolates from rubber were found to be distinct having close relation with the isolate infecting cyclamen. The isolates from primula and vaccinia were placed together depicting the closeness of these two.

Two species of *Colletotrichum* namely *C. gloeosporioides* and *C. acutatum* prevailing in Indian rubber plantations are responsible for Colletotrichum leaf disease in *Hevea* and they produce different symptoms: raised spots, anthracnose and small papery lesions. *C. acutatum* was found to be associated with the raised spot symptom, whereas *C. gloeosporioides* was responsible for the other two symptoms (Saha *et al.*, 2002). In rubber plantations, the damage caused by *C. acutatum* was more than that by *C. gloeosporioides* as the former infect immature leaves (Kumar *et al.*, 2002) thereby delaying the apical growth of the young plants. In this study, our objective was to establish a phylogeny of the two species of *Colletotrichum* from rubber with the same species infecting other crops based on the homology of the rDNA-ITS nucleotide sequences. Restriction analysis of the rDNA was found to be very useful to identify the *Colletotrichum* infecting rubber at the species level (Saha *et al.*, 2002). In the present study, all the isolates collected from rubber plants could be grouped into two based on the restriction profiles of the amplified rDNA spacers. The rDNA restriction length polymorphisms detected among the fungal isolates, were generated due to sequence variations in ITS regions, which are rapidly evolving regions of ribosomal DNA/rRNA gene sequence and often best studied for comparing species and closely related taxa (Sreenivasaprasad *et al.*, 1996; Freeman *et al.*, 2000).

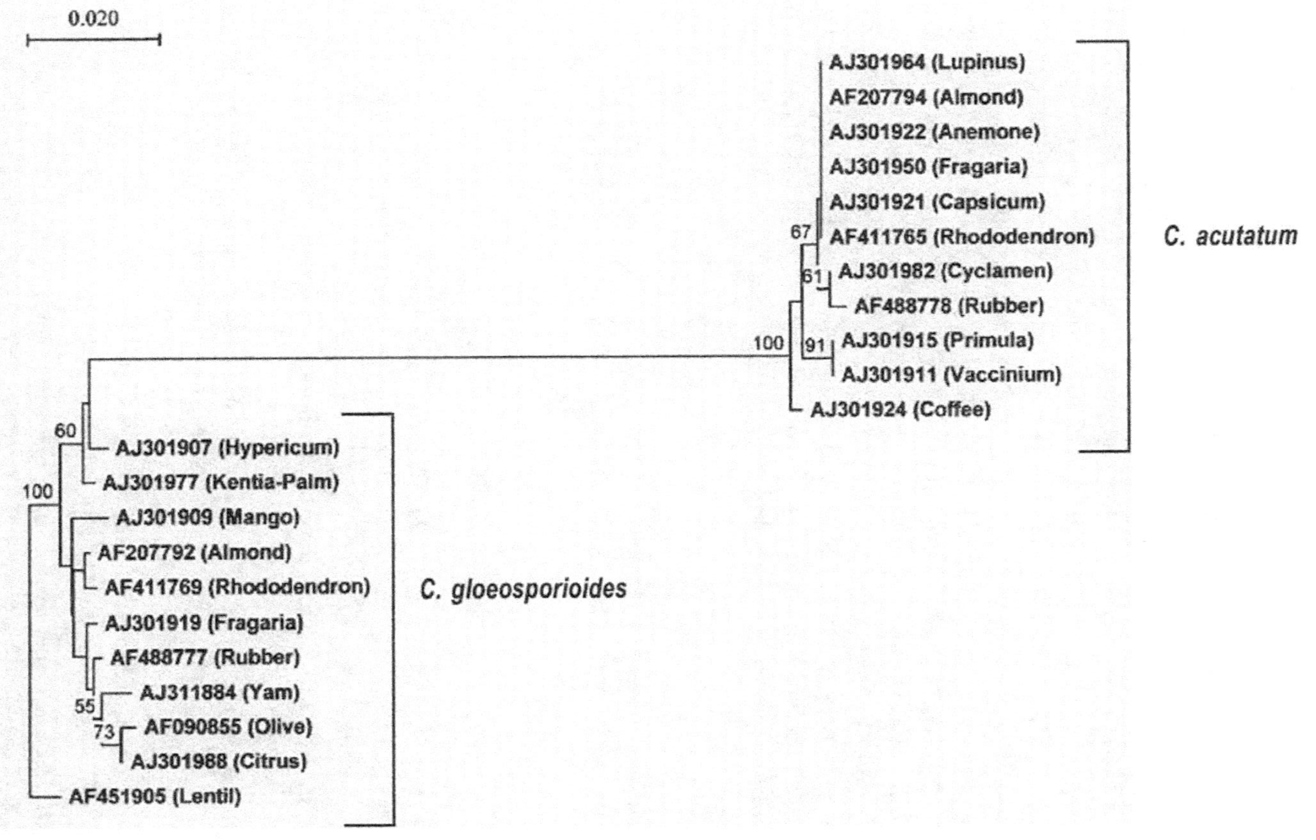

Figure 6 Neighbour-joining tree of the fungal isolates belonging to the species *C. gloeosporioides* and *C. acutatum* from different hosts including rubber reveals the relationships of the *Colletotrichum* isolates under study. These isolates form two major groups based on their ribosomal spacer sequences (ITS 1 - 5.8 S - ITS 2) representing two *Colletotrichum* species. The uniqueness of the rubber pathogens is also evident from the tree. Bootstrap values above 50% are indicated for the corresponding branches.

Nucleotide sequence analysis of the ITS 1 region showed a greater degree of inter-specific divergence compared to ITS 2 in the form of base substitutions as well as insertion/deletion. Among the base substitutions, transitions were found to occur at higher frequencies than transversions as noticed in any other genome which is considered to be a general property of DNA sequence evolution (Wakeley, 1996; Yang and Yoder, 1999). At the intra-species level, considerably more nucleotide variations were detected in ITS 1 than in ITS 2 regions of *C. gloeosporioides*. Therefore in *C. gloeosporioides*, ITS 1 sequence was proved to be more useful for understanding the phylogeny as reported earlier (Sreenivasaprasad *et al.*, 1996; Freeman *et al.*, 2000 Abang *et al.*, 2002). In the case of *C. acutatum*, ITS 2 was found to be more variable containing more number of nucleotide substitutions compared to ITS 1. Sequence alignment of the entire ITS region including the 5.8 S gene sequences from both the species infecting several crops was used to infer phylogenetic relationship among them. The *Colletotrichum* isolates infecting different hosts including rubber could clearly be separated into two groups belonging to two different species in the phylogram, which is supported by a bootstrap value of 100% indicating a high reliability. The *Colletotrichum* isolates infecting rubber appeared to be different from that of the other hosts based on the ITS sequence divergence.

Acknowledgements

We thank Dr. Arun Kumar, Assistant Mycologist for providing the fungal isolates and Dr. N. M. Mathew, Director, Rubber Research Institute of India for his encouragement to carry out this work.

References

Abang, M.M., Winter, S., Green, K.R., Hoffmann, P., Mignouna, H.D. and **Wolf, G.A.** 2002. Molecular identification of *Colletotrichum gloeosporioides* causing yam anthracnose in Nigeria. *Plant Path.* **51**: 63-71.

Bruns, T.D., White, T.J. and **Taylor, J.W.** 1991. Fungal molecular systematics. *Ann. Rev. Ecol. Syst.* **22**: 525-564.

Freeman, S., Minz, D., Jurkevitch, E., Maymon, M. and **Shabi, E.** 2000. Molecular analyses of *Colletotrichum* species from almond and other fruits. *Phytopath.* **90**: 608-614.

Green, S.J., Freeman, S., Hadar, Y. and **Minz, D.** 2004. Molecular tools for isolate and community studies of pyronomycete fungi. *Mycologia* **96**: 439-451.

Jeffries, P., Dodd, J.C., Jeger, M.J. and **Plumbley, R.A.** 1990. The biology and control of *Colletotrichum* species on tropical fruit crops. *Plant Path.* **39**: 343-366.

Jayasinghe, C.K. and **Fernando, T.H.P.S.** 1997. Growth at different temperatures and on fungicide amended media: Two characteristics to distinguish *Colletotrichum* species pathogenic to rubber. *Mycopath.* **143**: 93-95.

Johnston, P.R. and **Jones, D.** 1997. Relationships among *Colletotrichum* isolates from fruit-rots assessed using rDNA sequences. *Mycologia* **89**: 420-430.

Kumar, A., Jacob, C.K. and **Joseph, A.** 2002. Morphological and cultural characteristics of *Colletotrichum* isolates from *Hevea brasiliensis*. *Ind. J. Nat. Rub. Res.* **15**: 150-157.

Martin, M.P. and **Garcia-Figueres, F.** 1999. *Colletotrichum acutatum* and *Colletotrichum gloeosporioides* cause anthracnose on olives. *European. J. Plant Path.* **105**: 733-741.

Petch, T. 1906. Description of new Ceylon fungi. *Ann. Rep. Bot. Gardens, Peradeniya,* **3**: 1-10.

Saha, T., Kumar, A., Sreena, A.S., Joseph, A., Jacob, C.K., Kothandaraman, R. and **Nazeer, M.A.** 2000. Genetic variability of *Corynespora casiicola* infecting *Hevea brasiliensis* isolated from the traditional rubber growing areas in India. *Ind. J. Nat. Rub. Res.* **13**: 1-10.

Saha, T., Kumar, A., Ravindran, M., Jacob, C.K., Roy, B. and **Nazeer, M.A.** 2002. Identification of *Colletotrichum acutatum* from rubber using random amplified polymorphic DNAs and ribosomal DNA polymorphisms. *Mycol. Res.* **106**: 215-221.

Sreenivasaprasad, S., Mills, P.R. and **Brown, A.E.** 1994. Nucleotide sequence of the rDNA spacer 1 enables identification of isolates of *Colletotrichum* as *C. acutatum*. *Mycol. Res.* **98**: 186-188.

Sreenivasaprasad, S., Mills, P.R., Meehan, B.M. and **Brown, A.E.** 1996. Phylogeny and systematics of 18 *Colletotrichum* species based on ribosomal DNA spacer sequences. *Genome* **39**: 499-512.

Sutton, B.C. 1992. The genus *Glomerella* and its anamorph *Colletotrichum*. In: Colletotrichum – *biology, pathology and control*. (Eds.). Bailey, J.A., Jeger, M.J., CAB International, Wallingford, UK, 1-26.

Thompson, J.D., Higgins, D.G. and **Gibson, T.J.** 1994. CLUSTAL W: improving the sensitivity of progressive multiple sequence alignment through sequence weighting, positions-specific gap penalties and weight matrix choice. *Nucleic acids res.* **22**: 4673-4680.

Van de Peer, Y. and **De Wachter, R.** 1994. TREECON for windows: A software package for the construction and drawing of evolutionary trees for Microsoft Windows. *Computer application in Biosciences*, 10: 267-277.

Wakeley, J. 1996. The excess of transitions among nucleotide substitutions: New methods of estimating transition bias underscore its significance. *Tree* **11** : 158-163.

Waller, J.M. 1992. *Colletotrichum* diseases of perennial and other cash crops. In: Colletotrichum – *biology, pathology and control*, (*Eds.*). Bailey, J.A. and Jeger, M.J. CAB International, Wallingford, UK, 167-185.

White, T.J., Bruns, T., Lee, S. and **Taylor, J.** 1990. Amplifications and direct sequencing of fungal ribosomal RNA genes for phylogenetics. In: PCR protocol: A guide to methods and amplifications. (*Ed.*). Innis, M.A., San Diego, CA Academic Press, 315-322.

Yang, Z. and **Yoder, A.D.** 1999. Estimation of the transition/transversion rate bias and species sampling. *J. Mol. Evol.* **48**: 274-283.

28

Rapid and Species Specific Identification of Environmental Isolates of Mycobacteria by Pcr-Restriction Fragment Length Polymorphism Analysis (Pra) of Hsp-65 Gene.

Dinkar Sharma*, J. N. Srivastava*, V. D. Sharma, D. S. Chauhan, Deepti Parashar and V. M. Katoch.

Dept. of Microbiology and Molecular Biology, Central JALMA Institute for Leprosy and other Mycobacterial Diseases (ICMR), Tajganj, Agra - 282 001.

*Microbiology Lab., Dept. of Botany, D.E.I., Dayalbagh, Agra - 282 005.

ABSTRACT

PCR-Restriction fragment length polymorphism Analysis (PRA) of the hsp-65 (Heat shock protein) gene using Hae-III (*Haemophilous egyptius*) enzyme was applied to 14 mycobacterial isolates of representing 6 species biochemically identified comprising of *M.kansasii, M.marinum, M.avium, M.fortuitum, M.chelonae* and *M.smegmatis* for Indian isolates, to evaluate its potential as a rapid identification method. Isolates belonging to all the 6 species showed specific distinct patterns. This method, thus, appears to be relatively simpler and rapid identification method for mycobacteria.

KEY WORDS : Mycobacteria, hsp-65 KDa (Heat shock protein), PCR-RELP, rpoB, ICC (Indian Culture Collection),Hae III

Characterization of mycobacteria at species level is being done in most of the laboratories, based on the classical methods such as phenotypic and biochemical characteristics. These methods may take several days to weeks after getting desired growth. Sometimes these tests fail to give correct identification because of some limitations. Several other methods like HPLC or TLC and Gene Probe are also available to characterize mycobacteria with one or more limitation factors. 16S rRNA sequencing is another powerful technique to differentiate mycobacteria at species level, but it is not possible to perform at routine level in most of the laboratories. In contrast to these techniques, PCR-RELP targeting different gene sequences has been proved useful technique (1-8). PCR-RELP analysis of the hsp-65 gene offers an easy, rapid and inexpensive procedure to identify several mycobacterial species in a single experiment. Our primary aim was to evaluate this PRA method to Indian isolates of mycobacterial species.

Materials and Methods

Different species of mycobacteria such as *M.kansasii*, (Standard strain T-3, ICC-1342), *M.marium* (Std. Strain J-6, ICC-1666), *M.avium* (Std. Strain N-21, ICC-1622, ICC-1621, ICC-1623, ICC-1620), *M.fortuitum* (Std. Strain (N-2, ICC-1663, ICC-1314, ICC-1659, ICC-1661), *M.chelonae* (Std. Strain J-31, ICC-1312, ICC-1537) and *M.smegmatis* (Std. Strain N-18, ICC-1320, ICC-1327), deposited in the National Mycobacterial Repository Center, CJIL & OMD, Agra were included in this study.

Extraction of Nucleic Acids

DNAs were extracted by a physiochemical (Freeze/Boiling) procedure (Van, Embden *et al.*, 1993) using lysozymes and proteinase-K , from culture isolates.

PCR - Restriction Fragment Length Polymorphism Analysis (PRA)

PRA based on the amplification of a 439-bp fragment of the hsp-65 gene was performed with the primer Tb-ll (5'-ACCAACGATGGTGTGTCCAT-3') and Tb-12 (5'-GTTGTCGAACCGCATACCT-3') by method described earlier. (Telenti *et al.*, 1993).

RFLP Analysis of Amplicons

The amplified products (10μl) was digested with 5U of Hae-III for 2 hrs., according to the recommendation of the manufacturer and electrophoresed in 2% agarose gel in presence of ethidium-bromide at 1.2V/Cm for 4 hrs. Fragment sizes were estimated by GEL-Documentation (Bio-Red, USA) and compaired with appropriate controls run in parallel, standard strain and molecular weight marker.

Results and Discussion

The PRA method was selected in this investigation because of its ease and rapidity and it may help in speciate numerous species of mycobacteria with in a single experiment (1-7,10). This broader spectrum of the PRA method as compared to other molecular methods, is based on the amplification of a conserved hsp-65 gene present in all mycobacteria.. (3). The restriction analysis of a 439 bp fragment within this gene after Hae-III digestion was earlier reported to be highly effective for differentiating mycobacteria at species level (Telenti *et al.*, 1993).

The results obtained in present study showed distinct profiles for all 6 species, *M.kansasii, M.marinum, M.avium, M.furtuitum, M.chelonae* and *M.smegmatis* (3), (Figure 1, 2 and Table 1, 2).

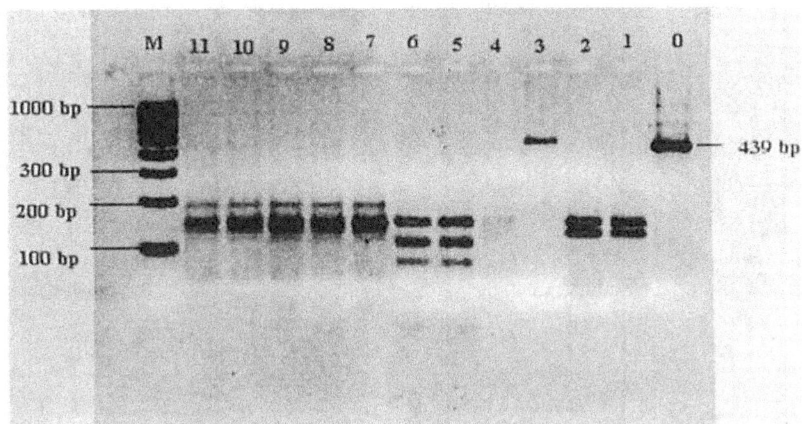

Figure 1 Gene amplification restriction analysis of environmental isolates (Slow growing) targeting 65 KDa gene region digested with Hae-III, Lane 0-uncut, Lane 1-2 – M.kansasii, Lane 5, 6-M.marinum, Lane 7-11 – M.avium, Lane 12 – Marker.

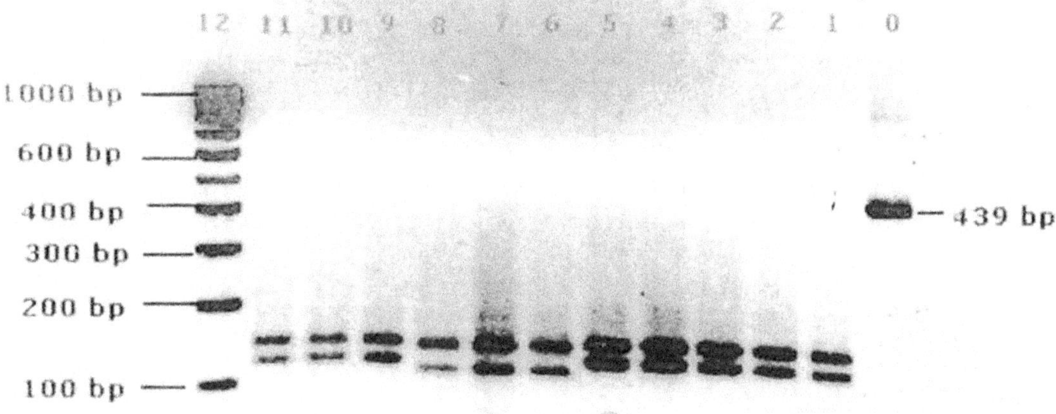

Figure 2 Gene amplification restriction analysis of environmental isolates (Rapid growing) targeting 65 KDa gene region digested with Hae-III, Lane 0 – uncut, Lane 1 - 5 – M.fortuitum, Lane 6 – 8 - M.chelonae, Lane 9-11 – M.smegmatis, Lane 12 – Marker.

Table 1 Fragment size generated for amplicons of 65 KDa from slow growing mycobacteria.

Lane No.	Strain No.	Species	No. of Bands	Band size (bp)
0	Uncut	–	1	439
1.	ICC-1342	M.kansasii	3	140,100,80
2.	T-3	M.kansasii Standard	3	140,100,80
3.				Result failure
4.				Result failure
5.	ICC-166	M.marinum	3	160,115,80
6.	J-6	M.marinum Standard	3	160,115,80
7.	ICC-1622	M.avium	3	190,140,130
8.	ICC-1621	M.avium	3	190,140,130
9.	ICC-1623	M.avium	3	190,140,130
10.	ICC-1620	M.avium	3	190,140,130
11.	N-21	M.avium Standard	3	190,140,130
12.	Marker			

Table 2 Fragment size generated for amplicons of 65 KDa from rapid growing mycobacteria.

Lane No.	Strain No.	Species	No. of Bands	Band size (bp)
0	Uncut	–	1	439
0	Uncut			
1.	ICC-1663	M. fortuitum	3	155, 150,100
2.	ICC-1314	M.fortuitum	3	155, 150, 100
3.	ICC-1659	M.fortuitum	3	155, 150, 100
4.	ICC-1661	M.fortuitum	3	155, 150, 100
5.	N-2	M.fortuitum Standard	3	155, 150, 100
6.	ICC-1312	M.chelonae	4	160,150,65,60
7.	ICC-1531	M.chelonae	4	160,150,65,60
8.	J-31	M.chelonae Standard	4	160,150,65,60
9.	ICC-1320	M.smegmatis	3	155,150,70
10.	ICC-1327	M.smegmatis	3	155,150,70
11.	N-18	M.smegmatis Standard	3	155,150,70
12.	Marker			

Conclusion

PRA, targeting hsp-65 gene using the Hae-III enzyme appears to be useful for the rapid identification of Indian isolates of environmental Mycobacteria belonging to *M.kansasii M.marinum, M.avium, M.furtuitum, M.chelonae.* (All pathogens) and M.smegmatis (Nonpathogen) to the species level by Microbiology Laboratory. It would be advisable that to analyze larger number of strain using Hae-III enzyme. The results need to be finally compaired with other PCR-RFLP technique specially based on rRNA (1-5), rpoB (4) 16s-rRNA which will confirm the final gold standard for comparison.

References

Devallois, A., Goh, K.S. and **Rastogi, N**. 1993 *"Rapid identification of mycobacteria to species level by PCR-retriction fragment polymorphism of an algorithm to differentiate 34 mycobactorial species". J clin microbial* 31:175-178.

Hans, A.B., Champ, G., Levy frehault V. 1989 *"Detection and identification of mycobacteria by amplification of mycobacteria DNA" Mol. Microbiol* 3: 843-849

Katoch, V.M. 1999 *"Development of species and strain specific amplified rDNA restriction (ARDRAD) technique(s) for clinicaliy important pathogenic mycobacteria" DBT project report* (BT/R&D/09/94).

Kim, B.J., Lee, K.H., Dark, B.N., Kim, S.J. and **Kook, Y.H.,** 2001 *"Differentiation of mycobacterial species by PCR-Restriction analysis of DNA (342 base pairs) of the RNA polymerase Gene (rpoB)" J Clin microbial;* 39 : 2102-2109.

Roth, A, Reischl, U.D.O., Streubel Naumann, L., Kroppenstedt, R.M., Habianicht, M., Fischer, M., and **Mauch, H.,** 2000 *"Novel diagnostic algorithm for identification of the 16s-23s rRNA gene spacer and restriction endonuclease." J Clin Microtriol;* 38 : 1094-1104.

Sharm, D. 2004 *"Identification and characterization of environmental isolates of mycobacteria (NTM) by PCR-RFLP assay targeting 65 –KDa antigen gene."* M.sc. dissertation (unpublished). Dayalbagh Educational Institute, Dayalbagh, Agra.

Telenti, A., Marchesi, F., Balz, M., Bally, F., Bottger and **Bodmer, T.** 1993 *"Rapid identification of mycobacteria to the species level by PCR and restriction enzyme analysis" J clin microbial,* 31: 175-178.

Van Soolingen D. Dettass PEW, Harmans PWM 1993 *"Comparison of various repetitive DNA elements of genetic marker for strain differentiation and epidemiology of M.tuberculosis." J. clin. Microbiol*, 31 : 1997-1995.

Van Embden, J.D.A., Cave, M.D., Crawford, J.T., Dele, J., Eisenach, K.D., Gicquel, B., Hermons, P., Matri, C., Medan, R., Shinnick, T.M., Small, P.M. 1993 *"Strain identification of M. tuberculosis by DNA finger printing : Recommended for standardized Methodology." J. Clin Microbiol* 1993; 31: 406-409.

Vaneechoutte, M., Been houwer, H.D., Ciacys, G., and Verschvaegen, G., de -Rouck, A., Palpe, N., Elaichoum, A. and **Portales, F.** 1993 *"Identification of Mycobacterial species by using amplified ribosomal DNA resrtriction analysis" J. clin microbial*, 31;2061-2065.

29

Pseudomonas fluorescens Mediated suppression of Black gram root rot incited by *Macrophomina phaseolina*

M.R.Kalaivani, R.Bhuvaneswari, K.Kavitha, S.Mathiyazhagan, V.Sendhilvel, V.Menaka and G.Chandrasekar

Deptt. of Plant Pathology, Center for Plant Protection Studies, Tamil Nadu Agricultural University, Coimbatore - 641 003

ABSTRACT

Twenty isolates of plant growth promoting fluorescent pseudomonads were isolated and evaluated against black gram root rot viz., *M. phaseolina*. Among these isolates, VPf6 and CPf3 recorded maximum percent inhibition of *M. phaseolina* mycelial growth 78.88 and 75.43 respectively. These isolates were characterized and identified by biochemical studies. Antibiotic viz., 2,4-DAPG (diacetylphloroglucinal) and phenazine was eluated from *P. fluorescens*, the isolate VPf6 and CPf3.It was found to be antifungal action to *M. phaseolina* at a concentration of 150-200 µl/well. The sensitivity of 2,4-DAPG and phenazine against *M. phaseolina* was maximum at pH 4.0 in PDA medium. The antibiotic (2,4-DAPG and phenazine) production was maximum when the bacteria grown in glycerol and ammonium sulphate as sole carbon and nitrogen source respectively. Siderophores play a vital role in the suppression of plant pathogens by creating competition for iron. All the bacterial strains produced siderophore in Chrome azurol S (CAS) plate assay method. The five isolates tested, among the isolates VPf6 and CPf3 produced the maximum intensity of HCN. *P. fluorescens* isolate VPf6 recorded the highest salicylic acid production followed by isolate CPf3. All the five isolates, such as VPf6, CPF3, DPf1, CPf5 and CBPf2 produced the lytic enzymes viz., â-1, 3 glucanase and protease under *in vitro* condition.

KEY WORDS : Biological control, *Pseudomonas fluorescens*, *Macrophomina phaseolina*, 2,4- diacetylphloroglucinal, phenazine, lytic enzymes, HCN, Siderophore.

Macrophomina *phaseolina* (Tassi) Goid., a serious soil borne pathogen, causes seed ling rot /dry rot in black gram(Vigna mungo (L.) Hepper) leading too extensive yield loss. The hazards in the fungicidal treatment for disease management by way of harmful residues and development of resistant strains have led to the use of biocontrol agents for disease management. Heterotrophic rhizobacteria of *Pseudomonas fluorescens* types have been successfully used for biological control of several plant pathogens (Howell and Stipanovic, 1979; Weller and Cook, 1983;Ganesan and Gnanamanicam, 1987). A primary mechanism of pathogen suppression is by the production of siderophore, HCN, lytic enzymes and antibiotics. Flurorescent *Pseudomonas* spp., produce secondary metabolites with antibiotic activities many of which have been implicated in suppression of soil-borne diseases like phenazine-1-carboxylic acid (PCA), 2, 4 – diacetylphlroglucinol (DAPG), oomycin – A, pyocyanin, pyoluteorin and pyrrolnitrin (Thomashow and Weller, 1996). In the present study, an attempt has been made to isolate plant growth promoting fluorescent pseudomonads, study their antibiotics, siderophores, HCN, lytic enzymes and exploit this property on laboratory scale for assessments of in vitro antagonism

Materials and Methods

Isolation and Characterization of Pseudomonas fluorescens

Twenty isolates of fluorescent pseudomonads were isolated from rhizosphere soil of black gram crops grown in different parts of Tamil Nadu (Table 1) using King's medium B (KMB) (King *et al.*, 1954). The isolates were characterized based on standard biochemical tests (Hildebrand *et al.*, 1992). The following tests were carried out: Gram reaction, pigment production (King's B), oxidase test, starch hydrolysis, KOH test, anaerobic growth and arginine dihydrogenase test. Results of these tests were scored either as positive or as negative. Grouping was made with the aid of a determinative scheme developed by earlier workers (Barrett *et al.*, 1986; Hildebrand *et al.*, 1992).

M.phaseolina was isolated from naturally infected black gram plants using potato dextrose agar (PDA) medium and mass–multiplied in sand maize medium at 19:1 w/w (sand:maize). Pathogenicity was tested in black gram plants and pathogen was maintained in PDA slants. For *in vitro* screening of fluorescent pseudomonads against *M. phaseolina*, the bacterial isolates were streaked on one side of a Petri dish (1 cm from the edge of the plate) with PDA medium and mycelial disc (8 mm diameter) of seven-day-old culture of *M.phaseolina* was placed on the opposite side of the Petri dish perpendicular to the bacterial streak (Vidhyasekaran *et al.*, 1997). The plates were incubated at room temperature (28 ± 2 C) for 4 days and the zone of inhibition was measured.

Extraction of 2, 4 Diacetylphloroglucinol (2, 4 DAPG) (Rosales et al., 1995)

Strains of P. fluorescens were grown in 100ml of Pigment Production broth (peptone 20g, Nacl-5g, KNo3-1g, glycerol-20 ml, pH-7.2 and Distilled water-1000 ml) for four days on a rotary shaker at 30°C. The fermentation broth was centrifuged at 3500 rpm for five min. in a tabletop centrifuge and the supernatant was collected. It was acidified to pH 2.0 with 1N HCl and then extracted with an equal volume of ethyl acetate. The ethyl acetate extracts were reduced to dryness in vacuo. The residues were dissolved in methanol.

Extractions and Isolation of Phenazine

The strains of *P. fluorescens* were grown in nutrient broth at 30°C on a rotary shaker. The cells were collected by centrifugation at 3500rpm for seven minutes. The pellets were suspended in pigment production broth and then incubated on a rotary shaker for four days at 30°C. The antibiotic phenazine-1-carboxylic acid (PCA) was isolated as per the procedure described by Rosales *et al.*, (1995).

The antibiotic was separated in to respective fractions after acidifying the culture filtrate with 1N HCl to pH 2.0 and then extracting the culture filtrates with an equal volume of Benzene (Phenazine). Then the benzene phase was extracted with 5 per cent NaHCO3. Phenazine-1-carboxylic acid was recovered from the bicarbonate layer while oxychlororaphine remained in the benzene layer. The bicarbonate fraction was extracted once again with benzene to recover phenazine from bicarbonate fraction. The antibiotic was air dried and dissolved in methanol.

Effect of 2,4 diacetylphloroglucinol and Phenazine on the growth of M. phaseolina and their sclerotia

PDA medium was poured in sterile Petri plate and allowed to solidify. A 5mm sterilized cork borer was placed inside the medium and wells were made in the Petri plates at the opposite ends. The pathogen *M. phaseolina* was cut with 5mm cork borer and placed in the center of the well. The antibiotic was poured into the wells at opposite ends. The effective dose for the inhibition of *M. phaseolina* was standardized by pouring different doses of antibiotics namely 100µl, 150µl, and 200µl/well. The mycelial growth and inhibition zone was recorded after incubation at 30^0C for 4 days. Tracing the surface area of inhibition in a tracing paper and then plotting it on the graph sheet measured surface area of inhibition.

The sclerotia of *M. phaseolina* were dipped in 2,4-DAPG and phenazine for 5 min and placed in PDA medium. Sclerotia dipped in sterile water served as control. The plates were incubated at 30^0C for 4 days and the sclerotial germination.

Detection of antibiotics using TLC

For the detection of phenazine, 10ml of samples were chromatographed on TLC aluminium sheets - Silica gel 60 F 254 (Merck, Germany). The solvents used were Isopropanol/ ammonia/ water (8:1:1). Plates were viewed under short wave length UV light (254nm). Rf values of the spots were calculated. The secondary metabolite, 2, 4 DAPG was detected through F254 coated TLC plates using 2, 4 DAPG obtained from Toronto Chemicals, Canada as the marker. For the detection of 2, 4 DAPG the plates were developed in acetonitrile / methanol/ water (1:1:1) and visualized by short wave length (254nm) and sprayed with diazotized sulphanilic acid.

Dissolve 50mg diazotized sulphanilic acid in 20ml 20% Na2CO3 solution. Diazotization is done by dissolving 25g sulphanilic acid in 125ml 10% sodium nitrite solution. Add the mixer drop by drop to 60 ml of 8M HCl in ice-cold condition. After 10 min, it was filtered under ice-cold condition. It was washed extensively with the cold water and then with ethanol followed by ether. The crystals were air dried and stored at 4°C.

Effect of 2, 4 DAPG and phenazine on the growth of M. phaseolina in different pH of the medium

The PDA medium was adjusted to pH 4.0, 6.0, 8.0 and 10.0 and autoclaved. 2,4-DAPG and phenazine isolated from *P. fluorescens* isolate VPf6 and CPf3 were added to the plates at the rate of 150 ml/well in PDA medium having the pH of 4.0, 6.0, 8.0 and 10.0 and incubated at 28±2°C for 4 days. The mycelial growth, inhibition zone and surface area of inhibition were recorded after 4 days of incubation.

Effect of different Carbon and Nitrogen sources on the production of 2, 4 DAPG and its antifungal action against M. phaseolina

The pigment production broths were substituted with different carbon sources namely dextrose, lactose, sucrose and glycerol and different nitrogen sources such as ammonium nitrate, ammonium sulphate, sodium nitrate and potassium nitrate. *P. fluorescens* isolate VPf6 and CPf3 were inoculated in the broth and incubated at 28±2°C for 72h. DAPG were extracted from the broth containing different carbon and nitrogen sources and were assayed for its antifungal action. DAPG were added to the plates containing PDA medium at the rate of 200 µl/well and incubated at 28±2°C for 4 days. The mycelial growth, inhibition zone and surface area of inhibition were recorded after 4 days of incubation.

Salicylic acid and IAA production

Salicylic acid production of the strains was determined as per the method of Meyer et al. (1992). The strains were grown in the standard succinate medium (Succinic acid- 4.0g, K2HPO4-6.0g, KH2PO4- 3.0g, (NH4)2 SO4-1.0g, MgSO47H2O-0.2g, distilled water-1000ml, pH 7.0) at 28±2°C for 48h. Cells were collected by centrifugation at 6000g for 5min. Four ml of cell free culture filtrate was acidified with 1N HCl to pH 2.0 and salicylic acid was extracted with equal volume of chloroform. Four ml of water and 5ml of 2M FeCl3 was added to the pooled chloroform phases. The absorbance of the purple iron salicylic acid complex, which was developed in the aqueous phase, was read at 527nm. A standard curve was prepared with salicylic acid in succinate medium and quantity of salicylic acid produced was expressed as mg/ml.

The bacterial strains were grown in trypticase soybroth (Animal peptone-15g, soypeptone-5g, NaCl-5.0g, Glycine-4.4g, Distilled water-1litre) with tryptophan (100 µl/ml) and incubated at 28±2°C. To one ml of cell free culture filtrate, 2ml of Salkowsky reagent (1ml of 0.5 M FeCl3 in 50 ml of 35 % perchloric acid) was added and incubated at 28 ± 2°C for 30 min. The absorbance was read at 530 nm. A standard was prepared using IAA and presence of IAA in culture filtrate was quantified as µg/ml (Gorden and Paleg, 1957).

Quantification of β-1, 3-glucanase and protease produced by P. fluorescens

The bacterial isolates were grown in nutrient broth for 3 days and incubated. Then the culture was centrifuged and the supernatant was taken as enzyme extract. The enzyme activity was colorimetrically assayed. Crude enzyme extract of 62.5 ml was added to 62.5 ml of 4 per cent laminarin and then incubated at 40°C for 10 min. The reaction was stopped by adding 375 ml of dinitro salicylic acid (dinitro salicylic acid (DNS) reagent was prepared by adding 300ml of sodium hydroxide (4.5%) to 880ml of a solution containing 8.8g of dinitro salicylic aid and 22.5g of potassium sodium tartarate) and heated for 5 min. on boiling water bath. The resulting solution was diluted with 4.5 ml distilled water and the absorbance was read at 500 nm. The crude extract preparation with laminarin with zero time incubation served as blank. The enzyme activity was expressed as mg equivalent of glucose/ min/ ml of culture filtrate.

The bacterial strains were grown in nutrient broth supplemented with glucose (5g/l) and casein (5g/l) and incubated in rotary shaker for 3 days. Then it was centrifuged at 10000 rpm at 4°C for 10 min. The supernatant was used as the enzyme source. The activity of protease was determined by the method of Anson, 1938. The reaction mixture contained 0.8ml of Hammerstein casein (6.0gl^{-1}, dissolved in 0.05M phosphate buffer pH 6.0) and 0.2ml of culture filtrate. It was incubated without shaking at 37°C for 15 min. The reaction was stopped by the addition of 0.5ml of ice-cold 10 per cent tricarboxyclic acid. The sample was then centrifuged at 1000 rpm for 10 min. One ml of the supernatant was used for the assay of tyrosine, according to Lowry *et al.*, (1951). One unit of enzyme activity was defined as the amount of enzyme required for the formation of 1 mmol of the product per minute of the reaction, under the standard assay conditions.

Siderophore Production

Production of siderophore by bacterial antagonist was assayed by plate assay. The tertiary complex Chrome azural S (CAS) / Fe^{3+} / hexadecyl trimethyl ammonium bromide served as an indicator. Forty eight hour old culture of the bacterial isolates were streaked on to the succinate medium (Succinic acid-4.0g, K_2HPO_4-3.0g, $(NH_4)_2 SO_4$. $7H_2O$-0.2 g, Distilled water-1 liter, pH-7.0) amended with indicator dye. To prepare one liter of blue agar, 60.5mg of chrome azurol S (CAS) was dissolved in 50ml of distilled water and mixed with 10ml of Iron (III) solution (1mM $FeCl3$. $6H2O$ in 10mM HCl). While constantly stirring this solution was slowly added to 72.9mg of hexa decyl trimethyl ammonium bromide (HDTMA) dissolved in 40ml of water. The resultant dark blue liquid was observed for the formation of bright zone with yellowish fluorescent colour in the dark colored medium. It was the indication of production of siderophore. The result was scored either positive or negative to this test (Schwyn and Neilands, 1987). The surface area of production was recorded by tracing the area of colour change in a tracing paper and there by plotting it on a graph sheet.

Detection of the Nature of Siderophore

The bacterial isolates of *P. fluorescens* were inoculated in 10 ml of King's B broth. It was incubated in a rotary shaker at 120 rpm for 48 h. The bacteria multiplied in the broths were used as the sample for the determination of the nature of siderophore.

Hydroxamate nature of siderophore examined by tetrazolium salt test. Instant appearance of a deep red colour by addition of siderophore sample to tetrazolium salt under alkaline conditions indicated the presence of hydroxamate (Snow, 1984). Carboxylate nature of siderophore detected by Vogel's chemical test where the disappearance of pink colour on addition of Phenolphthalein to siderophore sample under alkaline condition indicated carboxylate nature (Vogel, 1987).

Hydrogen Cyanide Production

Qualitative assay and Quantitative assay

HCN Production was determined by using the modified protocol of Miller and Higgins (1970). Bacteria were grown on Tryptic soy agar (TSA) (animal peptone–15.0g, soyapeptone-5.0g, sodium chloride-5.0g, glycine-4.4g, distilled water-1000ml). Filter paper discs soaked in picric acid solution (2.5g of picric acid, 12.5g of sodium carbonate, and 1000ml of distilled water) were placed in the upper lid of each Petri plate. Dishes were sealed with parafilm and incubated at 28°C for 48h. A change from yellow to light brown, brown or reddish brown of the discs were recorded as an indication of weak, moderate or strong production of HCN for each strains respectively.

Quantitative assay

Bacteria were grown on tryptic soy broth (TSB). Filter paper was cut into uniform strips of 10cm long and 0.5cm wide. The strips were saturated with alkaline picrate solution and placed inside the conical flasks in a hanging position. After incubation at 28°C for 48h, the sodium picrate present in the filter paper was reduced to reddish compound in proportion to the amount of hydrocyanic acid evolved. The colour was eluted from the filter paper by placing it on a clean tube containing 10ml of distilled water and the absorbance was measured at 625nm (Sadasivam and Manickam, 1992).

Results and Discussion

Characterization of P. fluorescens

Twenty isolates of *P. fluorescens* isolated from rhizosphere soil of black gram crop grown in different parts of Tamilnadu.The isolates were found to be fluorescent on KMB and all the tested isolates did not show positive result on starch hydralysis. Most of the bacterial isolates showed a positive response for Pigment production (Kings B), Oxidase test, KOH test, Arginine dihydrogenase. Hence the isolates were grouped under *P. fluorescens* (Table 1).

Table 1 Characterization of P.fluorescens and their efficacy on inhibition of
M. phaseolina

Isolates	Gram reaction	Pigment production (Kings B)	Oxidase test	Starch hydrolysis	KOH test	Anaerobic growth	Arginine dihydro-genase	Inhibition zone (mm)
CPf1	–	+	+	–	+	+	+	43.44[d]
CPf2	–	+	+	–	+	+	+	46.87[d]
CPf3	–	+	+	–	+	+	+	75.43[g]
CPf4	–	+	+	–	+	+	+	59.55[e]
CPf5	–	+	+	–	+	+	+	33.43[c]
VPf1	–	+	+	–	+	+	+	22.76[b]

Table 1 *Contd...*

Isolates	Gram reaction	Pigment production (Kings B)	Oxidase test	Starch hydrolysis	KOH test	Anaerobic growth	Arginine dihydro-genase	Inhibition zone (mm)
VPf2	−	+	+	−	+	+	+	42.54[d]
VPF3	−	+	+	−	+	+	+	36.44[c]
VPf4	−	+	+	−	+	+	+	16.77[a]
VPf5	−	+	+	−	+	+	+	23.64[b]
VPf6	−	+	+	−	+	+	+	78.88[g]
CBPf1	−	+	+	−	+	+	+	33.95[c]
CBPf2	−	+	+	−	+	+	+	60.22[f]
CBPf6	−	+	+	−	+	+	+	41.12[d]
DPf1	−	+	+	−	+	+	+	68.98[f]
DPf2	−	+	+	−	+	+	+	8.87[a]
DPf3	−	+	+	−	+	+	+	21.31[b]
TPf1	−	+	+	−	+	+	+	43.11
TPf2	−	+	+	−	+	+	+	15.36[a]
TPf3	−	+	+	−	+	+	+	12.39[a]

Symbol '+' denotes a positive result, symbol '-' denotes negative results Pf – P. fluorescens

In a column, means followed by the same letter do not differs significantly (P=0.05) by DMRT

Many of the fluorescent pseudomonads, predominantly *P. fluorescens*, were isolated from suppressive soil for the management of soil borne diseases (Mills *et. al.*, 1989; Weller and cook, 1986; Stephens *et. al.*, 1993). The present study indicates that the majority of the strains isolated from rhizosphere of black gram belonged to *P. fluorescens*.

After isolating and identifying the fluorescent pseudomonads, selecting an effective isolates is the first and foremost important step in biological control. *P. fluorescens* isolate VPf6 showed the maximum inhibitory effect on mycelial growth.

Among the twenty isolates of *P. fluorescens* tested for their efficacy in inhibiting the mycelial growth, five isolates, *P. fluorescens* CPf3, CPf5, VPf6, CBPf2 and DPf1 showed higher inhibitory effect on mycelial growth of *M. phaseolina* when compare to other tested isolateds. Among these isolates CPf3 and VPf6 exhibited the maximum inhibition of mycelial growth in vitro by recording a inhibition zone of mm (table1)

Effect of Phenazine on the growth of M. phaseolina

The phenazine isolated from *P. fluorescens* isolate VPf6 recorded 35.87 per cent inhibition of mycelial growth of *M. phaseolina* at 200ml concentration over untreated control (Table2). It was followed by 150 μl (30.95%). In case of isolate CPf3, the per cent inhibition of mycelial growth was 28. 66 per cent at 200 ml concentration followed by 26. 33 per cent at 150 μl concentration/well. The phenazine isolated from other isolates such, as DPf1, CPf5 and CBPf2 were not effective in inhibiting the mycelial growth of *M. phaseolina*.

Effect of 2, 4 DAPG on the mycelial growth of M. phaseolina

2,4 DAPG isolated from *P. fluorescens* isolate VPf6 recorded the least mycelial growth of 66.5mm at 200 µl concentration and 150 µl concentration/well. It was statistically on par with isolate CPf3, which recorded a mycelial growth of 68.0mm at 200 µl concentration and 150 µl concentration/well as against 91.00 mm in untreated control. Isolates such as CPf5 recorded the mycelial growth of 76.50mm and was on par with isolate DPf1 (Table 3). The 2, 4 DAPG isolated from isolate CPf5 was not effective in inhibiting the mycelial growth of *M. phaseolina*. Hence, in further studies on antibiotic production, the isolate such as VPf6 and CPf3 were selected.

Table 2 Antifungal action of phenazine against *M. phaseolina*

Anta-gonitic bacteria	Mycelial growth of pathogen (mm)*			Inhibition (mm)			Per cent inhibition of mycelial growth over control			Surface area of inhibition (mm^2)		
	100 µl/well	150 µl/well	200 µl/well	100 µl/well	150 µl/well	200 µl/well	100 µl/well	150 µl/well	200 µl/well	100 µl/well	150 µl/well	200 µl/well
VPf6	62.00[b]	58.00[b]	56.00[b]	14.00[a]	20.00[a]	21.00 [a]	31.00	34.55	35.77	66.0[a]	78.0[a]	76.0[a]
CPf3	68.00[b]	70.00[b]	69.00[b]	0.00[b]	0.00 [b]	0.00 [b]	23.22	23.26	28.33	0.00[b]	0.00 [b]	0.00 [b]
DPf1	90.00[a]	90.00[a]	90.00[a]	0.00 [b]	0.00 [b]	0.00 [b]	0.00	0.00	0.00	0.00 [b]	0.00 [b]	0.00 [b]
CPf5	90.00[a]	90.00[a]	90.00[a]	0.00 [b]	0.00 [b]	0.00 [b]	0.00	0.00	0.00	0.00 [b]	0.00 [b]	0.00 [b]
CBPf2	90.00[a]	90.00[a]	90.00[a]	0.00 [b]	0.00 [b]	0.00 [b]	0.00	0.00	0.00	0.00 [b]	0.00 [b]	0.00 [b]
Control	90.00[a]	90.00[a]	90.00[a]	0.00 [b]	0.00 [b]	0.00 [b]	-	-	-	-	-	-

Mycelial growth was recorded after 4 days of incubation at 28±2°C

* Values are mean of three replications. In a column, means followed by a common letter are not significantly different at the 5 % levels by DMRT.

Table 3 Antifungal action of 2,4-DAPG against *M. phaseolina*

Anta-gonitic bacteria	Mycelial growth of pathogen (mm)*			Inhibition (mm)			Per cent inhibition of mycelial growth over control			Surface area of inhibition (mm^2)		
	100 µl/well	150 µl/well	200 µl/well	100 µl/well	150 µl/well	200 µl/well	100 µl/well	150 µl/well	200 µl/well	100 µl/well	150 µl/well	200 µl/well
VPf6	64.50[b]	68.50[b]	66.50[a]	20.00[a]	22.00[a]	23.00[a]	28.33	24.86	22.87	59.0[a]	62.3[a]	63.5[a]
CPf3	64.00[b]	69.00[b]	68.50[b]	18.50[a]	20.60[a]	21.50[a]	28.88	23.33	23.33	50.4[a]	53.3[b]	55.0[a]
DPf1	74.40[ab]	76.50[a]	76.50[a]	12.20 [a]	13.60[a]	15.00[a]	17.33	15.11	13.88	29.0[b]	30.8[b]	32.0[b]
CPf5	73.00[ab]	74.50[a]	74.50[a]	13.50[a]	15.30[a]	16.20[a]	18.88	15.00	16.11	30.2[b]	32.5[b]	33.2[b]
CBPf2	90.00[ab]	90.00[a]	90.00[a]	0.00 [b]	0.00 [b]	0.00[b]	0.00	0.00	0.00	0.00[c]	0.00[c]	0.00[c]
Control	91.00[ab]	91.00[a]	91.00[a]	0.00 [b]	0.00 [b]	0.00[b]	-	-	-	-	-	-

Mycelial growth was recorded after 4 days of incubation at 28±2°C

* Values are mean of three replications. In a column, means followed by a common letter are not significantly different at the 5 % levels by DMRT.

Studies on antibiotics produced by *Pseudomonas* spp. are abundant since they are common inhabitants of rhizosphere. Antibiotics produced by different PGPR have a broad-spectrum activity (Thomashow *et al.*, 1997). The antibiotic phenazine and 2,4 DAPG isolated from the cell cultures of *P. fluorescens* isolate VPf6 and CPf3 effectively suppressed the growth of *M. phaseolina*. Phenazine and 2,4 DAPG were highly inhibitory to M.phaseolina. The results are in line with the findings of Bakker et al. (2002) that *P. putida* WCS 358r produced an antifungal compound phenazine–3-carboxylic acid (PCA) and 2, 4 DAPG which had both antifungal and antibacterial action.

Effect of 2, 4 DAPG on the germination of sclerotial fungi

The sclerotia of *M. phaseolina* were highly sensitive to 2,4-DAPG and phenazine produced by *P. fluorescens* isolate VPf6 and CPf3. It completely inhibited the sclerotial germination of all the sclerotial fungi, when sclerotia were treated for duration of 5 min and 10 min, indicating it's the broad spectrum of activity of antibiotics. But in untreated control 100 per cent germination of the sclerotia was observed irrespective of the pathogens tested (Table 4).

Table 4 Effect of 2, 4 DAPG and phenazine on the sclerotial germination of M.phaseolina

Isolates	*DAPG*		*Phenazine*	
	Germination (%) of sclerotia exposed to different duration		*Germination (%) of sclerotia exposed to different duration*	
	5 min	*10 min*	*5 min*	*10 min*
CPf3	0	0	0	0
VPf6	0	0	0	0
CPf1	12	3	11	3
CBPf2	14	2	12	4
DPf1	18	7	21	8
Control	100	100	100	100

Values are mean of three replications

Detection of antibiotics using TLC

In this study, the secondary metabolites responsible for antifungal activity were isolated and the metabolites were detected through thin layer chromatography. The secondary metabolite, 2, 4 DAPG was detected through F254 coated TLC plates using 2, 4 DAPG obtained from Toronto Chemicals, Canada as the marker. The 2, 4-DAPG isolated from P. fluorescens,the isolate VPf6 and CPf3 recorded the Rf value0.88. The results were in concomitant with the findings of Rosales et al. (1995). Since the gene responsible for the production of DAPG is highly conserved, the Rf value of DAPG from P. putida was similar to the Rf value of 2, 4 DAPG extracted from VPf6 and CPf3. Similarly, the Rf values of phenazine were 0.57 for isolate VPf6 and CPf3 isolate and for the standard phenazine visualized through fluorescent coated TLC plates under short wave UV light at 254nm.

Influence of pH on the antifungal action of 2, 4 DAPG and phenazine against M. phaseolina

The sensitivity of the antifungal action of 2, 4-DAPG from isolate VPf 6 and CPf 3 against *M. phaseolina* was the higher at pH 4.0 in the PDA medium. The inhibition of mycelial growth of *M.phaseolina* was 30.55 and 25.44 per cent over control at pH 4.0 followed by pH 6.0 and 8.0 (Table 5). The surface area of inhibition of *M.phaseolina* was the higher at pH 4.0 (104.5 mm^2). However, different pH of the medium (pH 4.0, 6.0 and 8.0) were not inhibitory to the growth of M. phaseolina in the absence of 2, 4 DAPG but in pH 10.0 the growth of *M.phaseolina* was 8.0 cm after 4 days of incubation.

Table 5 Influence of pH on the antifungal action of 2,4 DAPG against *M.phaseolina*

pH of the medium	Isolate VPf 6				Isolate CPf 3			
	*Mycelial growth (mm)**	*Per cent inhibition of mycelial growth over control*	*Inhibition zone (mm)*	*Surface area of inhibition (mm^2)*	*Mycelial growth (mm)**	*Per cent inhibition of mycelial growth over control*	*Inhibition zone (mm)*	*Surface area of inhibition (mm^2)*
pH4.0	60.60[b] (90.0)	30.55	26.60[a]	103.50[a]	68.00[b] (90.0)	25.44	23.00[a]	96.00[a]
pH 6.0	64.00[b] (90.0)	32.00	24.30[ab]	81.60[ab]	72.00[b] (90.0)	22.00	19.00[ab]	73.50[b]
pH 8.0	71.30[a] (90.0)	20.77	18.60[b]	51.80[b]	79.00[a] (90.0)	13.22	11.00[b]	54.00[c]
pH 10.0	68.60[a] (80.0)	14.25	22.00[ab]	68.30[ab]	78.00[a] (80.0)	3.50	10.00[b]	32.50[d]

* Mycelial growth was recorded after 4 days of incubation at 28±2°C

Values are mean of three replications. In a column, means followed by a common letter are not significantly different at the 5% levels by DMRT.

Values in parentheses are growth of pathogens in untreated control

The phenazine produced by isolate VPf 6 recorded the least mycelial growth of 66.6 mm at pH 4.0 with inhibition zone of 25.3mm with mean surface area inhibition of 92.6mm^2 (Table 6). But in case of isolate CPf 3 recorded a mycelial growth of 68.3mm at pH4.0 with an inhibition of 23.6mm with mean surface area inhibition of 64.3mm^2.

The antifungal action of 2, 4-DAPG and phenazine is a pH dependent factor. The antifungal activity of 2, 4-DAPG and phenazine was more at pH 4.0 to pH 6.0 against *M. phaseolina*. But, it declined at alkaline pH. These results indicated that the antibiotic was effective both in acidic to neutral pH. Mycelial growth of *P. ultimum* var. *sporangiiferum* was completely inhibited at pH 4.5, in the presence of 2, 4-DAPG at a concentration of 16mg/ml, whereas 2, 4-DAPG was not sufficient to completely inhibit the mycelial growth at pH 7.5 and 8.0 even at a concentration of 40mg/ml indicating that the efficacy was a pH dependent factor (de Souza *et. al.*, 2003).

Table 6 Influence of pH on the antifungal action of phenazine against *M.phaseolina*

pH of the *medium*	*Isolate VPf 6*				*Isolate CPf 3*			
	Mycelial growth (mm)[*]	*Per cent inhibition of mycelial growth over control*	*Inhibition zone (mm)*	*Surface area of inhibition (mm²)*	*Mycelial growth (mm)*[*]	*Per cent inhibition of mycelial growth over control*	*Inhibition zone (mm)*	*Surface area of inhibition (mm²)*
pH 4.0	65.60[c] (90.0)	28.11	25.30[a]	92.60[a]	68.30[b] (90.0)	25.22	23.60[a]	64.30[a]
pH 6.0	74.60[a] (90.0)	18.11	13.60[b]	54.30[a]	76.30[ab] (90.0)	16.22	15.00[b]	39.80[b]
pH 8.0	69.30[bc] (90.0)	24.00	15.60[bc]	59.30[a]	83.60[a] (90.0)	8.11	12.60[b]	41.00[b]
pH 10.0	72.60[ab] (80.0)	11.00	19.00[b]	61.80[a]	71.60[b] (80.0)	11.5	17.30[b]	56.50[ab]

[*] Mycelial growth was recorded after 4 days of incubation at $28\pm2°C$

Values are mean of three replications. In a column, means followed by a common letter are not significantly different at the 5 % levels by DMRT.

Values in parentheses are growth of pathogens in untreated control

The antifungal activity of Phenazine-1-carboxamide (PCN) and Phenazine-1-carboxylic acid (PCA) was 10 times higher at neutral pH suggesting that it may contribute to the superior biocontrol performance of *P. chlororaphis* strain PCL 1391 in tomato against *F. oxysporum* (Chin-A-Woeng *et al.*, 1998).

Effect of different carbon and nitrogen sources on the production of 2, 4 DAPG and its activity on the growth of M. phaseolina

Different carbon sources such as dextrose, lactose, sucrose and glycerol were used to assess the influence of nutrient sources on antibiotic (2,4-DAPG) production. The production of 2,4-DAPG by isolate VPf 6 and CPf3 was higher when glycerol was used as the sole carbon source (Table 7). The surface area of inhibition of *M.phaseolina* by 2,4-DAPG from VPf 6 extracted from glycerol based carbon source was 79.0 mm² followed by lactose (66.6mm²) and dextrose (56.5mm²).

Studies on the influence of various nitrogen sources on the production and antifungal action of 2,4-DAPG from *P. fluorescens* isolate VPf6 grown in ammonium sulphate as nitrogen source recorded maximum (26.11%) (Table 8) inhibition of the mycelial growth of *M. phaseolina* followed by potassium nitrate (26.66 %).But in case of isolate CPf3, significant reduction in the mycelial growth of *M. phaseolina* was recorded when potassium nitrate was used as nitrogen source. The mean surface area of inhibition of *M. phaseolina* was maximum from 2,4-DAPG extracted (77.3 mm²) from potassium nitrate amended PP broth (Table 8) followed by sodium nitrate (70.6 mm²). The constituents of the medium

Table 7 Influence of pH on the antifungal action of phenazine against *M.phaseolina*

	Isolate VPf 6					Isolate CPf 3				
Different carbon	Quality of 2,4 DAPG /100 ml of PP broth (µg/ml)	Mycelial growth (mm)[*]	PIOC	Inhibition n zone (cm)	Surface area of inhibition n (mm²)	Quality of of 2,4 DAPG /100 ml of PP broth (µg/ml)	Mycelial growth (mm)[*]	PIOC	Inhibition n zone (mm)	Surface area of inhibition n(mm²)
Dextrose	85	67.00[b]	27.66	21.00[a]	56.50[ab]	56	72.00[b]	22.11	29.30[a]	87.50[a]
Lactose	73	65.00[b]	29.88	21.00[a]	66.60[ab]	67	65.30[c]	27.55	26.30[a]	58.30[a]
Sucrose	66	69.30[b]	25.11	24.30[a]	45.60[b]	29	65.60[c]	28.22	18.30[b]	58.30[a]
Glycerol	87	62.30[b]	30.88	28.30[a]	79.00[a]	527	64.30[c]	28.66	28.30[a]	70.80[a]
Control		90.00[a]	0.00	–	–		90.00[a]	0.00		

* Mycelial growth was recorded after 72h of incubation at $28\pm2°C$

PIOC- Percent inhibition of mycelial growth over control

Values are mean of three replications. In a column, means followed by a common letter are not significantly different at the 5 % levels by DMRT

Table 8 Effect of different nitrogen sources on the production of 2,4 DAPG and its antifungal action against *M.phaseolina*

	Isolate VPf 6					Isolate CPf 3				
Different nitrogen sources	Quality of 2,4 DAPG /100 ml of PP broth (µg/ml)	Mycelial growth (mm)[*]	PIOC	Inhibition n zone (cm)	Surface area of inhibition (mm²)	Quality of of 2,4 DAPG /100 ml of PP broth (µg/ml)	Mycelial growth (mm)[*]	PIOC	Inhibition zone (mm)	Surface area of inhibition (mm²)
Ammonium nitrate	584	66.60[b]	23.88	23.30[a]	81.00[a]	482	68.30[b]	22.00	22.30[b]	63.30[ab]
Ammonium sulphate	88	64.60[b]	24.11	24.60[a]	73.50[a]	233	69.00[b]	21.22	21.00[b]	48.30[b]
Sodium nitrate	185	65.60[ab]	25.00	25.30[a]	66.50[a]	14	65.60[c]	25.00	24.00[b]	71.60[a]
Potassium nitrate	87	65.00[ab]	26.66	25.00[a]	58.50[a]	52	60.60[c]	30.55	28.30[a]	78.30[a]
Control		90.00[a]	0.00	–	–		90.00[a]	0.00	–	–

* Mycelial growth was recorded after 4 days of incubation at 28 ± 2 °C

PIOC- Percent inhibition of mycelial growth over control

Values are mean of three replications. In a column, means followed by a common letter are not significantly different at the 5 % levels by DMRT.

exerted a control over the yield of antibiotics produced by bacterial antagonists. The antifungal action of DAPG and phenazine against *M. phaseolina* by isolate VPf6 and CPf3 was higher when glycerol was used as the sole carbon source. The antibiotic production of *Pseudomonas* clusters and its in vitro antagonism against *P. ultimum* and *R. solani* in PDA medium was higher when glucose was used as the sole carbon source (Nielson *et al.*, 1998).

Among the different nitrogen sources tested, DAPG isolated from *P. fluorescens* isolate VPf6 grown in ammonium sulphate as nitrogen source recorded maximum inhibition of the mycelial growth of *M. phaseolina*. In case of isolate CPf3, the antifungal action was higher in potassium nitrate amended medium followed by sodium nitrate, ammonium nitrate and ammonium sulphate.

Siderophore Production

The bacterial strains produced siderophore in Chrome azurol S (CAS) plate assay method. All isolates produced yellow pigmentation in blue colored medium. The surface area of production was higher in isolate VPf6 ($185mm^2$) and isolate CPf3 ($92\ mm^2$) followed by isolate CPf5 ($72\ mm^2$), DPf1 ($56\ mm^2$) and isolate CBPf2 ($11\ mm^2$). (Table 9)

Siderophores play a vital role in the suppression of plant pathogens by creating competition for iron. In the present study, all the bacterial strains produced siderophore in Chrome azurol S (CAS) plate assay method. The isolates produced either hydroxamate type or carboxylate nature of siderophore. Siderophore production on CAS medium was different between the rhizobacterial isolates. Most isolates produced yellow halo surrounding the colonies. The production of fluorescent siderophores by *P. fluorescens*, which contributed to its antagonistic action against *P. ultimum* (Kloepper *et al.*, 1988).

Table 9 Testing of antagonistic bacterial strains for the production of siderophores and its nature

S.No	Rhizobacterial strains	Colour of siderophore pigment	Surface area production $(mm)^2$	Nature of siderophore	
				Hydroxamate	Carboxylate
1.	CPf3	Yellow	92^b	+	−
2.	CBPf2	Yellow	11^a	+++	−
3.	VPf6	Yellow	185^b	-	*
4.	DPf1	Yellow	56^c	+++	−
5.	CPf5	Yellow	72^d	+++	−

Values are mean of three replications. In a column, means followed by a common letter are not significantly different at the 5 % levels by DMRT.

+++ Appearance of deep red colour, + Appearance of light red colour, *Disappearance of pink colour, − Appearance of pink colour

Hydrogen Cyanide Production

P. fluorescens isolates were tested for HCN production, the intensity of HCN production was strong in isolate VPf6 and CPf3recording an absorbance value of 0.074 and 0.062 followed by isolate DPf1 that recorded an absorbance of 0.012 (Table 10).

In the present study,the five isolates tested,VPf6 and CPF3 produced the maximum intensity of HCN. However, no cyanide production was detected in CBPf1, CPF5 and DPF1 isolates. Mondal *et al.*, (1998) suggested that *P. fluorescens* (Rb-26) produced maximum HCN production, which in turn, showed stronger growth inhibition of *X. auxonopodis* pv. *malvacearum*.

Table 10 Hydrogen cyanide production by antagonistic bacteria

Antagonistic Bacteria	Qualitative assay	Quantitative assay (O.D value)
Vpf6	++	0.074[b]
Cpf3	++	0.062[b]
Cpf5	–	–
DPf1	+	0.012[a]
CBPf2	–	–

++ Produced HCN; - Absence of HCN

Values are mean of three replications. In a column, means followed by a common letter are not significantly different at the 5 % levels by DMRT.

Salicylic acid and IAA Production

P. fluorescens isolates were tested for salicylic acid production. Among them, isolate VPf6 and CPf3 recorded the highest salicylic acid production of 65.90 µg/ml and 62.34 µg/ml. The other isolates such as DPf1, CPf5 and CBPf2 produced 29.84, 25.79 and19.67 mg/ml of salicylic acid. (Table11). Indole Acetic Acid production by *P. fluorescens* strains was quantified. Among the isolates, VPf6 and CPf3 isolate recorded the maximum IAA production of 20.0 µg/ml and 14.0 µg/ml followed by CPf5, CBPf2 and DPf1isolates recorded 3.0,5.0 and 4.0 µg/ml (Table 11). The plants easily absorb salicylic acid of bacterial origin and they involve in triggering resistance in the host system. In the present study, *P. fluorescens* isolate VPf6 recorded the highest salicylic acid production followed by isolate CPf3. *P. fluorescens* strain WCS374 produced large quantity of salicylic acid under conditions of iron limitations (Leeman *et al.*, 1995).

The *Pseudomonas* spp. may increase plant growth by releasing phytohormones (Arshad and Frankenberger, 1991) or by producing gibberellin like substances (Brown, 1972). In the present study, Indole acetic acid productions by different PGPR strains were quantified. The isolate VPf6 recorded the highest IAA production followed by isolate CPf3 It has been established that fluorescent pseudomonad's produced plant growth promoters such as gibberellins, cytokinins and indole acetic acid, which can either directly or indirectly, modulate the plant growth and development (Dubeilovsky *et al.*, 1993; Glick, *1994).*

Table 11 Salicylic acid and IAA production by *P. fluorescens*

Antagonistic Bacteria	Salicyclic acid (μg/ml)	IAA production (μg/ml)[*]
Vpf6	65.90[a]	20.0[a]
Cpf3	62.34[a]	14.0[b]
Cpf5	29.84[b]	3.0[d]
DPf1	25.79[bc]	5.0[c]
CBPf2	19.67[bc]	4.0[c]

Values are mean of three replications. In a column, means followed by a common letter are not significantly different at the 5% levels by DMRT.

Lytic Enzyme Production

Quantification of β-1, 3 glucanase and protease

The β-1, 3 glucanase activity was higher in *P. fluorescens* isolate CPf3 recording an activity of 2741.84 nmol of glucose released /h/ml followed by isolate VPf6(2312.58 nmol of glucose released /h/ml) but isolate DPf1 recorded the lowest β 1,3 glucanase activity of 1118.80 nmol of glucose released /h/ml (Table 12).

Assay of protease activity revealed that the *P. fluorescens* isolate VPf6 recorded the highest protease activity of 385.88 U/ml followed by isolate CPf3 (220.78U/ml) (Table12).

Table 12 Quantification of β-1, 3 glucanase and protease production by *P. fluorescens* strains

S.No	Rhizobacterial strains	β1,3 Glucanase (μmol equivalent of glucose released/h/ml)	Protease (Units/ml)
1.	VPf6	2312.58[e]	395.88[a]
2.	DPf1	1118.80[c]	159.00[c]
3.	CPf5	1430.00[b]	129.83[e]
4.	CBPf2	1263.89[d]	128.76[d]
5.	CPf3	2741.84[a]	220.78[b]

Values are mean of three replications. In a column, means followed by a common letter are not significantly different at the 5 % levels by DMRT.

All the five isolates, such as VPf6, CPF3, DPf1, CPf5 and CBPf2 produced the lytic enzymes viz., β-1, 3 glucanase and protease under in vitro condition. But, the level of production of lytic enzymes varies with each and every isolate. It may be one of the reasons for in vitro antagonism. Alstrom (2001) reported that the bacterial strains in the red pigmented members of enterobacteriaeae possessed the ability to produce cellulase, protease (proteinase) and phospatase and some even produced chitinase. In the present study the extra cellular secretion of glucanase and protease also increase the antifungal action of the rhizobacteria in addition to the production of antibiotics, siderophore, SA and HCN.

References

Ahl, P., Voisard, C. and ***Defago, G.*** 1986. Iron bound siderophores, cyanic acid and antibiotics involved in suppression of *Thielaviopsis basicola* by *P. fluorescens* strain. *J. Phytopathology*, 116: 121-134.

Alström, S. 2001. Characteristics of bacteria from oilseed rape in relation to their biocontrol activity against *Verticillium dahliae. J. Phytopathology*, 149:57-64.

Anson, M. L. 1938. The estimation of pepsin, papain and cathepsin with haemoglobin. J. Gen. Physiol. 22: 79-89.

Arshad, M. and ***Frankenberger, Jr. W. T.*** 1991. Microbial production of plant hormones. In: *The rhizosphere and plant growth*. Keister, D.L, Cregan, P.B, (Eds.) Dordrecht, The Netherlands: Kluwer Academic Publishers, pp327–334.

Audenaert, K., Damme, A. Van., Cornelis, P., Cornelis, T. and ***Höfte, M.*** 2002a. Induced resistance by *Pseudomonas aeruginosa* 7NSK2: bacterial determinants and reactions in the plant. *Bulletin* OILB/SROP, 25: 223-226.

Barrett E.L., Solanes, R.E., Tang, J.S. and ***Palleroni, N.J.*** 1986. *Pseudomonas fluorescens biovar* V its resolution in to distint component groups and relationship of these to other *P. fluorescens* biovars to *P. putida* and to psychophilic pseudomonads associated associated with food spoilogy. *Journal of General Microbiology* 132:2709-2721.

Brown, M.E. 1972. Plant growth substances produced by microorganisms of soil and rhizosphere. *J. Appl. Bacteriol.*, 35: 443-451.

de., Thomas-Oates, J.E. and ***Lugtenberg, B.J.J.*** 1998. Biocontrol by phenazine-1-carboxamide-producing *Pseudomonas chlororaphis* PCL1391 of tomato root rot caused by *Fusarium oxysporum* f. sp. *radicis-lycopersici*. Mol. *Plant–Microbe Interact.*, 11: 1069-1077.

de Souza, J.T., Arnould, C., Deulvot,C., Lemaneau, P., Gianinzzi-Pearson, V. and ***Raaijmakers, J.M.*** 2003. Effect of 2,4 Diacetylphloroglucinol on *Pythium* : Cellular responses and variation in sensitivity among propagules and species. *Phytopathology*, 93: 966-975.

Dubeikovsky, A.N., Mordukhova, E.A., Kochethov, V.V., Polikarpova, F. Y. and ***Boronin, A. M.*** 1993. Growth promotion of black current soft wood cuttings by recombinant strain *P. fluorescens* BSP53 synthesizing an increased amount of indole-3-acetic acid. *Soil Biol. Biochem.*, 25: 1277-1281

Ganesan, P. and ***Gnanamanickam, S.S.*** 1987. Biological control of Sclerotium rolfsii sacc. in peanut by inoculation with *P. fluorescens. Soil Biol. Biochem.*, 19: 35-38.

Glick, B. R., 1994. The enhancement of plant growth by free living bacteria. *Can. J. Microbiol.*, 41: 109-117.

Gorden, S.A. and **Paleg, L.G.**1957.Quantitative measurement of Indole acetic acid. *Physiol. Plant Pathol.*, 10:347-348.

Hildebrand, D.C, Schroth, M.N. and **Sands, D.C.** 1992. Pseudomonas,In:Schaad NW (ed) Laboratory Guide for Identification of Plant Pathogenic Bacteria,2nd edition,American *Phytopatholoical Society*,St.Paul,MN.

Howell, C. R., and **Stipanovic, R. D.** 1979. Control of *Rhizoctonia solani* on cotton seedlings with *Pseudomonas fluorescens* and with a antibiotic produced by the bacterium. *Phytopathology* **70:** 712-715.

King's, B., Ward, M.K. and **Raney, D.E.** 1954. Two simple media for the demonstration of pyocyanin and fluorescin. *J. Lab. Clin. Med.*, **44:** 301-307.

Kloepper, J.W., Lifshitz, R. and **Schroth, M.N.** 1988. *Pseudomonas* inoculants to benefit plant protection. In: *ISI Atlas of Science*, Institute for Scientific information, Philadelphia, pp 60-64.

Leeman, M., van Pelt, J.A., den Ouden,F.M., Heinsbroek, M., Bakker, P.A.H.M. and **Schippers, B.** 1995. Induction of systemic resistance against *Fusarium* wilt of radish by lipopolysaccharides of *Pseudomonas fluorescens*. *Phytopathology* **85:** 1021–1027.

Lowry, O.H., Rosebrogh, N.J., Farr, A.L. and **Randall, R.L.** 1951. Protein measurement with Folin phenol reagent. *J. Biol. Chem.*, 193: 265-275.

Meyer, J.M., Azelvandre, P. and **Georges, C.** 1992. Iron metabolism in Pseudomonas: salicylic acid, a siderophore of *Pseudomonas fluorescens* CHA0. *Biofactors*, 4: 23-27.

Miller, R.L. and **Higgins, V.J.** 1970. Association of cyanide with infection of birds foot trefoil by Stemphylium rot. *Phytopathology*, 60: 104-110.

Mondal, K.K., Dureja, P., Singh, R. P., and **Verma, J .P.** 1998. Secondary metabolites of *P. fluorescens* in the suppression of bacterial blight of cotton induced by *Xanthomonas axonopodis* pv *malvacearum*. In: proc. of 7th Int. Congr. Plant Pathol. Edinburgh, Scotland, Abst No. 5.2.79.

Rosales, A.M., Thomashow, L., Cook, R.J. and **Mew, T.W**. 1995. Isolation and identification of antifungal metabolites produced by rice associated antagonistic *Pseudomonas* sp. *Phytopathology*, 85: 1028-1032.

Sadasivam, S. and **Manickam, A.** 1992. Biochemical methods for Agricultural Sciences, Wiley Eastern Ltd, New Delhi, p246.

Schwyn, B. and **Neilands, J.B.** 1987. Universal chemical assay for thedetection and determination of siderophore. *Anal. Biochem.*, 169: 47-56.

Snow, G.A. 1984. Mycobactin. A growth factor for *Mycobacterium johnei* II degradation and identification of fragments. *J. Chem. Soc.*, 25: 2588- 2596.

Thomashow, L.S., Bonsall, R.F. and ***Weller***, D. M. 1997. Antbiotic production by soil and rhizosphere microbes in situ. In: *Manual of Environmental Microbiology*, C.J. Hurst ed., A.S.M. Press, Washington. D.C. pp 493-500.

Thomashow, L.S. and ***Weller, D.M.*** 1996. Current concepts in the use of introduced bacteria for biological disease control; Mechanisms and antifungal metabolites. In: ***Plant Microbe Interactions.*** Vol I Stacey,G. and Keen,N. (Eds.) Chapman and Hall, New York pp 187-235.

Vidyasekaran, P. Sethuraman, K. Rajappan, K. and ***Vasumathi, K.*** 1997. Poweder formulations of *P. fluorescens* control pigeon pea wilt. Biological control 8:166-171.

Vogel, A. E. 1987. Class reactions (reactions for functional groups) In: Elementary Practical Organic Chemistry. CBS publisher, New Delhi pp 190-194.

Weller, D.M. and ***Cook, R.J.*** 1983. Suppression of take-all of wheat by seed treatments with fluorescent pseudomonads. *Phytopathol.* 73: 463-469.

30 | *Prediction and Confirmation of Triple Helix DNA Formation in Heat Shocked Yeast*

Chand Pasha and Venkateswar Rao. L

Department of Microbiology, Osmania University, Hyderabad - 500 007, India.

ABSTRACT

Antisense technology is the upcoming field of research for therapy and understanding the programmed up and down regulations of genes. After heat shock many genes are found to be down regulated and only few were expressed in *Saccharomyces cerevisiae*. Using poly purine templates for BLAST, various number of possible Triple helical structures were predicted. Frame shift assays using nondenaturing PAGE, UV spectra, melting complexes, DMSO foot printing and Rnase treatment were analyzed for triplex formation. Heat shock caused accumulation of triple helix structures in yeast. Heat shock proteins were reported and found to bind the triple helical structures. Double stranded DNA structure was retrieved from triple helix with the help of HSP. Natural existing junk DNA (Non coding DNA) and expression of few genes at particular stage of life can be explained using this knowledge of Triple helix DNA.

KEY WORDS : Triplex DNA, Heat shock, HSP, Antisense, BLAST

Although triplex formation was first described over 30 years ago [Felsenfeld, and Rich, 1957; Lipsett, 1963; Riley *et al.*, 1966], it has been realized since 1987 that the formation of these structures offers the possibility of designing agents endowed with considerable sequence-recognition properties (He!le' ne, 1991).

One outstanding problem with triplex formation is that, since hydrogen bonds are only formed with the purine strand of the duplex, it is generally restricted to targeting homopurine-homopyrimidine sequences. Two triple helix motifs have been characterized, depending on the orientation of the third strand, which can be positioned either parallel or antiparallel to the purine strand of the duplex. Parallel triplex formation is achieved using a pyrimidine-rich oligonucleotide and is characterized by the formation of T:AT and C:GC triplets [Moser, and Dervan, 1987; LeDoan, T., *et al.*, 1987; Thuong, N. T. and He! le' ne, 1993; De los Santos, *et al.*, 1989]. Alternatively, purine-rich oligonucleotides bind antiparallel to the purine strand of the duplex, generating G:GC, A:AT and T:AT triplets [Beal, and Dervan, 1991; Chen, 1991; Pilch, *et al.*, 1991].

Intermolecular DNA triplexes have pronounced sequence recognition properties and have been used successfully to decrease transcription from specific genes [Orson, *et al.*,1991; Birg, *et al.*, 1990; Kim and Miller, 1995]. It has been previously demonstrated that triplex formation at a promoter can block the binding of various transcription factors thereby inhibiting transcription. Another mechanism may be to prevent transcriptional elongation. Our cells and tissues are challenged constantly by exposure to extreme conditions that cause acute and chronic stress. Consequently, survival has necessitated the evolution of stress response networks to detect, monitor, and respond to environmental changes (Morimoto, and Santoro, 1998; Morimoto, *et al.*, 1990; Feige and van Eden, 1996; Baeuerle, 1995; Baeuerle, and Baltimore, 1996). Prolonged exposure to stress interferes with efficient operations of the cell, with negative consequences on the biochemical properties of proteins that, under ideal conditions, exist in thermodynamically stable states. In stressed environments, proteins can unfold, misfold, or aggregate. Therefore, the changing demands on the quality control of protein biogenesis, challenges protein homeostasis, for which the heat shock response through the elevated synthesis of molecular chaperones and proteases, repairs protein damage and assists in the recovery of the cell. In vitro studies revealed the formation of triple helix DNA by heat shock (Chudinov and Glazkov, 2002).

In eukaryotic genome functions, several genes are not yet discovered as they are not expressed, even the stability of triplex structures is low in comparison with that of double helical DNA. The need to increase the possibility for triplex targeting is critical for the success of this technique as a means for artificially enhancing the gene expression. There are now several areas of research that target the problem of stability. One possible strategy, with which this paper is concerned is to use compounds that bind spherically to triplex (not duplex) DNA, thereby facilitating the formation of double helix structures.

Materials and Methods

Culture Conditions

Microorganism

Saccharomyces cerevisiae - VS$_3$ was isolated from soil samples collected from hot regions near the Kothagudem Thermal Power Plant located in Khammam District, A.P, India. The organism was isolated, mutated by UV and identified as *Saccharomyces cerevisiae -VS$_3$* strain in our lab (Kiran Sree, N., 2000). It was maintained on Yeast extract Peptone Dextrose agar medium (YEPD) (1% Yeast Extract, 2% Peptone, 2% Glucose, 2% Agar-Agar, p^H 6.0).

Prediction of Triple helix DNA in Yeast

Query sequence GAGAAAAATGAA with poly purines was reported to form triple helix DNA (Chudinov and Glazkov, 2002) and used to BLAST (Search for short, nearly exact matches) against yeast genome data available with NCBI Gene bank. Results obtained (top 10) were checked for their function analysis from NCBI information.

Confirmation of Triple Helix DNA

Yeast Chromosome preparation and Heat shock

Thermotolorant yeast *Saccharomyces cerevisiae* VS3 strain was grown in YEPD broth for 48 hours at 30°C and 150 rpm. Yeast cells Washed with 0.5M PBS and were suspended in protoplast solution (O 0.5MPBS, 6M sorbitol, 2mM Beta Mercapto ethanol , 100U lyticase and 1μl of Rnase (50 ng/μl)). The mixture was incubated at 30°C and 50 rpm for 2 hours. Samples were collected at an interval of 10 minutes up to 2 hours and checked for protoplast conversion by methylene blue staining, spherical shape of protoplasts and hypotonic rupture of protoplasts. After 90% cells converted to protoplasts, heat shock was given at 50^{0}C for 45 minutes.

Nondenaturing PAGE Assay for Triplex Formation

Rnase and Lyticase treated yeast suspension was centrifuged at 22000 rpm and supernatant containing enzymes was removed.. The pellet was used for phenol extraction of DNA. The chromosomal DNA was suspended in 0.5M PBS and shocked at 50°C for 45 minutes. Then 5ml of the mixture was combined with 3 ml of 30% aqueous glycerol containing xylenecyanole and Bromphenol Blue. The products were resolved in 20 ×20 cm slabs of 4% gel (acrylamide/bisacrylamide 19:1 with Ethidium Bromide) in TAE at 24 °C for 16–18 h at 60 mA, 110 V. Bands were visualized by Trans illuminator in a gel documentation.

Dimethyl Sulfate (DMS) Footprinting

Heat shocked chromosome preparations (250μl) was purified by phenol extraction and treated with 10 μl of a 1:40 dilution of DMS in dimethyl sulfoxide for 3 min at 0°C. Reactions were stopped by DNA precipitation with ethanol. For both DNA cleavage of DMS-treated substrates and removal of crosslink, pellets were dried and resuspended in 100 μl of 0.1 M NaOH and heated to 90°C for 1 h. After an additional precipitation, samples were resuspended in 5 μl of loading buffer and loaded on an 8 M urea/4% poly acrylamide gel and used for electrophoresis under denaturing conditions.

Thermal Stability

Thermal stability of complexes was assessed by monitoring the UV absorption of the solution in Spectronic spectrophotometer while the temperature of DNA sample was steadily raised at 1°C/min from 25 to90°C. The concentration of DNA was 100 μg/10μl in TA buffer.

RNase Hydrolysis.

Heat shocked DNA (2ml) was treated with 5 μl of RNase solution (50 ng/μl) in a final volume of 50 μl at 37°C for 1 h. Reaction mixtures were extracted three times with phenol/chloroform, and products were precipitated.

Results and Discussion

Multistranded DNA structures have attracted a great attention in recent years, as they possibly play a substantial role in chromosomal DNA organization (Ivanov. *et al.*, 2002) regulating gene expression [Agazie, *et al.*, 1996; Soyfer, V.N. and Potoman, 1996; Vasquez, and Wilson, 1998; Praseuth, *et al.*, 1999). In BLAST analysis many of the genes (80%) forming triple helix DNA are not yet characterized. Remaining are expressed but down regulated after the heat shock.

Prediction of Triple helix DNA in Yeast.

Table 1 Probable Triple helix DNA forming Genes obtained by BLAST

BLAST Tool: Search for short, nearly exact matches.

Query sequence GAGAAAAATGAA

S.No	Gene Sequences Name	Chromosome	Score	E	Function
1.	gi|1015838|emb|Z49618.1|SCYJR118C	X	24	3.0	Unknown
2.	gi|476045|emb|X78993.1|SCRACII	70kD II.	24	3.0	given below
	1. Proliferating cell nuclear antigen,				
	2. Mitochondrial regulator of splicing				
	3. Constitutive acid phosphotase,				
	4. Gepressible acid phosphotase				
	5. Probable protein kinase,				
	6. Alpha mannosyl transferase				
	7. Unknown Protein				
	8. Glucose repression mediator protein				
	9. UV damage repair protein,				
	10. Alpha-aminoadipate reductase				
	11. transketolase,				
	12. Elongation factor EF-1-alpha				
	13. Small nuclear ribonucleoprotein,				
	14. Nuclear factor for cytB synthesis				
	15. Probable transfer RNA-Gly synthetase,				
	16. Ribosomal protein YmL36 fragment				
	17. Transcription factor TFIIIC ,				
	18. Unknown Protein				
	19. Regulator of carbon catabolite repression				
3.	gi|1360372|emb|Z73215.1|SCYLR043C	X	24	3.0	Un known

Table 1 *Contd....*

S.No	Gene Sequences Name	Chromosome	Score	E	Function
4.	gi|1360370|emb|Z73214.1|SCYLR042C	XII	24	3.0	Un known
5.	gi|4086|emb|X69881.1|SCORF12	unidentified DNA	24	3.0	Un known
6.	gi|1301875|emb|Z71310.1|SCYNL034W	XIV	24	3.0	Un known
7.	gi|536461|emb|Z35966.1|SCYBR097W	II	24	3.0	Un known
8.	gi|1420165|emb|Z74949.1|SCYOR041C	XV	24	3.0	Un known
9.	gi|6318278|gb|U14000.1|SCU14000 DNA alkylation damage resistance	Mms4P	24	3.0	
10.	gi|1079672|gb|U39205.1|SCU39205 1. Bts1p: Geranylgeranyl diphosphate synthase 2. Ald6p: Acetaldehyde dehydrogenase 3. Pdr12p: Multidrug resistance transporter	XVI, left arm, cosmid 8460	24	3.0	

Under microscope, spherical Protoplasts were observed. At 25th minute, 80% protoplast formation was noted. When these protoplasts were diluted in distilled water, decrease in protoplast count was noted due to hypotonic rupture. Protoplasts are stained by methylene blue and observed for loss of membrane bound oxidases (blue stained). Protoplasts were also stained by using geimsa stain and by dropping from height observed for chromosomes. Protoplasts (>80%) that were heat shocked were found to aggregate the chromosomal DNA. Chromosomal DNA isolated from protoplasts showed 16 bands indicating each chromosome as a separate band. After heat shock number of bands was decreased and meshed, indicating the formation of complexes. Heat shocked and DMS foot printed yeast showed less aggregates than only heat shocked yeast. Rnase cleaves single strand RNA/DNA. Rnase treatment to heat shocked yeast DNA (aggregates) yielded more mobility of bands confirming triple helical DNA formation by heat shock. (Figure 1).

Dimethyl sulfate (DMS) methylates G residues, creating an adduct that can be broken chemically. Typically, complexes are formed and treated with DMS, then the reaction is stopped. Triple helix DNA complexes were found to be protected from DMS mediated DNA cleavage (Church, and Gilbert, 1984).

Melting Complexes

Melting temperature of un shocked yeast DNA was 55^0C, whereas in heat shocked yeast DNA, it is 61^0C. Increased T_m (melting temperature) of heat shocked yeast Genomic DNA also confirms the formation of triple helix DNA. Heat shocked DNA was treated with Heat shock protein (HSP 70) for 20 minutes. After Hsp treatment DNA is phenol extracted and used for detection of melting temperatures. T_m (melting temperature) of Heat shocked and Hsp treated DNA is similar to un shocked yeast. Heat shocked and HSP treated Genomic DNA was found to have the T_m 54^0C. Heat shocked yeast chromosomes were reversed by Hsp 70 treatment (Figure 2).

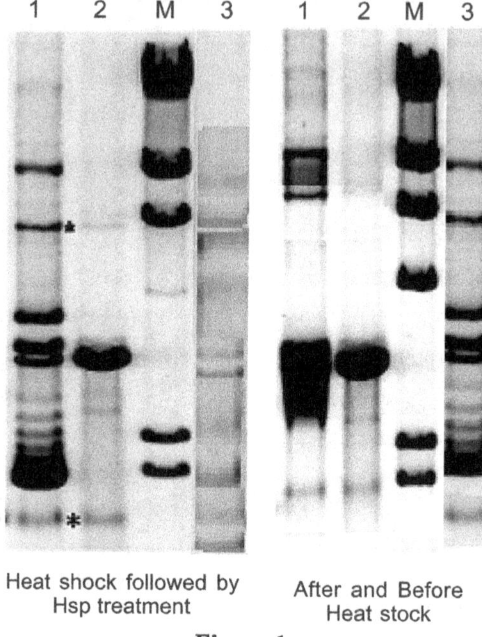

Figure 1

A :

1. Yeast chromosomes after heat shock and Hsp treatment.
2. Yeast chromosomes heat shocked and DMS treated, followed by Hsp treatment
   M) Molecular weight Marker
3. Heat shocked and Rnase treated Chromosomal DNA

B :

1. Yeast chromosomes after heat shock.
2. Yeast chromosomes after heat shock and DMS printing.
3. Unshocked Yeast chromosomes.

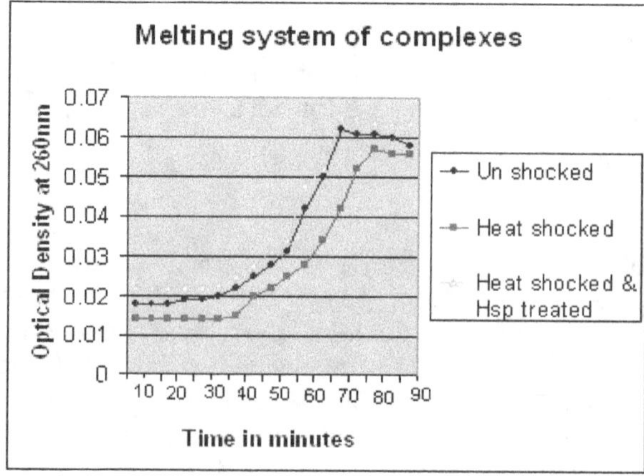

Figure 2

After heat shock, formation of Triple helix DNA was found which confirms earlier *invitro* reports (Kiran Sree, *et al.,* (2000). These triple helix structures may be responsible for blockage of genes after heat shock. Triple helix DNA was converted to double helix by the treatment of HSP 70. HSP70 can be used to convert inactive supra structures of DNA into the active double helix and subsequent invitro gene expression to characterize the gene.

Acknowledgements

We are grateful to "Department of Biotechnology, Ministry of Science and Technology, Government of India, New Delhi for providing financial assistance to complete the work.

References

Agazie, Y.M., Burkhokder, G.D., and Lee, J.S., 1996, *Biochemistry*. Triple Helical Nucleic Acids 361, 461–466.

Baeuerle, P.A. 1995. *Inducible gene expression. Vol. I. Environmental stresses and nutrients. Vol. II. Hormonal signals*. Birkhauser, Boston, MA.

Baeuerle, P.A. and **Baltimore, D.** 1996. NF-kappa B: ten years after.*Cell* 87 : 13-20.

Beal, P. A. and **Dervan, P. B.** 1991. Second structural motif for recognition of DNA by oligonucleotide-directed triple-helix formation. *Science* 251, 1360-1363

Birg, F., Praseuth, D., Zerial, A., Thuong, N. T., Asseline, U., Le Doan, T. and He! le' ne, C. 1990. Inhibition of simian virus 40 DNA replication in CV-1 cells by an oligodeoxynucleotide covalently linked to an intercalating agent.Nucleic Acids Res. 18, 2901-2908

Chen, F. M. 1991. Intramolecular Triplex Formation of the Purine· purine· pyrimidine Type. *Biochemistry* 30, 4472-4479

Chudinov and **Glazkov** 2002. *In vitro* Formation of Triple-Stranded DNA Structures in the Human a 1-Antitrypsin Gene. *Molecular Biology, 36, 1, 2002, 100–103.*

Church, G.M. and Gilbert, W. 1984. Genomic sequencing.*Proc. Natl. Acad. Sci. USA*, 81, 1991-1995.

De los Santos, C., Rosen, M. and **Patel, D.** 1989. NMR studies of DNA (R+)n. (Y-)n.(Y+)n triple helices in solution: imino and amino proton markers of T.A.T and C.G.C+ base-triple formation. *Biochemistry* 28, 7282-7289

Feige, U. and **W. van Eden** 1996. Infection, autoimmunity and autoimmune disease. *Stress Inducible Cellular Responses* 77: 359-373.

Felsenfeld, G. and **Rich, A.** 1957. *Biochim. Biophys. Acta*. 26, 457-468.

He! le' ne, C. (1991) *Eur. J. Cancer* 27, 1466-1471

Ivanov. S. A., Ya. I. Alekseev, and Gottikh, M.B. 2002. Pyrimidine Oligodeoxyxylonucleotides Form Triplexes with Purine DNA in Neutral Medium. *Molecular Biology, 36, 1, 2002, 131–139.*

Kim, H.G. and *Miller, D. M.* 1995. A model for recognition scheme between double stranded DNA and proteins. *Biochemistry* 34, 8165-8171

Kiran Sree, N., Sridhar,M., Suresh, K., Banat, I.M. and Venkateswar Rao, L. 2000. Isolation of thermotolerant, osmotolerant, flocculating, *Saccharomyces cerevisiae* for ethanol production. *Biores. Tech.* 72, 43-46

Le Doan, T., Perrouault, L., Praseuth, D., Habhoub, N., Decout, J. L., Thuong, N. T., Lhomme, J. and *He! le' ne, C*. 1987. Sequence-specific recognition, photocrosslinking and cleavage of the DNA double helix by an oligo-[alpha]-thymidylate covalently linked to an azidoproflavine derivative. *Nucleic Acids Res.* 15, 7749-7760

Lipsett, M. N. 1963. Interactions of poly c and guanine trinucleotide. *Biochem. Biophys. Res. Commun.* 11, 224-228

Morimoto, R.I. and *M.G. Santoro* 1998. Stress-inducible responses and heat shock proteins: new pharmacologic targets for cytoprotection. *Nature Biotech.* 16 : 833-838.

Morimoto, R.I., A. Tissieres, and *C. Georgopoulos* 1990. The stress response, function of the proteins, and perspectives. In *Stress proteins in biology and medicine* (ed. R.I. Morimoto, A. Tissieres and C. Georgopoulos), 1-36. Cold Spring Harbor Laboratory Press, Cold Spring Harbor, NY.

Moser, H. E. and Dervan, P. B. 1987. Sequence-specific cleavage of double helical DNA by triple helix formation. *Science,* 238, 645 ± 650

Orson, F. M., Thomas, D. W., McShan, W. M., Kessler, D. J., and *Hogan, M. E.* 1991. Oligonucleotide inhibition of IL2Rá mRNA transcription by promoter region collinear triplex formation in lymphocytes. *Nucleic Acids Res.* 19, 3435-3441

Pilch, D. S., Levenson, C. and *Shafer, R. H.* 1991. Structure, stability, and thermodynamics of a short intermolecular purine-purine-pyrimidine triple helix. *Biochemistry* 30, 6081-6087

Praseuth, D., Guieysse, A.L., and Helene, C. 1999. Triple helix formation and the antigne strategy for sequence-specific control of gene expression. *Biochim. Biophys. Acta,* 1489, 181–206.

Riley, M., Maling, B. and Chamberlain, M. J. 1966. *J. Mol. Biol.* 20, 359-389

Soyfer, V.N. and Potoman, V.N. 1996. *Triple-Helical Nucleic Acids,* Berlin: Springer-Verlag,

Thuong, N. T. and *He! le' ne, C.* 1993. Sequence-specific recognition and modification of double-helical DNA by oligonucleotides. *Angew. Chem.* 32, 666-690

Vasquez, K.M. and Wilson 1998. Triplex-directed modification of genes and gene activity *J. H. Trends Biochem. Sci.,* 1998, 23, 4–9.

Unit – IV

Microbial Biotechnology

31 Role and Control of Microorganisms Involved in Dental Caries and Periodontal Diseases - A Review

K.R. ANEJA*, PRANAY JAIN and RAMAN ANEJA

Department of Microbiology, Kurukshetra University, Kurukshetra - 136 119, India.
*e-mail: anejakr@rediffmail.com

ABSTRACT

Periodontal diseases and Dental caries are probably the most widespread gum and root infections in humans, affecting virtually all adults and children throughout the world. While the Periodontal diseases are marked by overwhelmed inflammatory response of the gingival tissue to the plaque accumulation in the form of Periodontitis and Gingivitis, the Dental caries, on the other hand, is the dissolution of the teeth enamel due to the plaque. Although bacteria are responsible for the onset and progression of the disease, other factors have been linked to increased susceptibility and include age, poor oral hygiene, tobacco smoking, stress, systemic diseases, decreased immunocompetence and genetic traits. Microbial diagnosis is significant in the treatment planning of these infections. *Streptococcus mutans* and anaerobic lactobacilli are common in dental caries, *Actinomyces* and *Peptostreptococcus spp.* in gingivitis, *Actinobacillus actinomycetemcomitans* and *Porphyromonas gingivalis* in periodontitis. A multi-factorial antimicrobial approach is necessary for the successful management of the destructive periodontal disease and for that of dental caries. Commonly used drugs in periodontal diseases and dental caries include, Metronidazole, Clindamycin, Amoxicillin and Ciprofloxacin. Soil provides a vast array of microbial isolates capable of producing secondary metabolites that can be used biotechnologically for the synthesis of novel antibiotics that would be effective in controlling the infectious microorganisms in the near future.

KEYWORDS : Periodontitis, Gingivitis, Dental caries, Antibiotics, Soil Microflora

The mouth supports a varied and dense population of microbial flora. Common bacterial diseases of the oral cavity are gingivitis, periodontitis, periodontosis, dental caries, abscesses associated with teeth, pharyngitis and tonsillitis. Of these, the two common diseases namely periodontitis and dental caries have affected mankind since the earliest of times, and studies from ancient India and China have indicated that periodontitis has been observed for many centuries (Boghani, 2000).

Periodontal disease is the general description applied to the inflammatory response of the gingiva and surrounding connective tissue to the bacterial plaque accumulations on the teeth (van Winkelhoff, 2003; Boghani, 2000). Based on the damage caused to the gums and teeth, periodontal disease is of three types: (i) The first type of the chronic periodontal disease referred to as *Marginal gingivitis* or *Chronic Marginal Gingivitis* is a mild condition involving a painless accumulation of blood (hyperemia) of the gingival margin without gross pus formation; (ii) The second type is referred to as *Ulcerative Gingivitis*, also known as *acute gingivitis*, *Vincent's infection*, and *Pyorrhea simplex*, is an acute inflammatory disease involving destruction of exposed gingival tissue; and (iii) The third type, *Periodontoclasia*, also known as *Periodontitis* and *Pyorrhea profunda*, is characterized by an extension of the inflammatory response into the deeper periodontal structures with a protracted destruction of the periodontal membrane (van Winkelhoff, 2003). The classification of periodontitis, as delineated by the American Academy of periodontology (Neville *et al.*, 2002) is given in Table 1. Tooth decay, also termed as dental caries is an infectious microbial disease that results in localized dissolution and destruction of the calcified tissues of the teeth (Ross and Holbrook, 1994).

Table 1 Classification of periodontitis*

Name of the disease	Subcategories
A. Chronic periodontitis	• Localized chronic periodontitis • Generalized chronic periodontitis
B. Aggressive periodontitis	• Localized aggressive periodontitis • Generalized aggressive periodontitis
C. Periodontitis as a manifestation of systemic diseases	• Associated with hematologic disorders • Associated with genetic disorders
D. Necrotizing periodontal diseases	• Necrotizing ulcerative gingivitis • Necrotizing ulcerative periodontitis
E. Abscesses of the periodontium	• Gingival abscess • Periodontal abscess • Pericoronal abscess
F. Periodontitis associated with endodontic lesions.	

* Modified from Neville *et al.*, (2002).

According to Neville *et al.*, (2002), Gingivitis is an inflammation limited to the soft tissues that surround the teeth. The primary types of gingivitis are: plaque related gingivitis, necrotizing ulcerative gingivitis, medication influenced gingivitis, allergic gingivitis, specific infection related gingivitis and dermatosis related gingivitis.

Gingivitis and periodontitis are chronic bacterial infections. As in other infections, the host bacterial interactions determine the nature and extent of the resulting disease. Pathogenic microorganisms may cause disease indirectly by stimulating a host response or directly by producing toxins and invading the tissues. Overwhelming evidences show that organisms present in microbial plaque constitute the primary and possibly, the only extrinsic etiologic agent participating in the causation of gingivitis and periodontitis (Loe *et al.*, 1965). At present, the prevalence of dental caries in India is approximately 60-65%, while the prevalence of periodontal disease is approximately 65-100% (Saini et. al., 2003).

Microbiological information of the subgingival plaque is being given emphasis in diagnosis and treatment planning in patients with severe periodontitis (van Winkelhoff, 2003). This approach is based on a number of facts as mentioned : Periodontitis is a bacterial disease and knowledge of the infectious agents can, in principle, improve the diagnosis and treatment of the disease and may contribute to the prognosis. Based on the microbiology of the subgingival plaque in the patient, periodontitis is considered as a collection of disease rather than one disease entity. Not all patients show a favourable treatment response upon mechanical periodontal treatment. Rationale use of antimicrobial agents is based on the knowledge of the causative agents of the disease.

A multifactorial antimicrobial approach is necessary for the successful management of destructive periodontal disease. Effective antimicrobial periodontal therapy aims to overwhelm periodontal pathogens with aggressive initial therapy and prevent previously suppressed pathogens from rising up a new through daily oral hygiene measures and frequent professional cleaning (Slots and Jorgensen, 2000).

This review deals with the various periodontal and dental caries microbiota, the mechanism involved in these microbial infections, their treatment and control by various antimicrobial agents and the significance of soil microflora in producing antibiotics which will be useful in treating these infections and also in overcoming the resistance developed amongst the pathogens to the known antibiotics.

Periodontal microbiota

Destructive periodontal disease is an infectious process, and hence, may not be managed by anti-infective intervention. Causes may include nonspecific bacterial plaque, as in gingivitis and mild forms of chronic (adult) periodontitis, or specific bacterial infections as in aggressive (early onset) periodontitis. Various microbial pathogens involved with periodontal disease and dental caries include aerobic and anaerobic bacteria, spirochetes, molds and yeasts and protozoa (Table 2) and these have been extensively discussed by Saini *et al.,* (2003), van Winkelhoff (2003), Schuster (1999), Socransky (1977) and Slots (1997).

Recent techniques that have been applied to detect periodontal pathogens are immunoassays (ELISA, immuno-flourescence), enzyme tests, DNA probes (whole genomic probes), anaerobic culture and polymerase chain reaction (PCR). In daily practice, the aerobic culture, staining, biochemical tests, the anaerobic culture technique and the PCR technique are used most frequently (van Winkelhoff, 2003).

Table 2 Microorganisms associated with Periodontal disease and Dental caries

Microorganisms	*Representattive species*
A. Bacteria Aerobic Gram positive cocci	Viridans Group Streptococci • *Streptococcus milleri group* • *Streptococcus mutans group*
Aerobic Gram positive bacilli	*Rothia dentocariosa*
Aerobic Gram negative coccobacilli	*Actinobacillus spp.* *Campylobacter spp.* *Capnocytophaga spp.* *Eikenella spp.*
Anaerobic Gram positive rods	*Actinomyces spp.* *Peptostreptococcus spp.* *Eubacterium spp.*
Anaerobic Gram negative rods	*Porphyromonas spp.* *Fusobacterium spp.* *Prevotella spp.* *Bacteroides spp.*
Anaerobic Gram negative cocci	*Veillonella spp.*
Spirochetes	*Treponema spp.*
B. Molds & Yeasts	*Candida albicans* *C. dubliniensis* *C. glabrata, C. parapsilosis* *Saccharomyces cerevisiae*
C. Protozoa	*Entamoeba gingivalis* *Trichomonas tenax*

Most subgingival pathogens are strict anaerobic Gram negative bacteria, but Gram positive strict anaerobic bacteria have also been implied. Of the various pathogens revealed, Streptococcus mutans and anaerobic lactobacilli are common in dental caries; *Actinomyces* and *Peptostreptococcus spp.* in gingivitis; *Actinobacillus actinomycetemcomitans* and *Porphyromonas gingivalis* in periodontitis (Saini et. al., 2003; van Winkelhoff, 2003). The number of different bacterial species that have been isolated from dental plaque is well over four hundred. Yet, the pathogens that have been associated with progression of periodontitis is limited to less than twenty (van Winkelfhoff, 2003). Table 3 reveals the association between various periodontal pathogens and progression of periodontal disease at various stages.

Systematic studies on the anaerobic bacterial flora of orodental infections reported from India are few. Saini *et. al.,* (2003) undertook the study to isolate and identify aerobic as well as anaerobic microbes from patients of adult periodontitis, gingivitis and deep seated dental caries and findings were compared against the normal healthy individuals. They found that aerobic/ anaerobic ratio was high in normal flora as compared to dental caries, gingivitis and periodontitis. Ninety seven per cent of orodental infections were polymicrobial and three or more microbes were found in 84 % cases of study groups as compared to 28 % in controls. The ratio of aerobes, facultative anaerobes to anaerobes decreased as the lesion progressed from normal gingiva to dental caries to gingivitis to periodontitis as shown in Table 4. Maximum number of anaerobes were isolated from the cases of periodontitis.

Table 3 Association between putative periodontal pathogens and progression of periodontal disease. *

Very Strong	Strong	Moderate	Early stage of investigation
Actinobacillus actinomycetemcomitans Porphyromonas gingivalis Spirochetes of acute necrotising gingivitis	Bacteroides forsythus Prevotella intermedia Eubacterium nucleatum Treponema denticola	Streptococcus intermedius Peptostreptococcus micros Fusobacterium nucleatum Campylobacter rectus	Selenomonas spp. Enteric rods Pseudomonas spp. Staphylococcus spp. Cytomegalovirus Epstein Barr virus type 1

* Modified from Slots, 1997; van Winkelhoff, 2003

Table 4 Aerobes, Facultative Anaerobes and Anaerobes isolated from gingivitis, periodontitis and normal flora. *

Isolates	Normal Flora	Gingivitis	Periodontitis
Streptococcus mutans	+	+	+
Non mutans streptococci	+	+	–
S. sanguis	+	+	+
S. mitior	+	+	+
Others			
Staphylococcus aureus	–	+	+
Coagulase negative staphylococci	+	+	+
Lactobacilli (Aerobic)	–	–	–
Actinobacillus actinomycetemcomitans	+	+	+
Klebseilla	+	+	+
Escherichia coli	–	–	+
Pseudomonas	+	–	+
Non fermenters	+	+	+
Yeast Cells	+	–	–
Gram positive bacilli			
Lactobacilli			
Actinomyces	–	+	+
A. naeslundii	+	+	+
A. viscosus	+	+	+
A. israeli	–	+	+
Propionibacterium	–	+	+
Bifidobacterium	+	+	+
Gram negative bacilli			
Prevotella melaninogenica	–	+	+
Prevotella intermedia	–	+	+
Porphyromonas gingivalis	–	+	+
Fusobacterium	+	+	+
Gram negative cocci			
Veillonella	+	+	+
Gram positive cocci			
Peptostreptococcus spp.	+	+	+

* Modified from Saini *et al.*, (2003)

The majority of subgingival bacteria are probably commensal bacteria and are considered as endogenous pathogens. This implicate that the prevalence of these bacterial species in the oral cavity is similar both in periodontal health and in disease. These species may accumulate in the subgingival area, and over time, may comprise significant part of the subgingival plaque. This may not be true for *Actinobacillus actinomycetemcomitans* and *Porphyromonas gingivalis*. These pathogens have characteristics of exogenous pathogens. This presumption is based on the finding that they have a low occurrence in periodontally healthy subjects (Saini *et al.*, 2003; Griffen *et al.*, 1998; van Winkelhoff, 2003).

Fungal species have gained considerable importance as opportunistic pathogens of immuno compromised individuals, particularly those infected with human immunodeficiency virus (HIV) (Hazen, 1995 ; Phelan *et al.*, 1997). The cultivable microflora from HIV positive patients with periodontal diseases have been found to be comprised of periodontopathic species similar to those found in HIV negative subjects with periodontitis, mainly Gram negative bacteria (Barr, 1995; Brady *et al.*, 1996). The presence of yeasts was a striking feature of the subgingival plaque samples from some AIDS patients in one study. Among the subgingival samples, *Candida albicans*, *C. dubliniensis, C. glabrata, C. parapsilosis* and *Saccharomyces cerevisiae* are the yeast recovered (Jabra-Rizk *et al.*, 2001).

Ghayoumi and coworkers (2002) have implicated *Dialister pneumosintes* as a candidate periodontal pathogen. They found the association of subgingival *D. pneumosintes* with demographic variables (age and gender) and the presence of subgingival periodontopathic *Bacteroides forsythus* and *Porphyromonas gingivalis*. Subgingival *D. pneumosintes* occurred with significantly higher prevalence in older individuals and was closely associated with subgingival *B. forsythus*. According to Ghayoumi *et al.*, (2002), *D. pneumosintes* may play an important role in the microbial complex responsible for destructive periodontal disease.

The Periodontal Disease and Dental Caries : Mechanism Involved

The most important new finding concerning periodontal disease is the realization that these clinical entities are really specific infections. These infections are unusual in that massive or even obvious bacterial invasion of the tissues is rarely encountered. Rather bacteria in the plaque touching the tissue elaborate various compounds such as H_2S, NH_3, amines, endotoxins, enzymes (such as collagenase) and antigens, all of which penetrate the gingiva and elicit an inflammatory response. This inflammatory response, although overwhelmingly protective, appears to be responsible for net loss of periodontal supporting tissue, and leads to the periodontal pocket formation, loosening of the teeth, and eventual tooth loss (van Winkelhoff, 2003).

Although bacteria are essential for the onset and progression of the disease, they are probably not enough. This concept is based on the fact that periodontal pathogens can occur in a subject without clinical signs of the disease. In order for the disease to develop, a susceptible host is required. The nature of susceptibility for destructive periodontal disease has long been an enigma. Several factors have been linked to increased susceptibility and these include age, poor oral hygiene, tobacco smoking, stress, systemic diseases, decreased immunocompetence (neutropenia) and genetic traits (gene polymorphisms). None of these factors by themselves are probably capable to produce the disease, but it seems likely that a set of risk factors is necessary to initiate the disease process. For this reason, periodontitis is considered as a multifactorial disorder (van Winkelhoff, 2003).

Establishing a cause and effect relationship between a single microorganism in oral flora and caries has been very difficult. Most of the investigators believe that development of caries of enamel is preceded by the formation of microbial plaque in the tooth (Madigan *et al.*, 2000). As dental plaque accumulates and acid products are formed, dental caries (tooth decay) is the usual result. Thus, tooth decay is an infectious disease caused by microorganisms. The role of the oral flora in tooth decay has now been well established through studies on germ free animals. The smooth surfaces of the teeth that are exposed to frequent cleaning by the tongue, cheek, saliva or toothbrush or to the abrasive action of food mastication are relatively resistant to dental caries. The tooth surfaces in crevices, where food particles can be retained are the sites where tooth decay predominates. Thus, the shape of the teeth is important (Madigan *et al.*, 2000).

Madigan and coworkers (2000) have found a correlation between the dental caries and sugars present in the mouth of a person. According to them, diets high in sugars are especially cariogenic because lactic acid bacteria ferment the sugars to lactic acid, which causes the decalcification of the enamel of the tooth. Once breakdown of the hard tissue has begun, proteolysis of the matrix of the tooth enamel occurs through the action of proteolytic enzymes released by bacteria. Microorganisms further penetrates into the decomposing matrix, but the later stages of the process may be exceedingly slow and are often highly complex. The structure of the calcified tissue also plays an important role in the extent of dental caries (Madigan *et al.*, 2000).

van Winkelhoff (2003) classified periodontitis on the basis of the occurrence of microorganisms in the host. Periodontitis associated with endogenous (commensal) pathogens be considered an *opportunistic infection* whereas periodontitis associated with exogenous pathogens may be considered a *true infection*. This view has clinical implications and allows a differentiated treatment strategy; endogenous pathogens may only need suppression, but exogenous pathogens should be eliminated in susceptible subjects.

Treatment and Control

A view to the past indicates that in ancient times, India was probably the most advanced country in dental health services in the world. In the old Ayurvedic literature of India, details of gum diseases and forms of treatments have been described. The practice of oral hygiene was included in daily rituals (D' Silva, 1999). Ancient scriptures such as Vedas, Puranas etc. proposed that the natural dentition could be preserved by appropriate periodontal treatment. Today, Ayurvedic treatment is still very popular amongst rural and uneducated people in India. For the treatment of chronic periodontitis, massaging of the gums with oils and ointments prepared from herbs is still advised. It has been reported that some of the Ayurvedic preparations have various ingredients which have analgesic, anti-inflammatory properties as well as being able to facilitate cleansing and mouth freshening (D' Silva, 1999).

Most periodontal treatment in present times is based on the modern allopathic concept of defining etiology of the diseases, analyzing signs and symptoms and executing treatment based on this concept. The research carried out in India is based on allopathic concept with special emphasis on Indian environment and Indian habits (Boghani, 1991).

Periodontal disease is probably the most widespread inflammatory disorder in humans, affecting virtually all adults and children throughout the world. Severity of periodontal disease ranges from mild gingival inflammation to advanced periodontal attachment loss, sometimes resulting in loss of teeth. Historically, a predominantly mechanical approach to treatment has been employed, and outcomes have often been disappointing (Slots and Jorgensen, 2000). Use of antimicrobial agents in periodontal treatment has often been haphazard and empirical. However, there is evidence that the addition of appropriate antimicrobial therapy to traditional periodontal treatment may substantially enhance clinical outcomes and reduce the need for costly surgical procedures.

Current antimicrobial periodontal therapy employs mechanical debridement performed with and without surgery, antibiotics and antiseptics (Slots and Jorgensen, 2000). A proficient cost effective antimicrobial periodontal treatment may offer populations who are currently unable to receive adequate care due to lack of financial resources, a practical and valuable means of maintaining a functional periodontum and dentition for a prolonged period (Slots, 1997).

New products and treatment modalities for the management of periodontal disease continue to offer a larger number of choices, many of which involve antimicrobials. Specific pathogenic bacteria play a central role in the etiology and pathogenesis of destructive periodontal disease. Under suitable conditions, periodontal pathogens colonize the subgingival environment and are incorporated into a tenacious biofilm. Successful prevention and treatment of periodontitis is contingent upon effective control of the periodontopathic bacteria. This is accomplished by professional treatment of diseased periodontal sites and patient performed plaque control. Appropriate antimicrobial agents that can be administered systemically (antibiotics) or via local delivery may enhance eradication or marked suppression of the subgingival pathogens (Slots and Jorgensen, 2000; Mombelli *et al.*, 2002).

The choice of antimicrobial therapy is empirical and based on the experience. Another approach is microbiological analysis of the subgingival microflora which establishes the presence and levels of periodontal pathogens. Consequently, the selected antimicrobial therapy can be based on known susceptibility profiles of the target microorganisms. Using this protocol, therapy failure with an empirically selected antibiotic can be avoided (van Winkelhoff, 2003).

In vitro susceptibility of periodontal and dental caries microorganisms to antimicrobial agents

Extensive studies have been carried out involving putative periodontal and dental caries pathogens to a vast array of antimicrobial agents. Wilson and coworkers in 1990 investigated the effects of Minocycline on subgingival plaque samples from patients with chronic periodontitis. The results implied that adequate reductions in the numbers of viable subgingival plaque bacteria are unlikely to occur after exposure to Minocycline at concentrations attainable in gingival crevicular fluid after systemic administration.

Hoshino and coworkers (1996) undertook a study to clarify the anti- bacterial effect of a mixture of Ciprofloxacin, Metronidazole and Minocycline, with and without the addition of Rifampicin, on bacteria taken from infected dentine of root canal walls. The efficacy was also determined against bacteria of carious dentine and infected pulps which may be the precursory bacteria of infected root dentine. The respective drug alone (10, 25, 50 and 75 μg/ml) substantially decreased the bacteria recovery, but could not kill all the bacteria. Bacteria taken from carious dentine and infected pulps were also sensitive to the drug combination. These results indicated that the bactericidal efficacy of the drug combination is sufficiently potent to eradicate bacteria from the infected dentine of root canals.

The study on the sensitiveness of slow growing anaerobic bacteria to antibiotics is delicate when the technical motives are considered that make it difficult to transpose the standard methods frequently used in microbiological laboratories. The main methods used to determine susceptibility to antibiotics are – Disk Diffusion test and antibiotics containing microdilution plates. Kamagate *et al.*, (2001) carried out a study to compare the effectiveness of each of these methods on severe anaerobes bacteria isolated in subgingival flora of patients suffering from developing periodontitis (rapidly progressive periodontitis, refractory periodontitis, active stage of adult chronic periodontitis). The observed bacteria were *Porphyromonas gingivalis, Prevotella intermedia, Fusobacterium nucleatum, Campylobacter rectus,* and *Peptostreptococcus micros.* The drugs used were Ampicillin, Tetracycline, Erythromycin and Metronidazole. The comparison of the MIC of each of these methods permitted to show a strict correlation in the results observed with these three methods, if only the growth of the severe anaerobes on agar medium did not exceed 72 hours.

Effects of three different types of short term applications (1-3 times during a week) of Chlorhexidine on the susceptibility of 863 clinical isolates of *Streptococcus mutans* and 53 isolates of *Streptococcus sobrinus* from 58 subjects were studied by Jarvinen and coworkers (1995). Chlorhexidine resistant isolates were not found either before or after the treatment. The MIC's to Chlorhexidine of all isolates of *S. mutans* were lesser or equal to 1 μg/ml and of *S. sobrinus* were also susceptible to Ampicillin, Penicillin, Cefuroxime and Tetracycline.

Buchman *et al.*, (2003) tested a hypothesis that the bacterial susceptibilities in aggressive periodontitis change upon administration of systemic antibiotics as adjunct to periodontal therapy. The periodontal pathogens investigated prior to and one year after periodontal therapy were tested sensitive to the antimicrobial agents. In aggressive periodontitis, changes in bacterial susceptibility upon the administration of systemic antibiotics were associated with the limited number of isolates tested following therapy.

Milazzo and coworkers (2003) compared the in vitro activity of Faropenem, an oral penem, with those of Penicillin, Cefotixin, Clindamycin, Erythromycin and Metronidazole against anaerobic pathogenic isolates involved in systemic infections. Faropenem show structural similarities with both Penicillins and Cephalosporins, and are characterized by a broad antibacterial spectrum, a potent Penicillin binding protein affinity and good β-lactamase stability. The organisms tested comprised *Porphyromonas gingivalis, Prevotella spp., P. melaninogenica, P. intermedia, Actinomyces sp. Fusobacterium nucleatum, Peptostreptococcus spp.* and *Bacteroides forsythus.* The antimicrobial properties of Faropenem were investigated by studying MIC's, MBC's, time-kill kinetics and post antibiotic effect. Faropenem was highly active against all the anaerobes tested and was bactericidal against both β-lactamase positive and negative anaerobes, with a maximum bactericidal effect at 10 X MIC at between 12 and 24 hrs. Faropenem may be useful for treating infections caused by periodontal bacteria or oral flora. Table 5 depicts the spectrum of various antimicrobial agents on selected dental pathogens.

Table 2.5 Spectrum of activity of antimicrobial agents against selected oral microorganisms *

	Viridans Group Streptococci	*Anaerobic Gram negative bacilli*	*Actino- bacillus spp.*	*Capnocy- tophaga spp.*	*Eikenella coorodans*	*Actinomyces spp.*
Penicillin	+/-	+/-	+/-	+	+	+
Amoxicillin	+/-	+/-	+/-	+	+	+
Amoxicillin Clavulanate	+/-	+/-	+/-	+	+	+
Oral Cepha- losporins**	-	-	+/-	-	-	+
Erythromycin	+/-	-	-	-	-	+
Azithromycin/ Clarithromycin	+/-	-	-	-	-	+
Ciprofloxacin/ Levofloxacin	+/-	-	+	+	+	-
Clindamycin	+	+/-	-	+	-	+
Metronidazole	-	+	-	-	-	-
Doxycycline/ Tetracycline	+/-	-	+	+	+	+

** Includes agents such as Cephalexin, Cefitixime, Cefluroxime axetil, Cefprozil.

* Modified from van Winkelhoff, 2003; Slots, 1996

Systemic Antibiotics

Empirical antibiotic therapy may be used for periodontal diseases with known microbial etiologies, such as acute necrotizing ulcerative gingivitis, that is caused by anaerobic organisms that can be eradicated by systemic Amoxicillin- Metronidazole combination therapy. However, even the most careful clinical examination cannot delineate the likely microbial pathogens in most types of periodontitis nor determine their susceptibility to various antibiotics. Microbiological analysis of subgingival plaque provides valuable information that will facilitate accurate diagnosis and optimal treatment planning. Slots (1996) suggested that therapeutic antibiotic therapy in periodontitis may be deferred until the results of culture, nucleic acid based diagnostic tests and antimicrobial susceptibility determination are available.

Systemic antibiotics should be used singly or in combination. Single agent antibiotic therapy in current periodontitis includes, Metronidazole, Clindamycin and Ciprofloxacin. Common combination therapy in periodontitis includes Metronidazole and Amoxicillin, and Metronidazole and Ciprofloxacin (van Winkelhoff, 2003).

Locally Delivered Antibiotics

Controlled release devices that contain Tetracycline-HCL, Doxycycline, Minocycline, Metronidazole for direct pocket placement are commercially available in various countries. In controlled release drug delivery, the antimicrobial agent is released during an extensive period of time by zero order drug release kinetics (Rams and Slots, 1996). Locally applied tetracycline fibres have been shown to have some clinical benefits up to six months following therapy when used in conjunction with scaling and root planning (Newman *et al.*, 1994).

Locally Delivered Antiseptics

Mechanical root debridement to remove dental calculus is important in periodontal therapy, but is frequently inadequate in curing severe periodontal infections (Slots, 1996). To augment mechanical debridement, topical antiseptics may be used to kill periodontal pathogens during initial therapy and to suppress their repopulation during maintenance therapy. To have therapeutic value, pharmaceuticals must be delivered at an effective concentration and for an adequate length of time (Slots, 1996).

Subgingival antimicrobials can be administered either by non controlled delivery for fast acting pharmaceuticals or by controlled drug delivery devices for slow acting agents. Povidone-iodine is able to kill microorganisms rapidly enough to overcome the drug-diluting effect of gingival crevicular fluid and is effective when delivered by subgingival irrigation (Nakagawa and Saito, 1990). Sodium hypochlorite is a strong oxidizing agent that possesses numerous attractive properties for antiseptic use, include rapid bactericidal action, ease of use and very low cost. Irrigation with dilute (0.5%) sodium hypochlorite solution has been shown to cause significantly greater and longer lasting reduction in plaque and gingivitis than irrigation with water (Lang and Brecx, 1986). In contrast, Chlorhexidine digluconate mouthrinses at concentration of 0.1 % to 0.2 %, which have a long history of use in controlling supragingival plaque and gingivitis (Lang and Brecx, 1986), may require controlled-release delivery to exert effective killing of subgingival microorganisms (Caufield *et al.*, 1987). Also, chlorhexidine tends to be more bactericidal for gram positive organisms populating supragingival plaque than for gram negative species inhabiting periodontal pockets (Slots, 1997).

Soil as a Source of Antibiotics

The soil remains the best economical source that harbours the microorganisms which produce antibiotics and soil all around the world are continuously screened for the isolation of new antibiotic producing microorganisms. Antibiotics are microbial products or their derivatives that kill susceptible microorganisms and inhibit their growth even when used at low concentrations, while harming the host as little as possible (Aneja, 2003).

Penicillin, the first antibiotic to be used therapeutically was discovered from a mold *Penicillium notatum* by the Scottish physician, Alexander Fleming in 1929 and a new medical era was born. With the discovery of a means of producing large quantities of penicillin, and the successful treatment of once fatal infectious diseases, a concerted effort was begun to search for more antibiotics. The search was expanded to include other fungi, as well as algae, animals, bacteria and plants. Although none of the antibiotics discovered since penicillin have had the same notoriety, many are just as important and many are more commonly used than penicillin, especially the over-the - counter antibiotics used as topical applications. Most have been bacterial metabolites, or more specifically, *Actinomycetes*, which are mycelia producing bacteria that were once thought to be fungi. Unlike those antibiotics isolated from fungi, some of those from Actinomycetes did not require intravenous injections nor were they taken orally. Some examples of such antibiotics include bacitracin and neosporin (Singh, 1998).

Most of the antibacterial agents act on bacterial wall synthesis or protein synthesis. Peptidoglycan is one of the major wall targets because it is only found in bacteria. Some other compounds target bacterial protein synthesis, because bacterial ribosomes (termed 70 S ribosomes) are different from the ribosomes of humans (80S) and other eukaryotic organisms. Similarly the one antifungal agent (griseofulvin) binds specifically to the tubulin proteins that make up the microtubules of fungal cells, these tubulins are different from tubulins of humans (Singh, 1998).

As the putative periodontal and dental caries pathogens are developing resistance against the commercially known antibiotics, the need of hour is to explore soil microflora for discovering new bacterial, fungal and actinomycetes isolates in order to synthesize novel antibiotics that can be used in the treatment and control of not only the periodontitis and dental caries, but also of other dreadful diseases for which there has been no optimal antibiotic yet discovered.

The Future Prospects

The last 25 years have brought unprecedented advances to our understanding of periodontal disease. In 1970, periodontitis was believed to effect most individuals over the age of 35 years, to progress steadily in an individual once initiated until teeth were lost, to be the primary cause of tooth loss in adults, to be caused by the bacterial mass accumulating on the tooth surface and subgingivally, and to involve the host in some fashion or another (Williams *et al.*, 1996). However, impressive research advances made in the recent years in the epidemiology of periodontal disease, the specific bacterial etiology of periodontal disease, and the immunoinflammatory mediators of periodontal tissue destruction have greatly altered our view of periodontal disease.

Impressive research into new ways to diagnose the periodontal disease is well underway. Investigators are seeking new ways to diagnose an individual's degree of risk for periodontal disease initiation susceptibility to disease progression, level of disease 'activity' and the likely response to treatment and recurrence of active disease. New diagnostic tests should greatly advance our ability to more accurately and specifically diagnose periodontal disease. The future also looks promising for new treatment strategies to slow or arrest periodontal disease progression. The bacterial specificity of periodontal disease etiology revealed since 1970 has logically led to the use of antibiotics in periodontitis treatment. In the late 1980's, the concept of locally delivering antibiotics to the periodontal pockets was introduced, and subsequent clinical trials have indicated that it is possible to reduce pocket depth and inflammation with tetracycline locally delivered to the periodontal pockets (Williams, 1996).

Likely, we have barely scratched the surface in studying the efficacy of locally delivery antimicrobial agents to alter the progression of periodontal disease. As new agents are developed and better delivery systems to the periodontal pockets are developed, the future should see a variety of antimicrobial agents available, which can slow periodontal disease progression. The future also holds promise for slowing periodontal disease progression by blocking inflammatory pathways important in periodontal tissue destruction. Clinical trials of Flubiprofen, Naproxen and Ketoprofen indicate that it is possible to slow periodontal disease progression with non steroidal anti-inflammatory drug which inhibit one destructive pathway. In addition, data from animal models indicate that chemically modified tetracycline as an inhibitor of collagenase can slow disease progression in animals (Williams, 1996). Again, we have likely only just begun to explore the wide range of molecular mediators of tissue destruction, which may be targeted for blocking and thereby slow or arrest periodontal disease progression.

Acknowledgement

Pranay Jain, working as a research scholar in the Department of Microbiology, Kurukshetra University, Kurukshetra is grateful to the honourable Vice Chancellor, Dr. A.K. Chawla for providing the opportunity to carry out research work.

References

Aneja, K.R. 2003. Experiments in Microbiology, Plant Pathology and Biotechnology. 4th ed. New Age International Publishers, New Delhi, 608 pp.

Barr, C.E. 1995 . Periodontal problems related to HIV-1 infection. *Adv. Dental Res.* 9: 147-151.

Boghani, C.P 1991. Periodontal education. *Indian Soc. Periodontol.* 15:1.

Boghani, C.P. 2000. Progress of periodontal research and practice in India. In: Progress of periodontal research and practice in Asian Pacific countries. (Eds.) Bartold, P.M. and Ishikawa, *L.Asian Pacific Society of Periodontology Bulletin.*

Brady, L.J., Walker, C., Oxford, G.E. Stewart, C., Magnusson, I. and **McArthur, W.** 1996. Oral diseases, mycology and periodontal microbiology of HIV-1 infected women. Oral Microbiol. Immunol. 6: 371-380.

Buchman, R., Muller, R.F., Van Dyke T.E. and **Lange, D.E.** 2003. Change of antibiotic susceptibility following periodontal therapy. A pilot study in aggressive periodontal disease. J. Clin. Periodontol. 30 : 222-229.

Caufield, P.W., Allen, D.N. and **Childers, N.K.** 1987. In vitro susceptibilies of suspected periodontopathic anaerobes as determined by membrane transfer assay. Antimicrob. *Agents Chemother.* 31: 1989-1993.

D' Silva 1999. The effect of neem gel on plaque and gingivitis. *J. Indian Soc. Periodontol.* 2: 63.

Ghayoumi, N., Chen, C., and **Slots, J.** 2002 . Dialister pneumosintes, a new putative periodontal pathogen. *J. Periodontol. Res.* 37 : 75-78.

Gilfillan, G.D., Sullivan, D.J., Haynes, K., Parkinson, T., Coleman, D.C. and **Gow, N.A.R** 1998. Candida dubliniensis: phylogeny and putative virulence factors. *Microbiology* 144 : 829-838.

Griffen, A.L., Becker, M.R., Lyons, S.R., Moeschberger, M.L. and *Leys, E.J.* 1998. Prevalance of Porphyromonas gingivalis and periodontal health status. *J. Clin. Microbiol.* 36: 3239-3242.

Haffajee, A.D. and *Socransky. S.S.* 1994. Microbial etiological agents of destructive periodontal disease. *Periodontol.* 2000 5: 78-111.

Hazen, K. 1995. New and emerging yeast pathogens. *Clin. Microbiol. Rev.* 8: 462-478.

Hoshino, E., Kurihara-Ando, N., Sato, I., and *Iwaku, M.* 1996. *In vitro* antibacterial susceptibility of bacteria taken from infected root dentine to a mixture of Ciprofloxacin, Metronidazole and Minocycline. Int. Endod. J. 29: 125-130.

Jabra-Rizk, M.A., Baqui, A.A.M.A., Kelley, J.I., Falkler, Jr., W.A., Merz, W.G. and *Meiller, T.F.* 1999. Identification of Candida dubliniensis in a prospective study of patients in the United States. *J. Clin. Microbiol.* 37: 321-326.

Jabra-Rizk, M.A., Ferreira, S.M.S. Sabet, M., Falkler, W.A., Merz, W.G. and *Meiller, T.F.* 2001. Recovery of Candida dubliniensis and other Yeasts from Human Immunodeficiency Virus associated periodontal lesions. *J. Clin. Microbiol.* 39: 4520-4522.

Jarvinen, H., Pienihakkinen, K., Huovinen, P. and *Tenovuo, J.* 1995. Susceptibility of Streptococcus mutans and Streptococcus sobrinus to antimicrobial agents after short term oral chlorhexidine treatment. Eur. J. Oral Sci. 103 : 32-35.

Kamagate, A., Kone, D., Coulibaly, N.T. and *Brou, E.* 2001. A comparative study of various evaluation methods of the antibiotic sensitivity of strict anaerobic bacteria of the subgingival flora. *Odontostomatol* Trop. 24 : 9-12.

Lang, N.P. and *Brecx, M.C.* 1986. Chlorhexidine digluconate- an agent for chemical plaque control and prevention of gingival inflammation. *J. Periodontal Res.* 21 : 74-89.

Loe, H., Thellade, E. and Jensen, S.B. 1965. Experimental gingivitis in man. J. Periodontol. 36: 177-187.

Madigan, M.T., Martinko, J.M. and *Parker, J.* 2000. Brock Biology of Microorganisms. 9th ed. Prentice Hall Inc. USA, 991pp

Michael, G., Jorgensen, D.D.S. and *Slots, J.* 2002. The ins and outs of periodontal antimicrobial therapy. J. California Dental Assoc. At Website: www. nih.nlm.gov.in

Milazzo, I., Blandino, G., Nicolitti, G. and ***Speciale, A.*** 2003. Faropenem, a new oral penem: antibacterial activity against selected anaerobic and fastidious periodontal isolates. *J. Antimicrob. Chemother.* 51: 721-725.

Mombelli, A., Schmid, B., Rutar, A. and ***Lang, N.P.*** 2002. Local antibiotic therapy guided by microbiological diagnosis. *Clin. Periodontol.* 29: 743-749.

Nakagawa, T and ***Saito, A.*** 1990. Bacterial effects on subgingival bacteria of irrigation with a povidone-iodine solution (neojodin). Bull. Tokyo Dent. Coll. 31:199-203.

Neville, B.W., Damm, D.D., Allen, C.M. and ***Bouquot, J.E.*** 2002. Oral and Maxillofacial Pathology. 2nd ed. Elsevier. pp 844.

Newman, M.G., Komman, K.S. and ***Doherty, F.M.*** 1994. A 6-Month multi-center evaluation of adjunctive tetracycline fiber therapy used in conjunction with scaling and root planning in maintenance patients: clinical results. *J. Periodontol.* 65: 685-691.

Phelan, J.A., Begg, M.D., Lamster, I.B., Gorman, J., Mitchell-Lewis, D., Bucklan, R.D. and ***el-Sadr, W.M.*** 1997. Oral candidiasis in HIV infection; predictive value and comparison of findings in injecting drug users and homosexual men. *J. Oral Pathol. Med.* 26: 237-243.

Rams, T.E. and ***Slots, J.*** 1996. Local delivery of antimicrobial agents in the periodontal pocket. *Periodontal* 2000 10: 139-159.

Ross, P.W. and ***W.P. Holbrook*** 1994. Dental Plaque. In: Clinical & Oral Microbiology. Blackwell Scientific Publishers, Glasgow.

Saini, S., Aparna, Khanna, P, Mahajan, A., Arora, D.R., Saini, O.P. and ***Gupta, Naveen*** 2003. Antibiotic susceptibility of bacterial isolates in gingivitis and periodontitis. J. Indian Dent. Assoc. 74: 216-222.

Schuster, G.S. 1999. Oral flora and pathogenic organisms. Infectious Disease Clinics of North America. 13: 757-773.

Singh, B.D. 1998. Biotechnology. 1st ed. Kalyani Publishers, New Delhi. pp 574.

Slots, J. 1996. Systemic antibiotics in periodontics. *J. Periodontol.* 67: 831-838.

Slots, J. 1997. Subgingival microflora and periodontal disease. *J. Clin. Periodontol.* 6: 351-382.

Slots, J. and **Jorgensen, M.G.** 2000. Efficient antimicrobial treatment in periodontal maintenance care. *J.Am. Dent. Assoc.* 131: 1293-1304.

Socransky, S.S. 1977. Microbiology of periodontal disease- present status and future considerations. *J. Periodontol.* 48: 497-504.

van Winkelhoff, A.J. 2003. Microbiology in Diagnosis and treatment planning in periodontics. Int..J. Dent. *Hygiene* 1: 131-137.

William, R.C., Beck, J.D. and **Offenbacher, S.N.** 1996. The impact of new technologies to diagnose and treat periodontal disease. A look to the future. *J. Clin. Periodontol.* 23: 299-305.

Wilson, M., O'Connor, B. and **Newman, H.N.** 1990. Effect of Minocycline on subgingival plaque bacteria. *J. Appl. Bacteriol.* 69: 228-234.

32 | Current Scenario on the Biocontrol Potential of *Trichoderma* for the Management of Plant Diseases

N. Mathivanan

Centre for Advanced Studies in Botany, University of Madras,
Guindy Campus, Chennai - 600 025,

ABSTRACT : *Trichoderma* spp. have been used widely to control various root and foliar diseases in agricultural, horticultural and plantation crops and also for management of post harvest fungal diseases. The important mechanisms suggested for biocontrol of plant pathogens are mycoparasitism, antibiosis, competition and induction of defense response in host plants. It has been established that the chitinase producing *Trichoderma* spp. can be effective biocontrol agents against fungal pathogens. *Trichoderma* spp. are mainly applied in the rhizosphere, whereas little attempts have been reported on its application for the protection of foliar diseases. *T. harazianum* has been reported to be an effective biocontrol agent for foliar pathogens, *Botrytis cinerea*. In general, combination of two or more disease control strategies such as fungal antagonists, organic ammendments, chemical fungicides provide high level control than that could be achieved individually.

KEY WORDS : Biocontrol, *Trichoderma*, antibiosis, fungal antagonists.

The genus *Trichoderma* is considered the most important among fungi because of its potential use in agriculture as biocontrol agent for the management of plant diseases (Elad, 2000; Mathivanan *et al.*, 2000a) and also for various applications in industries (Manoharachary, 2003). The species of *Trichoderma* are free living and common soil-inhabiting fungi surviving saprophytically by utilizing the soil organic matter. Weindling (1932) for the first time demonstrated the biocontrol potential of Trichoderma and later several other workers have established the role of *Trichoderma* in controlling different plant pathogens (Sivan and Chet, 1986; Harman *et al.*, 1989; Elad, 2000). Further *Trichoderma* spp. have been used widely to control various root and foliar diseases in agricultural, horticultural and plantation crops (Alagarsamy *et al.*, 1987; Mukhopadhyay *et al.*, 1992; Whipps, 1992; Thomas *et al.*, 1993; Rajan *et al.*, 2002; Mathivanan *et. al.*, 2000b) and also for the management of post harvest fungal diseases (Batta, 2004).

Members of the genus *Trichoderma* have been studied extensively as biocontrol agents (Lewis and Papavizas, 1991; Haran *et al.*, 1996a; 1996b; Hermosa *et al.*, 2000; Sharma and Chandel, 2003) as they efficiently antagonize and control a wide range of fungal pathogens viz., *Pythium, Phytophthora, Rhizoctonia, Sclerotium. Fusarium, Botrytis,* etc.

(Adams, 1990; Devaki *et al.*, 1992; Abada, 1994; Lo *et al.*, 1996; Batta, 2004). *Trichoderma* is also an excellent component for integrated disease management system (Vyas, 1994; Mathivanan *et al.*, 1997) as most species are tolerant to various fungicides (Kay and Stewart, 1994; Malathi and Sabitha Doraisamy, 2003).

Biocontrol Mechanisms in *Trichoderma*

The important mechanisms suggested for biocontrol of plant pathogens are mycoparasitism, antibiosis, competition and induction of defense response in host plants. There are several mechanisms attributed to the antagonistic activity of *Trichoderma* towards pathogenic fungi of which, antibiosis is considered the most important, in which *Trichoderma* directly attacks the plant pathogens by secreting extracellular hydrolytic enzymes viz., chitinase, β-1,3 glucanase, cellulase and proteases (Haran *et. al.*, 1996a; Balasubramanian, 2003). These enzymes hydrolyze the fungal cell wall components such as chitin, glucan, cellulose and proteins successfully limiting the growth of fungal pathogens (Lorito *et al.*, 1994; Carsolio *et al.*, 1999; Balasubramanian, 2003). Thus, the production of chitinase, β-1,3 glucanase, proteases and cellulases play a vital role in antagonism exerted by *Trichoderma* (Flores *et al.*, 1997; Balasubramanian, 2003).

It has been established that the chitinase producing *Trichoderma* spp. can be effective biocontrol agents against fungal pathogens (Tokimoto, 1982; Sivan and Chet, 1989; Kubicek *et al.*, 2001). *Trichoderma harzianum*, which produces an endochitinase and N-acetyl-β-glucosaminidase was demonstrated to be a potent biocontrol agent against several phytopathogenic fungi like *Rhizoctonia solani, Sclerotium rolfsii* (Cook and Baker, 1983) and *Pythium ultimum* (Peer and Chet, 1990). *T. harzianum* enzymes even degrade sclerotia of S. rolfsii (Elad *et al.*, 1982). Similarly *Trichoderma viride* solubilized the hyphae of *Sclerotinia sclerotiorum* by producing β-1,3-glucanase (Jones *et al.*, 1972). Inhibition of spore germination and germ tube growth are the common phenomenon exhibited by the chitinolytic enzymes of *T. harzianum* (Lorito *et al.*, 1993). *Trichoderma* spp. frequently attaches to the host hyphae by coiling, hooks or appressorium-like bodies and penetrates the host by secreting cell wall lytic enzymes (Benhamou and Chet., 1993; Sreenivasaprasad and Manibhushanrao, 1993; Inbar and Chet, 1995; Katragadda and Murugesan, 1996). Production of many distinct chitinolytic enzymes has been reported in *T. harzianum* (de la Cruz *et al.*, 1992; Inbar and Chet, 1995; Haran *et al.*, 1996b) and these enzymes are induced when *T. harzianum* is grown in liquid medium containing chitin or its oligomers as carbon source. *T. harzianum* produced higher levels of chitinase when grown on medium amended with cell wall preparations of *F. oxysporum*. The synthesis of hydrolytic enzymes and peptaibols antibiotics is triggered by cell wall of *Botrytis cinerea* and peptaibol act synergistically with chitinase and β-1,3-glucanase in the inhibition of fungal spore germination and hyphal elongation (Schirmbock, 1994). Viterbo *et al.*, (2001) isolated a novel endochitinase gene from *T. harzianum*, which exhibited antifungal activity against fungal pathogens.

Production of inhibitory substances or antibiotics that suppresses the sporulation and pathogen growth is one of the mechanisms in biocontrol of plant diseases. It has been reported that *Trichoderma* spp. produce a wide range of volatile and non-volatile antibiotic substances (Sivasithamparam and Ghisalberti, 1998; Vyas and Mathur, 2002) and two such compounds namely, trichodermin and viridin, produced by *Trichoderma* sp. inhibited pathogenic fungal growth at very low concentrations (Weindling, 1941). Dennis and Webster (1971) explained the fungi-toxic action of *Trichoderma* on pathogenic *Pythium*. The volatile and non-volatile substances produced by *Trichoderma* spp. effectively inhibited growth of *R. solani* (Roy, 1977), *S. rolfsii* (Upadhyay and Mukhopadhyay (1983) and *Thanatephorus cucumeris* (Dubey and Patel, 2001). Claydon *et al.*, (1987) observed the production of antifungal alkyl pyrones in the culture filtrates of *T. harzianum* and later Govindasamy and Balasubramanian (1989) isolated a phenol like secondary metabolite from the culture filtrate of *T. harzianum*, which inhibited the uredospore germination of *Puccinia arachidis*. Recently it was observed that the seedling diseases of cotton caused by *R. solani* were suppressed by *T. viride* due to mycoparasitism and antibiotic production (Howell *et al.*, 2000).

In addition to direct inhibition of pathogen growth, *Trichoderma* spp. can also induce systemic and localized resistance in host plants against a variety of plant pathogens (Ahmed *et al.*, 2000; Howel *et al.*, 2000; Lo *et al.*, 2000; Harman *et al.*, 2004). *Trichoderma* spp. prevent the infection process of pathogenic fungi such as *B. cinerea* by inhibiting or degrading the pectinases and other enzymes, essential for pathogens to penetrate the leaf surfaces (Zimand *et al.*, 1996). Not only *Trichoderma*, its metabolites can also induce resistance in host plants as reported by Faize *et al.*, (2003). They observed the induced resistance to scab disease in apple due to the chitinases of *Trichoderma atroviride*. According to Mukherjee *et al.*, (2004) two G-protein alpha subunits, TgaA and TgaB were responsible for the antagonism of *Trichoderma* against plant pathogens. The above recent findings have thrown more light on the biocontrol mechanisms of *Trichoderma* and further research with the help of molecular tools would bring more clarity in this area.

Control of Plant diseases by Trichoderma

Biocontrol of soil-borne plant pathogens by the addition of antagonistic microorganisms to the soil is a potent disease control strategy. *Trichoderma* has emerged as a viable biofungicide and could be an alternative or supplement to chemical fungicides as they perform very well against several plant pathogens in green house as well as in the field conditions (Sharma, 1994; Mathivanan *et al.*, 2000a). The biological control of *Fusarium* wilt of melon and cotton and crown rot of tomato by *T. harzianum* strain T-35 has adequately been demonstrated (Sivan and Chet, 1986). Many investigators have also reported that seed or soil treatment with isolates of *Trichoderma* spp. can reduce damping-off disease induced by *R. solani*. These reports suggested that a considerable increment in the control of *R. solani* was achieved through mycoparasitism (Harman *et al.*, 1980; Chet and Baker, 1981; Elad *et al.*, 1981). Similarly, Elad *et al.*, (1980) also showed significant, reduction of diseased tomato plants caused by naturally occurring *S. rolfsii*.

T. viride is an excellent biocontrol agent exhibiting wide spectrum of activity against several plant pathogens and this species is being used widely, mostly for the management of root diseases in different crops (Jeyarajan *et al.*, 1994; Mathivanan *et al.*, 2000; Theradi Mani and Juliet Hepziba, 2003). Seed treatment with different species of Trichoderma has been found successful in controlling nursery diseases in tomato, and cauliflower, wilt in chickpea (Mukhopadyay *et al.*, 1992) and root rot in mung bean (Kehri and Chandras 1991). Seed treatment followed by soil application of *T. viride* significantly reduced root diseases viz., damping-off, root rot, wilt and collar rot in cotton, eggplant and sunflower in natural and in pathogen-inoculated soils (Mathivanan *et al.*, 2000a). Similarly seed and soil application of *T. harzianum* reduced the pigeonpea wilt caused by *Fusarium udum* under field conditions (Prasad *et al.*, 2002).

Trichoderma spp. are mainly applied in the rhizosphere, whereas little attempts have been reported on its application for the protection of foliar diseases. Application of *Trichoderma* conidia to foliage, flower and fruit effectively controlled plant diseases in different crops (Lattorre *et al.*, 1997; Elad, 1994; Harman, 2000). *T. harzianum* has been reported to be an effective biocontrol agent for foliar pathogen, *B. cinerea* (Elad, 1994; Hjeljord *et al.*, 2001) and *Trichoderma stromaticatum* was found to be a mycoparasite to *Crinipellis perniciosa* on cocoa (Sanogo *et. al.*, 2002).

Trichoderma Formulations for the control of Plant diseases

Several commercial products of *Trichoderma* have been developed worldwide for the management of soil-borne diseases. Ramakrishnan *et al.*, (1994) developed a talc formulation of *T. viride* for seed treatment and this formulation could be used for three months with adequate population of the antagonist (Jeyarajan and Nakkeeran, 1996). The population of *T. viride* in a biofungicide product was maintained as 28×10^8 cfu/g even after 4 months of storage (Mathivanan *et al.*, 2000a). Manav and Singh (2003) observed that the survival of *T. harzianum* was high in wet formulations prepared in wheat straw and wheat bran than the dry formulation in charcoal. A liquid conidial formulation of *T. harzianum* developed with an invert emulsion based on coconut and soybean oils (water-in-oil formulation) significantly reduced the post harvest apple gray mold disease caused by *B. cinerea* (Batta, 2004). Presently a number of commercial formulations of *Trichoderma* are available for the management of various foliar and root diseases worldwide (Whipps and Lumsden, 2001) and some of them are listed in Table 1.

The formulations of *Trichoderma* tested in field conditions gave satisfactory control of plant pathogens. A wheat bran culture of *T. harzianum* was used for controlling *R. solani* and *S. rolfsii* under field conditions in several crops (Baker and Cook, 1974; Chet *et. al.*, 1979; Hadar *et al.*, 1979). Application of T-35 as a conidial suspension or a wheat bran peat preparation resulted in high colonization of both rhizosphere and non-rhizosphere soils (Sivan and Chet, 1986). Kapoor and Pardeep Kumar (2004) reported that soil application of a wheat bran formulation of *T. harzianum* reduced the white rot of pea caused by *Sclerotinia sclerotiorum*. Harman *et al.*, (1989) developed a solid matrix priming

Table 1 Commercial formulations of *Trichoderma* for the control of plant diseases

Product	Species	Target pathogen/ disease	Crop	Formulation
In India				
Antagon-TV	**T. viride**	Seed and soil-borne pathogens	Legume, cotton vegetable crops, sunflower	Talc powder
Biocure-F	*T. viride*	Root diseases	Various crops	Talc powder
SunAgroderma	*T. viride*	Root diseases	Various crops	Talc powder
Eco-fit	*T. viride*	Seed and soil-borne pathogens	Legume, cotton, vegetable crops, sunflower	Talc powder
Ecoderma	*T. viride*	Root diseases, sheath blight	Various crops	Talc powder
Bumika	*T. viride*	Talc powderRoot diseases	Various crops	Talc powder
In abroad				
Bio-Fungus	*Trichoderma* spp.	Most of the seed and soil-borne pathogens	Flowers, trees, strawberries, vegetables	Granules wettable powder
Binab T	*T. harzianum*, *T. polysporum*	Wilt, take-all root rot	Flowers, fruits, ornamentals, turf, vegetables	Wettable powder and pellets
RootShield	*T. harzianum*	Pythium, Rhizoctonia, Fusarium	Trees, shrubs, Plantations, vegetables	Granules, wettable powder
T-22G, T-22 Planter box	*T. harzianum*	Pythium, Rhizoctonia, Fusarium	Trees, shrubs, plantations, ornamentals, vegetables	Granules wettable
Trichodex	*T. harzianum*	Botrytis, Colletotrichum, Sclerotinia, Sclerotiorum	Cucumber, soybean, strawberry, sunflower, tomato	Wettable powder
Trichopel, Trichodowels, Trichoseal	*Trichodowels*, *T. viride*	Armillaria, Fusarium Pythium, Rhizoctonia	Various crops	–
Trichoderma	*Trichoderma* sp.	Rhizoctonia, Sclerotium, Pythium, Fusarium	Nursery and field crops	Incorporated into soil or potting medium

technique to improve the biological seed treatment of *T. harzianum* and they proved that the delivery of biocontrol agents through seed is a good strategy. Southern blight disease of tomato caused by S. rolfsii was significantly controlled by the application of mycelial formulation of *Trichoderma koningii* (Latunde-Dada, 1993). Granular formulations developed with biomass of *Trichoderma* spp. reduced the damping-off of eggplant caused by *R. solani* and these formulations also inhibited the saprophytic growth of pathogen in soil-less mix (Lewis and Larkin, 1997). Talc formulations of *T. viride* effectively controlled dry root rot caused by *Macrophomina phaseolina* (Alagarsamy *et al.*, 1987) and root diseases in cotton, eggplant, okra and sunflower with commensurate increase in yield (Mathivanan *et al.*, 2000a; Theradimani and Juliet Hepziba, 2003). Further, the talc formulation of *T. viride* that produced at Nagarjuna Agricultural Research and Developed Institute, Hyderabad with the support of Department of Biotechnology, Government of India has been supplied to the farmers to cover 10,783 ha area spread over eight districts in AP and an adjacent Bellary district in Karnataka (Mathivanan *et al.*, 2003).

Trichoderma in integrated disease management

In general, combination of two or more disease control strategies such as fungal antagonists, organic amendments, chemical fungicides provide high level control than that could be achieved individually. Application of antagonists directly suppressed the growth of pathogens and reduced their population in phyllosphere or rhizosphere. While the organic amendments increased the residential or introduced antagonists that again reduced the level of primary inoculum of soil pathogens as well as incidence of root diseases (Manibhushanrao, 1995). Baby and Manibhushanrao (1993) studied the population density of *Trichoderma longibrachiatum* and found significant increase in the colony forming unit of this antagonist when it was introduced in soil with gliricidia leaf and neem cake. The establishment and survivability of introduced biocontrol agents in different soils was good when applied along with organic materials because the latter served as immediate nutrient source for the bioagents. Soil application of *T. viride* along with organic materials such as wheat bran, rice bran, saw dust, farmyard manure significantly reduced the root diseases in sunflower (Mathivanan *et al.*, 1998). Application of *T. harzianum* in combination with neem cake and a fungicide, Ridomil in the collar region of tobacco plants significantly reduced the black shank disease caused by *Phytophthora parasitica* var. *nicotianae* and increased the plant growth (Narayanaswamy *et al.*, 2004).

Trichoderma spp. can be used in integrated disease management system with fungicides because they were tolerant to various fungicides and also exhibited complementary effect in controlling plant pathogens (Wokocha, 1990; Kay and Stewart, 1994; Mondal *et al.*, 1995; Malathi and Sabitha Doraisamy, 2003). An integrated use of chemical fungicides with *Trichoderma* was reported to give excellent control of dry root rot in soybean (Vyas, 1994). The combined application of carbendazim and *T. viride* was effective in controlling soil-borne root diseases of cotton under field conditions. The seed treatment with *T. viride* reduced the plant mortality up to 56 %, whereas, integration of Carbendazim and *T. viride* reduced plant mortality up to 72% (Mathivanan *et al.*, 1997). The loose smut of wheat caused by *Ustilago segetum* var. *tritici* was significantly controlled by seed treatment with *T. viride* and *T. harzianum* in combination with reduced dose of Vitavax (Singh and Maheshwari, 2001).

Strain improvement in *Trichoderma*

Development of mutants is one of the strategies in strain improvement programme of biocontrol agents. Papavizas *et al.*, (1982) obtained UV induced mutants of T. harzianum tolerant to Benomyl and they suppressed the growth of *R. solani* in soil more effectively than the wild strains. Several mutant strains of *Trichoderma* have been developed with enhanced biocontrol activity against plant pathogens. A significant increase in cellulase activity and rhizosphere competence was observed in mutant strains of *T. harzianum,* which was active against *P. ultimum* (Ahmad and Baker, 1987a; 1987b). The mycelial growth of *T. harzianum* parent was completely suppressed at 4 mm concentration of Bavistin, while the mutants tolerated up to 16 mm (Viji *et al.*, 1993). Three stable mutants of *Trichoderma* showed different colony morphology, fungicide tolerance and antagonistic potential against *U. segetum-tritici* as compared to their parent (Selvakumar *et al.*, 2000). Zaldivar *et al.*, (2001) developed a mutant strain of *Trichoderma* with enhanced production of chitinase, b-1,3 glucanase and protease. In addition to biocontrol potential, the above mutant also exhibited increased fungicide tolerance than the wild parent.

Protoplast fusion is an efficient tool for fungal biotechnology used for intraspecific, interspecific and intergeneric hybridization of filamentous fungi and also for the improvement of industrial strains (Ohnuki *et al.*, 1982; Lalithakumari, 2000). The production of fungal protoplasts has become an important technique with an increasing number of applications in pure and applied biological research. The optimal conditions were standardized for release and regeneration of protoplasts in *Trichoderma* (Stasz *et al.*, 1988; Tschen and Li, 1994; Mrinalini, 1997; Balasubramanian, 2003). Protoplast isolation and hybridization by protoplast fusion has been successfully carried out with *Trichoderma koningii* (Hong *et al.*, 1984), *Trichoderma reesei* (Toyama *et al.*, 1984), *T. harzianum* (Stasz *et al.*, 1988) and *T. longibrachiatum* (Mrinalini and Lalithakumari, 1996).

A strain of *T. harzianum* (1295-22) obtained by fusing protoplasts of the auxotrophic mutants of *T. harzianum*, T12 and T95 gave better control of *P. ultimum* in cotton and maize when used as a seed protectant (Sivan and Harman, 1991). Protoplast fusion was effective in inducing rhizosphere competence and biocontrol potential in *Trichoderma* spp. (Peer and Chet, 1990; Lalithakumari *et al.*, 1996). Integration of protoplast of T. *harzianum* and *T. longibrachiatum* yielded novel fusant strains with enhanced fungicide tolerance and biocontrol potential against sheath blight pathogens of rice (Lalithakumari *et al.*, 1996). Balasubramanian (2003), Prabavathy *et al.*, (2004) developed fusant strains by integration of protoplasts between T. *harzianum* and T. *reesei* and these fusant progenies exhibited high antagonistic activity against plant pathogenic fungi belonging to Oomycetes and Deuteromycetes.

Suggestions for future research

It has been known for many years that the biocontrol agents such as *Trichoderma* are artificially multiplied, formulated and applied in to foliar and soil environment to manage plant diseases. It is considered a safe and environment friendly approach and can be a viable alternative or supplement to conventional fungicides. Although *Trichoderma* spp. are reported to offer satisfactory control of plant pathogens under controlled conditions (Lifshitz *et al.* 1986; Sharma 1994), their efficacy might vary under field conditions due to abiotic factors prevailing in different places (Burpee 1990). Inconsistent performance will affect the reproducible results in field environment of different agroclimatic regions. Soil properties such as pH, temperature, moisture, and nutrient availability greatly influence the survival, establishment and colonization of the introduced organisms. In addition, they also have to face strong resistance/competition with the native microbial population in the soil. Apart from these, the recent review of Brimner and Boland (2003) on the non-target effects of fungal biocontrol agents has highlighted the hazards associated with the use of fungi as biocontrol agents for controlling plant diseases. Therefore it is absolutely essential to focus the future research on *Trichoderma* to address the above issues; a few are outlined hereunder and that should be pursued on priority basis.

- Isolation of *Trichoderma* spp. from different ecological conditions and screening them against important plant pathogens (wide spectrum of biocontrol activity, rhizosphere competence, adaptability to different environment, plant growth promotion, tolerance to fungicides, induction of resistance in host plant, etc. should be the criteria for selecting the biocontrol strain).

- Determination of the survival dynamics of *Trichoderma* in the introduced soils.

- Study on the impact of *Trichoderma* introduction on soil ecology, especially on soil microflora and its possible non-target effects.

- Improvement of *Trichoderma* strains through mutation/protoplast fusion/genetic engineering, etc. with greater biocontrol potential.

- Development of cost effective technology for mass multiplication and formulation with long shelf life.

- Systematic evaluation of formulated product of *Trichoderma* in multi-location and on-farm trails at farmers' fields for the management of plant diseases.

Conclusions

With the growing concern over the environmental pollution due to indiscriminate use of agrochemicals, it is absolutely necessary to exploit the potential of microorganisms as biocontrol agents for the management of various plant diseases. *Trichoderma* spp. are the maximum exploited biocontrol agents for the control of a wide range of plant pathogens causing several diseases in agricultural, horticultural and plantation crops. They compete with other microorganisms for nutrients and space and inhibit the growth of plant pathogens by secreting fungitoxic substances and cell wall degrading enzymes. In addition, they also induce systemic resistance in host plants. *Trichoderma* can be one of the key components in the integrated disease management system adjunct to conventional chemicals as they exhibit tolerance to them besides, they exhibit synergism with many fungicides in controlling plant pathogens. Mutation and protoplast fusion technologies can be exploited to develop superior strains of *Trichoderma* with enhanced biocontrol potential and rhizosphere competence. Finally, the systematic application of *Trichoderma* with no or judicial use of conventional chemicals would be an ideal and eco-friendly approach for the management of plant diseases, which will ultimately help in achieving the goal of increasing farm productivity.

References

Abada, K.A. 1994. Fungi causing damping-off and root rot on sugar beet and their biological control with *Trichoderma harzianum. Agriculture Ecosystem and Environment* 51: 333-337.

Adams, P.B. 1990. The potential of mycoparasites for biological control of plant diseases. *Annual Review Phytopathology* 28: 59-72.

Ahmad, J.S. and **Baker, R**. 1987a. Competitive saprophytic ability and cellulolytic activity of rhizosphere competent mutants of *Trichoderma harzianum. Phytopathology* 77: 358-362.

Ahmad, J.S. and **Baker, R.** 1987b. Rhizosphere competence of Trichoderma harzianum with altered antibiotic production on selected in vitro. *Canadian Journal of Microbiology* 77: 182-189.

Ahmed, A.S., Sanchez, C.P. and **Candela, M.E.** 2000. Evaluation of induction of systemic resistance in pepper plants (Capsicum annuum) to Phytophthora capsici using Trichoderma harzianum and its relation with capsidiol accumulation. *European Journal of Plant Pathology* 106: 817-824.

Alagarsamy, G., Mohan, S. and **Jeyarajan, R.** 1987. Effect of seed pelleting with antagonists in the management of seedling disease of cotton. *Journal of Biological Control* 1: 66-67.

Baby, U.I., and ***Manibhushanrao, K.*** 1993. Control of rice sheath blight through the integration of fungal antagonists and organic amendments. *Tropical Agriculture* 70: 240-244.

Baker, K.F. and ***Cook, J.R.*** 1974. *Biological Control of Plant Pathogens.* W.H. Freeman, San Francisco. 433.

Balasubramanian, N. 2003. *Strain improvement of Trichoderma spp. by protoplast fusion for enhanced lytic enzyme and biocontrol potential.* Ph. D., Thesis, University of Madras, Chennai, India.

Batta, Y.A. 2004. Postharvest biological control of apple gray mold by Trichoderma harzianum Rifai formulated in an invert emulsion. *Crop Protection* 23: 19-26.

Benhamou, N. and ***Chet, I.*** 1993. Hyphal interaction between *Trichoderma harzianum* and Rhizoctonia solani: Ultrastructure and gold cytochemistry of the mycoparasitic process. *Phytopathology* 83: 1062-1071.

Brimner, T.A. and ***Boland, G.J.*** 2003. A review of the non-target effects of fungi used to biologically control plant diseases. *Agriculture Ecosystem and Environment* 100: 3-16.

Burpee, L.L. 1990. The influence of abiotic factors on biological control of soilborne plant pathogenic fungi. *Canadian Journal of Plant Pathology* 12: 308-317.

Carsolio, C., Benhamou, N., Haran, S., Cortes, C., Gutierrez, A., Chet, I. and ***Herrera-Estrella, A.*** 1999. Role of the Trichoderma harzianum endochitinase Gene ech2 in mycoparasitism. *Applied Environmental Microbiology* 65: 929-935.

Chet, I. and ***Baker, R.*** 1981. Isolation and biocontrol potential of *Trichoderma hamatum* from soil naturally suppressive to Rhizoctonia solani. *Phytopathology* 71 : 286-290.

Chet, I., Hadar, Y., Elad, J., Katan, J. and ***Henis, Y.*** 1979. Biological control of soil-borne plant pathogens by *Trichoderma harzianum.* In: *Soil-borne Plant Pathogens.* (Eds. Schippers, B. and Gams, W.). 585-592. Academic Press, London, UK.

Claydon, N., Allan, M., Hanson, J.R. and ***Avent, A.G.*** 1987. Antifungal alkyl pyrones of *Trichoderma harzianum. Trans of British Mycol Soci* 88: 505-513.

Cook, R.J. and ***Baker, K.F.*** 1983. The Nature and Practice of Biological Control of Plant Pathogens. *American Phytopathological society,* St. Paul, Minnesota, USA.

De La Cruz, J., Hidalgo-Gallego, A., Lora, J.M., Benltez, T., Pintor-Toro, J.A., and ***Llobell, A.*** 1992. Isolation and characterization of three chitinases from *Trichoderma harzianum. European Journal of Biochemistry* 206: 859-867.

Dennis, L. and ***Webster, J.*** 1971. Antagonistic properties of species group of *Trichoderma* III hyphal interactions. *Trans Br. mycol Soc* 57: 363-369.

Devaki, N., Shankara Bhat, S.S., Bhat, S.G. and **Manjunatha, K.R.** 1992. Antagonistic activities of Trichoderma harzianum against Pythium aphanidermatum and Pythium myriotylum on tobacco. *Journal of Phytopathology* 136: 82-87.

Dubey, S.C. and **Patel, B.** 2001. Evaluation of fungal antagonists against Thanatephorus cucumeris causing web blight of urd and mung bean. *Indian Phytopath.* 54: 206-209.

Elad, Y. 1994. Biological control of grape grey mould by *Trichoderma harzianum*. *Crop Protection* 13: 35-38.

Elad, Y. 2000. Biological control of foliar pathogens by means of Trichoderma harzianum and potential modes of action. *Crop Protection* 19: 709-714.

Elad, Y., Chet, I. and **Katan, J.** 1980. Trichoderma harzianum, a biocontrol agent effective against *Sclerotium rolfsii* and *Rhizoctonia solani*. *Phytopathology* 70: 119-121.

Elad, Y., Chet, I. and **Henis, Y.** 1981. Biological control of Rhizoctonia solani in strawberry fields by Trichoderma harzianum. *Plant Disease* 60: 245-254.

Elad, Y., Chet, I. and **Henis, Y.** 1982. Degradation of plant pathogenic fungi by Trichoderma harzianum. *Canad. J. microbiol.* 28: 719-725.

Faize, M., Malnoy, M., Dupuis, F., Chevalier, M., Parisi, L. and **Chevreau, E.** 2003. Chitinases of Trichoderma atroviride induce scab resistance and some metabolic changes in two cultivars of apple. *Phytopathology* 93: 1496-1504.

Flores, A., Chet, I. and **Herrera-Estrella, A.** 1997. Improved biocontrol activity of Trichoderma harzianum by over-expression of proteinases-encoding gene prb1. Current Genetics 31: 30-37.

Govindasamy, V. and **Balasubramanian, R.** 1989. Biological control of groundnut rust, Puccinia arachidis by *Trichoderma harzianum*. J. Pl. Dis. *Protection* 96: 337-345.

Hadar, Y., Chet, I. and **Henis, Y.** 1979. Biological control of Rhizoctonia solani demping-off with wheat bran culture of *Trichoderma harzianum*. *Phytopathology* 69: 64-68.

Haran, S., Schikler, H. and **Chet, I.** 1996a. Molecular mechanisms of lytic enzymes involved in the biocontrol activity of *Trichoderma harzianum*. *Microbiology* 142: 2321-2331.

Haran, S., Schikler, H., Oppenheim, A. and **Chet, I.** 1996b. Differential expression of Trichoderma harzianum chitinases during mycoparasitism. *Phytopathology* 86: 981-985.

Harman, G.E. 2000. Myths and dogmas of biocontrol changes in perceptions derived from research on *Trichoderma harzianum* T22. *Plant Disease* 84: 377-393.

Harman, G.E., Chet, I. and ***Baker, R.*** 1980. *Trichoderma hamatum* seedling disease induced in radish and pea by *Pythium* sp. and *Rhizoctonia solani. Phytopathology* 70: 1107-1172.

Harman, G.E., Howell, C.R., Viterbo, A., Chet, I. and ***Lorito, M.*** 2004. *Trichoderma* species-Opportunistic, avirulent plant symbionts. *Nature Review of Microbiology* 2: 43-56.

Harman, G.E., Taylor, A.G. and ***Stasz, T.E.*** 1989. Combining effective strains of *Trichoderma harzianum* with solid matrix priming to improving biological seed treatments. *Plant Disease* 73: 631-637.

Hermosa, M.R., Grondona, I., Iturriaga, E.A., Diaz-Minguez, J.M., Castro, C., Monte, E. and ***Garcia-Acha, I.*** 2000. Molecular characterization and identification of biocontrol isolates of *Trichoderma* spp. *Applied Environmental Microbiology* 66: 1890-1898.

Hjeljord. L.G., Stensvarrd, A. and ***Tronsmo, A.*** 2001. Antagonism of nutrient-activated conidia of *Trichoderma harzianum* (*atroviride*) P1 against *Botrytis cinerea. Phytopathology* 91: 1172-1180.

Hong, S.W., Hah, Y.C., Park, H.M. and ***Cho, N.J.*** 1984. Intraspecific protoplast fusion in *Trichoderma koningii.* Korean J.Microbio 22: 103-110.

Howell, C.R., Hanson, L.E., Stipanovic, R.D. and ***Puckhaber, L.E.*** 2000. Induction of terpenoid synthesis in cotton roots and control of *Rhizoctonia solani* by seed treatment with *Trichoderma virens. Phytopathology* 90: 248-252.

Inbar, J. and ***Chet, I.*** 1995. The role of recognition in the induction of specific chitinase during mycoparasitism by *Trichoderma harzianum. Microbiology* 141: 2823.

Jeyarajan, R., and ***Nakkeeran, S.*** 1996. Exploitation of biocontrol potential of *Trichoderma* for field use. In: *Recent Development in Biocontrol of Plant Pathogens.* (Eds. Manibhushanrao, K., and Mahadevan, A.). 61-66. Today and Tomorrow's Printers and Publishers, New Delhi, India.

Jeyarajan, R., Ramakrishnan, G., Dinakaran, D. and ***Sridhar, R.*** 1994. Development of products of *Trichoderma viride* and *Bacillus subtilis* for biocontrol of root-rot diseases. In: *Bio-technology in India* (Ed. Dwivedi, B.K.). 25-36. Bioved Research Society, Allahabad, India.

Jones, D., Gordon, A.H. and ***Bacon, J.S.D.*** 1972. Co-operative action by endo- and exo-b-(1,3)-glucanase from parasitic fungi in the degradation of cell wall glucanase of *Sclerotinia sclerotiorum. J. Biochemistry* 140: 47-55.

Kapoor, A.S., and ***Pardeep Kumar.*** 2004. Evaluation of different organic substrates, viability of *Trichoderma harzianum* and management of white rot of pea. *Plant Dis. Res.*19: 55-56.

Katragadda, H., and **Murugesan, K.** 1996. Hyperparasitic potential of *Trichoderma harzianum* over *Fusarium oxysporum* f. sp. *vasinfectum*. *Indian J. Microbiol* 36: 145-148.

Kay, S.J. and **Stewart, A.** 1994. The effect of fungicides on fungal antagonists of onion white rot and selection of dicarboximide-resistant biotypes. *Plant Pathology* 43: 863-871.

Kehri, H.K. and **Chandra, S**. 1991. Antagonism of *Trichoderma viride* to *Macrophomina phaseolina* and its application in the control of dry root-rot of mung. *Indian Phytopathology* 44: 60-63.

Kubicek, C.P., Mach, R.L., Peterbauer, C.K., and **Lorito, M.** 2001. *Trichoderma*: From genes to biocontrol. *Journal of Plant Pathology* 83: 11-23.

Lalithakumari, D. 1996. Protoplasts-A biotechnological tool for plant pathological studies. *Indian Phytopath.* 49: 199-212.

Lalithakumari, D. 2000. *Fungal protoplast: A biotechnological tool.* Oxford & IBH Publishing Co., Pvt., Ltd., New Delhi, India. 184 p.

Lalithakumari, D., Mrinalini, C., Chandra, A.B. and **Annamalai, P.** 1996. Strain improvement by protoplast fusion for enhancement of biocontrol potential integrated with fungicide tolerance in *Trichoderma* spp. *J. Pl. Dis. Protection* 103 : 206-212.

Latorre, B.A., Agosin, E., San Martin, R. and **Vasquez, G.S.** 1997. Effectiveness of conidia of *Trichoderma harzianum* produced by liquid fermentation against Botrytis bunch rot of table grape in Chile. *Crop Protection* 16: 209-214.

Latunde-Dada, A.O. 1993. Biological control of southern blight disease of tomato caused by *Sclerotium rolfsii* with simplified mycelial formulations of *Trichoderma koningii. Plant Pathology* 42: 522-529.

Lewis, J.A. and **Larkin, R.P.** 1997. Extruded granular formulation with biomass of biocontrol *Gliocladium virens* and *Trichoderma* spp. to reduce damping-off of eggplant caused by *Rhizoctonia solani* and saprophytic growth of the pathogen in soil-less mix. *Biocontrol Science and Technology* 7: 49-60.

Lewis, J.A. and **Papavizas, G.C.** 1991. Biocontrol of plant diseases: The approach for tomorrow. *Crop Protection* 10: 95-105.

Lifshitz, R., Windham, M.T., and **Baker, R.** 1986. Mechanism of biological control of pre-emergence damping-off of pea by seed treatment with *Trichoderma* spp. *Phytopathology* 76: 720-725.

Lo, C., Liao, T.E. and **Deng, T.C.** 2000. Induction of systemic resistance of cucumber to cucumber green mosaic virus by the root colonizing *Trichoderma* spp. *Phytopathology* 90 (Supplementary): S47.

Lo, C.T., Nelson, E.B. and **Harman, G.E.** 1996. Biological control of turfgrass diseases with a rhizosphere competent strain of *Trichoderma harzianum*. *Plant Dis.* 80: 736-741.

Lorito, M., Harman, G.E., Hayerss, C.K., Broadway, R.M., Tronsmo, A., Woo, S.L. and **Pietro, A.** 1993. Chitinolytic enzymes produced by *Trichoderma harzianum*: Antifungal activity of purified endochitinase and chitobiosidase. *Phytopathology* 83: 302-307.

Lorito, M. Peterbauer, C.K., Hayes, C.K. and **Harman, G.E.** 1994. Synergistic interaction between fungal cell wall degrading enzymes and different antifungal compounds on spore germination. *Microbiology* 140: 623-629.

Lu, Z., Tombolini, R., Woo, S., Zeilinger, S., Lorito, M. and **Jansson, J.K.** 2004. In vivo study of *Trichoderma*-pathogen-plant interaction using constitutive and inducible green fluorescent protein reporter systems. *Applied Environmental Microbiology* 70: 3073-3081.

Malathi, P. and **Sabitha Doraisamy**. 2003. Compatible of *Trichoderma harzianum* with fungicides against *Macrophomina phaseolina*. *Plant Dis. Res.* 18: 139-143.

Manav, M. and **Singh, R.S.** 2003. Shelf life of different formulations of mutant and parent strain of *Trichoderma harzianum* at variable temperatures. *Plant Dis.Res.* 18: 144-146.

Manibhushanrao, K. 1995. Sheath Blight Disease of Rice. Daya Publishing House, Delhi, India. 101.

Manoharachary, C. 2003. Role of *Trichoderma* Pers. Ex. Fr. as a biocontrol agent. Abstract of the National Conference on Biological Control of Crop Diseases: Present Perspectives and Future Challenges. 2. Centre for Advanced Studies in Botany, University of Madras, Chennai, India.

Mathivanan, N., Bharati N. Bhat, Prabavathy, V.R., Srinivasan, K. and **Chelliah, S.** 2003. *Trichiderma viride*: Lab to land for the management of root diseases in different crops. In: *Innovative Methods and Techniques for Integrated Pest and Disease Management* (Eds. Mathivanan, N., Prabavathy, V.R. and Gomathinayagam, S.). 52-58. Centre for Advanced Studies in Botany, University of Madras, Chennai, India.

Mathivanan, N., Srinivasan, K. and **Chelliah, S.** 1997. Evaluation of *Trichoderma viride* and carbendazim for integrated management of root diseases in cotton. *Indian Journal of Microbiology* 37 : 107-108.

Mathivanan, N., Srinivasan, K., and **Chelliah, S.** 1998. Evaluation of organic materials as carrier for commercial product of *Trichoderma*. *Journal of Biological Control* 12 : 97-100.

Mathivanan, N., Srinivasan, K. and ***Chelliah, S.*** 2000a. Biological control of soil-borne diseases of cotton, eggplant, okra and sunflower by *Trichoderma viride*. *J. Pl. Dis. Protection* 107: 235-244.

Mathivanan, N., Srinivasan, K. and ***Chelliah, S***. 2000b. Field evaluation of biocontrol agents against foliar diseases of groundnut and sunflower. *J. Biological Control* 14: 31-34.

Mondal, G., Srivastava, K.D. and ***Agarwal, R.*** 1995. Antagonistic effect of *Trichoderma* spp. on *Ustilago segetum* var. *tritici* and their compatibility with fungicides and biocides. *Indian Phytopath.* 48: 466-470.

Mrinalini, C. 1997. *Strain improvement of enhancement and integration of antagonistic and fungicide tolerance potential in Trichoderma* spp. Ph. D., thesis, University of Madras. Madras, India.

Mrinalini, C and ***Lalithakumari, D.*** 1996. Protoplast fusion: A biotechnological tool for strain improvement of *Trichoderma* spp. In: *Recent Development in Biocontrol of Plant Pathogens*. (Eds. Manibhushanrao, K. and Mahadevan, A.). pp. 133-146. Today and Tomorrow's Printers and Publishers, New Delhi, India.

Mukherjee, P.K., Latha, J., Hadar, R. and ***Horwite, B.A.*** 2004. Role of two G-protein alpha subunits, TgaA and TgaB in the antagonism of plant pathogens by Trichoderma virens. *Applied Environmental Microbiology* 70: 542-549.

Mukhopadhyay, A.N., Shrestha, S.M. and ***Mukherjee, P.K.*** 1992. Biological seed treatment for control of soil-borne plant pathogens. *FAO Plant Protection Bull.* 40: 21-30.

Narayanaswamy, H., Pradeep, S., Chandrashekhar, S.Y., and ***Basavaraja, M.K.*** 2004. Integrated management of black shank disease of tobacco in field crop. *Plant Dis. Res.* 19: 66-67.

Ohnuki, Y., Etoh, Y. and ***Beppu, T.*** 1982. Intraspecific and interspecific hybridization of *Mucor pusillus* and *M. miehei* by protoplast fusion. *Agricultural Biological Chemistry* 46: 451-458.

Papavizas, G.C., Levis, J.A. and ***Abd-ElMoity, T.H.*** 1982. Evaluation of new biotypes of *Trichodera harzianum* for tolerance of benomyl and enchanced biocontrol capabilities. Phytopathology 72:126-132.

Peter, S. and ***Chet, I.*** 1990. *Trichoderma* protoplast fusion; a tool for improving biocontrol agents. *Can. J. Microbiol.* 36: 6-9.

Prabavathy, V.R., Mathivanan, N., and ***Lalithakumari, D.*** 2004. Strain improvement in Trichoderma by protoplast fusion for multienzyme production and enhanced biocontrol potential. Abstract of the *National Symposium on Recent Trends in Biology, Biodiversity and Biotechnology of Fungi*. 45-46. Centre for Advanced Studies in Botany, University of Madras, Chennai, India.

Prasad, R.D., Rangeswaran, R., Hegde, S.V.* and *Anuroop, C.P. 2002. Effect of soil and seed application of *Trichoderma harzianum* on pigeonpea wilt caused by *Fusarium udum* under field conditions. *Crop Protection* 21: 293-297.

Rajan, P.P., Sharma, Y.R.* and *Anandaraj, M. 2002. Management of root rot disease of black pepper with *Trichoderma* spp. *Indian Phytopath.* 55: 34-38.

Ramakrishnan, G., Jeyarajan, R.* and *Dinakaran, D. 1994. Talc-based formulation of *Trichoderma viride* for biocontrol of *Macrophomina phaseolina*. *J. Biological Control* 8: 41-44.

Roy, A.K. 1977. Parasitic activity of *Trichoderma viride* on sheath blight fungus of rice. *J. Pl. Dis. Protection* 84: 675-683.

Sanogo, S., Pomella, A., Hebbar, P.K., Bailey, B., Costa, J.C.B., Samuels, G.J.* and *Lumsden, R.D. 2002. Production and germination of conidia of *Trichoderma stromaticatum*, a mycoparasite of Crinipellis perniciosa on cocoa. *Phytopathology* 92: 1032-1037.

Schirmbock, M., Lorito, M., Wang, Y.L., Hayes, C.K., Arisan-Atac, I., Scala, F., Harman, G.E.* and *Kubicek, C.P. 1994. Parallel formation and synergism of hydrolytic enzymes and peptaibol antibiotics: Molecular mechanisms involved in the antagonistic action of *Trichoderma harzianum* against phytopathogenic fungi. *Applied Environmental Microbiology* 60:4364-4370.

Selvakumar, R., Srivastava, K.D., Rashmi, A., Singh, D.V.* and *Dureja, P. 2000. Studies on development of *Trichoderma viride* mutants and their effects on *Ustilago segetum tritici*. *Indian Phytopath.* 53:185-189.

Sharma, B.K. 1994. Efficacy of biocontrol agents for the control of chickpea stem rot. *Journal of Biological Control* 8:115-117.

Sharma, S.N.* and *Chandel, S.S. 2003. Screening of biocontrol agents in vitro against *Fusarium oxysporum f.* sp. *gladioli* and their mass multiplication on different organic substrates. *Plant Dis.Res.* 18:35-38.

Singh, D.* and *Maheshwari, V.K. 2001. Biological seed treatment for the control of loose smut of wheat. *Indian Phytopath.* 54:457-460.

Sivan, A.* and *Chet, I. 1986. Biological control of *Fusarium* spp. in cotton, wheat and muskmelon by *Trichoderma harzianum*. *Journal of Phytopathology* 116: 38-47.

Sivan, A.* and *Chet, I. 1989. Degradation of fungal cell walls by lytic enzymes of *Trichoderma harzianum*. *Journal of General Microbiology* 135: 675-682.

Sivan, A. and **Harman, G.E.** 1991. Improved rhizosphere competence in a protoplast fusion progeny of *Trichoderma harzianum*. *Journal of General Microbiology* 137:23.

Sivasithamparam, K. and **Ghisalberti, E.L.** 1998. In: *Trichoderma and Gliocladium Vol.* 1. (Eds. Kubicek, C.P. and Harman, G.E.). 139-191.Taylor and Francis, London.

Sreenivasaprasad, S., and **Manibhushanrao, K.** 1993. Efficacy of *Gliocladium virens* and *Trichoderma longibrachiatum* as biocontrol agents of groundnut root and stem rot diseases. *Internat. J. Pest Management* 39:167-171.

Stasz, T.E., Harman, G.E. and **Weeden, N.F.** 1988. Protoplast preparation and fusion in two biocontrol strains of *Trichoderma harzianum*. *Mycologia* 80: 141-150.

Theradi Mani, M. and **Juliet Hepziba, S**. 2003. Biological control of dry root rot of sunflower. *Plant Dis. Res.* 18: 124-126.

Thomas, J., Bhai, R.S., Vijayan, A.K. and **Naidu, R.** 1993. Evaluation of antagonists and their efficacy in managing root diseases of cardamom. *J. Biological Control* 7: 29-36.

Tokimoto, K. 1982. Lysis of the mycelium of *Lentinus edodus* caused by mycolytic enzymes of *Trichoderma harizanum* when the two fungi were in antagonistic state. *Tr. Br. mycol. Soc.* of Japan 23: 13-20.

Toyama, H., Yamaguchi, K., Shinmyo, A. and **Okada, H.** 1984. Protoplast fusion of *Trichoderma reesei* using immature conidia. *Applied Environmental Microbiology* 47: 363-368.

Tschen, J.S.M. and **Li, I.F.** 1994. Optimization of formation and regeneration of protoplasts from biocontrol agents of *Trichoderma* sp. *Mycoscience* 35:257-263.

Upadhyay, J.P. and **Mukhopadhyay, A.N.** 1983. Effect of non-volatile and volatile antibiotics of *Trichoderma harzianum* on growth of *Sclerotium rolfsii*. *Indian J. myco.* and *Plant Pathology* 13 : 232-233.

Viji, G., Baby, U.I. and **Manibhushanrao, K.** 1993. Induction of fungicidal resistance in *Trichoderma* sp. through UV irradiation. *Indian Journal of Microbiology* 33: 123-129.

Viterbo, A., Haran, S., Friesem, D., Ramot, O. and **Chet, I.** 2001. Antifungal activity of a novel endochitinase gene (chit36) from *Trichoderma harzianum* Rifai TM. *FEMS Microbiol. Lett.* 200 : 169-174.

Vyas, S.C. 1994. Integrated biological control and chemical control of dry root rot on soybean. *Indian J. mycol Pl. Path* 24: 132-134.

Vyas, R.K. and **Mathur, K.** 2002. *Trichoderma* spp. in cumin rhizosphere and their potential in suppression of wilt. *Indian Phytopath.* 55: 455-457.

Weindling, R. 1932. *Trichoderma lignorum* as a parasite of other fungi. *Phytopathology* 22: 837.

Weindling, R. 1941. Experimental consideration of the mould toxins of *Gliocladium* and *Trichoderma. Phytopathology* 31: 991-1003.

Whipps, J.M. 1992. Status of biological disease control in horticulture. *Biocontrol Sci. and Tech.* 2: 3-24.

Whipps, J.M. and **Lumsden, R.D.** 2001. Commercial use of fungi as plant disease biological control agents: Status and prospects. In: *Fungi as Biocontrol Agents: Progress, Problems and Potential.* (Eds. Butt, T.M., Jackson, C. and Magan, N.). 9-22. CABI Publishing, Wallingford, UK.

Wokocha, R.C. 1990. Integrated control of Sclerotium rolfsii infection of tomato in the Nigeria Savanna: Effect of *Trichoderma viride* and some fungicides. *Crop Protection* 9: 231-234.

Zaldivar, M., Velasquez, J.C., Contreras, I. and **Perez, L.M.** 2001. *Trichoderma aureoviride* 7-121, a mutant with enhanced production of lytic enzymes: Its potential use in waste cellulase degradation and/or biocontrol. *Electronic Journal of Biotechnology* 4: 1-9.

Zimand, G., Elad, Y. and **Chet, I.** 1996. Effect of *Trichoderma harzianum* on *Botrytis cinerea* pathogenicity. *Phytopathology* 86: 1255-1260.

33

The Need for Commercialization of Biocontrol Agents in India

H.B. Singh*, S. Srivastava and A. Singh
National Botanical Research Institute, Lucknow-226 001 *E.mail:hbsl@rediffmail.com

ABSTRACT

Biological control is becoming an integral part of plant disease management practices all over the world. In our country, the work on biological control was started in middle of the twentieth century. Since then a number of attempts have been made to control plant diseases by using formulations of bioagents. Recently, the amendments made in Insecticides Act 1968, made it mandatory that some bioagents be registered before they are commercialized. This paper presents an overview about the need and importance of registration of biocontrol agents in India.

KEY WORDS : Commercialization, Biocontrol agents in India.

Over the past century, the research has repeatedly demonstrated that phylogeneti cally diverse microorganisms can act as natural antagonists of various plant pathogens. The interactions between microorganisms and plant hosts can be complex. However, fewer isolates of microorganisms can suppress plant diseases under diverse growing conditions and fewer still have broad-spectrum activity against multiple pathogenic taxa. Nonetheless, intensive screening tests carried out all over the world have yielded numerous candidate organisms for commercial development. Screening is a critical step in the development of biocontrol agents. The success of all subsequent stages depends on the ability of a screening procedure to identify an appropriate candidate.

Many useful bacterial biocontrol agents have been found by observing zones of inhibition in Petri plates. However, this method does not identify biocontrol agents with other modes of action such as parasitism, induced plant resistance, or some forms of competition. Screening methods for parasitism include burying and retrieving propagules of the pathogen to isolate parasites.

For competition, methods include looking for microbes that quickly colonize sterilized soil and have the ability to exclude other organisms that attempt to invade the space, and looking for microbes that colonize the infection court. Primary screening for new biocontrol microbes is still performed and it seems likely that continued prospecting will be required to diversify the potential applications of biocontrol as well as replace more widely used biocontrol products should resistance develop.

Interest in biological control research continues reflecting the desire of multiple constituencies to develop sustainable methods for controlling plant diseases. Growing concerns about environmental health and safety have led to substantial regulatory changes in the past several years. In addition, the majority of growers continue to express interest in biologically-based pest management strategies of all types as central components of an IPM approach. Such market realities promote the development of biocontrol products. Still, the path to developing and applying effective biocontrol methods is a long one, fraught with a variety of difficulties. Scientific, regulatory, business management and marketing issues must all be handled effectively for a biocontrol product to be successful in the marketplace.

Need for commercialization

The agriculture, practiced as monoculture of crops by mankind, is highly vulnerable to a large number of diseases all over the world. On the one hand the demand for food is increasing continuously as a result of logarithmic growth in population and on the other there is increasing pressure on the farming community to increase the productivity. Though the productivity is increasing in relation to various crops in our country it is not in synchrony with the population growth. Besides the reducing fertility of the cultivable soil, various diseases of the crops caused by pathogens are the most important reason for inability of the farmers to meet the demands. Most of the time a farmer tries to manage the diseases by chemical means, which though controls the disease to some extent but also affects the human and cattle health due the problem of residual effects. There is wide recognition of the fact that biological control of plant diseases i.e. use of microbes as antagonists to arrest the growth of pathogen, is a more effective alternative to the chemical management. Although a number of reports are there for controlling the pathogen *in vitro* and *in vivo*, very few reports are there for large scale and successful application of biocontrol agents in disease management. And there are still fewer examples of commercialization of biological control formulations despite the origin of concept of use of biocontrol agents for increasing plant yield over 100 years ago.

Our research in the immediate years after the independence has been mainly basic. Only after the green revolution the research in agricultural disciplines started to move towards applied areas and the scientists started to categorize the effective researches on the basis of their reach to the farmers, how easily it could be applied in the field and the cost-benefit ratio etc. In our country and elsewhere N, P, K has been in use since ancient times through the use of cow dung, leftovers of the previous crops etc and the disease management practices were in vogue by using the simple techniques of roguing and burning the diseased plants, use of $CuSO_4$ etc. But only after the fungicides began to be produced on commercial scale did the production of crops burgeoned. Same is the case with biological control agents. The scientists have searched and found effective antagonist, which can control diseases, tested their effectiveness on microplots and developed the formulations. The next in line is commercialization and that too on large scale so that the more and more farmers can benefit from this new, effective and eco-friendly method of plant disease management.

Commercializing the biocontrol agents is also necessary to overcome the negative and harmful effects of the chemicals more prominent in our current scheme of disease management. This is more so in case of developing countries where the chemicals banned in most of the developed countries are still in use e.g. PCNB, a highly toxic fungicide is still used for disease management whereas it is banned in USA. The successful development and commercialization of the biocontrol fungi will effectively reduce the use of hazardous chemicals and ultimately it may eliminate them. This will help us to provide our future generations with safe and clean environmental conditions.

Problems of commercialization

The reasons for negligible commercialization of biocontrol fungi are a plenty. It is a long-term process. The rhizospheric competence of the formulation to be commercialized varies with changing soil conditions (Ahmad and Baker, 1987). After isolating a rhizosphere competent strain of an effective biocontrol agent/antagonist all the research must be directed towards the development and production of the strain. The field trials are necessary supplements for the development of a product, which need to be out under different soil conditions in different fields, on different crops and various pathogenic fungi in order to develop a broad-spectrum product, which can effectively control a large number of pathogenic fungi under different conditions.

Another constraint is the selection of a suitable substrate for the development of effective formulation. A number of substrates have been tried for the development of biocontrol formulations, which can be commercialized later on but a lot of them are food grains and they definitely can not be deemed as cost-effective. The economically viable method to use the agricultural wastes viz., tea, coffee waste, rive hay, distilled wastes from oil yielding plants etc for mass production of biocontrol agents is an alternative (Singh *et al.*, 2004). The most economical method for mass multiplication of biocontrol fungi for field application has recently been developed (Singh and Johri, 2000)

A number of commercialized products have been developed and are available in the world market for disease management but they are either location specific or disease specific or both. Not many products are available which are broad-spectrum in nature and used on large scale. As on today a large number of diseases are still not covered by the commercialized biocontrol formulations. Thus, there is the need from moving ahead with selecting the effective strains of biocontrol fungi and developing them as broad-spectrum formulations for use against a large number of soil-borne diseases in varying soil conditions. Like a large chunk of bioformulations commercialized all over the world are Trichoderma-based e.g. in our country, twenty-nine manufacturers are producing *Trichoderma*-based formulations.

Weller (1988) has reviewed the more important cases related with the failure or inconsistent performance of the biological control agents in the rhizosphere. Firstly the presence of low disease pressure for the test to be effective and the treatment favouring the increased damage from the non-target diseases. Secondly the root colonization effected by the introduced strain or the loss of ecological competence by the strain is variable and the production of antibiotics, wherever it is necessary for the effective action of antagonist, is either too late or too low in quantity that it does not prove to be effective for controlling the disease.

The more important reason for developing the biocontrol product and its commercialization is that funding for such projects is short-term and it is not feasible to sort out various problems associated with the commercialization. The ideal condition would be that the flow of funds should be continuous not like 2 or 3-year projects, which result only by ending the project at the isolation and testing stage e.g. The development of T-22 strain of *Trichoderma harzianum* by Harman and his associates took more than a decade and it was only in last decades of last century that the sale of its product, Root Shield, started to pick up from scratch (Harman, 2000). The other problems related to the institutes, technical limitations and unrealistic expectations have to be overcome if we are to give any thought to the development of commercial products for plant disease management.

The economics of commercialization is another important factor in the commercial deployment of the effective strains of the biocontrol fungi in the disease management. Primarily, a limited market pays the developmental and registration costs and mostly it is disease specific for a single crop. In most of the cases, either there is no established infrastructure for scale up and commercialization of biocontrol agents (Cook, 1993) or it is in infancy, which does not prove economically viable and there is no impetus for the development of such products. There is a need for significant changes in the infrastructure facilities for the use of biocontrol agents to be effective against one disease on one crop as has been achieved on sweet corn in USA (Mathre *et. al.*, 1999). Several species of *Trichoderma* have been used as plant disease control agents. Recently it has been reported that *Trichoderma* species ate opportunistic, avirulent plant symbionts which induce resistance in crops and thus help in plant disease protection as well as enhancement in yield (Harman *et al.*, 2004).

Success stories

The number of commercial biocontrol formulations is only around thirty (Table 1). The preliminary trials for diseases management using biocontrol formulations involve the mass production of biocontrol fungi on different substrates. Our group has developed a process for successful mass production of biocontrol fungi on agri-waste and filed a US patent for the process (Singh *et al.*, 1999). Our group has also filed a US patent on a novel strain of *Trichoderma harzianum* having growth promontory effect, potential to control a large number of soil-borne fungal pathogens and also nematode inhibitory effect (Kalra *et al.*, 2000). We have also successfully controlled collar rot disease of mints using *Trichoderma and Gliocladium* sp. (Singh and Singh, 2004). Different scientists have tried a number of substrates (approximately 25) for producing bioagents *en* masse

There has been a number of plant diseases controlled by using biocontrol formulations developed on these substrates e.g. wilt of chickpea, collar rot of *Mentha* spp., corm rot of gladiolus etc. in addition to promotion of growth in some ornamental crops, legumes and cucurbitaceous crops (Singh *et al.*, 2004)

A number of bioagents have been registered under Insecticides Act 1968. A list is given in Table 1.

Table 1. List of manufacturers registered for production and sale of bio-pesticides

Products based on	Name of the Manufacturer	Category	9(3) regular valid upto	9(3B) provisional Section
Trichoderma viride 1% WP (min. cfu 2 x 10^6/gm)	K.N. Bioscience Ltd., Hyderabad	Fungicide	—	February 2006
-do-	Pragathi Bio-fertilizer, Nellore (AP)	Fungicide	—	February 2006
-do-	M/s Green Tech. Agro Services Pvt. Ltd., Coimbatore	Fungicide	—	February 2006
-do-	M/s Vishwamithra Bio Agro Pvt. Ltd., Guntur	Fungicide	—	February 2006
-do-	M/s Sri Biotech (formerly Vermigreen Bio-fertilizer) , Hyderabad	Fungicide	—	March 2006
-do-	International Panacea Ltd., N. Delhi	Fungicide	—	September2004
-do-	Bio tech International Ltd., Delhi	Fungicide	—	December 2005
-do-	Agriland Biotech Ltd., Baroda	Fungicide	—	December 2005
-do-	M/s PJ Margo, Bangalore	Fungicide	—	April 2004
-do-	Margo Bio Controls Pvt. Ltd., Bangalore	Fungicide	Regular 9(3)	-
-do-	Multiplex Agri Care Pvt. Ltd., Bangalore	Fungicide	—	21.10.2005
Trichoderma viride 1.15% WP (min. cfu 2 x 10^6/gm)	Romvijay Bio tech Pvt. Ltd., Pondicherry	Fungicide	Regular 9(3)	
-do-	Sun Agro Industries, Delhi	Fungicide	—	April, 2004
-do-	T. Stanes and Co. Ltd., Coimbatore	Fungicide	—	April, 2004
Trichoderma viride 6% WP (min. cfu 2 x 10^8/gm)	Indore Bio tech Inputs and Research P. Ltd., Indore	Fungicide	—	21.2.2004

Table *Contd...*

Products based on	Name of the Manufacturer	Category	Section	
			9(3) regular valid upto	9(3B) provisional
Trichoderma harzianum 1% WP (min. 2 x 10^6/gm min)	M/s Green Tech. Agro Services Pvt. Ltd., Coimbatore Services Pvt. Ltd., Combatore	Fungicide	—	February 2006
-do-	M/s Vishwamithra Bio Agro Pvt. Ltd., Guntur.	Fungicide	—	February 2006
Trichoderma harzianum 1% WP (min. 2 x 10^6/gm min)	M/s Vermigreen Bio-fertilizer, Hyderabad	Fungicide	—	March 2006
-do-	Bio Pest Management Bangalore	Fungicide	--- —	March 2006
Trichoderma harzianum 0.5% WP (min. cfu 1x 10^9/gm)	Pest Control India Ltd, Mumbai	Fungicide	--- —	25.4.2004
Pseudomonas fluorescence 0.5% WP (min. cfu 2 x 10^8/gm)	K.N. Bioscience Ltd., Hyderabad	Fungicide	—	February 2006
-do-	Pragathi Bio-fertilizer, Nellore (AP)	Fungicide	—	February 2006
-do-	M/s Vishwamithra Bio Agro Pvt. Ltd	Fungicide	—	February 2006
-do-	M/s Vermigreen Bio-fertilizer, Hyderabad	Fungicide	—	March 2006
Pseudomonas fluorescence 1.25% WP (Min. cfu 2.5 x 10^8/g)	Sun Agro Industries India Porur, Chennai-16	Fungicide	—	21.8.2004
NPV *Helicoverba armigera* 2 % AS (min. POB 1x0^9/ml)	Bio Pest Management Bangalore	Insecticide	—	March 2006
-do-	Indore Bio tech Inputs and Research Pvt. Ltd., Indore	Insecticide	—	March 2006
-do-	Ganesh Bio Control Systems, Gondal	Insecticide	—	March 2006
NPV *Helicoverpa armigera* (min. cfu 1x 10^9/gm)	Bio Tech Indus. Ltd., New Delhi.	Insecticide	—	April, 2004
NPV *Helicoperpa armigera* 0.43% AS (min.cfu1x10^9/gm)	Pest Control India, Mumbai	Insecticide	—	25.4.2004

Table Contd...

Products based on	Name of the Manufacturer	Category	Section	
			9(3) regular valid upto	9(3B) provisional
NPV *Spodoptora litura* 0.5% AS (min. cfu 1x 10^9/gm) -do-	Pest Control India, Mumbai M/s International Pancera Biotech, New Delhi	Insecticide Insecticide	— —	25.4.2004 31.10.2004
Beauveria bassiana 1% WP (min. cfu 1x 10^8/gm)	T. Stanes and Co. Ltd., Coimbatore	Insecticide	—	8.5.2004
Bacillus thuringiensis var. Kurstaki 5% WP	Wockhardt Life Science Ltd., Mumbai	Insecticide	Regular 9(3)	—
Bacillus thuringiensis var. Kurstaki 7.5% WP	Ajay Bio Tech Ltd., Pune	Insecticide	—	13.1.2005
Bacillus thuringiensis var. Kurstaki 5.8% WP	Bio-tech International Ltd., New Delhi.	Insecticide	Regular 9(3)	—
Bacillus thuringiensis var. Israelensis WP	M/s Aventis Crop Science Ltd.	Insecticide	Regular 9(3)	—
Bacillus thuringiensis var. Kurstaki tech. & 2.5% AS	Fertilizer, Chennai	Insecticide	Regular 9(3)	—
Bacillus thuringiensis var. Israelensis 8% WP /5% WP	M/s Tuticorin Alkali Chemicals & Fertilizer, Chennai	Insecticide	Regular 9(3)	—

Registration requirements in India

Several biocontrol agents have come under Insecticides Act, therefore it is imperative that these may be registered before they are produced on commercial scale. It is important that better quality biofungicide product be registered according to the norms set by Central Insecticide Board. To produce biofungicide in laboratory is one thing and its large scale production after registration from CIB is another thing. It requires certain specified parameters. Following data is needed before a biofungicide product based on *Trichoderma* be registered under 9(3B) and 9(3)

A. Chemistry

    1. Analytical test report

    2. Shelf life

    3. Container Content compatibility study

B. Bioefficacy

    1. Laboratory Bioassay

    2. Field Trials (two location data)

    3. Data on non-target organisms

C. Toxicological studies

    (a) (For formulated products to be directly manufactured)

        1. Single dose oral- Rat (21 days) (Toxicity/Infectivity/Pathogenicity)

        2. Single dose oral- Mouse (21 days) (Toxicity/Infectivity/Pathogenicity)

        3. Single dose Pulmonary- Rat (14 days) (Toxicity/Infectivity/Pathogenicity)

        4. Single dose Dermal- Rabbit (21 days) (Toxicity/Infectivity/Pathogenicity)

        5. Single dose Dermal- Intraperitoneal (21 days) (Toxicity/Infectivity/Pathogenicity)

        6. Primary skin irritation

        7. Eye irritation

    (b) Human Safety Records

D. Environmental Toxicological Studies (For formulation only)

Non Target Vertebrates

    1. Toxicity to Chicken

    2. Toxicity to Pigeon

    3. Toxicity to Freshwater Fish

Future Prospects

There is a growing demand for sound, biologically-based pest management practices. Recent surveys of both conventional and organic growers indicate an interest in using biocontrol products, suggesting that the market potential of biocontrol products will increase in coming years. Applications of diverse biological control strategies have been successful in the greenhouse industry (where environmental design and control is greatest) and continue to increase. Funding for basic and applied research continues to be provided by different funding agencies in the country. Such investment will ensure that innovations in biological control research continue.

An upswing in commercial interests has also developed in the past few years, and prospects for increased growth are positive. Clearly, the future success of the biological control industry will depend on innovative business management, product marketing, extension education, and research. Increased demand for organic produce and participation in home gardening activities by pesticide-wary urban populations all over the world has enlarged the market for biocontrol products.

The field of plant pathology will contribute substantially to making the 21st century the age of biotechnology by the development of innovative biocontrol strategies. The development of such strategies will benefit a developing country like India which is rich in natural resources being made untenable by excessive and injudicious use of chemicals in agriculture.

Acknowledgement

The authors are grateful to Dr. P. Pushpangadan, Director, National Botanical Research Institute for his valuable encouragement and providing necessary facilities.

References

Ahmad, J.S. and **Baker, R.** 1987. Rhizosphere competence of *Trichoderma harzianum. Phytopathology.* 77: 182-189.

Chet, I., Viterbi, A., Shoresh, M. and **Harel, M.** 2004. Enhancement of plant disease resistance by th biocontrol agent *T. asperellum.*

Harman G.E. 2000. Myths and Dogmas of Biological Control: Change in Perceptions desired from Research on *Trichoderma harzianum* T-22. *Plant Disease.* 84 (4): 377-393.

Harman, G.E., Howell, Viterbo, A., Chet, I. and **Lorito, M.** 2004. *Trichoderma* species- opportunistic, avirulent, plant symbionts. *Nature Rveiews- Micorbiology.* 2: 43-56.

Kalra, A., Singh, H.B., Pandey, R., Patra, N. K., Kumar, S., Gupta, M. L., Dhawan, O.P. and *Kumar, S.* 2000. A process for the isolation of a novel strain of *Trichoderma harzianum* useful as nematode inhibitor, fungicide and growth promoter. Indian Patent Filed NF 101/2000.

Mathre, D. E., Cook, R. J. and *Callan, N. W.* 1999. From discovery to use: Traversing the world for commercializing bocontrol agents for plant disease Control. *Plant Disease.* 83 (11): 972-983.

Singh, A. and *Singh, H.B.* 2004. Control of collar rot in mint (*Mentha* spp.) caused by *Sclerotium rolfsii* using biological means. *Current Science.* 87 (3): 361-366.

Singh, H.B., Singh, A., Singh, S.P. and *Nautiyal, C.S.* 2004. Commercialization of biocontrol agents: the necessity and its impact on agriculture. In: Eco-agriculture with Bioaugmentation: An emerging concept [Eds. S.P. Singh and H.B. Singh]. Rohitashwa Printers, Lucknow, India.1-20.

Singh, H. B. and *Johri, J. K.* 2000. Gyamin Kshetra mein paudh rog niyantrana hetu jaivik niyantrana samagriva takniki nirman. National Symposium on Gramin Prodyodiki, Aug. 23-24, 2000. RRL Trinuvanantapuram, P. 16.

Singh, H. B, Kalra, A., Patra, N. K., Kumar, S. Pandey, R., Khanuja, S.P.S. and *Shashney, A. K.* 1999. A process for the preparation of novel growth media from distillation and other plant wastes for mass multiplication of biocontrol fungi. Indian Patent No. 235/D/99 US No. 09/281784 Dt. 31.3.99.

Singh, H.B., Singh, Sarita, Singh, A. and *Nautiyal, C.S.* 2004. Mass production, formulation and delivery systems of fungal and bacterial antagonistic organisms in India. In: Eco-agriculture with Bioaugmentation: An emerging concept [Eds. S.P. Singh and H.B. Singh]. Rohitashwa Printers, Lucknow, India. 53-69.

Weller, D. M. 1988. Biological control of soil borne plant pathogens in the rhizosphere with bacteria. *Ann. Rev. Phytopathol.* 26: 379-407.

34 | Evaluation of Mosquito Larvicidal Toxins in the Extra Cellular Metabolites of two Fungal Genera *Beauveria* and *Trichoderma*

SARDUL SINGH SANDHU* and PALLAPOTU VIKRANT

Fungal Biotechnology and Invertebrate Pathology Laboratory,
Dept. of Biological Sciences, R.D. University, Jabalpur, M.P., India-482001
*e-mail: sardulsinghs@yahoo.co.uk

ABSTRACT : The germ plasm of fungi is a rich source of biologically active novel compounds. Fungal secondary metabolites may act as potential agents for control of mosquitoes. Therefore, the secondary metabolites of *Beauveria bassiana*, *Beauveria nivea*, *Trichoderma harzianum* and *Trichoderma viride* were bio-assayed. Except *Trichoderma viride*, the secondary metabolites of all the other three fungal strains were found to be toxic to the III instar larvae of *Culex quinquefasciatus*. The LC_{50} (concentration required to kill 50% of the mosquito larvae) values were evaluated.

KEYWORDS : *Beauveria bassiana*, *Beauveria nivea*, *Trichoderma harzianum*, *Trichoderma*, secondary metabolites, toxic, *Culex quinquefasciatus*, LC_{50}

Mosquitoes are vectors of many human and animal diseases. *Culex* is described as Zoophagic as it takes its meal from animals as well as humans and can also be described as Ornithrophagic because it frequently feed on birds. During this process they act as vectors of human and animal diseases. Insecticides have been the only control measure for this medically important vector. Development of resistance to chemical insecticides and their deleterious effects on the ecology has led to a concept of replacing or supplementing them with biological control agents. This developed due to recent concern over the continued use of insecticides, insect resistance and particularly, their environmental impact. The naturally occurring entomopathogens are important regulatory factors in insect populations. The comparison of entomopathogens with conventional chemical pesticides is solely from the perspective of their efficacy and their cost.

The use of microbial control agents render safety for humans and other non-target organisms, reduction of pesticide residues in food, preservation of other natural enemies and increased bio-diversity in managed ecosystems (Lacey *et al.*, 2001). Many scientists (Balaraman *et al.*, 1979; Daoust and Roberts, 1982; Daoust and Roberts, 1983; Sandhu *et al.*, 1993 and Sandhu and Sharma, 1994) have earlier studied the use of fungal isolates for the control of mosquito larvae.

Therefore, in our present study we have evaluated the efficacy of extra-cellular metabolites of *Beauveria bassiana*, *Beauveria nivea*, *Trichoderma harzianum* and *Trichoderma viride* against the III instar larvae of *Culex quinquefasciatus* for their control.

Materials and Methods

Mosquito colonies and Fungal strains

The laboratory reared mosquito larval colonies of *C. quinquefasciatus* that were fed with dog biscuit and yeast powder (1:1) were used. The fungal strains *B. bassiana*, *B. nivea*, *T. harzianum* and *T. viride* that were maintained on sabouraud-dextrose agar (SDA) (Peptone 10 G, Dextrose 40 G, Agar 20 G and Distilled water 1000 mL) were used in the present study.

Extraction of extra-cellular metabolites

500 ml Erlenmeyer flasks containing 250 ml Richard's broth (KNO_3 10 gms; KH_2PO_4 5 gms; $MgSO_4.7H_2O$ 2.5 gms; $FeCl_3$ in trace; Sucrose 35 gms and Distilled water 1000 ml) were seeded with 5 mm discs that were separated from 10 days old actively growing culture on SDA medium at 28 ± 2^0C. Inoculated flasks were incubated at 28 ± 2^0C for 21 days.

Under aseptic conditions the metabolized growth medium was filtered through a pre-weighed Whatmann filter paper-I and was centrifuged at 4, 000 g for 15-20 min. The pellet was thrown and the supernatant was again filtered *in vacuo* by micro-filtration using sterile micro-filters, 0.45 mm, Minisart (Sartorius, Göttingen, Germany) and CFCF was finally obtained (Templeton and Walker, 1978).

Bioassay

Hundred III instar larvae were placed in 500 ml beakers containing sterilized tap water. The CFCF obtained from the above step was added in concentration gradients from 100 ml/ml to 500 ml/ml concentrations. Culture medium without the fungal metabolites served as control and sterilized tap water served as control over control. The experiment was replicated thrice. The beakers were then held at 28 ± 2^0C. All the bioassay cups were covered with sterilized muslin cloth and were maintained at 28 ± 2^0C. Observations on larval mortality were made after every 24 hours.

Statistical Analysis

The LC_{50} value was deduced using the probit regression analysis (Finney, 1971).

Results and Discussion

In the present study we have evaluated the mosquito larvicidal potential of the extra-cellular metabolites of four fungal strains *B. bassiana*, *B. nivea*, *T. harzianum* and *T. viride*. From the results presented (Table-1) it was evident that the extra-cellular metabolites of *T. viride* had no larvicidal toxins. But, in comparison the extra-cellular metabolites of all the other three fungal strains were larvicidal. It was observed that the larvae in the beginning became very active exhibiting typical movements. Gradually the physiological behavioral changes in the movements decreased and later on the larvae died. In the controls no such changes were observed. These changes may be attributed to the presence of some components in the extra cellular metabolites of the fungi that might be toxic to the mosquito larvae. The LC_{50} values were deduced and were found to be 631.0 ml/ml, 512.9 ml/ml and 398.1ml/ml respectively for *B. bassiana*, *B. nivea* and *T. harzianum* (Table 1).

Table 1 Efficacy of fungal metabolites against the III instar larvae of *Culex quinquefasciatus*

S.no.	Fungal Strain	LC_{50} (µl/ml)	Fiducial Limits (µl/ml)		χ^2 (n-2) Equation	Regression
			Lower	Upper		
1.	*Beauveria bassiana*	631.0	177.4	2242.6	0.4033	0.2808x + 2.946
2.	*Beauveria nivea*	512.9	68.8	3821.2	0.7023	0.1808x + 3.155
3.	*Trichoderma harzianum*	398.1	229.1	691.9	3.4164	0.559x + 2.0425
4.	*Trichoderma viride*		No mosquito larvicidal activity			

Prior to invasion of the haemocoel of the insects by the fungal spores, they must germinate on the surface of the hard external cuticle, and then penetrate the cuticle itself. Even before the growing hyphae reach the haemocoel, the cells that mediate the immune response, the haemocytes, become activated and aggregated beneath the site of hyphal penetration (Huxham *et al.*, 1989) thereby preventing the fungal pathogenesis. In cases such as these the presence of such myco-toxins elevates the potential of the fungal strains in the control of insect pests. These toxins can be classified in two groups: (i) low-molecular-weight compounds including cyclodepsipeptides (cyclic peptides of 6 or 7 amino acids) and (ii) higher-molecular-weight protein molecules (Vey *et al.*, 1986 and Mollier *et al.*, 1994). A number of cyclodepsipeptides have been purified and extensively studied (Vey *et al.*, 1986 and Kleinhauf and von Dohren, 1986). The higher-molecular-weight protein toxins were initially described as proteases in *Metarhizium anisopliae* and *Beauveria bassiana* (Kucera, 1980) but have not been purified or further characterized. Mollier *et al.* (1994) also reported cytotoxic activity in culture filtrates from the entomopathogenic fungus *Beauveria sulfurescens*.

References

Balaraman, K., Bheema Rao, U.S. and **Rajagopalan, P.K.** 1979. Isolation of *Metarhizium anisopliae, Beauveria tenella* and *Fusarium oxysporum* (Deuteromycetes) and their pathogenecity to *Culex fatigans* and *Anopheles stephensi. Indian Journal of Medical Research* 70: 718-722.

Daoust, R.A. and **Roberts, D.W.** 1982. Virulence of natural and insect-passages strains of *Metarhizium anisopliae* to mosquito larvae. *Journal of Invertebrate Pathology* 39: 107-117.

Daoust, R.A. and **Roberts, D.W.** 1983. Studies on the prolonged storage of *Metarhizium anisopliae* conidia; effect of temperature and relative humidity on conidial viability and virulence against mosquitoes. *Journal of Invertebrate Pathology* 41: 143-150.

Finney, D.J. 1971. Probit Analysis, 3rd Ed., *Cambridge University Press, London.*

Huxham, I.M., Lackie, A.M. and **McCorkindale, N.J.** 1989. Inhibitory effects of cyclodepsipeptides, destruxins, from the fungus *Metarhizium anisopliae*, on cellular immunity in insects. *Journal of Insect Physiology* 35(2): 97-105.

Kleinhauf, H. and **von Dohren, H.** 1986. Peptide antibiotics. In: "Biotechnology" (H. Pape and H.J. Rehm, Eds.), pp. 283-306. V.C.H. Verlagsgesellschaft, Federal republic of Germany.

Kucera, M. 1980. Proteases from the fungus *Meatrhizium anisopliae* toxic for *Galleria mellonella* larvae. *Journal of Invertebrate Pathology* 35: 304-310.

Lacey, L.A., Frutos, R., Kaya, H.K. and **Vails, P.** 2001. Insect pathogens as biological control agents: Do they have a future? *Biological Control* 21: 230-248.

Mollier, P., Lagnel, J., Quiot, J.M., Aioun, A. and **Riba, G.** 1994. Cytotoxic activity in culture filtrates from the entomopathogenic fungus *Beauveria sulfurescens. Journal of Invertebrate Pathology* 64: 208-213.

Sandhu, S.S. and **Sharma, M.** 1994. Larvicidal activity of fungal isolates *Beauveria bassiana, Metarhizium anisopliae* and *aspergillus flavus* against *Culex pipiens. Biological Control of Insect Pests* 1: 145-150.

Sandhu, S.S., Rajak, R.C. and **Sharma, M.** 1993. Bioactivity of *Beauveria bassiana* and *Metarhizium anisopliae* as pathogens of *Culex tritaeniorhynchus* and *Aedes aegypti*: effect of instar, dosages and time. *Indian Journal of Microbiology* 33(3): 191-194.

Templeton, G.E. and **Walker, H.L.** 1978. *In vitro* production of phytotoxic metabolite by *Colletotrichum gloeosporioides* f sp. *aeschynomene. Plant Science Letters* 13: 91-96.

Vey, A., Quiot, J.M. and **Pais, M.** 1986. Toxemie d'orgigine fongique chez les Invertebres et ses consequences cytotoxiques; etude sur l'infection a' *Metarhizium anisopliae* (Hyphomycete, Moniliales) chez les Lepidopteres et les Coleopteres. C. *Society of Biology* 180: 105-122.

35

Incidence and Biological Activity of Fumonisins and their Management

S.M. REDDY, S. GIRISHAM, B. VIJAYAPAL REDDY and K. NARASIMHA RAO
Department of Botany & Microbiology, Kakatiya University, Warangal – 506 009.

ABSTRACT : Incidence of species of *Fusarium* producing fumonisins and fumonisins contamination of foods, feeds and fodders was most common in different parts of the world. *Fusarium moniliforme, F. proliferatum, F. nygamei, F. anthophylum, F. napiforme, F. globosum, F. diamini* and *F. thapsium* are the fumonisins producing fungi. Different methods of isolation, detection and quantification of fumonsins are discussed. Biosynthetic pathway of fumonisins is outlined. Role of fumonisins in the cause of leucoencephalomalacia is also discussed. Different health problems associated with fumonisins intake are detailed. Finally strategies to be adapted in the containment of fumonisins and infestation of foods, feeds and fodders by *Fusarium* species are discussed.

KEYWORDS : *Fusarium,* fumonisins, leucoencephalomalaceia, maize, fusario toxicosis, management, natural incidence, biological activity.

Fumonisins or a novel toxic metabobites *of Fusarium verticelloides* and related species of *Fusarium* though toxicity of *F. moniliforme* to horses is reported as back as 1904 (Marasas *et al.* 1976) and other experimental animals such as monkeys, pigs, sheep and rats (Kirek *et al.* 1981) was reported, it is Wilson *et al.* (1985), who implicated naturally infested corn samples by *F. moniliforme* in the field and out break of Leucoencephalomalacia (LEM) and for hepatocarcinogenicity in rats. Marasas *et al.* (1979) and Marasas *et al.* (1984) recorded *F. moniliforme* as most prevalent fungus associated with human and animal dietary corn. Li *et al.* (1980) and Yang (1980) have also implicated *F. monilforme* infested corn and the incidence of human esophageal cancer in China. Gelderblom *et al.* (1984) could isolate and identify the cancer promoting activity of *F. moniliforme* (fusarin C), a mutagenic metabolite. Thiel *et al.* (1982) and Gelderblom *et al.* (1984) could detect moniliforme, zearalenone, deoxynivalenol and fusarinin in corn naturally contaminated with *F. moniliforme* from high esophageal cancer risk area of Tanskei, South Africa. However, Gelderblom *et al.* (1986) were of the opinion that fusarin elaborated by *F. moniliforme* is not involved in the hepatocarcinogenicity. It is, Gelderblom *et al.* (1988) who have isolated two compounds which have cancer promoting activity and identified them to be fumonisin B_1 and B_2. Bezuidenhout *et al.* (1988) have elucidated the structure of fumonisins. Presently, approximately 20 fumonisins analogs have been isolated and characterized (MacKentzie *et al.* 1998). Marasas *et al.* (1988) could demonstrate the leukoencephalomalacea in horse by intravenous injection of fumonisin B_1 isolated from *F. moniliforme* (MRC 826). Association of these toxins with porcine pulmonary edema in swine (Harrison *et al.* 1990) and cancer promoting activity in rats have been recorded (Reeder *et al.* 1992). Gelderblom *et al.* (1992) have further created special interest in the study of this group of fusarial toxins. Bullerman *et al.* (2001) and Riley *et al.* (2001) have excellently reviewed different aspects of fumonisins in particular and fusarial toxins in general. According to Abbas *et al.* (2000) aflatoxins are a chronic problem in the south eastern USA, while the fumonisins are a chronic problem worldwide and therefore, may represent a greater challenge to control. Marasas *et al.* (1984) aptly states that if the new era of research on food-borne diseases caused by mycotoxins was introduced by the aflatoxins in 1960, the twentieth century closed with worldwide attention focused on the fumonisins. The next chapter to be written on food-borne diseases caused by *Fusarium* mycotoxins may be called "Fumonisins – Food-borne fusarium toxins for the third millinium". An attempt, therefore, is made to review its worldwide incidence, biological activity of fumonisins and their management.

1. Chemical Structure

Fumonisins ($C_{34}H_{59}NO_{15}$) are diesters of propane-1,2,3-tricarboxylic acid (tricarboxylic acid) and 2-amino-12,16-dimethylpolyhydroxydicosanes, in which the C_{14} and C_{15} hydroxyl groups are esterfied with the terminal carboxyl group of tricarboxylic acid (Fig. 1). There are as many as six structurally related fumonisin analogues. Seo *et al.* (1999) reported N-acetyl derivatives of the type C fumonisins by *F. oxysporum*. Akiyama *et al.* (2000) have reported a novel FB_1 analogue. Branham and Plattner (1993) have reported fumonisin C_1 (FG) which lacks the amino and terminal methyl group (CH_3) of FB_1. Wang *et al.* (1998) reported that *F. moniliforme* which showed dimorphism produce FB_1.

Of the six known fumonisins, FA_1 and FA_2 the acetyl derivatives of FB_1 and FB_2 respectively are produced in the lowest yields in cultures of *F. moniliformi* and have the lowest toxicities. These two structural analogues and FB_4 do not occur under natural conditions.

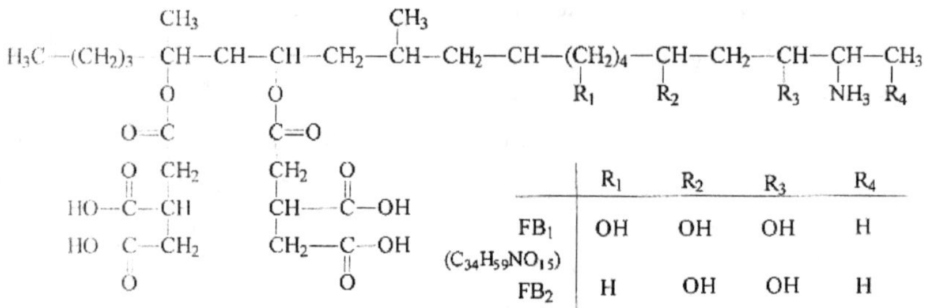

$$H_3C-(CH_2)_3-CH-CH-CH-CH_2-CH-CH_2-CH-(CH_2)_4-CH-CH_2-CH-CH-CH_3$$

with substituents at R_1, R_2, R_3, NH_3, R_4 and the methyl/ester groups as drawn.

	R_1	R_2	R_3	R_4
FB$_1$ ($C_{34}H_{59}NO_{15}$)	OH	OH	OH	H
FB$_2$ ($C_{34}H_{59}NO_{14}$)	H	OH	OH	H
FB$_3$ ($C_{34}H_{59}NO_{14}$)	OH	H	OH	H
FB$_4$ ($C_{34}H_{59}NO_{13}$)	H	H	OH	H
FA$_1$ ($C_{34}H_{61}NO_{15}$)	OH	OH	OH	CH$_2$CO
FA$_2$ ($C_{34}H_{61}NO_{14}$)	H	OH	OH	CH$_2$CO

Fig. 1 Chemical structure of different fumonisins produced by *F. moniliforme.*

2. Organisms

Fusarium moniliforme which is known to be a producer of fumonisins, was reported on corn earlier as *Oospora verticelloides* in 1881 in Italy. Subsequently Shelton (1904) incriminated this as a cause of mold-corn toxicosis of animal in USA and subsequently described as *F. moniliforme*. It is Gelderblom *et al.* (1988) who isolated fumonisins from cultures of *F. moniliforme* and conclusively proved it to be responsible for encephalomala-leionephropathy. Subsequently Ross *et al.* (1990) reported that *F. proliferatum* also produces this toxin. Thiel *et al.* (1992) and Nelson *et al.* (1992) reported that fumonisins are produced even by *F. nygamei* and *F. anthophylum* respectively. Klittich *et al.* (1997) reported that *F. napiforme* also produces fumonisins. Sydenham *et al.* (1997) added another species of *Fusarium*, *F. globosum* to fumonisins producers. Production of fumonisins by *F. polyphialium* and *F. oxysporum* (Abbas *et al.*, 1995) and *Alternaria alternata* (Chen *et al.*, 1992) require confirmation. Large number of isolates of representative species of both liseola section and other section of *Fusarium* are to be screened before reaching any decisive conclusions. Production of these toxins by *Alternaria alternata* if proved to be correct, gives ample scope of production of these toxins by other group of fungi.

3. *Biosynthesis*

Cladas *et al.* (1998) have tried to trace biosynthetic pathway of fumonisins through radiolabelled amino acids, water and molecular oxygen. They concluded that O_2 of tricarboxylic acid moieties of fumonisins were derived from water, while lipid back bone of hydroxyl fumonisins. Using ^{14}C isotope concluded that biosynthesis of fumonisins involves addition of methionine, glutamate, serine or alanine to the hydrocarbon backbone supporting the hypothesis that as opposed to being a modified lipid (Blackwell *et al.*, 1994). Fumonisins are the inhibitors of *de nov* spingolipid biosynthesis and alters the sphingolipid metabolism (Voss *et al.*, 1994).

4. *Isolation and Quantification*

The development of analytical methods for the quantification of naturally occurring levels of fumonisins was difficult because they do not possess any chromatophores and, therefore, do not absorb UV or visible light nor do they produce fluorescence. Alberts *et al.* (1990) described quantification of fumonisins by HPLC analysis of the malyl derivatives. But with this method of fumonisins detection in culture filtrate at level of 17.9 g/kg can only be detected. Shephard *et al.* (1990) developed HPLC technique which involves fluorescence detection of O-phtholdialdehyde (OPA) derivatives proved to be a suitable method and has been evaluated in international collaborative studies. Cawood *et al.* (1991) described the method of isolation and quantification of FB_3 and FB_4 and FA_1 and FA_2 in addition to known FB_1 and FB_2. Scott and Lowrence (1992) by employing 4-flouro-7-nitrobenzofurazan (4-flouro-7-nitrobenzo-2-oxo-1,3-diazole and naphthalem-2,3-dicarboxyaldehyde potassium cyanide) determined the amount of FB_1 in corn. Consequently, the methods of Sherphard *et al.* (1996) as modified by Sydenham *et al.* (1996) and Thiele *et al.* (1996) recommended for the quantitative analysis of FB_1 and FB_2 in corn based foods and feeds. Gelderblam *et al.* (1998) have made an attempt to isolate and purify fumonisins B_1 and B_2 from culture of *F. moniliforme*. Extraction of the fumonisins was achieved by methanol : Water (3:1) followed by a solvnt partitioning step using chloroform. Subsequent purification of the aqueous phase was subjected to amberlite aAD-2, silica gel, and reverse phase (C_{12}) columns yielding FB_1 and FB_2. Quantitation of fumonisins was also accomplished by liquid secondary ion mass spectrometer (Koof Macher *et al.*, 1991) post hydrolysis GC (Sydenham *et al.*, 1990), GC-MS (Plattener *et al.*, 1990), fast bombardment (Kor Macher *et al.*, 1991), HPLC (Thiel *et al.*, 1993) and TLC (Shelby *et al.*, 1994).

Stack and Eppley (1992) described a liquid chromatographic method after derivatizing with O-phtholdialdehyde and 2-mercaptoethanol. Holcomb *et al.* (1993) described method of fumonisins B_1 detection in rodent feed through acetonitrile : water (80 : 50) and clean up with tanden C_{18} sep-pak vac and strong anion-exchange (SAX) columns and eluting with acetic acid (1%) in methanol and quantified in a gradient elution HPLC using per column derivatization with FMOC and fluorescence detection. Margos and Richard (1994) employed ELISA and liquid chromatography for detection of fumonisins in milk. ELISA technique though simple and useful for routine analysis but was of low sensitivity. Several immunochemical methods for the determination of FB_1 have also been developed including ELISA with polyclonal antibodies (Azcona Olivera *et al.*, 1992b and Usleber *et al.*, 1994), monoclonal antibodies (Azcona Olivera, 1992a) and anti-idiotype antibodies (Chu *et al.*, 1995). Schneider *et al.*, (1995) have developed a competitive direct enzyme immunoassay (EIA) and an enzyme linked immunofilteration assay (ELISA) for detection of FB_1. Immunoaffinity columns which use monoclonal antibodies are commercially available and have detection limits at ppm level. Liquid chromatography of fumonisins following extraction with methanol-acetone and strong anion exchange followed by derivatization with naphthalene-2,3-dicarboxaldehyde was sensitive with detection limits of 5mg/ml. Sydenham *et al.* (1996) while comparing the sensitivity of HPLC and monoclonal antibody-based (MAB) competitive direct enzyme linked immunosorbent assay (CD ELISA) for detection of fumonisins in corn concluded that HPLC is more sensitive and reproducible than ELISA probably due to the lipid based matrix effect. Xie *et al.*, (1997) detected two naturally occurring structural isomers of partially hydrolyzed fumonisins B_1 in corn by on-line capillaryliquid chromatography and fast atom bambardment mass spectrometry. Karuna and Sashidharn (1999) have developed ion excange chromatography method for detection of fumonisins. Poling and Platner (1996) developed a rapid purification of fumonisins B_1 and B_2 by loading onto a log NH phase extraction cartridge and eluting fumonisins in acetic acid (5%) in methanol. The elutants were diluted by 1.5 times with water and loaded into a log to 18 cartridge. Some of the common methods of extraction, purification and detection of fumonisins are précised in Table 1. Isolation and characterization of fumonisins was reviewed by Merdith (2000). Kulisek and Hazebroek (2000) have recommended the use of phosphate buffered saline (PSB) as an extraction solution for fumonisins in maize in conjunction with ELISA as an alternative to methanol-water or acetonitrile-water. Ho *et al.* (2000) have developed a novel immunoassay for fumonisins using fumonisins liposome conjugates filled with flourophore (Sulfortodamine). Groland *et al.* (2001) while comparing different methods of extraction and clean up procedures for detection of fumonisins, felt that the use of acetonitrile and water (1:1) as an extraction solvent and immunoaffinity column for clean up proved to be better for recovery of fumonisins and chromatographic resolution when compared to methanol : water (3:1) and strong anion exchange (SAX) respectively.

Table 1 Analytical methods for detection of fumonisins mycotoxins

Method	Reference	Solvent System
Liquid chromatography		Methanol : Water (3:1)
		0.5% acetic acid in methanol
		Methanol, Sodium hydrogen phosphate
HPLC	Sydenham *et al.* (1992)	Acetonitrile : Water (1:1)
		Acetonitrile : Water (70:30)
		40% acetonitrile + 60% 0.05 M KH_2PO_4
HPLC	Sydennam *et al.* (1992) O-Phthaldialdehyde	Methanol : Water (3:1)
HPLC	Shephard *et al.* (1990)	Methanol : Water (3:1)
		Acetonitrile : Water : Aceticacid (47:53:0.5)
HPLC	Holcomb *et al.* (1993)	Precolumn derivatization with FMOC and flouroscence detection
HPLC	Thiel *et al.* (1993)	—
TLC and competitive immunoassay	Shelby *et al.* (1994)	—
TLC	Sydenham (1989)	Methanol : Water (3:1)
Cawood (1991)	Spray : 0.5% p-anisoldehyde	
Enzyme immunoassay	Usleber *et al.* (1994)	—
LC-ELISA	Stack and Epply (1992)	
OPLC		
Reverse phase chromatography	Poling and Platner (1996) Cawood *et al.* (1991)	Solid phase extraction column —
Ion exchange chromatoraphy	Karuna and Sashidharan (1999)	Purified Dowex - 1 Methanol : Water (3:1)
TLC-LSD		
LSIMS (Liquid secondary ion mass spectrometry)	Koof Macher *et al.* (1991)	
Post hydrolysis GC	Sydenham *et al.* (1990)	—
GC-MS	Plattner *et al.* (1990)	—

Thiel *et al.* (1996) reported that limits of detection to be 100 ppm for TLC 50 ppb for LC-MS and 200 ppb for ELISA. Xie *et al.* (1997) could detect isomers of partially hydrolyzed fumonisin B_1 by on line capillary liquid chromatography fast atom bombardment mass spectroscopy. Wilkes *et al.* (2001) have developed a elctrospray mass spectrometry for detection of fumonisins. Maragos *et. al.,* (2001) have developed fluorescence polarization method for detection of fumonisins in maize.

5. *Natural Incidence*

Ever since characterization of fumonisins of *F. moniliforme*, its natural incidence has gained paramount importance in view of its carcinogenicity and causal agent of leukoencephalomalaesia. *F. moniliforme* which is ubiquitous and reported to contaminate wide variety of agricultural commodities and elaborate fumonisins from different parts of the world. Incidence and toxigenic potentialities of species of *Fusarium* are précised in Table 2. Out of different commodies, corn is most susceptible and supports the elaboration of fumonisins B_1 and B_2. It is also reported that fumonisins contaminate animal feeds and fodders. Groves *et. al.,* (1999) recorded fumonisins B_1, B_2, B_3 in 19, 25 and 6% of the corn samples respectively collected from Shandong. Martins *et al.* (2001) reported fumonisins B_1 and B_2 in black tea and medicinal plants from Portugal. Oho *et al.* (2002) feel that the seed moisture of freshly harvested maize at crop harvest stage is critical for infestation and elaboration of fumonisins by *F. verticelloides*. Abbas *et al.* (1999) reported fumonisins in rice sheath rot caused by *F. proliferatum* both *in vitro* and *in vivo*.

Chulze *et al.* (1996) recorded *F. moniliforme* and *F. proliferatum* at different stages of ear development of corn. Fumonisins production was maximum during late stages of development of ear of corn. Corn ear damage caused by European corn borer increased the incidence of *F. moniliforme* and fumonisins production (Low *et al.* 1991). Shelby *et al.* (1994) recorded more amount of fumonisins in corn hybrids grown under stress conditions than grown under optimal conditions. Rice *et al.* (1994) recorded fumonisins in feccal samples of cattle, sheep and rats fed with diets mixed with *F. moniliforme*. Castella *et al.* (1999) could record FB_1 and FB_2 in maize from Spain. Yamashita *et al.* (1995) have reported the incidence of fumonisins in corn and different cereals from Philippines and Thialand respectively. Hennigen *et al.* (2000) recorded high level contamination of corn flour with fumonisins but could not find any difference in fumonisins level between brand and unbranded corn flour samples.

Table 2 Natural incidence of species of *Fusarium* and fumonisins in different agricultural commodities

Food	Species	Mycotoxins	Country	Reference
Corn	*F. moniliforme*	FB1	Argentina	Solovey *et al.,* (1999)
	F. proliferatum			Magnoli *et al.,* (1999)
	F. nygamai			Torres *et al.,* (2001)
				Pacin *et al.,* (2002)
	F. moniliforme	FB1	Brazil	Hirooka *et al.,* (1996)
	F. moniliforme	FB1	China	Groves *et al.,* (1999)
		FB2		
		FB3		
	F. moniliforme	FB1 & FB2	Croatia	Segvic and Pepeljnjak (2003)
	F. moniliforme	FB1	Egypt	Allah (1997)
		FB2		
	Fusarium spp.	Fumonisins	Europe	BrunoDoko *et al.,* (1995)
	F. verticilloides	FB1	France	Melcion *et al.,* (1998)
	F. proliferatum			
	Fusarium spp.	FB1	India	Bhatt *et al.,* (1997).
	F. moniliforme	Fumonisins	Iran	Shephard *et. al.,* (2002)
	F. proliferatum	FB1 & FB2	Italy	Pascale *et al.,* (1999)
	F. moniliforme	Fumonisins		
	F. moniliforme	Fumonisins	Korea	Mazzami *et al.,* (1999)
	F. moniliforme	FB1, FB2,	South	Sydenham *et al.,* (1992)
		FB3	Africa	Odhav *et al.,* (2001)
	Alternaria alternata	Fumonisins	Southern USA	Abbas *et al.,* (2000)
	F. verticilloides	FB1	Spain	Castella *et al.,* (1999)
	F. moniliforme	Fumonisins	Taiwan	Cheng *et al.,* (2000)
	F. moniliforme	Fumonisins	Venezula	Mazzami *et al.,* (2000)
	F. proliferatum	Fumonisins	USA	Desjardians *et al.,* (2000)
	F. graminarium	FB1		
	F. moniliforme	FB1		
	F. anthophilum		USA	Abbas *et al.,* (2000)
	F. dlamini			
	F. napiforme			
	F. nygamai			
	F. moniliforme	FB1 & FB2	Uruguayan	Maya & Silva (1997)
	F. moniliforme	FB1	Western Kenya	Kedera *et al.,* (1999)
	Fusarium spp.	FB1, FB2 & FB2	W.Africa Zambia	Albert *et al.,* (1990)

Food	Species	Mycotoxins	Country	Reference
Cereal grains	*Fusarium* spp.	Fumonisins	Europe	Visconti & Doko (1994)
	Fusarium spp.	Fumonisins	Italy	
	Fusarium spp.	Fumonisins	Canada	Campbell *et al.*, 2002
Rice	*Fusarium* spp.	Fumonisins	USA	Abbas *et al.*, (2000)
	Fusarium spp.			Desjardilus *et al.*, (2000)
	F. moniliforme			
	F. proliferatum			
Wheat	*Fusarium* spp.	Fumonisins	USA	Desjardius *et al.*, (2000)
				Castella *et al.*, (1999)
Poultry feed	*F. nygamai*	FB1, FB2,	Argentina	Magnoli *et al.*, (1999)
	F. proliferatum	FB3		
	F. moniliforme			
Sorghum millet	*F. moniliforme*	Fumonisins	Africa	Nelson *et al.*, (1992)
	F. subglutinans		Asia	Viconti and Doko (1994)
	F. napiforme			
	F. nygamai			
Agricultural	*F. moniliforme*			
commodities	*F. proliferatum*	FB2	Japan	Akiyama *et al. m* (2000)
Swine corn	*F. moniliforme*	Fumonisins	Spain	Castella *et al.*, (1999)
Poultry feed	*F. proliferatum*			
Soybean feed				
Horse feed peas				
Barley wheat				
Black Tea	*F. moniliforme*	FB1 & FB2	Portugal	Martins *et al.*, (2001)
Medicinal				
plants				

6. Biological Activity

The involvement of food-borne *Fusarium* mycotoxins in hemorrhage feed-refusal, emetic, estrogenic, neurotoxic and pulmonary diseases in animals and the association of food-borne *Fusarium* mycotoxin with ATA, SG and EC in humans have been reviewed by several workers. It is much more difficult to establish relation between human diseases and food-borne *Fusarium* toxins.

Outbreak of encephalomalacia due to the consumption of a mouldy corn has been known for several decades. Neurotoxic and hepatotoxic symptoms may occur singly or in combination in horses. Feed refusal, lameness, atoxia, oral and facial paralysis and recumbency begins with in a few days after initial consumption of moldy corn and rapidly followed by seizers and morbidity. Histopathology of liver shows centrilobular necrosis, fibrosis and bile duct proliferation with increased mitosis, acute inflammation and fatty degeneration in fumonisins fed horses. Marasas *et al.,*(1988) and Kellerman *et al.,* (1990) have experimentally induced leukoencaphalomalacia in horses both by intravenous injection and oral feeding of fumonisins. Haschek *et al.,* (1992) experimentally induced pulmonary edema by administering higher doses of fumonisns, while at lower doses slowly progressive hepatic disease was most predominent. Gelderblom *et al.,* (1992) induced liver lesions in rats with fumosins B_1 and B_2. Wang *et al.,* (1991) and Wang *et al.,* (1992) discussed the mechanism of toxicity of fumonsins. According to them fumonisins are the inhibitors of *de nova* sphingolipid biosynthesis and alters the lipid metabolism as depicted in Fig. 2.

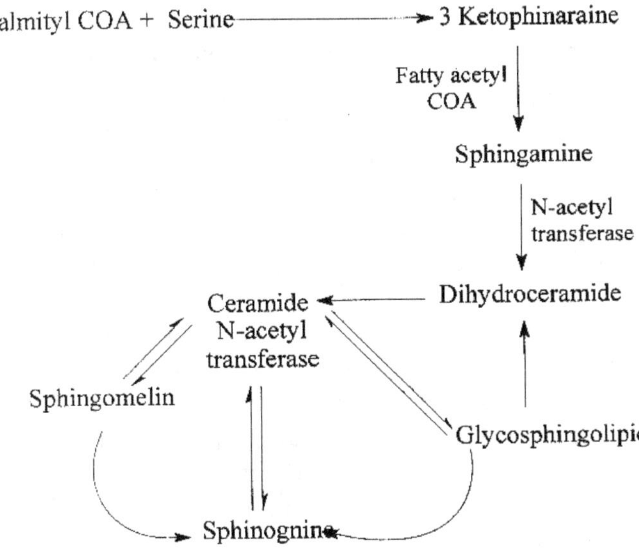

Figure 2   Mechanism of inhibition of sphingolipid synthesis by fumonisin B_1

Wang *et al.,* (1991) feels that fumonisins inhibit the conversion of sphingamine to N-acyl-(c)-sphinganines. This is further confirmed by the observations that fumonisins inhibited the activity of sphingosine N-acyltransferase ceramide sulphase in rat liver microsomes and reduced the conversion of sphingosine to ceramide by intact hepatocytes. The increase in intracellular spingamine coupled with the concomitant decrease in ceramides may induce apotopsis.

Rats fed with *F. moniliforme* infested feed when examined revealed development of hepatic nodules, cholantiofibrosis or cholangiocarcinomas with in 6 months. They also showed development of liver tumors and primary hepatocellular carcinoma and chronic interstitial nephritis in kidneys. No lesions were observed in the esophagus, heart or fore-stomach of fumonisins treated rats. Fumonisins exert strong effect on point mutation events with primary action on tumor promotion and selection of initiated cells on human cells caused antiproliferative effect after increased apopotic cell death as opposed to decreased cell proliferation (Wolf, 1994). Li *et al.*, (1989) have recorded a close correlation between the consumption of fermented corn and unfermented corn pancakes and the incidence of elevated stomach cancer mortality in rurallinque country in shandong province in China. Buch *et al.*, (1996) could induce leukoencephalomalacia and haemorage in the brain of rabbits with fumonisins. Bane *et al.*, (1992) have tried to correlate fumonisins contamination of feed and mysterious swine disease. Vass *et al.*, (1990) could find a correlation between hepatotoxicity and FB_1 and FB_2 content of aqueous extract and methanolic extracts of *F. moniliforme*. Weiblug (1993) have studied the effect of FB_1 and FB_2 on broiler chick. Colvin and Harrison (1992), Osweiler *et al.*, (1992) and Colvin *et al.*, (1993) have recorded porcicine pulmonary edema (PPE) and hydrothoracea in swine. Some of the mycotoxicoses attributed to the involvement of fumonisins and *Fusarium infestation* are precised in Table 3. Rosiler *et al.*, (1998) reported the out break of leukoencephalomalacia in Mexico due to consumption of fumonisins contaminated corn. On the other hand, You *et al.*, (1988) feel that stomach cancer increased by 30% among those consumed corn pancakes atleast daily. Groves *et al.*, (1999) feel that fumonisins contamination does not increase the risk of gastric cancer. Chatterjee and Mukherjee (1994) recorded significant reduction in the viability and phagocytic potential of macrophages for chicken peritonial exudate cells. Javed *et al.*, (1992) recorded lesions in the liver, kidneys, heart and lungs and subsequent death of broiler chicks. Gelderblom *et al.*, (1996) feel that FB_1 causes lipid peroxidation. Gelderblom *et al.*, (1997) feel that fumonisins disturb membrane components, the fatty acid storage pool and the accumulation of long-chain fatty acids within the cell which could eventually lead to the disintegration of membrane structure and cell death apart from inhibition of sphingolipid biosynthesis. FB_1 has also been found to effect the synthesis of cellular lipids by altering the incorporation of palmitic acid (Gelderbolm *et al.*, 1996). Increased levels of polyunsaturated fatty acids are associated with lipid peroxidation in FB_1 exposed primary hepatocytes suggests their indirect role in carcinogenicity. Gelderblom *et al.*, (1994) feel that prolonged exposure to fumonisins B_1 is essential for initiation of cancer rather than short exposure.

Marassas (1995) felt that the contamination of corn with fuminosins presents a real threat to human and animal health in Africa. Henry and Bosch (2001) have elegantly discussed the relationship between fumonisins incidence and mycotoxicosis epidemiology.

Table 3 Mycotoxicosis of animals and man due to fumonisins

Animal	Country	Symptoms	Mycotoxin	Reference
Rat (world wide)	South Africa	Equine leukoencephalomalacia (LEM) Porcine pulmonary edima (PPE)	FB1, FB2 and FB3	Sydenham *et. al.,*(1992)
Rat	—	Foetotoxic	FB1	Lebepe-Mazur *et al.,* (1995)
Rat	USA	Hepatocarcinogenio esophagel cancer hepatocellular carcinoma	FB1 moniliformin	Gelderblom *et al.,* (1991)
Rat	USA	Liver tumours	FB1	Gelderblom *et al.,* (1991)
Rat		Skin tumor (carcinogen)		
Rat		Kidney or liver apoptosis		
Rat	—	Foetotoxic	FB1	Lebepe-Mazur *et al.,* (1995)
Rat	USA	Hepatotoxicity, Nephrotoxicity	FB2, FB2	Cawood *et al.,* (1994)
Rat	South Africa	Carcinogenicity, Neurotoxicity	FB1, FB1	Bezuidenhout *et al.,* (1988) Jockson *et al.,* (1996)
Rat	Hungary	Reproductive females	FB1	Rafai (1991)
Horse		Leukoencephalomalacia	FB1	Marasas *et al.,* (1988)
Horse	Brazil	Leukoencephalomalacia	Fumonisins	Mallmann *et al.,* (1999)
Scoine Horse	—	Pulmonary edema encephalomalacia	Fumonisins	Bezuidenhout *et al.,* (1988)
Pig	—	Hepatic lesion syndrome & immunodepression	Fumonisins	Ballarini (2002)
Pig	—	Cardiovascular changes Cardiovascular changes, heptotoxic and nephrotoxic	FB_1 FB_1	Mattur *et al.,* (2001)
Horses and donkeys	Mexico, USA, Egypt, South Africa	Brain tissue necrosis Leukoencephalomalacia (Liquefactive necrosis of the white matter of brain)	Fumonisins	Gelderblom (1988)
Pig	Spain	Pulmonary edema syndrome lung disease	Fumonisins	Harrison *et al.,* (1990) Marasas *et al.,* (1988)
	Italy	DNA damage in fibroblasts (heart condition)	FB_1	Galvano *et al.,* (2002)
Humans	South Africa China	Cancer of Oesophagus	FB1, FB2	Sydenham *et al.,* (1990) Yang (1980)
Broiler chicks	—	Lesions in the liver, kidney heart and lungs	FB1	Javed *et al.,* (1993)
Soybean and Tomato	—	Phytotoxic	FB1, FB2	Abbas *et al.,* (1992)
Corn and Tomato seedling	—	Phytotoxic (reduction of root and shoot length)	FB1	Lamprecht *et. al.,* (1994)

Fumonisins are non-mutagenic according to the *Salmonella* test and non-genotoxic according to the DNA-repair assay with *Escherichia coli* (Gelderblom *et al.*, 1996) and do not induce unscheduled DNA synthesis in primary hepatocytes (Norred *et al.*, 1990). FB_1 mimics genotoxic carcinogens in both cancer initiation and promotion (Gelderblom *et al.*, 1994a). FB_1 induces Y-glutamyl transpeptidase (GGT) and the placental form of glutathione-S-transferase (STP).

It is established that FB_1 induces completely different toxic effects in different animal species. Mallir *et al.*, (2001) reported that fumonisins B_1 to be heptotoxic and nephrotoxic in milk fed calves. Galvano *et al.*, (2002) reported that fumonisin B_1 caused DNA damage in human fibroblasts. Moodley *et al.*, (2001) reported fumonisins B_1 as an etiologic agent of preeclampsia in pregnant women of S. Africa. Hepatoxicity and hepatocarcinogenicity in rats suppress both growth and foetal bone development (Lebepe-Mazur *et al.*, 1995). Prolonged exposure to FB_1 affected kidney in rats (Gelderblom *et al.*, 1992). Lemmer *et al.*, (1999) have investigated histopathology and gene expression changes in rat liver fed by B_1. Fumonisins induced apoptosis in human erythroleukamia cell lines (Minervini *et al.*, 2004). Fumonisins caused early apoptosis in rats (Moon *et al.*, 2000). Seefelder *et al.*, (2003) reported fumonisin B_1 induced apoptosis in cultured cells of human proximal tube cells. Bohni (2000) reviewed the role of fusarial toxins in animal health. Animals can be commonly treated against the chronic affects of mycotoxins by strengthening their immune system, in conjunction with the beneficial microorganisms or enzymes. Vitamin and carotenoid supplementation, short term administration of proteins or amino acids and addition of antioxidant trace elements such as selenium also promotes immune system.

FB_1 is an inhibitor of sphinganine N-acyltransferase and the increase in the sphingamine / sphingosine ratio in the urine or serum and has been proposed as biomarker to evaluate exposure to fumonisins (Qiu and Liu, 2001). Males are more sensitive than females to FB_1 in the disruption of sphingolipid metabolism (Zin and Liu, 2001). Howard *et al.*, (2001) feel that FB_1 inhibits ceramide synthesis and caused carcinogenesis in 2 year rodent. Foreman *et al.*, (2001) observed some subtle neurological signs in horses suffering from fumonisins toxicosis. Constable *et al.*, (2000) observed consistent cardiovascular changes in fumonisins treated horses with sphingosine mediated calcium-channel blackade of heart and vasculature. Delongchamp and Young (2001) feel that fumonisins increase age specific tumour incidence. Fumonisins are probably first example of apparently non-genotoxic (non-DNA reactive) agent producing tumours through a mode of action involving apotoic necrosis, atrophy and consequent regeneration (Cohen *et al.*, 2001).

Alteration in the n-6 fatty acid profile and decrease in the free cholesterol membrane associated levels which resulted in higher phosphatidyl-choline and cholesterol ratio suggesting a more rigid membrane structure FB_1 fatty acid storage pool and the accumulation of long-chain fatty acids with in the cell which could lead to the disintegration of membrane structures and eventually result in cell death (Gelderblom *et al.*, 1997).

Phytotoxicity

FB_1 and FB_2 are phytotoxic and cause damage to weed and crop cultivars of soybean and tomato. The primary centre of toxicity is plasma lemma or tonoplast. Abbas *et al.,* (1992) recorded light dependent cytoplasmic degeneration and chloroplast disruption. FB_1 is more toxic than FB_2 and FB_3 to corn or tomato seedlings and cause reduction in shoot and root length (Lamprecht *et al.,* 1994). Abbas *et al.,* (2000) reported the necrosis, growth inhibition and death of *Datura strmarium* seedlings by fumonisin B_1. Zanno and Vouro (1999) recommended fumonisins as herbicide to kill *Striga hormanthica*.

Management of Fumonisins

Galvano *et al.,* (2001) discussed elegantly the strategies to be adapted to contain fumonisins contamination of food grains. Fumonisins are generally concentrated on the outer pericarp layer of corn kernels (Syndenrham *et al.,* 1993b). Syndenham *et al.,* (1994) could reduce the fumonisins level very significatly by sorting out fine particals of less than 3 mm size. Syndenham *et al.,* (1993) reduced the fumonisins level in corn by treating with 0.1 m calcium hydroxide for 24 hrs at 25°C. Norred *et al.,* (1991) could reduce the fumonisins level in corn by ammoniation. Detoxification of fumonisins in maize by ammonia could reduce to 30% (Chourasia, 2001). Bhat (1997) recorded variation in fumonisins B_1 level in hybrids and hybroid of maize. Thampson (1996) reported the inhibition of growth of *F. moniliforme* in the presence of butylated hydroxyanisole or carvacrol (10). Juglal *et al.,*(2002) reported that clove oil (engenol) followed cinnamonthymol to be inhibitory to the production of fumonisins by *F. verticelloides.* Yamgish *et al.,* (2001) could produce mutant of *F. proliferatum* which produced little or no fuminosins and can be used as a competetor for fumonisins producing strains and thus minimise fumonisin contamination. Bacon *et al.,* (2001) suggested that the exploitation of *Trichoderma* sp. and *Bacillus subtilis* in the control of fumonisin B_1 production by *F. verticelloides.* Castello *et al.,* (2001) recorded reduction of fumonisin B_1 in extruded and baked corn based foods with sugars to the extent of 92.1%. Reynoso *et al.,* (2002) reported that antioxidants like PP and BHA were effective in arresting the growth of *F. verticilloides* and *F. proliferatum* at three a_w 0.99, 0.98 and 0.95. Trihydroxy butyrophenone and butylatedhydroxytoluene (BHT) were less effective either in single or combination. Use of combined mixture of antioxidants (BHA and PP) could be good strategy to minimize the entry of fumonisins into animal feed and human food. Charlubruns (2003) suggested crop management is the main strategy to control fumonisins by growing of adapted hybrids, proper plant nutrition, irrigation and insect control or use of transgenic hybrids help in reducing contamination of fumonsin B_1. Shim-Wan *et al.,* (1999) feel that strategies that target the regulatory elements of nitrogen metabolism may be effective of reducing the risk in fumonisins contamination in food. Beekrun *et. al.,* (2003) feel that naturally occurring phenolics of plants such as chlorophorin, iroka, maakiarin, benzoic acid, caffeic acid, ferulic acid and vanillic acid are effective inhibitors of fumonisins production by *F. verticilloides.* Ono *et al.,* (2002) feel that the decreased seed moisture of freshly harvested maize predrying of storage grains can control fungal growth and fumonisin contamination. King and Scott (1981), Shelby (1994) and Visconti (1996) have advocated cultivation of resistant varieties of maize in relation to geographical location in order to minimize fumonisin contamination.

Future Research

The incidence of fumonisins producing fungi are more wide spread under varied climatic conditions than visualized. Though large number of reports are available on the contamination of maize at different stages from production to consumption, the field incidence and meteorological and agronomic practices will favour infestation of species of *Fusarium* and fumonisins contamination. Location of fumonisins in pericap needs to be examined through intensive and extensive survey both at pre- and post-harvest stage. Inspite of accurate and sensitive method of detection of fumonisins are available, a need is felt for developing a simple and sensitive method by which fumonisins can be detected easily and accessable to common seed research laboratories. Pharmacokinetic studies need to be strengthened. Only preliminary reports are available for reducing the level of fumonisins. A simple and more effective method of management of these mycotoxins is the need of the hour.

Efforts should be focused on the molecular genetics of fumonisins producing fungi, mechanism of their action, difference in the metabolism and pharmacokinetics, immunobased and physicochemical techniques for their quantification and analysis of risk involved in the exposure of man and domestic animals to fumonisins and associated regulation for the control of fumonisins contamination need to be formulated. Similarly plant molecular biotechnology techniques to breed mycotoxin resistant cereals and nut producing cultivars should be evolved.

Acknowledgements

Thanks are due to CSIR, New Delhi (S.No. 38 (0953) 99-EMR-11) Department of Microbiology, Kakatiya University for financial assistance and facilities respectively.

References

Abbas, H.K., Cart Wright, R.D., Xie, W., Mirocha, C.J., Richard, S.L., Dvonak, T.J., Sciumbato, G.L. and **Shier, W.J.** 1999. Mycotoxin production by *Fusarium proliferatum* isolates from rice with *Fusarium* sheath rot disease, *Mycopathologia*, 147 : 97-104.

Abbas, H.K. and **Ocams, C.M.** 1995. First report of production of fumonisin B_1 by *Fusarium polyphialicum* collected from seeds of *Pinus strobes*. *Plant Dis.*, 79 : 642.

Abbas, H.K., Ocamb, C.M., Xie, W., Mirocha, C.J. and **Shier, W.T**. 1995. First report of fumonisin B_1, B_2 and B_3 production by *Fusarium oxysporum* var. *redolens*. *Plant Dis.*, 79 : 968.

Abbas, H.K., Paul, R.N., Boyette, C.D. and **Duke, S.O.** 1992. Physiolgoical and ultrastructural effects of fuminosins on Jimsonweed leaves. *Can. J. Bot.*, 70 : 18214-1833.

Abbas, H.K., Smeda, R.J., Gerwick, B.C. and **Shier, W.** 2000. Fumonisin B_1 from the fungus *Fusarium moniliforme* causes contact toxicity in plants : evidence from studies with biosyntheticaly labeled toxin. *J. Natural Toxins*, 9(1) : 85-100.

Akiyama, H., Kikuchi, Y., Narita, N., Suzuki, M., Goda, Y., Takatori, K., Ichinoe, M. and *Toyoda, M.* 2000. Fumonisin production by *Fusarium moniliforme* and *Fusarium proliferatum* isolated from several agricultural commodities and Japanese soil and detection of new fumonisins. *J. Food Hygi. Soc. Japan*, 41 : 30-37.

Alberts, J.F., Gelderblom, W.C.A., Thiel, P.G., Marasas, W.F.O., Van Schalkywyk, D.J., and *Behred, Y.* 1990. Effects of temperature and incubation period on the production of Fumonisin B_1 by *F. moniliforme*. *Appl. Environ. Microbiol.*, 56 : 1729-1733.

Azcona-Olivera, J.I., Abouzied, M.M. Platter, R.D. and *Pestka, J.J.* 1992. Production of monoclonal antibodies to mycotoxins fumonisins B_1, B_2 and B_3. *J. Agri. Food Chem.*, 40 : l531-534.

Azcona-Olivera, J.I., Abouzied, M.M., Plattner, R.D., Norred, W.P. and *Pestka, J.J.* 1992b. Generation of antibodies reactive with fumonisins B_1, B_2 and B_3 by using cholera toxin as the carrier-adjuvant. *Appl. Environ. Microbiol.*, 58 : 169-173.

Bacon, C.W., Vates, I.E., Hinton, D.M., Meredith, F., Allaben, W.T., Bucher J.R. and *Howard, P.C.* 2001. Biological control of *Fusarium moniliforme* in maize. *Envir. Health Perspe.*, 2001, 325-332.

Ballarini, G. 2002. Fumonisin mycotoxicosis in pigs *Rivista-di-suinicoltura*, 43 : 71-80.

Bane, D.P., Neumann, E.J., Hall, W.F., Harin, K.S. and *Slife, R,L.* 1992. Relationship between fumonisin contamination of feed and mystery swine disease. *Mycopathologia*, 117 : 121-124.

Beekrun, S., Govinden, R., Padayachoe, T. and *Odhav, B.* 2003. Naturally occurring phenols : a detoxification strategy for fumonisin B_1. *Food additives and Contaminants*, 20 : 490-493.

Bezuidenhout, S.C., Gelderblom, W.C.A., Gorst-Allman, C.P., Horak, R.M., Marasas, W.F.O., Spiteller, G. and *Vleggaar, R.* 1988. Structure elucidation of the fumonisins, mycotoxin from *Fusarium moniliforme*. *J. Chem. Soc. Chem. Commun.*, 743-745.

Bhat, R.V., Shetty, P.H.l, Amruth, R.P. and *Sudershan, R.V.* 1997. A food borne disease outbreak due to the consumption of moldy sorghum and maize containing fumonisins mycotoxins. *Clinical Toxicology*, 35 : 249-255.

Blackwell, B.A., Edward, O.E., Apsimon, J.W. and *Fruchier, A.* 1994. Relative configuration of the C-10 to C-16 fragment of fumonsin B_1. *Tetrahedron Letters*, 36 : 1973-1976.

Bohni, J. 2000. Fusariotoxins and their importance in animal nutrition. *Ubersichten-Zur-Tierernahrung*, 28 : 95-132.

Branham, B.E. and *Plattner, R.D.* 1993. Isolation and characterization of a new fumonisin from liquid cultures of *Fusarium moiliforme*. *Journal of Natural Products*, 56 : 1630-1683.

Bruno Doko, M., Sytlvie Rapior, Angelo Visconti and ***Johanne E. Schjoth*** 1995. Incidence and levels of fumonisins contamination in maize genotypes grown in Europe and Africa. *J. Agri. Food Chem.*, 429-434.

Bruns, H.A. 2003. Controlling aflatoxin and fumonisin in maize by crop management. *Journal of Toxicology. Toxin Reviews*, 22 : 153-173.

Buch, G., Foulkes, D.W., Kurono, H. and ***Mitchell, G.F.*** 1996. Total synthesis of racemic aflatoxin B_1. *J. Am. Chem. Soc.*, 87 : 882.

Bullerman, L.B., Labbe, R.G. and ***Garcia, S.*** 2001. *Fusarium.* Guide to food borne pathogens, 87-98.

Cladas, E.D., Sadikova, K., Ward, B.L., Jones, A.D., Winter, C.K. and ***Gilchrist, D.G.*** 1998. Biosynthetic studies of fumonisin B_1, and AAL toxins. *J. Agric. and Food Chem.*, 46 : 4734-4743.

Campbel, H., Choa, T.M.,m Bigier, B. and ***Underhill, L.*** 2002. Comparison of mycotoxin profiles among cereal samples from eastern Canada. *Candian J. Bot.*, 526-532.

Castelo, M.M., Jackson, L.S., Hanna, M.A., Reynolds, B.H. and ***Bullerman, L.B.*** 2001. Loss of fumonisin B_1 in extruded and baked corn based foods with sugars. *J. Food Sci.*, 66 : 416-421.

Castella, G., Bragulat, M.R. and ***Cabanes, F.J.*** 1999. Fumonisin production by *Fusarium* species isolated from cereals and feeds in spain. *J. Food Protec.*, 62 : 811-813.

Cawood, M.E., Gelderblom, C.W.A., Vleggaar, R., Behrend, Y., Thiel, P.G. and ***Marasas, W.F.O.*** 1991. Isolation of the fumonisin mycotoxins: A quantitative approach. *J. Agri. And Food Chem.*, 39 : 1958-1962.

Cawood, M.E., Gelderblom, W.C.A., Alberts, J.F. and ***Snyman, S.D.*** 1994. Interaction of ^{14}C-labelled fumonisin B. Mycotoxins with rat hepatocyte culture. *Food Chem. Toxicol.*, 32 : 627-632.

Chatterjee, D. and ***Mukherjee, S.K.*** 1994. Contamination of Indian maize with fumonisin B_1 and its effects on chicken macrophage. *Lett. Appl. Microbiol.*, 18 : 251-253.

Chen, J., Mirocha, C.J., Xie, W., Hogge, L. and ***Olson, D.*** 1992. Production of the mycotoxin fumonisin B_1 by *Alternaria alternata* f. sp. Lycopersici. *Appl. Environ. Microbiol.*, 58 : 3928-3931.

Chourasia, H.K. 2001. Efficacy of ammonia in detoxification of fumonisin contaminated corn. *Ind. J. Exper. Biol.*, 39 : 493-495.

Chu, F.S. and ***Li, G.Y.*** 1994. Simultaneous occurrence of fumonisin B_1 and other mycotoxins in moldy corn collected from the people's republic of China in regions with high incidences of esophageal cancer. *Appl. Environ. Microbiol.*, 60 : 847-852.

Chu, F.S., Huang, X. and **Maragos, C.M.** 1995. Production and characterization of anti-diotype and anti-anti-idiotype antibodies against fumonisin B_1. *J. AOAC Int.,* 43 : 261-267.

Chulze, S.N., Ramirez, M.L., Farnochi, M.C., Pascale, M., Visconti, A. and **March, G.** 1996. *Fusarium* and fumonisins occurrence in argentinian corn at different corn maturity stage. *J. Agric. Food Chem.,* 44 : 2797-2807.

Cohen, S.M. Bidlack, W.R., Dragan, Y., Goldsubrthy, T., Hard, G., Howard, P.C., Riley, R. and **Voss** 2001. Appotosis and its implications for toxicity, carinogencity and risk : fumonisin B_1 as an example. *Food Add. and Cont.,* 18: 187-201.

Colvin, B.M., Cooley, A.J. and **Beaver, R.W.** 1993. Fumonisin toxicosis in Swine : Clinical and Pathologic findings. *J. Vet. Diagn. Invest.,* 5 : 232-241.

Colvin, B.M. and **Harrison, L.R.** 1992. Fumonisin-induced pulmonary edema and hydrothorax in swine. *Mycopathologia,* 117 : 79-82.

Constable, P.D., Smith, G.W., Foreman, J.H., Waggoner, A.L., Eppley, R.M., Tumbleson, M.E., Benzon, G.J. and **Haschek, W.M.** 2000. Equine leukoencephalomalacia : magnetic resonance imaging of affected brains and cardiovascular toxicity following intravenous fumonisins B_1 administration to horses. Fumonisins risk assessment workshop, January, 10-12, College Park, Maryland USA.

Delongchamp, R.R. and **Young, J.F.** 2001. Tissue chinganine as a biomarker of fumonisin induced apoptosis. *Food Additives and Contaminants,* 18 : 187-201.

Desjardins, A.E., Manandhar, H.K., Plattner, R.D., Manandhar, G.G., Polling, S.M. and **Maragos, C.M.** 2000. *Fusarium* species fron Nepalese rice and production of mycotoxins and gibberellic acid by selected species. *App. Environ. Microbiol.,* 66 : 1020-1025.

Foreman, J.H., Constable, P.D., Waggnor, A.L., Eppley, R.M., Smoth, G.W., Tumbleson, M.E. and **Haschek, W.M.** 2001. Equine leukoencephalomalacia: Characterization of the clinical neurological diseases in horses following intravenous fumonisins B_1 administration to horses. *Food Add. And Conta.,* 18: 182-210.

Galvano, F., Campisi, A., Russo, A., Galvano, G., Palumbo, M., Renis, M., barcellona, M.L., Perezpolo, J.R. and **Vanella, A.** 2002. DNA damage in astrocytes exposed to fumonisin B_1. *Neuro Chemical Research,* 27 : 345-351.

Galvano, F., Piva, A., Ritieni, A. and **Galvano, G.** 2001. Dietary strategies to counteract the effects of mycotoxins : A review. *J. Food Prot.,* 64 : 120-131.

Galvano, F., Russo, A., Cardile, V., Galvano, G., Vanella, A. and **Renis, M.** 2002. DNA damages in human fibroblasts exosed to fumonisins B_1. *Food & Chem. Toxicology,* 40 : 25-31.

Gelderblom, W.C.A., Cawood, M.E., Sayman, D., Vieggar, R. and **Marasas, W.F.O.** 1993. Structure-activity relationships of fumonisins in short-term carcinogenesis and cytotoxicity assays. *Food Chem. Toxicol.,* 31 : 407-414.

Gelderblom, W.C.A., Cawood, M.E., Shyman, D. and *Marasas, W.F.O.* 1994. Fumonisins 13, dosimetry in relation to cancer initiation in rats liver. *Carcinogenesis,* 15 : 209-214.

Gelderblom, W.C.A., Jaskiewicz, K., Marasas, W.F.O., Thiel, P.G., Horak, R.M., Vleggaar, R. and *Kriek, N.P.J.* 1998. Fumonisins novel mycotoxins with cancer-promoting activity produced by *Fusarium moniliforme. Appl. Environ. Microbiol.,* 54 : 1806-1811.

Gelderblom, W.C.A., Jaskiewicz, K., Marasas, W.F.O., Thiel, P.G., Horak, R.M., Vieggaar and *Kriek, N.P.J.* 1997. *Appl. Environ. Microbiol.,* 54: 1806-1811.

Gelderblom, W.C.A., Jaskiewicz, K., Marasas, W.F.O., Thiel, P.G., Horak, R.M., Vleggaar, R. and *Kriek, N.P.J.* 1988. Fumonisins – Novel mycotoxins with cancer promoting activity produced by *Fusarium moniliforme. App. Environ. Microbiology,* 54 : 1806-1811.

Gelderblom, W.C.A., Kriek, N.P.J., Marasas, W.F.O., and *Thiel, P.G.* 1991. Toxicity and carcinogenicity of the *Fusarium moniliforme* metabolite, fumonisin B_1 in rats. *Carcinogenesis,* 12 : 1247-1251.

Gelderblom, W.C.A., Marasa, W.F.O., Steyo, P.S., Thiel, P.G., Vandermerwe, K.J., Vanrooyam, P.H., Vleggaer, T. and *Wessels, P.L.* 1984. *J. Chem. Soc. Chem. Comm.,* 122.

Gelderblom, W.C.A., Marasas, W.F.O., Veggeato, R., Thiel, P.G. and *Cawood, M.E.* 1992. *Mycophathologia,* 117 : 11-16.

Gelderblom, W.C.A., Smuts, C.M., Abel, S., Snyman, S.D., Cawood, M.E., Vander Westhuizen, L. and *Swanevelder, S.* 1996. Effect of fumonisin B_1 on protein and lipid synthesis in primary rat hepatocytes. *Food. Chem. Toxicol.,* 34 : 361-369.

Gelderblom, W.C.A., Smuts, C.M., Abel, S.D., Snyman, L., Vander West-Thuzen, L., Huber, W.W. and *Sworelder, S.* 1997. Effect of fumonisin B, on the levels and fatty acid composition of selected lipids in rat liver *in vivo. Food Chem Toxicol.,* 34 : 361-369.

Gelderblom, W.C.A., Thiel, P.G., Jaskiewicz, K. and *Marasas, W.F.O.* 1986. Investigations on the carcinogenicity of fusarin C – a mutagenic metabolite of *Fusarium moniliforme. Carcinogenesis,* 7 : 1899-1901.

Groves, F.D., Zhang, L., Chang, Y.S., Frank Ross, P., Howard, C., Norred, W.P., You, W.C. and *Franmeni, J.E.* 1999. *Fusarium* mycotoxins in corn and corn products in a high-risk area for gastric cancer in shandeng province, China. *J. AOAC Int.,* 82 : 657-661.

Harrison, L.R., Colvin, B.M., Green, J.T., Newmar, L.E. and *Cole, J.R.* 1990. Pulmonary edema and hydrothorax in swine produced by fumonisins B_1 a toxic metabolic of *Fusarium moniliforme. J. Vet. Diagn. Invest.,* 2 : 217-221.

Haschek, W.M., Waggoner, A.L., McAllister, M.M., Hsiao, S.H., Smith, G.W., Foreman, J.H., Tumbleson, M.E., Eppley, R.M. and **Constable, P.D.** 1992. Abstracts of the papers presented at the workship. *Food. Addit. Contamin.*, 18, 203.

Hirooka, B.Y., Xamagushi, M.M., Aoymma, S., Sugiura, Y. and **Veno, Y.** 1996. The natural occurrence of fumonisins in Brazilian corn kernels. *Food Add. And Conta.*, 13 : 173-183.

Ho, J.A. and **Durst, R.M.** 2000. *Anal. Chim. Acta*, 2000, 414, 61-69.

Holcomb, M., Thompson Jr, H.C. and **Han Kins, C.J.** 1993. Analysis of fumonisin B_1 in rodnt feed by gradient elution HPLC using precolumn derivatization with FMOC and fluorescence detection. *J. Agric. Food Chem.*, 41: 764.

Howard, P.C., Eppley, R.M., Stack, M.E., Warbritton, A., Voss, K.A., Lorentzen, R.J., Kovach, R.M. and **Bucci, T.J.** 2001. Fumonisin B_1 carcinogencity in a two year feedily study using F_{344} rats and $B6C3F_1$ mice. *Environ. Health Perspecti*, 109 : 277-282.

Jackson, L.S., Hlywka, J.J., Kannaki, R., Sen Thil, B., Lloyd, Bullerman and **Musser, S.M.** 1996. Effect of time, temperature and pH on the stability of Fumonisin B_1 in an aqueous model system. *J. Agric. Food. Chem.*, 44 : 906-912.

Jaskiewiez, K., Marasas, W.F.O. and **Taljaard, J.J.F.** 1987. Hepatitis in vervet monkeys course by *Fusarium moniliforme*. *J. Comp. Pathol.*, 97: 281-291.

Javed, T., Bennett, G.A., Richard, J.L., Dombrink, K.M.A., Cote, L.M. and **Buck, W.B.** 1993. Mortality in broiler chicks on feed amendel with *Fusarium proliferatum* culture material or with purified fumonisins B_1 and monilformin. *Mycopathologia*, 123 : 171-184.

Javed, T., Bunte, R.M., Bennett, G.A., Richard, I.L., Dombrink-Kurtzman, M.A., Cote, L.M. and **Buck, W.B.** 1992. Comparative pathologic changes in broiler chicks on feedamended with *Fusarium prolifiratum* culture material or purities fumonisins B_1 and moniliformin. In Abstracts of 106th AOAC International Annual Meeting 1992 Cincinatti.

Juglal, S., Govinden, R. and **Odhav, B.** 2002. Spices oils for the control of co-occurring mycotoxin producing fungi. *J. Food. Prote*, 65 : 683-687.

Karuna, R. and **Sashidhar, R.B.** 1999. Use of ion-exchange chromatography coupled with TLC-laser scanning densitometry for the quantitation of fumonisin B_1. *Talonta*, 50 : 381-389.

Kellerman, T.S., Marcasas, W.F.O., Thiel, P.G., Gelderblom, W.C.A., Cawood, M., Coetzer, J.A.W. and **Odersteport.** (1990). *J. Vet. Res.*, 57 : 269.

King, S.B. and **Scott, G.E.** (1981). Genotypic differences in maize to kernel infection by *Fusarium moniliforme*. *Phytopathology*, , 71 : 1245-1247.

Klittich, C.J.R., Leslie, J.F, Nelson, P.E. and **Marasas, W.F.O.** (1997). *Fusarium thapsinum* (*Gibberella thapsina*) a new species in section Liseola and from sorghum. *Mycologia*, , 89 : 643-652.

KoofMacher, W.A., Chiarelli, M.P., Jr. Lay, J.O., Bloom, J., Holcomb, M. and **McManus, K.J.** 1991. Characterization of the mycotoxin fumonisins : Comparison of thermospray, fast atom bombardment and electrospray mass spectrometry. *Rapid Commun. Mass Spectrum*, 5 : 463-468.

Kriek, N.P.J., Kellerman, T.S. and **Marasas, W.F.O.** 1981. A comparative study of the toxicity of *Fusarium verticillioides* (*F. moniliforme*) to horses, primates, pigs, sheep and Rats, Onderstepoort, *J. Vet. Res.*, 48 : 129-131.

Kulisek, E.S. and **Hazebroek, J.P**. 2000. *J. Agri. Food Chem.*, 48 : 65-69.

Abbas, H.K., Mirocha, C.J., Meronuk, R.A., Pokorny, J.D., Gould, S.L. and **Kommedahl, T.** 1988. Mycotoxins and *Fusarium* spp. Associalted with infected ears of cornin Minnesota. *Appl. Environ. Microbiol.*, 54 : 1930-1933.

Lamprecht, S.C., Marasas, W.F.O., Alberts, J.F., Cawood, M.E., Gelderblom, W.C.A., Shephard, G.S., Thiel, P.G. and **Calitz, F.J.** 1994. Phytotoxicity of fumonisins and TA toxin to corn and tomato. *Phytopathology*, 84 : 384-391.

Lebepe-Mazur, S., Bal, H., Hopmans, E., Murphy, P. and **Henrich, S.** 1995. Fumonisin B_1, is fetotoxic in rats. *Vet. Hum. Toxicol*, 37 : 126-130.

Lemmer, E.R., Gelderblom, W.C., Shephard, E.G., Abel, S., Seymour, B.L., Cruse, J.P., Kirsch, R.E., Marasas, W.F.O. and **Hall, P.M**. 1999. The effects of dietary it on overload on fumonisin B_1 induced cancer promotion in the rat liver. *Cancer Letters*, 146 : 207-215.

Li, J.Y., Ershow, A.G., Chen, Z.J., Wacholder, S., Li, G.Y., Guo, W., Li, B. and **Blot, W.J.** 1989. A case-control study of cancer of the esophagus and gastric cardia in Linxian. *Int. J. Cancer*, 43 : 755-761.

Li, M., Lu, S., Ji, C., Wang, Y., Wang, M., Chen, S. and **Tian, G.** 1980. Experimental studies on the carinogenicity of fungus-contaminated food from Linxian country. In Genetic and Environment factor in experimental and human cancer Gelboi, H.V., Ed., Japan Sciences Society Press, Tokyo, pp.139-148.

Low, W.C., Adler, A. and **Edinger, W.** 1991. Moniliformin and the European corn borer. *Mycotoxin Research*, 7 : 71-76.

MacKentzie, S.E., Savard, M.E., Blackwell, B.A., Miller, J.D. and **Apsimah, J.W.** 1998. Isolation of a new fumonisin from *F. moniliforme* grown in liquid culture. *J. Nat. Prod.*, 61 : 367-369.

Magnoli, C.E., Saenz, M.A., Chiacchiera, S.M. and **Dalcero, A.M. 1999**. Natural occurrence of *Fusarium* species and fumonisin-production by toxigenic strains isolated from poultry feeds in Argentina. *Mycopathologia*, 145 : 35-41.

Mallmann, C.A., Santurio, J.M. and **Dilkin, P.** 1999. Equine leukoencephalomalacia-associated with ingestion of corn contaminated with fumonisin B_1. *Revsta-de-Microbiolgia*, 30 : 249-252.

Maragos, C.M., Jolley, M.E., Plattner, R.D. and **Nasir, M.S.** 2001. Flourescence polarization as a means for determination of fumonisins in maize. *J. Agric. Food Chem.,* 49 : 596-602.

Maragos, C.M. and **Richard, J.L.** 1994. Quantification of stability of fumonisins B_1 and B_2 in milk. *J. AOAC Int.,* 77 : 1162-1167.

Marasas, W.F.O. 1995. Fumonisins : Their implications for human and animal health. *Nat. Toxins,* 3 : 193-198.

Marasas, W.F.O., Kellerman, T.S., Gelderblom, W.C.A., Coetzer, J.A.W., Thiel, P.G. and **Vander Lugt, S.S.** 1976. Leukoencephalomalacia in a horse induced by fumonisin B isolated from *Fusarium moniliforme.* Onderstepoort. *J. Vet. Res.,* 55 : 197-203.

Marasas, W.F.O., Kellerman, T.S., Piennar, J.G. and **Naude, T.W.** 1976. Leukoencephalomalacia of mycotoxicosis of equidae caused by *F. moniliforme* Sheldon. Onderstepoort. *J. Vet. Res.,* 43 : 113-122.

Marasas, W.F.O., Kriek, N.P.J., Wiggins, V., Steyn, P.S., Towers, D.K. and Hastic, T.J. 1979. Incidence, geographical distribution and toxigenicity of *Fusarium* species in South Africa corn. *Phytopathology,* 69 : 1181-1185.

Marasas, W.F.O., Nelson, P.E. and **Tousoum, T.A.** 1988. Reclassification of two important moniliformin-producing strains on *Fusarium.* NRRL 6l022 and NRRL 6322. *Mycologia,* 80 : 407-410.

Marasas, W.F.O., Nelson, P.E. and **Toussoum, T.A.** 1984. Toxigenic *Fusarium* species : identity and mycotoxicology, Pennsylvania state Univ., University Park, Pennsylvania, pp. 328.

Martins, M.L., Martins, H.M. and **Bernardo, F.** 2001. Fumonisins B_1 and B_2 in black tea and medicinal plants. *J. Food Prote.,* 64 : 1268-1270.

Mathur, S., P.D. Constable, Eppley, R.M., Tumbleson, M.E., Smith, G.W., Tranquilli, W.S., Morin, D.E. and **Haschek, W.M.** 2001. Fumonisin B_1 increase serum sphinganine concentration but does not alter serum sphingoshe concentration or induce cardiovascular changes in milk-fed calves. *Toxicological Sci.,* 60 : 379-384.

Maya, S.P. and **Silva, G.E.** 1997. Fumonisin levels in Uruguayan corn products. *J. AOAC Int.,* 80 : 825-828.

Mazzami, C., Berges, O., Luzon, O., Barrientes, V. and **Quijoda, P.** 1999. *Fusarium moniliforme,* fumonisins and *Aspergillus flavus* in kernels of corn hybrids in Guarcio state, 2000. *Rev. de Lafac. De.Agro.Uni.del-zulia,* 17 : 185-195.

Merdith, F.I. 2000. *Methods Enzymol.,* 311 : 361-373.

Minervini, F., Formelli, F. and **Flynn, K.M.** 2004. Toxicity and apoptosis induced by the mycotoxins nivalenol, deoxynivalenol and fumonisins B_1 in a human erythroleukemia cell line. *Toxicology in vitro,* 18 : 21-28.

Moodley, J., Moudley, D., Reddy, L., Dutton, M.F. and **Chuturgoon, A.A.** 2001. Fumonisin B_1 : Anb aetiological role in pre-eclampsia. *J. Obst. and Gyne.,* 21 : 599-600.

Moon-Woosung, Kim-Jootteon, Kang-Myoung Jae, Lee-Dohg Geun, Moon-WS, Kim J.H., Kang, M.J. and **Lee, D.G.** 2000. Early ultrastructural changes in apoptosis induced by fumonisin B_1 in rat liver. *Yonsei Medical Journal,* 41 : 195-204.

Nelson, P.E., Plattner, R.D., Shackelord, D.D. and **Desjardins, A.E.** 1992. Fumonisins B_1 production by *Fusarium* sp. other than *F. moniliforme* insection Liseola and by some related species. *Appl. Envin. Microbiol.,* 984-989.

Norred, W.P., Bacon, C.W., Riley, R.T., Voss, K.A. and **Meredith, F.I.** 1999. Screening of fungal species for fumonisin production and fumonisin-like disruption of Sphingo lipid biosynthesis. *Mycopathologia,* 146 : 91-98.

Norred, W.P., Plattner, R.D.,Vesonder, R.F., Hayes, R.F., Bacon, C.W. and **Voss, K.A.** 1990. Effect of *F. moniliforme* metabolites on unscheduled DNA synthesis (UDS) in rat primary hepatocytes. *Toxicologist,* 10 : 165.

Norred, W.P., Voss, K.A., Bacon, C.W. and **Rirrlery, R.T.** 1991. Effectiveness of ammonia treatment in detoxification of fumonisin – contaminated corn. *Food Chem. Toxicol.,* 29 : 815-819.

Odhav, B., Permaul, K., Rama Sunder, S., Padayachee, T., Mohanlal, V. and **Reddy, L.** 2001. Incidence and biocontrol of some mycotoxins in South Africa. Biological control of fungal and bacterial plant pathogens (ed. Blad, Y. and Freeman, S.), 113-136.

Oho, E.U.S., Sasaki, e.y., Hashimoto, E.H., Hara, L.N., Correa, B., Itano, E.N., Sugiura, T., Ueno, Y. and **Hirroka, E.Y.** (2002). Post harvest stage of corn : effect of beginning moisture content on mycoflora and fumonisin contamination. *Food add. and contam.,* 19: 1081-1090.

Osweiler, G.D., Ross, P.F., Wilson, T.M., Nelson, P.E., Witte, S., Carson, T.L., Rice, L.G. Rice and **Nelson, H.A.** 1992. Characterization of epozootic of pulmonary edema in swine associated with fumonisins in corn screening. *J. Vet. Invest.,* 4 :53-59.

Pacin, A.M., Boca, R.T., Gonzalez, H.H.L., Resnik, S.L., Burak, R., Broccoci, A.M., de Souza, J.C. and **de Souza, J.C.** Natural occurrence of mycotoxins and mycoflora of Argentinien popcorn.

Plattner, R.D., Norred, W.P., Bacon, C.W., Voss, K.A., Peterson, R., Shackelford, D.D. and **Weisleder, D.** 1990. A method of detection of fumonisin in corn samples associated with field cases of equine leukoencephalomalacia. *Mycologia,* 82 : 698-702.

Poling, M. and **Plantner, R.D.** 1996. Rapid purification of fumonisin B_3 and B_4 with solid phase extraction columns. *J. Agric. Food Chem.,* 44 : 2792-2796.

Qiu, M. and **Liu, X.** 2001. Determination of sphinganine, sphigosine and salso ratio in urine of humans exposed to dietary fumonisin B_1. *J. Food Add. Contamin.*, 18 : 263-269.

Reynoso, M.M., Torres, A.M., Ramirea, M.L., Rodriguez, M.I., Chuize, S.N. and **Magan, N.** 2002. Efficacy of antioxidant mixtures on growth, fumonisin production and hydrolytic enzyme production by *Fusarium verticilliodes* and *F. proliferatum in vitro* on maize-based media. *Mycological Research*, 106 : 1093-1099.

Rheeder, J.P., Marasas, W.F.O., Thiel, P.G., Sydenham, E.W., Shephard, G.S. and **Van Schlakwyk, D.J.** 1992. *Fusarium moniliforme* and fumonisins in corn in relation to human esophages cancer in Transkei. *Phytopathology*, 82 : 353-357.

Rice, L.G. and **Ross, P.F.** 1994. *J. Foos Protection*, 57 : 536-540.

Riley, R.T., Bacon, C.W., Meredith, F.I., Torres, O., Saenzde Tyada, Wang, E. and **Murrill Jr.A.H.** 2001. A summary of on going studies in the central highlands of Guatemala food additives and contaminants. 13 : 197-198.

Rosiler, M.R., Bautista, J., Fuetes, V.O. and **Ross, F.** 1998. *Zentralbl. Veterinarmed*, A45 : 299-302.

Ross, P.F., Rice, L.G., Osweiler, G.D., Nelson, P.E., Richard, J.L. and **Wilson, T.M.A.** (1990). Review and update of animals toxicoses associated with fumonisin contaminated feeds and production of fumonisins by *Fusarium* isolates. *Mycopathologia*, 117 : 109-114.

Schneider, E., Usleber, E. and **Martlbauer, E.** 1995. Rapid detection of fumonisins B_1 in corn-based food by competitive direct dipsick enzyme immunoassay / enzyme linked immunofiltration assay with integrated negative control reaction. *J. Agri. Food Chem.*, 43 : 2548-2552.

Scott, P.M. and **Lawrence, G.A.** 1992. *J. AOAC Int.*, 75 : 829-834.

Seefelder, W., Humpf, H.U., Schwerdt, G., Freudinger, R. and **Gekle, M.** 2003. Induction of apoptosis in cultured human proximal tubule cells by fumonsins and fumonisin metabolites. *Toxicology and Applied Pharmacology*, 192 : 146-153.

Segvic, M. and **Pepeljnjak, S.** 2003. Distribution and fumonisin B_1, production capacity of *Fusarium moniliforme* isolated from corn in croatia *Periodicu-Biologerum*, 275-279.

Seo, J.A. and **Lee, Y.W.** 1999. Natural occurrence of the "C" series of fumonisins in moldy corn. *Appl. Environ. Microbiol.*, 65 : 1331-1334.

Shelby, R.A., White, D.G. and **Bauske, E.M.** 1994. Differential fumonisins B_1 production in maize hybrids. *Plant Dis.*, 78 : 582-584.

Shephard, G.S., Thiel, P.G., Stockenstrom, S. and **Sydenham, E.W.** 1996. World wide survey of fumonisin contamination of corn and corn-based human staff. *J. AOAC Intern.*, 79 : 671-687.

Shephard, G.S.E.W., Sydenham, Thiel, P.G. and **Gelderblom, W.C.A.** 1990. Quantitative determination of fumonisins B_1 and B_2 by high performance liquid chromatography with fluorescence detection. *J. Liquid Chromatogr.,* 13 : 3077-3080.

Shim-Won, B.O., Woloshuk, C.P. and **Shim, W.B.** 1999. Nitrogen repression of fumonisin B_1, biosynthesis in *Gibberella futikuroi.* *FEMS-Microbiology Letters,* **177** : 109-116.

Sle, W., Mirocha, C.J. and **Chen, J.** 1997. Detection of two naturally occurring structural isomers of partially hydrolyzed fumonisin B_1 in corn by one-line capillary liquid chromatography-fast atom bombardment mass spectrometry. *J. Agri. And Food Chemistry,* 45 : 1251-1255.

Solovey, M.M.S., Somoza, C., Cano, G., Pacin, A. and **Resnik, S.** 1999. A survey of fumonisins, deoxynivalenol, zearalenone aflatoxins contamination in corn-based food products in Argentina. *Food Addi. And Conta.,* 16 : 325-329.

Stack, M.E. and **Epply, R.M.** 1992. Liquid chromatographic determination of Fumonisins B_1 and B_2 in corn and corn productions. *J. AOAC Int.,* 75 : 834-837.

Sydenham, E.W. The chromatographic determination of *Fusarium* toxins in maize associated with human / oesophageal cancer. M.Sc. thesis, University of Capte Town, South Africa, 1989.

Sydenham, E.W., Gelderblom, W.C.A., Thiel, P.G. and **Marasas, W.F.O.** 1990. Evidence for the natural occurrence of fumonisins B_1, a mycotoxin produced by *F. moniliforme* in corn. *J. Agric. Food Chem.,* 38 : l285-290.

Sydenham, E.W., lShephard, G.S., Thiedl, P.G., Marasa, W.F.O., Rheeder, L.E.P., Peralta Sanhueze, Gonzalez, H.H.L. and **Resnik, S.L.** 1993. Fumonisins in Agrenitina Field-trail corn. *J. Agri. Food Chem.,* 41 : 891-895.

Sydenham, E.W., Pieter, G., Thiel, Walter, F.O., Marasas, S.G., Shephard, D.J., Van Schalkwyk and **Klaus R. Koch.** 1990. Natural occurrence of some *Fusarium* mycotoxins in corn from low and high esophageal cancer prevalence areas of the transkei southern Africa. *J. Agric. Food Chem.,* 38 : 1900-1903.

Sydenham, E.W., Shephard, G.S., stockenstrom, S., Rheeder, J.P., Marasas, W.F.O. and **Vander Merwe, M.J.** 1997. Production of fumonisin B_1 analogues and related compounds by *Fusarium globosum*, a newly described species from corn. *J. Agric. Food Chem.,* 45 : 4004-4010.

Sydenham, E.W., Vander Westhuizen, L., Stockenstrom, S., Shephard, G.S. and **Thiel, P.G.** 1994. Fumonisin-contaminated maize : physical treatment for the partial decontamination of bulk shipments. *Food. Addtl. Contamin.,* 11 : 25-32.

Sydenham, M., Walter, Marasas, F.O., Gorden, S., Shephard, Pieter, G. Thiel and **E.X. Hirooka.** 1992. Fumonisin concentration in Borazilian feeds associated with field outbreaks of confirmed and suspected animal mycotoxicoses. *J. Agri. Food Chem.,* 40 : 994-997.

Syndenham, E.W., Shephard, G.S., Thiel, P.G., stockenstrom, S., Snijman, P.W. and **Van Schalkhyk, D.J.** 1996. Liquid chromatographic determination of fumonisin B_1, B_2 and B_3 in corn AOAC – IUPAC collaborative study. *J. AOAC Int.,* 79 : 688-696.

Thiel, P.G., Meyer, C.J. and **Marasas, W.F.O.** 1982. Natural occurrence of moniliformin together with deoxynivalenol and zearalenone in Transkeian Corn. *J. Agri. and Food Chem.,* 30 : 308-312.

Thiel, P.G., Marasas, W.F.O., Sydenham, E.W., Shephard, G.S., Gelderblom, W.C.A. and **Nieulwenhuis, J.J.** 1999. Survey of fumonisin production by *Fusarium* species. *Appl. Environ. Microbiol.,* 57 : 1089-1093.

Thiel, P.G., Marases, W.F.O., Sydenham, E.M., Shephard, G.S. and **Gelderblom, W.C.A.** 1992. The implications of natural occurring levels of fuminosins in corn for human and animal health. *Mycopathologia,* 117 : 3-9.

Thiel, P.G., Sydenham, E.W., Shephard, G.S. and **Van Schlkwyk, D.J.** 1993. Study of the reproducibility characteristics of liquid chromatographic method for the determination of Fumonisins B_1 and B_2 in corn IUPAC collaborative study. *J. AOAC Int.,* 76 : 361-366.

Thiel, P.G., Syndenham, E.W. and **Shephard, G.S.** (1996). The reliability and siognificance of analytical data on the natural occurrence of fumonisins in food. In Jackson, L., De Varies, J.W. and L.B. Bullerman (Eds.), Fumonisins in Food, Plenum Press, New York, 145-151.

Torres, A.M., Reynoss, M.M., Rojo, F.G., Ramirez, M.L. and **Chulze, S.N.** (2001). *Fusarium* species (section Liseola) and its mycotoxins in maize harvested in northern Argentina. *Food Add. And Conta.,* 18 : 836-843.

Usleber, E., Straka, M. and **Terplan, G.** 1994. Enzyme immuno assay for fumnonisin B_1 applied to Corn-bakd food. *J. Agric. Food Chem.,* 42 : 1392-1396.

Vass, K.A., Platter, R.D., Bacon, C.W. and **Norred, W.P.** 1990. *Mycopathologia,* 112 : 81.

Velaziquez, C. and **Monserrak, C.** 1999. Estefania Pena Nuria Sala, Ramon Canela. Linking extract and purification of maize samples for fumonisins analysis. *Food Add Conta,* 10 : 125-128.

Visconti, N. and **Doko, M.B.** 1994. Survey of fumonisin production by *Fusarium* isolated from cereals in Europe. *J AOAC,* 77 : 546-550.

Voss, K.A., Chamberlian, W.J., Bacon, C.W., Herbert, R.A., Walters, D.B. and **Norred, W.P.** 1994. *Fund. Appl. Toxicol.*

Wang, E., Norred, W.P., Bacon, C.W., Riley, R.T. and **Mlerrill, A.H.** 1991. Inhibition of sphingolipid biosynthesis by fumonisin : implications for diseases associated with *Fusarium moniliforme. J. Biol. Chem.,* 266 : 14486-14490.

Wang, E., Ross, F., Wilson, T.M., Riley, R.T. and **Morril, A.H.** 1992. Increase in serum sphingosine and sphinganine and decrease in complex sphigolipids in ponie given feed containing fumonisins mycotoxin produced by *F. moniliforme*. *J. Nutrition*, 122 : 1706-1716.

Weiblug, T.S., Ledoux, D.R., Bermudez, A.J., Turk, J.R., Pottinghaus, G.E., Wang, E. and **Jr. Merrill, A.H.** 1993. Effects of feeding *Fusarium moniliforme* culture material, containing know levels of fumonisin B_1 on the young broiler chick. *Poult. Sci.*, 72 : 456-466.

Wilkes, J.G., Lay, O.J., Trucksess, M.W. and **Pohland, A.E.** 2001. Electrospray mass spectrometry for mycotoxin detection and purity analysis. *Mycotoxins Protocols*, 37-48.

Wilson, T.M., Nelson, P.E. and **Knepp, C.R.** 1985. Hepatic neoplastic nodules, adenofibrosis and cholangiocarcinomas in male fisher G_4 rats fed corn naturally contaminated with *F. moniliforme*. *Carcinogenesis*, 6: 1155-1160.

Wilson, T.M., Rose, P.F., Owens, D.L., Rice, L.G., Green, S.A., Jenkins, S.J. and **Nelson, H.A.** 1992. Experimental reproduction of ELEM. A study to determine the minimum toxic dose in ponies. *Mycopathologia*, 117 : 115-120.

Wolf, G. 1994. Mechanism of the mitogenic and carcinogenic action of fumonisin B_1 a mycotoxin. *Nutr. Rev.*, 52 : 246-247.

Xie, W., Mirocha, C.J. and **Chen, J.** 1997. Detection of two naturally occurring structural isomers of partially hydrolyzed fumonisin B_1 in corn by one line capillary liquid chromatography atom bombardmnent mass spectrometry. *J. Agri. and Food Chemistry*, 45 : 1251-1255.

Yamagish, D., Visconti, A., avantaggiato, G. and **Ouchi, S.** 2001. A genetic approach to fumonisin biosyfnthesis in *F. prolifleratum*. *Bulletin of the Institute of Comprehensive Agricultural Sciences*, Kinki University, No. 9 : 13-28.

Yamashita, A., Yoshizawa, T., Aiura, Y., Sachez, P.C., Dizon, E.I. and Arim, R.H., Sardjono 1995. *Fusarium* mycotoxins and aflatoxins in corn from Southeast Asia. *Biosci. Biotech. Biochem.*, 59 : 1804-1807.

Yang, C.S. 1980. Research on esophageal cancer in China : A review. *Cancer Res.*, 40 : 2633-2644.

Zomo, M.C. and **Vurro, M.** 1999. Effect of fungal toxins on germination of *Striga hermonthica* seeds. *Weed Res.*, 39 : 15-20.

36 · Prospecting A Hearbal Drug Folk Medicine: Phansomba A Medicinal Mushroom

Vaidya, J.G.*, Bhosle, S.R., Lamrood, P.Y. and Bapat, Gauri

Department of Botany, University of Pune - 411 007. India

* Communicating author: jitendra@unipune.ernet.in

ABSTRACT

Mankind, during his evolution has developed a good weapon against ailments, diseases and disorders. 'The traditional medicinal system,' as called, is comprised of a variety of elements from the environment. Different civilizations settled globally have focussed mainly on plant parts and its products, hence called herbal drugs. Fungi were also included in these systems, specifically macro fungi generally termed as 'Mushrooms'. 'King of the Herbs' or 'Herb of Immortality', is a mushroom *Ganoderma* (Reishi, Ling zhi – Chinese, Mannentake - Japnese) no wonder it captures the top position in Traditional Chinese Medicine. Although Chinese, Japanese and other south eastern countries emphasize in using mushroom in regular diet and medicine system, even in India people use mushrooms in diet and as medicine. This article is a case study of such a fungus, a mushroom, a folk medicine from Western Ghats of Maharashtra, known as 'Phansomba.' The exploration of research carried out for more than a decade and future voyage in exploiting the folk drug and steps in opening up the drug to world by commercializing it.

KEY WORDS : Phansomba, *Phellinus, Ganoderma, Termitomyces.*

Human being had always strived to live and fight against the natural calamities. This character has helped in very fast evolution as compared to other living organisms. In his journey of evolution man has always taken help of his surrounding elements and conditions. This has made him to sustain through the toughest path of the evolution. The most important use of the environment and its elements by mankind was against variety of ailments, diseases and disorders. Thus came up the system of curing ailments/medicine. Different civilizations developed their own system of medicine. There are many ancient systems, which are still in use and some have modified during this period while some got extinct. The Greek medicine system in Europe, Yunani in gulf, Ayurveda in India and TCM (Traditional Chinese Medicine) in China are the most well established systems.

These systems have been authenticated by scientists world wide. Use of various plant parts are the core of these systems so the products are generally 'Herbal' products.

The Traditional Chinese Medicine has listed many plants and the paramount position has been allotted to a mushroom, called 'King of the Herbs', (Willard, 1990). In Chinese they call it as 'Reishi' or Mannentake. It was thought to cure a dead man and so was precious. The use of fungus as medicine is not only described in Chinese but also in all other medicine systems, even in Ayurveda there are notes about the medicinal properties and use of mushrooms (Vaidya & Lamrood, 2000).

(a)   *Medicinal Mushrooms.*

During the last 3-4 decades scientists around the world, especially from the southeastern countries have concentrated on isolating, purifying and characterizing the compounds from mushrooms. This deals with isolating, purifying and characterization of the compounds, and also screening the compounds for any kind of bioactivity. Many mushrooms (edible and non-edible and also poisonous) have been screened for their active principles. There are more than 2000 species of edible mushrooms, and 200 species are cultivated, whereas only 30-40 species are under commercial cultivation, these are mainly edible mushrooms. Few medicinal mushrooms like, *Lentinus edodes* (Shiitake), *Ganoderma lucidum* (Reishi), *Cordyceps* sp., *Grifiola frondosa*, *Coriolus versicolor* etc, are commercially cultivated (Table 1. & 2.). Lentinan, a polysaccharide purified from the mushroom *Lentinus edodes* is a best example of a compound commercialized and is an approved anticancer drug in China.

Ganoderma mushroom and its products are sold as ancient medicine under the trade name Reishi or Ling Zhi in China, Japan, Korea, Thiland, Malyasia and other Asian countries (Hobbs, 1996; Willard, 1990; Chang and Buswell, 1999; Chiu. *et.al.*, 2000; Hseu *et. al.*, 1996). The Reishi is even sold in Indian market with trade name Ganotherapy (therapy used with the help of powder of *Ganoderma* in the form of capsules or tea bags or ointment etc.) in some parts of western India. The use of ganotherapy is becoming more and more popular for curing the chronic diseases, which are not easily curable by allopathic treatment.

(b)   *Indian Folk Use of Mushrooms/Fungi as Medicine.*

Mushrooms come under the minor forest product category. The mushrooms not only support a tribal economically but also serve as a good nutrient food during the adverse conditions. The entire family including children hunt these mushrooms in rainy season. Amongst this ethnobotanical treasure, forest or wild fungi have important place in a tribal's life and has crucial role to play in the diet and medicine of those people who are well acquainted with their habitat and have experience (Harsh *et. al.*, 1993 and 1996).

Table 1 World production of mushrooms.*

Name of Mushroom	1981 Fresh wt × 10^3		1990 Fresh wt × 10^3		1997 Fresh wt × 10^3	
	Metric tons	%	Metric tons	%	Metric tons	%
Agaricus bisporus						
A. bitorquis	900	71.6	1,420	37.8	1,955.9	31.8
Lentinus edodes	180	14.3	393.0	10.4	1,564.4	25.4
Pleurotus spp.	35.5	2.8	900.0	23.9	875.6	14.2
Auricularia spp.	10.0	0.8	400.0	10.6	485.3	7.9
Volvariella volvacea	54.0	4.3	207.0	5.5	180.8	3.0
Flammulina velutipes	60.0	4.8	143.0	3.8	284.7	4.6
Tremella spp.	-	-	105.0	2.8	130.5	2.1
Hypsizygus spp.	-	-	22.6	0.6	74.2	1.2
Pholiota spp.	17.0	1.3	22.0	0.6	55.5	0.9
Grifola frondosa	-	-	7.0	0.2	33.1	0.5
Others	1.2	0.1	139.4	3.7	518.4	8.4
Total	1,357.2	100	3,763.0	100	6,158.4	100
Increasing %	-	72.5	25.4			

*Source: Chang, 1999

Many tribes collect edible and medicinal mushrooms for their day today consumption. Amongst the few edible mushrooms, *Termitomyces* is very popular. Several species of *Termitomyces* are consumed not only in India but in many other countries as well. The *Termitomyces* is very prized delicacy collected both for the house consumption and for sale in the local market. There are few other mushrooms which are also consumed as food and medicine viz., *Agaricus, Astraeus, Auricularia, Calvatia, Geastrum, Lycoperdon, Ganoderma, Phellinus, Podabriella* etc.

About the importance of higher fungi as food and medicine in folklore is very little known hence present study attempts to provide an insight into the scenario of ethno mycology of India.

This ethno mycological treasure of India can be broadly divided in to 4 major parts as :

1. Western Ghats and East Maharashtra

2. Madhya Pradesh i.e. Bastar Region

3. North India including Himalayas

4. East and North Eastern parts of India

Table 2 Country wise production of mushrooms (metric tons)*.

Name of the mushroom	Name of the country								Total	%
	China	Japan	Rest of Asia	North America	Latin America	Europe	Rest of Europe	Africa		
Agaricus bisporus	384400	-	68400	425300	51600	875000	115200	36000	1955000	31.8
Lentinus edodes	1397000	115300	47400	3600	300	500	300	-	1564400	25.4
Pleurotus spp.	760000	13300	88400	1500	200	6200	5800	200	875600	14.2
Auricularia spp.	48000	-	5300	-	-	-	-	-	485300	7.9
Volvariella volvacea	12000	-	60800	-	-	-	-	-	180800	3.0
Flammulina spp.	15000	10900	25700	-	-	-	-	-	284700	4.6
Tremela spp.	13000	-	500	-	-	-	-	-	130500	2.1
Hypsizygus marmoreus	2100	7200	100	-	-	-	-	-	74200	1.2
Pholiota nameko	3100	24500	-	-	-	-	-	-	55500	0.9
Grifola frondosa	2000	3100	-	-	-	-	-	-	33100	0.5
Hericium erinacium	800	-	-	-	-	-	-	-	-	-
Coprinus comatus	500	-	-	-	-	-	-	-	520800	8.4
Others	515100	2900	800	400	-	200	100	-		
Total	3972900	368000	297400	430800	52100	881900	121400	36200		
Per cent	64.5	6.0	4.8	7.0	0.8	14.3	2.0	0.6	6160800	100

*Source: Chang, 1999

Variety of fungi are regularly used by tribal people in India viz. of *Phellinus* spp., *Ganoderma* spp., *Xylaria* spp., *Bovista* spp., *Geastrum* spp., *Astraeus* spp., *Cyathus* spp., *Calvatia* spp., *Phallus impudicus*, *Scleroderma* spp., *Xylaria polymorpha*, *Micriporous xanthopus*, *Lycoperdon* spp., *Bovista* spp., *Agaricus campestris* etc. (Harsh *et. al.*, 1993 and 1996).

(c) ***'Phansomba.'***

In India, especially in Western Ghats many species of *Phellinus* are routinely used as a folk medicine to cure teeth, tongue, and throat related ailment, to stop excessive salivation in case of children, against diarrhoea. (Andhalkar, 1988; Vaidya, 1995; Vaidya and Bhor, 1991; Vaidya and Lamrood, 2000). The fungus is well known by the name Phansomba, Phanas-alambi, and Phanas-banda. Several authors viz., Bakshi, (1971; 1976), Chopra *et. al.*,(1956), Desai, (1927), Dymock *et. al.*, (1890). Gogate, (1972), Khory, (1887), Kirtikar, (1918), Nadkarni, (1954), Nanal, (1929), Sathe, (1982), Sawant, (1974), Singh, (1974), Shah, (1928), described the medicinal properties of this mushroom. But there is no previous study carried out to investigate the biochemical composition of the mushroom.

Many a times it is also found that samples of *Ganoderma*, knowingly or unknowingly are being adulterated with the species of *Phellinus* and sold as Phansomba. *Ganoderma* individually is mainly used by the tribal to cure tumerous growth in throat, inflammation of knee, cataract, asthma etc. Surprisingly their claims were proved to be positive using samples of *Ganoderma*. (Vaidya and Bhor, 1991; Vaidya and Rabba, 1993; Vaidya and Lamrood, 2000)

Our investigation have shown that many of the specimens collected as adulterated Phansomba belong to species of *Ganoderma* and have reported same active principles like Terpenoids, Ergosterol, as well as germanium, lectins, polysaccharides with that of Reishi.

The present report is a compilation of group work carried out over the years (1989 – till date) on various aspects like survey, collection, identification, pharmacognostic studies, biochemical studies with respect to specific bioactive molecule, anticancer, antimicrobial activity, toxicity studies, isolation of novel compound(s), Phase I clinical trial etc. and the results gathered from these various studies strongly revealed that the *Phellinus* species that are used in Western Ghats are equally competent with that of Korean Sang-Hwang or Reishi and has a much wider perspective to work on different aspects.

2. Previous Work

The Phansomba is a folk drug restricted to the western ghats of Maharashtra; there are reports of folk use of *Phellinus*, worldwide including in India. Various *Phellinus* species have been worked out in variety of aspects like culture isolation and characteristics, physiology, biochemical aspect, pharmacognosy, natural resource of compounds etc.

(a) *Foreign*

This mushroom is popularly known as Sang-Hwang in Korea; Song Gen in China and Meshimakobu in Japan. In recent years, from the cultured mycelium product (PL2 and PL5 strains that are similar to ATCC 26710) "meshima capsules" have been developed as a medicine in Korea and it was approved as a medicine manufactured by Korean New Pharmaceutical Co., in 1997 (Mizuno, 2000). This Korean mushroom is described as Sangmogi (mulberry ear) in Decoction chapter of Oriental Medical Thesaurus, Korea's old medical book, and as Sangshin, Sanghwanggo or Sang-i in Chinese medical books including Botanical List of Medical Herbs by Ming dynasty's Lee Shi-jin, Encyclopedia on Chinese Medicines, Encyclopedia on Oriental Medicine, etc. The medicinal literature in China such as Shin-nou-honzokyo (Imperial Medicine Handbook), Honzo-komoku (Medicine handbook), Chuuyaku-daijitenn (Great Dictionary of Chinese Medicines), Chuugoku-yakuyo-shinnkinn (Chinese Medicinal Fungi), and Toyoigaku-daijitenn (Great Dictionary published in 1967, cited in Mizuno, 2000) describes the efficacy of this mushroom as – it is used for the treatment of palsy, gonorrhea and abdominal pain, urination disorders, stomach problems, hematuria, gripes, penis ache, prolapsus ani, melena, over work, tural disorder, vitiated blood, lymphatic tumor, metrorrhagia, nose bleed, facial and diarrhoea. While, in Korea the mushroom is used mainly to treat various cancers e.g. cancers in the digestive system such as stomach, gullet, duodenum, colon and rectum and as an immune modulating/boosting medicine. (Mizuno, 2000; Kwon, 2000, 2001; http://www.nonkong.co.kr / english/sang.html).

There are many reports of isolation of pure compounds; Such as Triterpenes from *Phellinus gilvus* (Ahmad *et al.*, 1976), 7-phenylheptane-3-ones from *P. tremulae* Serck-Hanssen *et al.*, and isolation of benzyl dihydroflavones from *P. igniarius* (Mo *et al.*, 2003). The most exploited species of *Phellinus* are *P. ignarius, P. linteus, P. gilvus, P. pini, and P. pomaceus.*

Table 3 Comparisons of Medicinal Mushroom Phansomba and Sang-Hawang

Medicinal Properties	Phansomba[#]	Sang – Hawang[*]
Known	Tongue, teeth, throat related , diseases, excessive salivation, dysentery, diarrhoea	Treatment of palsy, gonorrhea, abdominal pains, urination disorders, stomach problems, hematuria, nose bleed, facial and diarrhoea
Literature	Same as above, anticancer, anti - inflammatory, anti-oxidant	Anti-tumor, anti-cancer, anti-diabetic.
Bioactivity/ Clinical Trials	"Mukhapak" diseases, 60 patients treated successfully as well as for edema and ulcerations on mucus membrane of buccal cavity, anticancer activity using SiHa cell lines, toxicity using *Allium* root tip assay	Trials on mouse and rats for anti-tumor, anti-cancer, anti diabetic activity, Cell toxicity using A549, SR-OV-3, SK-OV-3, Sk-MEL-2, HCT-15 and XF-498 cell lines derived from humans
Pharmacognostic studies	Ash, moisture, minerals, metals, crude fibers, total carbohydrates, total proteins, amino acids, organic acids, phenols	Moisture, crude protein, dietary fibers, carbohydrates, crude ash, K_{cal}, sodium, arsenic and lead
Biochemical studies	Lectins, Terpenoids, Ergosterol, Alkaloids, germanium, Polysaccharides.	Polysaccharides, carbohydrate analysis and polysaccharide fraction (KP).

Vaidya, (1995); Vaidya and Bhor, (1991); Andhalkar (1988)

* Mizuno (2000); Kwon, (2000; 2001) and http://www.nonkong.co.kr / english/sang.html

(b) *Indian.*

Bakshi (1976) has worked on the pathogenic and taxonomic aspect of the fungus in the north part of India. Recently Ajith & Janardhanan (2001) reported from Kerala, the use of *Phellinus linteus* and *P. rimosus* to cure mumps and also confirmed the anti-oxidant and anti-inflammatory activity of the studied samples. Screening of potential fungal candidate for preventing cancer has also been carried out by Jananrdhanan (2003).

(c) *Contribution of this team:*

Vaidya and Bhor (1991) started the work in the western ghats of the Maharashtra and they identified the Phansomba as a species of the genus *Phellinus*. Vaidya & Rabba (1993) studied the taxonomic details of the Phansomba samples.

(i) *Sample Collection and Taxonomy:*

More than 300 samples of *Phellinus* only from *Artocarpus*, more than 100 samples from other host trees, and a collection of 35 different species of Ganodermatoid fungi (table 4a and 4b) were collected.

Collections of specimens (basidiomata) were carried out from various localities of Western Ghats. Spore prints were obtained from fresh basidiocarps on cellophane paper and maintained in airtight plastic bags. For detailed microscopic examinations temporary slides were made by cutting free hand sections passing through all the parts of the basidiocarp. The sections were treated with 10% KOH solution and stained with 1% Phloxine and Cotton Blue after washing in distilled water. Lacto glycerin was used as a mounting medium for semi permanent slide. Permanent slides were prepared in P.V.A. (Polyvinyl Alcohol) medium as per Omar *et.al.,* (1979) method. Spores were isolated from a block of tube layers as per technique described by Steyaert (1972). All specimens were deposited in the Herbarium at the University of Pune, Department of Botany. Identification of *Phellinus* samples was done by using the key of Larsen and Cobb-Poulle, (1990) and Gilbertson & Ryvarden, (1986). Similarly, for *Ganoderma* the keys of Steyaert, (1972, 1980); Ryvarden, (2000); Ryvarden and Johansen, (1980); Gilbertson and Ryvarden, (1986) and Gottlieb and Wrigth (1999 a and b)

(ii) *Culture Isolation and Cultural Studies:*

Fresh tissues from the samples were used as and when available for the isolation of culture. In some cases of *Phellinus* old tissue was used for isolation and in that case successive pretreatments of tap water, NaOCl (5%), Alcohol and sucrose were given prior to inoculation. Potato dextrose agar and 2% Malt extract were used for isolation and maintenance of the cultures.

With the special medicinal mushroom culture facility, more than 20 different isolates have been isolated in pure culture from the species of *P. meriillii, P. sublinteus, P. gilvus, P. loydii* and isolates from *Ganoderma multiplicatum, G. stipitatum, G. poonensis, G. testateum, G. perzonatum,* and *G. praelongum* were obtained in pure form.

Growth rate and biomass yield on the above said culture media have been carried using standard methods. For growth rate cultures were grown on 90mm diameter plates containing the respective culture media and the growth was measured till the culture reaches the maximum (90mm). The growth rate was calculated in terms of average growth per day (mm per day) (table 5). For the biomass yield, flasks containing 100ml culture broth were inoculated and were incubated on shaker at 27°C. the yield was calculated on 15th day, 21st day and 28th day. The average yield per liter was calculated (table 5).

The collected samples were powdered using commercial mill (used for grinding spices and herbal products, general mesh size 80). The powder was further used for studying physicochemical parameters, pharmacognostic parameters, biochemical analysis and extraction of bioactive compounds.

Table 4a Phansomba Sample Survey (1984 – 2003)

Bhor (1984)	Rabba (1994)	Lamrood & Desh Pande (2002)	Lamrood, Sawant & Bhosle (2002-03)
P. badius	**P. fastuosus**	**P. llovdii**	**P. merilli**
P. chinchonensis	P. adamantinus	P. merrillii	P. fastuosus
P. durrissimus	P. calcitratus	Host other than	P. lloydii
P. gilvus	P. carteri	ARTOCARPUS	P. sublinteaus
P. linteus	P. conchatus	P. adamantinus	Bark and
P. merrillii	P. merrillii	P. aureobrunneus	wood of
P. pectinatus	**Host other than**	P. badius	Jack fruit tree
P. pachyphioeus	**ARTOCARPUS**	P. coffeatoporus	
P. robiniae	P. mcgregori	P. crocatus	
P. senex	P. sanctigeorgii	P. griseoporus	
P. sublinteus	P. melanoderms	P. lamaensis	
Bark and	P. sublintus	**P. linteus**	
wood of	P. torulosus	P. microcystideus	
Jack fruit tree	P. caryophyllii	P. microcystideus.	
	P. swieteniae	P. minutiporus	
	P. pectinatus	P. orientalis	
	P. callimorphus	P. pappianus	
	P. tropicalis	P. pectinatus	
	P. umbrinellus	P. poonensis sp. nov.	
	P. callimorphus	P. rickii	
	P. chryseus	P. sublinteus	
	P. shaferi		
	P. glaucescens		
	P. membranaceus		
	P. johnsonianus		
	P. ribis		
	P. sanjianli		

Table 4 b Adulterations used as Phansomba Sample

BHOR (1984)	BHOSLE (2002)
G. applanatum	G. lucidum
G. lucidum	G. orbiformum
	G. applanatum
	G. capense
	G. curtisii
	G. sessiliformae
	G. testaceum
	G. poonensis sp. nov.

Table 5 Growth rate and biomass of *Phellinus* cultures

Strains	Malt Extract agar Growth rate (cm day^{-1})	Malt extract broth Biomass (gm L$^{-1)}$)
Phellinus Strains		
P. merillii	0.45 – 0.6	63.12
P. sublinteus	0.32 – 0.38	41.053
P. gilvus	0.75 – 0.9	WP
P. loydii	0.2 – 0.23	WP
Ganoderma Strains		
G. stipitatum	0.706	116.6
G. poonensis	0.743	132.2
G. multiplicatum	0.597	122.6
G. testaceum	0.674	107.8
G. perzonatum	0.612	125.4
G. praelongum	1.059	121.4

(iii) *Study of Physico Chemical Parameters:*

Moisture content: One gram powder was kept in a previously weighed crucible and kept in oven at 80°C, the crucible was weighed at regular intervals till the weight was constant. Then the difference was calculated and moisture content was calculated in terms of percentage (table 6).

Table 6 Chemical composition of Phansomba samples powder extracts in percentage

Name of the extract	*Phellinus* (15 samples)	*Ganoderma* (10 samples)
Moisture content	5 – 15	6 – 10
Ash	2 – 3.5	2 – 4.4
Crude fibers	8 – 33	18 – 53
Ether extract	2 – 6	1 – 3
Alcohol extract	4 – 14	2 – 6
Aqueous extract	8 – 18	4 – 9
Alkaline extract	20 – 50	24 – 60

(*Source :* Vaidya and Bhor, 1991, Lamrood and Vaidya, 2003, and Bhosle 2004, unpublished data)

Ash content: One gram dried powder was taken in a pre weighed crucible, kept in furnace and ignited at 600°C for 2hrs, the furnace was cooled and the crucible was weighed again (the ash was checked for presence of black carbon particles, if present the crucible was re-ignited for one more hour). The ash content was calculated in terms of percentage (table 6).

Crude Fiber: was analysed using the method of Daniel (1991), the calculated crude fiber content are represented in terms of percentage (table 6).

Extractive Values: were analyzed using method described in Kokate (1997), the calculated values of different solvents are represented in terms of percentage (table 6).

Mineral and Metal Analysis: metal analysis was carried out using ***(AAS)*** Atomic Absorption Spectrophotometer (Varian and Perkin Elmer). The samples were processed prior to analysis. The samples were digested using diacid method; to one gram of sample powder 10ml of diacid mixture (HNO_3: HCl; 4:1) was added. The samples were then heated at 60°C till the reaction stops and further heated at 100 – 110°C till the solution gets cleared (3-4hrs). The solution was then filtered and diluted to 100ml and analysed under the AAS. The results are represented in terms of ppm (table 7).

General biochemical analysis: the different parameters studied under this topic are total carbohydrate content (Phenol-Sulphuric acid method); proteins (Lowry method); total free amino acid content (Sadasivam Manikam, 1997), phenol content (Sadasivam Manikam, 1997) (table 8)

General pharmacognostic study: This included qualitative analysis of amino acids, organic acids, phenols, sugars and terpenoids (table 9). This data is comparison between the samples of *Phellinus merrillii* and *P. fastuosus*, which has been reproduced from Vaidya & Bhor (1991). Similar kind of work is under progress for the samples of Phansomba (*Phellinus* and *Ganoderma*).

Bioactive compounds: apart from these studies extraction of secondary compounds (table 10) like Polysaccharide fractions (Mizuno, 1999), Terpenoids, viz Sesquiterpenes, Diterpenes and Triterpenes (Harborne, 1984), Ergosterol (Bhosle & Vaidya, 2002) and Lectins (Lis and Sharon, 1972) was done.

Table 7 Mineral and metal content in the samples (ppm).

Name of the metal	*Phellinus* (15 samples)	*Ganoderma* (10 samples)
Copper	0.01 – 1.3ppm	
Ferrous	*0.5 – 6ppm*	
Magnesium	0.9 – 7.9ppm	
Mangenese	0.06 – 0.77ppm	
Calcium*	1000 – 1130ppm	Work under progress
Lead*	trace	
Arsenic*	trace	
Cobalt	trace	

Table 8 Quantitative estimation of phansomba samples (percentage).

Estimation of	*Phellinus* (15 samples)	*Ganoderma*(10 samples)
Polyphenols	7 – 40%	0.05 – 2.5%
Total Carbohydrates	17 – 34%	11 – 36%
Proteins	9 to 17%	3.2 – 8.5%,
Total free Amino acids	0.43 to 2.17%.	0.5 – 4%

Table 9 Qualitative estimation of powder of phansomba samples (Vaidya & Bhor 1991).

Estimation of	*Phellinus fastuosus* (5 samples)	*Phellinus merrillii* (6 samples)
Free amino acids	B- alanine	—
	Aspartic acid	Aspartic acid
	Arginine	Arginine
	L-cystein	L-cystein
	Methionine	Methionine
Bound amino acids	L-cystein	Glutamic acid
	L-cystine	L-cystine
	Leucine	Leucine
Lipids	ND	ND
	Erythritol	Atranorin
	Lecanoric acid	Lecanoric acid
Organic acids	Salazinic acid	Salazinic acid
	Socellic acid	Socellic acid
	—	Erythrin

Table 9 Contd....

Estimation of	*Phellinus fastuosus* (5 samples)	*Phellinus merrillii* (6 samples)
Phenols	Catechol	Catechol
	Gallic acid	Gallic acid
	P-hydroxybenzoic acid	P-hydroxybenzoic acid
	4-methylresorcinol	2-methylresorcinol
	Salicylic acid	Orcinol
	Vanillic acid	Phloroglucinol
	—	Salicylic acid
Sugars	Arabinose	Fructose
	Galactose	Raffinose
	Maltose	Maltose
Terpenoids	Present	Present

ND = Not detected

Table 10 Comparative account of analysis of "Phansomba" samples

Name of the Test	*Phellinus*	*Ganoderma*
Lectins	*Phellinus badius,*	
P. membranaceous	Mild in *G. stipitatum*	
Alkaloids	Negligible	Negligible
Terpenoids	Variety of Diterpenes,	Variety of Diterpenes,
	Sesquiterpenes and Triterpenes	Sesquiterpenes and Triterpenes
Ergosterol	< 0.02%	0.001 – 0.04%
Germanium	0.2 – 0.5 ppm	WP
Polysaccharide	Qualitative	Qualitative

WP = Work in progress

(iv) *Antimicrobial Studies :*

Different extracts viz. water extracts, alcohol extracts, ethyl acetate extracts along with the polysaccharide fractions, Sesquiterpenes, Diterpenes and Triterpenes from 15 samples of *Phellinus* and 10 samples of *Ganoderma* were tested for their antimicrobial activity against human pathogenic microorganisms. The method used was well diffusion method (Barry, 1986).

Phellinus adamantinus; *P. coffeatoporus*; *P. merrillii*; *P. sublinteus* and *P. minutiporus*- strong activity against *E.coli* (724; 739); *Klebsiella pneumoniae* (432); *Proteus mirabilis* (1429) and *Candida ablicans* (3017; 1637), *Phellinus linteus* and *P. lloydii*- strong activity against *E.coli* (724; 739); *Klebsiella pneumoniae*; *Proteus mirabilis* and *Candida ablicans*, *Phellinus* rickii strong activity against *E.coli*; *Klebsiella pneumoniae*; *Proteus mirabilis*; *Candida ablicans*, *Phellinus fastuosus* - strong activity against *E.coli*; *Candida ablicans*, *Phellinus pappianus* - strong activity against *E.coli*; *Klebsiells pneumoniae*, *Phellinus merrillii* - strong activity against *E.coli*, *Phellinus badius* - moderate activity against *E.coli*; *Klebsiells pneumoniae* and *Candida ablicans*, *Phellinus griseoporus* and *P. crocatus* - moderate activity against *E.coli* and *Candida ablicans*, *Phellinus aureobruneus* - moderate activity against only *E.coli*. The water extracts of *Ganoderma lipsiense* showed maximum inhibition against *E.coli* 724. Triterpenoid extracts and Polysaccharide fractions of *Ganoderma* and *Phellinus* were compared with the standard, commercial antibiotics like Flucanazol, Tetracycline, Netilmicin, Ciprofloxacin, Amikasin, Ceptazidime, Cefotaxime, Ceftrixone, using *E.coli* 739, *Klebsiella pneumonia* 432, *Proteus* mirabilis 1429 and *Candida albicans* 3017 as test microorganisms. The results revealed that total terpenoid extract of *Ganoderma stipitatum* and *Phellinus adamantinus P. fastuosus, P. linteus, P. lloydii, P. merrillii, P. minutiporus* and *P. sublinteus* as well as Polysaccharide fraction of *Ganoderma stipitatum, G. poonensis, G. multiplicatum, G. multicornum, G. perzonatum, G. praelongum, G. chalceum* and *G. lucidum* var. *lucidum* and *Phellinus adamantinus, P. merrillii, P. minutiporus* and *P. sublinteus* showed strong activity as compared to standard antibiotics.

On the basis of the experimental evidence in this investigation, it appears that *Phellinus* could be a useful source for exploiting 'Antimicrobial compounds'. The work is in progress for the activity test against other microorganisms as well as the partial purification of the active compound in this laboratory. Recently, crude terpenoids (Diterpene, Sesquiterpenes and Triterpenes) and polysaccharide fractions were also tested for their antimicrobial action and the results are encouraging.

(v) *Novel Compound Study*

A methanolic extract of *Phellinus merrillii* was fractionated and three compounds were purified. The compounds are not named yet and further work on structural determination is going on but it was revealed that two compound are a straight chain molecule while one have aromatic rings attached to a sugar base (correct identification of these compounds is yet to be completed) seven compounds have been purified from the species *Phellinus lloydii*.

(vi) *Toxicity Studies*

Different concentrations of cold-water extract of medicinal mushroom *Phellinus merrillii* were tested for their toxicity by using *Allium* assay. Treated *Allium cepa* root tips did not show any chromosomal aberrations at a dose of 0.029 gm to 2.000 gm. However, decrease in mitotic index; root length and percent germination was observed with increasing concentration (from lower concentration to higher concentrations) whereas cell size was observed to be increasing. Thus, the results observed in *Allium* indicate no serious toxicity issues, indicating that the compounds in question may be safe in human therapy, and that the encouraging results indicate the great value of further study. Recently different concentrations of hot water extract were tested for their toxicity and EC50 was found to be 16.3 gm., which clearly indicates that the sample may be safe in human therapy.

(vii) *Ergosterol Analysis*

A new modified method was employed for Ergosterol Analysis (unpublished data) for 21 species of *Ganoderma*, the percentage ergosterol was found to be in range of 0.065 to 0.349, a range reported for *Ganoderma lucidum* (Reishi) is 0.03 – 0.04% (Mizuno, 2000).

(viii) *Anti Cancer Study*

Preliminary experiments to check anticarcinogenic activity using human cervical carcinoma cells (SiHa cells) were carried out with two species of Phansomba, *Phellinus merrillii.* and *P. fastuosus*. The study suggests that the hot and cold water extracts as well as hot and cold 50% ethanol extracts of these species show anticarcinogenic activity. Our studies revealed that cold and hot water extracts of *Ph. merrillii*, and cold and hot ethanol extracts of both species have high cytotoxic effect on SiHa cells. *Ph. merrillii* cold and hot water extracts demonstrated higher cytotoxic effect than those from *Ph. fastuosus*, while cold and hot ethanol extracts from both species entirely inhibited SiHa cell growth at a concentration of 500 ìg. (Bhonde *et. al.*, 2002).

3. Future Plans

(a) Collection of many more samples from the Western Ghats, obtaining them and preserving in pure culture form and developing a culture bank for Phansomba samples.

(b) Development of solid state fermentation for commercial scale, extraction of polysaccharides and other active principles by using the solid state fermentation.

(c) Development of novel antibiotic from the biomass.

(d) Isolation of active compounds from the Phansomba samples, purification characterization and structure identification.

(e) Anticancer, anti-diabetic and antimicrobial analysis of the pure compounds.

4. Potential Prospects Of Phansomba As Herbal Medicine:

As the work progresses, more uniqueness of this drug will emerge. As we collect more samples the authenticity of the drug will be strengthened. This will also facilitate in erecting not only a specimen or culture bank, but the data generated will also be helpful for research in the same field or other fields also. The cultural techniques will help to generate new strains with better quality and quantity of bioactive compounds. The biomass production technique in forthcoming period will enable to produce the compounds at least for the research, and if possible a commercial scale protocol has also been visualized. The physicochemical, or the biochemical work suggests that the drug has more amount of fiber (table 6) which is mainly contributed by the large molecules of polysaccharides, both branched and unbranched, some of which are known as anti-cancerous (Wasser and Weis 1999). The composition of which will be ascertained by the qualitative study of the isolated polysaccharide fraction. The compositions will at least reveal the nature of the polysaccharide and which may help in purification and characterization of the compound. The mineral and metal content show traces of toxic metals, well in the regulations put for the herbal drugs. It also shows that the clinically useful metals like germanium and selenium are in adequate quantity (table 6) and is equivalent to the general polypore mushroom content of Reishi, (Mizuno, *et. al.*, 1988). The extractive values (table 6) were useful especially during the extraction of terpenoids and the crude fibers, even during the quantification of ergosterol. The alkaline extractive value suggests that the alkali soluble polysaccharides are present in more quantity than that of the water soluble.

The in house technique developed for bioactivity testing like, antimicrobial or lectin assay, or *Allium* root tip assay for toxicity testing has helped in understanding the potential of the drug. The results have proven that this herbal drug comprises qualities like anti bacterial, anti candidial, lectin activity, which is also supposed to play an important role in anti cancer activity, and even the crude extracts have the cancer prevention activity against the Human cervical carcinoma. The purification of the active compounds as well the testing of the unknown purified compounds will open a new era.

These tests are inadequate to completely judge the drug, furthermore assays, tests are required, especially like anti-diabetic anti tumor / anti cancer. This may be done using animal trials and or using cell lines. Though the aspects like dose designing and therapy development have not been sighted because after analyzing toxicity and other activities, animal studies and human trials are the main stages which requires more facilities, group work and database development.

Over all the 'Phansomba' does open a new chapter in field of ethnomycology, a complete study of an organism possessing folk identity as a drug, covering the general biochemistry to isolation of novel compounds, from testing the crude fraction against bacteria and *Candida* to the human cervical carcinoma.

References

Ahmad, Saboor; Hussain, Ghausia; Razaq, Shamim. 1976. Triterpenoids of *Phellinus gilvus. Phytochemistry.* 15(2): 2000.

Ajith T.A., and *Janardhanan K.K.* 2001. Antioxidant and anti-inflammatory activities of methanol extract of *Phellinus* rimosus (Berk.) pilat. *Indian Journal of Experimental Biology,* 39, 1166-1169.

Andhalkar R.V. 1988. The uses of Phansomba in management of Mukhpaka (Sarvasara). M.D. (Ayu.). Thesis, University of Pune.

Bakshi, B. K. 1971. Indian Polyporaceae. Published by, Indian council of Agriculture Research, New Delhi, India.

Bakshi B. K. 1976. Forest Pathology – Principles and Practices in Forestry. Forest Research Institute, Dehradun, pp. 400.

Barry, A. L. 1986. Procedure for testing antimicrobial agents in agar media, Antibiotics in laboratory medicine, Second edition, published by: Williams and Wilkins, U.S.A. p. 1-26.

Bhonde R. R. Lamrood P.Y. and *Vaidya J.G.* 2002. Anricarcinogenic activity of two samples of Phansomba, *Phellinus merrillii* (Murr.) Ryv. and *Ph. fastuosus* (Lev.) Ryv., on SiHa cell lines. Int J Med Mushr, 4, 133-138.

Bhosle, S.R., and *Vaidya. J.G.* 2002. "Semi-micro determination of Ergosterol from some Aphyllophorales" Poster presented at National Conference on 'Current Trends in Plant Sciences' held at Department of Botany, University of Pune, during 26-28th March 2002.

Chang S.T. 1999. World production of cultivated edible and medicinal mushrooms in 1997 with emphasis on *Lentinus edodes* (Berk.) Sing. in China. *Int. J. med. mushr.,* 1, 291-300.

Chang S.T. and *Buswell, J.A.* 1999. *Ganoderma lucidum* (Curt.: Fr.) P.Karst. (Aphyllophoromycetideae)- A Mushrooming Medicinal Mushroom. *Int. J. Med. Mush.* 1: 139-146.

Chiu. et. al. 2000. Developmental Plasticity of Hong Kong Lingzhi as a Response to the Environment. Abstract from ISMS Congress 2000, May 15 - 19, Maastricht, The Netherlands. URL:

Chopra, R.N.; Nayar, S.L., and Chopra L.C. 1956. Glossary of Indian medicinal plants. ICAR Pub., New Delhi,0 pp 9 and 27.

Daniel M. 1991. Methods in Plant Chemistry and Economic Botany. Kalyani Publishers, New Delhi, INDIA. Pp 210.

Desai V.G. 1927. Aushadhi sangrah - medicinal plants from India: Identity, characters and uses. Shree Gajanan Book Depot, Dadar, (Bombay) Mumbai, India, pp 285-289.

Deshpande G.G. 2002. Some live standing tree diseases caused by Aphyllophorales from Campus of Pune University and some other places from Pune City. M.Phil. Thesis, University of Pune.

Dymock, W., Wardern, C.J., and Hooper, D. 1890. Pharmachographia Indica - a history of the principal drugs of vegetable origin. Part III. Education Society Press, Byculla, (Bombay) Mumbai, pp 629-635.

Gilbertson R.L. and Ryvarden L. 1986. North American Polypores. Published by Fungiflora, Oslo, Norway.

Gogate V.M. 1972. Dravyagunvigyan Ayurvedic materia medica. Maharashtra Vidyapeeth Granth Nirmiti Mandal. Continental Prakashan, Pune, India.

Gottlieb A.M. and Wrigth J.E. 1999 a. Taxonomy of Ganoderma from the southern South America: subgenus *Ganoderma. Mycol. Res.* 103(6), 661-673.

Gottlieb A.M. and Wrigth J.E. 1999 b. Taxonomy of Ganoderma from the southern South America: subgenus *Elfvingia*. Mycol. Res. 103(10), 1289-1298.

Harborne J.B. 1984. Phytochemical Methods. Chapman and Hall, London, II edition.

Harsh N.S.K., Rai B.K., and Tiwari D.P. 1993. Uses of *Ganoderma ludidum* in Folk Medicine. *Indian J. Trop. Biod.* 1, 324-326.

Harsh N.S.K., Tiwari C.K., and Rai B.K. 1996. Forest Fungi in the Aid of Tribal Women of Madhya Pradesh. *Sustainable Forestry*, 1, 10-15.

Hobbs C. 1996. Medicinal mushrooms: An exploration of tradition, healing and culture. Botanica Press, Santa Cruz, CA

Hseu R. S., Monocalvo J. M., Wang H. F. 1996. Molecular methods for *Ganoderma lucidum* complex systematic. 1996 Taipei International *Ganoderma* Research Conference p11-12. 1996 Taipei, Taiwan, R.O.C.

Janardhanan, K.K. 2003. Search for Cancer Chemoprevantive Agents from Mushrooms – Studies on tropical South Indian Mushrooms. Paper presented in National Symposium on Prospecting of Fungal Diversity and Emerging Technologies, held at Agharkar Research Institute, Pune, INDIA, February 6-7th 2003.

Jordan M. and David C. 1995. Biology of Higher Fungi in Encyclopedia of Fungi of Britain and Europe.

Khory R.N. 1887. Bombay Materia Medica and their therapeutics. Pri Raninas Union Press, pp 563-565.

Kirtikar Col. K.R. 1918. Indian medicinal plants. Part II. Pub. Sudhir Nath Basu, Bhuwaneshwar Ashram, Bhandurgun, Allahabad, pp 1394 - 1396.

Kokate C.K. 1997. Practical Pharmacognosy. Vallabh Prakashan, New Delhi, INDIA pp186.

Kwon H. J. 2000. Medicinal Effects of *Phellinus linteus.*

Kwon H. J. 2001. Rare, Unknown but Marvelous Medicinal Mushroom: Sang-Hwang, *Phellinus linteus*; Mulberry Yellow)

Larsen M. J., and Cobb-Poulle L. A. 1990. *Phellinus* (Hymenochaetaceae). A survey of the world texa. *Synopsis Fungorum* 3, Fungiflora, Oslo, Norway.

Lis H., and Sharon N. 1972. Soybean agglutinin. Method of Enzymology, Ginsberg V. ed., New York, 38, 360-368.

Mizuno T. 2000. Development of an antitumor response modifier from *Phellinus linteus* (Berk. Et Curt.) Teng (Aphyllophoromycetideae) (Review). *Int. J. Med. Mushr.* **2**, 21-34.

Mizuno Takahsi. 1999. The Extraction and Development of Antitumor- Active Polysaccharides from Medicinal Mushrooms in Japan (Review). *Int. J. Med. Mushr.* **1**: 9-29.

Mizuno, T., Ohtahara S. and Li J. 1988. Mineral Composition and Germanium Content of Several Medicinal Mushrooms. *Bull. Fac. Agr. Shizuoka Univ.* 38: 37-46.

Mo, Shun Yan; He, Wen Yi; Yang, Yong Chun; Shi, Jian Gong. 2003. Two Benzyl Dihydroflavones from *Phellinus ignarius. Chinese Chemical Letters*, 14(8): 810-813.

Nadkarni, K.M. 1954. Indian Materia Medica, Vol. I 3rd edit. Popular Book Depot, (Bombay) Mumbai, pp51-52 and 202.

Nanal, P.G. 1929. Sarthbhavpraksh (Marathi and Sanskrit), 1st edit. Pune, India, 206.

Omar M.B., Bolland L. and Heather W.A. 1979. A permanent mounting medium for fungi. *Bulletin of the British Mycological Society.* 13 (1), 231-232.

Ryvarden L. 2000. Studies in Neotropical Polypores 2: a Preliminary Key to Neotropical species of *Ganoderma* with a laccate pileus. *Mycologia.* 92(1), 180-191.

Ryvarden L. and Johansen I. 1980. A preliminary Polypore Flora of East Africa. Published by Fungiflora, Oslo, Norway.

Sadasivam, S and Manikam, A. 1997. Biochemical Methods. 2nd Edn. New Age International (P) ltd. Publishers. Pp 256.

Sathe, K.N. 1982. Gharguti Aushadhe. Ayurved Bhawan, Kakawati, (Bombay) Mumbai, India, 225 pp.

Sawant, S.Y. 1974. Medicinal Plants of Maharashtra. Continental Prakashan, Pune, India, pp 1-23.

Serck-Hanssen, Klaus; Wikstrom, Camilla. 1978. Novel 7-phenylheptane-3-ones from the fungus *Phellinus tremulae. Phytochemistry.* 17(9): 1678-1679.

Shah, B.C. 1928. Nighantu adarsh. Part II. A treatise on vegetable materia medica. Ahmedabad. pp. 622-625.

Singh, D. 1974. Unani Dravyagunadarsh. Vol II. Unani Pharmacopoeia, Ayurved and Tibbi Academy, Lucknow, U.P. India. pp. 204, 920.

Stamets P. 2000. Growing gourmet and medicinal mushrooms. 3d ed. Ten Speed Press, Berkeley, CA.

Steyaert R.L. 1972. Species of *Ganoderma* and related Genera, mainly of the Bogor and Lieden Herbaria. *Personia* 7(1), 55-118.

Steyaert R.L. 1980. Study of some *Ganoderma* species. *Bull. Jard. Bot. Nat. Belg.* **50**, 135 - 186.

Vaidya J.G. 1991. Phanas-alombe a True Mushroom or Phansomba. In: Proceedings of National Symposium on Mushrooms. (ed. Nair, M.C.), pp. 216-219.

Vaidya J.G. 1995. Biology of the Fungi. 1st ed. Satyajeet Prakashan. Kothrud, Pune – 38.

Vaidya, J.G. and Bhor, G.L. 1991. Medicinally important wood-rotting fungi with special emphasis on Phansomba. *Deerghayu Int.* 7, 16-19.

Vaidya, J.G. and Lamrood, P.Y. 2000. Traditional Medicinal Mushrooms and Fungi of India. *Int. J. Med. mushr.* **2:** 209-214.

Vaidya, J.G. and Rabba, A. S. 1993. Fungi in folk medicine. *Mycologist.* 7, 131-133.

Wasser S. P. and Weis A. L. 1999. Medicinal properties of substances occurring in higher Basidiomycetes mushrooms: Current perspectives (review). *Int. J. med. mushr.* **1**, 31-62

Wasser S.P. 2002. http://www.mushworld.com

Willard, T. 1990. Reishi Mushroom: Herb of Spiritual Potency and Medical Wonder. Sylvan Press, Washington. pp. 169.

Mass Production of *Trichoderma harzianum* on Various Indigenous Agricultural Products and Byproducts

Pratibha Sharma and S.K. Sain

Division of Plant Pathology, Indian Agricultural Research Institute, New Delhi-12, INDIA
Corresponding author Email: psharma032003@yahoo.co.in

ABSTRACT : Among the semi-solid and liquid substrates media evaluated for mass production of *Trichoderma harzianum,* sorghum grain and *Agaricus bisporus* mushroom extract (20%) was found to be the best. These media supplemented with K_2HPO_4, mancozeb and sucrose showed increased biomass production, spore concentration and spore viability. The total biomass production remained increased up to 40 days of incubation, however, the maximum spore concentration (6.7×10^{10} and 7.2×10^{10} g^{-1}) and spore viability (98.2 and 98.9 %) was observed after 25 and 20 days of incubation in both the lab formulations, respectively at 25 $\pm 2^0$C. The spores of *T. harzianum* remain viable up to 12 months at 5^0, 20^0 and 30 ± 5^0C storage in both the bioformulations, although at 5^0 C the spore viability remain higher (62.1-65.8%) as compared to 20^0 and 30 ± 5^0C after 12 month storage. A seed treatment dose of *T. harzianum* @ 2-4 g kg^{-1} seeds could be recommended for cauliflower and tomato seed treatment, which eventually resulted in, improved seed germination, plant vigour index and provide protection against soil borne diseases.

KEYWORDS : Mass production, agricultural products, byproducts, chemicals, *Trichoderma harzianum*

The biocontrol of plant pathogenic fungi was first described in 1926's, since then various kinds of biocontrol products have appeared in commercial market and several attempts are being made to standardize techniques for mass production of biocontrol agent by solid and liquid state fermentation with little or more effectiveness and shelf life. The largest technological impediment to the use of biocontrol agents is the lack of proper technology and facilities for biologically and economically – effective production of biocontrol agents which should show high spore concentration and long shelf-life in commercially available biocontrol products under field conditions. For the popularization of use of bioformulation the first and foremost step to be considered is the mass production of a biocontrol agent on a suitable medium which results in biomass with excellent shelf life even at adverse storage conditions with sufficient amount of viable spores.

Therefore, attempts were made to develop a mass production medium for *Trichoderma harzianum* using indigenous agricultural products and byproducts under the NATP CGP/399 entitled "Development of bioformulation from the improved strains of *Trichoderma* spp. for cauliflower and tomato"

Materials and Methods

Biocontrol fungal culture

The stock cultures of *T. harzianum* and *T. viride* isolates available in Biocontrol Laboratory, Division of Plant Pathology, IARI, New Delhi were used. The cultures were sub-cultured and maintained on PDA and *Trichoderma* Specific Medium (TSM) slants. For the experiment these cultures were grown in Pertiplates containing TSM medium.

Substrates for biomass production

In order to find out an inexpensive and suitable medium for the mass multiplication of *T. harzianum*, the following substrates were tested:

Semi-solid state multiplication :

1. Grains - sorghum, rice, wheat, pearl millet, and cluster bean.
2. Oil cakes- neem, coconut, groundnut, mustard, cotton seed, and sesame.
3. Grass straw- *Eragostis tenella* (L) P. Beauv. ex.Schult
4. Byproducts- rice bran, wheat bran, sugarcane bagasse, sugarcane molasses and refuge of *Agaricus bisporus* mushroom and its stipes at the packing.

Liquid state multiplication :

1. Liquid media containing: potato dextrose broth, TSM, oatmeal broth, coconut water, mushroom extract (5,10 & 20%) and sugarcane molasses (2, 4 & 6%).
2. Chemicals: $CaCO_3$ (0.4g L^{-1}), K_2HPO_4 (0.18g L^{-1}), sucrose (0.2gL^{-1}) and mancozeb (0.2g L^{-1}).

For the semi-solid substrate multiplication, pre-soaked dried and semidried substrates were autoclaved at 121^0C at 15 psi for 1 hr. For liquid state multiplication oatmeal, potato dextrose, TSM, mushroom extract and molasses broth media were prepared and sterilized. Flasks filled with these substrates were inoculated with spore suspension (10^6 ml^{-1}) of 7 days old *T. harzianum* (which showed superior performance in controlling plant diseases amongst 12 isolates of *T. harzianum* and *T. viride,*) (Sharma, 2004) cultures and incubated at 25 ± 2^0C for 20 days in BOD incubator. After the incubation period liquid substrates were filtered through Whatman No. 42 filter paper and the mycelium and spores thus obtained were air-dried, and powdered separately. Whereas, each of the semi-solid substrates was air-dried directly, powdered and sieved through 80-mesh sieve. Observation on total biomass production, spore concentration and spore viability were recorded in each substrate separately using Haemocytometer and dilution plate technique, respectively.

Effect of chemical supplements on biomass production rate

Further, one substrate among each semi-solid and liquid substrates found best for multiplication was used to determine the effect of low cost chemicals as a supplement on mass multiplication, spore production and spore viability. The $CaCo_3$, K_2HPO_4, sucrose and mancozeb were added to the specified medium alone and in their combinations. These media filled flasks were inoculated with *T. harzianum* isolate, incubated at 25 ± 2^0C in BOD incubator and the final observation was taken after 20 days of incubation.

Effect of incubation period on mass multiplication

The substrate (liquid and semi-solid) was found to be best for multiplication, was used to determine the effect of incubation period on total biomass, spore production and spore viability of *T. harzianum* in the final lab formulation. The sterilized and chemical supplemented liquid and semisolid substrate were inoculated with *T. harzianum* in the flasks and incubated at 25 ± 2^0C in BOD for different periods viz. 5, 10, 15, 20, 25, 30, 35 and 40 days. The observations on total biomass, spore concentration and spore viability were recorded at above mentioned time intervals.

Shelf life of lab bioformulations at various storage temperatures

The powdered bioformulation from *T. harzianum* in best suitable substrates (liquid and semi-solid) was stored at 5^0, 10^0 and 25^0 C to determine the shelf life of the bioformulation. The spore viability in terms of shelf life was measured at one-month interval to up to one year by dilution plate technique.

Effect of application doses on plant vigour index

To find out the suitable dose of biocontrol agent bioformulations for seed treatment of tomato (Pusa Sheetal) and cauliflower (Pusa Synthetic) seeds were treated with different amount of bioformulation (10×10^6 spore ml^{-1}) viz. 2, 4, 6, 8, and 10 g kg^{-1} seeds. The seeds were subjected for their germination and seedling vigour in 12" plastic pots filled with sterilized potting substrates. Fifteen seeds were placed in each pot for each treatment separately, incubated at 25 ± 5^0C in green house and were watered regularly. The germination of seeds of both the crops was recorded after 7 days and percent germination was calculated. Root and shoot length was measured 15 days after seed sowing and the vigour index was calculated by using the formula (Abdul Baki and Anderson, 1973):
Vigor Index = Mean root length + Mean shoot length x Germination (%)

For comparison of variance in data recorded during this study were subjected to analysis of variance using CRBD with three replications. Factorial RBD and arcsign transformed data were also used where it applicable.

Results and Discussion

Effect of various agricultural products and byproducts on biomass of *T. harzianum;*

In this study, among various indigenous agricultural products and byproducts tested, *T. harzianum* grew profusely onallmost all of the substrates except the rice grain and sugarcanes bagasse, on which *T. harzianum* showed partial colonization. However, among the semisolid substrates the maximum biomass and spore production were observed on sorghum grains (450.9 g kg^{-1} and 7.2 x 10^9 spores g^{-1}, respectively) followed by mushroom byproduct (426.1 g k^{-1} and 6.8 x 10^9 spores g^{-1}, respectively), pearl millet (392.5 g kg^{-1} and 6.1 x 10^9 spores g^{-1}, respectively) and groundnut cake (392.0 g kg^{-1} and 6.2 x 10^8 spores g^{-1}, respectively) (Table-1). The percent spore viability in all the dried preparations remained at par in these treatments, however, it was maximum in sorghum grains (95.4%) followed by mushroom substrate (94.3%), pearl millet (93.7%) and cluster bean (90.8%), *Eragostis tenella* (90.8%).

Table 1 Effect of solid and semi-solid agricultural products and byproducts on mass multiplication of *Trichoderma harzianum*

Substrate	Total biomass (g kg^{-1} dry substrate)	Spore production (spore g^{-1} dry substrate)	Per cent viable spores
Cluster bean grains	375.1	4.6 x 10^7	90.8 (72.34)
Sorghum grains	450.9	7.2 x 10^9	95.4 (71.61)
Rice grains	195.1	2.9 x 10^3	85.2 (67.37)
Wheat grains	257.2	4.5 x 10^5	88.2 (69.91)
Pearl millet grains	392.5	6.1 x 10^9	93.7 (75.46)
Rice bran	223.4	3.1 x 10^5	86.2 (68.19)
Wheat bran	280.6	4.9 x 10^8	89.9 (71.47)
Neem cake	353.7	4.8 x 10^8	87.5 (69.31)
Ground nut cake	392.0	6.2 x 10^8	90.6 (72.15)
Sesemum cake	375.2	5.5 x 10^8	89.2 (70.81)
Cotton seed cake	275.6	2.1 x 10^8	84.3 (66.66)
Mustered cake	362.1	4.9 x 10^8	87.8 (68.70)
Coconut cake	361.8	3.7 x 10^8	84.2 (66.58)
Mushroom byproducts	426.1	6.8 x 10^9	94.3 (76.19)
Sugarcane molasses	341.5	4.2 x 10^9	90.1 (71.66)
Sugar cane baggase	205.4	2.3 x 10^6	87.3 (69.12)
Eragostis tenella straw	326.1	6.5 x 10^8	90.8 (72.24)
CD (p=0.05)	22.099		5.440

*The data in parenthesis are arsine transformed prior to analysis.

Among the liquid substrate media the overall spore production was higher and total biomass production was lesser in all substrates as compared to semisolid substrate media for mass production. All liquid substrates showed its higher spore concentration at 9th and 10th dilution with the highest in mushroom 20 % extract (5.6 x 10^{10} spore g^{-1}) followed by molasses 6% (5.3 x 10^{10} spore g^{-1}), mushroom 10% extract (4.2 x 10^{10} spore g^{-1}), and molasses 4% (3.4 x 10^{10} spore g^{-1}) (Table-2). The maximum total biomass was recorded in mushroom 20% extracts (9.22 g L^{-1}) followed by mushroom extract at 10 and 5 % with the production of 9.16 and 9.12 g L^{-1}, respectively. All the dried bioformulations from liquid media showed higher spore viability with the highest in the 20% mushroom extracts (95.2 %) followed by mushroom extract 10%, molasses 6% and mushroom 5% with the percentage of 94.8, 93.7 and 93.2, respectively.

Table 2. Effect of liquid agricultural products and byproducts on mass multiplication of *Trichoderma harzianum*

Substrate	Total biomass (g kg^{-1} dry substrate)	Spore production (spore g^{-1} dry substrate)	Per cent viable spores
Potato dextrose broth	8.01	6.7 x 10^9	88.3 (70.00)
TSM broth	8.08	6.9 x 10^9	88.9 (70.54)
Oat meal broth	7.52	4.5 x 10^9	87.3 (69.04)
Coconut water	8.25	7.3 x 10^9	92.3 (73.89)
Mushroom extract 5 %	9.12	2.1 x 10^{10}	93.2 (74.88)
Mushroom extract 10 %	9.16	4.2 x 10^{10}	94.8 (76.82)
Mushroom extract 20 %	9.22	5.6 x 10^{10}	95.2 (77.34)
Molasses broth 2 %	8.29	1.8 x 10^{10}	91.5 (73.05)
Molasses broth 4 %	8.32	3.4 x 10^{10}	92.8 (74.55)
Molasses broth 6 %	9.04	5.3 x 10^{10}	93.7 (75.46)
CD (p=0.05)	1.664		2.593
*The data in parenthesis are arsine transformed prior to analysis.			

Kousalya *et. al.*, (1990) and Singh *et. al.*,(2001) have evaluated semi-solid substrates for the mass multiplication of *Trichoderma* isolates which showed less spore concentration and viability as compared to this present isolates and present bioformulations. Backman and Rodriquez-Kabana (1975) observed that *T. harzianum* exhibit better growth and delivery when formulated in diatomaceous earth with 10% molasses for the control of *Sclerotium rolfsii* in peanut.

Effect of chemical supplements on biomass of *T. harzianum* in best suitable medium

One medium among each semi-solid (sorghum grains) and liquid substrate media (mushroom extract 20%) found to be the best for mass multiplication were further tested to determine the effect of chemicals as supplements on total mass production, spore concentration and spore viability. Among the four chemicals viz. $CaCO_3$, K_2HPO_4, mancozeb and sucrose which added to the media alone and in their different combinations, resulted in enhanced ability in semi solid and liquid medium to increase total biomass, spore concentration and spore viability except the $CaCO_3$ and its combination with others, which showed slight decrease in mass multiplication in both the media.

The maximum total biomass, spore production and spore viability percentage in sorghum grain was with the K_2HPO_4 + sucrose + mancozeb combination (526.8 g k^{-1}, 5.8 x 10^{10} spore g^{-1} and 98.2%, respectively) followed by sucrose + mancozeb (510.7 g k^{-1}, 4.2 x 10^{10} spore g^{-1} and 97.1%, respectively), sucrose +K_2HPO_4 (488.4 g k^{-1}, 3.8 x 10^{10} spore g^{-1} and 97.2 %, respectively) and mancozeb + K_2HPO_4 (484.3 g k^{-1}, 9.1 x 10^9 spore g^{-1} and 97.8%, respectively) (Table 3). Similar trend of results was observed in liquid mushroom (20%) extract medium supplemented with above-mentioned chemicals. The highest total biomass, spore concentration and spore viability was observed with the supplement of K_2HPO_4 + sucrose + mancozeb combination (10.52 g L^{-1}, 6.2 x 10^{10} spore g^{-1} and 98.9 %, respectively).

Effect of incubation period on mass multiplication

Data in table 4 revealed that the spore concentration and spore viability both in sorghum grains and mushroom extract (20%) medium increased up to 25 and 20 days of incubation period, respectively, thereafter it decreased up to 40 days of incubation. Although, the mass production in both the media significantly increased up to 25 days of incubation thereafter, it increased in descending order up to 40 days. The present observation conclude that the 20-25 days incubation period is best suitable for mass multiplication in both the medium which resulted in maximum spore concentration, spore viability and biomass as well.

Sangle *et. al.,* (2003) has also reported the maximum spore concentration (8 x 10^9 spores g^{-1}), spore viability (92.5%) and total biomass (370.8 g kg^{-1} substrate) at 15 and 25 days of incubation, respectively at 27 ± 1^0C. However, Singh *et al.* (2001) had reported maximum conidial viability and spore concentration in 15 and 30 days old *T.virens* cultures on sorghum grain substrate, respectively.

Shelf life of lab bioformulations at various storage temperatures

A perusal of data presented in Table 5 indicates that over a period of 10 months at 5^0C temperature, spore viability in sorghum based bioformulation gradually reduced from 95.4 to 80.6 %. However, in mushroom extract based bioformulation it decreased from 95.8 to 85.2%. The sudden decline in spore viability was observed after 10 months. Although, the decrease in spore viability at 25^0 and 30 ± 5 ^{0}C was higher than that of 5^0C temperature storage.

Table 3 Effect of chemical supplements on semi-solid (sorghum grain) and liquid (mushroom extract) substrate on mass multiplication of *Trichoderma harzianum*

Suppliments g L^{-1} medium	Sorghum grains			Mushroom extracts (20%)		
	Total biomass (g kg^{-1} dry substrate)	Spore production (spores g^{-1} dry substrate)	Per cent viable spores	Total biomass (g L^{-1} dry substrate)	Spore production	Per cent viable spores
CaCo$_3$ 4g	445.6	6.8 x 10^9	89.5 (71.09)	8.15	7.1 x 10^9	89.9 (71.47)
Sucrose 0.2g	469.8	8.6 x 10^9	96.2 (78.96)	9.59	8.9 x 10^9	96.8 (79.69)
K$_2$HPO$_4$ 0.18g	455.9	8.2 x 10^9	96.7 (79.53)	9.28	8.8 x 10^9	97.2 (80.37)
Mancozeb 0.2g	471.2	1.2 x 10^{10}	96.9 (79.85)	9.82	2.5 x 10^{10}	97.5 (80.90)
CaCo$_3$ + sucrose	457.8	8.1 x 10^9	91.6 (73.15)	9.28	8.4 x 10^9	92.3 (73.78)
CaCO$_3$ + K$_2$HPO$_4$	450.3	7.9 x 10^9	90.2 (71.76)	9.24	8.1 x 10^9	90.9 (72.44)
CaCO$_3$ +mancozeb	461.2	8.5 x 10^9	91.3 (72.84)	9.41	8.5 x 10^9	92.4 (74.00)
Sucrose +mancozeb	510.7	4.2 x 10^{10}	97.1 (80.19)	10.41	4.8 x 10^{10}	98.2 (82.29)
Sucrose+K$_2$HPO$_4$	488.4	3.8 x 10^{10}	97.2 (80.37)	10.25	4.1 x 10^{10}	98.3 (82.51)
K$_2$HPO$_4$+mancozeb	484.3	9.1 x 10^9	97.8 (81.47)	10.160	9.5 x 10^9	98.5 (82.73)
CaCO$_3$+sucrose+K$_2$HPO$_4$	482.9	3.1 x 10^{10}	94.2 (76.06)	10.02	3.5 x 10^{10}	95.2 (77.34)
K$_2$HPO$_4$+sucrose+mancozeb	526.8	5.8 x 10^{10}	98.2 (82.29)	10.52	6.2 x 10^{10}	98.9 (83.93)
Control	450.9	7.2 x 10^9	95.4 (77.61)	9.22	5.6 x 10^{10}	95.8 (78.17)
CD(p=0.05)	21.974		2.211	1.990		2.270

*The data in parenthesis are arsine transformed prior to analysis.

Table 4 Effect of incubation period on mass multiplication of *Trichoderma harzianum*

Incubation period (in days)	Sorghum grains			Mushroom extracts (20%)		
	Total biomass (g kg^{-1} dry substrate)	Spore production (spores g^{-1} dry substrate)	Per cent viable spores	Total biomass (g L^{-1} dry substrate)	Spore production	Per cent viable spores
5	296.8	2.5	69.3 (56.35)	5.2	3.6	70.6 (57.17)
10	450.3	4.2	85.4 (67.54)	8.3	4.4	87.4 (69.21)
15	507.8	5.3	92.6 (74.21)	9.4	5.4	93.5 (75.23)
20	526.8	5.	98.2 (82.29)	10.52	6.2	98.9 (83.98)
25	542.5	6.7	95.3 (77.48)	11.4	7.2	96.8 (79.86)
30	554.1	5.9	92.7 (74.32)	12.3	6.9	94.3 (76.19)
35	566.3	5.2	90.4 (71.95)	12.9	5.8	90.8 (72.34)
40	568.4	5.0	82.8 (65.05)	13.2	5.2	84.6 (66.89)
CD (p=0.05)	30.795	0.845	1.765	2.285	0.915	2.218

*The data in parenthesis are arsine transformed prior to analysis.

Table 5 Effect of storage period and temperature on shelf life of bioformulation

Storage period (in month)	Spore viability on sorghum substrate at different temperature			Spore viability on mushroom extracts at different temperature		
	5⁰C	25⁰C	30± 5⁰C	5⁰C	25⁰C	30± 5⁰C
0	95.4(77.61)	95.4(77.61)	95.4(77.61)	95.8(78.47)	95.8(78.17)	95.8(78.17)
1	95.1(77.21)	93.5(75.23)	91.2(72.74)	95.4(77.61)	94.5(77.75)	91.8(73.36)
2	94.6(76.56)	92.3(73.89)	87.2(69.04)	95.1(77.21)	93.8(75.58)	87.9(69.44)
3	93.5(75.35)	90.4(71.95)	83.4(65.96)	94.5(76.44)	92.4(74.00)	84.4(66.74)
4	92.3(73.89)	89.7(71.23)	80.2(63.58)	93.8(75.58)	90.6(72.15)	80.8(64.01)
5	91.2(72.74)	88.9(70.54)	76.1(60.73)	92.6(74.21)	88.8(70.45)	77.0(61.34)
6	90.2(71.76)	84.2(66.58)	70.2(56.91)	91.5(73.05)	85.5(67.62)	72.5(57.48)
7	89.2(70.81)	80.4(63.72)	66.1(54.39)	90.1(71.76)	81.4(64.45)	66.9(54.88)
8	85.5(67.62)	77.2(61.48)	61.7(51.77)	89.7(71.23)	78.2(62.17)	62.5(52.24)
9	82.2(65.05)	70.8(57.29)	56.6(48.79)	86.3(68.28)	72.5(58.37)	57.8(49.49)
10	80.6(63.87)	62.2(52.06)	51.2(45.69)	85.2(67.37)	64.6(53.49)	53.0(46.72)
11	76.4(60.94)	50.2(45.11)	48.3(44.03)	79.5(63.08)	52.2(46.26)	49.4(44.66)
12	62.1(51.41)	40.5(39.51)	38.4(38.29)	65.8(54.21)	43.7(41.38)	40.2(39.35)
CD (p=0.05)	Storage period =2.56; Temperature=3.52; Storage period x Temperature= 3.29			Storage period =2.88; Temperature=3.89; Storage period x Temperature= 4.92		

*The data in parenthesis are arsine transformed prior to analysis.

Effect of application doses on plant vigour index

To determine the optimum amount of *T. harzianum* mass preparation for seed treatment, different doses viz. 2, 4, 6, 8, and 10 g kg^{-1} seeds of cauliflower and tomato were evaluated in green house conditions. Results presented in Figure 1a & 1b as bar and line diagrams indicated that the significantly higher seed germination and plant vigour index in both the crops of cauliflower and tomato was observed when seeds were treated with powdered bioformulation of *T. harzianum* @ of 2,4,6 and 8 g kg^{-1} seeds.

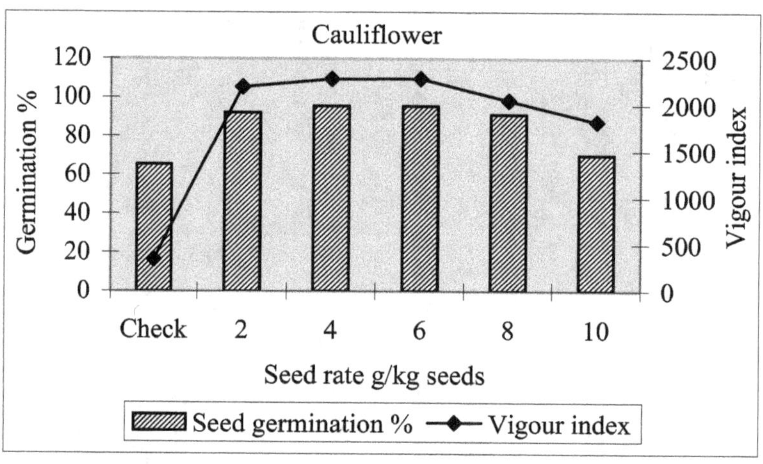

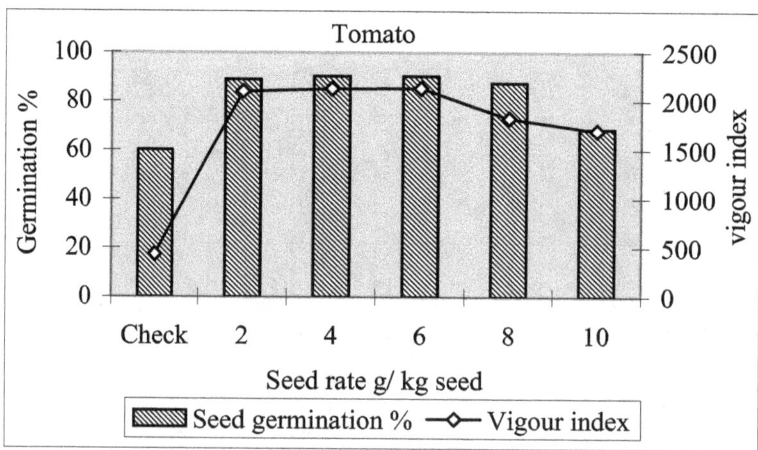

Figure 1 a, b. Effect of doses of bioformulation on seed germination and plant vigour index of tomato and cauliflower.

The results of doses 2, 4, 6, and 8 g Kg^{-1} for seed germination and vigour index were statistically at par. However, the maximum seed germination and vigour index was observed with the seed treatment dose @ 4 and 6 g kg^{-1} seeds in both the crop. Whereas, with the seed treatment with 10 g kg^{-1} rate, seed germination was decreased. Sharma and Sain (2004) and Sharma *et. al.,* (2003), reported that seed treatment from *T. harzianum* and *T. viride* (10^8 conidia ml^{-1}) suspension successfully controlled, *Pythium aphanidermatum* damping off and *Sclerotinia sclerotiorum* rot of cauliflower and tomato crop. Sharma (2002a, 2002b, 2002c and 2003) have also reported that use of *T. harzianum* was compatible with recommended doses of thiram (sulfur fungicide), captaf, iprodione (dicraboximide), carbendazim and mancozeb (bisdithiocarbamate) @2 g/kg seed treatment (4×10^6 c.f.u. ml^{-1}) and @ 2 g/lit water for seedling dip showed a good control of nursery diseases of vegetable corps like cauliflower, cabbage, tomato, chilli and brinjal.

In conclusion the laboratory and green house studies conducted during the present investigation clearly indicated that the sorghum grains (semi solid)and mushroom extract (liquid)along with some low cost chemicals like mancozeb, sucrose and K_2HPO_4 could be one of the best inexpensive and most suitable medium for mass production of the *T. harzianum* which resulted in significant higher biomass multiplication (77.5 %) at 20th day after incubation, 16.8 % as compared to sorghum grain alone, 60.7% as compared to wheat bran with higher spore concentration and spore viability. The seed treatment dose @ 2 - 4 g kg^{-1} seeds of the *T. harzianum* bioformulation could be the best recommended dose for cauliflower and tomato crops against soil borne diseases and increasing vigour index.

Acknowledgements

The authors thank NATP CGP-III for providing the funds in the form of project for conducting this research work.

References

Abdul Baki, A.A. and **Anderson J.D**. 1973. Vigor determination in soybean seed by multiple criteria. *Crop Science.* 13: 630-633.

Backham, P.A. and **Rodriquez-Kabana, R.** 1975. A system for growth and delivery of biological control agents to the soil. *Phytopathology.* 65: 819-821

Kousalya, G., Jeyrajan, R. and **Gangadharan, K.** 1990. Mass multiplication of Treichoderma spp. *J. Biol. Control.* 4: 70-71

Sangle, U.R., Bambawale, O.M., Nashim Ahmad and **Singh, S.K.** 2003. Substrate and temperature requirements for sporulation of sub-tropical isolates of *Trichoderma* spp. *Ann. Pl. Protec. Sci.* 11: 192-195

Sharma, Pratibha. 2002a. *Biological Management of Vegetable Nursery Diseases.* National Symposium on Agriculture in Changing Global Scenario held at I.A.R.I., New Delhi from Feb, 21-23. 407

Sharma, Pratibha. 2002b. Use of bioagents with pesticides in plant disease management. *New Agriculturist.* 13: 54-60

Sharma, Pratibha. 2002c. Use of improved *Trichoderma harzianum* isolates against major soil borne diseases of vegetable crops. In: 2nd International Agronomy Congress on Balancing Food and Environmental Security A Continuing Challenge, New Delhi, Nov. 26-30, 2002, Extended summaries Vol. 2: 1014-1016.

Sharma, Pratibha. 2003. Use of biocontrol agents in sustainable vegetable production, In: Proceedings of International Seminar on Downsizing Technology for Rural Development at Bhubaneswer Vol.I 193-202

Sharma, Pratibha. 2004. In: NATP CGP/III Project report "Development of bioformulation from improved strains of *Trichoderma* spp. for cauliflower and tomato". -20

Sharma, Pratibha and Sain, S.K. 2004. Induction of systemic resistance in tomato and cauliflower by *Trichoderma* spp. against stalk rot pathogen, *Sclerotinia sclerotiorum* (Lib) de Bary. *J. Biol. Control.* 18: 21-28

Sharma, Pratibha, Sain, S.K. and James, S. 2003. Compatibility Study of *Trichoderma* isolates With Fungicides Against Damping-off of Cauliflower and Tomato caused by *Pythium aphanidermatum. Pesticide Res. J.* 15: 133-138

Singh, G., Singh, U.S. and Mukhopadhyay, A.N. 2001. Mass production of spore powder of *Trichoderma virens* and its viability on coated lentil seeds. *Ann. Pl. Protec. Sci.* 9: 16-18

Singh, R.S., Singh, H.V., Puneet Singh and Jaspal Kaur. 2001. A comparison of different substrates for mass production of *Trichoderma. Ann. Pl. Protec. Sci.,* 9: 248-253

38 | Anatomical and Histochemical changes in the leaves of *Vigna unguiculata* (cowpea) due to inoculation with AM fungi and *Rhizobium*

Lakshmipathy, R., Balakrishna, A.N., Bagyaraj, D.J., Rajanna, M.D.* and Sharath, S.

Dept. of Agricultural Microbiology, University of Agricultural Sciences, GKVK Campus, Bangalore-560 065.

*Deptt. of Botany, University of Agricultural Sciences, GKVK Campus, Bangalore-560 065.

ABSTRACT

The anatomical and histochemical changes in the leaves of *Vigna unguiculata* due to inoculation with *Rhizobium* sp. and arbuscular mycorrhizal fungi *Glomus mosseae* and *Glomus fasciculatum* were studied. The epidermis, mesophyl tissue and vascular bundle thickness of cowpea leaves increased because of individual inoculation with AM fungi and *Rhizobium*. There was a significant increase in thickness of epidermis, mesophyll tissue and vascular bundles of cowpea leaves due to combined inoculation of *Rhizobium* plus *Glomus mosseae* compared to all other treatments. The same was true for the concentration of nucleic acids, proteins and polysaccharides in the leaf tissue.

KEYWORDS : *Rhizobium, G. mosseae, G. fasciculatum, cowpea*

Among the plant microbe symbiotic interactions, *Rhizobium* plays an important role with leguminous plants in improving growth by fixing atmospheric nitrogen. Mycorrhizal association is known to improve plant growth through improved mineral uptake, production of plant growth promoting substances, resistance to water stress, transplanting shock, protection against root pathogens, and synergistic interactions with beneficial soil microorganisms (Jeffries 1987, Bagyaraj, 1996). Research in the last few decades have brought out their potential application in agriculture and forestry with nursery raised transplanted crops (Burgess *et. al.*, 1993; Bagyaraj, 1992). Synergistic interaction between AM fungi and *Rhizobium* enhancing nodulation, percent AM root colonization and plant growth is well documented (Din and Moawad, 1998). Manjunath and Bagyaraj (1994) reported that dual inoculation of AM fungi and Rhizobium increased AM root colonization, N_2 content in some forage legumes and grain legumes. Shockley *et. al.*, (2004) suggested that legumes could be improved by the use of proper AM species and/ or *Rhizobium* strains.

A few studies have been made on the anatomical and histochemical modifications in the roots and leaves following mycorrhizal colonization. Ling-lee *et. al.*, (1977) studied the anatomy, ecology and physiology of eucalyptus ectomycorrhizae and also the histochemical distribution of polysaccharides in eucalyptus mycorrhiza. Doss *et. al.*, (1988) observed striking differences between mycorrhizal and non-mycorrhizal finger millet plants regarding size of roots, diameter of the root cap cells and carbohydrate content of the leaves. Krishna *et al.*, (1981) observed an increase in leaf thickness, size of the midrib veins, the mesophyll cells and number of plastids because of mycorrhizal colonization in the roots of finger millet. The leaves of mycorrhizal plants also contained higher amounts of insoluble polysaccharides and insoluble proteins. Gracy *et al.*, (2002) observed an increase in the concentration of RNA and protein in the roots and leaves of *Simarouba glauca* because of dual inoculation with *Glomus mosseae* and mycorrhizae helper bacterium (MHB) *Bacillus coagulans*. But information on the anatomical and histochemical changes brought out in the host by *Rhizobium* and mycorrhizal colonization is scanty (Chilvers *et. al.*, 1987; Gianinazzi and Gianinazzi - Pearson, 1992). The present work reports on the study of anatomical and histochemical changes in the leaves of *Vigna unguiculata* under the influence of single and/or dual inoculation with *Rhizobium* and AM fungi.

Material and Methods

Glomus fasciculatum and *Glomus mosseae* used in the study were obtained from the culture collection of AM fungi maintained in the glasshouse of the Department of Agricultural Microbiology, University of Agricultural Sciences, Bangalore using sand: soil (1:1 v/v) as the substrate and Rhodes grass (*Chloris gayana*) as the host. The AM fungal inoculum was added at the rate of 12,500 infective propagules per pot based on the most probable number estimation (Porter, 1979). *Rhizobium* was grown in modified YEMA broth based liquid biofertilizer medium for 5 days. Cowpea seeds were surface sterilized with 1% NaOCl for 1 minute and then washed thoroughly with water were treated with the liquid *Rhizobium* inoculant @ 3ml/kg seeds. Seeds treated with *Rhizobium* were sown at the rate of two seeds per pot containing 2.5 kg unsterilized soil: sand (2:1 v/v) mixture. The treatments, viz. uninoculated control, inoculated with *Rhizobium* (R), *G. mosseae* (G.m), *G. fasciculatum* (G.f), G.m+R and G.f + R, were replicated 4 times. The experiment was conducted in a glasshouse by following completely randomized design (CRD).

Leaf samples for anatomical and histochemical analysis were collected 90 days after sowing. Leaf sample of 0.5cm size including midrib region were collected from the photosynthetically active third leaf from top. Leaf samples collected were fixed in carnoy's B fixative (6:3:1 ethyl alcohol: chloroform: acetic acid) for 1 hour, dehydrated using n-butanol series and embedded in paraffin wax at 56^0c, serial sections of 10μm (leaf) and $12\ \mu$m (root) were taken and subjected to the following procedures.

Staining procedures

Test for insoluble protiens was carried out with mercuric bromophenol blue (MBB) in absolute alcohol as outlined by Mazia *et. al.,* (1953). The insoluble proteins appeared blue. For nucleic acids, hydrated sections were stained with 1% aqueous toluidine blue (TB) at 4.1 pH for 10 min as outlined by Feder and O'Brain (1968). The nucleic acid containing sites appeared blue (for DNA) and green (for RNA). For polysaccharides, hydrated sections were kept in per-iodic acid for 10-15 min and washed with water. They were kept in Schiff's reagent for 10-15 min as outlined by Jensen (1962). The polysaccharides appeared pink.

The leaf sections were also subjected to micrometry to measure the anatomical changes in the tissue. Appropriate control tests were conducted for each staining procedure.

Results and Discussions

Leaf structure

Considerable differences in leaf structures of cowpea on the basis of thickness of upper epidermis, palisade layer, spongy layer and vascular bundles were observed between the inoculated and uninoculated plants (Table 1). The thickness of the upper epidermis increased due to inoculation of *Rhizobium* and AM fungi.

Table 1 Anatomical changes in the leaves of cowpea due to inoculation with *Rhizobium* and AM fungi.

Treatments	Epidermis		Mesophyll tissue		Vascular bundle thickness (μm)
	Lower (μm)	Upper (μm)	Palisade (μm)	Spongy (μm)	
Uninoculated control	1.66	2.78	6.35	2.54	83.0
Rhizobium (R)	1.92	3.32	6.64	10.23	120.50
G. mosseae (G.m)	1.24	3.32	12.45	11.88	113.80
G.fasciculatum (G.f)	1.92	2.07	10.22	8.56	87.98
R + G.m	3.04	3.72	14.01	12.56	132.23
R + G.f	2.58	1.47	11.95	10.38	119.36
LSD @ 0.05%	0.06	0.06	0.16	0.19	11.55

Cells of the upper epidermis were largest in plants inoculated with *Rhizobium* plus *G. mosseae* (3.72 μm) and smallest in uninoculated plants (2.78 μm). Same trend was noticed in case of lower epidermis. The length of the mesophyl cells in the palisade layer was longer in the plants inoculated with *Rhizobium* plus *G.mosseae* (14.01 μm) which was followed by plants treated with *G. mosseae* alone (12.45 μm). The uninoculated plants had the smallest length of palisade cells, which measured 6.35 μm. Meosphyll cells with their increase in size have an extended cell wall surface exposed to intercellular air spaces, which would improve the gaseous excange between outside and the photosynthetic tissue.

The cell size in the spongy layer was larger in the leaves of plants treated with *Rhizobium* plus *G.mosseae* (12.56 µm). and it was 11.88 and 10.23 µm in plants singly inoculated with *G. mosseae* and *Rhizobium* respectively and smallest in uninoculated plants (5.24 µm). Considerable increase in the size of vascular bundles in *G. mosseae* plus *Rhizobium* treated plants reflects the higher rate of translocation of water and mineral nutrients compared to *G.mosseae* alone treated and uninoculated plants. This may be due to increased number of plastids and increased photosynthetic activities in the leaves of inoculated plants. Increased thickness of leaves, size of midrib veins, mesophyll cells, motor cells and number of plastids in leaves of fingermillet because of AM inoculation was reported by Doss *et. al.,* (1988). Similarly, Skoog and Schmitz (1972), Kinden and Brown (1975) observed the enlargement and better differentiation of vascular and other tissues in mycorrhizal plants. The results of the present study upholds the earlier observations made by other workers.

Metabolite Localization

Intense pink, blue and green + blue colour was taken by polysaccharides, proteins and nucleic acids respectively in the leaves of cowpea inoculated with AM fungi and *Rhizobium* respectively compared to uninoculated plants (Table 2). Enhanced concentration of nucleic acids and proteins in the leaves of finger millet and *Simarouba glauca* because of AM fungal inoculation has been reported earlier (Doss *et. al.,* 1968; Gracy *et. al.,* 2002). Maximum concentration of nucleic acids, proteins and polysaccharides occurred in the leaf tissue of cowpea plants co-inoculated with *G.mosseae* plus *Rhizobium*. Gracy *et. al.,* (2002) reported that co-inoculattion of *G. mosseae* + the MHB, *B. coagulans* resulted in higher concentration of RNA and proteins in the leaves of *S. glauca*. Higher concentration of nucleic acids, proteins and polysaccharides are indicative of increased metabolic activities, such as increase in the number (through mitosis) and size of the cells through elongation (Shah and Jain, 1981).

Table 2 Histochemical changes in the leaves of cowpea due to inoculation with *Rhizobium* and AM fungi.

Treatments	Total soluble polysaccharides	Nucleic acids	Proteins
Uninoculated control	+	+	+
Rhizobium (R)	+++	+++	+++
G.mosseae (G.m)	+++	++	+++
G.fasciculatum (G.f)	++	++	++
R + G.m	++++	++++	++++
R + G.f	+++	+++	++

+ = Poor, ++ = Average, +++ = Medium, ++++ = Rich

References

Bagyaraj, D. J. 1992. Vesicular arbuscular mycorrhiza: Application in agriculture. In: *Methods in Microbiology* (Eds. Norris, J.R., Read,D.J. and Verma, A.K.). 359-374. Acad. Press, London, U.K.

Bagyaraj, D. J. 1996. VA mycorrhizal fungi, nitrogen fixing bacteria and their interaction in improving plant productivity. In: *Use of mycorrhiza and nitrogen fixing symbionts in revegetating degraded sites.* FORSPA Publications, Bangkok, pp. 63.

Burgess, T. L., Malajczuk, N. and ***Grove, T.S.*** 1993. The ability of 16 ectomycorrhizal fungi to increase growth and phosphorus uptake of *Eucalyptus globulus Labill* and *E. diversicolor F. Plant and Soil* 53: 155-164.

Chilvers, G. A., Lapeyrie, F. F. and ***Horan, D. P.*** 1987. Ectomycorrhizal Vs endomycorrhizal fungi within the same root system. *New Phytology.* 107: 441-448.

Doss, D. D., Bagyaraj, D. J. and ***Joshi, S.*** 1988. Morphological and histochemical changes in the roots and leaves of finger millet *Eleusine coracona* colonized by VA mycorrhizae. *Proc. Indian Nat. Sci. Acad.* 54: 291- 293.

Feder, N. and ***O' Brien, T. P.*** 1968. Plant micro technique - Some principle and new methods. *American Journal of Botany.* 55: 123-142.

Gianinazzi, S. and ***Gianinazzi-Pearson, V.*** 1992. Cytology, histochemistry and immunocytochemistry as tools for studying structure and function in endomycorrhizas. *Methods in microbiology.* 24: 109-139.

Gracy, L. Sailo, Bagyaraj, D. J. and ***Syamsunder Joshi*** 2002. Anatomical and Histochemical changes in the roots and leaves of *Simarouba glauca due* to inoculation with an arbuscular mycorrhizal fungus and a mycorrhizal helper bacterium. *J. Plant Biology.* 29: 321-326.

Jeffries, P. 1987. Use of mycorrhizae in agriculture. CRC. *Critical Review of Biotechnology.* 5: 319-348.

Jensen, W. A. 1962. *Botanical Histochemistry.* Nowman Publishers, San Francisco, pp. 252.

Kin den, D. A. and ***Brown, M. F.*** 1975. Electron microscopy of VAM of yellow poplar. I. Characterization of endophytic structures by scanning electron stereoscopy; *Canadian J. Microbiology.* 21: 989-993.

Krishna, K. R., Suresh, H. M., Syamsunder Joshi and ***Bagyaraj, D. J.*** 1981. changes in the leaves of finger millet due to VA mycorrhizal infection. *New phytology.* 87: 717-722.

Ling lee, M., Ashford, A. E. and *Chilvers, G. A*. 1977. A histochemical study of polysaccharide distribution in *Eucalyptus* mycorrhizas. *New Phytology.* 78: 329-335.

Mazia, D., Brewer, P. A. and *Alfert, M.* 1953. The cytochemical staining and measurement of protein with mercuric bromo-phenol blue. *Biological Bulletin.* 104: 56-67.

Porter, W. M. 1979. The most probable number method for enumerating infective propagules of vesicular arbuscular mycorrhizal fungi in soil; *Arid Soil Res. Rehabilita.* 4: 515-519.

Skoog, F. and *Schmitz, R. V.* 1972. Cytokinins. In: *Plant Physiology,* VIB: *Physiology and Development - The Hormones.*

Shah, C. K. and *Jain, B. K.* 1981. Histochemical study during foliar and floral differentiation in some monocotyledons I. Rice (*Oryza sativa L.*). *Indian J. Plant Physiology.* 24: 321-324.

39 Studies on Nodulation and Effect of Rhizobium Inoculation on Green Gram [*Vingna radiata* (L.) Wilczek]

K.V. Mallaiah and P.N.L. Chandrika
Deptt. of Microbiology, Acharya Nagarjuna University, Nagarjunanagar-522 510, Guntur Dt. A.P.

ABSTRACT

The nodulation pattern of eight cultivars of green gram [*Vigna radiata* (L) Wilczek] commonly cultivated in Andhra Pradesh was studied. The number of days taken for appearance of nodules on different cultivars varied from 6 to 15 days. The nodules recorded per plant were highest (57) in cv ML-267 and least (18) in cv LGG 410. *Rhizobium* strains were isolated from all the eight cultivars and identity confirmed by their nodulating ability on homologous hosts and other cultural and biochemical tests. The colonies were white or cream coloured and highly mucilaginous. The colony diameter ranged from 4.5 mm to 12.0 mm in 48 hours. The isolate from cvML-267 was relatively slow grower and that from cv LGG-450 was relatively fast grower. The isolates showed good growth on medium without NaCl, and growth was much reduced on medium with 1% or more concentration of NaCl and showed only traces of growth at 2% NaCl concentration. Inoculation tests with cv ML-267 showed that green gram responds well for *Rhizobium* inoculation, and nodulation was relatively more in seed inoculation method than soil inoculation or seedling inoculation methods. The nitrogen content of seed and yield significantly increased in inoculated plants when compared to control plants without *Rhizobium* inoculation. Total biologic yield and harvest index were also significantly more in inoculated plants compared to control plants.

KEY WORDS : Green gram, *Rhizobium*, inoculation, nodulation, yield

Green gram [*Vigna radiata* (L) Wilczek] is an important pulse crop of India, cultivated throughout the country. It occupies an area of 2.8 million ha. occupying about 12% of the total pulse crop acreage in India. It is cultivated mainly as a rainfed crop during kharif and also in rabi season as a second crop after harvesting of paddy in rice fallows. A large number of cultivars of the crop are cultivated in different parts of the country, and the new cultivars of the crop are released for their selected characters like disease resistance, drought resistance and yield. The nodulation pattern on eight cultivars of green gram, commonly cultivated in Andhra Pradesh and effect of *Rhizobium* inoculation on ML-267 cultivar of the crop was studied.

Materials and Methods

Authentic seed samples of eight cultivars of green gram viz. LGG-407, LGG-410, LGG-450, LGG-460, ML-267, MGG-295, PDM-54 and WGG-2 were obtained with the courtesy of Regional Agricultural Research Station (ARAS), Lam, Guntur. All the eight cultivars were raised in the field plots in the botanical experimental garden of our University. For studying the nodulation characteristics, five plants of each cultivar were carefully uprooted at weekly intervals, brought to the laboratory, washed under running tap water and observations were recorded on the number of nodules per plant, size, shape and colour of the nodules, distribution on tap root or lateral roots and occurrence in clusters or individuals etc.

Rhizobium sp. was isolated from fresh healthy nodules of all the eight cultivars on Yeast Extract Mannitol Agar (YEMA) medium with 0.1% congo red, and their identity as *Rhizobium* sp. was confirmed by characterization tests and ability to nodulate the homologous hosts.

To study the cultural characteristics, the *Rhizobium* isolates were spot inoculated on Petriplates with YEMA medium, incubated for 48 h at room temperature and observations were recorded on colony diameter, elevation, colour, shape, texture and mucilage. Three replicates were maintained for each isolate. The progress of growth of the eight isolates of *Rhizobium* was studied by inoculating them on YEMA medium and recording the colony size at 24 h intervals up to 144 h. The effect of Sodium chloride (NaCl) on growth of *Rhizobium* isolates was studied by inoculating the YEMA medium plates supplemented with different concentrations of the salt. The different concentrations used were 0, 0.05%, 0.1%, 0.2%, 0.3%, 0.5%, 1.0%, 2.0% and 3.0%. The growth was recorded after one week.

The effect of *Rhizobium* inoculation on growth, nitrogen content and yield of green gram was studied using cultivar ML-267, which is a short duration variety, extensively cultivated in the region. *Rhizobium* strain isolated from the same cultivar was used in this study. The inoculation of *Rhizobium* to green gram was carried out in three different methods viz. seed inoculation, soil inoculation and seedling inoculation, under sterile conditions. Uninoculated controls were also maintained for the purpose of comparison. For raising the plants surface sterilized seeds and sterilized soil in sterile pots were used. Observations were recorded on days to nodulation, days to flowering, days to maturity, number of nodules per plant at vegetative, reproductive and harvesting stages, seed yield per plant and also nitrogen content of the plant at different growth stages. Nitrogen content of the nodules, shoot and seeds was estimated by microkjeldahl method given by AOAC (1978). The seed yield of five randomly selected plants was recorded in grams, and the average value of the five plants was taken as seed yield per plant in grams. The total dry weight of the plant including economic yield (i.e. seed yield) was taken as the biological yield of the plant in grams. The total dry weight of the plant was calculated by taking the weight of the plants dried in hot air oven. The harvesting index, defined as the proportion of economic yield to total biological yield (Donald, 1965), was calculated by using the formula harvest Index (%) = Economic yield / Total biological yield $\times$ 100.

Results and Discussion

Nodulation characteristics: The nodulation characteristics of green gram cultivars raised in the field plots are given in the Table 1. The number of days taken for the appearance of nodules on different cultivars varied much from 6 to 15 days. The nodules appeared quite early by 6 days in the cv. MGG-295 and by 7th day in WGG-2. The nodules appeared by 10th day in cv. PDM-54, by 12th day in LGG-407, by 13th day in LGG-450 and 14th day in case of LGG-410 and LGG-460. In ML-267, the nodules appeared very late on 15th day. The number of nodules per plant began to increase with age and maximum number was recorded by 45th day when the plants are in full bloom, the nodules began to senesce as the pod formation started and shriveled and dropped by harvesting time. The maximum number of nodules recorded per plant (average of 5 plants) varied from 18 in LGG-410 to 57 in ML-267. In MGG-295, on which nodulation started quite early, the average number per plant was 35, while in ML-267, on which nodulation started very late, the number was 57, the highest among all the cultivars. Thus it is clear that there is no correlation between the time taken for the appearance of nodules and maximum number of nodules formed.

Table 1. Nodulation characteristics of green gram cultivars raised in field plots

Green gram cultivar	First appearance of nodules (days)	Max. no. nodules formed	Nodule size (mm)	Nodule shape	Nodule colour	Nodule distribution (tap root/lateral or both)	Nodule occurrence (singles/groups of both)
LGG-407	12	30	1.5	Round to irregular	Light to dark brown	Both	Both
LGG-410	14	18	1.5	Round to irregular	Light to dark brown	Both	Both
LGG-450	13	23	4.0	Round to irregular	Light to dark brown	Both	Both
LGG-469	14	30	3.0	Round to irregular	Light to dark brown	Both	Both
ML-267	15	57	5.0	Round to irregular	Light to dark brown	Both	Both
MGG-295	06	35	1.5	Round to irregular	Light to dark brown	Both	Both
PDM-54	10	30	2.0	Round to irregular	Dark brown	Both	Both
WGG-2	07	30	2.0	Round to irregular	Dark brown	Both	Both

There was little variation in the shape of the nodules formed on different cultivars. Majority of the nodules were round in shape but, a few were lobed or irregular. The size of the nodules varied from 1 to 5 mm in diameter. The average size was relatively more i.e. 5 mm in ML-267 and 4 mm in LGG-450. The nodules were very small, from below

1 mm to 2 mm with an average of about 1.5 mm in LGG-407, LGG-410 and MGG-295. The colour of the nodules varied from light to dark brown with white lines in all the cultivars. In ML-267, the nodules were light to dark brown without any clear white lines. In all the other cultivars the colour of the nodules was typically light brown and a few are dark brown. The white lines are prominent on dark brown nodules than on the light brown nodules.

The pattern of nodule distribution was the same on all the cultivars. The nodules were found on both tap root as well as lateral roots. In general, the first formed laterals have relatively more nodules than the others. The nodules mainly occurred as isolated individuals, but formation in clusters was also observed on all the cultivars.

Even though all cultivars showed nodulation in natural soil, the number of nodules formed per plant showed wide variation, and that the different cultivars of the same crop showed wide variation in nodulation was reported also for other legume crops like black gram (Vaishya and Gajendragadkar, 1982; Mathur *et. al.*, 1998; Reddy and Mallaiah, 2002), Bengal gram (Bapat *et. al.*, 1977; Kenjale, *et. al.*, 1980)), red gram (Singh *et al.*, 1979), soybean (Narula *et. al.*, 1980) and other crops. Other characters of nodules like shape, colour and distribution did not show much variation among the cultivars of green gram, and Ali *et. al.*, (1997) basing on their studies on legume root nodule morphology, reported that "legume root nodules have a recognizable morphology, which is generally characteristic and uniform throughout the population of a susceptible host".

Table 2 Colony characters of *Rhizobium* isolates from green gram cultivars on YEMA medium after 48 h. of incubation

Rhizobium Isolate from cultivar	Colony characters after 48 h. of incubation						
	Size (mm)	Shape	Colour	Elevation	Mucilage	Margin	texture
LGG-407	4.5	Irregular	white	Slightly elevated	Thick mucilaginous	Wavy	Smooth
LGG-410	10.0	Irregular	White	Slightly elevated	Thin transparent watery	Wavy	Smooth
LGG-450	10.0	Irregular	Cream	Elevated	Slightly mucilaginous	Wavy	Smooth
LGG-460	9.0	Irregular	Cream	Highly elevated	Thick mucilaginous	Wavy	Smooth
ML-267	4.5	Irregular	White	Slightly elevated	Thick mucilaginous	Wavy	Smooth
MGG-295	10.0	Irregular	White	Elevated	Slightly mucilaginous	Wavy	Smooth
PDM-54	12.0	Round drop like	White	Elevated	Thick mucilaginous drop like	Slightly Wavy	Smooth
WGG-2	7.5	Round	White	Slightly elevated	Mucilaginous	Entire	Smooth

Cultural characteristics of Rhizobium isolates:

The colony characteristics of the 8 isolates of *Rhizobium* from the 8 green gram cultivars on YEMA medium after 48 hours of incubation at room temperature are given in the Table 2. The colony diameter of the different isolates ranged from 4.5 (isolates from LGG-407 and ML-267) to 12.0 mm (isolate from PDM-54) after 48 hours of incubation. The colonies of the various isolates were circular to irregular in shape, smooth in texture, and slightly elevated with raised centre and gradual sloping towards margins. The colonies of the isolates from LGG-450 and LGG-460 were cream coloured while the other six isolates were white in colour. The colonies of all the 8 isolates were mucilaginous indicating that the isolates produce extracellular slime. However, the amount of mucilage production showed variation. The colonies of the isolate from LGG-410 were thin, transparent and watery, while the isolates from LGG-407, LGG-460, ML-267 and PDM-54 were thickly mucilaginous. Of these, the colonies of the isolate from PDM-54 appear conspicuously as thick drops of mucilage.

The progress of the growth of the different isolates on YEMA medium, measured as colony diameter at 24 h intervals up to 144 hours showed much variation (Table 3). The colony diameter after 24 h of incubation ranged from 4.0 mm (ML-267, LGG-407 and LGG-410) to 10.0 mm (PDM-54), and after 144 h it ranged from 5.5 mm ((ML-267) to 20.0 mm ((LGG-450). The isolates from ML-267 and LGG-407 did not show much growth after 1st 24 hour period. The isolates from LGG-460, PDM-54 and WGG-2 though showed fast growth initially showed fast decrease in growth with increasing incubation time. The isolates from LGG-410, LGG-450 and MGG-295 showed relatively fast growth up to 72 hours or more and then decreased.

Table 3 Growth progress of *Rhizobium* isolates from green gram cultivars on YEMA medium

Rhizobium Isolate	Colony diameter of *Rhizobium* isolates on YEMA medium (in mm) after					
	24 h.	48 h.	72 h.	96 h.	120 h.	144 h.
LGG-407	4.0	4.5	5.5	6.5	7.0	7.5
LGG-410	4.0	10.0	16.0	17.0	17.0	17.0
LGG-450	9.0	12.0	15.0	18.0	19.0	20.0
LGG-460	8.5	11.0	12.5	13.5	14.5	15.0
ML-267	4.0	4.5	5.0	5.5	5.5	5.5
MGG-295	5.5	10.0	14.5	16.5	17.5	17.5
PDM-54	10.0	12.0	14.0	15.0	15.5	16.0
WGG-2	6.5	8.5	9.5	10.0	10.5	11.0

The root nodule bacteria of *Vigna* species is generally considered as slow growers, and treated as belonging to the genus *Bradyrhizobium* (Jordan, 1984), however, the present study shows that the isolates of green gram are relatively fast growers, and show much variation in growth pattern. Hence, they are referred to as *Rhizobium* rather than *Bradyrhizobium* in the present study.

Effect of NaCl on growth of Rhizobium isolates:

All the isolates of *Rhizobium* from green gram showed good growth on medium without the salt (Table 4) indicating that they have no special requirement of the salt for growth. Except for the isolate from LGG-407, the growth of the isolates on medium containing 0.05% to 0.5% salt the growth is substantially more than on medium containing no salt, which shows that salt at low concentration stimulates the growth of *Rhizobium* isolates from green gram. The growth was very much reduced on medium containing 1% salt concentration, indicating that increased salt concentration inhibits the growth. The growth of the isolates was only a trace on medium containing 2% salt concentration, and none on medium containing 3% salt concentration. Graham and Parker (1964) reported that most strains of *Rhizobium* do not grow on medium containing 2% NaCl. Thamizchelvan *et al.* (1991) reported that fast growers tolerate up to 3% NaCl but tolerance is restricted to 1% only in case of slow growers. The green gram isolates, though relatively fast growers showed good growth only up to 1%.

Table 4 Growth of *Rhizobium* isolates on YEMA medium with different concentrations of Sodium Chloride (NaCl) after one week

Rhizobium Isolate from	Colony diameter of YEMA medium (in mm) with NaCl concentration (%)								
	0	0.05	0.1	0.2	0.3	0.5	1.0	2.0	3.0
LGG-407	13.0	10.2	10.0	13.0	10.0	11.0	3.0	Trace	–
LGG-410	7.0	14.0	17.0	15.0	13.0	11.0	8.0	Trace	-
LGG-450	9.5	11.0	15.0	12.5	8.5	20.2	10.0	Trace	-
LGG-460	7.0	9.0	15.0	11.0	8.0	14.0	6.0	Trace	-
ML-267	5.5	5.0	5.5	6.0	5.0	5.0	2.0	Trace	-
MGG-295	8.5	12.0	17.5	12.5	9.0	20.0	10.0	Trace	-
PDM-54	10.5	17.0	16.0	13.0	10.0	20.0	10.0	Trace	-
WGG-2	16.5	10.0	11.0	9.0	12.0	20.0	12.0	Trace	-

Effect of Rhizobium inoculation on green gram :

Growth : The seeds readily germinated in both inoculated and control pots and seedlings emerged by 4th day. In control plants flowering started by 35th day and all plants flowered by 37 days. *Rhizobium* inoculation did not show any discernible effect on initiation of flowering and, irrespective of the method of inoculation flowering started by 35th day (Table 5). The number of days taken to maturity of the pods, taken as blackening of the pods ranged from 63 days in control plants and one or two days more in inoculated plants.

Nodulation : The nodules first appeared on seed inoculated plants by 11th day and it took 14 days to appear in soil inoculated plants and 15 days on seedling inoculated plants. The number of nodules increased with age of the plants up to reproductive stage.

Table 5 Effect of *Rhizobium* sp. isolate (ML-267) inoculation on growth and nodulation of green gram cv. ML-267

Sl. No.	Method of inoculation	Days to nodulation	Days to nodulation	Days to nodulation	Av. No. of Nodules/ Plant at		
					Veg. stage	Rep. stage	Harv. stage
1.	Control (without inoculation)	nil	35	63	nil	nil	nil
2	Seed inoculation	11	35	65	38.8	76.0	1.4
3	Soil inoculation	14	35	64	22.4	49.4	0.8
4	Seedling inoculation	15	35	64	29.2	69.6	0.8

Table 6 Effect of *Rhizobium* sp. isolate (ML-267) inoculation on nitrogen content of nodules, shoot and seeds of green gram cv. ML-267

Sl. No.	Method of inoculation	N-content of shoot (%) at			N-content of nodules Rep. stage	N-content of seeds (%)	Protein content of seeds (%)
		Veg. stage	Rep. stage	Harv. stage			
1..	Control (without inoculation)	0.85	1.20	0.70	–	2.5	15.62
2.	Seed inoculation	1.97	3.01	1.77	4.99	5.01	31.31
3.	Soil inoculation	1.11	2.01	1.08	3.12	3.99	24.93
4.	Seedling inoculation	1.10	2.01	1.00	3.97	4.12	25.75

At vegetative stage i.e. around 25 days when the plants are vigorously growing, the nodules developed per plant varied from 37 to 41 with an average of 38.8 per plant in seed inoculated plants. The average number of nodules per plant was 29.2 in seedling inoculation and 22.4 in soil inoculation. The number of nodules observed was the maximum at reproductive stage i.e., actively flowering period at the age of 40 days. At this stage the average number of nodules on seed inoculated plants was 76, and it was 69.6 on seedling inoculated plants and 49.4 in soil inoculation method (Table 5). At the harvesting stage the number of nodules per plant greatly reduced and only one or two were found attached to the plant. That the nodulation was highest at flowering stage was also reported for black gram (Lalman and Singh, 1989), cowpea (Garg *et. al.*, 1982) and bengal gram (Garg *et. al.*, 1981).

Nitrogen content : The nitrogen content of the nodules at reproductive stage, that of shoot at three different stages of growth (vegetative, reproductive and harvesting stages) and of fresh seeds soon after harvesting was estimated (Table 6). The nitrogen content of the nodules, in terms of percent dry weight, was highest (4.99%) in seed inoculated plants followed by that of seedling inoculated plants (3.97%) and soil inoculated plants (3.12%).

The nitrogen content of the shoot showed much variation both with respect to the growth stage and type of *Rhizobium* inoculation. In control plants, the nitrogen content was lowest when compared to that of inoculated plants at all stages of growth. The nitrogen content of the shoot at vegetative, reproductive and harvesting stages was 0.85%, 1.20% and 0.70% respectively in controls, showing that the nitrogen content gradually increased from vegetative to reproductive stage and sharply decreased by harvesting stage. The same trend was observed in *Rhizobium* inoculated plants also, but with higher nitrogen content. The nitrogen content of the shoot in seed inoculated plants was 1.97%, 3.01% and 1.77% respectively at the three stages of plant growth, and it was relatively more than in soil inoculation and seedling inoculated plants.

The nitrogen content of the seeds at harvesting stage was significantly higher in *Rhizo-bium* inoculated plants when compared to uninoculated controls. The percent nitrogen content of the seeds from seed inoculated plants was 5.01% while that in control was 2.5%. The nitrogen content of the seeds from seedling inoculated plants was 4.12% and that of soil inoculated plants was 3.99%.

That the major effect of nodulation is the increase of nitrogen content of the plant and seed was well established in literature, and it is the major reason for the use of legumes as green manure crops (Allen and Allen, 1981; Tilak, 1991; Subba Rao, 1999).

Yield : The seed yield per plant varied from 4 to 13 grams. On average the yield was only 4.8 g in control plants and it was highest of 11.9 g in seed inoculated plants. In soil and seedling inoculations, the seed yield was 9.1 g and 9.9 g respectively (Table 7).

Table 7   Effect of *Rhizobium* sp. (ML-267 isolate) inoculation on seed yield, biologic yield and harvest index of green gram cv. ML-267

Sl. No.	Method of inoculation	Seed yield per plant (g)	Percent increase over control	Total biological yield	Harvest index (%)
1.	control	4.8	–	10.6	45.28
2.	Seed inoculation	11.9	247.91	19.9	59.79
3.	Soil inoculation	9.1	189.57	16.1	56.52
4.	Seedling inoculation	9.9	202.50	17.5	56.57

The total biological yield, estimated as the total dry weight of plants at harvesting, was highest in seed inoculated plants (19.9 g) and least (10.6 g) in control plants. In soil and seedling inoculations the biological yield was 16.1 g and 17.5 g respectively.

The harvesting index was also highest in seed inoculated plants with 59.79% and it was 56.57%, 56.52% and 45.28% in seedling inoculated, soil inoculated and control plants respectively, gradually reducing in that order.

The economic and biological yield of green gram is significantly higher in inoculated plants than in control, and it clearly shows that *Rhizobium* inoculation, irrespective of the method of inoculation, is having positive effect on growth and yield of green gram.

References

Ali, S.S., Ramesh, R. and **Gupta, N.** 1997. *Rhizobium* – plant interaction in determination of legume root nodule morphology. *J. Indian bot. Soc.,* 76: 185-190

Allen, O.N. and Allen, E.K. 1981. *The Leguminosae – A source book of characters, uses and nodulation. Mac Millan Publishers Ltd., London, U.K.* p 812

AOAC. 1978. "Official and tentative methods of analysis" *Association of Official Agricultural Chemists, Washington DC.*

Bapat, N., Singh, L., Vaishya, U.K. and Dubey, J.N. 1977. Variation in symbiotic effectiveness in Bengal gram (*Cicer arietinum* L.) *Indian J. Microbiol.,* 17: 163-166

Garg, F.C., Bansal, R.K. and Tauro, P. 1981. Variation in nodular nitrogenase activity of winter legumes during various growth phases. *Indian J. Microbiol.,* 21: 74-75

Garg, F.C., Beri, N. and Tauro, P. 1982. Nodular nitrogenase activity during various growth phases of cowpea (*Vigna unguiculata*). *Indian J. Microbiol.,* 22: 289-290

Graham, P.H. and Parker, C.A. 1964. Diagnostic features in the characterization of root nodule bacteria of legumes. *Plant and Soil,* 20: 383-396

Jordan, D.C. 1984. The Genus *Bradyrhizobium.* In *"Bergey's manual of systematic Bacteriology,* Volume-I (Eds. Kreig and Holt). *Williams & Wilkins Publication, Baltimore,* pp 242-244

Kenjale, L.D., Shinde, P.A. and Utikar, P.G. 1980. Varietal response to *Rhizobium* inoculation in gram (*Cicer arietinum* L.) under field conditions. *Indian J. Microbiol.,* 20: 144-146

Lalman and Singh, B. 1989. Effect of SO_2 fumigation on growth and nodulation of *Vigna mungo. J. Indian Bot. Soc.,* 68: 89-94

Mathur, V.L., Laxmikant and Joshi, A. 1998. Study of nodulation, leghaemoglobin content and NRA in black gram [*Vigna mungo* (L) Hepper]. *Legume Research,* 21: 221-224

Narula, N., Lakshmi Narayana, K. and Tauro, P. 1980. Field evaluation of *Rhizobium* japonica, and soybean varieties by acytylene-reduction method. *Indian J. Microbiol.,* 20: 298-301

Reddy, A.S. and Mallaiah, K.V. 2002. Studies on the effect of *Rhizobium* inoculation on black gram. In *"Bioinoculants for sustainable agriculture and forestry"* (Eds. S.M. Reddy et al.), *Scientific Publishers, Jodhpur, India,* pp93-98

Singh, H.P., Baghel, S.S. Singhania, R.A. and Pareek, R.P. 1979. Variety-strain interaction between pigeon pea (*Cajanus cajan* L.) and *Rhizobium* sp. *Indian J. Microbiol.*, 19: 170-172.

Subba Rao, N.S. 1999. *Soil Microbiology. Oxford & IBH Publishing Co. Pvt. Ltd.*, New Delhi,

Tamizchelvan, P., Manimaran, R. and Santha Guru, K. 1991. Isolation, characterization and evaluation of rhizobia from tree legumes. *Indian J. Microbiol.*, 31: 209-210

Tilak, K.V.B.R. 1991. *Bacterial fertilizers, Published by ICAR, New Delhi*, 66pp

Vaishya, V.K. and Gajendragadkar, G.R. 1982. Effect of rhizobial inoculation on nodulation and yield of different genotypes of Urid [*Vigna mungo* (L) Hepper]. *Indian J. Microbiol.*, 22:132-134

40 | *Soil Microbiology and Biotechnology*

Nair L.N.

38/18, Vinayak Apartments, Prabhat Road, Lane No. 10, Pune-411004.
e-mail : lizzienair@hotmail.com

ABSTRACT : Soil microbiology includes metabolic activities of its living population (microorganisms) and their role in energy flow and cycling of nutrients associated with primary productivity. Biological activities bring about changes in the physical and chemical composition of the soil. Maximum population of microorganisms occur in areas where plants grow compared to virgin soil. One of the important sites of microbial activity is the region of plant roots – rhizosphere. Root exudates stimulate the various microorganisms of the rhizosphere. Rhizosphere effect is most pronounced with bacteria, significant with fungi and less significant with actinomycetes. Antogonism amongest microorganisms is a common phenomenon in soil. Bacteria, actinomycetes and fungi of some cultivated plants were analysed. The actinomycetes isolated mostly were spices of Streptomyces, *Micromonospora* and *Streptosporangium*.

Soil biotechnology offers potential solutions for increasing crop productivity. One such process is biological nitrogen fixation by *Rhizobium*-legume symbiosis. For most of the plants mycorrhizae are the chief organisms involved in nutrient uptake there by increasing crop productivity.

KEYWORDS : Microorganisms, interaction, biotechnology, mycorrhizae, *Rhizobium*.

Soil microbiology cannot be called a pure discipline as its parentage can be traced through Bacteriology, Mycology, Soil science, Biochemistry and Plant Pathology. Among these, Soil sciences, Microbiology and Chemistry of biological systems are the three disciplines, which are closely woven into the fabric of Soil Microbiology.

Soil has a complicated system formed by the interactions between a physical, chemical matrix with variable composition and a living population with all microbial forms. Within the range of physical and chemical conditions the soil biological communities are sustained. The main thrust of Soil Microbiology is the metabolic activities of its living population (microorganisms) and their role in the energy flow and cycling of nutrients associated with primary productivity. The activities are mainly controlled by environmental factors, which may be favourable or unfavourable for the growth and metabolism of these organisms.

Short History

It is not possible to review the vast history of Soil Microbiology here. Hence only some milestones are mentioned. The title of "Father of Soil Microbiology" should go to the Dutch lens grinder Antono Van Leeuwenhoek (1676) who reported occurrence of small animalcules in water and microorganisms in decaying plant material. However, the designation goes to Winogradasky (1856-1953) in recognition of his contributions like nitrifying bacteria wherein it was shown that microorganisms could use inorganic substrates as sources for growth and energy. The discovery of fungal root associations was first shown in Orchidaceae by Reissek (1847) and later by Pfeffer (1877). However, it was Frank (1885) who coined the term "Mycorrhizen" and later distinguished (1887) the 'Ectotrophisch' and 'Endotrophisch' types. Louis Pasteur's (1830-1900) contribution on microbial fermentation delineated anaerobic metabolism.

Hiltner (1904) claimed that root exudates made the soil around roots a favourable medium for the activity of soil microorganisms and coined the term rhizosphere. Taylor and Lochhead (1985) made comparative studies of rhizosphere of different plants keeping soil as control. Vagnerova *et. al.,* (1960), Rovira (1965), Odunfa and Oso (1978) are some of the workers on Soil Microbiology. The pioneer workers in Actinomycetes were Waksman in USA and Krassilnikov in USSR and that of fungi include Parkinson and Clarke (1964) and several others. Lynch's (1983) contribution to Soil Microbiology and Biotechnology are noteworthy. Studies on Actinomycetes and root mycoflora is carried out by Nair and Nair (1986, 1989, 1990) and microflora of mangroves by Nair *et. al.,* (1991).

General Aspects

Soil functions as a unique ecosystem to the microbial inhabitants and plays a pivotal role in the welfare of mankind. Man depends on soil for his food and to certain extent the productivity of soil determines his living standards. Soil is the natural medium in which plants grow, multiply and die and thus provide a perennial source of organic matter. This organic matter is recycled by microorganisms and serve as nutrients for the growth of plants. Plants in turn serve as food for man and animals. The fertility of soil is the ability of soil to provide organisms living in it with food and other necessary conditions for life and multiplication.

Soil environment is not a static one, but a constantly changing medium activated by the interplay of a vast number of organisms both micro- and macroscopic forms of life.

Biological activities play an indispensable role in the formation of available nutrients from complex compounds and also bring about changes in the physical and chemical composition of soil. The physical condition of soil is improved through the activities of certain organisms. These include the growth of plant roots, the passage of soil through the digestive tract of earthworms and the development of fungal hyphae interwoven around soil particles to form aggregates. The surface of these hyphae also acts as changed surface on which clay particles can bind. Ultimately there will be an increased soil aggregation, which improves soil structure. This increases aeration, infiltration of water and provide easier paths through which plant roots can grow leading to good plant growth. Earthworms feed on organic matter and soil and the casts they leave behind is rich in Ca, N_2, C, and P and has several enzymes.

Soil fertility is increased by several biochemical processes such as N_2 fixation, humus formation, cellulose and chitin degradation etc. Various other transformations of inorganic phosphates, Fe, and S, also take place in soil mediated by microorganisms. A biological equilibrium is attained by antagonism, predation and parasitism.

Soil Profile

In V.S. of virgin soil (Fig. 1) several bands known as horizons are visible. These horizons differ in their structure, texture and composition.

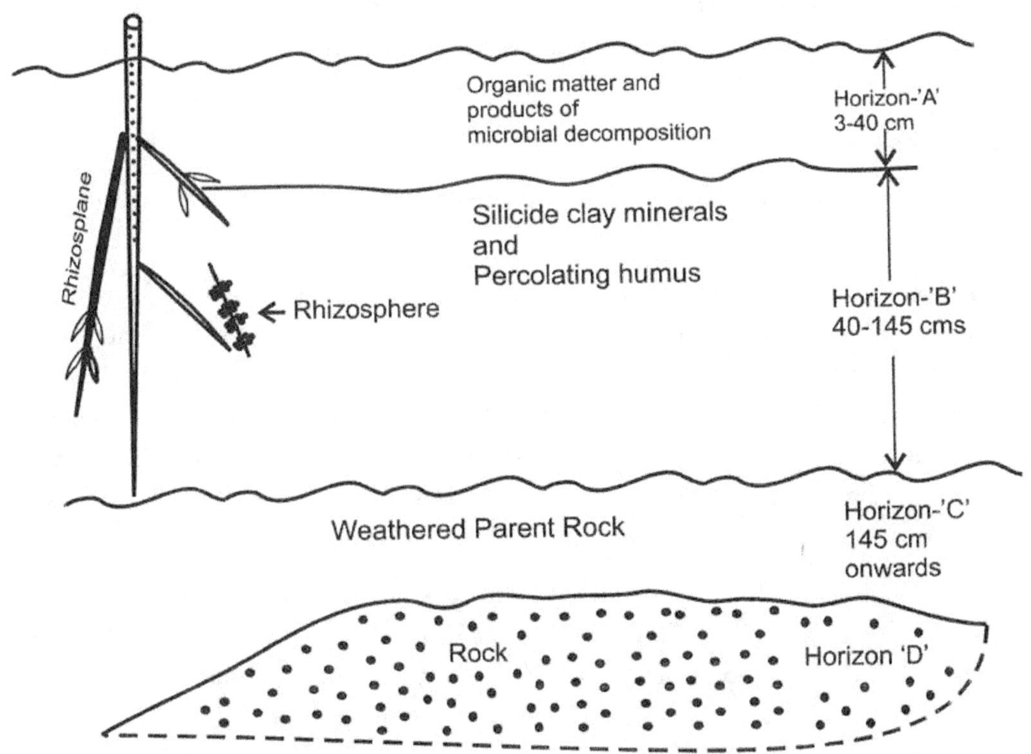

Figure 1   V.S. of soil profile

Horizon 'A' which is the uppermost one (3-35 to 40 cms) is formed and differentiated by the accumulation of organic matter and its decomposition products through microbial degradation. 'A' horizon is the surface soil and designates the stratum subjected to marked leaching. It is also a layer of great biological concern, as roots, small animals and microorganisms are most dense in this region. In this zone the concentration of organic matter is highest and serves as the reservoir for microbial food. Aerobic microorganisms are maximum here.

Horizon 'B'- 40-145 cms is made up of accumulations of silicate clay minerals and percolating humus particles. The 'B' horizon or the sub-soil underlying the 'A' horizon has little organic matter, few plant roots and sparse microflora. Iron and Aluminium compounds often accumulate here. 'A' and 'B' horizon represent true soil.

Horizon 'C' – 145 cms onwards at the bottom of true soil is 'C' horizon, which consists of parent rock material of soil proper. In this layer organic material is very less and there is little or no life noted. Individual horizons and profiles differ from one another in their thickness and chemical constitution.

Horizon 'D' is rock.

Soil is made up of five major components. i. Mineral matter, ii. Water, iii. Air, iv. Organic matter, v. Living population.

(i) ***Mineral matter :*** The physical properties of soil depend on the mineral fraction, which consists of particles of varying sizes. The largest ones being stones and gravel with a diameter of 2 mm or more to clay particles of 0.002 mm with sand 0.05 mm and silt particles 0.02-0.05 mm. Silt exert less influence on physical chemical and biological properties of soil. Sand affect movement of water and air. The texture of soil is determined based on the content of sand silt and clay.

(ii) ***Water :*** Moisture levels vary at different areas since water is derived from atmosphere precipitation. Part of the water moves with gravitational pull and is present in the larger pores of soil where air is present. Water replaces air from the pores. This forms the soil solutions, which contains inorganic salts, which are required by plants. However, the downward movement of water removes it from the zone of microbes and plant roots, and it becomes inaccessible to microbes and plants.

(iii) ***Air*** *:* Aeration and moisture are directly related since the portion of pore space which is not occupied by water is filled by air. This air constitutes soil atmosphere, though not similar to the atmosphere. In the subterranean atmosphere oxygen is consumed by the microorganisms and plant roots releasing carbon dioxide.

(iv) ***Organic matter (Chemical composition)*** *:* The three main components which provide nutrients for plant growth are, i. Organic matter, ii. Derivatives of parent rock material and iii. Clay fraction. These are released into soil solution and absorbed by plant roots. The organic fraction of soil is the potential source of C, N, P and S for plant growth. Microbiological decomposition of organic matter is an essential step to release the bound nutrients present in organic residues and make them in available form. The carbon, nitrogen, sulphur and phosphorus cycles are completed by the activity of various microorganisms. The inorganic minerals come from rock (sand and silt) and made available to plants by mineral decomposition.

(v) ***Living population*** *:* It is fallacious to imply that true soil is the mineral soil and that organic matter and microorganisms are intrusions. Soil organisms occur in a bewildering array of habitats. There is an unusually complex physical and chemical framework of environment provided by soil, which profoundly affect the activity and interaction of organisms. This response of organisms to the environment helps in achieving the overall homeostasis of the ecosystem. A small volume of soil contains a vast assemblage of minute organisms of diverse form and function. Their small size, however, precludes the morphological complexity obtained by plants and animals. The enormous numbers of living things that inhabit soil are obscured because of their minute size and opacity of soil. All groups of microorganisms exist in soil. Because of their presence, soil is transformed from an inert mass of mineral and organic matter to a dynamic living system.

Maximum population of microorganisms occur in areas where plants grow compared to virgin soil. An organism in its natural habitat is considered to be the product of that since it has got adjusted to the environmental conditions prevailing there (e.g. virgin soil) –population.

In contrast to this there exist in specialized habitats organisms having a broad ecological amplitude and are considered as cosmopolitan ones and is not indicative of a particular habitat.

The living population consists of Bacteria, Actinomycetes, Fungi, Algae, Protozoa and viral particles.

Rhizosphere

Even though microorganisms are distributed in all types of soil, they vary qualitatively and numerically from place to place. One of the important sites of microbial activity is the region of plant roots. There is an unique environment under the influence of plant roots, referred to as the rhizosphere- a term coined by Hiltner (1904). The rhizosphere is divided into two areas the inner rhizosphere at the very root surface referred to as rhizoplane and the outer rhizosphere embracing the soil adjacent to the root surface. The biochemical interaction between microorganisms and roots are most pronounced in the inner rhizosphere or rhizoplane than in outer rhizosphere. This is because the root exudates are more concentrated on the surface of roots. The term rhizosphere effect indicates the overall influence of plant roots on soil microorganisms.

The most striking influence the root exerts in the rhizosphere is the stimulation of various types of microorganisms. This stimulation is referred to as rhizosphere effect, which is due to root exudates. The weird and peculiar ecology of the root region (rhizoplane and rhizosphere) has root exudates which, with their tempting constituents containing amino acids, sugars, nucleotides, flavonones, vitamins etc stimulate the growth and multiplication of microorganisms. They also have CO_2 exhaled by the rootlets that search for crevices for the vital fluid. Rejected transient roots, parts of ephemeral existence such as root hairs, non-cambial feeder rootlets, and root caps are also present in the rhizosphere. Root exudation is a process involving several pathways and biochemical mechanisms. The chemicals and elaborated metabolites are released on the surface or into the root environment.

Leakages are compounds of low molecular weight that diffuse into the apoplast, cell wall and plasma and via apoplast move to the root surface and leak directly from epidermal or cortical cells.

Secretions are compounds, which cross the membrane barriers as a result of expenditure of metabolic energy.

Mucilages are the groups contributing to organic material, (i) Those originating in root cap cells and secreted by Golgi bodies, (ii) Polysaccharide hydrolysates of primary cell walls between the epidermis and sloughed root cap cell. (iii) Those secreted by epidermal cell with only primary cell walls: include root hairs. (iv) Those produced by bacterial degradation of walls of old dead cells.

These substances stimulate the growth of more intensive microflora in the rhizosphere region. This is a distinctive microbial complex, which differs from that of root free soil. These organisms, which require vitamins, amino acids etc. for maximum growth are known as nutritionally fastidious organisms.

The rhizosphere effect is found to be most pronounced with bacteria, significant with fungi and less significant with actinomycetes. Minimum effect is on protozoa and algae. Nematodes are more in rhizosphere than in non–rhizosphere. The plant root system starts building up a unique rhizosphere microflora at the time when its first root enters into the soil. The microflora keeps on increasing as the plants grow more and more and occupy larger area in the soil. The maximum microflora is observed when the plant is in its active growing or metabolic state. This is because it excretes maximum quantity of metabolites into soil from roots. This secretion decreases with the flowering stage and by this time cellulolytic forms start appearing. The old and sloughed off cells of the roots serve as the main source of carbon and energy for the cellulolytic microorganisms along with root exudates. This leads to a qualitative difference in the microflora.

Besides the age of the plant, the region of the root also is important for supporting a better population of microorganisms. As the root tip region has greater metabolic activity, it is probable that more exudates of biologically reactive substances will be produced in this region. The crown zone has more sloughed off root tissue and hence supports a good number of microorganisms dominated by cellulolytic forms. Several changes are brought about by the root system of plants to which the microorganisms respond. They include lowering the concentration of certain mineral nutrients due to absorption, partial desiccation of soil due to absorption of water, increase in carbohydrates followed by CO_2 release and contribution to microbial food in the form of root exudates and dead and sloughed off root parts. Root exudates influence the proliferation and survival of root infecting pathogens in soil. Comparatively the pathogens dominate the rhizosphere and rhizoplane regions. Spores of many pathogenic fungi have been shown to germinate by the stimulus provided by root exudates of the susceptible cultivars of the host plants.

Rhizosphere microflora varies in different plants. Antagonism amongst microorganisms is a common phenomenon in soil. It embraces competition, antibiosis, predation and parasitism. Microbial requirements are changed due to foliar application of nutrients by spray, addition of inorganic amendments etc.

Bacteria, actinomycetes and fungal flora of some cultivated plants, like Citrus, Cotton, Cabbage and five Mangroves were analyzed. The actinomycetes isolated were mostly species of *Streptomyces* and one of *Micromonospora* (Figs. 1-8) and *Streptosporangium*. These actinomycetes were used for antagonistic studies against human pathogenic bacteria (Fig. 3) and plant pathogenic fungi. Most of the fungi showed inhibition of growth (Fig. 4) (Nair 1984, Hossein 1985, Rao 1987).

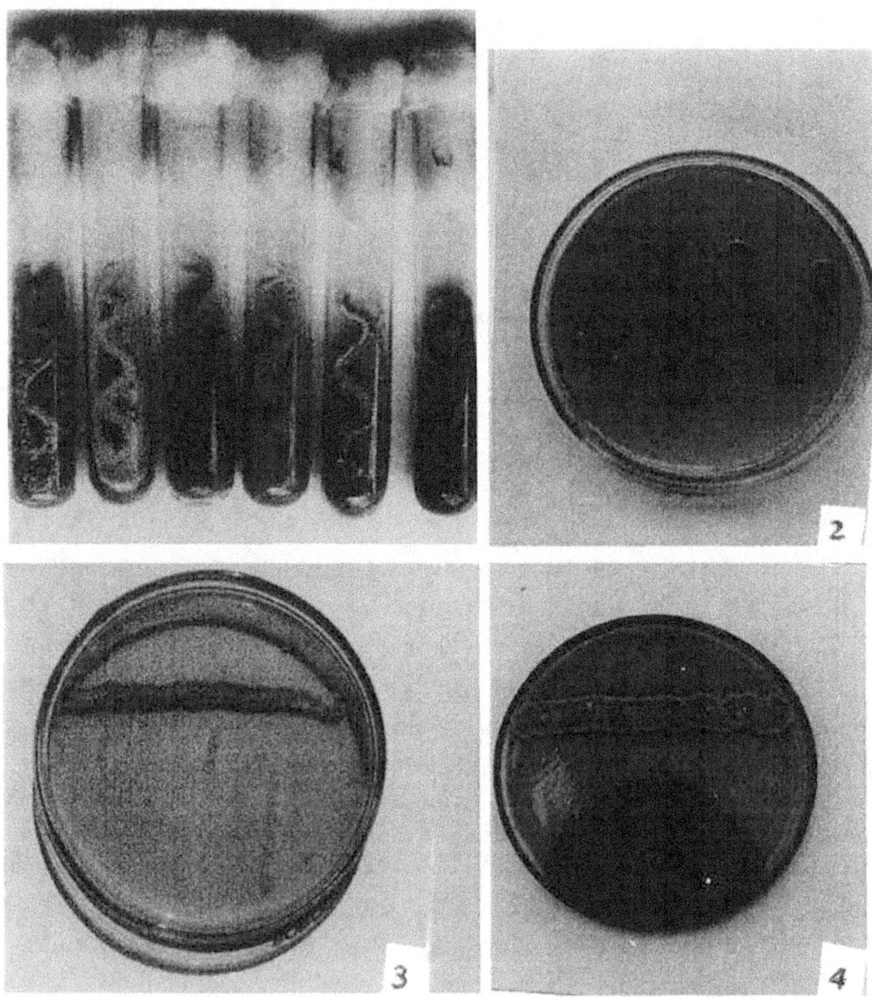

Figs. 1-4. Actinomycetes from soil. Fig. 1. *Streptomyces* spp. and last tube of
Micromonospora x ½ N.S. Fig. 2. *Micromonospora* note pigment x
Fig. 3 *Micromonospora* antagonism to pathogenic bacteria x ½ N.S.
Fig. 4. *S. griseobrunnens* antagonism to *Curvularia* x ½ N.S.

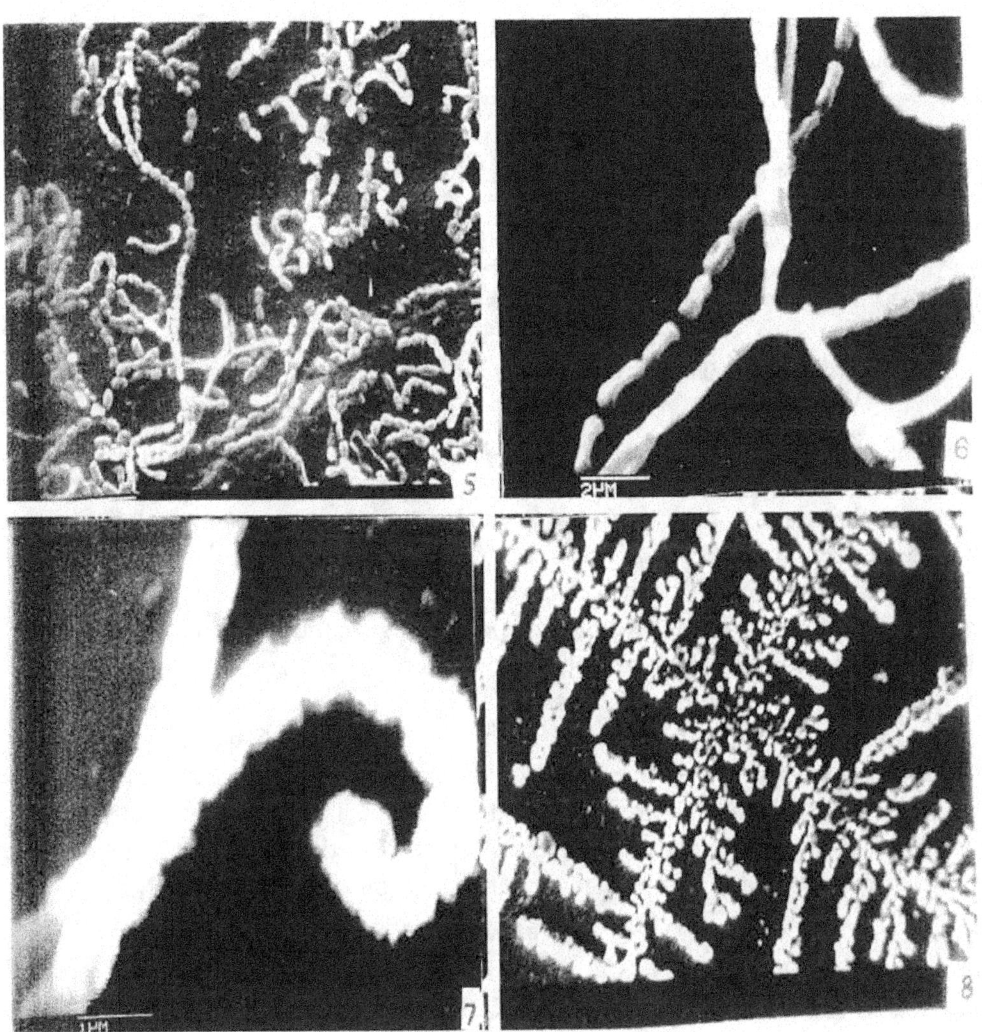

Figs. 5-8. SEM pictures of Actinomycete spore chains. Fig. 5. *Streptomyces fulvoviridus* Fig. 6. *S. achromogenes*. Fig. 7. *S. rutgerensis*. Fig. 8 *Micromonospora*

Fungal flora showed that the soil mycoflora was dominated by species of *Aspergillus* (45 species) as is characteristic of tropical flora. There was a succession of fungi. During the early stages of growth of cultivated plants and during rainy season in mangroves. Primary saprophytic colonizers were the Zygomycetous fungi, which dominated the flora, disappeared later when their position was taken up by Deuteromycetous fungi. In turn their teliomorphs appeared as secondary colonizers, which were mostly cellulolytic Deuteromycetous forms. Maximum members were of Deuteromycetous fungi. Members of Coelomycetes appeared at a latter stage. During winter maximum number of organisms were isolated and as summer approached the number reduced. The correlation of growth of organisms was compared with physical and chemical composition of soil.

While isolating organisms in pure cultures and studying their activities certain negative points should be born in mind. In rich soils there is a diversity of life and hence no single nutrient supplepment (provided in the laboratory) is expected to supply all the requirements of so many different microorganisms like autotrophs and heterotrophs. Because of the complexity of soil microorganisms various methods are involved in the isolation of pure cultures under controlled conditions on synthetic media in laboratories. The use of consortia (defined mixed cultures) of bacteria, actinomycetes and fungi for the study of more complex processes like decomposition of xenobiotic compounds has become very common since these cultures exhibit some of the more complex interactions reflective of the soil ecosystem. The attempts made with pure cultures or a mixture of cultures are inadequate since they defy extrapolation to the more complex situation in soil. The conditions like stresses or edaphic factors under which the organism operates its metabolic activities cannot be matched in the defined cultural conditions in a laboratory. As a result of this, the microorganisms themselves change phenotypically or even genotypically to the artificial condition in the laboratory. In soil also there can be mutations taking place and during isolation there is a possibility that this mutant which occurs as a miniscule portion in soil population may grow *in vitro* than majority of strains. Thus, it is a genetic variant; possible that will be studied in the laboratory. Therefore data obtained from such cultures will not provide a complete picture of the process occurring in soil.

The realm of soil microbiology should include studying the properties of the microorganisms themselves and the impact of variability of edaphic factors on these organisms. In short soil microbial ecology should include an evaluation of the behavior of organisms in their natural habitats.

Soil Biotechnology

For increasing crop productivity biotechnology offers potential solutions. For this, it should be possible to optimize microbiological processes that deliver nutrients to plant roots. One such process is biological nitrogen fixation by *Rhizobium* - legume symbiosis.

Nitrogen fixation

Nitrogen is the mineral element most commonly limiting in crop production. Though this element is in abundance in the atmosphere plants are not able to assimilate the molecular form.   There are free living nitrogen fixers like *Beijerinkia, Azotobacter, Clostridium* etc. in soil. Biological nitrogen fixation by the *Rhizobium* -legume symbiosis can be made to contribute more to future production since it is possible to genetically enhance biochemical efficiency and competitiveness in soil. In the third world where fertilizer availability and cost is a major handicap biological organisms can make immense contributions. *Azolla* and free living Cyanobacteria are being widely used. Legume- *Rhizobium* endosymbiosis is one of the most highly evolved associations between a prokaryotic *Rhizobium* and the eukaryotic plant.

Nitrogen fixation is a faculty reserved to only a few prokaryotes. Atmospheric N_2 is a triple bonded element N =N. The reduction of triple bonded elemental nitrogen to assimilable form is catalysed by nitrogenase. Nitrogenase purified from diverse systems like the anaerobic *Klebsiella pneumoniae*, the aerobic *Azotobacter vinelandii*, the oxygenic photoautotroph *Anabaena cylindrica* and the symbiotic *Rhizobium*, all have essentially similar biochemical features. *In vivo* this enzyme nitrogenase is repressed by ammonia in all systems except the marine cyanobacterium *Anabaena* where nitrate is a more potent repressor (Bottomley *et. al.*, 1979).

Pulses are significant in agricultural systems since they provide grains rich in proteins and herbage, which also adds to soil fertility. This is so, since they have an advantage of harbouring two symbionts, the rhizobia and the vesicular arbuscular mycorrhizal fungi. The eukaryotic fungus provides excess phosphorus required for nitrogen fixation along with water and other nutrients.

Biological nitrogen fixation is as energy expensive to the legume is as fertilizer nitrogen to the farmers. These nitrogen fixers can be grown *in vitro* and hence are available to farmers in the form of Azorhiza mixed with carrier material.

Another aspect is, they can be isolated and the compatible strains can be induced *in vitro* in tissue culture raised seedlings, which can be hardened. *Rhizobium* entered callus cells and rhizoids of *Phaseolus* produced nodules (Figs. 9-12) and fixed nitrogen as evaluated by acetylene reduction assay. Nitrogenase activity of *Rhizobium* tissue culture associations were 2.5×10^5 m moles per gram fresh weight of callus per hour at 5 days incubation and increased to 3.1×10^5 m moles per gram per hour to 4.5×10^5 m moles per gram per hour at 10 and 15 days of incubation respectively.

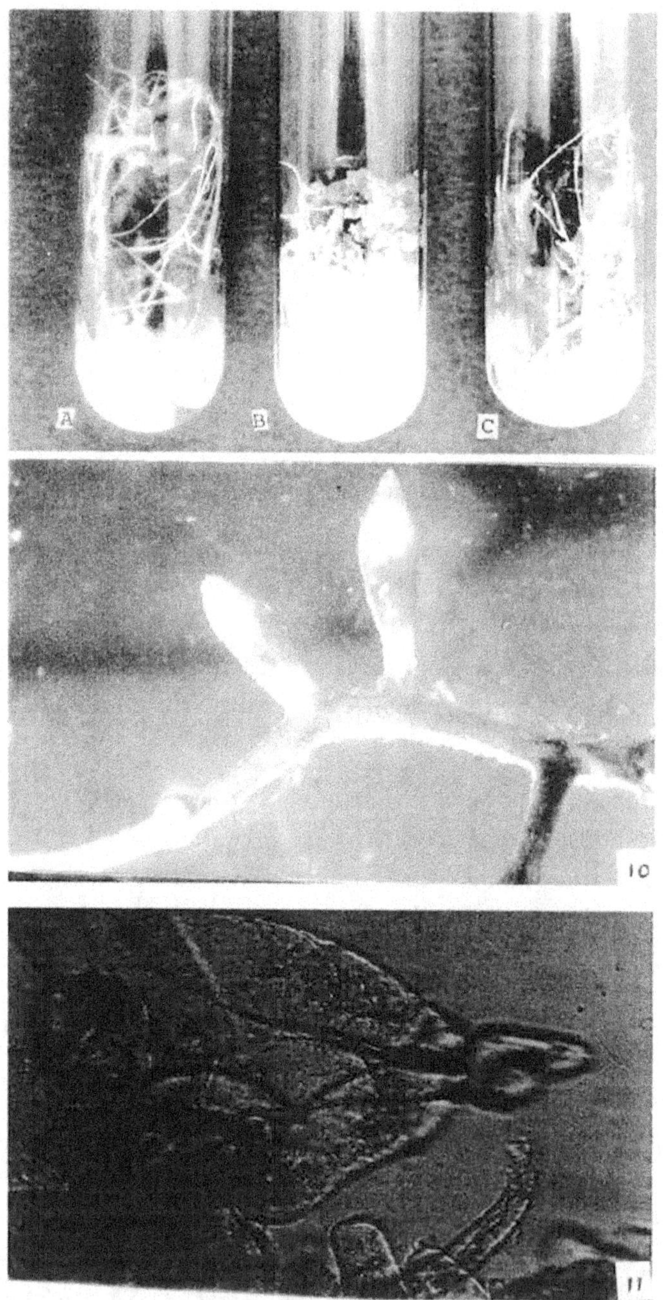

Figs. 9-11. Dual axenic cultures of *Phaseolus- Rhizobium* calli from A) cotyledon, B) Plumule and C) Radicle. Note profuse rhizogenesis x N.S. Fig. 10. Nodule like structures x 100. Fig. 11 Rhizobium in cells of callus x 450.

Mycorrhizae

Mycorrhizae are of two main categories; the ectomycorrhiza and endomycorrhiza. Ectomycorrhizae are restricted to few woody shrubs and trees of temperate tropical and sub-tropical trees. On the contrary endomycorrhiza other wise known as vesicular arbuscular mycorrhiza (VAM) or arbuscular mycorrhiza (AM), are characterized by and arbuscules and/or vesicles are ubiquitous and occur over a broad ecological range on a wide variety of host plants. They are obligate biotrophs and cannot grow in the absence of a host hence they cannot be grown in axenic cultures.

For most of the plants mycorrhizae are the chief organisms involved in nutrient uptake. The VAM fungi are influenced by the root exudates of plants, derive carbohydrates prepared by the host and in turn influence the absorption of soil derived nutrients like phosphorus, copper, zinc, calcium, nitrogen etc. and supply them to the host. Phosphorus ions are absorbed faster by mycorrhizal hyphae. They mop up rock phosphates more thoroughly dissociating them chemically at the particle surface (Hayman, 1983). They occur in terrestrial ecosystems. Such responses are more pronounced inhabiting more than 90% of vascular plants (Sylvia, 1994), in nutrient deficient soils (Jalali and Thareja, 1985). Mycorrhizal fungi produce phosphatases, phytases and organic acids that act upon nutrients and make them available to plants and accelerate nitrogen fixation.

VAM fungi produce antibiotics; utilize surplus carbohydrates and other nutrients present in root exudates thus preventing growth and attack by root pathogens. Jalali and Jalali (1991) are of the opinion that it is quite likely that VAM fungus elicits a resistance mechanism by host, which may inhibit subsequent attack by pathogens.

VAM fungi have a unique ecological position in the rhizosphere compared to that of other microorganisms since they are partly inside and partly outside the roots. The internal parts of the fungus does not have to encounter competition with other microorganisms and hence can have a more functional biomass having greater effect on the plant.

VAM fungal associations with plant roots help them in surviving in several environmental stresses like low soil fertility, temperature fluctuations, drought, salinity, transplantation stress and other natural stresses like heavy metal toxicities, organic acids, pH, root pathogens etc. It increases the longevity of feeder roots of plants by providing deterrent to root infection. The metabolic activities like photosynthesis, respiration and enzyme activities of plants are altered due to VAM symbiosis (Dugassa *et. al.*, 1996). This will alter the biochemical composition of cells leading to changes in the composition of root exudates and rhizosphere microflora. Mycorrhizas are of great significance in low input agricultural systems of tropical acidic infertile soil. Since they increase the efficiency of growth and productivity of agricultural crops mycorrhizal inoculation and management is essential. This has been proved by the application of compatible strains of VAM and *Rhizobium* (Tripartite symbiosis) in fields of pigeon pea. They are useful in *ex vitro* establishment of plantlets.

Though there are multibeneficial aspects, the use of VAM fungi is not yet a common procedure mainly because they cannot be grown in *in vitro* like *Rhizobium*. There are other difficulties like selection of the fungus, farmers being unable to handle the inoculum and lack of standard quality and low cost production.

However, VAM inoculum is grown on rootstocks of selected plants and supplied on a small scale to farmers and researchers.

Enzymes

The microorganisms present in soil have the ability to produce a wide variety of enzymes, which help them to degrade organic matter. One such enzyme is cellulase produced by species of fungi like *Aspergillus, Penicillium, Trichoderma, Sporotrichum* etc. and a large variety of bacteria and actinomycetes. Cellulase complex includes several enzymes in order to carry out hydrolysis of cellulose, which is the main constituent of plants. The cellulolytic enzymes are C or b- 1 4 exoglucanase, Cx or b- 1 4 endoglucanase, or CMCase and b glucosidase or cellobiase. These are adaptive enzymes produced in presence of the substrate cellulose. All these three enzymes may not be produced by one organism. Different organisms collectively contribute to the degradation of cellulose. A large number of other enzymes like proteases, xylanases, pectinases etc. are also produced by soil microorganisms.

Industrial Microbiology

Microorganisms are exploited for fermentation and other allied industries. During fermentation modifications of a chemical structure is brought about by enzymes of microbes. The organisms include *Lactobacillus* in lactic acid production or actinomycetes and fungi involved in antibiotic production (secondary metabolites) or primary metabolites like organic acids (Citric acid). e.g. Production of penicillin by *Penicillium chrysogenum*, gentamycin by *Micromonospora*, streptomycin by *Streptomyces* spp., citric acid by *Aspergillus niger* etc. All these organisms are isolated from soil.

The germplasm from soil should be genetically engineered to meet the challenges in the 21st century. However for obtaining genetically engineered organisms and for utilizing them for the correct procedures, one needs knowledge of its basic characteristics.

References

Bottomley, P.J., van Baalen, C., Grillo, C.F. and **Tabita, F.R.** 1979. Synthesis of nitrogenase and heterocysts by *Anabaena* sp. CA in the presence of high levels of ammonia. *J. Bacteriol.* 140 (3): 38-43.

Dugassa, G.D. Altin, N. and **Schoenbeck, F.** 1996. Effect of arbuscular mycorrhizas, (AM) on health of *Linum usitatissimum*, L. infected by fungal pathogens. *Plant soil,* 185: 171- 182.

Hayman D.S. 1983. The physiology of Vesicular arbuscular mycorrhizal symbiosis. *Can. J. Bot.* 61: 944-963.

Hossein R. 1985. Microflora of *Citrus*. Ph.D. thesis, University of Pune.

Jalali, B.L. and **Jalali, I.** 1991. Mycorrhiza in plant disease control. In *Handbook of Applied Mycology*. I soil and plants; Eds. D.K. Arora *et al*. Marcel Dekker Inc. NY.

Jalali, B.L. and **Thareja, M.L.** 1985. Plant growth responses to VA mycorrhizal inoculation of soils incorporated with rock phosphate. *Ind. Phytopath.* –38 : 306-310.

Lynch, J.M. 1983. *Soil Biotechnology. Microbiological factors in crop productivity.* Oxford: Blackwell Scientific Publications.

Nair V.S. 1984. Microflora of Cotton and Cabbage with special reference to antagonism. Ph.D. thesis, University of Pune.

Nair L.N. and **Nair V.S.** 1986. Studies of two species of *Streptomyces. Ind. J. Micro.*, 26: 101- 104.

Nair V.S. and **Nair L.N.** 1989. Root mycoflora of cotton. *Geophytology* 19(1) : 30-33.

Nair V.S. and **Nair L.N.** 1990. Antagonism in Cabbage microflora. *Vegetos* 3 (1):118.

Nair L.N., Rao V. and Chaudhary S. 1991. Microflora of *Avicennia officinalis. Proc. Sympo. Sig. Mangr.* 52-55.

Odunfa, V.S.A. and Osa, B.A. 1978 Bacterial population in rhizosphere soils of cowpea and Sorghum. *Rev. Eco. Biol. Sol.* 15(4): 413-420.

Parkinson, D. and Clarke, J.H. 1964. Studies on fungi in the root regions III. Root surface fungi of three species of *Allium. Plant and Soil.* 20: 166-174.

Rao V.P. 1987. *Microflora of Mangrove mud.* Ph.D. thesis, University of Pune.

Rovira, A.D. 1965. Plant root exudates and their influence upon soil microorganisms. In *Ecology of soil borne plant pathogens,* Eds. K.F. Baker and W. Snyder. 170 – 180

Sylvia, D.M. 1994. Vesicular arbuscular mycorrhizal fungi: *In: Methods of soil analysis part 2: Microbiological and biochemical properties*: R. W. Weaver *et al.* (eds.) Adison WI. Soil. Sci. Soc.

Taylor, G.S. and Lochhead 1985. Qualitative study of soil microorganisms:II A survey of bacterial flora   of soils differing in feasibility. *Canad. J. Res.* (C) 16: 162-173.

Vangernova, K., Macura, J. and Catska, V. 1960. Rhizosphere microflora of wheat-I composition and properties of bacterial flora during the first stage of growth. *Folia Microbiol.* (Prague) 5: 295-310.

 # Development of Cellulase less Mutants of *Pleurotus ostreatus* for biobleaching and biopulping in Paper and Rayon Industries

Ch. Vijaya, M. Vijaya and M.A. Singaracharya
Dept. of Microbiology, Acharya Nagarjuna University, Guntur - 522 001.
*Dept. of Microbiology, Kakatiya University, Warangal - 506 009, A.P.

ABSTRACT

Among the 75 basidiomycetous fungi screened, 25 organisms were positive in production of lignolytic enzymes. Out of these 25 organisms, seven were positive in the production of high lignolytic and low cellulolytic activities. *Pleurotus ostreatus* was finally selected for the present study. Chemicals like EDTA, TCA, NaN_3 and KSCN and heavymetals like Cu, Mg, Zn completely suppressed both the enzymes in their different concentrations. Veratryl alcohol was identified as a useful chemical for biobleaching as it increased the lignolytic production and completely suppressed the cellulolytic production of *P. ostreatus*. The exposure of UV rays for 20 minutes and X-rays for 0.4 seconds were useful in the development of cellulase less (cel) mutants for biobleaching of unbleached kraft pulp (UKP) in Paper and Rayon industries.

KEY WORDS : *Pleurotus ostreatus,* veratryl alcohol, biobleaching, ligninases, cellulases, cel mutants, UV-rays, X-rays.

In many parts of Andhra Pradesh number of wood based industries like paper and rayon are well established. During the manufacturing process residual lignin in kraft pulp is highly modified by alkaline condensation reactions during pulping and gives the pulp a characteristic dark brown colour (Katagiri *et. al.,* 1995). This residual lignin is commercially removed by bleaching with chlorine based chemicals, which are costly and toxic (Ramakrishna, 2003). They release toxic effluents which are difficuplt to degrade, are carcinogenic and mutagenic (Ander *et. al.,* 1997). Number of researchers (Kirk and Yang, 1979; Michael *et. al.,* 1991; Wuyep *et. al.,* 2003) tried to replace these chlorinated chemicals by non-toxic, biological and efficient ligninase producing white rotp fungi for bleaching and pulping of unbleached kraft pulp (UKP).

The selection of white rot fungi for delignification of UKP is a difficult job. There are some strains of white rot fungi which are very good lignin degraders, but these fungi degrade the cellulose fibrils which are very important for the production of paper, rayon etc. (Kamra, 1998). Therefore, the biobleaching due to delignification is neutralized by the high dry matter loss (cellulose) during fungal treatment. This makes the treatment uneconomical, undesirable and unacceptable.

Farell (1986) viewed that the existing powerful lignin degrading microbes can be further improved either by mutation, protoplast fusion or by recombinant DNA technology and aimed at enhancing the production of lignolytic enzymes, while suppressing the production of cellulolytic enzymes. Thus the cel mutants of white rot fungi can cause sufficient modification of bleaching of UKP with the use of energy from other substances like hemicellulose, causing very less damage to the cellulose fibrils (Johnsrud and Eriksson, 1985). In this regard, we felt that it would be worthwhile to explore for the development of cel mutants for improving the bleaching of UKP by limiting the loss of energy from cellulose. Among the 75 isolates *Pleurotus ostreatus* was selected which is a high lignolytic and poor cellulolytic white rot fungus. An attempt was made using physical and chemical mutagens to transform this as a cel mutant.

Material and Methods

Microorganism : Seventy five white rot fungi were screened from Andhra Pradesh, for positive lignolytic activity by guaiacol plate assay technique. (Plate 1) (Rama Krishna and Singaracharya, 2003). Among these fungi 25 showed the lignolytic activity (Table 1). Further, these 25 organisms were screened for high lignolytic and low cellulolytic activities (Table 2), out of which seven organisms were selected as useful fungi for the present study. All these fungal strains were maintained on 2% malt extract agar slants (pH 4.5) (Christinson 2000). Myecelial pieces of $2mm^2$ of 6-7 day old plate culture were used as inoculum for liquid cultures.

1(a)   1(b)

Figure 1(a, b) *Guaiacol plate assay :* Fig. 1A - No zone indicating absence of ligninase activity. Fig.1B – Brown zone around the colony formed due to guaiacol oxidation indicating the presence of ligninase activity.

Extraction of Enzymes: For quantitative estimation of enzymes, selected fungal isolates were inoculated in to 50 mL of malt broth after sterilisation and incubated at room temperature ($30^0C \pm 2$). After incubation the cultures were harvested by filteration through whatman No.1 filter paper, dialysed and used for estimation of enzymes.

Table 1 List of basidiomycetous fungi for lignolytic activity

Fungi	Ligninases	Fungi	Ligninases
Agaricus bisporus	+	Agaricus campestris	−
Aleurodiscus amorphous	−	Agaricus rodmani	−
Armillaria mellea	+	Bolbitius vitellinus	−
Auricularia auricula	+	Boletus edulis	−
Caldesiella sp.	+	Cantharellus flocosus	−
Calvatia cyathiformis	−	Clavaria sp.	−
Cantharellus cibrarius	+	Clavaria vermicularis	−
Coriolopsis occidentalis	+	Coprinus comatus	−
Corticium rolfsii	+	Coprinus fimetarius	−
Corticium sp.	−	Daedalea quercina	−
Fomintopsis sp.	−	Fomes fomentarius	−
Gaestrum triplex	+	Ganoderma applanatum	−
Ganoderma lucidum (a)	+	Ganoderma lucidum	−
Hygrophorus sp.	−	Grifola berkeleyi	−
Inonotus sp.	−	Hydnum repandum	−
Irpex lactus	+	Irpex sp.	−
Laetiporus sulphureus	+	Lactarius volemus	−
Lenzites betulina	+	Lentinus edodes	−
Lepiota morgani	−	Lenzites sepiaria	+
Lepiota rachodus	−	Phaeolus sp.	−
Lycoperdon perlatum	+	Phallus impudicus	−
Pleurotus cornucopiae	+	Phallus ravenelli	−
Pleurotus eryngii	+	Phellinus igniaricus	+
Pleurotus ostreatus	+	Phellinus sp.	−
Pleurotus servinus	−	Phlebia sp.	−
Polyporus brumalis	+	Pluteus atromarginatus	−
Stream ostrea	+	Podaxis pistillaris	−
Streum sanguinolentum	−	Polyporus sp.	−
Thelephora sp.(a)	−	Polyporus squamosus	−
Thelephora sp.(b)	−	Poria obliqua	−
Thelephora terrestris	+	Porodisculus sp.	+
Trametes versicolor	+	Pycnoporus cinnabarinus	−
Tremella frondosa	+	Schizophyllum commune	−
Tyromyces sp. (a)	+	Solenia candida	−
Tyromyces sp. (b)	−	Trametes sp.	−
Vaginata sp.	−	Typhula inearnata	−
Vararia sp.	−	Volvariella volvacea	−

Influence of Chemicals: To study the effect of various chemicals on the production of cellulolytic and lignolytic enzymes EDTA, TCA, NaN_3 and KSCN (20, 50 & 100 mM), Cu, Mg and Zn (5, 10 and 20 mM), veratryl alcohol (12.4, 24.8, 37.2, 49.6 and 62.0 mM) were added to 50 ml of malt broth before sterilisation. Malt broth without chemicals was used as control.

Isolation of cel mutants: After cultivating the fungus on the malt agar plates for 3 days at 37°C the cultures were irradiated with UV light and X-rays. In UV irradiation, the plates were exposed for 0, 10, 20, 30, 40, 50 and 60 minutes at an intensity of 83 $mWcm^{-2}$. In X-ray irradiation the plates were exposed for 0, 0.4, 0.8 1.2, 1.6, 2.0 and 2.4 seconds at 40 mA and 70 KV. After exposure all the plates were further incubated at 37°C for 4 days. After sufficient growth, approximately 2 mm^2 of the fungal mycelium from the spreading edge of each plate was transferred into a separate malt broth flasks (40 ml). Flasks were incubated for 7, 11 and 14 days and cellulolytic (C_x and C_1) and lignolytic (LiP and Lac) enzymes were estimated in the culture filtrate.

Assay of cellulolytic and lignolytic enzymes: Among the cellulases, C_x activity was assayed by viscosity measurements as suggested by Reese *et al* (1950). Ostwald Fenske viscometers made up of pyrex glass were used. The reaction mixture consisting of 15 ml 0.8% Carboxy Methyl Cellulose (CMC), 5.0 ml of enzyme and 1.0 ml of Mc Iivaines buffer at pH 5.5 was used and the loss of viscosity was measured after every 10 minutes and relative cellulase activity was calculated. Relative enzyme activity (REA) = $1000/t_{50}$ where, t_{50} is time taken for 50% loss of viscosity.

C_1 activity was estimated by DNS method (Miller 1959). 3.5 ml of 0.8% CMC and 1.0 ml of MC Iivaines buffer and 0.5 ml of enzyme were incubated at room temperature for 6 h. From this reaction mixture 0.2 ml was taken and added with 3 ml of 3, 5-dinitro salicylic acid (DNS) and placed in a boiling water bath for 15 minutes. 2 ml of 20% sodium - potassium tartarate was added while the tubes were hot and cooled immediately to room temperature in running tap water. The optical density was measured at 575 nm and activity was expressed in terms of mg/mL of reducing sugars liberated in 6 h.

Lignin peroxidase (LiP) was assayed in an assay mixture consisting of 25 mM sodium tartarate (pH 2.5), 2 mM veratryl alchol, 0.4 mM H_2O_2, 50 - 275 L of enzyme in a total volume of 0.5 ml. The reaction was initiated by addition of H_2O_2 and oxidation of veratryl alcohol to veratraldehyde was determined by an increase in absorbance at 310 nm (Crawford and Crawford 1976).

Laccase (Lac) was measured according to the method of Kirk and Farell (1987). 50 mM sodium acetate buffer (pH 4.5) and 2 mM guaiacol were used in total volume of 2.5 ml reaction mixture. The activity was measured at 440 nm and enzyme activity was expressed as the amount of tetraguaiacol formed per minute/mL of enzyme extract.

Poly acrelamide gel electrophoresis (PAGE): Poly acrelamide gel electrophoresis (Coll *et. al.,* 1993) of samples was performed to demonstrate the increased ligninase activity in presence of various concentrations of veratryl alcohol (Fig. 2) and with different UV and X-ray exposures (Fig. 3 & 4).

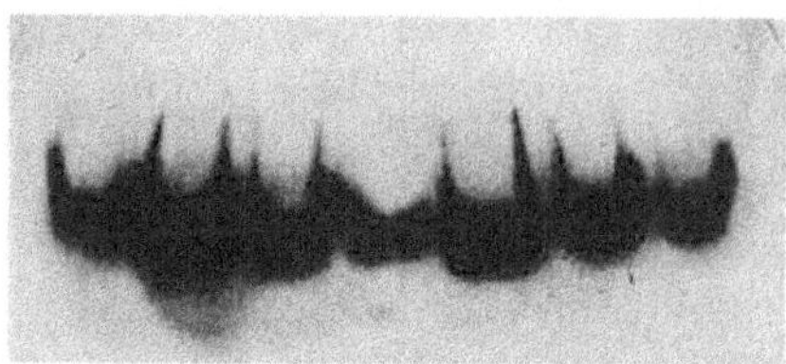

Figure 2 PAGE of Laccase activity in presence of various concentrations of veratryl alcohol. Lane 1 – Control; Lane 2 – 12.4 mM, Lane 3 – 24.8 mM; Lane 4 – 37.2 mM; Lane 5 – 49.6 mM; Lane 6 – 62.0 mM.

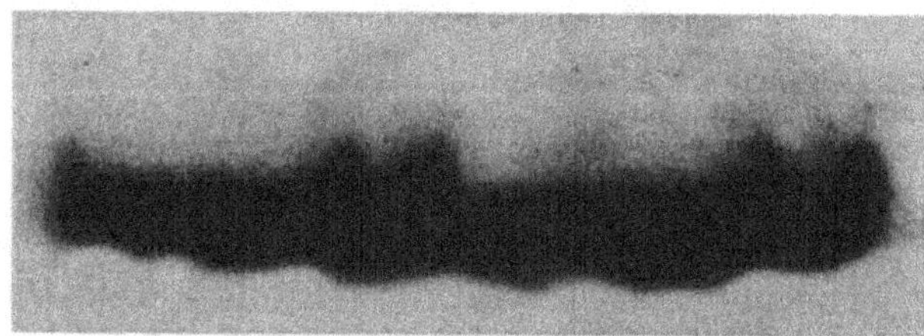

Figure 3 PAGE of Laccase activity in sample exposed to U.V rays (min). Lane 1 – Control; Lane 2 – 10, Lane 3 – 20, Lane 4 – 30, Lane 5 – 40, Lane 6 – 50

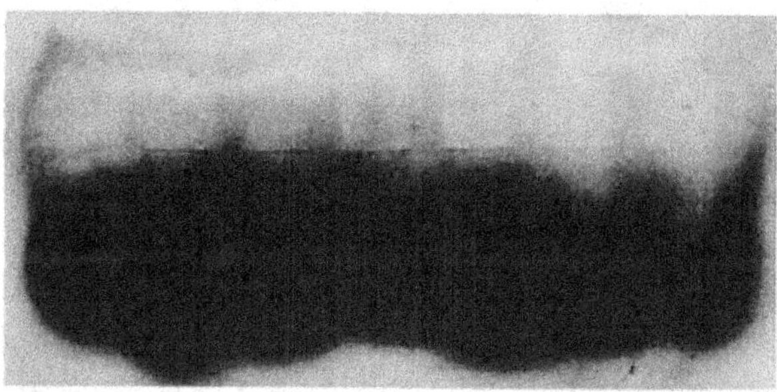

Figure 4 PAGE of Laccase activity in sample exposed to X rays (sec). Lane 1-Control; Lane 2 – 0.4, Lane 3 – 0.8, Lane 4 – 1.2, Lane 5 – 1.6, Lane 6 – 2.0, Lane 7-2.4.

Results and Discussion

From the seven selected white rot fungi for biobleaching, *P. ostreatus* was preferred for further studies based on its high lignolytic and moderate cellulolytic activities (Table-1). In addition *Stereum ostrea* and *Corticium rolfsii* also showed high lignolytic and poor cellulolytic activity. *Lenzitis betulina* and *Theliophora terrestris* secreted meagre amounts of lignolytic enzymes.

Table 2 Screening of six white rot fungi for high lignolytic and low cellulolytic activities in malt broth ater 7 days of incubation

Fungi	Cx	C_1	Lip	Lac
Pleurotus ostreatus	20	42	90	92
Lenzitis betulina	100	30	20	30
Lycoperdon perlatum	50	47	80	88
Theliophora terrestreus	66	83	25	32
Stereum ostrea	33	53	84	84
Corticium rolfsii	25	61	82	81

C_x = REA (Relative enzyme activity).
C_1 = mg/ml of reducing sugars liberated in 6 h.
Lip = Lignin peroxidase – 0.01 increase in OD equal to one unit of enzyme.
Lac = Laccase – 0.01 increase in OD equal to one unit of enzyme.

Table 3 Influence of different chemicals on cellulolytic (c_x & c_1) and lignolytic (lip & lac) enzymes by *P. ostreatus* after 11 days of incubation

Chemical	Concentration	Cx	C_1	Lip	Lac
Control	–	40	40	30	31
Cu	5	33	23	12	10
10	50	26	16	16	
20	33	16	21	20	
Mg	5	100	27	32	40
10	100	23	12	16	
20	111	20	16	10	
Zn	5	83	46	21	14
10	66	26	25	23	
20	83	20	35	17	
EDTA	20	0	20	1	2
50	0	40	3	2	
100	0	30	10	2	
TCA	20	67	53	6	6
50	50	60	3	2	
100	45	40	2	4	
NaN$_3$	20	0	10	1	1
50	0	10	0	0	
100	0	6	0	0	
Veratryl alcohol	12.4	16	51	90	92
	24.8	16	49	93	91
	37.2	0	47	104	102

C_x = REA (Relative enzyme activity).
C_1 = mg/ml of reducing sugars liberated in 6h.
Lip = Lignin peroxidase – 0.01 increase in OD equal to one unit of enzyme.
Lac = Laccase – 0.01 increase in OD equal to one unit of enzyme.

Chemicals (EDTA, TCA, NaN_3), metal ions (Cu, Mg and Zn), and veratryl alcohol were tried for suppression of cellulase and promotion of ligninases. Except veratryl alcohol all the chemicals decreased both the enzymes with the increased concentrations (Table-2). However, veratryl alcohol increased substantially the lignolytic production and decreased cellulolytic activity.

Among the different concentrations of veratryl alcohol tested, 37.2mM concentration was found to be very useful and ideal for promotion of lignolytic and complete suppression of cellulolytic activities (Table 3) (Fig.2).

The UV and X-rays were used for development of cel mutants of *P. ostreatus* and the variations that were obtained with different exposures on the production of C_x, C_1, Lip and Lac are presented in Table 4 and Fig 6.

Table 4 Influence of X-rays and uv rays on cellulolytic (c_x & c_1) enzymes by *P. ostreatus*

Days of incubation	Time of exposure		C_x		C_1	
	A (min)	B (sec)	A	B	A	B
7	0	0	66	40	34	42
	10	0.4	66	20	26	36
	20	0.8	33	28	26	44
	30	1.2	40	33	32	40
	40	1.6	33	20	32	46
	50	2.0	45	28	34	44
	60	2.4	40	20	38	46
11	0	0	28	28	52	48
	10	0.4	20	17	52	44
	20	0.8	18	25	54	46
	30	1.2	25	28	52	42
	40	1.6	17	25	57	44
	50	2.0	28	20	54	40
	60	2.4	28	22	54	40
14	0	0	22	25	10	22
	10	0.4	18	15	24	6
	20	0.8	17	22	22	20
	30	1.2	18	25	22	4
	40	1.6	15	22	24	12
	50	2.0	22	18	10	4
	60	2.4	25	20	14	10

A = UV rays, B = X-rays, C_x = REA (Relative enzyme activity), C_1 = mg/ml of reducing sugars liberated in 6h.

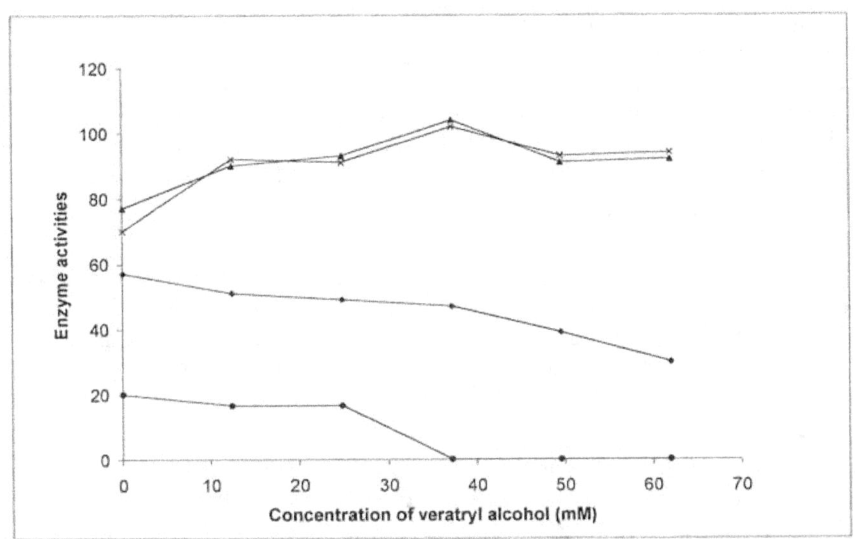

Figure 5 Influence of veratryl alcohol on cellulolytic (Cx & C_1) and lignolytic (Lip & Lac) enzymes by *P. ostreatus* after 14 days of incubation.

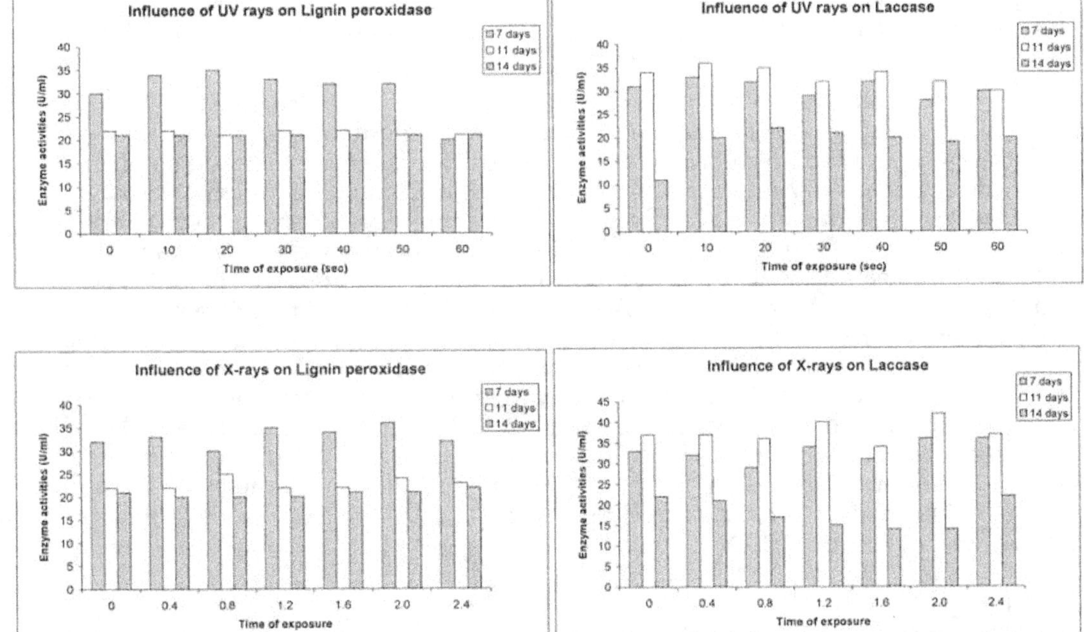

Figure 6 Influence of UV rays and X rays on Lignolytic enzymes (Lip and Lac)

From the data it is evident that 11 days incubation of the mutated organism showed the desired trends of decreased cellulase and increased ligninase productions. The subsequent incubation (14 days) resulted for decreasing activities of both the enzymes. The 20 minutes exposure of UV rays substantially increased the production of laccase (35 U/ml) and considerably decreased the cellulase (C_X) activity. No useful production trends for the bleaching purpose was recorded in the production of lignin peroxidase and cellulase (C_1) activity with UV irradiation.

An exposure to X-rays for 0.4 seconds substantially reduced the cellulase (C_x and C_1) production and increased the laccase activity. Hence, the *P. ostreatus* culture exposed to 0.4 seconds can be designated as cel mutant and can be successfully used for further commercialization of bleaching process. However, 0.4 to 2.4 seconds exposures of X-rays did not develop any desired cel mutants. Only marginal improvement in the laccase production was noticed (36-42 U/ml) with 2.0 seconds exposure.

The role of veratryl alcohol in promoting the activity of lignolytic enzymes was well established (Harvey *et al.*, 1985 and Warishi *et al.*, 1991). Veratryl alcohol as a secondary metabolite of most white rot fungi has been reported to increase the initial rate of lignin degradation to CO_2 (De Jong *et al.*, 1994). Brendly *et al.*, (1986) proposed that veratryl alcohol probably functions via an induction type of mechanism. It has also been demonstrated that veratryl alcohol is responsible for the proper turnover of the enzyme cycle enabling improved oxidation of non phenolic substrates (Koduri and Tien, 1994).

Palmieri *et al.*, (2000) studied the role of metal ions, Cu, Mg and Zn which showed positive effect on lignolytic enzymes. However, in our study the influence of these metal ions on lignolytic activity was negative. Similarly it was observed that chemicals like EDTA, TCA, NaN_3 proved to suppress the cellulase and lignolytic enzymes (Frederick *et al.*, 1991).

The first genetic experiment to produce cel mutants were carried out by UV light irradiation of spore suspension of white rot fungus *Polyporus adustus* (Eriksson and Goodell, 1974). Eriksson *et al.*, (1983) isolated spontaneous acid cel, xylanase positive and phenol oxidase positive mutants of *Sporotrichum pulverlantum* and compared them with parent strains. However, these mutants have low capacity of lignin degradation as compared to wild types. Johnsrud and Eriksson (1985) developed new cellulase deficient strains with increased ability to degrade lignin by cross breeding of selected and mutated homokaryotic strains of *Phanerochaete chrysosporium*. Alic *et al.*, (1987) recorded phototrophic strains of *P. chrysosporium* from crosses between two auxotrophic strains of lignin degrading basidiomycetes. Kamra *et al.*, (1998) while working on biodelignification of lignocellulosic feeds with white rot fungi, strengthened the concept of improvement of strains for increasing digestibility of substrate by cel mutants that produced enhanced phenol oxidase. These studies indicate the chemical application of veratryl alcohol at 37.2 mM concentration ideal for application to obtain cellulose minus strains. Further, the UV rays (20 minutes) and X-rays (0.4 seconds) were considered to be useful in development of cel mutants of *P. ostreatus* for biobleaching of UKP.

Acknowledgements

The authors are thankful to Prof. A. Sadanandam, Head, Department of Botany for encouragement. One of the authors (Ch. Vijaya) is greatful to Dr. R. Mallikarjuna Reddy, Coordinator, Biotechnology, Jawahar Bharathi Degree College, for his valuable suggestions.

References

Alic, M., C. Letzring and **Gold, M.H.** 1987. Mating system and basidiospore formation in the lignin degrading Basidiomycetes *Phanerochaete chrysosporium*. *Appl. Environ. Microbiol.*, 53: 1464-1469.

Ander, P., Eriksson, K.E., Kolar, M.C. and **Kringstad, K.P.** 1997. Studies on the mutagenic properties of bleaching effluents. *Sven. Paperstidn.*, 80: 454-459.

Brendly, N.D. Faison, Kent Kirk, T. and **Robert Farell, L.** 1986. Role of veratryl alcohol in regulating ligninase activity in *Phanerochaete chrysosporium*. *Appl. Env. Microbiol.*, 52: 251-254.

Christon, J.H. 2000. Manual of Environmental Microbiology, ASM Press, Washington DC, 62-70.

Coll, M.P., Fernadex-Abalos, J.M., Villanueva, J.R., Santamaria, R. and **Perez, P.** 1993. Purification and characterization of a phenol oxidase (Laccase) from the lignin-degradating basidiomycete PMI (CECT 2971). *Appl. Environ. Microbiol.*, 59: 2607-2613.

Crawford, D.L. and **Crawford, R.L.** 1976. Microbial degradation of lignocellulose : The lignin component. *Appl. Env. Microbiol.*, 31: 714-717.

De Jong, Field, J.A. and **de Bont, J.A.M.** 1994. Aryl alcohol in the physiology of white-rot fungi. *FEMS Microbiol. Rev.*, 13 : 153-188.

Eriksson, K.E. and **Goodell, E.W.** 1974. Pleotrophic mutants of the wood rotting fungus *Polyporus adustus* lacking cellulase, mannase and xylanase. *Can. J. Microbiol.*, 20: 371-378.

Eriksson, K.E., Johnsrud, S.C. and **Ualbander, L.** 1983. Degradation of lignin and lignin model compounds by white rot fungus *Sporotrichum pulverulentum*. *Archives in Microbiology*, 135 : 161-168.

Farell, R.C. 1986. Third International Conference on Biotechnology in pulp and paper industry, Stockholm, 61-63.

Frederick, C., Michel, J.R., Balachandra Dass, S., Eric A. Grulke and **Adinarayana Reddy, C.** 1991. Role of Manganese peroxidases and lignin peroxidases of *Phanerochaete chrysosporium* in the decolourisation of kraft bleach plant effluent. *Appl. Env. Microbiol.*, 57: 2368-2375.

Harvey, P.J., Schoemaker, H.E. and **Palmer, J.M.** 1985. Veratryl alcohol as a mediator and the role of radical cations in lignin biodegradation by *Phanerochaete chrysosporium* FEBS 3298. Elsevier Science Publishers, 195 : 242-246.

Johnsrud, S.C. and **Ericksson, K.E.** 1985. Cross-breeding of selected and mutated homokaryotic strains of *Phaerochaete chrysosporium* K-3 : New cellulase deficient strains with increased ability to degrade lignin. *Appl. Microbiol. & Biotechnol.,* 21 : 320-327.

Kamra, D.N. 1998. Biodelignification of lignocellulosic feeds with white rot fungi for the feeding of ruminants. In "Fungi in Biotechnology", CBS Publishers, Anil Prakash (ed.), New Delhi.

Katagiri, N., Tsutsumi, Y. and **Nishida, T.** 1995. Correlation of brightening with cumulative enzyme activity related to lignin biodegradation during biobleaching of kraft pulp by white rot fungi in the solid state fermentation system. *Appl. Env. Microbiol.,* 61 : 617-622.

Kirk, T.K. and **Yang, H.H.** 1979. Partial delignification of unbleached kraft pulp with lignolytic fungi. *Biotechnol. Lett.,* 1 : 347-352.

Kirk, T.K. and **Farell, R.L.** 1987. Enzymatic "Combustion" : the microbial degradation of lignin. *Ann. Rev. Microbiol.,* 41 : 465-505.

Koduri, R. and **Tien, M.** 1994. Kinetic analysis of lignin peroxidase explanation for the mediation phenomenon by veratryl alcohol. *Biochemistry,* 33: 4225-4230.

Miller, G.L. 1959. Use of dinitrosalicylic acid reagent for the determination of reducing sugars. *Analyt. Chem.,* 31: 426-428.

Palmieri, G., Giardina, P., Balnco, C., Fontanella, B. and **Sannia, G.** 2000. Copper induction of laccase isoenzymes in the lignolytic fungus *Pleurotus ostreatus. Appl. Env. Microbiol.,* 66: 920-924.

Ramakrishna, G. 2003. A biotechnological approach to the biodegradation of lignin by white rot fungi. Ph.D. thesis, Kakatiya University, A.P.

Reese, E.T., Siu, R.G.H. and **Levinson, H.S.** 1950. The biological degradation of soluble cellulose derivatives. *J. Bacteriol.,* 59: 485-497.

Warishi, H., Jin Haung, H., Brain Dunford and **Michael H. Gold** 1991. Reactions of lignin peroxidase compounds I and II with veratryl alcohol. *Biological Chemistry,* 266: 20694-20691.

Wuyep, P.A., Khan, A.V. and **Nok, A.J.** 2003. Production and regulation of lignin degrading enzymes from *Lentinus suarrosulus* (mont). *Singer and Psathyrella atroumbonata Pegler. African J. Biotech.,* 2: 444-447.

42 | *In Search of New Antimicrobial Compounds from Plants : Polyalthia longifolia*

J. Annapurna, M. Marthanda Murthy and J. Madhusudan Rao

Microbiology Laboratory, Organic Chemistry Division, Indian Institute of Chemical Technology, Hyderabad - 500 007,

ABSTRACT

Polyalthia longifolia or Devadaru, belongs to the family annonaceae. The plant is well known for its medicinal values. Its leaves and bark are used in several ailments. The bark is used in skin diseases, fever, diabetes, hypertension, helmithiasis and is febrifuse. Several compounds such as alkaloids, terpenoids, sterols and steroid glycosides have been isolated from different parts of this plant. Proanthocyanidin along with β-sistosterol and leucocyanidin were isolated from stem bark. In view of its importance of biological activity of its metabolites, autobioassay guided fractionation of seeds of *P. longifolia* has resulted in isolation and characterisation of diterpenoids. The two clerodane diterpenoids 16á-hydroxy-cleroda - 3,13 (14) z diene- 15,16 -olide (1) and 16-Oxo-cleroda - 3, 13(14) E-diene - 15 oic acid (2) isolated from the hexane extract of the seeds of *Polyalthia longifolia*, have demonstrated significant antibacterial and antifungal activities.

KEY WORDS : Antibacterial, antifungal, *Polyalthia longifolia*, seeds, diterpenoids

Throughout the ages, Nature has provided humans with sources of the essentials of life, including food, medicines, and raw materials for the manufacture of clothing and shelters. In particular, higher plants have been the source of medicinal agents since earliest times, and today they continue to play a dominant role in the primary health care of about 80% of the world's population. Natural products, and medicinal agents derived therefrom, are also essential features in the health care systems of the remaining 20% of the population residing mainly in developed countries, with more than 50% of all drugs in clinical use having a natural product origin (Balandrin *et. al.*, 1993). Of the world's 25 best-selling pharmaceutical agents, 12 are natural product-derived (Neil and Lewis, 1993), and natural products continue to play an important role in drug discovery programs of the pharmaceutical industry and other research organizations (Kinghorn and Balandrin, 1993). Research into the chemical and biological properties of natural products over the past two centuries has not only yielded drugs for the treatment of many human ailments, but has provided the stimulus for the development of modern synthetic organic chemistry and the emergence of medicinal chemistry as a major route for the discovery of novel and more effective therapeutic agents.

Plants are natures "Chemical factories", providing the richest source of chemical compounds on Earth. Although only about 10,000 secondary plant metabolites have been chemically defined, it is estimated that the total number of plant chemicals may amount to 400,000 or more. During their evolution from the Devonian period, many plant taxa have evolved highly sophisticated defense systems, largely a complex array of defense chemicals produced by the plants themselves.

In the recent past, there has been a global trend towards revival of interest in the medicinal plants. One of the reasons is that herbal medicines are believed to be comparatively safer than most of the modern drugs and the other belief is, they are economical and environmentally better suited for local conditions. The introduction of number of in vitro test systems, which specifically analyze the molecular targets in screening of plants, has motivated to investigate plants. The current problem of resistance reflects the Darwinian selection whereby these drugs kill susceptible bacteria but allow resistant ones to survive. Concern emerged in the late 1990's that resistance was accumulating without the discovery of new antibiotic classes.

As mentioned earlier, plants have formed the basis for the treatment of diseases throughout the ages and continue to be major source of primary health care for about 80% of the world's population. Sophisticated plant-based traditional medicine systems have been in existence for thousands of years in countries such as china (Chang and But, 1986) and India (Kapoor, 1990) and medicinal plants are used extensively in African traditional health systems. Numerous phytomedicines are registered and extensively used in Europe, and more than 600 botanical items have been officially recognized in various editions of the United States Pharmacopoeia (Tyler, 1993), though current regulations prohibit most from being marketed as drugs. In Australia, the traditional therapeutic use of native plants by indigenous people was not recorded in written form and the extent of that use and knowledge is only now becoming evident.

Of the 119 plant-derived drugs commonly used in one or more countries, 74% were discovered as a result of chemical studies directed at the isolation of the active constituents of plants used in traditional medicine. Well-known examples include the cardiac glycosides from *Digitalis purpurea* L., the antihypertensive agent and tranquilizer, reserpine, from the East Indian snakeroot, *Rauvolfia serpentina* (L.) Bentham ex Kurz; the antimalarial agent, quinine, from Cinchona spp.; and the analgesics, codeine and morphine, from *Papaver somniferum* L. Secondary metabolites isolated from medicinal plants have also served as precursors or models for the preparation of effective agents through semi-syntheses or lead-based total syntheses. Examples include the anticancer agent, etoposide, a semi-synthetic derivative of epipodophyllotoxin isolated from *Podophyllum* spp., and anticholinergic drugs modeled on the belladonna alkaloids (e.g., atropine) isolated from *Atoropa belladonna* L. and other medicinal plant species (Balandrin *et. al.*, 1993).

Of the estimated 250,000 currently known higher plant species, very little is known about their secondary metabolites; this is particularly true for tropical flora, which constitute over 60% of this estimated number. Even less is known about the far more abundant (though taxonomically relatively unexplored) insect and microbial worlds (Gentry, 1993), as well as the biologically rich and enormously diverse marine environment (Norse, 1994). Given the rapid destruction of tropical habitats, especially the rainforests, and the degradation of some marine ecosystems, this lack of knowledge is alarming. Considering that the 119 drugs mentioned above were isolated from only about 90 plant species, the potential for drug discovery from plants and other natural sources is enormous, but little time remains to explore this rapidly diminishing resources. It is also worth considering the potential of higher plants to yield novel molecules towards the development of new antibiotics.

Screening programme is directed towards the discovery of new leads from plants. Several strategies have been employed including the following by earlier researches. In certain circumstances, higher plants produce antibiotic compounds (phytoalexins) in response to infection (Haenen, 1985; Watson et. al., 1985). Such compounds have been examined for their antimicrobial and antifungal actions. However, there is no compelling reason to support that plant anti-infective agents are active against human or veterinary pathogens; and in searching for antimicrobial agents it is technically easier to screen for natural products rather than for phytoalexins. Additionally the quantity of phytoalexins produced in comparison to the amounts of constitutive agents is often small, even in infected plants (Mitscher et. al., 1987). This approach does have the attraction that phytoalexins are highly active compounds produced in response to infection.

The compounds isolated to date belong to different classes of chemical structures. Recent examples include alkaloids (Verpoorte, 1987: Bhattarcharyya et. al., 1985), Chromans (Mitscher et. al., 1985b), flavonoids (Mitscher et. al., 1985c), antharaquinoes (Ognuntimein, 1987), phloroglucinols (Marthuenda and Garcia, 1985), saponins (Bader et. al., 1987), lactones (Mares, 1987). These substances differ in their structure from the clinically used antibiotics. However, further work on potency, toxicity and the development of resistance is obviously warranted. Hence it is appropriate to continue research on medicinal, ethnobotanical and aromatic plants in search of new compounds.

The annonaceae is a large family of tropical and subtropical trees and shrubs comprising about 120 genera and more than 200 species. The phytochemistry of this family has been reviewed by Leboeuf et al.(1982). Their review details the various alkaloids, carbohydrates, lipids, amino acids, proteins, polyphenols, essential oils terpenes, and aromatic compounds typically found in these plants. Many species in this family are used traditionally for various purposes. However most of the previous phytochemical studies have not attempted to explain these uses through testing of biological activities

Bioactivity directed isolation studies are strongly suggested in the search for potentially bioactive secondary metabolites in plants. Many such activities directed towards fractionation lead to the isolation, structure elucidation of extremely bioactive compounds. The purpose of this paper is to isolate bioactivity directed fractionation and chemistry of this rapidly growing class of natural compounds.

Polyalthia longifolia belongs to Annoaceae family. Annonaceae species mainly contained a group of compounds called acetogenins which are possessing anticancer as well as insecticidal properties. *P. longifolia* is an ornamental tree and is widely distributed in India. Several compounds have been isolated from different parts of the plant. Aporphine and azafluorene type alkaloids have been isolated from the leaves of *P. longifolia* (Jossang *et al.*, 1982; Wu *et al.*, 1990), Proanthocyanidine trimer (Wu,1989) and clerodane diterpernoids, sistosterols (Phadnis *et al.*, 1988) have been isolated from this plant. Proanthocyanidin trimer (Agarwal and Misra, 1979) have been reported from the bark of this plant. Flavonoid and steroidal glycosides have been reported from the leaves. Clerodane type diterpenoids with antifeedant properties have been isolated from the leaves of this plant. Enthalimanes have been reported from the stem bark of this plant. Several compounds have been isolated from stem, stem bark and leaves of the plant Polysaccharide has been isolated from the seeds of this plant (Haq *et al.*, 1974). Recently we reported the isolation of clerodane type diterpenoids from the seeds of *P. longifolia* which has not been reported earlier, instead of terpenoids, alkaloids and steroids have been isolated. The methanol extract of stem parts of *P. longifolia* showed significant cytotoxic activity. Bioassay guided fractionation of the methanol extract of stem parts of this plant yielded aporphine alkaloid liriodemine.

These are some of the compounds isolated by earlier workers and their structures are shown below.

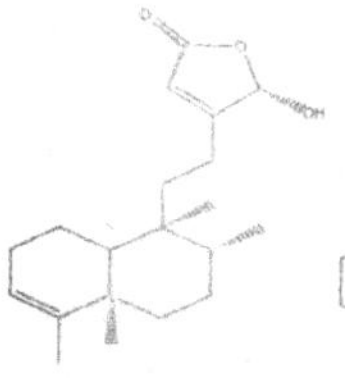

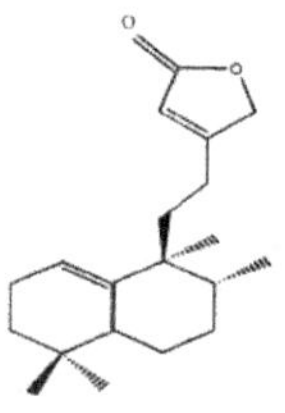

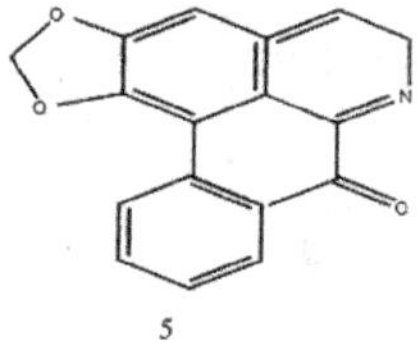

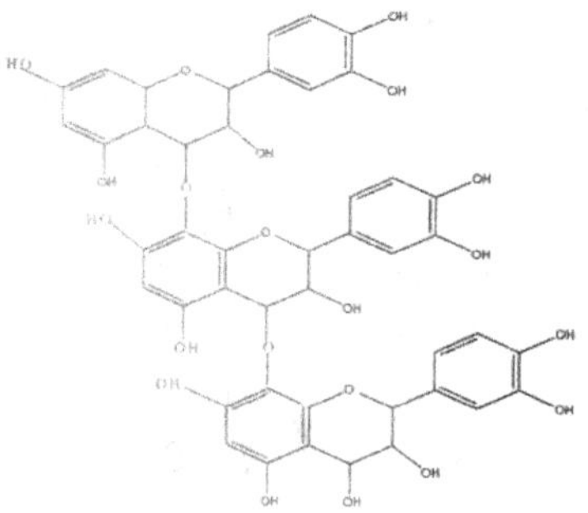

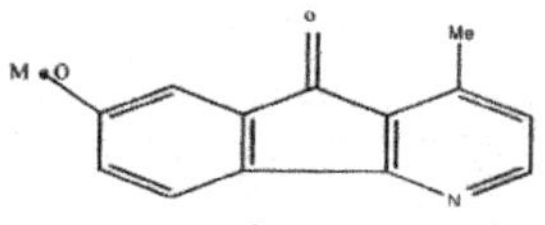

1. Clerodane diterpenoids 2. Clerodane diterpenoids 3. Ent halimane diterpenoids
4. Proanthocynadin   5. Aporphine alkaloid   6. Azafluorene alkaloid

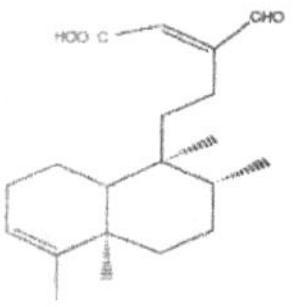

In our pursuit to isolate novel antimicrobial compounds, we have earlier worked on many medicinally/ethno pharmacological important plants (Annapurna *et. al.,* 1983, 1989, 1999, 2004, Annapurna and Iyengar, 2000, Marthnda Murthy *et. al.,* 2003, 2004,) In our earlier study the leaves have shown good antimicrobial activity which was reported in 1983.

Materials and Methods

Plant and microorganisms

Polyalthia longifolia Thw. (Annonaceae) (Sastri, 1969) seeds were collected from IICT campus, Hyderabad in August 2001 and authenticated by a toxonomist Prof. Prabhakar, Department of Botany, Osmania University, Hyderabad, India. It is also known as false asoka now cultivated throughout India in gardens and avenues. *Bacillus Subtilis, Escherihia coli* and *Pyricularia oryzae* were used as test oragninsms.

Extraction Procedure

Seeds were ground finely and extracted successively with solvents of increasing polarity such as hexane and methanol. The solvents were evaporated under vaccuum and the residue was used for testing. Since nonpolar solvent hexane gave highest activity (Table 1) further bioassay guided fractionation was carried out with hexane extract. It was fractionated over silica gel column using hexane ethylacetate, stepwise elution resulting in seven pooled fractions (Table 2). All the fractions were evaluated for antimicrobial activity against gram positive and gram negative bacteria and fungi. The fractions which were active were further purified to isolate pure compounds. These two compounds were charachterised by spectroscopic methods and designated as cleredane diterpenoids.

Table 1 Antimicrobial activity of crude extracts of *Polyalthia longifolia*

Solvents	Inhibition Zone in mm		
	Bacillus subtilis	*E.coli*	*Pyricularia oryzae*
Hexane	34	35	15
Methanol	25	26	10
Gentamycin	26	28	–

Table 2 Antimicrobial activity of different pooled fractions

Fractions No.	Name of Microbes Tested Inhibition Zone in mm		
	B.subtilis	E.coli	Pyricularia oryzae
1.	NIL	NIL	NIL
2.	+ +	+ + +	NIL
3.	26	28	20
4.	30	32	30
5.	30	32	30
6.	22	24	15
7.	24	26	NIL

Antimicrobial Testing

All the media and standard antibiotics used were from Hi Media, Bombay, India. Bacterial strains were grown on nutrient agar and suspended in Mueller Hinton broth and fungal strains in sabouroud broth. Two-fold serial dilutions were employed to determine MIC values (Jones et. al., 1984). One mg of the sample in 1 ml of 10% acetone water (1000 mg/ml) was transferred aseptically to 9 ml of broth (1st tube, 100 µg/ml) and subsequently 5 ml from the 1st tube was added to 5 ml of 2nd tube and so on to get serial dilutions of concentrations 100-1 µg/ml. To the above tubes, 30 µl of suitably diluted inoculum grown at 35 + 2 °C for 20 hr (approximately 10^7 CFU/ml) were added and the culture tubes, along with controls, were incubated at 37 °C for 24 hr. The results (average of triplicates) are presented in mg/ml along with standards. The lowest concentration of antimicrobial agent that resulted in complete inhibition of microorganisms represents MIC and anti fungal activity by paper disc method (Bauer and Kirby ,1996).

Autobioassay

Paper chromatogram was run in hexane ethylacetate mixture to 15 cms from the point of application and air dried completely. Seeded agar was prepared with 24 hr old culture at the above said concentration of cells and poured aseptically for setting. After solidification the dried paper was kept on the seeded agar plate and covered. This biogram was kept in refrigerator for one hour for the compound diffusion. After an hour the paper was removed from the plate with sterile foreceps under laminar hood and incubated for 24 hr in incubator at 35° C. The inhibition zone corresponds to the presence of the active compound present at that rf values. The zones of inhibition are shown in the autobiogram showing two active compounds in the present hexane extract.

Purification

Further purification was been done using column chromatography with silica gel as stationary phase and hexane ethylacetate as mobile phase. The two active fractions characterised by spectroscopic studies and the details of the structures are shown having marked antibacterial and antifungal activities.

Tested materials

16á-hydroxy-cleroda - 3,13 (14)z diene- 15,16 -olide *(1)* and 16-Oxo- cleroda – 3, 13(14) E-diene – 15 oic acid *(2)*.

Results and Discussions

The two compounds shown below were isolated for the first time from the seeds of *P. longifolia* which have shown to be having antibacterial activity. The details of the spectral data are shown below along with structures.

16-hydroxy-cleroda - 3,13 (14)z diene- 15,16 -olide (Phadnis et al., 1988) colourless sticky mass, [a]D + 16.67°(MeOH), MS m/z : 318. ^{1}H NMR (200 MHz, CDCl3): d 5.99 (s, 16 H), 5.84 (s, 14H), 5.19(s, 3H), 1.59 (18-CH3), 1.01 (19-CH3), 0.82 (17-CH3), 0.78 (20-CH3) ^{13}C NMR (50 MHz, CDCl3): d 170.99 (C-13), 170.13 (C-15), 144.37 (C-4), 120.42 (C-3), 117.16 (C-14), 98.76 (C-16), 46.53 (C-10), 38.73 (C-5), 38.20 (C-9), 36.76 (C-6), 36.41 (C-8), 34.91 (C-11), 29.70 (C-1), 26.81 (C-7), 21.35 (C-12), 19.91 (C-19), 18.32 (C-2), 18.18 (C-20), 17.96 (C-18), 15.98 (C-17).

16-Oxo- cleroda - 3, 13(14) E-diene – 15 oic acid (2) pale yellow semi solid, [a]D –53.06° (MeOH), MS m/z : 318. ^{1}H NMR (200 MHz, CDCl3) d 9.55 (H-16), 6.5 (H-14), 5.23 (H-3), 1.60 (18-CH3), 0.99(19-CH3, 0.83 (17-CH3), 0.69 (20-CH3), ^{13}C NMR (50 MHz, CDCl3): d 194.20 (C-16), 170.16 (C-15), 157.43 (C-13), 144.23 (C-4), 133.83 (C-14), 120.73 (C-3), 46.61 (C-10), 39.37 (C-9), 38.28 (C-5), 37.07 (C-11), 36.34 (C-6), 36.07 (C-8), 27.66 (C-7), 26.8 (C-2), 19.90 (C-19), 19.2 (C-12), 18.17 (C-1), 18.00 (C-20), 18.00(C-18), 15.85 (C-17).

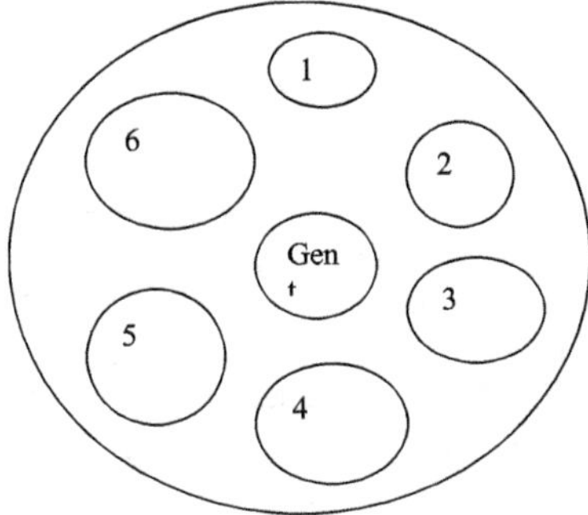

Figure 1 Inhibition zones of hexane extract

The Fig. 1 shows concentrations of 5, 10, 15, 20, 25 and 30 mg/ml represented from 1-6 inhibition zones. From the resuts it is evident that the inhibition zones are not directly proportional to the concentrations used in the present study. To get proportional zones the compound has to be diluted many fold. The crude hexane extract was tested by paper disc method at different concentrations to know the concentration range expand to be used for MIC determinations. Even at 5 µg/ml the activity is marked and hence MIC was carried out in the range of 1-200 µg on different microorganisms.

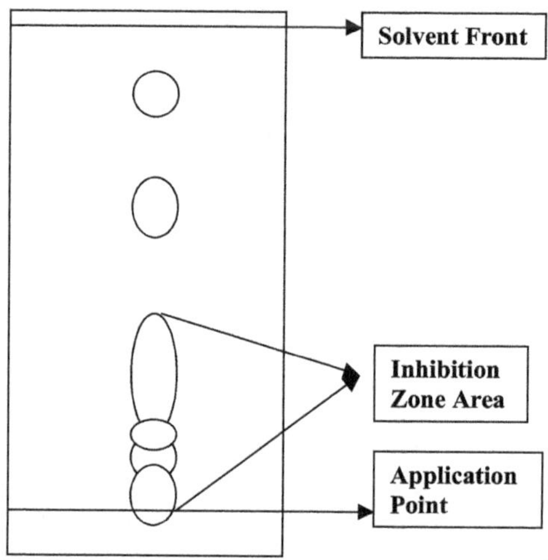

Figure 2   Visualization of autobiogram

The above Fig. 2 shows the visualization of the chromatogram after selecting a good solvent system for good separation for the purification of the active compound for structure elucidation after attaining the Rf values. In the present figure the visualisation has been done using sulfuric acid.

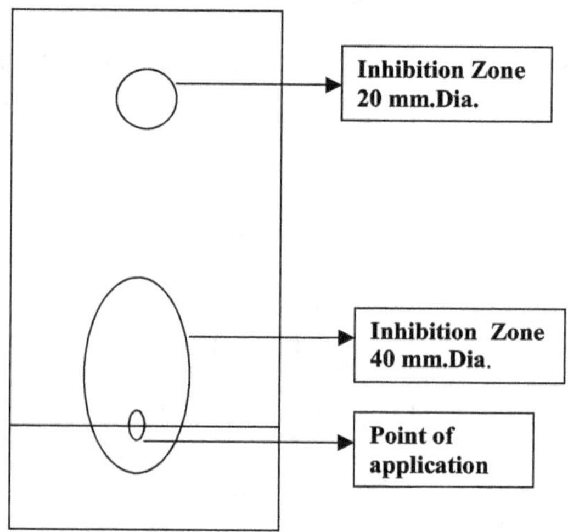

Figure 3   Autobiogram

The Fig. 3 shows two inhibition zones having Rf values 0.3800 and 0.8220 indicating the presence of active compounds. Accordingly the active compounds were purified by column chromatography followed by testing of all the fractions.

The Table 1 shows the activity of hexane and methanol extracts of the seeds of *P. longifolia* compared with standard gentamycin. Hexane extract showed highest activity.

Table 2 shows the number of fractions collected and poooled according to their rf values into 7 fractions. All the fractions are tested for the antimicrobial activity. The activity was tested against *Bacillus subtilis* and *E.coli* and *Pyricularia oryzae*. Fraction nos. 3 and 4 have shown highest activity at 39=35 mm inhibition zone. Accordingly these were further purified to isolate compounds responsible for antimicrobial activity.

The nonpolar extracts of seeds *P. longifolia* demonstrated antimicrobial activity in the present study. The polar solvent fraction methanol has not shown remarkable activity as non polar hexane fraction. Antimicrobial activity of the hexane fraction was compared with standard gentamycin and nystatin. The MIC was determined and the results are reported in Table 3. Our results show that hexane fraction has broad-spectrum antimicrobial activity against Gram-positive and Gram-negative bacteria and fungi such as *Candida albicans* and *Saccharomyces cereviseae*.

Table 3 Antibacterial activity of compound 1 and 2 of *P. longifolia* seeds

Microorganism	*Minimum inhibitory concentrations µg/ml*		
	1	*2*	*Gentamycin*
Gram positive			
Arthrobacter citreus	6.25	12.50	6.25
Bacillus cereus	3.125	12.50	6.25
B. licheniformis	3.125	25.0	3.125
B. polymyxa	1.56	12.5	3.125
B. pumilus	1.56	12.5	6.25
B. subtilis	1.56	6.25	6.25
Clostridium sp.	3.125	6.25	3.125
Staphylococus aureus	6.25	12.25	25.00
Streptococcus sp.	12.50	25.00	25.00
Gram negative			
Escherichia coli	0.78	1.56	6.25
Klebsiella aerogenes	1.56	1.56	3.125
Pseudomonas aeruginosa	0.78	3.125	3.125
P. putida	3.125	3.125	6.125
Salmonella typhimurium	0.78	1.56	3.125
Sarcina lutea	1.56	3.125	6.25
Nocardia sp.	3.125	6.25	6.25

In general, the activities against most of the tested bacterial cultures used have shown good activity when compared with the standard gentamycin. The activity against Gram-positive and Gram-negative bacteria was in the range of 1.00-3.13 µg/ml. The activity was more significant against Gram-negative than Gram-positive bacteria. The most sensitive Gram-negative bacteria were two *E.coli* strains with MIC values of 3.13 and 6.25 µg/ml, followed by *Klebsiella*, *Pseudomonas* and *Salmonella* having 6.25 mg/ml MIC values. *Sarcina lutea* was inhibited at 50 µg/ml concentration. Among Gram-positive bacteria, most of the strains of *Bacillus* have MIC values of 12.5 µg/ml. *Staphylococcus* was inhibited at 25 µg/ml while *Streptococcus* and *Arthrobacter* at 200 µg/ml and 100 µg/ml, respectively.

Fungal cultures were not very sensitive when compared with standard nystatin (Table 4). The MIC for *Candida albicans* and *Saccharomyes cereviseae* were 25 and 50 µg/ml, respectively, whereas nystatin inhibited the above fungi at 1.56 and 3.13 µg/ml. From the above results it is evident that the active principle from *P. longifolia* seed extract is more active against bacteria than fungal cultures. The active fraction is nonpolar in nature.

Table 4 Antibacterial activity of compound 1 and 2 of *P. longfolia* seeds

Fungi	*Minimum inhibitory concentrations µg/ml*			
	1	*2*	*Nystatin*	*Dithane M 45*
Yeasts				
Candida albicans	12.5	25	1.56	ND
Saccharomyces cerevisiae	25	5	3.13	ND
Plant pathogens				
Pyricularia oryzae	50	25	ND	25
Altenaria alternata	50	50	ND	50
Colletotrichum gloeosporioides	50	25	ND	50
Drechslera specifera	50	25	ND	25
Fusarium sp.	100	50	ND	50

The ethanobotanical and Ayurvedic uses of this plant signifies the presence of important compounds used for different ailments such as cutaneous affections, dropsy, leprosy, scabies, intermittent fever, jaundice, ulcers, etc. (Sastri 1969). Further purification and structure elucidation of active compounds are in progress.

Conclusions

Compounds (1) and (2) were isolated from hexane extract for the first time from the seeds of *P. longfolia*. These compounds demonstrated very good activity against gram positive and gram negative bacteria. Of the two compounds, compound (1) is highly active against gram negative bacterial such as *E.coli*, *K. aerogenes*, *Pseudomones* sp. and *S. lutea* with 0.78-1.5 μg of MIC values. Amongst gram positive bacteria *Bacillins* sp. are highly active with less than 2 μg of MIC. In general compound (1) is highly active against gram negative bacteria and compound (2) against all the bacteria tested. Compound (1) exhibited more activity than standard gentamycin. Compound (2) showed good activity against fungi than compound (1) with MIC values of 25 to 50 µg. The activity is compared with fungicide Dithane M-45. *Candida* and *Saccharomyces* recorded MIC values of 12.5 –25 μg which is comparable with standard Nystatin.

The prehistoric man first turned to mother nature for remedies and nature did not disappoint him. Since then the relationship between humans and plants has been very intimate throughout the development of civilization. Virtually every pharmacological class of drug includes a natural product prototype. Other sources of drug discovery such as synthetic compounds, microbes etc were identified much later and their importance cannot be underestimated.

Acknowledgements

We express our gratitude to Prof. M. Prabhakar, Taxonomist, Osmania University, Hyderabad and Director IICT for their cooperation.

References

Annapurna, J., Sakti Mitra, Iyengar, D.S., Nagabushan Rao, S. and **Bhalerao, U.T.** 1983. Antimicrobial activity of leaf extracts of *Polyalthia longifolia*. Phytopath.Z. 106:183-185

Annapurna, J., Iyengar, D.S. and **Bhalerao, U.T.** 1989. Antimicrobial activity of leaf extracts of *Enterolobium saman*. *Indian Drugs* 26: 272-274.

Annapurna, J., Bhalerao, U.T. and **Iyengar, D.S. 1999.** Antimicrobial activity of *Saraca asoca* leaves. *Fitoterapia*, 70: 80-82

Annapurna, J. and **Iyengar, D.S.** 2000. Antimicrobial activity of *Millingtonia hortensis* leaf extract. *Pharm. Biol*, 38: 157-160

Annapurna, J., Chowdary, I.P., Lalitha, G., Ramakrishna, S.V., and **Iyengar, D.S.** 2004. Antimicrobial activity of *Euphorbia nivulia* leaf extract, Pharmaceutical Biology (In press).

Agrawal, S. and *Misra, K.,* 1979 Current Sci. 48: 141.

Balandrin, M.F., Kinghorn, A.D. and *Farnsworth, N.R.* 1993. In: Human Medicinal Agents from Plants. Ed. A.D. Kinghorn and M.F. Balandrin, ACS Symposium Series 534, Washington, DC, p.2.

Bhattacharya, P., Chakarabarty, P. K. and *Chowdhury, B.K.* 1985. Glycozolidol, an antibacterial carbazole alkaloid from *Glycosmis pentaphylla. Phytochemistry* 24: 882-883.

Bader, G., Binder, K., Hiller, K., and *Ziegler-Bohme, H.* 1987. The antifungal action of triterpene saponins of *Solidago virgaurea* L. *Pharmazie* 42:140-146.

Bauer, A.W. and *Kirby* 1966. WMM. *Am. J. Clin. Pathol.* 45:493-495.

Berdy, J.H. 1989. In: Progress in Industrial Microbiology Ed. M.E. Bushell and U. Grafe, Elsevier, Amsterdam, vol. 27, p.3.

Chang, H.M. and *But, P.P.H.* 1986. Pharmocology and Applications of Chinese Materia Medica, Singapore, World Scientific Publishing Co.

Gentry A. 1993. In: Human medicinal agents from plants Ed by A.D., Kinghorn and M.F. Balandrin, ACS symposium series 534, Washington DC, 1993, p.13-21.

Haenen, J.M. 1985. Phytoalexins: antibiotic substances from higher plants. *Pharm Inter* 6: 194-196

Haq, Q.N., Khuda, A.M.H. and *Kaimuddin, M.* 1974. *Bangladesh J. Sci. Ind. Res.* 9: 101- 115.

Jones, R.N., Barry, A.L., Gaven, T.L. and *Washington, J.A.* 1984. In: *Lenette EH,*

Ballows, A., Havsle, W.J. and *Shadomy, H.J.* eds., *Manual of Clinical Microbiology,* Washington DC, American Society for Microbiol. 972-977.

Jossang, A., Leboeuf, M. and *Cave, A.* 1982. *Tetrahedron Letters,* 23: 5147

Kapoor, L.D 1990. *CRC Handbook of Ayurvedic Medicinal plants.* CRC Press, Boca Raton FL, 208-217.

Kinghorn, A.D., and *Balandrin, M.F.* (Eds) 1993. *Human medicinal agents from plants,* ACS symposium series 534, Washington D.C., p.48.

Leboeuf, M., Cave, A., Bhaumik, O.K., Mukherjee, B., and *Mukherjee, R.* 1982 *Phytochemistry,* 21: 2783-2813.

Mares, D. 1987. Antimicrobial activity of protoanomonin, a lactone from ranunculaceous plants. *Mycopathologia,* 98: 133-140.

Marhuenda, E. and *Garcia, M.D.* 1985. Mise an evidence desproprietesantimicrobiennes des sommites fleuries d' ononis natrix. Identification de lacide ferulique. *Plant Med. Phytother.* 19: 163-172.

Marthanda Murthy, M., Subramanyam, M., Giridhar, K.V. and *Annapurna Jetty* 2004. Antimicrobial activities of bharangin and bharangin monoacetate from *Premna herbaceae* Roxb. *Journal of Enthanopharmacology, (Under review)*

Marthanda Murthy, M., Subramanyam, M., Hima Bindu, M. and *Annapurna, J.* 2003. Antimicrobial activity of clerodane diterpenoids from *Polyalthia longifolia* seeds. *Fitoterapia* (In press)

Mitscher, L.A., Darke, S. Gollapudi, S.R. and *Okunte, K.* 1987. A modern look at folkloric use of anti-infective agents. *J. Nat. Prod.* 50:1025-1040.

Mitscher, L.A., Gollapudi, S.R., Darke, S. and *Oburn, D.S*. 1985a. Antimicrobial agents from higher plants: Two dimethylbenzisochromones from *Karwinskia humboldtiana. Phytochemistry* 24 : 1681-1683.

Mitscher L.A., Gollapudi S.R., Khanna I.K., Drake S.D., Hanumaiah T., Ramaswamy T. and *Rao K.V.J.* 1985b. Antimicrobial agents from higher plants; activity and stuctural revision of flemiflavanone-D from *Fleiminga stricta. Phytochemistry.* 24: 2885-2887.

Neill, M.O. and *Lewis, J.A.,* 1993. *Human Medicinal Agents from Plants*, Ed A.D. Kinghorn and M.F. Blandrin, ACS Symposium Series 534, Washington, DC, 48.

Kumar Singh and *Mahendra Sahai* 1995. *Phytochemistry;* 38: 189 -193.

Norse E.A. 1994. (Ed.), *Global marine biological diversity. A strategy for building conservation into decision making.* Island press, Washington D.C.

Oguntimein, B.O. 1987. The terpenoids of *Annona reticulata* II. *Fitoterapie.* 58: 411-413.

Phadnis, A.P., Patwardhan, S.A., Dhaneshwar, N.N., Tavale, S.S. and *Row, T.N.G.* 1988. *Phytochemistry.* 27: 2899 - 2993.

Sastri, B.N. 1969. Ed. *Wealth of India, Raw Materials*, Council of Scientific & Industrial Research, New Delhi. 8:188-192.

Sivarajan, V.V. and *Balachandran, I.* 1994. *Ayurvedic Drugs and Their Plant Sources.* 448-449, New Delhi, IBH Publishing Co. Pvt. Ltd

Tyler , V.E. 1993. *Human Medicinal agents from Plants*, Ed. A.D. Kinghorn and M.F. Balandrin, ACS Symposium Series 534,Washington, DC. p.25.

Verpoorte, R. 1986. Antimicrobially active alkaloids, Actual Chim, Ther 13, 195-209. *Chem. Abstr.* 106: 2598.

Watson, D.G., Rycroft, D.S., Freer, I.M., and *Brooks, C.J.W.* 1985. Sesquiterpenoid phytoalexins from suspended callus cultures of *Nicotiana tabacum. Phytochemistry* 24: 2195-2200.

Wu. Y- C. 1989. Heterocycles. 29: 463 - 467.

Wu, Y-C, Duh, C.-Y., Wang, S-K., Chen, K-S. and *Yang, T-H.J*. 1990. *Nat Prod.* 53: 1327-1332.

43 Production of β-Glucosidase from Penicillium rubrum and Aspergillus sydowi

Buddolla Viswanath* and B. Rajasekhar Reddy

Dept. of Microbiology, Sri Krishnadevaraya University, Anantapur-515 003, A.P.
e-mail : Buddolla@yahoo.co.in

ABSTRACT

The aim of the present study was to investigate the production of β- glucosidase in liquid medium (Dextrose Peptone Agar) under shaking conditions on a laboratory scale by two fungal cultures *Aspergillus sydowi* and *Penicillium rubrum* isolated from the soil. During the course of the growth of fungal cultures on liquid medium, content of extracellular protein and activity of β – glucosidase in the cultural filtrate were monitored at regular intervals over a period of 5 weeks. Cultures of *Aspergillus sydowi* exhibited peak β–glucosidase activity at 4th week interval, whereas in case of *Penicillium rubrum* it was observed at 3rd week interval. β-glucosidase activity of *Penicillium rubrum* was higher than that of *Aspergillus sydowi*.

KEY WORDS : *Aspergillus sydowi, Penicillium rubrum*, β-glucosidase, Dextrose peptone Agar.

Cellulose is the most abundant carbohydrate polymer in the biosphere. Cellulose bio conversion through enzymatic hydrolysis to obtain sugars, which can be fermented to produce liquid fuels, has received global attention during the last few years (Mehta, 2004). β–Glucosidase is an essential component of cellulase complex responsible for the final step in the cellulose degradation, namely hydrolysis of cellobiose to glucose that means they have maximum activity toward cellobiose. β–glucosidases can catalyze the hydrolysis of the β–glucosidic linkages of aryl and alkyl β–glucosides, β–linked oligoglucosides and several other oligosaccharides with the release of glucose usually in the β–configuration (Stuart *et al.* 1997). These enzymes are widely used in various biotechnological processes. In the flavor industry, β-glucosidases are the key enzymes in the enzymatic release of aromatic compounds from glucosidic precursors present in fruits and fermented products (Riou *et al.* 1998). The cellulase complex of *Trichoderma* species has been extensively studied and considered to have commercial importance. *Trichoderma* species excrete large amounts of endo-exoglucanases while producing comparatively small amounts of β–glucosidase. In view of the importance of β–glucosidase to the efficient enzymatic saccharification of cellulose and its poor production by *Trichoderma* sp. its search in other organisms attracts a great deal of attention. Therefore, production of β–glucosidase by *Aspergillus sydowi* and *Penicillium rubrum* was studied.

Materials and Methods

The fungal cultures of *Aspergillus sydowi* and *Penicillium rubrum* isolated from soils, were used in the present study and maintained on Dextrose peptone agar. Spore suspension was prepared by adding 2 ml of sterile distilled water to the freshly (7 days) grown slants of the above cultures on Dextrose peptone agar. The density of spores in the suspension was quantified with the help of a haemocytometer. This spore suspension was used to inoculate 50 ml of Dextrose peptone broth amended with 0.5% cellulose into 250 ml Erlenmeyer flasks, at spore density of 2×10^6. pH of the media was adjusted to 5.0. The flasks were incubated on an orbitary shaker with shaking at 200 rpm at 30^0C. At desired intervals, 250 ml Erlenmeyer flasks with growing culture of *Penicillium rubrum* and *Aspergillus sydowi* were withdrawn during the course of different experiments. Each sample was monitored for fungal growth, pH, protein and β–glucosidase activities.

The cultures in the flasks were aseptically filtered through pre-weighed Whatman No-1 filter paper to separate mycelial mat and culture filtrate. The filter paper along with mycelial mat was dried at 70^0 C in an oven until constant weight and this weight was recorded. Difference between the weight of the filter paper bearing mycelial mat and weight of only filter paper represented biomass of fungal mat. Fungal growth was expressed in terms of mg/flask.

An aliquot of these *Penicillium rubrum* and *Aspergillus sydowi* culture filtrates with appropriate dilution was used for estimation of soluble protein content according to Lowry *et al.* (1951). Bovine serum albumin was used as protein standard.

β–glucosidase activity in the culture filtrate of *Penicillium rubrum* and *Aspergillus sydowi* was determined according to the method of Herr (1979). B-D-Glucosidase activity was measured in assay mixture containing 0.2 ml of 5 mM p-nitrophenyl β-D-glucopyranoside (PNPG) dissolved in 0.05 M citrate buffer pH 4.8 and 0.2 ml of diluted enzyme solution with appropriate controls. After incubation for 30 min at 50^0C, the reaction was stopped by adding 4 ml of 0.05 M NaOH-glycine buffer (pH-10.6). The yellow coloured p-nitrophenol liberated was determined at 420 nm.

Results and Discussion

Growth of both the fungal cultures increased till 3rd week and then declined in case of *Aspergillus sydowi* but in case of *Penicillium rubrum* increased till 4th week (Fig 1). Maximum dry weight of biomass of *Penicillium rubrum* recovered at the end of 4th week incubation was 2.416 grams as against 2.883 grams by *Aspergillus sydowi* at 3rd week interval. It appears that the growth of *Penicillium rubrum* was slower than *Aspergillus sydowi*.

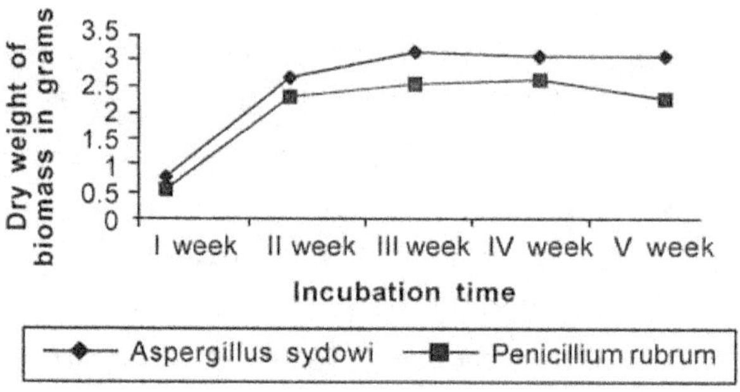

Fig 1 Growth of fungal cultures

It is found that during the course of growth of both *Aspergillus niger* and *Penicillium rubrum* pH changes occurred in the culture broth (Fig 2).

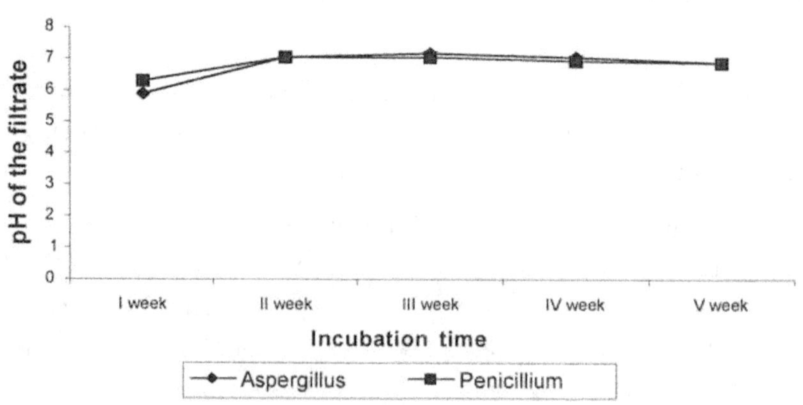

Fig 2 Changes in the pH of broths of fungal cultures

At the end of 28 day interval maximum of about 360 μgrams/ml of extracellular protein was recovered from the filtrate of *A. sydowi* whereas 369 μgrams/ml of extracellular protein was recovered in the filtrate of *P. rubrum* at 14 day interval. Fig 3 shows the differences in the production of extracellular protein in the filtrate on weekly intervals.

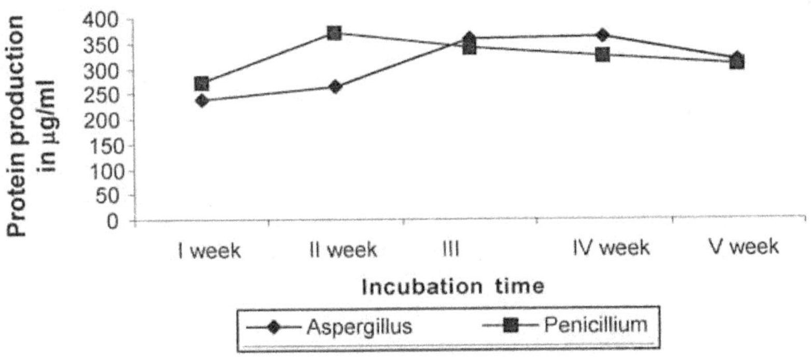

Fig 3 Secretion of extracellular protein in the filtrate of Fungal cultures

β– glucosidase was more in the filtrate of 4th week cultures of *Aspergillus sydowi* with 0.718 nmoles/ml/min but in case of *Penicillium rubrum* peak production of β– glucosidase was observed in the 3rd week filtrate with 1.03 nmoles/ml/min (Fig 4).

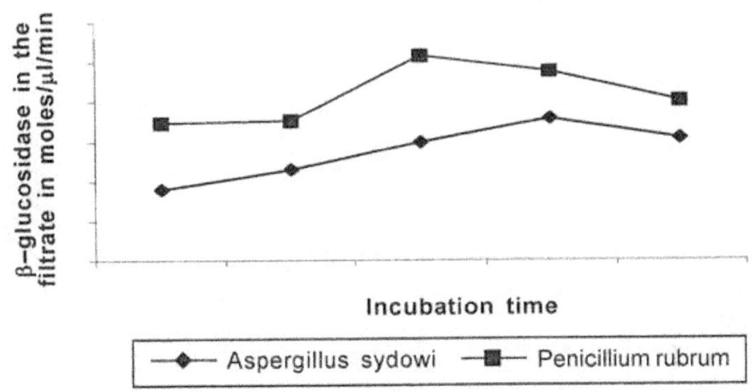

Fig 4 Production of β–glucosidase in the filtrate of fungal cultures

Production of β–glucosidase was very less in both *Penicillium rubrum* and *Aspergillus sydowi* when compared with other works of Freer and Detroy, (1985), Gokhale *et al.*, (1988) Stuart *et al.*, (1997)) Narasimha (2002). This may be due to the use of glucose rich medium, Dextrose peptone agar for their cultivation. Because, the addition of a high concentration of glucose in the medium repressed β–glucosidase expression (Freer and Destroy, 1985). Further work is in progress to get large amount of β–glucosidases by decreasing the glucose concentration in the medium.

<table><tr><td>**44**</td><td># *Ecofriendly Management of Fungal Biodeteriogens Using Floral Extracts*</td></tr></table>

Arun Arya and Vishal K. Muliya

Department of Botany, Faculty of Science, The Maharaja Sayajirao University of Baroda, Vadodara - 390 002

e-mail : aryaarunarya@rediffmail.com

ABSTRACT

Museum objects being organic in nature are attacked by fungi, bacteria and actinomycetes. Considering the environmental health hazards efforts are on to find out ecofriendly alternatives. During recent years, the antimicrobial principles of higher plants referred as antibiotics, phytocides are used to control fungal organisms. Essential oils are important in this connection as they can be used as fumigant fungicides. These being lipophilic can easily penetrate deeper through living tissue unbarred by the selective permeability of the cell membrane. Effect of nine flower extract was observed on two different fungi. Flower extract of *Ageratum conyzoids, Parthenium hysterophorous* and *Vinca rosea* were found effective against two biodeteriogens, *Myrothecium roridum* and *Chaetomium globosum*.

KEY WORDS : *Myrothecium roridum, Chaetomium globosum,* floral extract, ecofriendly control.

Present environmental constraints would restrict the use of many prevalent synthetic fungicides. As such there is a pressing need for pesticides of plant origin because of their ecofriendly nature. Higher plants have been recognized to release volatile substances which keep the air remarkably free from pathogenic micro-organisms (Mahadevan 1982).

Flowers are having volatile substance which attract insect for pollination, attract human and become symbol of love, affection and feelings. Carnation flower (*Dianthus caryophyllus* L. Caryophyllaceae) is a plant yielding volatile oil and used as cardiac stimulant, diaphroretic and vermifuge. The effect of characteristic aroma of floral buds of clove is having very excellent antibacterial activity. Chemically essential oils are composed of organic compounds usually unsaturated hydrocarbons called terpenes. The essential oils secreted as a result of known metabolic process had been observed to be accumulated in the form of oily droplets in some cells or spaces in the plant tissues. Presence of essential oil acts as protection in a number of plants against the plant parasites and against the depredation of animals (Raymond *et. al.*, 1951, Dikshit *et. al.*, 2004). In the present communication antifungal efficacy of essential oil / flower extract against two fungal biodeteriogens i.e. *Myrothecium roridum* and *Chaetomium globosum* is reported.

Materials and Methods

Isolations from miniature paintings kept in Baroda Museum and Picture gallery resulted in pure culture of *Myrothecium roridum*. Isolations from stamp papers obtained from a private collection resulted in culture of *Chaetomium globosum*. This ascomycetous fungus is characterized by perithecia with coiled hairs. These two organisms were used as test organisms.

Fresh flowers were collected locally as well from Botanical garden of Maharaja Sayajirao University of Baroda. Hundred gram flowers were washed in running tap water for 5 min. Then they were crushed in mixer for 10 min and then filtered with cheese cloth. Filtrate was centrifused at 2000 rpm for 20 min to remove microscopic debris and finally crystal clear supernatant was taken and it was diluted to make the final volume to 100 ml. Thus each ml of solution is having extract of one gram.

The technique used to evaluate antifungal property was Poisoned food technique (Nene and Thapliyal 1979). The solution of floral extract was mixed with PDA at three different percentage 2%, 5% and 10% v/v and then it was sterilized by autoclaving at 121 C temperature. Then it was transferred to sterile Petriplates in aseptic condition. When medium became solidified and cool, fungus was inoculated in the center. The colony diameter was measured after seven days and percentage inhibition was calculated. The results are depicted in Table 1.

Table 1 Percentage inhibition of two fungi *in vitro* in aqueous floral extract

Flower	*Myrothecium roridum*			*Chaetomium globosum*		
	2%	5%	10%	2%	5%	10%
Mirabilis jalapa L.	01	00	0.5	00	00	01
Ageratum conyzoids Linn.	02	05	20	02	00	00
Calotropis gigentia R. Br.	00	00	10	05	02	02
Mimusops elangi L.	05	00	10	05	05	20
Tegetus erectus L.	01	05	08	01	02	05
Nyctanthes arbor tristis L.	02	05	05	02	05	08
Parthenium hysterophorus A. Gray	02	05	10	02	05	05
Ocimum sanctum L.	00	00	00	00	00	00
Vinca rosea L.	05	06	10	03	06	08

Results and Discussion

Rio Earth Summit 1992 in Brazil concluded that mankind is threatened by drastic global environmental changes triggered by our activities. There is a need for having sustainable development approach. We need to investigate and develop alternative strategies for different problems. The green plants may be explored as important natural resource as "natural chemotherapeutants."

Problems in Museums and Libraries and their Ecofriendly Alternatives

Museum contains various articles of organic nature like paper, wood, cloth, ethnological objects, animal and birds etc. Miniature paintings displayed in Baroda Museum were attacked by silver fish and fungi. The damage by insects is quite common on cloths and cloth made objects. Different objects stored in store-well get damaged during rainy season. Increase in humidity is one such reason.

It is a common practice in Oriental Institute of Baroda to keep dried rhizome powder of *Acorus calamus* (Sweet flag) in small cloth packets. It is well known that *A. calamus* is having insecticidal activity and it can replace use of Napthelene balls. It is cheap, effective and simple to handle, with no adverse effect on stored grains, domestic animals and human beings (Virmani *et. al.*, 1977). It is toxic to wide range of biting and sucking insects. The botanical pesticides can be used to substitute the synthetic chemicals. These can be used in cupboards or filter papers can be dipped in botanicals and then kept in direct contact with materials.

It is clear from Table 1 that there is maximum inhibition of *M. roridum* by 10% conc. of *Ageratum conyzoids*, which is up to 20%. It is followed by 10% conc.of *M. elangi*, where 10% inhibition of *Chaetomium* was observed. Earlier at 60 h, *A. conyzoids* showed maximum inhibition (32.20%) of *Rhizoctonia solani* (Gautam *et. al.*, 2003). Patnaik *et. al.*, (1996) reported that the essential oil of *Ageratum* sp. had fungicidal activity against many fungi like *Fusarium solani, F. oxysporum, Phomopsis vexans, Candida albicans* etc. Flower extracts of *V. rosea* and *P. hysterophorus* showed comparatively good inhibition of both the biodeteriorating fungi in comparison to some other flowers. Naidu and John (1981) observed that leaf extract of *P. hysterophorus* inhibited the growth of *R. solani*. It was observed that in certain cases colony diameter was more in comparison to control.

References

Dikshit, A., Shahi S.K., Pandey, K.P. Patra M. and **Shukla A.C.** 2004. Aromatic plants-a source of natural chemotherapeutants. *Science Letters.* 27(5&6): 145-164.

Gautam, K., Rao, P.B. and **Chauhan S.V.S.** 2003. Efficacy of some botanicals of the family compositae against *Rhizoctonia solani* Kuhn. *J. Mycol. Pl. Pathol.* 33(2) : 230-235.

Mahadevan, A. 1982. *Biochemical aspects of disease resistance* part I. Performed inhibitory substances: *Prohibitins.* Today and tomorrows Printers and Publishers, New Delhi, India 425.

Naidu, V.D. and **John V.T.** 1981. *In vitro* inhibition of rice fungal pathogens by extracts of higher plants. *Int. Rice Res. News.* 6 : 12

Nene, Y.l. and **Thapliyal, P.N.** 1979. *Fungicide in plant disease control.* Oxford and IBH Pub. Co., New Delhi 507.

Patnaik, S. Subramayam V.R. and **Kole C.** 1996. Antibacterial and antifungal activity of ten essential oils in vitro. *Microbios* 86 (349): 237-346

Raymond E.K. and **Donal, F. Other.** 1951. *Encyclopedia of Chemical Technology.* The Inter Science Encyclopedia, Inc. New York, 9: 569.

Virmani O.P, Srivastava G.N. and **Datta S.C.** 1977. Essential oil of *Acorus calamus. Ind. Purf* 21: 103-119.

45 Management of Aflaroot and Dry Root Rot of Groundnut using *Pseudomonas fluorescens*

A.B. Ade and L.V. Gangawane

Soil Microbiology and Pesticides Laboratory, Deptt. of Botany, Dr. Babasaheb Ambedkar University, Aurangabad - 431 004 M.S.

ABSTRACT

Pseudomonas fluorescens from the rhizosphere was found to be more effective in the management of aflaroot and dry root rot of groundnut. In addition it was possible to transfer the antagonistic gene in *Rhizobium* nodulating groundnut. The transconjugants were also very effective in controlling these diseases, hence the transgenic *Rhizobium* can control the disease and also helps in N- fixation. This is more ecofriendly when compared with other methods of management of crop diseases.

KEYWORDS : *Pseudomonas fluorescens, Aspergillus flavus, Aspergillus niger*, aflaroot of groundnut, dry root rot of groundnut

Groundnut (*Arachis hypogaea* L) is an important oil seed crop of India covering nearly half of the area under oil seeds. The groundnut is also known for its nitrogen fixing capacity though the root nodule bacteria and forms an important member in many crop rotation all over the country. In Maharashtra it is grown during Rabi and Kharif wherever irrigation facilities are available. The total production of groundnut is nearly 7.5 million tones. However this crop suffers from aflaroot disease caused by *Aspergillus flavus* (Mehan and Chohan, 1974) which produces aflatoxin in the kernels. In addition *Aspergillus niger* is also responsible to cause dry root rot (Raju and Krishnamurthy, 2000) of groundnut.

Now a days use of microbes for biological control has been advocated. In the present work the efforts were made to manage these diseases with the antagonistic bacterium *Pseudomonas fluorescens*.

Materials and Methods

Isolation of *Pseudomonas* from the rhizosphere and soil of groundnut was done by using King's medium (1954). The identification of *Pseudomonas* was done by holding a plate under UV light. The colonies showing fluorescence on the King's medium were picked up on the slants ands preserved at 4C. The pathogens from the infected groundnut were isolated on Martin's Rose Bengal agar medium. Antagonism of *Pseudomonas fluorescens* was tested against the *Aspergillus flavus* and *Aspergillus niger*. *Pseudomonas fluorescens* isolates were inoculated in the nutrient broth. After 3 days of incubation period the culture filtrates were filtered through the cintered glass filters and were used for studying antagonism against the pathogens. The *Aspergillus flavus* and *Aspergillus niger* spore suspension were plated on Czapek Dox agar medium. The wells (8 mm diameter) were prepared with the help of cork borer. The agar wells were filled with the culture filtrates (0.1 ml) and the plates were placed in laboratory for observation of antagonistic gene around the well. Simultaneously the filter paper disks dipped in the culture filtrate were also placed on the seeded agar plates and antagonistic zones were observed after three days. (Saxena *et. al.,* 1995, Ramesh, 2000, Charita Devi *et. al.,* 2003). The antagonistic gene was transferred into *Rhizobium* nodulating groundnut. The transfer of the gene was done through horizontal gene transfer (Leaderberg, 1987). Of the 100 transconjugant Rhizobia 12 were antagonistic to *Aspergillus flavus* and *Aspergillus niger*.

Results and discussion

Antogonistic effect of *Psedomonas fluorescens* against *Aspergillus flavus* and *Aspergillus niger* is presented in Table 1.

Table 1 Antagonism of *Pseudomonas fluorescens* against *Aspergillus flavus* and *Aspergillus niger* causing aflaroot and dry root rot of groundnut.

Pseudomonas fluorescens isolates	Inhibition zone against *Aspergillus flavus* (in mm)	Inhibition zone against *Aspergillus niger* (in mm)
PF-1	5	3
PF-2	5	3
PF-3	3	3

Application of rhizobial transconjugants each showing antagonism against *Aspergillus flavus* and *Aspergillus niger* were selected for *in vivo* studies. The healthy seeds of groundnut cv JL-24 were inoculated with pathogen and transconjugants (TC). The percentage control efficacy (PCE) was calculated (Samoucha *et. al.,* 1987). The results in table 2 indicated that PCE was highly increased due to treatment of rhizobial transconjugants. In case of aflaroot disease TC-98 was more effective followed by TC-30 and TC-64 in decreasing manner while in case of dry root rot TC-77 was more effective than that of TC-89, TC-68 and TC-93. The plasmid profile of antagonistic rhizobial transconjugant was also observed on Agarose Gel Electrophoresis. The gene transafer in *Mesorhizobium ciceri* for nodulation in chickpea have been very well explained by Chimote and Kashyap (2003).

Table 2 Percentage Control Efficacy (PCE) of Rhizobial Transconjugants on aflaroot and dry root rot of groundnut cv JL-24 caused by *Aspergillus flavus* and *Aspergillus niger* respectively.

Treatment	Percentage Control Efficacy (PCE)
T1- *Aspergillus flavus* only	13.30
T2 AF + TC-30	80.00
T3 AF + TC -98	86.00
T4 AF + TC -64	20.00
T5 AF + TC -46	33.33
C.D. 0.05	29.29
T1 *Aspergillus niger* only	20.00
T2 AN + TC-77	86.60
T3 AN + TC-89	81.60
T4 AN + TC-68	60.00
T5 AN + TC-93	46.66
C.D. 0.05	41.62

Abbreviation :

AF= *Aspergillus flavus*

AN= *Aspergillus niger*

References

Charita Devi, M. Reddy, M.N. and **Eswara Reddy, N.P.** 2003. Biocontrol efficacy of *Pseudomonas* spp. against *Macrophomina phaseolina* the incitant of root rot of groundnut (*Arachis hypogaea* L.). In: *Proc. of National Symposium on Plant Pathogen diversity in relation to plant health*, Jan 16-18, p 49.

Chimote, Vivek and **Kashyap L.R.** 2003. Site specific mutagenesis in nodulation genes of *Mesorhizobium ciceri* strain MC 18-7 using *in vitro* transposition in a hetero logous cloned DNA. *Indian J. Genet.* 63(1) : 1-4.

Leaderberg J. 1987. Genetic recombination in bacteria: A discovery account. *Annual Review of Genetics* 21:23-46.

Mehan, V.K. and **Chohan, J.S.** 1974. Effect of filtrates of aflatoxin producer and non producer isolates of *Aspergillus* on different crop plants. *Ind. J. Mycol. Pl. Pathol.* 4: 75-76.

Raju Ram Bhadra, M. and ***Krishnamurthy, K.V.M.*** 2000. Efficacy of *Trichoderma* spp. in the management of collar rot of groundnut caused by *Aspergillus niger* van Teighem. *Indian J. Plant Prot.* 28(2) 177-199.

Ramesh K.R. 2000. Inhibition of *Rhizoctonia solani* causal agent for collar rot of teak seedlings by fungicides and biocontrol agents in *in vitro* condition. *Indian Forester* 126(3) : 284-288.

Samoucha, Y. Hingal Shoffer, U. and ***Gisi, U.*** 1987. Effect of disease intensity and application types as efficacy and synergy of fungicide mixture against *Phytophthora infestance. Phytopathology* 120 :44-52.

Saxena Deepak, Patel, K.R., Patel, K.A., Rao, K.K. and ***Mehta M.H.*** 1995. Role of fluorescent *Pseudomonas* in controlling plant diseases. *Indian J. Mycol. Pl. Pathol.* 25 (1&2): 131.

Evaluation of Cluster Bean Germplasm for *Xanthomonas axonopodis pv. cyamopsidis*

Krishnappa M.,* Chakravarthy C.N. and Thippeswamy B.

Department of P. G. Studies and Research in Applied Botany, Kuvempu University, Jnana Sahyadri, Shankaraghatta - 577 451, Shimoga (District), Karnataka,
*Corresponding author

ABSTRACT : *Xanthomonas* is a genus of phytopathogenic bacteria that attack a variety of hosts with economic importance including beans, citrus, rice, grape, and cotton. Cluster bean is an annual legume. The crop is affected by a number of diseases. Among them bacterial blight caused by *Xanthomonas axonopodis* pv. *cyamopsidis* was a devastating disease in major cluster bean growing districts in Karnataka during kharif 2002 and 2003. During the field survey, infected leaf, stem and seed samples were collected and tested for target bacteria. Also 100 different genotypes of cluster bean seeds were tested, which were received, from Main Pulses Research Station, Gujarat. Among them 60 genotypes yielded the bacterial pathogen with an incidence of 2 to 40 %. Plants were raised from the naturally infected seeds in the experimental field, to know the transmission of pathogen. Based on the morphological, biochemical, pathogenicity test and hypersensitive reaction, the pathogen was confirmed as *Xanthomonas axonopodis* pv. *cyamopsidis*.

KEY WORDS : Cluster bean seeds, screening, transmission, *Xanthomonas axonopodis* pv. *cyamopsidis*.

Cluster bean (*Cyamopsis tetragonoloba* (L.) Taub.) commonly called as guar, is a grain legume crop that is high in protein and contains 19 to 43% galactomannan gum, is used as an emulsifier, in cosmetics and pharmaceuticals. India exports the guar gum to USA and earns Rs. 170 million annually (Yadava, 1999). The production of Cluster bean is limited by a destructive bacterial blight caused by *Xanthomonas axonopodis* pv. *cyamopsidis* (Patel and Patel, 1958), that occurs in most cluster bean growing districts in Karnataka state. The disease has been reported from Brazil (Almeida *et al.*, 1992), Madison (Undersander *et al.*, 1991), Arizona (Mihil and Alcorn, 1985) and United states (Orellana *et al.*, 1965). In India Gupta, (1978) reported the yield loss up to 58 per cent and also Gandhi and Chand (1985) reported the bacterial blight is the most destructive pathogen affecting seed quality and reduced the yield in cluster bean. Present study is concentrated on screening the different germplasm and also study the transmission of pathogen.

MATERIALS AND METHODS

Collection of seed samples

Field survey was carried out during kharif 2002 and 2003 in major cluster bean growing districts in Karnataka. During the field survey infected leaf, stem and seed samples were collected and also one hundred different cluster bean genotype seeds were received from Main Pulses Research Station, Sardar krushinagar, Gujarat for testing the target pathogen.

Isolation of the pathogen from infected plant materials

Infected stems and leaves were washed thoroughly with tap water. The cut ends of the stem were dipped in 10-ml sterile distilled water in a test tube and observed for exudation of milky ooze. The cut ends of the leaf were also observed under compound microscope for bacterial oozing. The suspension from both stems and leaves was serially diluted on YDC agar medium (Yeast extract 10 g, Dextrose 20 g, Calcium carbonate 20 g, Agar 20 g and Distilled water 1000 ml). The plates were incubated at 30°C for 48-h. Suspected colonies were sub-cultured on YDC medium and were subjected to various tests for confirmation.

Detection of *Xanthomonas axonopodis* pv. *cyamopsidis* in cluster bean seeds

A total of 100 different Cluster bean genotypes were tested by the direct plating method. Four hundred seeds (in replicates) from each sample were assayed for the presence of the pathogen. The seeds were surface disinfected in sodium hypochlorite (1%) for 2 min. rinsed thoroughly in sterile water, blot dried and plated on YDC agar plates and incubated for 72 h at 30 °C (Schaad and Kendrick, 1975). Suspected colonies of *X. axonopodis* pv. *cyamopsidis* appearing yellow, mucoid and shiny that developed around the seeds were transferred to fresh YDC plates. After 24 h, pure colonies were transferred to YDC slants and stored at 4 °C for identification.

Characterization of the bacterial isolates

Bacterial colonies isolated from seeds, stems and leaves were characterized based on their cultural, morphological, biochemical, pathogenicity tests and hypersensitive reaction. For all the tests 24 - 48 h old cultures grown on YDC medium were used (Carmen, 1992).

Bio-chemical tests

All bacterial isolates were tested by Gram staining, oxidase test (Kovacs, 1956), levan formation, starch hydrolyzing test, esterase activity, action on litmus milk, gelatin hydrolyzing test, arginine hydrolyzing tests and nitrate reduction tests (Lelliot and Stead, 1987) were carried out for the confirmation of *X. axonopodis* pv. *cyamopsidis*.

Hypersensitive reaction

The 24 h old bacterial culture was infiltered into leaf mesophyll tissue of two months old Tobacco (*Nicotiana tabacum* var. *xanthi*) by using hypodermic syringe. Plants inoculated with distilled water served as control. All the inoculated plants were incubated at $28 + 2°$ C in green house to observe the reaction (Klement *et. al.*, 1964).

Pathogenicity test

Twenty four hour old bacterial culture [10^8 cfu] was sprayed as a fine mist using a low pressure sprayer to healthy 20 days old susceptible Pusa Navbahar plants. The plants were incubated at $28 + 2°$ C in a green house and observed for blight symptoms

Transmission studies

Ten different cluster bean genotypes which were showing higher per cent incidence of X. *axonopodis* pv. *cyamopsidis* in direct plating methods were selected for transmission studies. Four hundred seeds from each of the selected samples were sown separately in 20X10-m experimental plot. The seedlings were observed for expression of symptoms at regular interval. The infected plant parts were taken and subjected for oozing test then incubated on YDC plates. The seeds collected from the experimental plots were taken and subjected for direct plating for the recovery and confirmation of the transmission.

Results and Discussion

Field survey

The bacterial blight was prevalent in 50 % of the inspected fields. The incidence of the disease varied from 10 to 65 % in various cultivars of cluster bean. The highest incidence of bacterial blight was found in Pusa Navbahar (65 %) in Hassan field, whereas the lowest (10%) were observed in HG-75 in Chikmagalure field. In Haryana, Gandhi and Chand (1985) have reported disease intensity of 71 % in Pusa Navbahar and 40 % in HG-75. This is due to variations in environmental factors in the field conditions in Karnataka.

Cluster bean affected by X. *axonopodis* pv. *cyamopsidis* was initiated from the leaf tip. In several cases, it occurred at more than one site on the same leaf, developing into 'V' shaped lesions, which rapidly spread to the entire lamina giving a totally blighted appearance. The symptoms on the stem were linear brownish black streaks and later the stem split open longitudinally. In advanced stages of infection, the stem cracked and upper shoot bent downward while still remaining attached to the main stem. In some cases whole plants were blighted.

Characterization of bacterial isolate

All the bacterial isolates were characterized on the basis of bio-chemical tests. These isolates were yellow pigmented, gram-negative, aerobic, small rod shaped bacteria with rounded ends, singly or in short chains, 0.8 × 1.7m in size, capsulated and motile with a single polar flagellum. The bacteria hydrolyzed starch, leaven negative, hydrolyzed tween-80, oxidase negative, litmus test positive, arginine test negative, liquefied gelatin, they produced the acid from glucose and did not reduce the nitrate. The results of the biochemical tests were compared with Burgeys manual *X. axonopodis* pv.*cyamopsidis* test results (Krieg and Holt, 1984). The bacteria grow well in yeast dextrose calcium carbonate agar, yeast glucose calcium carbonate agar, nutrient dextrose agar and nutrient agar. On these media, bacterial colonies were mucoid, shiny and spherical with varied shades of yellow.

Hypersensitive reaction

The 24-h old bacterial cultures induced brown localized necrosis within 48-h of inoculation in tobacco plant leaf.

Pathogenicity test

The 24-h old culture induced typical blight symptoms as 'V' shaped on the leaf in blight susceptible Pusa Navbahar variety. Bacterial isolates from seeds, produced similar symptoms as observed in naturally infected plants during pathogenicity test. Hence the bacterial pathogen was confirmed as *X. axonopodis* pv. *cyamopsidis.*

Detection of *Xanthomonas axonopodis* pv. *cyamopsidis* in cluster bean seeds

Incubated plates showed yellow mucoid bacterial growth around the seeds. Such seeds were considered as infected. Colonies from the seeds, were isolated and subjected to biochemical, hypersensitive and pathogenicity tests to confirm the *X. axonopodis* pv. *cyamopsidis.* Out of one hundred different cluster bean seeds tested, only 60 genotypes yielded the bacterial pathogen with an incidence of 2-40% (Table 1). Srivastava and Rao (1963) have reported that the seed infection ranges from 3 to 95% in different varieties of cluster bean. Bacterial blight is the most destructive pathogen affecting the seed quality and yield loss in cluster bean (Gandhi and Chand, 1985 and Gupta, 1978). Seeds are the carriers of inoculum (Parashar and Sharma, 1984).

Table 1 Incidence of *Xanthomonas axonopodis* pv. *cyamopsidis* in different Cluster bean genotypes seeds

Sl. No.	Variety	*Per cent of infection on YDC Plates
01.	GAUG-8819	DN
02.	GAUG-8832	06
03.	GAUG-9002	DN
04.	GAUG-9003	DN
05.	GAUG-9005	DN
06.	GAUG-9007	DN
07.	GAUG-9008	DN
08.	GAUG-9009	02
09.	GAUG-9012	04
10.	GAUG-9013	DN
11.	GAUG-9106	DN
12.	GAUG-0109	08
13.	GAUG-9406	40
14.	GAUG-9702	DN
15.	GAUG-9703	04
16.	GAUG-9712	09
17.	GAUG-9802	02
18.	GAUG-9803	23
19.	GAUG-9807	DN
20.	GAUG-9808	30
21.	GAUG- 9809	05
22.	GAUG-9814	03
23.	GAUG-9901	04
24.	GAUG-9902	20
25.	GAUG-9907	18
26.	GAUG-9910	DN
27.	GAUG-9911	20
28.	GAUG-9912	10
29.	GAUG-9914	DN
30.	GAUG-9915	36
31.	GAUG-21	30
32.	GAUG-26	15
33.	GAUG-60	25
34.	GAUG-69	08
35.	GAUG-73	05

Table 1 Contd....

Sl. No.	Variety	*Per cent of infection on YDC Plates
36.	GAUG-0001	04
37.	GAUG-0010	10
38.	GAUG-0013	06
39.	GAUG-0101	DN
40.	GAUG-0102	DN
41.	GAUG-0103	02
42.	GAUG-0104	10
43.	GAUG-0105	12
44.	GAUG-0106	DN
45.	GAUG-0107	03
46.	GAUG-0108	DN
47.	GAUG-0110	DN
48.	GAUG-0111	DN
49.	GAUG-0112	10
50.	GAUG-0113	04
51.	GAUG-0114	12
52.	GAUG-0115	08
53.	HGS-880	15
54.	HGS-885	08
55.	HGS-365	14
56.	HGS-847	10
57.	HGS-855	15
58.	HGS-857	05
59.	HGS-859	04
60.	HGS-867	10
61.	HGS-870	10
62.	HGS-875	06
63.	HGS-881	DN
64.	HGS-884	08
65.	HGS-891	02
66.	HGS-899	07
67.	HGS-903	DN
68.	HGS-905	DN
69.	RGS-936	12
70.	RGC-986	10
71.	RGC-1002	DN
72.	RGC-1017	DN

Table 1 *Contd....*

Sl. No.	Variety	*Per cent of infection on YDC Plates
73.	RGC-1019	DN
74.	RGC-1020	03
75.	RGC-1022	DN
76.	RGC-1024	04
77.	RGC-1025	DN
78.	RGC-1026	04
79.	RGC-1027	02
80.	RGC-1028	DN
81.	RGC-1029	DN
82.	GAUG-1030	DN
83.	RGC-1031	DN
84.	RGC-1033	04
85.	RGC-1038	02
86.	RGC-1047	04
87.	RGC-1055	DN
88.	RGC-1059	DN
89.	RGC-1066	DN
90.	RGC-1074	06
91.	RGM-112	06
92.	RGM-113	DN
93.	RGM-114	DN
94.	RGM-115	DN
95.	CAZG-50	DN
96.	CAZG-60-1	DN
97.	CAZG-90-2	04
98.	CAZG-97	04
99.	CAZG-9819	DN
100.	CAZG-9820	DN

Data based on 100 seeds.

DN = Did not notice

Transmission studies

Among 100 different genotypes, 10 different genotypes which were showing higher per cent of infection on YDC agar plate were selected for transmission studies. Percent transmission varied in different genotypes. GAUG-9406 (30%), GAUG-9808 (26%), GAUG-9902 (18%), GAUG-9907 (15%), GAUG-9911 (20%), GAUG-9915 (28%), GAUG-21 (21%), GAUG-60 (18%), HGS-880 (14%) and HGS-855 (13%). Recovery of the pathogen from the seeds of experimental plot plants was done, infection range was 10 - 32% (Table 2). Transmission studies showed that the plants are highly susceptible to bacterial blight after flowering stage (39th day old) and the pathogen perpetuated in the plant till harvest. Heavy spotting of the pods proved seed infection. The primary inoculation of the disease was carried in the infected seeds, which transmitted the disease (Srivastava and Rao, 1963).

Table 2 Seed transmission and recovery of pathogen in cluster bean seeds

Variety	No. of seeds germinated*	Infection on YDC plates (%)	Average % of seed to seedling transmission	Recovery of the pathogen in %
GAUG-9406	360	40	30	32
GAUG-9808	371	30	26	26
GAUG-9902	382	20	18	16
GAUG-9907	388	18	15	13
GAUG-9911	383	20	20	15
GAUG-9915	378	36	28	22
GAUG-21	372	30	21	21
GAUG-60	368	25	18	19
HGS-880	389	15	14	10
HGS-855	386	15	13	12

*Data based on 400 seeds.

X. axonopodis pv. *cyamopsidis* is responsible for causing bacterial blight of cluster bean, with recent emphasis on the intensive cultivation of cluster bean for it's high protein and galactomannan gum. This disease has become wide spread in Karnataka and appears almost every season, causing considerable yield loss depending on the variety and environmental conditions. Our findings are the evidence that, pathogen is carried in the seeds both internally and externally. Detection of the pathogen in the seeds and its survival in nature is important from the point of understanding the disease cycle and in controlling the disease.

Acknowledgement

The authors are grateful to Dr. S.B.S. Tikka, CCPI and Research Scientist, Main Pulse Research Station, Gujarat Agricultural University, Sardar Krushinagar, Gujarat, for providing one hundred different genotypes of cluster bean seeds. Thanks to the Chairman, Department of Applied Botany for providing facilities.

References

Almeida, I.M.G., Beriam, L.O.S., Malavolta, V.A., and **Ambrosan, E. J.** 1992. Isolation and characterization of *Xanthomonas campestris* pv. *cyamopsidis* in Brazil and reaction of guar genotypes to the bacterium. *Summa-phytopathologica,* 18: 255-261.

Carmen, N. M. 1992. *Seed Bacteriology Laboratories Guide.* Danish government institute of seed pathology for developing countries, Ryvanges Alle 78, DK-2900 Hellerup, Copenhagen, Denmark.

Gandhi, S. K. and **Chand, J. A.** 1985. Yield losses in guar due to bacterial blight caused by *Xanthomonas campestris* pv. *cyamopsidis. Indian Phytopath.,* 38 : 516-519.

Gupta, D.K. 1978. Yield losses in guar (*Cyamopsis tetragonoloba*) caused by bacterial blight *(Xanthomonas cyamopsidis,). Ind. J. Mycol. and Pl. Pathol.,* 8 : 27.

Klement, Z., Farkas, G.L. and **Lovrekovich, L.** 1964. Hypersensitive reaction induced by pytopathogenic bacteria in the tobacco leaf, *Phytopathol.* 54: 474-477.

Kovacs, N. 1956. Identification of *Pseudomonas pyocyanea* by the oxidase reaction, *Nature.* 178: 703.

Krieg, R. and **Holt G. Jhon.** 1984. *Bergeys manual of systemic bacteriology.* Vol-1 (ed), Williams and Wilkins, Baltimore, USA.

Lelliot. and **Stead.** 1987. *Methods for the diagnosis of bacterial diseases of plants, Methods in plant pathology,* Vol-2 ed, T.F. prece Backwell Scientific Publications, 216.

Mihil, J. D. and **Alcorn, S.M.** 1985. Bacterial blight of Guar in Arizona. *Plant Disease.* 69: 811.

Orellana, R.G., Thomas. and **Kinman, M. L.**1965. A bacterial blight of Guar in the United state, *F.A.O. Plant Prot. Bull.,* 13: 9-13.

Parashar, R. D. and **Sharma D.D.K.** 1984. The detection of *Xanthomonas campestris* pv. *cyamopsidis* in guar seed lots. *Phytopathological notes.* 37: 353-355.

Patel, A.J. and **Patel, M. K.** 1958. A new bacterial blight in *Cyamopsis tetragonoloba* (L). Taub. *Curr. Sci.,* 27: 258- 259.

Schaad, N. W. and **Kendrick, R.** 1975. A qualitative method for detecting *Xanththomonas campestris* in crucifer seed. *Phytopathol.* 65: 1034-1036.

Srivastava, D.N. and **Rao, Y.P.** 1963. Seed transmission and control of bacterial blight of guar *Cyamopsis tetragonoloba* (L). Taub. *Ind. Phytopathol.* 16: 390-391.

Undersander, D. J., Putnam, D. H., Kaminski, A. R., Kelling, K.A., Doll, J. D., Oplinger, E. S. and **Gunsolus, J.L.** 1991. *Alternative field crops manual.* University of Wisconsin, Madison, WI-53706

Yadava, N. D. 1999. *Technique and management of field crop production.* Agrobios Publ. Jodhpur. 443-452.

47   *Biotechnological Aspects of Thermophilic Fungal Glucoamylases*

Pardeep Kumar and T. Satyanarayana*
Dept. of Microbiology, University of Delhi, South Campus, Benito Juarez Road,
New Delhi - 110 021,
*Phone : 091-11-24102008, Fax:091-11-26885270, e-mail : tsnarayana@vsnl.net

ABSTRACT

Thermostable microbial amylases have found a number of commercial applications because of their overall inherent stability. Glucoamylase is a typical fungal enzyme and mainly used in the production of high glucose and fructose syrups from starch. Thermophilic fungi secrete high enzyme titres active at elevated temperatures, and therefore, are attracting increased attention as sources of thermostable glucoamylases. A number of thermophilic fungal glucoamylases have been studied and efforts are being made to enhance their activity and stability at high temperatures to suit industrial processes. The main focus of this review is on the production, purification, characteristics and applications of thermophilic fungal glucoamylases.

KEY WORDS : Thermophilic moulds/fungi, glucoamylase, thermostable, starch, saccharification

Starch is one of the most abundant biopolymers and an important constituent of human and animal diets. Availability of starch as renewable energy source has greatly encouraged research on starch-hydrolyzing enzymes. Amylases are the most widely used thermostable enzymes in starch industry. Glucoamylase is particularly important in the production of glucose, which is subsequently used to produce crystalline dextrose, dextrose syrups and high fructose corn syrups.

Glucoamylase is an exo-enzyme that hydrolyzes a-1,4 glycosidic linkages of a-glucans from the non-reducing end, and also a-1,6 linkages, though at a very slow pace, releasing b-D glucose as the sole end product in theoretically 100% yield. The enzyme is rare in prokaryotes and is a typical fungal enzyme. Thermostability being the desired property of glucoamylases, the search for thermophilic organisms is gaining momentum, as the cellular components (enzymes, proteins and nucleic acids) of thermophilic organisms are also thermostable (Jakob, 1989; Niehaus *et al.*, 1999; Demirijan *et al.*, 2001; Haki and Rakshit, 2003). The benefits of elevated process temperatures include higher reaction rates and minimum risk of contamination by mesophiles. Thermophily in fungi is not as extreme as in bacteria. The highest temperature range at which fungi have been reported to thrive is 45-55°C (Nevalainen and Te'o, 2003). *Thermomyces lanuginosus* (Mangallam *et. al.*, 1977; Rao *et. al.*, 1981; Jensen *et. al.*, 1988; Duo-Chaun, *et. al.*, 1998; Haasum *et. al.*, 1991; Mishra and Maheshwari, 1996; Singh *et. al.*, 1996), *Rhizomucor pusillus* (Kanlayakrit *et. al.*, 1987), *Thermomucor indicae-seudaticae* (Kaur and Satyanarayana, 2001; Kumar and Satyanarayana, 2003) and *Talaromyces dupontii* (Tamura *et. al.*, 1981) are some of the known glucoamylase producing thermophilic moulds. In this article, an attempt has been made to emphasize on the production, characteristics and potential applications of glucoamylases of thermophilic moulds.

Substrates

A number of substrates, essentially starchy or related nature, including starch, glycogen, amylose, amylopectin, dextrins, maltodextrins, maltose, isomaltose, dextran, panosan and oligo-, di- and poly-sachharides, have been reported to be hydrolysed by glucoamylases (Pandey, 1995). Starch is the second most abundant biopolymer after cellulose, and is a major storage product of many economically important crops such as wheat, rice, maize, tapioca, and potato. Starch of cereals forms an important constituent of the human diet, and a large proportion of the food consumed by world's population originates from them (Van der Marrel *et. al.*, 2002). Natural starch is insoluble in cold water and is present in the form of microscopic granules. The shape and diameter of the granule depends on the botanical origin. The size of commercially important starch sources ranges from 2-30 (maize starch) to 5-100 μm (potato starch) (Robyt, 1998).

Starch is a polymer of α-D glucopyranose units linked to one another through glycosidic bonds. Glycosidic bonds are stable at high pH but hydrolyze at low pH. At one end of the polymeric chain is present a free aldehyde group known as reducing end. Starch is made up of two high molecular weight compounds, amylose (15–25%) and amylopectin (75–85%). Amylose is a straight chain/linear polymer of glucose subunits joined together by a-1,4 linkages, having a degree of polymerization of 10^2 to 10^5, and molecular weight ranging between several thousands to half a million. Amylopectin, on the other hand, is a branched polysaccharide (molecular weight up to 1-2 million) and forms the bulk of starch molecule. It consists of short a-1,4 linked linear chain of 10-60 glucose units and a-1,6 linked side chains with 15-45 glucose units. The average number of branching points is 5%, and varies with the botanical origin. Its total degree of polymerization is approximately two million linked units (Lee, 1991; Zobel and Stephan, 1995; Hermasson and Svegmark, 1996). Starch granules are organized into amorphous and crystalline regions. In root and tuber starches, amylopectin forms the crystalline region, while the amorphous region is occupied by amylose. In cereal starches amylose is, however, complexed with lipids and forms a weak crystalline structure.

Starch hydrolyzing enzymes

The ability to utilize starch as carbon and energy source is widely distributed among plants, animals and microorganisms. In order to hydrolyze starch in different environmental niches, microorganisms produce different starch hydrolyzing enzymes with different physicochemical properties (Antranikian, 1992). Among microbes, filamentous fungi are capable of secreting high titres of enzymes that are generally active at temperatures well above their growth optima, and therefore, are attracting increased attention as sources of thermostable amylolytic enzymes suitable for industrial applications.

Amylolytic enzymes (a-glucanases) hydrolyze the glycosidic linkages in various a-glucans. Various starch-hydrolyzing enzymes are: α-amylase, β-amylase, glucoamylase, cyclodextrin glycosyl transferase, α-glucosidase, isoamylase, pullulanase and amylopullulanase. Depending upon their mode of action amylases can be divided into two categories: endoamylases and exoamylases. Endoamylases are dextrinizing or liquefying amylases that randomly hydrolyze the α–1,4 linkages, while the branching points (α–1,6 linkages) remain unattacked, resulting in the formation of linear and branched oligosaccharides of varying chain lengths (dextrins), e.g. α–amylases. Exoamylases, on the other hand, are saccharifying or saccharogenic amylases that hydrolyze polysaccharides from the non-reducing end successively, resulting in short end products. One type cleaves each bond to produce solely β–glucose (glucoamylases), and another type breaks every alternate bond to produce maltose (β-amylases). Debranching enzymes, isoamylase, pullulanase, and amylopullulanase cleave a–1,6 glycosidic linkages in starch.

Glucoamylase (1,4-α-D-Glucan Glucanohydrolase, E.C.3.2.1.3)

Glucoamylase, also known as amyloglucosidase or γ–amylase, is a typical fungal enzyme. It is an extracellular exo-acting enzyme that removes glucose units from the nonreducing end of the starch molecule in a stepwise manner and changes the anomeric configuration of the resultant glucose molecule, thereby producing β-D glucose as sole end product. It cleaves α–1,4 linkages of α-glucans from the non-reducing end, and also α–1,6 and the rare α–1,3 linkages, but at a very slow pace. It preferentially hydrolyzes polysaccharides with a high molecular weight. The rate of the reaction decreases with the decreasing chain length of the dextrin substrate, and maltose is attacked most slowly (Gerhartz, 1990). Prolonged incubation time and high concentration of starch (35-40%) usually cause reversion of the reaction catalyzed by glucoamylase. The major reversion products include maltose, isomaltose and some oligosaccharides (Antranikian, 1992).

Glucoamylase structure

Glucoamylases are essentially the sole members of glycoside hydrolase family 15 (Coutinho and Henrissat, 1999). The majority of glucoamylases are multidomain enzymes consisting of a catalytic domain connected to a starch-binding domain by an O-glycosylated linker region (Sauer *et. al.*, 2000). All glucoamylase catalytic domains have $(a,a)_6$ barrel structures, with six inner a-helices surrounded by six outer ones, and a central conical well shaped active site (Coutinho and Reilly, 1997). In eukaryotes a peripheral thirteenth a-helix is present (Fig. 1). The starch-binding domain (SBD) consists of eight β-strands organized into two β-sheets forming a twisted β-barrel structure (Sorimachi *et. al.*, 1996; Jacks *et. al.*, 1995). Most glucoamylases produced by filamentous fungi have SBDs connected to the C-termini of their catalytic domains by O-glycosylated linkers of variable lengths, with O-glycosylation extending up to thirty residue belt (Coutinho and Reilly, 1997). *Aspergillus* glucoamylase used industrially is of 65.79 kDa, with the catalytic domain (1-480) being 50.456 kDa, the linker (481-508) of 3.454 kDa, and the SBD (509-616) of 11.88 kDa. N- and O-glycosylation of the enzyme varies among strains, and can add a further 12-15% to the total molecular mass (Aehle, 2004). Commercially used *Aspergillus* glucoamylase is a mixture of GA I and GA II. GA I can bind starch granules much better, since it contains the SBD, but in industrial practice this confers no advantage, since the two forms encounter soluble dextrin molecules, which they hydrolyze at the same rate (Meagher and Reilly, 1989; Meagher *et. al.*, 1989).

Mechanism of glucoamylase action

The widely accepted mechanism of hydrolysis involves proton transfer to the glycosidic oxygen of the scissile bond from a general acid base catalyst, formation of an oxocarbonium ion, and a nucleophilic attack of water assisted by a general base catalyst (Konstantinidis and Sinnott, 1991; Tanaka *et. al.*, 1994). The final step involves the addition of hydroxyl ion to the carbonium ion intermediate. The OH group is added in the β-configuration hence liberating β-D-glucose as the end product. The conserved tryptophan residues of the starch-binding domain (SBD) are essential for its attachment and interaction with the substrate (Juge *et. al.*, 2002). It is commonly suggested that the SBD not only binds onto raw-starch but also disrupts a-glucan interchain binding at the surface of the starch granule. This behavior further facilitates the hydrolysis of amylose and amylopectin chains by the exo-acting catalytic domain that releases glucose from the nonreducing ends of the polymer chains (Southall, *et. al.*, 1999). SBD increases the activity of the glucoamylase towards starch granules by around 100-fold (Svensson *et. al.*, 1982).

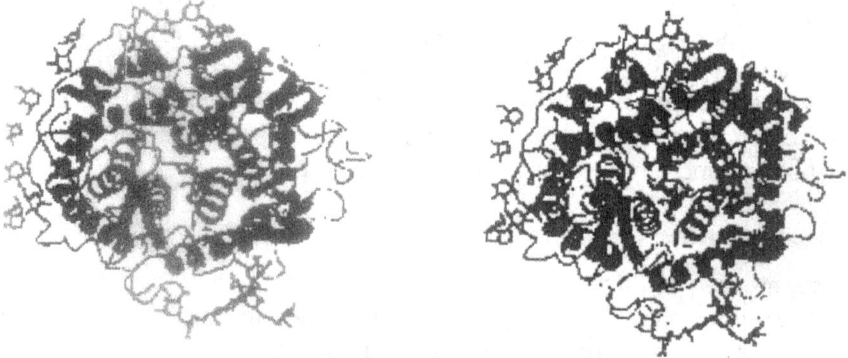

Figure 1 Stereo view of tertiary structure of *Aspergillus awamori* var. X-100 glucoamylase (Aleshin *et. al.*, 1994).

Industrial starch hydrolysis

Hydrolysis of starch using dilute acids was a common practice, but more recently it has been replaced by enzymatic treatment, because of low glucose yields, formation of large amount of salts, and the need to use corrosion resistant equipments, in the former method. Also the use of enzymes being environmentally safe offers a bonus. The benefits of using enzymes in industrial processes lie in their specificity and efficiency resulting in the formation of fewer byproducts, less toxic wastes and reduced handling problems. The starch industry is one of the largest users of thermostable enzymes for the hydrolysis and modification of this useful raw material in to a variety of products (Haki and Rakshit, 2003).

The complete hydrolysis of starch requires the coordinate action of several enzymes, such as a-amylase, b-amylase, glucoamylase, debranching enzymes (pullulanases), and α-glucosidases (Sunna et. al., 1997). The operating range of glucoamylase, in terms of pH, temperature and compatibility with other enzymes should be improved to maximize the efficiency of its action in food processing operations (James and Lee, 1997).

The enzymatic conversion of starch includes gelatinization, which involves the dissolution of starch granules, thereby forming a viscous suspension; liquefaction, involving partial hydrolysis and loss in viscosity; and saccharification, that results in the production of glucose and maltose via further hydrolysis (Haki and Rakshit, 2003). Both liquefaction and saccharification are carried at high temperature, and hence thermostable enzymes are of utmost importance in starch hydrolysis. One extremely valuable advantage of conducting biotechnological processes at elevated temperatures is reducing the risk of contamination by common mesophiles. Other benefits of elevated process temperature include, higher reaction rates due to a decrease in viscosity and an increase in the diffusion coefficient of substrates and higher process yield due to increased solubility of substrates and products, and favourable equilibrium displacement in endothermic reactions (Mozhaev, 1993; Krahe et. al., 1996).

Gelatinization

Gelatinization is achieved by heating starch with water, and starch is water-soluble only at high temperature. Heating of starch suspension above a critical temperature results in a multistage process called gelatinization, which involves tangential swelling of the amorphous regions of the granule, disruption of the readily ordered structures, and eventually opening of the crystal structure as the polymer chain becomes increasingly hydrated. During this process, amylose leaches out of the granule and makes up the continuous gel phase outside the granule, which in turn increases the viscosity of starch slurry. Gelatinization greatly increases the chemical reactivity of inert starch granules towards amylolytic enzymes, and is widely used for enhancing starch hydrolysis, and in the manufacture of sugar syrups (Hermasson and Svegmark, 1996; Oates, 1997). For hydrolysis of starch to proceed immediately after gelatinization, among other things avoiding a lot of cooling time, the enzyme has to be thermostable.

Conventional starch hydrolysis

It is a two-step process involving liquefaction and saccharification and requires a combined action of different amylolytic enzymes including á-amylase, â-amylase, glucoamylase and pullulanase (Hyun and Zeikus, 1985a, b, c). The properties of enzymes used in the process determine the operating conditions for starch processing. During liquefaction, aqueous slurry of starch (30-40 % DS) is gelatinized (105 °C, 5 min) and partially hydrolysed (95 °C, 2 h) by highly thermostable á-amylase to about dextrose equivalents (DE) 5-10. The reaction is carried out at pH 6.0-6.5 and in most cases Ca^{2+} ions (generally 50 ppm) are required for maintaining the activity and stability of the enzyme. Liquefaction is followed by saccharification, where glucoamylase is used along with pullulanase (60 °C, pH 4.0-4.5, 48 h) to yield greater than 95-96 % glucose, or with â-amylase (55 °C, pH 5.0- 5.5, 72 h) to yield around 80-85 % maltose.

Fleming (1968) classified glucoamylases into two groups, one converts starch and β-limit dextrins completely into glucose, and the other converts starch and β-limit dextrins into 80% and 40% glucose, respectively. Applicability of glucoamylase in starch industries can be benefited from thermostability and enhanced activity in the neutral pH range (Reilly, 1999). Glucoamylase alone is used to improve the efficiency of conversion of starch to glucose, but due to its less affinity towards the branching points (α-1,6 glycosidic bonds), it is used in combination with pullulanase to hasten the glucose formation, or with â-amylase to increase the maltose level in high maltose syrups (Norman, 1982). Due to pH variation, large amounts of salts have to be removed from the product stream by ion exchangers. Apart from the first step all other steps are time consuming, leading, in most cases, to reverse reactions and lower yields. Therefore, undesirable products like branched oligosaccharides, panose, isopanose and isomaltose are also formed.

The ideal process

For replacement of traditionally used enzymes, it is most important to find more hermostable enzymes with unique properties. Thermostable enzymes active at the same pH and temperature range can be applied in a one step starch hydrolysis process, avoiding temperature and pH adjustments. The enzymes that simultaneously hydrolyze α-1, 4- and α-1, 6-linkages can significantly enhance the starch saccharification process. The enzymes with simultaneous α-1, 4 and α-1, 6 bond hydrolyzing capabilities are novel kinds of pullulanases, known as amylase-pullulanases (Kim and Kim, 1995) and amylopullulanases (Saha and Zeikus, 1989a). Overproduction of such enzymes can be achieved by employing genetic techniques and optimization of fermentation processes (Antranikian *et. al.*, 1987a,b; Vihinen and Mantsala, 1989). The use of enzymes with these properties will significantly lower the cost of sugar production.

Glucoamylase producing thermophilic moulds

Glucoamylases are produced by a wide variety of microorganisms. Filamentous fungi, however, constitute the major source of glucoamylases among all microorganisms (Pandey *et. al.*, 2000). *Aspergillus* and *Rhizopus* are mainly used for commercial production of glucoamylases. Because of greater stability and activity at elevated temperatures, the glucoamylases of thermophilic moulds are of particular interest.

Degradation of starch usually requires cooperation of several enzymes because only a few enzymatic activities can degrade it alone, and thus, starch-degrading microbes usually have several amylolytic enzymes. *Humicola grisea, Talaromyces emersonii* and *Thermomyces lanuginosus* produced a mixture of amylases containing α–amylase, glucoamylase and α-glucosidase (Jackson and Seidman, 1985; Vanacker *et. al.*, 1990; Bunni *et. al.*, 1992). Other thermophilic mould strains investigated for their thermotostable glucoamylases include, *Humicola lanuginosa* (Taylor *et. al.*, 1978), *Thermomyces lanuginosus* (Hassum *et. al.*, 1991), *Thermomucor indicae-seudaticae* (Kumar and Satyanarayana, 2003; Kaur and Satyanarayana, 2004), and *Myrothecium* sp. M1 (Malek and Hossain, 1994). Bacterial glucoamylases from *Clostridium* sp. G005 (Ohnishi *et. al.*, 1992), *Methanococcus jannaschii* (Bult *et. al.*, 1996) and *Thermoanaerobacterium thermosaccharolyticum* (Ducki *et. al.*, 1998) have also been studied for their thermostability. These enzymes have approximately 40% sequence identity to glucoamylases from Aspergilli.

Glucoamylase production

In order to obtain optimum yield of an enzyme, development of suitable medium and culture conditions is obligatory (Narang and Satyanarayana, 2000). Several thermophilic moulds including *Humicola lanuginosa, H. grisea, Rhizomucor pusillus, Talaromyces dupontii, Thermomucor indicae-seudaticae* and others have been reported to produce glucoamylase in both submerged as well as solid-state fermentations.

Submerged fermentation (SmF)

Submerged fermentation involves the growth and fermentation by microorganisms of liquid substrate under continuous agitated conditions, ensuring uniform distribution of nutrients. Under submerged conditions fungal growth occurs in the form of spore balls floating in the liquid medium. SmF is widely used and in most cases preferred over SSF because it is less problematic, heat transfer is much better and homogeneity is much better too (Pandey *et. al.*, 2001). The optimized cultural conditions for glucoamylase production by some thermophilic moulds are presented in Table 1.

Thermomyces lanuginosus, a thermophilic fungus has been shown to produce several amylases. Chadha *et. al.*, (1997) optimized the culture conditions for the production of amylases by *T. lanuginosus* in submerged fermentation. The mould produced a-amylase (18.4 U ml^{-1}) and glucoamylase (11.2 U ml^{-1}) optimally, in the medium containing rice flour (2% w/v) and corn steep liquor as carbon and nitrogen sources, at pH 5.5 and 50 °C after 72 h of incubation at 150 rpm.

Jensen *et. al.*, (1987) compared the production of extracellular starch hydrolyzing enzyme from *T. lanuginosus* in four different cultivation systems. The highest extracellular amylase production was attained in static cultures. Static conditions favored α–amylase synthesis, while glucoamylase production was higher in shake flasks.

Table 1 Optimized cultural conditions for glucoamylase production by some thermophilic moulds.

Organism	Optimum Temp. (°C)	Carbon source	Nitrogen source	Agitation (rpm)*	Incubation Time (days)	References
Humicola grisea	42	Potato starch	–	130	2	Campos and Felix, 1995
H. grisea var. *thermoidea*	45	Maltose	–	Static	6	Tosi *et al.*, 1993
Thermomyces lanuginosus	50	Soluble starch	L-asparagine	–	2	Rao *et al.*, 1981
T. lanuginosus	50	Glucose/ starch	Yeast extract	220	4	Rubinder *et al.*, 2002
T. lanuginosus ATCC34626	–	Soluble starch starch	L-asparagine	–	–	Nguyen *et al.*, 2002
T. lanuginosus IISc	50	*Potato soluble starch*	L-asparagine	–	*4*	*Mishra and Maheshwari, 1996*
T. lanuginosus	50	Rice floor	Corn steep liquor	150	3	Chadha *et al.*, 1997
Thermoascus aurantiacus	40	Starch	Sodium nitrate	–	6	Ohno *et al.*, 1998
Thermomucor indicae-seudaticae	40	Sucrose	Yeast extract	250	2	Kumar and Satyanarayana, 2003
Aspergillus fumigatus	30	Oat	–	Static	6	DaSilva and Peralta, 1998

*rpm (revolutions per minute)

– not recorded

Thermomucor indicae-seudaticae secreted glucoamylase optimally in a synthetic medium containing starch, in shake flasks as well as in laboratory fermenter. The highest enzyme production (30,000 UL^{-1}) was reported when 5×10^6 spores from 3 day old culture of *Thermomucor indicae-seudaticae* were inoculated into 50ml synthetic medium containing 5% starch, and incubated at 40°C for 48h. A 1.7 fold increase in enzyme production was attained when the mould was grown in a laboratory fermenter (Kaur and Satyanarayana, 2004).

Rubinder *et. al.,* (2002) studied the effect of different parameters on amylase production by a mutant strain [III (51)] of *T. lanuginosus* using a Box-Behnken design. The regression models computed showed significantly high R squared value for α-amylase and glucoamylase activities, suggesting that they are appropriate for predicting relationships between corn flour, soybean meal and pH.

Thermoascus aurantiacus IFO 31693 was identified as an extracellular amylase producing thermophilic mould. The addition of 2 to 4% starch as carbon source and 0.5% sodium nitrate as nitrogen source were effective for enzyme production. The organism was incubated in 2% starch medium for six days at 40°C (Ohno *et. al.*, 1998).

Solid-state fermentation (SSF)

A general opinion about the choice of fermentation method for the production of any microbial product would normally be SmF, unless there are particular reasons for using SSF including favourable economics, if the desired product is formed only in SSF or to follow the government regulations to utilize organic solid wastes as substrates.

Solid-state fermentation involves the growth and fermentation by microorganisms, especially fungi, on moist, water insoluble, and solid substrate in the absence or near absence of free flowing water. The solid substrate not only supplies the nutrients but also serves as an anchorage to the microbial cells. The water content is quite low with the necessary moisture existing in the absorbed or complex form within the solid material so that the microorganism is almost in constant contact with gaseous oxygen in the air.

Solid-state fermentation (SSF) offers numerous advantages over submerged fermentation (SmF) including, high volumetric productivity and concentration of the products, less effluent generation, simple fermentation equipments, and others. There are several factors (physical, chemical and biochemical), which one needs to consider for any SSF process (Pandey *et. al.*, 2001).

The major factors which affect microbial growth and activity in SSF include selection of suitable microorganism and substrate, pretreatment of the substrate, particle size (inter particle space and surface area) of the substrate, water content and water activity (a_w) of the substrate, relative humidity, type and size of inoculum control and temperature of the fermenting matter/removal of metabolic heat, period of cultivation, maintenance of uniformity in the environment of SSF, and the gaseous atmosphere (oxygen consumption rate and carbon dioxide evolution rate).

Fungal enzymes were extensively produced by SSF technique in United States, pioneered by Takamine (1914). The ideal solid substrate is one that provides all the necessary nutrients to the microorganism for their optimum function. Filamentous fungi have been largely exploited for various products in solid-state fermentation. The main advantages of SSF are the ease with which fungi can grow on complex natural solid substrates like agro-industrial residues without any chemical transformation. A variety of substrates have been employed for the production of amylases: wheat bran (Pandey, 1990; Jaleel *et. al.*, 1992), wheat straw, Cassava (Soccol *et. al.*, 1994), rice bran (Pandey, 1993) and Copra waste (Pandey, 1995).

Fungi cannot directly utilize these macromolecules, and therefore, secrete different enzymes to hydrolyze these macromolecules into smaller metabolizable compounds (Mitchell *et. al.*, 1992). Although this technique has been in practice since ancient times, little progress has been made as compared to SmF processes. The major difficulties associated with SSF include heat build up, difficulty in determining growth, uncontrolled pH and moisture content, problems associated with scale up etc. Gowthaman *et. al.,* (2001) reviewed fungal solid-state fermentation in detail highlighting the importance of nutritional factors in the solid substrate and the cultivation techniques like tray system, packed column reactor, rotating drum bioreactor, and several other configurations.

Kumar and Satyanarayana (2004) reported glucoamylase production by *Thermomucor indicae-seudaticae* in solid-state fermentation. Greater enzyme production was observed with substrate to moisture ratio of 1:2.5, using wheat bran as the substrate at 40 °C and pH 7.0. The enzyme titres were high when 10g wheat bran was taken in 250 ml Erlenmeyer flasks, and a decline was observed at higher substrate levels. A very high enzyme titre (392.3 Ug^{-1}DMB) was recorded when wheat bran was supplemented with 2% cotton oil seed cake. *H. lanuginosa* secreted amylolytic enzymes optimally after four days of incubation at 45 °C (Chawla, 1996). The substrate to moisture ratio of 1:2.5 and a$_w$ in the range of 0.92 to 0.8 favoured enzyme production. The enzyme titres under optimal conditions in trays containing wheat bran were slightly lower than those in flasks.

Rhizomucor pusillus NH-139 produced one raw starch-digesting glucoamylase on solid wheat bran medium. The optimum temperature for growth was 45°C (Kanlayakrit *et. al.,* 1987).

Improvement in production and properties of glucoamylases

Properties of an enzyme used in biotechnological processes determine the fermentation conditions and ultimately the cost of the product. There is a continuous search for enzymes that are active at the desired temperature and pH range, which can significantly hasten the production rate and lower the production costs. Enzymes with desired properties can be obtained either by screening new isolates from different environmental niches or by using classical and /or molecular methods of mutations. While searching for new isolates, the target should be extreme environments, especially hot pools, which are excellent sources of a number of bacterial genes encoding economically relevant enzymes. Improvement of microbial strains for the overproduction of desired products has been the hallmark of all commercial fermentation processes. The principal goals of strain improvement are the reduction of variation between batches and the improvement of production levels or the specific enzyme activity of the product. Conventional strain improvement is achieved by classical mutation and random selection, or genetic recombination (Parekh *et al.,* 2000). Classical mutation involves the exposure of a given population to radiations or chemicals, used as mutagens, and the random screening of the surviving population for the desired characteristics.

Rubinder *et. al.,* (2002) reported approximately 7- and 3-fold higher specific activities of α-amylase (190 Umg^{-1}protein) and glucoamylase (105 Umg^{-1}protein), respectively, as compared to the wild type parent strain, when the thermophilic fungus *Thermomyces lanuginosus* was subjected to three cycles of mutagenesis, using NTG and UV radiations.

Amylase hyper-producing, catabolite resistant, recombinant strains were produced by intraspecific protoplast fusion of *T. lanuginosus* strains, using well-characterized, morphological, and 2-deoxy-D-glucose resistant markers. The fusant heterokaryons exhibited enhanced amylase activities as compared to the parental strain. Diploids derived from heterokaryons segregated to stable haploid recombinant strains. In the haploid strain, approximately 5-fold higher specific activities of a-amylase and glucoamylase than the parent strain were recorded (Rubinder *et. al.*, 2000).

In contrast to random method of mutation, screening and selection, molecular techniques offer a number of possible ways to specifically direct strain improvement, once gene encoding the activity of interest has been cloned. Filamentous fungi are of great interest as efficient expression hosts for a wide range of valuable gene products originating from other organisms (Nevalainen and Te'o, 2003). However, industrially exploited production hosts should have GRAS status (Generally Regarded As Safe) or otherwise a long history of safe use in industry. Glucoamylase encoding genes have been cloned and sequenced from several filamentous fungi including *Aspergillus niger* (Boel *et. al.*, 1984), *A. awamori* (Innis *et. al.*, 1985), *Rhizopus oryzae* (Ashikari *et. al.*, 1986), *A. oryzae* (Hata *et. al.*, 1991a, b), *Neurospora crassa* (Stone *et. al.*, 1993), and *A. terreus* (Ventura *et. al.*, 1995). Some of the thermophilic fungal glucoamylase encoding genes have also been cloned. The gene encoding a thermostable glucoamylase from *Talaromyces emersonii* was cloned and heterologously expressed in *Aspergillus niger*. This glucoamylase gene encoded a protein that contained 618 amino acid residues with a calculated molecular mass of 62.827 kDa. The expressed enzyme showed high specific activity towards maltose, isomaltose, and maltoheptaose, having 3-6 fold elevated K_{cat} as compared to *A. niger* glucoamylase. *T. emersonii* glucoamylase showed significantly improved thermostability with a $t_{1/2}$ of 48h at 65°C in 30% (w/v) glucose in comparison with *A. niger* glucoamylase (Nielsen *et. al.*, 2002).

Humicola grisea var. *thermoidea* was transformed using a bleomycin-phleomycin resistance gene linked to regulatory sequences from *A. nidulans* (Allison *et. al.*, 1992). The transformants were obtained at a frequency of 0.52-2 per mg of plasmid DNA. Extra copies of *Humicola* gla 1 gene were introduced into the genome of several *Humicola* strains by transformation. Some transformants produced almost 3-fold more glucoamylase than the parent strain.

Thielavia terrestris, a soil-borne thermophilic fungus was found to produce extracellular glucoamylase in culture filtrates. The corresponding glaA gene was cloned and expressed in *Aspergillus oryzae* under the control of an *A. oryzae* alpha-amylase promoter and an *A. niger* glucoamylase terminator. The coding region contained five introns. Based on the amino acid sequence, the glucoamylase was 65% identical to *Neurospora crassa* glucoamylase. Sequence comparisons suggested that the enzyme belongs to the glycosyl hydrolase family 15. The 75-kDa recombinant glucoamylase showed a specific activity of 2.8 Umg^{-1} protein with maltose as substrate. With maltotriose as a substrate, the enzyme was optimally active at pH 4.0 and 60°C. The enzyme was stable at 60°C for 30 min. The

K_m and K_{cat} of the enzyme for maltotriose were determined at various pHs and temperatures. At 20°C and pH 4.0, the enzyme had a K_m of 0.33 ± 0.07 mM and a K_{cat} of (5.5 ± 0.5) × 103 min^{-1} for maltotriose. The temperature dependence of K_{cat}/K_m indicated an activation free energy of 2.8 kJ/mol across the range of 20-70°C. Overall, the enzyme derived from this mould exhibited properties comparable with its homologue derived from mesophilic fungi (Rey et. al., 2003).

Protein engineering by using modeling strategies or more recently by directed evolution (Kolkman and Stemmer, 2001) might be used to modify the properties of enzyme protein of interest (MacCabe et. al., 2002). It involves the premeditated change of amino acids and is usually based on the known 3-D structure of a given protein and its biochemically established mechanism. Protein engineering involves *in vitro* mutagenesis, the alteration of one or a small number of known amino acids of the target protein through the use of short oligonucleotide primers coding for the change that permits modification of catalytic or other properties. In site-specific mutagenesis, a detailed knowledge of the three-dimensional structure of the enzyme is required. Site directed mutagenesis has been used to generate thermostable enzymes by three approaches: a) removal of Asn-Gly bonds subject to deamination (Chen et. al., 1994a, b) elimination of thermolabile aspartyl bonds and c) replacement of Gly residues in helices with Ala residues. The other properties that can be altered include substrate specificity, activity in alkaline and acid solutions and oxidative stability (Leisola et. al., 2002).

While protein engineering relies on established knowledge from studied genes and proteins, directed evolution explores either natural or mutated gene pools in order to select desired properties. Enzymes are often highly specialized for specific biological functions within the context of a living organism. For industrial processes on the other hand enzyme should be stable over long periods of time. Directed evolution work around these problems by creating gene libraries and applying mutagenesis/recombination and/or gene and domain shuffling techniques (Gibbs et. al., 2001; Kolkman and Stemmer, 2001), for isolating the right DNA encoding protein with desired properties (Arnold, 1996; Farinas et. al., 2001; Joern et. al., 2002). The transformation frequencies with filamentous fungi and even yeast are not advanced enough to allow effective screening for enzyme evolution, and therefore, existing molecular evolution programmes are mainly carried out in *E. coli* with which transformation frequencies and robotized screening methods are highly developed to enable compilation of mutant libraries and effective high throughput screening of gene products (Nevalainen and Te'o, 2003).

Bakir et. al., (1993) studied the catalytic and pH properties of *A. awamori* glucoamylase by cassette mutagenesis where nine single amino acid mutations were made in the enzyme active site. The Glu179® Asp mutation expressed in yeast resulted in large decrease in K_{cat}, however K_m remained unchanged, thus confirming the involvement of this residue in catalytic function. Other mutations, Asp176®glu and Glu 180®Asp indicated the involvement of these residues in substrate binding or structural integrity. Insertions of Asp or Gly between residues 176 and 177 resulted in complete loss of activity.

The two forms of *Aspergillus niger* glucoamylase GAI (from the N-terminus, catalytic domain + linker + starch binding domain) and GAII (catalytic domain + linker) were shuffled to make RGAI (SBD + linker + catalytic domain), RGAIDL (SBD + catalytic domain) and RGAII (linker + catalytic domain). The results indicated that; a) movement of SBD to N-terminus or replacement of the native SBD affects soluble starch hydrolysis, b) SBD location significantly affects insoluble starch binding and hydrolysis, and c) insoluble starch hydrolysis is imperfectly correlated with its binding by the SBD (Cornett *et. al.*, 2003).

Liu and Wang (2003) constructed twelve mutations to improve the thermostability of glucoamylase from *Aspergillus awamori*. The thermal unfolding of the catalytic domain followed a putative hierarchical behavior. The unfolding of the13 a-helices obeyed the random order mechanism, in which the a-helices 8,1,11 unfolded more rapidly. The disulfide bonds engineered to lock the a-helix 11 on the surface of the catalytic domain dramatically increased the thermostability. Substituting G396 and G407 with Ala residues slightly increased thermostability, but their specific activity and catalytic efficiency were reduced.

Purification and characterization of glucoamylases of thermophilic moulds

With better knowledge and purification of enzymes the number of applications has increased many fold, and with the availability of thermostable enzymes a number of new possibilities for industrial processes have emerged. Purification is a necessary prerequisite for fully understanding the nature and mechanism of enzyme action.

Isolation of a protein from the biological environment requires a series of purification steps, each step removing some of the impurities and bringing the product closer to the final specification (Kaul and Mattiasson, 1993). Different techniques including column fractionation, ion exchange, and hydrophobic interaction and gel chromatography have been used to purify glucoamylases. The choice of procedure for enzyme purification depends on their location. Initial processes include crude fractionation, clarification, concentration of crude enzyme using processes such as centrifugation, ultrafiltration (Grootegoed *et. al.,* 1973; Tsvetkov and Emanuilova, 1989) and salt precipitation (Yamane *et. al.,* 1973; Krishnan and Chandra, 1983). The concentrate is then further purified using high resolution techniques based on chromatographic and electrophoretic separations. It is designed to remove aggregates, degradation products and to prepare a solution suitable for the final formulation of the purified enzyme (Kaul and Mattiasson, 1993).

Fungal glucoamylases are generally glycoproteins having 5 to 20% carbohydrate content in the form of glucose, glucosamine, mannose and galactose (James and Lee, 1997). The carbohydrate part of the enzyme is believed to be responsible for the higher thermostability of the enzyme. Removal of carbohydrate moiety reduces the stability and activity of the enzyme (Saha and Zeikus, 1989b). The molecular mass of fungal glucoamylases ranges from 26.85kDa to 112kDa and the optimal pH of most glucoamylases fall within the range of 4.5 to 5.5 (Fogarty, 1983; Vihinen and Mantsala, 1989). Most glucoamylases are stable on the acid side of pH 7.0, only a few have been reported to be stable at or above neutral pH. Glucoamylase of thermophilic mould *Thermomucor indicae-seudaticae* is optimally active pH 7.0 (Kumar and Satyanarayana, 2003), and that of *Humicoila lanuginosa* is stable up to pH 11.0 (Taylor *et. al.*, 1978). The methods used for purification of glucoamylases are species specific, since one method used to purify glucoamylase from one organism does not necessarily work for purification of other glucoamylases (James and Lee, 1997). The important characteristic features of some thermophilic fungal

Table 2 Characteristics of glucoamylases produced by thermophilic moulds.

Organism	Optimum Temp. (°C)	Optimum pH	K_m/V_{max}	pI	M.W.	Carbo-hydrate (kDa)	References content (%)
Humicola grisea	60	6.0	0.14/--	8.4	74	5.0	Campos and Felix, 1995
H. grisea var. thermoidea	55	5.0	--/--	–	63	1.8	Tosi *et al.*, 1993
Scytalidium thermophilum	70	5.5	0.210/62	8.3	75	9.8	Aquino *et al.*, 2001
Thermomyces lanuginosus	70	–	--/--	–	57	10-12	Rao *et al.* 1981
T. lanuginosus ATCC34626	70	4.6-6.6	0.8/--	4.1	75	3.27	Nguyen *et al.*, 2002
T. lanuginosus IISc	–	–	0.04/660	–	–	11	Mishra and Maheshwari, 1996
Thermomucor indicae-seudaticae	60	7.0	0.4/143	8.2	42	9-10.5	Kumar and Satyanarayana, 2003
Rhizomucor pusillus NH-139	65	–	--/--	–	68	–	Kanalayakrit *et al.*, 1987
Aspergillus fumigatus	65	4.5-5.5	0.1/161	–	42	23	Da Silva and Peralta, 1998

glucoamylases are shown in Table 2.

Fagerstrom and Kalkkinen (1995) characterized glucoamylase of *Trichoderma reesei* having pH and temperature optima of 5.5 and 70 °C, respectively. The enzyme was slightly inhibited by the action of b-cyclodextrin, while remained unaffected by EDTA and metal ions. K_m and V_{max} values were 0.11 mg ml^{-1} and 28.5 s^{-1}, respectively. The amino acid sequencing showed about 60% identity with extensively studied *A. niger* glucoamylase.

T. lanuginosus produced both a-amylase and glucoamylase, and they did not exhibit any synergistic effect like those of *Aspergillus* sp. (Abe *et. al.*, 1988). Glucoamylase of *T. lanuginosus* affected up to 76% conversion of soluble starch or raw potato starch to glucose in 24 h, indicating that it was independent from any catabolite repression (Mishra and Maheshwari, 1996). A thermostable glucoamylase of 57 kDa was purified from the thermophilic fungus *Thermomyces lanuginosus* (Rao *et. al.*, 1981). It was optimally active at 70 °C.

The *Humicola* glucoamylase of 63 kDa exhibited pH and temperature optima of 5.0 and 55 °C, respectively (Campos and Felix, 1995). The enzyme had a single catalytic site for starch and maltose, as both were hydrolyzed by the same catalytic site at the same rate. Another form of glucoamylase (M.W. 74 kDa; pI 8.4) produced by the same fungus was optimally active at 60°C and pH 6.0 and was remarkably insensitive to end product inhibition. It retained 65% activity in presence of 950 mM glucose.

The glucoamylase of *Thermomucor indicae-seudaticae* was optimally active at 60 °C and at pH 7.0. The purified enzyme had a molecular mass of 42kDa with a pI of 8.2. It is a glycoprotein with 9-10.5% carbohydrate content, with a $t_{1/2}$ of 12h at 60°C and 7h at 80°C. The K_m and V_{max} values for soluble starch hydrolysis at 60°C were 0.40 mg mL^{-1} and 143 mmol mg^{-1} protein min^{-1} (Kumar and Satyanarayana, 2003).

Glucoamylase of a thermotolerant mould *Neosartorya fischeri* M-1 was most active in the temperature range of 55-60°C and exhibited the highest activity at pH 4.0 to 4.4. The K_m and V_{max} values of the enzyme for amylopectin were 2.4 gL^{-1} and 12mg glucoseL^{-1}min^{-1}, respectively. The molecular mass of the enzyme, as estimated by the gel filtration method was 32kDa (Hang and Woodams, 1993).

Rhizumucor pusillus NH-139 produced only a single form of raw-starch-absorbable, raw starch-digesting glucoamylase. The electrophoretically homogenous preparation of glucoamylase had a molecular weight of 68,000Da and its optimal temperature on gelatinized and raw corn starch was 65°C and 50°C. However, this raw-starch-digesting glucoamylase, unlike other glucoamylases, could not completely hydrolyze glycogen but hydrolyzed it to the extent of 80% as glucose. (Kanlayakrit *et. al.*, 1987).

A thermostable glucoamylase from *Aspergillus fumigatus* was purified to homogeneity. It was a glycoprotein with 23% carbohydrate content and an apparent molecular mass of 42 kDa. The enzyme was optimally active at pH 4.5-5.5 and 65°C and preferentially attacked polysaccharides, such as starch, glycogen, and amylopectin rather than maltose and maltotriose. The K_m and V_{max} of soluble starch hydrolysis at 40°C and pH 5.0 were 0.1mg mL^{-1} and 161 mmol glucose equivalents liberated min^{-1} mg^{-1} protein, respectively. The enzyme was not affected by 500 mM D-glucose and retained about 80% of its original activity in the presence of 1000mM of this sugar (DaSilva and Peralta, 1998).

Ohno *et. al.*, (1998) purified two forms of amylases from the culture filtrate of *Thermoascus aurantiacus* when inoculated into the medium containing 2% starch. One (amylase [2]-b) was obtained as partially purified protein (MW, 75,000) and the other (amylase [22]) as an electrophoretically homogeneous protein (MW, 56,000). Amylase[2]-b and [22] were identified as being exo-and endo-type enzymes respectively, with optimum pH of 5.0 and 4.5.The optimum temperatures were 65 and 70°C.

The purified amylolytic enzymes (á-amylase and glucoamylase) of *Thermomyces lanuginosus* ATCC 34626 had molecular masses of 61 and 75 kDa, respectively (Nguyen *et. al.*, 2002). Their pI values were 3.5-3.6 and 4.1-4.3. Optimum pH ranges for á- amylase and glucoamylase activities were 4.6-6.6 and 4.4-5.6, respectively. Both purified enzymes had temperature optima at 70°C.

Industrial Applications

Glucoamylase has got wide applications in the food and fermentation industries, in the processes of starch saccharification, brewing and distillation. It has got 14% distribution and sales (second position after proteases) on worldwide distribution and sales of industrial enzymes (Pandey, 1995), and is mainly used in the production of glucose syrups, high fructose corn syrups and ethanol (Vihinen and Mantsala, 1989; Jensen and Olsen, 1999). Glucoamylase is considered to be less stable than a-amylase, as it rapidly unfolds above 60°C. Hence, it is used industrially at 55-60°C in soluble form. This temperature is sufficiently high to prevent microbial contamination. The conversion of starch to sugars is one of the most important biotechnological processes with an annual output of over 8 million tons (Lee, 1991). Glucoamylases are also used in commercial baking to improve the flour quality, retard dough staling and improve dough for efficient machinability. They have also been used to enhance bread crust colour, bleach flour and to improve the quality of high fiber baked products (James and Simpson, 1996).

Conclusions and Future Perspectives

Glucoamylase is an extremely important industrial enzyme used in the enzymatic conversion of starch into high glucose and fructose syrups. Fungal glucoamylases are most widely used in industrial starch saccharification, however its activity towards α-1,6 linkages is only 0.2% of that for α-1,4 linkages, which adversely affects the yield. Also most glucoamylases are unstable at elevated temperatures and are active in the acidic pH range. Development of thermostable glucoamylases, capable of performing industrial saccharification at elevated temperatures near neutral pH would be of significant importance to starch processing industry. A little success has been achieved through protein engineering towards increased thermostability of fungal glucoamylase, and much is expected in the near future. Besides thermostability, protein engineering of glucoamylases should be targeted towards improving their ability to hydrolyze raw starches, broad substrate specificity and stability at wider pH range, for making them suitable for industrial applications. Alternatively, screening of thermophilic organisms and subsequent cloning and expression of the target genes in a suitable host could also be attempted for obtaining glucoamylases with the desired properties. Comparison of gene sequences of different thermostable glucoamylases provides important information in understanding the role of individual amino acids and would be very helpful in future glucoamylase engineering.

References

Abe, J., Nakajiima, K., Nagano, H., Hizkuri, S. and Obata, K. 1988. Properties of the raw starch digesting amylase of *Aspergillus* sp. K-27: a synergistic action of glucoamylase and a-amylase. *Carbohydrate Research.* 175: 85-92.

Aehle, W. 2004. Industrial Enzymes. In: *Enzymes in Industry.* (Ed. Aehle, W.). pp.101-257. WILEY-VCH Verlag GmbH and Co. KgaA, Weinheim, Germany.

Aleshin, A.E., Firsov, L.M. and Honzatko, R.B. 1994. Refined structure for the complex of acarbose with glucoamylase from *Aspergillus awamori* var. *X-100* to 2.4 A° resolution. *J. Biological Chem.* 269: 15631-15639.

Allison, D.S., Rey, M.W., Berka, R.M., Armstrong, G. and Dunn-Coleman, N.S. 1992. Transformation of the thermophilic fungus *Humicola grisea* var. *thermoidea* and overproduction of *Humicola* glucoamylase. *Current Genetics* 21:225-229.

Antranikian, G. 1992. Microbial degradation of starch. In: *Microbial Degradation of Natural Products.* (Ed. Winkel, G.). pp. 27-51. VCH, Weinheim, Germany.

Antranikian, G., Herzberg, C. and Gottschalk, G. 1987a. Conditions for the overproduction of a-amylase, pullulanase, and a-glucosidase by a new *Clostridium* isolate. *Appl. Microbiol. Biotechnol.* 27: 75-81.

Antranikian, G., Herzberg, C. and Gottschalk, G. 1987b. Production of thermostable a-amylase, pullulanase and a-glucosidase in continuous culture by a new *Clostridium* isolate. *Appl. Microbiol. Biotechnol.* 53: 1668-1673.

Aquino, A.C., Jorge, J.A., Terenzi, H.F. and Polizeli, M.L. 2001. Thermostable glucose-tolerant glucoamylase produced by the thermophilic fungus *Scytalidium thermophilum. Folia Microbiologia* (Praha). 46: 11-16.

Arnold, F.H. 1996. Directed Evolution: Creating Biocatalysts For The Future. *Chemical Engineering Science.* 51: 5091-5102.

Ashikari, T., Kinchi-Goto, N., Tanaka, Y., Shibano, Y., Amachi, T. and Yoshizumi, H. 1986. *Rhizopus* raw-satrch-degrading glucoamylase: Its cloning and expression in yeast. *Agric. Biological Chem.* 50: 957-964.

Boel, E., Hansen, M.T., Hjort, I., Hoegh, I. and Fill, N.P. 1984.Two different types of inverting sequences in the glucoamylase gene from *Aspergillus niger. EMBO Journal* 3:1581-1591.

Bult, C.J., White, O., Olsen, G.J., Zhou, L., Fleischmann, R.D., Sutton, G.G., Blake, J.A., FitzGerald, L.M., Clayton, R.A., Gocayne, J.D., Kerlavage A.R., Dougherty, B.A., Tomb, J.F., Adams, M.D., Reich, C.I., Overbeek, R., Kirkness E.F., Weinstock, K.G., Merrick, J.M., Glodek, A., Scott, J.L., Geoghagen, N.S.M. and Venter, J.C. 1996. Complete genome sequence of the methanogenic archaeon, *Methanococcus jannaschii. Science* 273: 1058-1073.

Bunni, L., Coleman, D.C., Mc Hale, L., Hackett, T.J. and Mc Hale, A.P. 1992. cDNA clone and expression of a *Talaromyces emersonii* amylase encoding genetic determinant in *E. coli. Biotechnol. Lett.* 14: 1109-1114.

Campos, L. and Felix, C.R. 1995. Purification and charcterization of a glucoamylase from *Humicola grisea. Appl. Environmental Microbiology* 61: 2436-2438.

Chadha, B.S., Singh, S., Vohra, G. and Saini, H.S. 1997. Shake culture studies for the production of amylases by *Thermomyces lanuginosus. Mycologia* 67: 608-615.

Chawla, M. 1996. Amylolytic enzymes of the thermophilic mould *Humicola lanuginosa.* M. Sc. Dissertation, University of Delhi South Campus, New Delhi, India.

Chen, H.M., Ford, C. and Reilly, P.J. 1994a. Substitution of asparagine residues in *Aspergillus awamori* glucoamylase by site-directed mutagenesis to eliminate N-glycosylaion and inactivation by deamination. *Biochemical Journal* 301: 275-281.

Chen, H.M., Bakir, U., Reilly, P.J. and Ford, C. 1994b. Increased thermostability of Asn182®Ala mutant *Aspergillus awamori* glucoamylase. *Biotechnology and Bioengineering* 43: 101-105.

Cornett, C.A.G., Fang, T.Y., Reilly, P.J. and Ford, C. 2003. Starch-binding domain shuffling in *Aspergillus niger* glucoamylase. *Protein Engineering* 16:521-529.

Coutinho, P.M. and Henrissat, B. 1999. Carbohydrate –Active Enzymes: an integrated database approach. In: *Recent Advances in Carbohydrate Bioengineering* (Eds. Gilbert, H.J., Davies, G., Henrissat, B. and Svensson, B.). pp. 3-12. The Royal Society of Biochemistry, Cambridge.

Coutinho, P.M. and Reily, P.J. 1997. Glucoamylase structural, functional and evolutionary relationships. *Proteins* 29: 334-347.

DaSilva, W.B. and Peralta, R.M. 1998. Purification and characterization of a thermostable glucoamylase from *Aspergillus fumigatus. Can. J. Microbiol.* 44: 493-497.

Demirijan, D., Moris-Varas, F. and Cassidy, C. 2001. Enzymes from extremophiles. *Current Opinion in Chemical Biology* 5: 144–151.

Ducki, A., Grundmann, O., Konermann, L., Mayer F. and Hoppert, M. 1998. Glucoamylase from *Thermoanaerobacterium thermosaccharolyticum*: Sequence studies and analysis of macromolecular architecture of the enzyme. *J. General Appl. Microbiol* 44: 327-335.

Duo-Chaun, L., Yi-Jung, Y., You-Liang, P. and Chong-Yao, S. 1998. Purification and characterization of extracellular glucoamylase from the thermophilic fungus *Thermomyces lanuginosus. Mycol. Res.*102: 568-572.

Fagerstrom, R. and Kalkkinen. 1995. Characterization, subsite mapping and partial amino acid sequence of glucoamylase from the filamentous fungus *Trichoderma reesei. Biotechnol. Appl. Biochem.* 21: 223-231.

Farinas, E.T., Bulter, T. and Arnold, F.H. 2001. Directed enzyme evolution. *Current Opinion in Biotechnology* 12: 545-551.

Fleming, I.D. 1968. *Starch and its Derivatives.* 4th edition (Ed. Radley, J.A.). pp. 498-527. Chapman and Hall Ltd., London.

Fogarty, W.M. 1983. Microbial amylases. In: *Microbial enzymes and Biotechnology.* (Ed. Fogarty, W.M.). pp.1-92. Appl. Science Publishers, London.

Gerhartz, W. 1990. *Enzymes in Industry–Production and Applications.* V.C.H. Weinheim, Germany. pp. 113-148.

Gibbs, M.J., Nevalainen, K.M.H. and Bergquist, P.L. 2001. Degenerate oligonucleotide gene shuffling (DOGS): A method for enhancing the frequency of recombination with family shuffling. *Gene.* 271: 13-20.

Gowthaman, M.K, Chundakkadu, K. and Moo-Young, M. 2001. Fungal solid-state fermentation – an overview. In: *Applied Mycology and Biotechnology.* Vol. 1. (Eds. Khachatourians, G.G. and Arora, D.K.). pp. 305-352. Agriculture and Food production, Elsevier Science, The Netherlands.

Grootegoed, J. A., Lauweres, A. M. and Heinen, W. 1973. Separation and partial purification of extracellular amylase and protease from *Bacillus caldolyticus. Archives of Microbiology* 90: 223-232.

Haasum, I., Erickson, S. H., Jensen, B. and Olesen, J. 1991. Growth and glucoamylase production by a thermophilic fungus *Thermomyces lanuginosus* in a synthetic medium. *Appl. Micobiol: Biotechnol.* 34: 656-660.

Haki, G.D. and Rakshit, S.K. 2003. Developments in industrially important thermostable enzymes: a review. *Bioresource Technol.* 89: 17-34.

Hang, Y.D. and Woodams, E.E. 1993. Characterization of glucoamylase from *Neosartorya fischeri. Lebensmittel-Wissenschaft und-Technologie* 26: 483-484.

Hata, Y., Kitamoto, K., Gomi, K., Kumagai, C., Tamura, G. and Hara, S. 1991a. The glucoamylase cDNA from *Aspergillus oryzae* its cloning, nucleotide sequence and expression in *Saccharomyces cerevisiae. Agric.Biological Chem.* 55: 941.

Hata, Y., Tsuchiya, K., Kitamoto, K., Gomi, K., Kumagai, C., Tamura, G. and Hara, S. 1991b. Nucleotide sequence and expression of the glucoamylase–encoding gene (*glaA*) from *Aspergillus oryzae. Gene.* 108: 145- 150.

Hermasson, A.M. and Svegmark, K. 1996. Developments in the understanding of starch functionality. *Trends in Food Science and Technology.* 7: 345-353.

Hyun, H.H. and Zeikus, J.G. 1985a. General biochemical characterization of thermostable pullulanase and glucoamylase from *Clostridium thermohydrosulfuricum*. *Appl. Environ. Microbiol.* 49: 1168-1173.

Hyun, H.H. and Zeikus, J.G. 1985b. Regulation and genetic enhancement of glucoamylase and pullulanase production in *Clostridium thermohydrosulfuricum*. *J. Bacteriol.* 164: 1146-1152.

Hyun, H.H. and Zeikus, J.G. 1985c. Simultaneous and enhanced production of thermostable amylase and ethanol from starch by co-cultures of *Clostridium thermosulfrogenes* and *Clostridium thermohydrosulfuricum*. *Appl. Environ. Microbiol.* 49: 1174-1181.

Innis, M.A. et. al., 1985. Expression, glycosylation and secretion of an *Aspergillus* glucoamylase by *Saccharomyces cerevisiae*. *Science.* 228: 21-26.

Jacks, A.J., Sorimachi, K., LeGal-Coeffet, M.F., Williamson, G., Archer, D.B. and Williamson, M.P. 1995. ^{1}H and ^{15}N assignment and secondary structure of the starch-binding domain of glucoamylase from *Aspergillis niger*. *European J. Biochem.* 233: 568-578.

Jackson, L. E. and Seidman, M. 1985. Enzymatic hydrolysis of granular starch directly to glucose. *European Patent Application* 0171218.

Jakob, K. 1989. Thermophilic organisms as sources of thermostable enzymes. *TIBTECH.* 7: 349-353.

Jaleel, S.A., Srikanta, S. and Karanth, N.G. 1992. Production of fungal amyloglucosidase by solid state fermentation - influence of some parameters. *J. Microbiol. Biotechnol* 7: 1-8.

James, J.A. and Lee, B.H. 1997. Glucoamylase: Microbial Sources, Industrial Applications And Molecular Biology-A Review. *J. Food Biochem.* 21:1-52.

James, J.A. and Simpson, B.K. 1996. Application of enzymes in food processing. *Critical Rev. Food Sci. Nutrition* 36: 437-463.

Jensen, B. and Olsen, J. 1999. Amylases and their industrial potential. In: *Thermophilic moulds in Biotechnology* (Eds. Johri, B. N., Satyanarayana, T. and Olsen, J.). pp. 115-137. Kluwer Academic Publishers, The Netherlands.

Jensen, B., Olesen, J. and Allermann, K. 1987. Effect of media composition on the production of extracellular amylases from the thermophilic fungus *Thermomyces lanuginosus*. *Biotechnol. Lett.* 9: 313-316.

Jensen, B., Olsen, J. and Allermann, K. 1988. Purification of extracellular amylolytic enzymes from the thermophilic fungus *Thermomyces lanuginosus*. *Can. J. Microbiol.* 34: 218-223.

Joern, J.M., Meinhold, P. and Arnold, F.H. 2002. Analysis of shuffled gene libraries. *J. Mol. Biol.* 316:643-656.

Juge, N., Le Gal-Coeffet, M.F., Furniss, C.S.M., Gunning, A.P., Kramhoft, B., Mirris, V.J., Williamson, G. and Svensson, B. 2002. The starch binding domain of glucoamylase from *Aspergillus niger*: overview of its structure, function, and role in raw-starch hydrolysis. *Biologia, Bratislava* 57: 239-245.

Kanlayakrit, W., Ishimatsu, K., Nakao, M. and Hayashida, S. 1987. Characteristics of raw starch digesting glucoamylase from thermophilic fungus *Rhizomucor pusillus. J. Fermentation Technol.* 65: 379-385.

Kaul, R. and Mattiasson, B. 1993. Secondary purification. *Bioseparation* 3:1-26.

Kaur, P. and Satyanarayana, T. 2001. Partial purification and characterization of glucoamylase of thermophilic mould *Thermomucor indicae-seudaticae. Indian J. Microbiol.* 41: 195-199.

Kaur, P. and Satyanarayana, T. 2004. Production and starch saccharification by a thermostable and neutral glucoamylase of a thermophilic mould *Thermomucor indicae-seudaticae. World J. Microbiol. Biotechnol.* 20: 419-425.

Kim, C.H. and Kim, Y.S. 1995. Substrate specificity and detailed characterization of bifunctional amylase-pullulanase enzyme from *Bacillus circulans* F-2 having two different active sites on one polypepetide. *European J. Biochem.* 227: 687-693.

Kolkman, J.A. and Stemmer W.P. 2001. Directed evolution of proteins by exon shuffling. *Nature Biotechnology* 19: 423-428.

Konstantinidis, A. and Sinnott, M.L. 1991. The interaction of 1-fluoro-D-glucopyranosyl fluoride with glucosidases. *Biochemical J.* 279:587-593.

Krahe, M., Antranikian, G. and Markel, H. 1996. Fermentation of extremophilic microorganisms. *FEMS Microbiology Reviews* 18: 271–285.

Krishnan, T. and Chandra, A.K. 1983. Purification and characterization of a-amylase from *Bacillus licheniformis* CUMC 305. *Appl. Environ. Microbiol.* 46: 430-437.

Kumar, S. and Satyanarayana, T. 2003. Purification and kinetics of a raw starch-hydrolyzing, thermostable and neutral glucoamylase of a thermophilic mold *Thermomucor indicae-seudaticae. Biotechnol. Progress.* 19 : 936-944.

Kumar, S. and Satyanarayana, T. 2004. Production of thermostable and neutral glucoamylase by a thermophilic mould *Thermomucor indicae-seudaticae* in solid-state fermentation. *Indian J. Microbiol.* 44: 53-57.

Lee, B.H. 1991. Bioconversion of starch wastes. In: *Bioconversion of Waste Materials to Industrial Products.* (Ed. Martin, A.M.). pp. 265-291. Elsevier Science Publ. Co., New York.

Leisola, M., Jokela, J., Pastinen, O., Tarunen, O. and Shoemaker, H. 2002. Industrial use of enzymes.http://www.hut.fi/Units/Biotechnology/Kem-70.415/ INDUSTRIAL_USE_OF_ENZYMES.DOC

Libby, C.B., Cornett, C.A.G., Reilly, P.J. and Ford, C. 1994. Effect of amino acid deletions in the O-glycosylated region of *Aspergillus awamori* glucoamylase. *Protein Engineering*. 7:1109-1114.

Liu, H.L. and Wang W.C. 2003. Protein engineering to improve the thermostability of glucoamylase from *Aspergillus awamori* based on the molecular dynamics simulations. *Protein Engineering*. 16:19-25.

MacCabe, A.P., Orejas, M., Tamayo, E.N., Villanueva, A. and Ramon, D. 2002. Improving extracellular production of food-use enzymes from *Aspergillus nidulans*. *J. Biotechnol*. 96: 43-54.

Malek, S.A.S. and Hossain, Z. 1994. Purification and Characterization of a thermostable glucoamylase from *Myrothecium* isolate. *J. Appl. Bacteriology* 76: 210-215.

Mangallam, S., Subrahamanyam, A. and Gopalkrishna, K.S. 1977. Preliminary observations on the production of amyloglucosidase by thermophilic fungus. *Cur. Sci* 46: 16.

Meagher, M.M. and Reilly, P.J. 1989. Kinetics of the hydrolysis of di- and trisaccharides by *Aspergillus niger* glucoamylases É and ÉÉ. *Biotechnol. Bioeng*. 34: 689-693.

Meagher, M.M., Nikolov, Z.L. and Reilly, P.J. 1989. Subsite mapping of the *Aspergillus niger* glucoamylase É and ÉÉ with malto- and isomaltooligosaccharides. *Biotechnol. Bioeng*. 34: 681-688.

Mishra, R.S. and Maheshwari, R. 1996. Amylases of thermophilic fungus *Thermomyces lanuginosus*: Their purification, properties, action on starch and response to heat. *J. Biosciences* (Bangalore) 21: 653-672.

Mitchell, D.A., Targonski, Z., Rogalski, J. and Leonowicz, A. 1992. *Solid substrate cultivation*. (Eds. Doelle, H.W., Mitchell, D.A. and Rolz, C.E.). pp. 29-40. Elsevier App. Sci., London.

Mozhaev, V. 1993. Mechanism-based strategies for protein thermostabilization. *Trends in Biotechnol*. 11:88-95.

Narang, S. and Satyanarayana. T. 2001. Thermostable a-amylase production by an extreme thermophile *Bacillus thermoleovorans*. *Lett. Appl. Microbiol* 32:31-35.

Nevalainen, H. and Te'o V.J.S. 2003. Enzyme Production in Industrial Fungi-Molecular Genetic Strategies for Integrated Strain Improvement. *Appl. Mycol. Biotechnol*. 3:241-259.

Nguyen, Q. D., Rezessy-Szabo, J. M., Claeyssens, M., Stals, I. and Hoschke, A. 2002. Purification and characterization of amylolytic enzymes from thermophilic fungus *Thermomyces lanuginosus* strain ATCC 34626. *Enzyme and Microbial Technol.* 31: 345-352.

Niehaus, F., Bertoldo, C., Kahler, M. and **Antranikian, G.** 1999. Extremophiles as a source of novel enzymes for industrial applications. *Appl. Microbiol. Biotechnol.* 51: 711-729.

Nielson, B.R., Lehmbek, J. and Frandsen, T.P. 2002. Cloning, heterologous expression, and enzymatic characterization of a thermostable glucoamylase from *Talaromyces emersonii. Protein expression and purification* 26: 1-8.

Norman, B.E. 1982. A novel debranching enzyme for application in the glucose syrup industry. *Starch/Starke* 10: 340-346.

Oates, C.G. 1997. Towards and understanding of starch granule structure and hydrolysis. *Trends in Food Sci.Technol.* 8: 375-382.

Ohnishi, H., Kitamura, H., Minowa, T., Sakai, H. and Ohta, T. 1992. Molecular cloning of a glucoamylase gene from a thermophilic *Clostridium* and kinetics of the cloned enzyme. *European J. Biochem.* 207: 413-418.

Ohno, N., Fukuda, H., Hongxian, W., Kasamura, M., Shinoyama, H. and Fujii, T. 1998. Amylases Produced by a Thermophilic Fungus, *Thermoascus aurantiacus*, and Some of Their Properties. *J. Fermentation and Bioeng.* 85: 458.

Pandey, A. 1990. Improvements in solid state fermentation for glucoamylase production. *Biological wastes* 34: 11-19.

Pandey, A. 1993. The production of glucoamylase by *Aspergillus niger* NCIM 1245 in solid state fermentation. *Process Biochem.* 28: 305-309.

Pandey, A. 1995. Glucoamylase research: an overview. *Starch/Starke* 42: 439-445.

Pandey A., Nigam P., Soccol C.R., Soccol V.T., Singh D. and Mohan R. 2000. Advances in microbial amylases. *Biotechnol. Appl. Biochem.* 31: 135-152.

Pandey, A., Soccol, C.R., Rodriguez-Leon, J.A. and Nigam, P. 2001. *Solid state fermentation in biotechnology: fundamentals and applications.* Asiatech publishers, Inc.New Delhi.

Parekh, S., Vinei, V.A. and Strobel, R.J. 2000. Improvement of microbial strains and fermentation processes. *Appl. Microbiol. Biotechnol.* 54: 287-301.

Rao, V.B., Sastri, N.V. and Subba Rao, P.V. 1981. Purification and characterization of a thermostable glucoamylase from the thermophilic fungus *Thermomyces lanuginosus. Biochemical . J.* 193: 379-387.

Reilly, P.J. 1999. Protein engineering of glucoamylase to improve industrial properties: A Review. *Starch/Starke* 51: 269-274.

Rey, M.W., Brown, K.M., Golightly, E.J., Fuglsang, C.C., Nielsen, B.R., Hendriksen, H.V., Butterworth, A. and Xu, F. 2003. Cloning, heterologous expression, and characterization of *Thielavia terrestris* glucoamylase. *Appl. Biochem. Biotechnol.* 111(3): 153-166.

Robyt, J.F. 1998. *Essentials of carbohydrate chemistry.* Springer, New York.

Rubinder, K., Chadha, B.S., Singh, N. and Saini, H.S. 2000. Amylase hyper-producing haploid recombinant strains of *Thermomyces lanuginosus* obtained by intraspecific protoplast fusion. *Can. J. Microbiol.* 46 (7): 669-673.

Rubinder, K., Chadha, B.S., Singh, N., Saini, H.S. and Singh, S. 2002. Amylase hyperproduction by deregulated mutants of the thermophilic fungus *Thermomyces lanuginosus. J. Industrial Microbiol. Biotechnol.* 29(2): 70-74.

Saha, B.C. and Zeikus, J.G. 1989a. Novel highly thermostable pullulanase from thermophile. *TIBTECH* 7: 234-238.

Saha, B.C. and Zeikus, J.G. 1989b. Microbial glucoamylases: biochemical and biotechnological features. *Starch/Starke* 41(2): 57-64.

Sauer, J., Sigurskjold, B.W., Christensen, U., Frandsen, T.P., Mirgorodskaya, E., Harrison, M., Roepstroff, P. and Svensson, B. 2000. Glucoamylase: structure/ function relationship and protein engineering. *Biochimica et Biophysica Acta* 1543: 275-293.

Singh, C. B., Arvind, S. S. and Singh, S. H. 1996. Mutagenesis for hyperproduction of extracellular amylases by *Thermomyces lanuginosus. Acta Microbiology Polonica.* 45:31-36.

Soccol, C. R., Iloki, I., Marin, B. and Raimbault, M. 1994. Comparative production of a-amylase, glucoamylase and protein enrichment of raw and cooked cassava by *Rhizopus* strains in submerged and solid-state fermentations. *J. Food Sc. Technol.* 31: 320-323.

Sorimachi, K., Jacks, A.J., Le Gal-Coeffet, M.F., Williamson, G., Archer, D.B. and Williamson, M.P. 1996. Solution structure of the granular starch binding domain of glucoamylase from *Aspergillus niger* by nuclear magnetic resonance spectroscopy. *J. Mole. Biol.* 259: 970-987.

Southall, S.M., Simpson, P.J., Gilbert, H.J., Williamson, G. and Williamson, M.P. 1999. The starch-binding domain from glucoamylase disrupts the structure of starch. *FEBS Letters.* 447: 58-60.

Stone. P.J., Makoff, A.J., Parish, J.H. and Radford, A. 1993. Cloning and sequence analysis of the glucoamylase gene of *Neurospora crassa. Curr. Genetics.* 24: 205-211.

Sunna, A., Moracci, M., Rossi. M. and Antranikian, G. 1997. Glycosyl hydrolases from hyperthermophiles. *Extremophiles.* 1: 2-13.

Svensson, B., Pedersen, T.G., Svendsen, I., Sakai, T. and Ottesen, M. 1982. Characterization of two forms of glucoamylase from *Aspergillus niger*. *Carlsberg Res. Communi.* 47: 55-69.

Takamine, J. 1914. Enzymes of *Aspergillus oryzae* and the applications of its amylolytic enzymes to the fermentation industry. *Industrial Engi. Chem.* 6 : 824-828.

Tamura, M.K., Shimizu, M.H. and Tago, M. 1981. *Highly thermostable glucoamylase and process for its production.* United State patent No. 4, 247, 637.

Tanaka, Y., Tao, W., Blanchard, J.S. and Hehre, E.J. 1994. Transition state structures for the hydrolysis of a-D-glucopyranosyl fluoride by retaining and inverting reactions of glycosylases. *J. Biological Chem.* 263: 79-89.

Taylor, P.M., Napier, E.J. and Fleming, L.D. 1978. Some properties of a glucoamylase produced from the thermophilic fungus *Humicola lanuginosa*. *Carbohydrate Res.* 61: 301- 308.

Tosi, L.R.O., Terenzi, H.F. and Jorge, J.A. 1993. Purification and characterization of an extracellular glucoamylase from the thermophilic fungus *Humicola grisea* var. *thermoidea*. *Can. J. Microbiol.* 39: 846-852.

Tsvetkov, V.T. and Emanuilova, E.I. 1989. Purification and properties of heat stable a-amylase production from *Bacillus brevis*. *Appl. Microbiol. Biotechnol.*31: 246-248.

Van der Maarel, M.J.E.C., Van der Veen, B., Uitdehaag, J.C.M., Leemhuis, H. and Dijkhuizen, L. 2002. Properties and applications of starch- converting enzymes of the a-amylase family. *J. Biotechnol.* 94: 137-155.

Vanacker, P., Bacle, B., Vidal, G. and Lacoste, L. 1990. Recherche de nouvelles activites saccharifiantes thermostable chez les champignons filamenteux. *Can. J. Microbiol.* 36: 625-630.

Ventura, L., Gonzalez-Candelas, L., Perez-Gonzalez, A. and Ramon, D. 1995. Molecular cloning and transcriptional analysis of the *Aspergillus terreus* gla1 gene encoding a glucoamylase. *Appl. Environmental Microbiol.* 61: 399-402.

Vihinen, M. and Mantsala, P. 1989. Microbial amylolytic enzymes. *Critical Reviews in Biochem. Mol. Biol.* 24:321-418.

Yamane, K., Yamaguchi, K. and Maruo, B. 1973. Purification and properties of a cross reacting material related to a-amylase and biochemcial comparison with the parent a-amylase. *Biochimica et Biophysica Acta* 295:323-340.

Zobel, H.F. and Stephan, A.M. 1995. Starch: structure, analysis and application. In: *Food Polysaccharide and Their Applications* (Eds. Stephan, A. M. and Marcel D.). Inc., New York, Basel, Hongkong.

Impact of Fertilizers on VAM Fungi - A Review

ASHOK AGGARWAL, SEEMA SHARMA, VIPIN PARKASH, ANIL GUPTA and R. S. MEHROTRA

Botany Department, Kurukshetra University, Kurukshetra, Haryana, 136119.

ABSTRACT

Vam fungi contribute to phosphorus nutrition of the plant by enhancing phosphorus uptake from the soil. Agricultural practices like application of fertilizers influence the soil and the rhizosphere microflora. Some fertilizers influence the microbial activity, however some fertilizers have inhibitory effect on specific microorganisms. Phosphate fetilization is well known to reduce VAM colonization of host plant roots. When ammonium nitrogenous fertilizer is used instead of nitrate, it is responsible for suppression of root colonization and spore numbers. Optimum benefit in terms of plant growth can be obtanied from mycorrhizal symbiosis at a low fertilizer level out put.

KEY WORDS : Mycorrhiza, phosphorus uptake, fertilizers, colonization.

Soil is the surface layer of earth on which the human beings depend for their existence. Soil consists mainly of mineral, organic matter, water and gaseous phases besides hosting a dynamic population of microorganisms. Plant roots also provide an ecological niche for many of the microorganisms that are present in the soil. One of the important members of functional soil microbial community is the 'mycorrhiza' that forms symbiotic relationships with the roots of the higher plants. Natural ecosystems are undisturbed habitats, where the diversity of VAM species is maintained on the basis of natural phenomena of ecosystem i.e. survival of the fittest. Mycorrhizae are believed to protect the environmental quality by enhancing beneficial biological interactions. It has been well established that vesicular arbuscular mycorrhizae contribute to the phosphorus nutrition of plant by enhancing phosphorus uptake from the soil. (Draft and Nicolson, 1966, Gray, 1971, Mosse *et. al.*, 1973; Gerdemann, 1974; Johnson, 1976; Trappe, 1981; Reid, 1984). The mycorrhizal mediated phosphorus uptake takes place in three phases, uptake by the fungal hyphae in the soil system, its movement to the hyphae inside the root cortex through entry points and its subsequent release to the host plant. The symbiont takes up phosphate from the soil in the form of orthophosphate ions from the labile pool and its translocation to the root takes place by cytoplasmic streaming of polyphosphate granules in the vacuoles of the mycorrhizal endophyte. These dynamic microbial populations are influenced by various factors for example, the fertility level, moisture content, soil air, temperature, organic matter, cultural agricultural practices.

In agriculture we use various practices to increase the crop yield. Agricultural practices are known to influence the soil and the rhizosphere microflora. One of the agricultural practices includes application of fertilizers. Fertilizers improve the microbial activity, as the nutrients are made readily available. Some fertilizers however, have inhibitory effect on specific microorganisms for example, addition of nitrates inhibits the activity of free-living nitrogen fixing bacteria like *Azotobacter.* (Eggleton, 1934; Clark, 1939; Parr and Bose, 1948; Waksman, 1952; Lochhead, 1958; Venkateson and Rangaswamy, 1964; Bagyaraj and Rangaswamy, 1967a). Direct influence of fertilizers on the microbial population in the soil as well as on the rhizosphere of crop plants have been reported by various workers (Peterseva and Rogacheva, 1958; Louw and Webley, 1959 a, b; Sundar Rao and Chayanulu, 1961; Bagyaraj and Rangaswamy, 1967b). Fertilizers are being applied to increase the productivity of crops. Among the many fertilizers used, Nitrogen and Phosphatic fertilizers are important ones. Prakash *et. al.,* (1976) observed higher frequency values and higher fungal population from the rhizosphere soil amended with Urea. Katznelson (1965) reported unpredictable effects of soil treatment on the rhizosphere microflora while Papavizas and Davey (1961) pointed out that application of fertilizers had little effect on the number of microorganisms in the rhizosphere. According to Samtsevich and Borisova (1961) there was little effect of fertilizers on the microbial population in uncropped soil and wheat rhizosphere (based on pot experiments) but in the field experiments the effects depend on time of the year. While Davey and Papavizas (1960) and Bagyaraj and Rangaswamy (1967a) reported significant changes in the microbial population of soil and rhizosphere as a result of treatment of fertilizers. Wallander and Nylund (1989) observed that fungal biomass was negatively correlated with N fertilization. The fungal biomass decreased at increased N-levels. Noguchi (1992 a, b, c) studied the effect of organic fertilizers on soil microorganisms and concluded that fertilizers affect the growth as well as the activity of soil microorganisms. Lugauskas *et. al.,* (1997) investigated the amount of microorganisms and composition of fungi in soddy podzolic, sandy pasture soil fertilized with various rates and ratios of NPK fertilizers and demonstrated that composition of fungal species was limited if the concentration of any fertilizer in the soil was reduced. *Penicillium, Mortierella* and *Mucor* were the dominant species in the soil and *Trichoderma* species was observed frequently in the soil. Louw and Webley (1959 a, b) Sunder Rao and Chayanulu (1961), Bonnier and Jacques (1966), Bagyaraj and Rangaswamy (1967a), Gorina (1967), Shivappashetty and Rangaswamy (1968), Venkata Rao and Sadasivaiah (1968), Mishra (1971 a, b, 1972) also carried out work on similar lines and reported different effects on soil microorganisms as a result of different application of fertilizers.

Phosphate fertilization is well known to reduce VAM colonization of host plant roots (Daft and Nicholson, 1969; Khan, 1972; Mosse *et. al.,* 1973; Hayman, 1975; Son and Smith, 1988; Daniels Hetrick, 1984; Thompson, 1994; Koide and Mingguang, 1990; Sreenivasa and Bagyaraj 1989a; Harini Kumar and Bagyaraj, 1989; Saif, 1986; Guttay and Dandurand, 1989 and Thingstrup and Jakobsen, 1998 and Jakobsen, 1998). High phosphorus level in soil or P addition is known to reduce VAM colonization of roots and sporulation (Sreenivasa and Bagyaraj, 1989a). Bagyaraj and Powell (1985) have also observed depressed VAM development at recommended phosphorus level due to higher availability of phosphorus. Shukla and Vanjare (1990) reported the effect of nitrogen and phosphorus on VAM population of oilseed crops (Sesamum, Sunflower, and Mustard) and revealed adverse effects of increasing level of nitrogen and phosphorus on VAM fungi. Devi and Sitaramaiah (1990) reported decrease in mycorrhizal colonization with corresponding increase in the concentration of phosphorus. McGonigle *et. al.,* (1990) reported decline in arbuscular colonization of field grown maize root with addition of 400Kg. P/hectares. In pot culture study, Jones and Sreenivasa (1992) observed increase in per cent root colonization of sunflower with increase in phosphorus levels up to 50% of recommended dose of phosphorus, which decreased thereafter. Studies have indicated that addition of an excess of readily available phosphorus eliminates beneficial mycorrhizal effects (Sreenivasa *et. al.,* 1993). Mehrotra and Baijal (1994) observed that percentage of mycorrhizal infection and number of hyphal entry points increased with increasing level of nitrogen and phosphorus fertilizers up to $N_{40}P_{60}$Kg./h, but the same was lowest at $N_{60}P_{80}$ Kg./h level of fertilization application to sunflower plant. Miranda and Harris (1994a) studied that external mycelial growth, root colonization and their relationship were reduced at supra-optimal phosphorus concentration. Martensson and Carlgren (1994) found that after five years of heavy phosphorus fertilization, 45Kg P/ha/year, the spore frequencies were reduced, moderate additions of phosphorus fertilizers i.e. 5 and 15Kg P/ha/year did not affect spore frequencies whereas at zero phosphorus fertilized treatment had highest spore number in the field soil. According to Miranda and Harris (1994b) phosphorus levels of 37.3 mg Kg.$^{-1}$ and above decreased germination and hyphal growth of *Scutellospora heterogama* and *Glomus etunicatum.* Chandrashekara *et. al.,* (1995) while studying the response of two sunflower genotypes to VAM mycorrhizal inoculation and phosphorus levels found that percent root colonization and chlamydospores count decreased with increasing phosphorus levels. Siqueira *et. al.,* (1998) reported that heavy application of soluble phosphorus at planting (in coffee crop) time might reduce mycorrhiza formation, fungus sporulation and the crop's mycorrhizal dependency and thus mycorrhizal effectiveness towards coffee crop. Miller and Jackson (1998) reported that VAM colonization of lettuce decreased with use of chemical inputs including pesticides and phosphorus and nitrogen fertilizers. According to Ryan and Ash (1999) addition of phosphorus markedly reduced VAM colonization in clover and shoot phosphorus concentration had a strong negative linear correlation with VAM colonization. Olsson *et. al.,* (1999) observed that Dazomet application strongly suppressed colonization of the linseed roots by AM fungi and in plots with no Dazomet application, root colonization by the AM fungi was observed more. The result of the study conducted by Leu *et. al.,* (2000) clearly indicated a negative impact of high phosphorus and micronutrient on development of both internal mycorrhizal root colonization and extraradical hyphae of *Glomus intraradices* in *Zea mays.* In studies conducted at Citrus Research and Education Center, University of Florida, Florida, U.S.A., *Poncirus trifolia* × *Citrus sinensis* and *Citrus aurantium* were grown in peat-perlite medium with four levels of Superphosphate and Rock phosphate and inoculated with *Glomus intraradices.* They found that root colonization by VAM and growth responses were less with Superphosphate than with Rock phosphate amended soil (Graham and Timmer, 1985). Between two sources of phosphorus i.e. Rock phosphate and Superphosphate. Rock phosphate was found to encourage greater root colonization compared to Superphosphate. Thalkappian *et. al.,* (2000) observed deleterious effect of higher phosphorus fertilizers on percentage root colonization of *Glomus fasciculatum* in Cassava plantations in alluvial soil of coastal Tamil Nadu. Khalid and Sanders (2000) reported the inhibitory effect of phosphorus fertilizers on VAM fungi of field grown barley. Nogueira and Cardoso (2000) studied that in Soybean, increased phosphorus rates decreased root colonization, active external mycelium and total external mycelium by *Gigaspora margarita* and *Glomus intraradices.* Titus and Leps (2000) demonstrated that fertilizers had the greatest effect on AM levels of *Holcus lanatus, Molinia caerulea, Potentilla erecta, Prunella vulgaris* and *Ranunculus auricomus* by decreasing the level of *Holcus* hyphae and vesicles, *Potentilla* vesicles, *Prunella* hyphae and *Ranunculus* hyphae and vesicles respectively. Boddington and Dodd (2000) found that addition of phosphorus fertilizer (10mg P Kg.$^{-1}$) reduced mycorrhizal colonization by *Acaulospora, Gigaspora* and *Glomus* compared with the addition of organic matter (10mgP Kg.$^{-1}$). Sharma (2000) investigated the effect of two fertilizers i.e. Urea and Superphosphate on mycorrhizal fungi of sunflower crop. Both the fertilizers showed inhibitory effect on soil microorganisms but the degree varied with the change of concentration. In the present investigation different species of *Glomus* were inhibited at higher concentrations of Superphosphate. There was also decrease in root colonization and VAM spores with increasing Urea and Superphosphate concentration In present study application of Superphosphate had more deleterious effect on mycorrhizal spore number than on percentage root colonization. Al-Karaki (2002) conducted a field experiment to determine the effects of arbuscular mycorrhizal fungi (AMF) and soil phosphorus (P) on bulb yield, benefit and cost analysis, and P use efficiency in garlic (*Allium sativum* L.). They observed AMF-root associations were highest when soil P was lowest and decreased with increasing P application. Treatments included application of the AMF *Glomus mosseae* inoculum to the planting furrows after treatment with different levels of P (0, 20, 40, and 60 kg P/ ha. Corkidi *et. al.*, (2002) investigated the effects of nitrogen (N) fertilization on arbuscular mycorrhizas at two semiarid grasslands with different soil properties and N-enrichment history (Shortgrass Steppe in Colorado, and Sevilleta National Wildlife Refuge in New Mexico) their findings supported the hypotheses that fertilization with N alters the balance between costs and benefits in mycorrhizal symbioses and AM fungal communities from N fertilized soils are less beneficial mutualists than those from unfertilized soils. Podeszfinski *et. al.*, (2002) conducted a two-year field trial to evaluate the impact of fertilization and arbuscular mycorrhizae on the in situ establishment of turfgrasses in low and high phosphorus content experimental sites. Organic or chemical fertilizers were applied on Kentucky Bluegrass and Creeping Bentgrass plots inoculated with either *Glomus mosseae, G. aggregatum* or *G. intraradices*. Soil surveys for indigenous Glomales populations revealed the predominance of *Gigaspora margarita* and *Scutellospora calospora* species at the high and low P sites, respectively. Both turfgrass species showed higher root colonization levels, up to 60%, at the low phosphorus soil site. Huat *et. al.*, (2002) assessed the effects of fertilizer and vesicular-arbuscular mycorrhizal colonization on growth and rate of photosynthesis in *Azadirachta excelsa* seedlings in a pot experiment. They observed that fertilizer had the strongest effect on growth, biomass and photosynthetic rate, and leaf nutrient concentrations of *A. excelsa* seedlings. Plants supplied with slow release fertilizer gave the most vigorous growth and higher rates of photosynthesis among the fertilizer treatments. These results demonstrated the potential of using slow release fertilizer in promoting seedling growth. Azcon *et. al.*, (2003) performed experiment aimed to determine the combined effects of N (1, 5 or 9 mM) and P (0.1 or 0.5 mM) fertilization on mycorrhizal and non-mycorrhizal lettuces and observed that these fertilizer levels were inhibitory to AM-intra and extra radical colonization. Prakash *et. al.*, (2004) also noticed decrease in mycorrhizal colonization in *Sesamum indicum* with increased concentration of rock phosphate.

In contrary to deleterious effect, fertilizers can have no or stimulatory effect on VAM population. Cooke *et. al.*, (1992) pointed out that fertilization of *Acer saccharum* in a Quebec hardwood forest soil with a base cation mixture did not appear to affect either the condition or the incidence of VAM. Awasthi *et. al.*, (1996) reported that application of fertilizers to peach seedlings increased the spore number and root colonization, highest spore number and root colonization were observed with the combination of nitrogen and phosphorus. Olsen *et. al.*, (1999) found that phosphorus addition might not reduce VAM colonization of roots if the inoculum potential of the soil of *Capsicum annum* is high. The level of colonization of roots by VAM fungi decreases with increased phosphorus availability, which is not as marked as thought. A reasonable degree of colonization was observed at available phosphorus levels well above those required for maximum yield.

Mechanism Involved in Low VAM Root Colonization with Increased Fertilizers Application

The two mechanisms have been proposed to explain phenomenon of low VAM root colonization with high doses of fertilizer applications are that soil phosphorus acts directly upon VAM fungus in the soil or indirectly through the increased phosphorus status of the host plant. The reduced root colonization and extra radical hyphal development at high soil phosphorus indicates that the soil phosphorus was inhibitory to VAM fungi.

Regarding the mode of action of phosphorus fertilizers on VAM fungi, Sutton (1973) described the VAM mycorrhizal formation processes as consisting of three phases : 1. a lag phase during which spore germination, germ tube growth and initial penetration occurs, 2. a rapid growth phase coinciding with the development of external mycelium, multiple infection and extensive spread of the fungus within the root and 3. a stable phase during which the proportion of mycorrhizal to non-mycorrhizal roots remains approximately constant. Schwab *et. al.,* (1983) observed that major effect of phosphorus on VAM formation occurred during the rapid growth phase. Another explanation for the suppression of root colonization by VAM fungi is consistent with the well-known fact that high phosphorus in roots depressed VAM fungal colonization (Koide and Mingguang, 1990) by deficiencies of soluble carbohydrates (Jakobsen and Rosendahl, 1990; Thompson *et. al.,* 1990). The primary effect of phosphorus on VAM formation seems to be a reduction in growth of external hyphae caused by high soil phosphorus concentration which can be detected during the lag phase of growth of the fungus. Abbott *et. al.,* (1984) reported that reduction in the length of external hyphae occurred at increasing additions of phosphate is of a greater magnitude than its effect in decreasing the length of internal hyphae of the colonized root. This reflects the decrease in spore number as these are formed by the external hyphal system.

Other Factors Responsible for Low VAM Colonization with Fertilization

Miller *et. al.,* (1995) found that the reduction in colonization occurs to greater extent in the roots growing in a fertilized zone than in those outside this zone. Tuyev *et. al.,* (1990) reported that types of nitrogen fertilizers play a significant role in the regulation of microbial population. Nitrogen fertilizers caused changes in taxonomic structure of microbial populations and in their biochemical activity leading to formation of different metabolites. It may be further informed that efficiency of VAM fungi however, could depend on the form of nitrogen present in the fertilizer. Urea being highly soluble in water is subject to rapid leaching and also changes to ammonia. When ammonium (relatively immobile) instead of nitrate (highly mobile) form of nitrogenous fertilizers is used as a nitrogen source, it might be responsible for suppression of root colonization and spore number. Pereia *et. al.,* (1996) reported that mycorrhizal colonization of *Senna multijuga, S. macrantheram, Melia azadirachta* and *Jacaranda mimosifolia* inoculated with *Glomus etunicatum* and *Gigaspora margarita* at 4mM of NO_3^- and NH_4NO_3 was reduced after 120 days and there was more accumulation of calcium and potassium in plants treated with NH_4^+ compared with those treated with NO_3-. High mineral nutrition especially of nitrogen is well known to reduce and finally inhibit mycorrhizal development. Wallander and Nylund (1989) observed that mycorrhizal infection is negatively correlated with the N-concentration of the shoot at different N-concentration in the growing media. Ammonia in the media has more inhibiting effect than NO_3 because the roots more easily take it up. Some authors have concluded that nitrate have more inhibiting effect than Ammonium (Rudawska, 1986). According to Hayman (1970) plots of wheat without nitrogen and other fertilizers, consistently contained more spores than plots with fertilizers. Valentine *et. al.*, (2002) also studied the effects of arbuscular mycorrhizal (AM) colonisation by *Glomus mosseae* on growth and photosynthesis of cucumber (*Cucumis sativus* L. cv. Telegraph Improved) plants supplied with low phosphorus (0.13 mM P) nutrient solutions containing nitrate (NO3-), ammonium (NH4+) or a NO3-/NH4+ mixture, with the same total (4 mM) N concentration. Mycorrhizal colonization varied with N-form, plants supplied with NO3- only had the highest total AM colonization and highest proportion of arbuscules, with no differences between the other two N treatments. Root systems of NH4+-fed AM plants had the lowest number of vesicles per unit root length compared with AM plants grown in NO3- or NO3-/NH4+. In the presence of NH4+ either on its own or in combination with NO3-, both AM and uninoculated plants had higher leaf N concentrations than NO3—fed plants

Guttay and Dandurand (1989) reported that at lowest soil phosphorus level, NPK fertilization increased the mycorrhizae and at the highest soil phosphorus levels it decreased mycorrhizae. Bevege (1972), Hetrick (1984) have found that nitrogen inhibited mycorrhizal colonization at high levels of phosphorus and increased colonization at moderate levels of phosphorus.

Effects on Low Mycorrhizal Colonization on Mineral Nutrition

Mycorrhizal fungi increase the volume of soil exploited by plants. As a result, root colonization often results in enhanced uptake of relatively immobile metal micronutrients, such as Cu, Zn and Fe (Faber *et. al.*, 1990; Leu *et. al.*, 2000). High phosphorus levels in soil inhibit mycorrhizal development. Uptake of micronutrients by mycorrhizal plants is related to internal and external development of AM fungi, levels of phosphorus also influence nutrient acquisition. Lambert and Weindensaul (1991) concluded that high phosphorus inhibition of mycorrhizal activity was the main reason for decreased Cu and Zn uptake. Jamal *et. al.*, (2002) observed that mycorrhizal fungi enhance Zn and Ni uptake as a result of increased phosphorus uptake by mycorrhizal fungi in phosphorus deficient soils of Pakistan. Azcon *et. al.*, (2003) observed that high availability of nitrogen and phosphorus in soil reduced the content of micro and macronutrients in AM plants (lettuce). At the lowest phosphorus application (0.1mM) the AM colonization increased nutrient acquisition. Highest application of nitrogen and phosphorus to the soil reduced the uptake of N, P, K, Mn and Zn in AM compared to non-mycorrhizal lettuce plants. Their results demonstrated some negative effects of high N and P application in soil on the acquisition of N, P, K, Fe, Mn and Zn for mycorrhizal plants.

Conclusion

The need for continued use of fertilizers in increasing the agricultural productivity to meet the growing demands of the people cannot be overemphasized. It is now becoming clear that the ultimate "sink" of the fertilizers applied in agriculture and other areas is soil. Soil, being the storehouse of multitudes of microbes, in quantity and quality, receives the chemicals in various forms and acts as the scavenger of the harmful substances. The fertilizers reaching the soil in significant quantities alter the ecological balance. This may affect the overall microbial population, of which some of them may be selectively inhibited or killed. The fertilizers might alter the physiological conditions prevailing in the rhizosphere, which in turn alters the rhizosphere mycoflora both quantitatively and qualitatively. The chemicals used should avoid serious injuries to the great variety of microbes whose functions are vital to the crop-producing power of the soil. Factors that can contribute to their survival and performance of essential activities are microbial adaptability, adequate soil management and use of the minimum effective dosage of the fertilizers. Sustainability of soil-plant systems requires good development and function of mycorrhizal symbiosis. Applications of easily soluble fertilizers especially phosphorus fertilizers increase the fraction of easily available phosphorus in arable soils, but this is negatively correlated with presence of VAM fungi. Finally, it can be said that higher concentrations of the fertilizers are not required in increasing plant growth as these result in greater degree of soil disturbance. The optimum benefit in terms of plant growth can be obtained from mycorrhizal symbiosis at a low fertilizer input. It is suggested that the low concentrations of the fertilizers must be applied which would have both an economic and environmental impact.

Acknowledgements

Authors, Vipin Parkash and Seema Sharma are thankful to CSIR New Delhi and Kurukshetra University, Kurukshetra for financial assistance respectively.

References

Abbott, L.K., Robson, A.D. and **De Boer, G.** 1984. The effect of phosphorus on the formation of hypha in soil by the vesicular-arbuscular mycorrhizal fungus, *Glomus fasciculatum. New Phytologist*, 97 : 437-446.

Al-Karaki, G. N. 2002. Benefit, cost, and phosphorus use efficiency of mycorrhizal field-grown garlic at different soil phosphorus levels. J. *Pl. Nutrition.* 25(6):1175-1184.

Awasthi, R.P., Godara, R.K. and **Kaith, N.S.** 1996. Effect of fertilizer and bio-fertilizer on spore number and root colonization of Peach seedlings. *Journal of Hill Research*, 9 (1): 28-32.

Azcon, R., Ambrosano, E. and **Charest, C.** 2003. Nutrient acquisition in mycorrhizal lettuce plants under different phosphorus and nitrogen concentration. *Plant Science*, 165(5): 1137-1145.

Bagyaraj, D.J. and **Powell, C. L. I.** 1985. Effect of vesicular-arbuscular mycorrhizal inoculation and fertilizer application on the growth of Marigold. *New Zealand J. Agri. Res.* 28: 169-173.

Bagyaraj, D.J. and **Rangaswamy, G.** 1967a. Effect of fertilizers on the microflora of soil and the rhizosphere of ragi (*Eleusine coracana*). *Ind. J. Microbiol.* 7 : 29-38.

Bagyaraj, D.J. and **Rangaswamy, G**. 1967b. Studies on the effect of foliar nutrient sprays on the rhizosphere microflora of *Eleusine coracana*. *Mysore J. Agri. Res.* 1: 176-185.

Bevege, D.I. 1972. Vesicular arbuscular mycorrhizae of *Araucaria* : Aspects of their ecology and physiology and role in nitrogen fixation. Ph.D. Thesis, Univ. of New England, Armidle, New South Wales.

Boddington, C.L. and **Dodd, J.C.** 2000. The effect of agricultural practices on the development of indigenous arbuscular mycorrhizal fungi II. Studies in experimental microorganisms. *Plant and Soil*, 218 (1/2): 145-157.

Bonnier, C. and **Jacques, B.** 1966. Inoculation of alfa-alfa seeds in direct contact with various phosphate fertilizers. *Bull. Rech. Agron. Gem. Boux.*, 1 : 3-12.

Chandrashekara, C.P., Patil, V.C. and **Sreenivasa, M.N.** 1995. Response of two sunflower (*Helianthus annuus* L.) genotypes to VA mycorrhizal inoculation and phosphorus levels. *Biotropia*, 8: 53-59.

Clark, F.E. 1939. Effect of soil amendments upon bacterial populations associated with roots of wheat. *Trans. Kanas. Acad. Sci.*, 42: 91-96.

Cooke, M.A., Widden, P. and **O'-Halloran, I.** 1992. Morphology, incidence and fertilization effects on the vesicular-arbuscular mycorrhizae of *Acer saccharum* in a Quebec hardwood forest. *Mycologia*, 83(31): 422-430.

Corkidi, L., Rowland, D. L., Johnson, N.C. and **Allen, E. B.** 2002. Nitrogen fertilization alters the functioning of arbuscular mycorrhizas at two semiarid grasslands. *Plant and Soil.* 240(2): 299-310.

Daft, M.J. and **Nicolson, T.H.** 1966. The effect of *Endogone* mycorrhizae on plant growth. *New Phytologist*, 65 : 343-350.

Daniels Hetrick, B.A. 1984. Ecology of VA mycorrhizal fungi. In: *VA mycorrhiza*. (Eds. Powell, C.L.L. and Bagyaraj, D. J.), CRC Press Inc, Boca Raton, Florida, pp. 35-55.

Davey, C.B. and **Papavizas, G.C.** 1960. Effect of decomposing organic soil amendments and nitrogen on fungi in soil and bean rhizosphere. *Trans. Internat. Congress Soil Sci. 7th Congress (Madison Wisc.) Com.*, **III:** 551-557.

Devi, G.U. and ***Sitaramiah, K.*** 1990. Effect of Superphosphate, rock phosphate as sources of phosphorus in combination with *Glomus fasciculatum* on root colonization, growth and chemical composition of blackgram. In: *Current Trends in Mycorrhizal Research* (Eds. Jalali, B.L., Chand, H.), TERI, New Delhi and HAU, Hisar, pp. 144-145.

Eggleton, W.G.E. 1934. Studies on the microbiology of grassland soil. Part I. General Chemical and microbial features. *J. Agric. Sci.* 24: 416-434.

Faber, B A., Zasoski, R. J., Barau, R. G. and ***Uriu, K.*** 1990. Zinc uptake by corn affected by vesicular arbuscular mycorrhizae. *Plant and Soil*, 129: 121-131.

Gerdemann, J.W. 1974. Mycorrhizae. In: *The plant root and its environment.* (Ed. Carson, E.W.), University Press of Virginia, Charlottesville, 205-217.

Gorina, E.I. 1967. Species composition of bacterial dissolving mineral phosphate of calcium. *Izuakad Nauk. Turkm. S.S.R. Biol. Nauk.,* 2: 19-24.

Graham, J.H. and ***Timmer, L.W.*** 1985. Rock phosphate as source of phosphorus for vesicular-arbuscular mycorrhizal development and growth of citrus in a soil less medium. *J. American Soc. Horticulture Sci.* 110 (4) : 459-492.

Gray, L.E. 1971. Physiology of vesicular-arbuscular mycorrhizae. In: *Mycorrhizae,* (Ed. Haeskaylo, E.), Misc. Publ., US For. Serv., Washington, DC. pp. 1189.

Guttay, A.J.R. and ***Dandurand, L.M.C.*** 1989. Interaction of the vesicular-arbuscular mycorrhizae of maize with extractable soil phosphorus levels and nitrogen-potassium fertilizers. *Biology and Fertility of Soils*, 8: 307-310.

Harinikumar, K.M. and ***Bagyaraj, D.J.*** 1989. Effect of cropping sequence, fertilizers and farmyard manure on vesicular-arbuscular mycorrhizal fungi in different crops over three consecutive seasons. *Biology and Fertility of Soils*, 7 (2): 173-175.

Hayman, D.S. 1970. *Endogone* spore numbers in soil and vesicular-arbuscular mycorrhiza in wheat as influenced by season and soil treatment. *Tr. Br. mycol. Soc.* 54 (1): 53-63.

Hayman, D.S. 1975. The occurrence of mycorrhiza in crops as affected by soil fertility. In: *Endomycorrhizas,* (Eds. Sanders, F.E., Mosse, B. and Tinker, P.B.), Academic Press, London, 495-509.

Hetrick, A.D. 1984. Ecology of VA mycorrhizal fungi. In: *VA Mycorrhiza* (Eds. Powell, L. and Bagyaraj, D.J.), CRC Press, Boca, Raton, Florida, 35-55.

Huat, O. K., Awang, K., Hashim, A. and ***Majid NM.*** 2002. Effects of fertilizers and vesicular-arbuscular mycorrhizas on the growth and photosynthesis of *Azadirachta excelsa* (Jack) Jacobs seedlings. *Forest Ecology and Management.* 158(1-3): 51-58.

Jakobsen, I. 1998. Mycorrhizal fungi a biological resource in plant nutrition. In: Phosphorus balance and utilization in agriculture towards sustainability. Proceeding of a seminar 17-19 March, 1997. *Kungl.-skogs-och-Lantbru ksakademiens-TidsKrift*, 137 (7) : 113-118.

Jakobsen, I. and ***Rosendahl, L.*** 1990. Carbon flow into soil and external hyphae from roots of mycorrhizal cucumber plants. *New Phytologist*, 115 : 77-83.

Jamal, A., Ayub, N, Usman, M. and ***Khan, A. G***. 2002. Arbuscular mycorrhizal fungi enhance zinc and nickel uptake from contaminated soil by soybean and lentil. *Internat. J. of Photoremediation*, 4(3): 205-221.

Jones, N.P. and ***Sreenivasa, M.N.*** 1992. Response of Sunflower to the inoculation of VA mycorrhiza and phosphate solubilizing bacteria in black Clayey Soil. *J. Oilseeds Res.* 10: 86-92.

Katznelson, H. 1965. *Ecology and soil borne plant pathogens.* (Eds. K.F. Baker and W.C. Snyder), Univ. of Calif. Press, Berkeley, 187-209.

Khalid, A. and ***Sanders, F.E.*** 2000. Effects of vesicular-arbuscular mycorrhizal inoculation on the yield and phosphorus uptake of field grown barley. *Soil Biology and Biochemistry*, 32 (11/12): 1691-1696.

Khan, A.G. 1972. The effect of VA mycorrhizal associations on growth of cereals: 1. Effect on maize growth. *New Phytologist*, 71 : 613-619.

Koide, R. and ***Mingguang, L***. 1990. On host regulation of the vesicular-arbuscular mycorrhizal symbiosis. *New Phytologist*, 114: 59-74.

Leu, A., Hamel, C., Hamilton, R. I., Ma, B. L. and ***Smith, D. L.*** 2000. Acquisition of Cu, Zn, Mn and Fe by mycorrhizal maize (*Zea mays* L.) grown in soil at different P and micronutrient levels. *Mycorrhiza*, 9:331-336.

Lochhead, A.G. 1958. The soil bacteria and growth-promoting substances. *Bacterial Review*, 22: 145-153.

Louw, H.A. and ***Webley, D.H.*** 1959a. The bacteriology of root region of oat plant grown under controlled pot culture conditions. *J. Appl. Bacteriology*, 22: 216-226.

Louw, H.A. and ***Webley, D.H.*** 1959b. A study of soil bacteria dissolving certain mineral phosphate fertilizers and related components. *J. Appl. Bacteriology*, 22: 227-233.

Lugauskas, A., Repeckiene, J., Salina, O. and ***Vasiliauskiene, V.*** 1997. Spread of microorganisms in pasture soil fertilized with various rates of NPK fertilizers. *Zemdirbyste- Mokslo- Darbai*, 59: 193-208.

Martensson, A.M. and ***Carlgren, K.*** 1994. Impact of phosphorus fertilization on VAM diaspores in two Swedish long-term field experiments. *Agriculture, Ecosystems and Environment*, 47: 327-334.

McGonigle, T.P., Evans, D.G. and **Miller, M.H.** 1990. Effect of degree of soil disturbances on mycorrhizal colonization and phosphorus absorption by maize in growth chamber and field experiments. *New Phytologist*, 116 : 629-636.

Mehrotra, V.S. and **Baijal, U.** 1994. Effect of vesicular-arbuscular mycorrhiza on the growth and nutrition of sunflower plant at different levels of added N and P fertilizers. *Pro. Nat. Acad. Sci. India*, 64 (B) III: 299-304.

Miller, M.H., McGonigle, T.P. and **Addy, H.D.** 1995. Functional ecology of vesicular arbuscular mycorrhizas as influenced by phosphate fertilization and tillage in an agricultural ecosystem. *Critical Reviews in Biotechnology*, 15 (3/4): 241-255.

Miller, R.L. and **Jackson, L.E.** 1998. Survey of vesicular arbuscular mycorrhizae in Lettuce production in relation to management and soil factors. *J. Agri. Sci.* 130 (2): 173-182.

Miranda, J.C.C. de. and **Harris, P.J.** 1994a. The effect of soil phosphorus on the external mycelium growth of arbuscular mycorrhizal fungi during the early stages of mycorrhiza formation. *Plant and Soil*, 166: 271-280.

Miranda, J.C.C.de. and **Harris, P.J.** 1994b. Effects of soil phosphorus on spore germination and hyphal growth of arbuscular mycorrhizal fungi. *New Phytologist*, 128 : 103-108.

Mishra, R.R. 1971a. Effect of certain chemical fertilizers on the rhizosphere mycoflora of *Oryza sativa*. I. Nitrogenous fertilizers: Ammonium nitrate and Urea. *Mycopath. Mycol. Appl.*, 44 (2): 168-176.

Mishra, R.R. 1971b. Effect of certain chemical fertilizers on the rhizoshpere mycoflora of *Oryza sativa* Linn. II. Potassium fertilizer – Potassiumsulphate. *Ibid*, 45 (2): 119-124.s

Mishra, R.R. 1972. Effect of certain chemical fertilizers on the rhizosphere mycoflora of *Oryza sativa* Linn. III Phosphate fertilizers – Superphosphate. Ibid, 46(1): 97-102.

Mosse, B., Hayman, D.S. and **Arnold, D. J**. 1973. Plant growth responses to vesicular-arbuscular mycorrhiza: V. Phosphate uptake by three plant species from P deficient soils labeled with 32P. *New Phytologist*, 72: 809-815.

Noguchi, K. 1992a. Organic fertilizers and soil microorganisms I. *Agriculture and Horticulture*, 67: 6, 673-677.

Noguchi, K. 1992b. Organic fertilizers and soil microorganisms II. *Agriculture and Horticulture*, 67: 7, 785-790.

Noguchi, K. 1992c. Organic fertilizers and soil microorganisms III. *Agriculture and Horticulture*, 67: 8, 883-887.

Nogueira, M.A. and **Cardoso, E.J.B.N.** 2000. External mycelium production by arbuscular mycorrhizal fungi and growth of soybeans as a function of phosphorus fertilizer rate. *Review Brasileira de Ciencia do Solo*, 24 (2): 329-338.

Olsen, J.K., Schaefer, J.T., Edwards, D.G., Hunter, M.N., Galea, V.J. and **Muller, L.M.** 1999a. Effects of mycorrhizae, established from an existing intact hyphal network on the growth response of Capsicum (*Capsicum annuum* L.) and tomato (*Lycopersicon esculentum* Mill.) to five rates of applied phosphorus. *Australian J.Agri. Res.* 50: 223-237.

Olsson,P. A., Thingstrup, I., Jakobsen, I. and **Baath, E.** 1999. Estimation of the biomass of arbuscular mycorrhizal fungi in a linseed field. *Soil Biol. Biochem.* 31(13): 1879-1887.

Papavizas, G.C. and **Davey, C.B.** 1961. Extent and nature of the rhizosphere of lupinus. *Plant and Soil, 14:* 215-236.

Parr, C.H. and **Bose, R.D.** 1948. Phosphate manuring of legumes. *Indian Farming,* 5: 156-162.

Pereira, E.G., Siqueira, J.O., Vale, F.R. and **Moreira, F.M.S**. 1996. Influence of mineral nitrogen on growth and mycorrhizal colonization of tree seedling. *Pesquisa Agropecuaria Brasileira,* 31 (9): 653-662.

Peterseva, A.N. and **Rogacheva, P.V.** 1958. The effect of bacterial fertilizer on the root microflora of various plants. *Dokl. Uses. Akad. Sel'sokhoz Nauk, Im.* VI. *Lemina,* 3: 20-25.

Podeszfinski, C., Dalpe, Y. and **Charest, C**. 2002. In situ turfgrass establishment: 1. Responses to arbuscular mycorrhizae and fertilization. *J.Sustainable Agri.* 20(1):57-74.

Prakash, A., Tandon, V. and **Sharma, N. C.** 2004. Effect of rock phosphate and VAM inoculation on growth and nutrient uptake in Sesamum indicumL. *Physiology and Molecular Biology of Plants,* 10(1): 137-141.

Prakash, D.M., Mashkoor A., Shabbir A and **Khan A. M.** 1976. Studies on rhizosphere mycoflora of egg plant in relation to soil amendments. *Acta Botanica Indica,* 4: 68-70.

Reid, C.P.P. 1984. Mycorrhizae: A root soil interphase in plant nutrition. In: *Microbial-Plant Interactions.* Am. Soc. Agron., Spec. Publ. 47, Madison, Wisc, 28-50.

Rudawska, M. 1986. Sugar metabolism of ectomycorrhizal Scots pine seedling as influenced by different nitrogen forms and levels. In: *Physiological and Genetical Aspects of Mycorrhizae.* Proc. Ist. Eur. Symp. on Mycorrhizae, INRA, Paris, 389-394.

Ryan, M. and **Ash, J.** 1999. Effects of phosphorus and nitrogen on growth of pasture plants and VAM fungi in SE Australian soils with contrasting fertilizer histories (Conventional and biodynamic). *Agriculture, Ecosystems and Environment,* 73: 51-62.

Saif, S.R. 1986. Vesicular-arbuscular mycorrhizae in tropical forage species as influenced by season, soil texture, fertilizers, host species and ecotypes. *Angewandte-Botanik*, 60 (1-2): 125-139.

Samtaevich, S.A. and ***Borisova, V.N.*** 1961. Effect of fertilizers on the root-microflora of winter wheat. *Microbiologia (U.S.S.R.)*, 30: 1033-1041.

Schwab, S.M., Menge, J.A. and ***Leonard, R.T.*** 1983. Comparison of stages of vesicular-arbuscular mycorrhiza formation in sudan grass grown at two levels of phosphorus nutrition. *Amer. J. Botany*, 70: 1225-1232.

Sharma, D. 2000. Effect of some soil fungicides and chemical fertilizers on the soil mycoflora with special reference to mycorrhizal fungi of sunflower. Ph. D. Thesis. Kurukshetra University, Kurukshetra.

Shivappashetty, K. and ***Rangaswamy, G.*** 1968. Studies on the effect of heavy doses of phosphate fertilizer application on the soil and rhizosphere microflora of sunhemp (*Crotolaria juncea* L.). *Mysore Journal of Agricultural Science*, 2: 257-266.

Shukla, B.N. and ***Vanjare, N.*** 1990. Extraction of vesicular-arbuscular mycorrhizae spores and effect of fertilizers on their population in oilseeds cropped soils. In: *Current Trends in Mycorrhizal Research* (Eds. Jalali, B.L. and Chand, H.), Tata Energy Research Institute, New Delhi, pp. 33-34.

Siqueira, J.O., Junior, O.J.S., Aylas, W.W.F. and ***Guimaraes, P.T.G.*** 1998. Arbuscular mycorrhizal inoculation and Superphosphate application influence plant development and yield of coffee in Brazil. *Mycorrhiza*, 7: 293-300.

Son, C.L. and ***Smith, S.E.*** 1988. Mycorrhizal growth responses: Interaction between photon irradiance and phosphorus nutrition. *New Phytologist*, 108 : 305-314.

Sreenivasa, M.N. and ***Bagyaraj, D.J.*** 1989a. Suitable form and level of phosphorus for mass production of VA Mycorrhizal fungus, *Glomus fasciculatum*. *Zentralbl. Mikrobiol.*, 144: 33-36.

Sreenivasa, M.N. and ***Bagyaraj, D.J.*** 1989b. Use of pesticides for mass production of vesicular-arbuscular mycorrhizal inoculum. *Plant and Soil*, 119: 127-132.

Sreenivasa, M.N., Krishnaraj, P.U., Gangadhara, G.A. and ***Manjunathaiah, H.M.*** 1993. Response of Chilli to the inoculation of an efficient vesicular-arbuscular mycorrhizal fungus. *Scientia Hortic.*, 53 : 45-52.

Sunder Rao, W.V.B. and ***Chayanulu, M.V.*** 1961. Rhizosphere bacterial flora of wheat and berseem. *Ind. J. Microbiology*, 1: 9-16.

Sutton, J.C. (1973). Development of vesicular-arbuscular mycorrhiza in crop plants. *Canadian J. Bot.* 51: 2487-2493.

Thingstrup, I. and ***Jakobsen, I.*** 1998. Effect of soil P levels on mycorrhiza functioning in Phosphorus balance and utilization in agriculture – towards sustainability. Proceeding of a seminar 17-19 March 1997. *Kungl.-skogs-och-Lantbruksakademiens-Tidskrift*, 137 (7) : 121-123.

Thalkappian, P., Sivasaravanan, A. and ***Sundaram, M.D.*** 2000. Effect of phosphorus levels on the mycorrhizal colonization, growth, yield and nutrient uptake of cassava (*Manihot esculenta* Crantz.) in alluvial soils of coastal Tamil Nadu. *Mycorrhiza News*, 11(4): 15-17.

Thompson, B.D., Robson, A.D. and ***Abbot, L.K.*** 1990. Mycorrhizas formed by *Gigaspora calospora* and *Glomus fasciculatum* on subterranean clover in relation to soluble carbohydrate concentrations in roots. *New Phytologist*, 114 : 217-225.

Thompson, J.P. 1994. What is the potential for management of mycorrhizae in agriculture? In: *Management of Mycorrhizas in Agriculture, Horticulture and Forestry.* (Eds. Abbott, L.K. and Malajuk, N.), Kluwer Academic Publishers, Dordrecht/Boston, London, pp. 191-200.

Titus, J.H. and ***Leps, J.*** 2000. The response of arbuscular mycorrhizae to fertilization, mowing and removal of dominant species in a diverse oligotrophic wet meadow. *Amer. J. Bot.* 87 (3): 392-401.

Trappe, J.M. 1981. Mycorrhizae and productivity of arid and semi-arid range lands. In: *Advances in food producing systems for arid and semi-arid lands.* Academic Press, London, 581-599.

Tuyev, N.A., Chernyaeva, I .I. and ***Sviridova, O.V.*** 1990. Role of nitrogen fertilizers in the transformation of humus by microorganisms. *Agrokemia-es-Talajtan*, 39 (3-4) : 551-554.

Valentine, A. J., Osborne, B. A., and ***Mitchell, D. T.*** 2002. Form of inorganic nitrogen influences mycorrhizal colonization and photosynthesis of cucumber. *Scientia Horticulturae.* 92(3-4): 229-239.

Venkatarao, B.V. and ***Sadashivaiah, T.*** 1968. Studies on nitrogen metabolization through phosphate fertilizing of a legume in the Bangalore red soil. *Mysore J. Agricultural Sci.* 2: 251-256.

Venkatesan, R. and ***Rangaswamy, G.*** 1964. Qualitative and quantitative studies of actinomycete population from non-rhizosphere soil: Rhizosphere of rice and stubble root surface. *Indian J. Microbiol.* 4 : 110-120.

Waksman, S.A. (1952). *Soil Microbiology.* John Wiley and Sons. Inc., New York.

Wallander, H. and ***Nylund, J.E.*** 1989. Nitrogen nutrition and mycorrhizal development. *Agriculture, Ecosystem and Environment*, 28: 547-552.

49 | Management of Charcoal Rot of Pigeonpea Through Protoplast Fusion Between Carbendazim Sensitive and Resistant Isolates of *Macrophomina phaseolina*

WADIKAR M. S., and KAMBLE S. S.*

Department of Botany, Deogiri College, Aurangabad.
*Mycology and Plant Pathology Research Laboratory, Department of Botany, Shivaji University, Kolhapur.

ABSTRACT

There was variation in the minimum inhibitory concentration (MIC) of carbendazim against *Macrophomina phaseolina* both on agar plates and on pigeonpea plants. Isolates MP-2 was most sensitive. The mutant isolate EMSMP-1 was more resistant and showed MIC 400 mg/ml. The protoplasts of both the isolates were obtained by using enzymatic method and their fusion was carried out in PEG. The fusants were obtained on agar plates. Fusants were inoculated on pigeonpea plants. The fusants were unable to infect pigeon pea plants.

KEY WORDS : MIC, Carbendazim, *Macrophomina phaseolina*, Protoplasts, Fusants, PEG.

Pigeonpea (*Cajanus cajan* (L.) Mill sp.) is one of the most extensively consumed pulses in our country. It is grown all over India. This crop is subjected to many diseases caused by various fungi. Chief among them is the charcoal rot caused by *Macrophomina phaseolina* (Tassi.) Goid. The disease is managed by using Bordeaux mixture, Carbendazim and other effective fungicides.

Protoplast fusion is an important alternative approach to the construction of fungal strains with novel gene combinations. Protoplast recombination can be used to blend several desirable properties together in a single biotype. In the present investigation carbendazim sensitive and resistant isolates causing charcoal rot of pigeonpea were fused together by protoplast fusion. The fusants were again inoculated on pigeonpea plants.

Material and Methdos

Samples of pigeonpea exhibiting charcoal rot symptoms were collected from different districts of Maharashtra State Viz. Aurangabad, Beed, Osmanabad, Latur, Ahmadnagar, Jalna, Nanded, Jalgaon and Akola. Eleven isolates of *Macrophomina phaseolina* were obtained from the samples on Czapek-Dox Agar medium. Carbendazim sensitivity was tested by food poisoning technique using Czapek-Dox-Agar medium and also on *Cajanus cajan* (L.) Millsp. For the (*in vivo*) studies *Cajanus cajan* plants were grown in earthern pots and were kept in glass house. The plants were then inoculated by pouring 100ml mycelial suspension of *Macrophomina phaseolina* in the earthern pots. The percent infection was recorded after various incubation periods.

Carbendazim resistant isolate EMSMP-1 (MIC 400 mg/ml) was obtained by treating sensitive MP-2 mycelial suspension with 0.1% EMS solution. This resistant isolate showed reduction in pathogenicity due to induction of carbendazim resistance.

Protoplast Fusion of *Macrophomina phaseolina* Sensitive and Resistant Isolates

A. *Isolation of Protoplast :*

100 mg young mycelia of *Macrophomina phaseolina* sensitive and resistant to carbendazim were collected aseptically from 24hr. old cultures by suction filteration and rinsed with sterile distilled water and transferred to 100 ml conical flasks containing 5mg Novozym 234, a lytic enzyme in 0.6 M mannitol. The flasks were kept under shaken conditions for 3 hrs. at 100 rpm. The release of protoplasts was examined under light microscope. After maximum release, they were collected by centrifugation at 5000 rpm for 5min.

B. *Fusion of Protoplasts :*

Viable and purified protoplasts of *Macrophomina phaseolina* sensitive and resistant to carbendazim isolates were used for fusion studies. Protoplasts were fused by following the method of Stasz *et. al.,* (1988). Polyethyleneglycol (PEG) was used as the fusogen. Different concentrations of PEG viz 30, 40 and 50% with 10 ml. Tris-HCL and 10ml $CaCl_2$ were used for the fusion of protoplasts. The fusion mixtures were serially diluted and plated on the non-selective media (PDA with osmotic stabilizer) and selective media (media amended with carbendazim) and checked for regeneration. The regenerated fusants were inoculated on pigeonpea plants as mentioned earlier.

Results and Discussion

It was observed that there was variation in minimum inhibitory concentration (MIC) of carbendazim against *Macrophomina phaseolina* both on agar plates and on pigeon pea plants. MIC of carbendazim on agar plates ranged from 80 to 140 mg/ml and ED50 from 27.81 to 67.26 mg/ml, while it was 65 to 127 mg/ml and ED50 23.18 to 56.74mg/ml on pigeonpea plants. (Tables 1 & 2). Isolate MP-2 was the most sensitive (MIC 80 mg/ml *in vitro*; 65 mg/ml *in vivo*).

Table 1 MIC of carbendazim against *Macrophomina phaseolina* causing charcoal rot on pigeonpea (*in vitro*)

Sr. No.	Isolate	Locality	Data characteristics to dose response curve				
			Regression				
			Const.	Coeff.	Corel Coeff.	ED50 (μg/ml)	MIC (μg/ml)
1	MP – 1	Osmanabad	-0.3022	0.0011	0.1896	56.34	100.00
2	MP – 2	Nilanga	-0.1554	0.0011	0.1672	27.81	80.00
3	MP – 3	Parbhani	-0.2776	0.0014	0.2387	67.26	125.00
4	MP – 4	Maliwada	-0.3191	0.0013	0.2313	62.64	131.00
5	MP – 5	Beed	-0.2489	0.0009	0.1890	58.48	140.00
6	MP – 6	Jalgaon	-0.1554	0.0011	0.1672	27.81	80.00
7	MP – 7	Nanded	-0.2421	0.0008	0.1862	57.88	135.00
8	MP – 8	Latur	-0.1855	0.0009	0.1959	44.53	100.00
9	MP – 9	Ahmednagar	-0.3146	0.0009	0.1959	44.03	130.00
10	MP – 10	Jalna	-0.2489	0.0009	0.1890	58.48	140.00
11	MP – 11	Akola	-0.2577	0.0011	0.1838	45.31	95.00

- ED50 = carbendazim concentration causing 50% reduction in radial growth.
- MIC = minimum inhibitory concentration.

Table 2 MIC of carbendazim against *Macrophomina phaseolina* causing charcoal rot on pigeonpea (*in vivo*)

Sr. No.	Isolate	Locality	Data characteristics to dose response curve				
			Regression				
			Const.	Coeff.	Corel Coeff.	ED50 (µg/ml)	MIC (µg/ml)
1	MP – 1	Osmanabad	-0.3041	0.0013	0.1916	56.74	80.00
2	MP – 2	Nilanga	-0.1325	0.0006	0.1179	31.09	65.00
3	MP – 3	Parbhani	-0.2147	0.0009	0.1532	37.76	79.00
4	MP – 4	Maliwada	-0.3567	0.0025	0.3069	46.39	116.00
5	MP – 5	Beed	-0.1812	0.0007	0.1518	39.88	120.00
6	MP – 6	Jalgaon	-0.2674	0.0014	0.2140	48.69	100.00
7	MP – 7	Nanded	-0.1295	0.0009	0.1393	23.18	66.00
8	MP – 8	Latur	-0.2568	0.0015	0.2243	42.25	93.00
9	MP – 9	Ahmednagar	-0.2553	0.0010	0.1850	50.11	105.00
10	MP – 10	Jalna	-0.2262	0.0008	0.1718	53.16	127.00
11	MP – 11	Akola	-0.2225	0.0016	0.18942	27.57	70.00

- ED50 = carbendazim concentration causing 50% reduction in radial growth.
- MIC = minimum inhibitory concentration.

Results in Table 3 show that there was no development of resistance through sp. However, UV, SA and EMS treatment to MP-2 mycelial suspensions yielded 10, 17 and 24 mutants. Mutants isolate EMSMP-1 was the resistant isolate showing MIC 400 mg/ml. This is because a change at the target site of carbendazim might be attributed in the induction of resistance in the sensitive isolates of *Macrophomina phaseolina*. This view was supported by many workers (Fuchs, *et. al.*, 1977; Brown, 1971; Dekker and Gielink, 1979). They have developed resistant mutants of *Phytophthora megasperma* f. sp. *medicaginis* to pimaricin and pyrazaphos which act as inhibitor of sterol biosynthesis.

Table 3 Development of carbendazim resistance in *Macrophomina phaseolina* through various treatments

Treatment	Colony Character	No. of Mutant	Concentration	Resistance factor	Growth in mm	Time (Minutes)
Ultraviolet	Black	10	240 ppm (0.1%)	3	17-26	30
Sodium Azide (SA)	Faint black	17	320 ppm (0.1%)	4	35-42	30
Ethyl-methyl sulphonate (EMS)	Whitish black	24	400 ppm (0.1%)	5	60-75	30

Since protoplast fusion is now regarded as an effective technique in genetic manipulation, the application of which in the present investigation justifies considerable optimism for future scope. The carbendazim sensitive MP2 and resistant EMSMP-1 were chosen as the two parents for protoplast fusion. The standardization of techniques for the protoplast isolation, regeneration and fusion were adopted according to some earlier Indian workers (Annamalai and Lalitha Kumari, 1992; Vijaypalini, 1995; Elaverasan, 1996; Mrinalini, 1997; Karpagam, 1994).

The fusants were inoculated on pigeonpea plants but all fusants were non-pathogenic i.e. fusants were unable to infect pigeonpea plants. There seems transfer of fungicide resistance ability and pathogenicity between strains by protoplast fusion. According to Lalithakumari (1994) of the two fungicide resistant mutants of *Venturia inaequalis*, one was resistant to penacozole designated as PRM and another resistant to carbendazim. The fusants were screened on a range of PDYEA amended with penaconazole, with carbendazim and with both penaconazole and carbendazim. The fusants grew on the medium amended with both the fungicides. Penaconazole + carbendazim resistant fusants exhibited either parental and biparental characters. Fusants with biparental characters were assumed to be recombinants. The fusants were slow growing and on further subcultuning. They segregated into fast growing and slow growing colonies.

Protoplast fusion carried out between sporulating carbendazim sensitive strain and non-sporulating carbendazim resistant mutant of *Venturia inaequalis* showed spore formation and tolerance to carbendazim (Lalithakumari, 1994).

Thus protoplast fusion provided ample information about the transfer of fungicide resistance in plant pathogens and it is an important fact for build up of resistance in the field even after the discontinuation of stress chemical.

Acknowledgement

The authors thank Prof. D. Lalithakumari, Ex-Director, CAS in Botany, University of Madras, Chennai for providing Laboratory facilities during the course of investigations.

References

Annamalai, P. and **Lalithakumari, D.** 1990. Decreased activity of Drechslera oryzae field isolate to endifenphos. *Indian Phytopathol.* 43 (4): 553-558.

Brown, J. P. 1971. Susceptibility of the cell walls of some yeasts to lysis by enzymes *Helix pomatia. Can. J. Microbiol.* 17: 205-208.

Dekker, J. and **Gielink, A. J.** 1979. Acquired resistance to pimaracin in *Cladorsporium cucumarinum* and *Fusarium oxysporum* f. sp. *narcissi* associated with decreased virulence. *Neth. J. Plant. Pathol.* 85 : 67-73.

Elavarsan, A. 1996. *Genetics of the stability and survival of fitness of Rhizoctonia solani (Kuhn) laboratory mutants resistant to benzimidazole and dicarboximide fungicides.* Ph. D. Thesis, University of Madras, Madras, India.

Fuchs, A., De Raig, S. P., Thyl, J. M. van and **De varies, F. W.** 1977. Resistance to trifornia; a non existent problem. *Neth J. Pl. Pathol.* 836: 189-205.

Karpagam, S. 1994. *Variations in protoplast regenerated isolates of Trichoderma harzianum.* M. Phil. Dissertation, University of Madras, India.

Lalithakumari, D. 1994. Resistance to sterol biosynthesis inhibitors (Baycor and sisthane) against *Venturia inaequalis*, ICAR Report, New Delhi.

Mrinalini, C. 1997. *Strain improvement for enhancement and integration of biocontrol potential and fungicide tolerence in Trichoderma* spp. Ph. D. Thesis. University of Madras, India (81-83).

Revathi, R. and **Lalithakumari, D.** 1992. *Venturia inacqualis.* A novel method for protoplast isolation and regeneration. Z. *Pfanzenkrankh. Pflanzensch.* 100: (211-219).

Stasz, T.E., Harman, G.E. and **Weeder, N.F.** 1988. Protoplasts preparation and fusion in two biocontrol strains of *Trichoderma harzianum. Mycologia* 80: 141-150.

Vijayapalani, P. 1995. *Biochemical, physiological and molecular aspects of penconazole and carbendazim resistance in mutants and protoplast fusants of Venturia inaequalis (Cooke) Wint,* Ph. D. Thesis, University of Madras, India. (209-213).

50 *Treatment of Distillery Effluents*

Jayashree Deshpande

Dept. of Environment Science, S.B.E.S. College of Science, Aurangabad 431 001 M.S.

ABSTRACT

There are nearly 166 distilleries in India with production of 500 million litres of alcohol. It is also estimated that about 12 to 14 litres of waste water is released for every litre of alcohol. The distillery waste water is also known as spent wash, stillage or vinase. However, most of the distilleries are not treating waste water because of non-availability of a sure process under Indian conditions. The following review is, therefore, discussed in relation to treatment of distillery effluent by various methods.

KEY WORDS : Sludge, *Aspergillus* spp., bioremediation

In India nearly 500 million litres of alcohol is produced per year from about 166 distilleries. For every litre of alcohol about 12 -14 litres of waste water is released. This distillery waste water is also known as spent wash, vinase or stillage. These are from the molasses based distillery. The population equivalent of distillery waste water in India is nearly 7 times more than the entire population. Most of the distilleries are not treating waste water because of non-availability of a sure process under Indian conditions (Handa and Seth, 1990).

Farooqui *et. al.,* (1998) studied the characteristics of granular sludge developed on cane sugar mill waste, including the effect of temperature on methanogenic activity of the sludge granules, yield coefficient of methanogenes, adaptation of the sludge to different wastes, the effect of unfed storage of granular sludge at room temperature (25-30°C) on its methanogenic activity and physical characterization of the sludge.

Sludge from UASB reactors operating at three different distilleries were analysed for total solids, volatile suspended solids, polymers, sedimentation characteristics, methanogenic activity and methanogenic population. The sludge from Associated Distillery, Hissar showed highest methanogenic activity even though its volatile suspended solid content was lowest ($49.9 \, g \, L^{-1}$) while lowest activity was found in sludge from Panipat Distillery having highest volatile suspended solid contents ($97.5 \, g \, L^{-1}$). Acetoclastic methanogenic population was also higher in the sludge from Hissar distillery. The sedimentation characteristics of the sludge are dependent on their proximate composition (Sharma and Singh, 2000).

Recent studies by Senthil Kumar *et. al.,* (2001) on population studies of sugar mill effluent showed that most of the physico-chemical parameters like colour, odour, total solids, COD, BOD, alkalinity and fluoride were found to exceed the ISI prescribed permissible values while pH, phosphate and sulphate were found within permissible limits. The concentrations of toxic metals such as Cd, Cu, Fe, Hg, Mn, Ni, Pb and Zn were also determined by inducting plasma emission spectrophotometry (ICP-AIS). The results showed that Fe, Hg, Mg, Mn and Pb content were higher than the permissible limits.

Distillery Wastewater Treatment

It is essential to release any industrial effluent only after treatment for controlling pollution. These are physico-chemical and biological treatment processes. Environmentalists have done some researches on the treatment of different industrial wastes and many industries are using these methods of releasing the treated effluents in the environment. Juwarkar *et. al.,* (1987) mixed the paper mill wastewater with chlorination and hypochlorite wastewater at various proportion and found that SAR of wastewater was reduced from 18.8 to 5.3. This treated wastewater was used to irrigate soil samples. All the soil samples equilibrated with PMW and crossed the ESP limit alkali soil. Similarly soils of loamy sand, sandy loam, sandy clay loam and loam texture with good drainage did not exhibit salinity problem on the water and had electrical conductivity of 2000 to 3000 $^\wedge$ mhos/cm. Saxena *et. al.,* (1986) treated the tannery effluents with chlorides and carbonates of sodium, ammonia, calcium, sodium dichromate, myrobalam extract and sulphuric acid. These treated effluents were used for irrigation purposes in dry areas. The effluents stimulated the growth of many crop plants but also had toxic chemicals which retarded the germination.

Kumar *et. al.,* (1997) noted that anaerobically digested molasses spent wash is a dark brown recalcitrant effluent which has high COD and high pollution potential. Bacterial enrichment cultures were set up to obtain isolates capable of distillery effluents decolonization and bioremediation. Two-gram negative aerobic bacterial cultures isolated from soil samples were collected on Associated Distillery Ltd., Hissar, India. Both the cultures grew well on medium containing distillery waste supplemented with glucose as readily available C source. These bacteria gave maximum decolourisation (36.5% and 32.5%) and COD decrease (41.0% and 39.0%) after 8 days. Again, Kumar *et. al.,* (1997) studied the soil samples from the same distillery containing cane molasses. These were incubated with diluted anaerobically digested spent wash and a facultative anaerobic pure bacterial culture. The bacterial culture L2 achieved 31% decolourisation and 56% COD decrease in 7 days. By this, they showed that such culture of bioremediation of digested spent wash offers a realistic approach for decreasing the pollutant potential of the recalcitrant waste before disposal.

Reboll *et. al.,* (2000) studied the use of sewage wastewater on Citrus trees irrigation during three different seasons. Many biochemical changes in Citrus leaves were observed when compared with the irrigation of normal ground water but the soil elements were within the optimal ranges. Rohella *et. al.,* (2001) studied the removal of colour, turbidity and COD from pulp and paper mill effluents using polyelectrolytes for pollution control and its safe disposal. The effluent was treated with 0.2 ml/L of Rishyly to 80 L. for the removal of colour, turbidity and COD. It was noted that this method is techno-economically viable. Decolourisation of textile effluent was tried by Sharma and Arora (2001), as many dyes and pigments are used in processing of textiles. Many of these dyes are potential hazards and some of them are even carcinogenic. They used the methods such as coagulation, ion exchange, membrane filtration and surface adsorption. These methods reduced BOD, COD, foul odour and taste.

Biological treatment of industrial effluents has been observed to be more useful in removal of colour, oil etc. There are some reports available on this aspect. Ramana *et al.* (2000) used *Acetobacter xylimim* for the production of cellulase membrane in the wastewater. The carbon and nitrogen sources were added in the wastewater. The bacterium was able to utilize a wide range of protein and nitrogen sources. Sodium dodecyl sulphate polyacrylamide gel electrophoresis (SDS-PAGE) analysis of pellicle proteins revealed electrophoretic bands of molecular masses in the range of 116-20 Kda. Furthermore, the strain can be useful for removal of various nitrogenous and carbon substrates present in wastewater. Srinivasan and Murthy (2000) used *Trematis versicolor* for removal of the wastewater colour. This organism is a white rot fungus and has the ability to degrade lignin. The fungus can produce non-specific enzyme system which oxidizes the recalcitrant compounds present in the effluent. The fungus could decolourise the effluent from pulp plant and dyeing industry. Kouser and Singara Charya (2002) applied four fungi (*Aspergillus niger*, *Curvularia lunata*, *Fusarium oxysporum* and *Mucor mucedo*) isolated rom textile and dye contaminated soils for colour removal of textile and dye wastewater. The physical, chemical and biological characteristics were analysed and related to the process of decolourization. *Aspergillus niger* and *Mucor mucedo* were resistant in the soils and also efficient in the decolourization and enzyme production. Other two fungi were not so efficient in colour removal and also enzyme secretion. Chauhan and Thakur (2002) treated the pulp and paper mill effluent by *Pseudomonas fluorescens*. This bacterium degraded 4 chlorophenol through an oxidative route as indicated by oathring cleavage. The strain produced significant reduction in colour (75%), phenol (66%), chemical oxygen demand (79%) and lignin (45%). However, there was increase in chloride content (92%) after 15 days in bioreactor. Similar observations were made by Jeyaramraja *et. al.,* (2001) when they used *Aspergillus fumigatus* to declorize the paper mill effluent in a bioreactor.

Researches on mechanical and chemical treatment of distillery effluents can be revealed as below. Gadre and Godbole (1986) used two bench scale upflow anaerobic filters for anaerobic digestion of waste. The results showed that there was substantial reduction in pollution load with simultaneous production of biogas. Essentially the in-built buffering capacity of anaerobic filter was more useful. Routh and Dhaneshwar (1986) also showed that anaerobic treatment of spent wash in immobilized cell reactor was more efficient than conventional methods. The advantage was first acidification and second methanogenesis phase. Ramteke *et. al.,* (1989) used pyrochar from paper mill sludge for removal of colour of distillery effluent. They observed that granular and powdered activated pyrochar can remove colour and COD. Berchmans and Vijayvalli (1989) used electro-chemical oxidation technique for removal of BOD and decolourisation. For this, performance of hybrid anaerobic reactor using distillery effluent was assessed by Bardiya et. al., (1995). The process involved was diphasic in nature with hybrid reactor design for methane digester. Optimum biogas production of 10 kg/m^3/day in hybrid reactor. Patil and Kapadnis (1995) studied the decolourisation of spent wash pigment melanodin by chemical and biological methods. Spent wash from an anaerobic digester was treated with hydrogen peroxide, calcium oxide and soil bacteria, at 144 hours of incubation with treatment at varied concentration of hydrogen peroxide, the maximum decolonisation and COD reduction was 98.67 and 88.40% respectively. Vaidyanathan et. al.,(1995) made an attempt to determine the bio-kinetic coefficients for a two stage anaerobic digester in a bench scale model using distillery spent wash and application of these coefficients for the design of digester. The distillery spent wash treatment by anaerobic filter has also been studied by Ilyas et. al., (1998). They concluded that conservation of fresh water and preservation of environment, imparting greater efficiency and stability to the reactor. Handa and Seth (1990) showed that among the various treatment alternatives, anaerobic digestion appeared to be most attractive and produces a needful by-product biogas.

Daryapurkar and Chakrapani (1999) showed that yeast sludge is most important waste stream generated from alcohol distillery for which no useful applications are still practiced in India. Bhagatkar (1999) studied the aerobic bio-oxidation of post anaerobic distillery effluents. The study evaluated the biological kinetic parameters like K, KS, Y, Kd, Umax for an activated sludge system using post distillery effluents. A bench scale continuous reactor of 6-9 litres was also developed. Akunna and Clark (2000) reported the performance of granular bed anaerobic baffled reactor used in the treatment of whisky distillery wastewater. It was found to be very effective. It also combined the advantages of baffled reactor system and up-flow of anaerobic sludge blanket systems. Acidogens were mostly non-granular while methanogenes were granular. The aerobic composting of spent wash process is completed with 20-22 days while press mud and spent wash absorption ratio is 1 : 1.7 for fresh press mud.

During this biotechnological era, biological treatment of wastewaters is gaining much importance. Seenappa and Rao (1995) used earthworm *Eudrilus eugeniae* for conversion of distillery waste into organic manure. Similarly Joshi and Kapadnis (1997) gave fungal and microbial biomass treatment to distillery waste. This helped in preparation of compost fertilizer. The fungal biomass of *Aspergillus* sp. and *Phialomyces* was developed on spent wash to remove the colour. The decolourised spent wash was then enriched with nitrogen content by growing mixed culture of *Azotobacter* sp. This compost fertilizer is used as other agro based waste products like press mud. A good compost fertilizer was obtained within 45 days. Interestingly Singh and Srivastava (1999) removed   the basic dyes from aqueous solutions by chemically treated *Psidium gujava* leaves. The leaves were treated with formaldehyde and sulphuric acid. 2 g/L leaves were sufficient for removal of methylene blue. Gonzalez et. al., (2000) used white rot fungus, *Trametes* sp. to detoxify the distillery wastewater. Use of thermotolerant yeasts, *Saccharomyces cerevisae* and *Kluyveromyces marxianus* has also been made. Molasses pigment solution treated with *Chlorella saccharophila* and *C. vulgaris* and removed the colour to 30-33%.

References

Akunna, J.C. and **Clark, M.** 2000. Performance of a granular-bed anaerobic baffled reactor (GRABBR) treating whisky distillery wastewater. *Bioresource Technol.* 74(3): 257-261.

Baradiya, M.C., Hashiya, R. and **Chandna, S.** 1995. Performance of hybrid reactor for anaerobic digestion of distillery effluent. *J. Assoc. Environ. Management* 22: 237-239.

Berchmans, L. and **Vijayvalli, R.** 1989. Electrochemical oxidation as a tool for pollution control: Effect of anodic oxidation on the treatment of distillery effluent. *Indian J. Environ., Health.* 30(4): 309.

Bhagatkar, M. 1999. Aerobic bioxidarion of post anaerobic distillery effluents. *J. Indian Assoc. Environ. Manage.* 26(3): 177-182.

Chauhan, N. and **Thakur, I.S.** 2002. Treatment of pulp and paper mill effluent by *Pseudomonas fluorescens* in fixed film bioreactor. *Pollution Res.* 21(4): 429-434.

Daryapurkar, R. and **Chakrapani, D**. 1999. Challenging problems in distillery waste treatment - Indian scenario. *J. Indian Assoc. Environ. Manage.* 26(3): 141-149.

Farooqui, I.H., Khursheed, A. and **Siddiqui, R.H.** 1998. Characterization of granular sludge developed on cane sugar mill waste. *Indian J. Environ. Health.* 40(1): 58-66.

Gadre, R.V. and **Godbole, S.H**. 1986. Treatment of distillery waste by upflow anaerobic filter. *Indian J. Environ. Health.* 28(1): 54-59.

Gonzalez, T., Terron, M.C., Yague, S., Zapico, E., Galletti, G.C. and **Gozalez, A.E.** 2000. Pyrolisis / gas chromatography / mass spectrometry monitoring of fungal - biotreated distillery wastewater using *Trametes* sp. l-62n (CECT 20197). *Rapid Communi. in Mass Spectrometry*, 14(15): 1417-1424.

Handa, B.K. and **Seth, R.** 1990. Waste management in distillery industry. *J. Indian Assoc. Environ. Managet.* 18: 44-54.

Ilyas, M., Isa, M.H., Farooqui, I.H., Khursheed, A. and **Siddiqui, R.H.** 1998. Treatment of distillery spentwash by anaerobic filter - A water conservation approach. *Indian J. of Environ. Health*, 40(2): 153-159.

Jeyaramraja, P.R., Anthony, T., Rajendran, A. and **Rajkumar, K.** 2001. Decolourization of paper mill effluent by *Aspergillus fumigatus* in bioreactor. *Pollution Res.* 20(3): 309-312.

Joshi, R.D. and **Kapadnis, B.P**. 1997. Pretreatment of distillery wastewater using fungal and microbial biomass to use it in preparation of compost fertilizers. Abstract, *National Seminar on Biomass Productivity and utilizations* (6-8 Feb. 1997).

Kousar, N. and **Singara Charya, M.A.** 2002. Decolourisation of textile and dye amended soils by fungi. *Indian J. Environ. Health.* 44(1): 65-70.

Kumar, V., Wati, L., Nigam, P., Banat, I.M., McMullan, G., Singh, D. and ***Marchant, R.*** 1997. Microbial decolourisation and bioremediation of anaerobically digested molasses spent wash effluent by aerobic bacterial cultures. Microbios. 89(359): 81-90.

Patil, N.B. and ***Kapadnis, B.P.*** 1995. Decolourisation of melanodin pigment from distillery spent wash. *Indian J. Environ. Health.* 37(2): 84-87.

Ramana, K.V., Tomar, A. and ***Singh L.*** 2000. Effect of various carbon and nitrogen sources on cellulose synthesis by *Acetobacter xylinum. World J. Microbiol. Biotechnol.* 16(3): 245-248.

Ramteke, D.S., Wate, S.R. and ***Moglie, C.A.*** 1989. Comparative adsorption studies of distillery waste on activated carbon. *Indian J. Environ. Health.* 31(1): 17-24.

Reboll, V., Cerezo, M., Roig, A., Flors, V., Lapena, L. and ***Garcoa, P.*** 2000. Influence of waste water vs ground water on young Citrus trees. *J. Sci. Food Agri.* 80(10): 1441-1446.

Rohella, R.S., Choudhary, S., Manthan, M. and ***Murty, J.S.*** 2001. Removal of colour and turbidity in pulp and paper mill effluents using polyelectrolytes. *Indian J. of Environ. Health.* 43(4): 159-163.

Routh, T. and ***Dhaneshwar, R.S.*** 1986. Anaerobic treatment of distillery spent wash. *Indian J. Environ. Health.* 29(2): 105-117.

Saxena, R.M., Kewal, P.F., Yadav, R.S. and ***Bhatnagar, A.K.*** 1986. Impact of tannery effluents on some pulse crops. *Indian J. Environ. Health.* 28(4): 345-348.

Seenappa, C. and ***Jagannath Rao, C.B.*** 1995. Conversion of distillery waste into organic manure using earthworm *Eudhlus eugeniae. J. Indian Assoc. Environ. Manage.* 22: 244-246.

Senthil Kumar, R.D., Narayanswamy, R. and ***Ramakrishnan, K.*** 2001. Pollution studies on sugar mill effluent - physico-chemical characteristics and toxic metals. *Pollution Res.* 20(1): 93-97.

Sharma, J.K. and ***Arora, M.K.*** 2001. Decolourisation of textile effluent. *Pollution Res.* 20(3): 453-457.

Singh, D.K. and ***Srivastava, B.*** 1999. Removal of basic dyes from aqueous solutions by chemically treated *Psidium gujava* leaves. *Indian J. Environ. Health.* 41(4): 333-345.

Srinivasan, S.V. and ***Murthy, D.V.S.*** 2000. Removal of colour from wastewater using *Trametes versicolor. J. Indian Assoc. Environ. Manage.* 27(3): 260-264.

Vaidyanathan, R., Meenambal, T. and ***Gokuldas, K.*** 1995. Biokinetic coefficients for the design of two stage anaerobic digester to treat distillery waste. *Indian Journal of Environmental Health.* 37(4): 237-242.

51 | *Biocontrol Potential of AM Fungi for the Management of Plant Parasitic Nematodes*

Sampat Nehra and P.C.Trivedi

Deptt. of Botany, University of Rajasthan, Jaipur - 302 004

ABSTRACT

Arbuscular mycorrhizal fungi are potential biofertilizers and bioprotectors, enhancing plant growth, yield, soil fertility and control of diseases. Plant parasitic nematodes cause disease in many economically important plants and huge lossess are incurred. Nematodes and AMF are intimately associated in feeder roots and their interaction may have antagonistic, neutral or synergistic effect on plants. Effect of various antagonistic species Viz. *Meloidogyne* spp., *Rotylenchulus reniformis*, *Tylenchulus semipenetrans*, *Heterodera* spp., *Pratylenches* spp. were studied in detail and it was found that population of nemotodes is reduced due to AMF and plant growth is enhanced.

KEY WORDS : Bio control, AM Fungi, nematodes, interaction, control

Arbuscular mycorrhizal fungi (AMF) are commonly known as AMF, which has a more appropriate meaning. It refers to the presence of intracellular mycelium, vesicles and arbuscles-that are formed inside the root during different developmental stages.

AMF occur in roots of most Angiosperms, Pteridophytes, Bryophytes and Gymnosperms. These mycorrhizas are the most commonly occurring groups. AMF is plays a very important role in soil. It provides organic link between the root and the bulk of the soil. The fungus provides the plant with minerals, especially phosphorus, the plant supplies the fungus with photosynthetic sugars (Norris *et. al.*, 1991;1992). The bi-directional transport leads to enhanced plant growth. These fungi are potential biofertilizers and bioprotectors to enhance plant growth, yield, act as soil conditioners (soil fertility) and play a vital role in sustainability of the rapidly degrading environment (Norris *et al.*, 1994; Nehra, 2004 b). Mycorrhizal symbionts produce extraradical hyphae that may extend several centimeters into the soil and exude organic materials that cause soil particles to adhere and help to create small aeration, water percolation as well as stability (Tisdall, 1991).

Plant-parasitic nematodes : The word nematode (Greek nema=thread, eidos=likeness, resembling) is derived from the Greek word nema. The terms nema and nematology were introduced by Cobb (1932). The nematodes may be defined as triploblastic, bilaterally symmetrical, unsegmented, pseudocoelomate organisms. The plant parasitic nematodes are popularly called eelworms, phytonematodes, phytohelminths or simply plant nematodes. All plants are susceptible to attack by one or more species of plant parasitic nematodes. The major damage done by nematodes is caused by their feeding activity on plant root system. Plant parasitic nematodes are obligate parasites. The amount of damage caused by plant parasitic nematodes is related to many variables including the nematode species, the size of the nematode population, the susceptibility of the host plant, various environmental factors and the presence of other organisms contributing to the total damage inflicted upon the crop.

There are numerous estimates of the economic importance of nematodes in crop production on a worldwide and individual country basis, but precise values can not be determined. It was because of nematodes 'small size and hidden way of life' and lack of definite information on their occurrence and pathogenecity. The assessment of the society of the nematologists committee on crop losses indicated annual losses in the United states due to plant parasitic nematodes to the tune of $ 1,038, 374,300 in the field crops, $225,145,900 in fruit and nut crops, $266,989,100 in vegetable crops and $598,176,34 in ornamental crops (Anon,1971). In a world wide survey conducted by Sasser (1989) the ten most important genera of plant parasitic nematodes were *Meloidogyne, Pratylenchus, Heterodera, Ditylenchus, Globodera, Tylenchulus, Xiphinema, Radopholus, Rotylenchulus* and *Helicotylencus*. This order of importance of various genera was found to be fairly representative for most regions of the world. Estimated overall annual yield loss on the world's major crops due to damage by plant parasitic nematodes is 12.3%.

The 20 crops that represent a miscellaneous group important for food or export value were reported to have an estimated annual yield loss of 14%. Losses for the 40 crops in developed countries average 8.8% compared with 14.6% for developing countries. Global crop losses due to nematodes on 21 crops, 15 of which are life sustaining, were estimated at $ 77 billion annually based on 1984 figures and prices. The United states portion alone is $ 5.8 billion. These figures are staggering and the real figure, when all crops throughout the world are considered, probably exceeds $ 100 billion annually.

Several workers have attempted to assess crop losses caused by plant parasitic nematodes in India. Van Berkum and Seshadri (1970) were the first to have calculated theses losses in India in terms of money. They estimated annual losses due to 'ear cockle' caused by *Anguina tritici* on wheat, amounting to $ 10 billion, due to *Pratylenchus coffeae* on coffee to $ 3 billion and due to 'Molya 'disease caused by *Heterodera avenae* in Rajasthan province alone to $ 8 million. Paruthi and Bhatti (1981) reported the loss in wheat yield due to *Anguina tritici* ranged from 1 to 9 per cent. The yield of okra, tomato and brinjal suffered 90.9, 46.2 and 2.3 per cent losses, respectively due to *Meloidogyne incognita* infestation @ 3-4 larvae/g soil under field conditions (Bhatti and Jain,1977). Many important nematode pests of importance are potato cyst nematode *Globodera rostochinensis* in Nilgiris, the citrus nematode *Tylenchulus semipenetrans* in citrus crops, the burrowing nematode, *Radopholus similis* in banana, the reniform nematode, *Rotylenchulus reniformis* in cotton, maize,finger millet, cow pea and black gram respectively.

Losses inflicted by nematode pests continue to increase and are becoming a limiting factor in maximizing forage yields (Jain and Hasan,1995). The potential role of AMF in the biological control of plant parasitic nematodes is very important because of perceived urgency to develop and adopt environmentally safe and economic and efficient method for managing nematodes. Plant parasitic nematodes and AMF occupy the same ecological niche of most agro-ecosystem but they have opposite effect on the root system and plant growth. AMF acts as obligate symbiont enhancing plant growth, the nematode as obligate parasite take out vital nutrients from the roots reducing plant growth. Though largely based on green house experiments recent research has indicated that AMF has the potential as biocontrol agent when both groups of organisms occur simultaneously in the root or rhizosphere of the same plant. The plants heavily colonized with mycorrhizal fungi are able to grow well inspite of the presence of damaging levels of plant parasitic nematodes (Hasan and Jain,1991; Jain and Hasan, 1994a; Trivedi,1995; Nehra *et. al.*, 2003 ; Nehra 2002; 2003; 2004 a; Nehra and Trivedi, 1999, 2004a, b).

The nature of interaction between these two groups of organisms, however, differs in terms of their effect on plant growth or yield in terms of each organisms effect on the other. For example many workers have demonstrated a reduction in nematode population or diseases incidence, but cases where nematode population remain unaffected or even increased under the influence of AMF are not uncommon.

Table 1 Effect of VAM fungi on plant growth and Nematodes in forage crops.

Host	Nematode species	Effect in mycorrhizal plants	References
Alfalfa	*Meloidogyne hapla*	Improved plant growth, reduced nematode numbers and adult	Grandison and Cooper, 1986
White clover	*Meloidogyne hapla*	Improved plant growth, fewer nematode numbers	Cooper and Grandison,1986
Soybean	*Meloidogyne incognita*	Increased yield, fewer galls	Schenck *et. al.*, 1975
	Heterodera glycine	Increased yield, decreased reproduction	Francl and Dropkin,1985
Cowpea	*Meloidogyne incognita*	Improved plant growth, fewer nematode and galls	Jain and Hasan,1988 Jain and Sethi,1988
	Heterodera cajani	Improved plant growth, reduced larval penetration and reproduction	Jain and Sethi,1987,1988
Berseem	*Tylenchorhynchus vulgaris*	Improved plant growth, nematode population unaffected	Hasan and Jain, 1991 b
	Meloidogyne arenaria	Improved plant growth, nematode population increased	Kassab and Taha,1990
Stylo	*Meloidogyne incognita*	Improved plant growth, fewer galls	Jain and Hassan,1988
Oat	*Meloidogyne incognita* *Meloidogyne hapla*	Improved plant growth, fewer nematodes	Sikora and Schoenbeck, 1975
Sorghum	*Pratylenchus zeae* *Tylenchorhynchus vulgaris*	Improved plant growth, nematode population low	Jain and Hassan,1986

Since plant-parasitic nematodes and AMF are intimately associated in feeder roots, it is logical to consider an interaction between these two groups of organisms in terms of their combined effect on plant growth. Interactions can be quantified by measuring the effect on fungal colonization of roots or sporulation and nematode attraction to roots, penetration or subsequent development and reproduction. Plant responses to concomitant infection can be assessed by determining the influence of interaction on plant growth or yield, by either stimulation from mycorrhizal development or suppression from nematode infection; the latter effect is emphasized in review article. The nature of interaction varied to antagonistic, neutral or synergistic

Table 2 Possible effects of interactions between plant parasitic nematodes and VA Mycorrhizae (After Hussey and Roncadori, 1982).

Types of interaction	Components	Effect on components
Neutral	Fungus	Root infection or sporulation not altered
	Host	Mycorrhizal stimulation of vegetative growth or yield not altered, nematode suppression of vegetative growth or yield not offset
	Nematode	Attraction to roots, penetration, development and reproduction not altered
Positive	Fungus	Root infection or sporulation increased
	Host	Nematode suppresssion of vegetative growth or yield offset
	Nematode	Attraction to roots, penetration, development and reproduction suppressed
Negative	Fungus	Root infection or sporulation suppressed
	Host	Vegetative growth or yield response to mycorrhizae suppressed
	Nematode	Attraction to roots, penetration, development and reproduction increased

Where no changes are evident, the interaction is categorised as neutral. An interaction is rated positive if AMF offset nematode damage to plants; since the response of a plant to the endophyte is rarely enhanced in the presence of plant parasitic nematodes. The variation in the nematode population or disease incidence in the mycorrhizal plants was found to be due to the varying host-nematode-symbiont combinations. Besides this other factors like (1) the sequence in which the plants are being infected with the nematodes relative to the time of AMF infection; (2) the infection levels of both nematode and fungus and; (3) soil fertility may also influence the interaction. So each nematode-mycorrhizal fungus-plant combination may be unique.

The antagonistic effect of AMF on plant parasitic nematodes could have either a physiological or a physical basis. AM fungi may-

1. Improve plant vigor and growth to offset yield losses normally caused by nematodes.
2. Physiologically alter or reduce root exudates responsible for chemotactic attraction of nematodes or
3. Directly retard nematode development or reproduction within the root tissue.

An interaction could have a physical basis if the endophytes and endoparasitic nematodes in the root compete for nematode activities. A negative interaction occurs when plant growth or yield of dually infected plants is less than that of unchallenged mycorrhizal plants or if root colonization or sporulation by the mycorrhizal fungus is suppressed. Although the presence of endophyte in the cortex is not expected to enhance nematode damage, the symbionts could increase nematode reproduction.

Nematode-Mycorrhiza Interaction

As discussed earlier AM fungi-nematode-plant combination may be unique. To generalise nematode resistance mechanism by the mycorrhizal plants will be extremely difficult. The primary effect of AM fungi has generally been to increase host resistance even in the presence of damaging levels of nematodes. During recent past improved 'P' nutrition by the mycorrhizal plants resulting into better plant vigor and growth largely has been considered as one of the major mechanism of the host tolerance against nematodes (Hussey and Roncadori,1982; Schoenbeck,1979). However, some studies have also shown that AM fungi may effect host response and nematode activities differently from 'P' nutrition (Cooper and Grandison,1986; Grandison and Cooper,1986; Smith,1987; Smith *et. al.*,1986b). Sikora (1979) observed that the prior presence of AM fungi *Glomus mosseae* has resulted into an increased plant resistance against *Meloidogyne* spp. and suggested that fungal symbiont may have exerted its influence of the nematode by:

1. Altering root attractiveness
2. Reducing larval penetration
3. Impeding larval development
4. Retarding giant cell formation.

He has further emphasized that these effects most probably result from complex physiological changes associated with mycorrhizal infection rather than caused by direct competition between the two organisms. Later, hypothesis proposed by Hussey and Roncadori (1982) to explain AM fungal effect and nematode parasitism and activities generally have been considered to have either physical or physiological basis as mentioned earlier. Further, Smith (1987) and Brundret (1991) while supporting the Hussey and Roncadori (1982) hypothesis and considering them as base discussed the following mechanisms in detail regarding increased tolerance or resistance to nematodes by AM fungi which may be due to-

1. Increased root growth and functions
2. Nutritional effect other than 'P'
3. Alteration in root exudation
4. Competition for host photosynthates
5. Production of nematistatic compounds
6. Parasitism of eggs.

They also suggested that probably AM fungi do not directly interact with plant parasitic nematodes inspite of their proximity in root tissues but more likely, they alter the host either physically or physiologically.

1. Resistance

Horsfall and Diamond (1957) reported that disease incidence is greater whenever the level of sugars in host plants is low which agrees with our findings since largest quantity was recorded in mycorrhizal plants. It was also observed by Batemann and Miller (1966) that resistant plants had more sugar content. It therefore appears that high sugar content in mycorrhizal plants some how affects the resistance of plants to nematodes.

Mycorrhizal plants also had higher amino acid content compared to control confirming the observations of Nemee and Meredith (1981). Further, the mycorrhizal plants had higher concentrations of phenylalanine and serine which are known to reduce the growth and reproduction of root-knot nematodes (Krishnaprasad,1971; Parvatha Reddy,1974). Thus the presence of increased quantities of sugars, amino acids like phenylalanine, serine and phosphorus may each or collectively play a role in suppressing the development of *M. incognita* in plants and thus indirectly affect the host nematode relationship.

2. AMF and nutrient uptake

The effect of AMF and 'P' fertility on host response to nematode parasitism and on nematode infection, development and reproduction are summarized in Table 3.

AMF have been shown to enhance the uptake of Ca, Cu, Mn, S and Zn in addition to 'P' and increase water uptake (Gerdemann, 1968; Harley and Smith,1983; Hayman,1982). Thus AM fungi may increase host tolerance by increasing uptake of key nutrients that would be deficient in non-mycorrhizal nematode infected plant.

3. Sequence and inoculation level of AMF

Prior establishment of mycorrhiza mitigated the deleterious effects of nematode on plant growth characters to a considerable extent (Suresh and Bagyaraj,1984; Hussey and Roncadori,1982; Sharma and Trivedi,1994; Nehra and Trivedi,1999; 2004 a ;b; Nehra, 2002; 2003).

Table 3 Interaction of AM fungi (Myc) and phosphorus with plant-parasitic nematodes and the effect on host response and nematode infection, development and reproduction. (After Smith,1987)

Host	Fungus	Nematode P effect	Myc	Interaction Component				
				Host	Infection	Development	Reproduction	Reference
Alfalfa	Four species mixture*	*Meloidogyne hapla*	Myc	+	+	–	nd	Cooper &
			P	+	+	+.0	nd	Grandison, 1986
Bean	*Glomus etunicatum*	*Meloidogyne incognita*	Myc	+	+nd	nd	nd	Jain & Sethi,
			P	+	nd	nd	+.–	1988
Clover	Four species mixture*	*Meloidogyne hapla*	Myc	+	+.–	–	nd	Baltruschet *et. al.,*
			P	+	+	+	nd	1973
Cotton	*Glomus fasciculatum*	*Meloidogyne incognita*	*Myc*	+.0	–	nd	–	Hayman, 1982
			P	nd	nd	nd	nd	
	Glomus intraradices	*Meloidogyne incognita*	Myc	+	–	–	–.0x	Jain, 1986
			P	+	+	+	–.0x	
	Gigaspora margarita	*Meloidogyne incognita*	Myc	+	+	nd	0	Ingham, 1989
			P	–	+	nd	0	
Onion	*Glomus fasciculatum*	*Meloidogyne hapla*	Myc	+	nd	nd	+	Grandison &
			P	nd	nd	nd	nd	Cooper, 1986
	Glomus fasciculatum	*meloidogyne hapla*	Myc	0	0	0	Ox	Harley & Smith.
			P	nd	0	0	Ox	1983
Potato	*Glomus etunicatum*	*Globodera rostochinensis*	Myc	+	+	nd	–	Jain & Sethi.
			P	nd	nd	nd	nd	1987
Soybean	*Glomus Fasciculatum*	*Heterodera glycines*	Myc	+	nd	nd	–.0x	Carling *et. al.,*
			P	nd	nd	nd	nd	1996
	Glomus ambisporum	*Heterodera glycines*	Myc	+	0	0	+.0	Atilanto *et. al.,*
			P	+	0	nd	0	1981

Table 3 Contd...

Host	Fungus	Nematode P effect	Myc	Interaction Component				
				Host	Infection	Development	Reproduction	Reference
	Gigaspora margarita	*Heterodera glycines*	Myc P	+ +	0 0	0 nd	+.0 0	Atilanto *et. al.,* 1981
Tamarillo	Four species mixture*	*Meloidogyne incognita*	Myc P	+ –.+	– +	– +	nd nd	Bagyaraj *et. al.,* 1979
Tomato	*Glomus margarita*	*Meloidgyne incognita*	Myx P	0 +	0 +	0 0	0 +.0x	Jain & Hasan. 1986
	Gigaspora mosseae	*Meloidogyne incognita*	Myc P	0 +	0 +	0 0	0 +.0x	Jain & Hasan. 1986
	Gigaspora fasciculatum	*Rotylenchulus reniformis*	Myc P	0 nd	0 nd	0 nd	0 nd	Hasan & Jain. 1981
	Four species mixture*	*Meloidogyne hapla*	Myc P	+ +	+.– +	– +	nd nd	Baltruschat *et. al.,* 1973.

Here += increase, - =decrease, nd= not determined

⁺component effect in interaction: Host=change in plant growth or yield when compared with unchal-lenged plant at lowest P level.

Infection= Change in number of nematodes infecting a root system when compared with a nonmycorrhizal root system at the lowest P level.

Reproduction= Change in nematodes or eggs per plant or gram root when compared with a nonmycorrhizal root system at the lowest P level.

* *Gigaspora margarita, Glomus mosseae, Glomus fasciculatum, Glomus etunicatum.*

X eggs per ovipositing females were recorded in addition to eggs per plant and (or) gram root.

Number of juveniles per 100 cm^3 soil.

4. AMF-Nematode-Host Combinations

Selection of host is of great importance in testing for nematode-mycorrhiza interactions. It was noted that tree and vine fruits, onion and legumes have been preferred hosts for model systems because of their greater dependence on AM fungi compared with hosts in Graminae and Solanaceae (Harley and Smith,1983; Hayman,1982). Similarly for selection of a nematode species, its feeding habit and importance as a plant pathogen should be kept in mind. Sedentary endoparasites suit for these type of interaction study because of their greater economic importance, ease of propagation, immobility of the female and proximity of the permanent feeding site to fungal structures in roots. AMF have been shown to differ in their ability to stimulate host growth, colonise roots and tolerate higher soil 'P' levels (Gerdemann, 1968; Harley and Smith,1983; Hayman,1982).Therefore selection of a AM Fungus is also important in a particular host-nematode symbiont interaction study.

Effect on Nematode Population

The potential of AM fungi to alleviate nematode induced plant stress has been widely investigated because of their ability to increase root growth and nutrient absorption (Smith, 1987). The inhibitory influence of AM fungi on phytonematodes is presented in Table 4. Many workers have demonstrated a reduction in nematode population or disease incidence (Bagyaraj *et. al.*, 1979; Cooper and Grandison, 1986; Grandison and Cooper, 1986; Jain and Hasan, 1988; Kassab and Taha, 1990; Sikora and Schonbeck, 1975; Sitaramaiah and Sikora, 1982; Sharma,1992; Nehra and Trivedi, 2004 a, b) but cases where nematode population remain unaffected (O' Bannon *et. al.*, 1974; Cason *et. al.*, 1983), or even increased under the influence of mycorrhiza (Atilano *et. al.*, 1981; Kassab and Taha,1990) are not uncommon.

(a) Effect of AMF on Meloidogyne spp.

Baltruschat *et. al.*,(1973) were the first to show that tobacco plants pre-inoculated with AMF, *Glomus mosseae* were less susceptible to root-knot nematode infection, they reported 75% reduction in the number of *M.incognita* larvae that developed into adults on tobacco plants pre-inoculated with *G.mosseae* compared with uninoculated control. Interaction between plant parasitic nematodes and plant pathogenic fungi occur when certain threshold levels of one or both components are present in the rhizosphere of the host plant. In a survey of soybeans in Florida, Schenck and Kinloch (1974) found that spore counts of endomycorrhizal fungi were consistently low when associated with high population densities of root-knot nematodes. Saleh and Sikora, (1984) observed that number of eggs of *M.incognita* were reduced at 55 % or greater colonization by *G.fasciculatum* but 50 % colonization significantly inhibited nematode development (Smith *et. al.*, 1986a).

Schenck *et. al.*, (1975) using different levels of nematode population found that at lower levels *M.incognita* stimulated spore production by the mycorrhizal fungus, *Endogone herogama* in soybean. At very high nematode levels the spore production decreased. Nematodes are checked if mycorrhiza are present before they are able to infect the peach plants (Hussey and Roncadori,1982).

Sikora and Schonbeck (1975) found that significantly fewer (75%) *M. incognita* and *M. hapla* juveniles developed into adults in tobacco, oats and tomatoes pre-inoculated with *G. mosseae*. Schenck *et al.,* (1975) conducted green house studies involving soybean inoculations with *M. incognita* and three AMF. They found that the symbionts made plants, tolerant to *M. incognita*. Roncadori and Hussey (1977) reported that *Gigaspora margarita* offsets the damage caused by *M.incognita* in cotton cv. stonville 213. Sikora, 1979) demonstrated that *M. incognita* larval development was impeded in tomato plants inoculated with *Glomus mosseae* when measured 8 to 16 days after nematode introduction. The slowdown in development may be responsible for the reduced size of galls often seen in mycorrhizal plants. He also reported that distribution of *M. incognita* galls on mycorrhizal tomato plants was negatively correlated with *G. mosseae* levels in the root system. The majority of galls were formed at the periphery of the root system where fungal levels were found to be significantly lower than in internal regions of the root systems. A significant decrease was detected in the number of larvae that penetrated mycorrhizal tomato roots, compared with non-mycorrhizal roots. The nematode population decrease found may be partially due to changes in root attractiveness brought about by the presence of fungal symbiont in the root or by the inability of a significant number of larvae to penetrate roots populated by mycorrhizal fungus (Sikora and Schonbeck,1975). Besides extensive histological studies showed that although, *G. mosseae* did not penetrate giant cells produced by *M. incognita* on tomato, the presence of AMF in the root retards giant cell formation. The giant cells produced on mycorrhizal tomato plants were smaller, fewer in number and contained less nuclei than giant cells of the same age produced on non-mycorrhizal plants. In addition, the cytoplasm of the giant cells produced on mycorrhizal plant was densely structured while on control it tended to be strongly vacuolated, with decreased auxin levels in the root (Dehne *et al.,*1978). Mycorrhiza may impede larval growth and giant cell formation.

Kellam and Schenck, (1980) also suggested that AMF decreased giant cell development. Thompson and Hussey (1981) reported that the presence of additional amount of phosphorus in tomato roots provided by AMF resulted in reduction of root-knot population. Strobel *et al.,* (1982) reported that *Gigaspora margarita* decreased number of *M. incognita* eggs produced per gm of peach root growth at a low P level, by increasing P fertility level negated the beneficial function of the fungus.

Pre-inoculation of *Piper nigrum* cv. *panniyar* colonizing with *G. fasciculatum* or *G. etunicatum* significantly reduced the root-knot index (*Meloidogyne incognita*) by 32.4 and 36 % respectively and also the growth of piper plants improved (Mukerji, 1999). The number of galls formed by *Meloidogyne incognita* on tomato was significantly lower in AM inoculated tomato plants. However, the AM colonization did not prevent the penetration by the larvae (Suresh *et al.,* 1985). Percentage infection of tomato by *G. fasciculatum* was reduced when *M. incognita* was inoculated 7 days before, or at the same time as the fungus, but not when inoculated 7 days after the fungus (Suresh and Bagyaraj, 1984), pre-inoculation with AMF 4 weeks before addition of nematodes is often necessary for extensive colonization and establishment by the fungus. Kellam and Schenck (1980) observed that the number of galls of *M. incognita* on soybean was also reduced by *G. macrocarpus* even though root systems of AMF plants were larger. Grandison and Cooper (1986) found that number of *M. incognita* per gram, root were decreased by *G. margarita* in both susceptible and resistant cultivars of Alfalfa. They also suggested that nematodes were absent from cortical tissue with more than 10% AMF colonization. Cooper and Grandison (1986) also found that the number of *M. hapla* on tomato or clover decreased more when AM fungi were inoculated 4 weeks before the nematode.

The ability of mycorrhizal plants to grow well despite infection by nematodes is generally considered to be the principal affect of mycorrhizal fungi or the interaction of host plants with parasitic nematodes (Hussey and Roncadori, 1982). Cooper and Grandison (1987) observed that AMF infection on Tomarillo influenced resistance to *Meloidogyne incognita*. Carling *et al.*, 1996 found that AM fungi made peanut plant more tolerant to the nematode and offset the reduction in growth caused by *M. arenaria* at the two lower 'P' levels. However, AM fungi and added 'P' increased the number of galls and egg production in *M. arenaria*, thereby increasing peanut susceptibility to nematode attack and had a minimal effect on AM colonization. Furthermore, a comparison of mycorrhizal and P-fertilized non-mycorrhizal plants of similar size showed that the latter are more susceptible to nematode attack indicating the likely involvement of factors other than 'P' nutrition in the interactions (Smith, 1987, 1988). On the other hand, mycorrhizal inoculations of Tamarillo (*Cyphomandra betacea*) against root-knot nematode, *M. incognita* could not be duplicated by adding 'P' and fertilizer was not therefore due merely to improved 'P' nutritions of the host (Cooper and Grandison, 1987).

Heald *et al* (1989) showed that *M. incognita* suppressed the growth of non-mycorrhizal *Cucumis melo* plants by 84% as compared to 21% in AM inoculated plants at 50 mg/g P. A similar trend was observed in soil with 100 mg/g Phosphorus. Pre-inoculation of *G. fasciculatum* in tomato var. Pusa ruby significantly reduced the number of galls/plant, eggmasses per plant and eggs per eggmass due to *M. incognita*. They also observed that the effect of AM fungi on reducing root-knot nematode was not same in the field and in the pots.

The time of AM inoculations, whether before, simultaneously or after inoculations with the pathogens or nematode, greatly affects the efficacy of AM fungi in controlling nematodes, nematode population was reduced up to 58% in the pots and only upto 36% in the field (Sharma and Bhargava, 1993). Studies conducted by several workers demonstrate that pre-inoculations of AM fungus suppressed nematode reproduction and development in roots to a greater degree when compared to plants inoclated simultaneously with both the organisms (Jain and Sethi, 1987; Suresh and Bagyaraj, 1984; Taha and Abdel-Kader, 1990).

Smith *et. al.,* (1986 a) in the field and microplot studies reported that fewer juveniles of *M. incognita* penetrated cotton roots infected with *Glomus intraradices* but the number of nematodes after 120 days was not different from control plants. Smith *et. al.,* (1986b) found that when AMF was pre-inoculated 4 weeks before *M. incognita*, reduced total eggs produced and number of eggs per gm of root but number of eggs per female were not affected. Pre-inoculation of *Acaulospora laevis* in Black pepper cv. Panniyar-1 reduced the root-knot index (0.75) followed by phorate treated pot (1.00). Pre-inoculation of *Glomus mosseae* or *Gigaspora margarita* showed root-index of 1.25. The pots where *Meloidogyne incognita* alone was inoculated the root-knot index was 3.5. Thus, pre-inoculation of AMF reduced significantly the root-knot index (Anandraj *et. al.,*1990; Nehra and Trivedi, 1999; 2004 b; Nehra; 2002; 2003; 2004 a). Singh *et. al.,* (1990) conducted a field experiment where *Glomus fasciculatum* was pre-inoculated in nursery bed and monitored for 4 weeks. The AMF treated tomato seedlings were transplanted in *M. javanica* plot and simultaneously half the recommended dose of phorate (3 kg a.i./ha) applied in the plot resulted in reduction of incidence of root-knot disease. These findings indicated that AM fungi can be well integrated into the existing practice of management of root-knot nematode. Application of AM fungi e.g. *G. fasciculatum* followed by *M. incognita* inoculation was more effective in reducing nematode infestation than the simultaneous application of AMF and nematodes or nematodes before AMF (Jain *et al.,* 1998, Bhagwati *et al.,* 2000, Nehra and Trivedi, 1999 ;2004 b; Nehra; 2002; 2003; 2004a). The report published by Sundarababu *et. al.,* (1995) revealed that Su-babul plants which received Arbuscular Mycorrhiza 21 days prior to *M. incognita* were able to offset the ill effect caused by the latter and had more vigorous plant growth. The gall formation and nematode population was significantly lower by the early establishment of AMF. Sundaram and Arangarasan (1996) pre-inoculating AMF at two doses (59 and 109) in tomato cv. Co-3 observed that *Glomus fasciculatum* inoculated at 10 gm level produced maximum benefit followed by *G. mosseae, Gigaspora margarita* and *Acaulospora laevis* in increasing yield and reducing *M. incognita* infection particularly least average gall number (10) compared wth maximum (135.3) in control and uptake of nutrients.

Rotylenchulus reniformis

Sitaramaiah and Sikora (1982) showed that inoculations of tomato transplants with *Glomus fasciculatum* significantly reduced root penetration by juveniles and the development of *Rotylenchulus reniformis* compared to control. Development of gelatinous matrix was delayed and fewer eggs per eggmass were produced on inoculated plants. Pre and simultaneous inoculation of *Glomus fasciculatum* in cotton cv. Cocker carolina queen reduced *Rotylenchus reniformis* juvenile penetration, egg mass production and population build up on mycorrhizal inoculated plants than control, (Sitaramaiah and Sikora, 1982). Similar results were also obtained in ragi (Jothi and Sunderbabu, 1998).

Tylenchulus semipenetrans

VAM offsets a loss in plant growth caused by nematodes is seen in work reported by O'Bannon *et. al.,* (1979). *Tylenchulus semipenetrans,* a sedentary semi endoparasite was less severe on rough lemon in the presence of *G. mosseae.* Although infected seedlings grew less than seedlings inoculated only with *G. mosseae.* They had significantly more growth than seedlings inoculated only with *T. semipenetrans,* providing another example of mycorrhizal symbionts increasing plant tolerance to nematodes.

VAM and Cyst Nematodes

Jain and Sethi, (1988) reported that *G. fasciculatum* decreased cyst production and reproduction of *Heterodera cajani.* First generation cyst nematodes (*Heterodera glycines*) were significantly fewer on soybeans colonized by an isolates of *G. fasciculatum* but other VAM had no effect (Francl and Dropkin,1985).

Pratylenchus spp.

Hussey and Roncadori (1978) reported that cotton roots colonized by *Gigaspora margarita* induced significantly less *Pratylenchus brachyurus* population per gram of root than non-mycorrhizal plants. Lower nematode reproduction was caused either by the symbiont altering the cortex to make it an unfavourable food source for the nematode or by the symbiont competing with the nematode for space in the cortex. Enhanced growth occurred in mycorrhizal plants and were unaffected by *P. brachyurus.*

Radopholus similis

O'Bannon and Nemec, (1979) reported that rough lemon seedlings inoculated with *Glomus etunicatus* before inoculation of *Radopholus similis* exhibited nearly 28 times more dry weight at the end of the test than seedlings inoculated with nematode alone, indicating that VAM *G. etunicatus* increases tolerance of rough lemon to *Radopholus similis.* According to Umesh *et. al.,* (1988) number of *Radopholus similis* in banana roots were significantly lower when the VAM (*Glomus fasciculatum*) was applied simultaneously or seven days prior to nematode inoculation. Increased phenol, lignins, total aminoacids, reducing and total sugar and higher N,P,K,Ca and Mg in Mycorrhizal inoculated banana root were associated with reduced reproduction of *Radopholus similis.*

Table 4 Inhibitory influence of AM Fungi on Phytonematodes

Host	AM species	Phytonematode species	Reference
Alfalfa	*Gigaspora margarita*	*Meloidogyne incognita*	Grandison & Cooper,1986
Banana	*Glomus fasciculatum*	*Radopholus similis*	Umesh *et. al.*,1988
Black pepper	*Acaulospora laevis*	*M. incognita*	Anandraj *et. al.*,1990
	Glomus fasciculatum	*M. incognita*	Sivaprasad *et. al.*,1990
Bean	*G. etunicatum*	*M. javanica*	Olivera & Zambolim,1987
Berseem	*Glomus* spp.	*M. arenaria*	Kassab & Taha,1990
	Glomus fasciculatum	*M. incognita*	Jain *et. al.*,1988
Bush bean	*G.mosseae*	*Rotylenchulus reniformis*	Sitaramaiah & Sikora,1982
Chick pea	*G. mannihotes*	*Meloidogyne javanica*	Deiderichs, 1987
Clover	*Glomus* sp.	*M. hapla*	Cooper & Grandison,1986
Cotton	*Gigaspora margarita*	*M. incognita*	Roncadori & Hussey, 1977
	Glomus mosseae	*Pratylenchus brachyurus*	Hussey & Roncadori, 1978
	G. fasciculatum	*M. incognita*	Saleh & Sikora,1984
	G. intraradices	*M. incognita*	Smith *et. al.*, 1986 c
Cow pea	*G.fasciculatum*	*Heterodera cajani*	Jain & Sethi,1988
	G. spp.	*Tylenchulus vulgaris*	Jain & Hasan,1988
		M. incognita	
	G. fasciculatum	*M. incognita*	Jain & Sethi, 1988
Grape	*G. fasciculatum*	*Meloidogyne arenaria*	Atilano *et. al.*, 1982
Maize	*G. fasciculatum*	*Pratylenchus zeae*	Jothi and Sundarababu,1997

Contd.....

Host	AM species	Phytonematode species	Reference
Oat	*G.mosseal*	*Meloidogyne hapla* *M.incognita*	Sikora and Schonbeck, 1975
Peach	*G.margarita*	*M.incognita*	Strobel *et al.*,1982
Rough Lemon	*G.mosseae*	*Tylenchulus semipenetrans*	O'Bannon *et al.*, 1979
	G.etunicatus	*Radopholus similis*	O'Bannon & Nemec,1979
Soybean	*G.mosseae*	*M.incognita*	Kellam & Schenck, 1980
Subabul	*G.macrocarpus*	*M.incognita*	Sundarababu *et al.*,1995
Tobacco	*Endogone giganter*	*Heterodera solanacearum*	Fox and Spasoff,1972
	G.mosseae	*M.incognita*	Baltruschat *et al.*,1973
	G.mosseae	*M.hapla,* *M.incognita*	Sikora and Schonbeck, 1975
Tomato	*G.mosseae*	*M.hapla,*	Sikora and Schonbeck, 1975
	G.fasciculatum	*M.incognita* *M.javanica*	Bagyaraj *et al.*,1979
	G.fasciculatum	*Rotylenchulus reniformis*	Sitaramaiah & Sikora, 1982
	G.fasciculatum	*M.incognita*	Suresh and Bagyaraj, 1984
	G.fasciculatum	*M.incognita*	Suresh *et al.*,1985
	Glomus spp.	*M.hapla*	Cooper and Grandison,1986
	G.fasciculatum	*M.arenaria*	Sundaram & Angarason, 1996
	G.etunicatum	*M.incognita*	Bhagwati *et al.*, 2000

Mechanism of Disease Control

The management of crop diseases caused by root pathogens has become one of the challenging research areas in plant pathology. Increasing knowledge and concern about the environmental consequences of pesticide applications have aroused interest in alternative methods of plant protection. The roots of most plant species live in symbiosis with certain soil fungi by establishing what are known as mycorrhizas. The fungus biotrophically colonises the root cortex and develops an extramatrical mycelium that helps plants to acquire mineral nutrients from the soil (Harley & Smith, 1983). It has been recognised that mycorrhizal symbiosis plays a key role in nutrient cycling in the ecosystem and also protects plants against environmental and cultural stress (Barea and Jeffries, 1995).

Generally AM fungi cause few changes in root morphology, but the physiology of the host plant changes significantly. The improved potential for mineral uptake from the soil accounts for changes in the nutritional status of host tissues, which, in turn, changes, the structure and biochemical aspects of root cells. This can alter membrane permeability and thus the quality and quantity of root exudation. Altered exudation induces changes in the composition of micro-organisms in the rhizospheric soil, which could be known as 'mycorrhizosphere'. The net effect of these alterations is a tolerant plant better able to withstand environmental stress and plant diseases (Lindermann, 1988). To evaluate the influence of AM fungi on disease incidence and development, the variable factors, namely plant pathogens, the symbiotic fungus and environmental conditions have been considered. Mostly, the host mediates the interactions between the pathogen and the symbiont. The characterisation of these interactions should therefore include information on the machanisms involved . Because interactions vary with specific host-symbiont pathogen combination, it is difficult to make generalisation on the effect of AM fungi on diseases.

In general following factors play major role in reducing nematode population and enhancing plant growth-

1. AM fungi improve plant nutrition and by doing so may aid the host in compensating for damage caused by parasitic nematodes,thereby increasing plant tolerance to these pathogens.

2. AM fungi may increase root growth, expand the absorptive capacity of the root system for nutrient and water cellular processes in roots(Hayman, 1982).

3. AM fungi reduced nematode infection and development on several hosts in spite of larger root system on mycorrhizal plants (Cooper & Grandison,1986; Grandison & Cooper,1986; Smith *et. al.*,1986a, Sharma 1992). It may be due to unfavourable conditions for nematodes developed on root system due to AM colonization. Changes in hormones, amino acids and cell permeability in roots have been attributed to mycorrhizal symbiosis (Hayman, 1982).

4. Mycorrhizal root colonization has been shown to affect root exudation (Gerdemann,1968; Harley & Smith, 1983; Hayman,1982). These changes could alter chemotactic attraction of nematodes to roots or directly retard nematode development within root tissues.

5. AM fungi may increase changes in the post-infectional nematode-host interaction by altering nematode reproduction and development.

6. Direct competition for space may account for reduced nematode infection on mycorrhizal root system since endoparasitic nematodes occupy similar root tissues as AM fungi. However, results of several studies do not support this hypothesis.

7. Increased tolerance or resistance to nematodes by AM fungi may be due to competition for host photosynthates, production of nematistatic compound and parasitism of eggs.

AM fungi do not directly interact with plant parasitic nematodes inspite of their proximity in root tissue. More likely, AM fungi alter the host either physically or physiologically, and thus indirectly affect the host nematode relationship (Smith,1987).

A possible mechanism of interaction between AMF and nematodes is given in Fig.1.

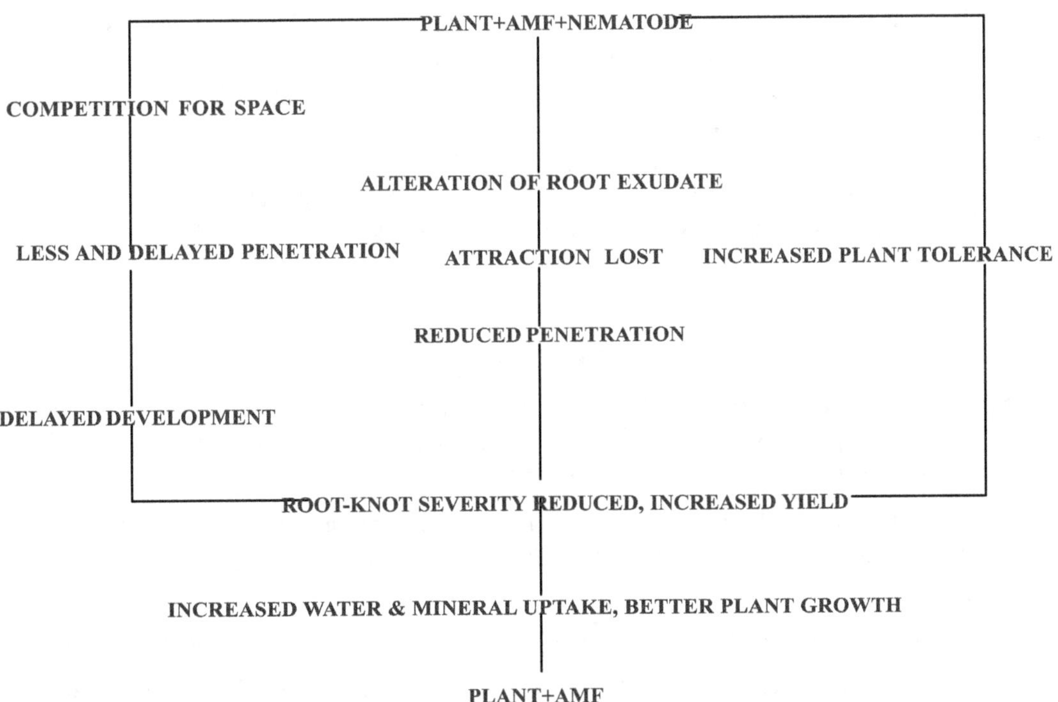

Figure 1 A diagrammatic scheme of the mechanism of interaction of root-knot nematode and AMF

Conclusions

The potential role of AMF as biocontrol agent for nematode infection has received considerable attention. It has been observed that the plants colonised with AMF generally grow well inspite of the presence of damaging levels of plant parasitic nematodes. Following aspects need special attention for the future work of AMF-nematode interaction study:

1. Considerable efforts are needed to select native efficient AMF for each crop for getting maximum benefits out of these fungi.
2. Positive results of green house experiment should be verified under field conditions to assess potential utilization of AMF as biocontrol agent.
3. Because of obligate nature of AMF these can not be cultured on large scale. Studies on priority basis should be undertaken to standaraize technique for mass scale multiplication and application.
4. It would be valuable to establish germplasm bank within a country, perhaps at one University and on large research institutes to help streamline the choice of endophytes for field trials.
5. Efforts should also be made to enhance the field of AMF inoculum with least contaminants, longer shelf life and easy transportation.

The existence of variability in the interrelationship between AMF and plant parasitic nematodes, in different combinations, is unique and an understanding of these interactions needs intensive study. Studies should determine whether AMF can be of sufficient value to warrant the development of technology for using these organisms. AMF should be used with other methods of nematode control as a part of integrated pest management strategies.

References

Anandraj, M., Ramana, K.V. and ***Sarma, Y.R.*** 1990. Interaction between vesicular arbuscular mycorrhizal fungi and *Meloidogyne incognita* in black pepper. *Proc. IInd Nat.Conference on Mycorrhiza*, Bangalore.

Anonymus, 1971. Estimated crop losses due to plant parasitic nematodes in the United States: Suppl. *J. Nematol. Special Publi.*, No.1 (USA):7.

Atilano, R.A., Menge, J.A. and ***Van Gundy S.D.*** 1981. Interaction between *Meloidogyne arenaria* and *Glomus fasciculatum* in grape. *J. Nematol.*13:52-57.

Atilano, R.A., Menge, J.A. and ***Vangandy, S.D.*** 1982. Interaction between *Meloidogyne arenaria* and *Glomus fasciculatum* in Grapes. *J. Nematol.* 13(1): 51-57.

Bagyaraj, D.J., Manjunath, A. and ***Reddy, D.D.R.*** 1979. Interaction of vesicular arbuscular mycorrhizal fungi with root-knot nematodes in tomato. *Plant Soil.* 51: 397-403.

Bagyaraj, D.J. 1984. Biological interactions with VA mycorrhizal fungi. In: *VA Mycorrhiza*, eds C.L. Powel, and D.J. Bagyaraj, CRC Press, Boca Raton, Florida, USA, 131-153.

Bagyaraj, D.J. 1992. Vesicular- arbuascular mycorrhiza- Application in agriculture. Methods in Microbiology.24: 359-373.

Baltruschet, H., Sikora, R.A. and ***Schoenbeck,*** F. 1973. Effect of VAM-mycorrhiza (*Endogone mosseae*) on the establishment of *Thielaviopsis basicoloa*, in *Meloidogyne incognita* in tobacco. Abstr. *IInd Intern. Cong. Plant Path. Minneapolish*, No. 661.

Barea, J.M. and ***Jeffries, P.*** 1995. Arbuscular mycorrhizas in sustainable soil plant systems. In: *Mycorrhiza: Structure, function, Molecular biology and Biotechnology.* (Eds) Varma, A. and Hock B., Berlin: Springer-Verlag. 521-560.

Barlett, E.M. and ***Lewis, D.H*** 1973. Surface phosphate activity of mycorrhizal roots of beech. *Soil Biol. Biochem.* 5: 246-257.

Bateman, D.F. and ***Miller, R.L.*** 1966. Pectic enzymes in tissue degradation. *Ann. Rev. Phytopathol.* 4: 119-146.

Bhagwati, B, Goswami, B.K. and ***Singh, C.S.*** 2000. Management of disease complex of tomato caused by *Meloidogyne incognita* and *Fusarium oxysporum* f.sp.*lycopersici* through bioagent. *Indian J. Nematol.* 30(1): 16-22.

Bhatti, D.S. and Jain, R.K. 1977. Estimation of loss in okra, tomato and brinjal yield due to *Meloidogyne incognita. Indian J. Nematol.* 7: 37-41 pp.

Brundret, M. 1991. Mycorrhiza in natural ecosystems. In: *Ecological Research* (Eds. Begon *et. al.,*) vol.21, Academic Press, London, New York, 171-313.

Carling, D.E., Roncadori, R.W*.** and ***Hussey, R.S. 1996. Interactions of arbuscular mycorrhizae, *Meloidogyne arenaria*, and phosphorus fertilization on peanut. *Mycorrhiza* 6:9-13.

Clarkson, D.T. 1985. factors affecting mineral nutrient acquisition by plants. *Ann. Rev. Plant Physiol.* 36:77-115.

Cason, T.K.M., Hussey, R.S. and ***Roncadori, R.W.*** 1983. Interaction of vesicular-arbuscular mycorrhizal fungi and phosphorus with *Meloidogyne incognita* on tomato. *J. Nematol.* 15: 410-417.

Cobb., N.A. 1932. The English Word Nema. *J. Amer Med. Assoc.* 98: 75.

Cooper, K. and Grandison, G.S. 1986. Interaction of vesicular-arbuscular mycorrhizal fungi and root-knot nematode as cultivars of tomato and white clover susceptible to *Meloidogyne hapla. Ann. Appl. Biol.* 108:555-565.

Cooper, K.M. and Grandison, G.S. 1987. Effect of vesicular-arbuscular mycorrhizal fungi on infection of tamarillo by *Meloidogyne incognita* in fumigated soil. *Plant Disease.* 71: 1101-1106.

Dehne, H.W., Schonbeck, F. and Baltruschat, H. 1978. Investigation on the influence of endotrophic mycorrhiza on crop diseases. III chitinase activity and ornithine cycle. *J. Phytopathol.* 95: 666-678.

Diederichs, C. 1987. Interaction between five endomycorrhizal fungi and nematode *Meloidogyne javanica* on chickpea under tropical conditions. *Tropical Agricultural.* 64: 353-355.

Ek, M., Ljungquist, P.O. and Stenstrom, E. 1983. Indole-3-acetic acid Production by mycorrhizal fungi determined by gas chromatography-mass spectrometry. *New Phytol.* 94:401-403.

Fox, J.A. and **Spasoff, L.** 1972. Interaction of *Heterodera solanacearum* and *Endogone gigantea* on tobacco (Abstr.). *J. Nematol.* 4:224-225.

Francl, L.J. and **Dropkin, V.H.** 1985. *Glomus fasciculatum* a week pathogen of *Heterodera glycines. J. Nematol.* 17: 470-475.

Gerdemann, J.W. 1968. Vesicular- arbuscular mycorrhiza and plant growth. *Ann. Rev. Phytopathol.* 397-418.

Grandison, G.S. and **Cooper, K.M.** 1986. Interaction of vesicular arbuscular mycorrhizae and culticars of alfalfa susceptible and resistant to *Meloidogyne hapla,* J. Nematol. 18: 141-149.

Graham, J.H. and **Syvertsen, J.P.** 1984. Influence of vesicular-arbuscular mycorrhiza on the hydraulic conductivity of roots of two citrus root stocks. *New Phytol.* 97: 227-284.

Harley, J.L. 1940. A study of the root system of beech in wood/ and soils with special reference to mycorrhizal infection. *J. Ecol.,* 28: 107-117.

Harley, J.L. and **Smith, S.E.** 1983. Mycorrhizal Symbiosis. Academic Press, New York, USA. pp 483.

Hasan, N. and **Jain, R.K.** 1991a. Biological control of nematodes in forage crops-current status and future prospects. In: *Recent Advances in Nematology* (Eds Dwivedi and Hussain). Biov. Res. Soc. Allahabad, India. 291-232.

Hasan, N. and **Jain, R.K.** 1991b. Parasitic nematodes and vesicular arbuscular mycorrhizal (VAM) fungi associated with berseem (*Trifolium alexandrium* L.) in Bundelkhand region. *Indian J. Nematol.* 17: 184-188.

Hayman, D.S. 1982. The physiology of vesicular-arbuscular endomycorrhizal symbiosis. *Can. J. Bot.* 6: 944-963.

Heald, C.M., Bruton, B.D. and **Davis, R.M.** 1989. Influence of *Glomus intraradices* and soil phosphorus on *Meloidogyne incognita* infecting *Cucumis melo. J. Nematol.* 21(1): 69-73.

Horsfall, J.G. and *Diamond, A.E.* 1957. The Disease Plant, In: *Plant Pathology- An Advanced Treatise* (ed. J.G.Horsfall and A.F.Diamond), Academic Press, New York. 1-16.

Hussey, R.S. and *Roncadori, R.W.* 1977. Interaction of *Pratylenchus brachyurus* and an endomycorrhizal fungus on cotton. *J. Nematol.* 9: 270-271.

Hussey, R.S. and *Roncadori, R.W.* 1978. Interaction of *Pratylenchus brachyurus* and *Gigaspora margarita* on cotton. *J. Nematol.* 10:16-20p.

Hussey, R.S. and *Roncadori, R.W.* 1982. Vesicular arbuscular mycorrhizae may limit nematode activity and improve plant growth. *Plant Dis.* 66: 9-14.

Jain, R.K. and *Hasan, N.* 1988. Role of vesicular arbuscular mycorrhizal (VAM) fungi *Glomus fasciculatum* and root-knot nematode *Meloidogyne incognita* on Stylo (*Strlosanthes hamata*). *Proc. 1st AAN Symp.* AMU, Aligarh, India. 12-13.

Jain, R.K. and *Hasan, N.* 1994 a. Vesicular arbuscualr mycorrhizal fungi, a potential biocontrol agent for nematodes. In: *Vistas in Seed Biology*, vol. 2 (eds. Singh T. and P.C. Trivedi), Printwell, Jaipur, India. 49-60.

Jain, R.K. and *Hasan, N.* 1995. Nematode problems in forage crops. In: *Nematode Pest Management- An Appraisal of Eco-friendly Approaches* (Swarup *et. al.,*) Nematological Soc. India, IARI, New Delhi, India, 217-227.

Jain, R.K. and *Hasan, N.* 1986. Association of vesicular arbuscular mycorrhizal fungi and plant parasitic nematodes with forage sorghum (*Sorghum bicolor L. Sorghum Newsletter.* 29: 84.

Jain, R.K. and *Sethi, C.L.* 1987. Pathogenicity of *Heterodera cajani* on cowpea as influenced by the presence of VAM fungi *Glomus fasciculatum* or *G. epigaes. Indian J. Nematol.* 17(2): 165-170.

Jain, R.K. and *Sethi, C.L.* 1988. Influence of endomycorrhizal fungi *Glomus fasciculatum* and *G.epigaes* on penetration and development of *Heterodera cajani* on cowpea. *Indian J. Nematol* 18: 89-93.

Jain, R.K., Hasan, N., Singh, R.K. and *Pandey, P.N.* 1998. Influence of the endomycorrhizal fungus, *Glomus fasciculatum* on *Meloidogyne incognita* and *Tylenchorhynchus vulgaris* infecting Berseem. *Indian J. Nematol.* 28 (1) : 48-51.

Jain, R.K. and Hasan, N. 1988. Role of vesicular arbuscular mycorrhizal (VAM) fungi and nematode activities in forage production. *Action Bot. India.* 16: 84-88.

Jothi, G. and Sundarababu, R. 1997. Studies on the management of root lesion nematode vesicular-arbuscular fungi, *Meloidogyne incognita* and soil fertility on peach. *Phytopathology.* 72: 690-694.

Kassab, A.S. and Taha, A.H.Y. 1990. Interactions between plant parasitic nematodes, vesicular-arbuscular mycorrhiza, rhizobia and nematicide on egyptian clover. Ann. *Agric. Sci.* 35:509-520.

Kaur, J., Gupta, R.P., Pandher, M.S., Holly, R.S. and ***Gupta, S.K.*** 1997. Synergistic effect of VAM and *Rhizobium* on soybeans. *Proc. Third Agricultural Science Congress. Punjab Agricultural University, Ludhiana.* 47.

Kellam, M.K. and ***Scenck, N.C.*** 1980. Interaction between a vesicular- arbuscular mycorrhizal fungus and root-knot nematode on soybean. *Phytopathology.* 70: 293-296.

Krishna prasad, K.S. 1971. *Effects of amino acid and plant growth substance on tomato and its root-knot nematode Meloidogyne incognita.* M.Sc. (Agric) Thesis, University of Agricultural Sciences, Bangalore,

Lindermann, R.G. 1988. Mycorrhizal interactions with the rhizosphere effect. *Phytopathology.* 78: 366-371.

Marx, D.H. and ***Bryan, W.C.*** 1971. *Far. Sci.* 17: 37-41.

Mukerji, K.G. 1999. Mycorrhizae in control of plant pathogens: Molecular Approaches, In: *Biotechnological Approaches in Biocontrol of Plant Pathogens,* eds.K.G.Mukerji, B.P. Chambola and R.K. Upadhyay, Kluwer Academic/ Plenum Publishers, New York, USA, 135-156.

Nemec, S. and ***Meredith, F.I.*** 1981. Amino acid content of leaves in mycorrhizal and non-mycorrhizal citrus root stocks. *Ann. Bot.* 47: 351-358.

Nehra, S. and ***Trivedi, P.C.*** 1999. Nematode diseases in ginger and their management. *In : Biotechnology and Plant Pathology-Current Trends* (eds. P.C.Trivedi). 281-299.

Nehra, S. 2002. Application of VAM fungi and organic amendment in the management of *Meloidogyne incognita* infecting ginger. *J. Indian Bot. Soc,* 81:93.

Nehra, S. 2003. Biological control of root-knot nematode infecting ginger using nematophagous fungi and AMF. *XXVI conf. Indian Bot. Soc, Jamia Hamdard,* N.Delhi, pp 29-31.

Nehra, S., Pandey, S. and ***Trivedi, P.C.***.2003. Interaction of arbuscular Mycorrhizal fungi and different levels of root-knot nematode on ginger. *Indian Phytopath.* 56(3) 297-299.

Nehra, S. 2004a. Integrated Management of Plant Parasitic Nematodes. In: Plant Diseases (eds. Sampat Nehra). Aavishkar Publishers. 367-399.

Nehra, S. 2004b. Arbuscular Mycorrhizal Fungi (AMF): Taxonomy and Multiplication. *In : Plant Diseases* (eds. Sampat Nehra). Pointer Publisher, Jaipur. 316-334.

Nehra, S. and ***Trivedi, P.C.*** 2004 a. Isolation, Multiplication and Screening of AM Fungi for the management of nematode *Meloidogyne incognita* infecting bajra.. *Proc. 91st. Indian Sci. Cong. (Plant Sci.),* Chandigarh, pp 14-15.

Nehra, S. and ***Trivedi, P.C***. 2004 b. Fungal biotechnology in the management of root-knot nematode infecting ginger. *Indian Phytopath. Soc. (Central zone) Meet. and Nat. Sem. Jaipur.* 42-43.

Norris, J.R., Read, D.J. and **Verma, A.K.** 1991. *Methods in Microbiology.* 23: Academic Press , London.

Norris, J.R., Read, D.J. and **Verma, A.K.** 1992. *Methods in Microbiology.* 24: Academic Press , London.

Norris, J.R., Read, D.J. and **Verma, A.K.** 1994. *Techniques in Mycorrhizal Research.* Academic Press , London.

O'Bannon, J.H. and Nemec, S. 1979. The response of citrus seedlings to a symbiont *Glomus etunicatum*, and a pathogen *Radopholus similis. J. Nematol.* 11: 270-275.

O'Bannon, J.H., Insena, R.N., Nemec, S. and **Vovlas, N.** 1979. The influence of *Glomus mosseae* on *Tylenchulus semipenetrans* infected and uninfected citrus lemon seedlings. *J. Nematology.* 11: 247-250 .

Paruthi, I.J. and **Bhatti, D.S.** 1981. Estimation of losses in wheat yield due to *Anguina tritici* and seed gall infestation in market grains. *Indian J. Nematol.* 11: 15-18.

Parvatha Reddy, P. 1974. *Studies on the action of amino acids on the root-knot nematode Meloidogyne incognita,* Ph.D.thesis, University of Agricultural Sciences, Bangalore, India.

Raj, J., Bagyaraj, D.J. and **Manjunath, A.** 1981. Influence of soil inoculation with vesicular-arbuscular mycorrhiza and a phosphate dissolving bacterium of plant growth and P uptake. *Soil Biol. And Biochem.* 13:105-108.

Roncadori, R. and **Hussey, R.** 1977. Interaction of the endomycorrhizal fungus *Gigaspora margarita* and root-knot nematode on cotton. *Phytopathology.* 67: 1507-1511.

Saleh, H. and **Sikora, R.A.** 1984. Relatioship between *Glomus fasciculatum* root colonization of cotton and its effect of *Meloidogyne incognita. Nematol.* 30: 230-237.

Sasser, J.N. 1989. *Plant parasitic Nematodes, The farmers hidden enemy.* Dept. Pl. Patho. NCSU, Raleigh, USA, 115.

Schenck, N.C. and **Kinloch, R.A.** 1974. Pathogenic fungi, parasitic nematodes and endomycorrhizal fungi associated with soybean roots in Florida. *Plant Dis. Rep.* 58: 169-173.

Schenck, N.C., Kinloch, D.H. and **Dickson,** D.W. 1975. Endomycorrhiza (eds. Sanders *et. al.,*) Academic Press, New York, 607-617.

Schenck, N.C., Kinloch, R.A. and **Dickson, D.W.** 1975. Interaction of endomycorrhizal fungi and root-knot nematode on soybean. In : *Endomycorrhiza* (eds. F.E.Sanders, B. Mosse and P.B. Tinker, Academic Press, London, 607-617.

Schonbeck, F. 1979. Endomycorrhiza in relation to plant disease. In: Soil-borne Plant Pathogens (eds. Schippers B & W Gams), Academic Press, London, 271-280.

Sharma, W. 1992. Studies on biochemical alterations resulting *in Meloidogyne* sp. infection in *Abelmoschus esculentus* and their control. Ph. D.thesis, University of Rajasthan, Jaipur.

Sharma, W. and **Trivedi, P.C.** 1994. Interaction of root-knot nematode with VAM on *Abelmoschus esculentus* and their effect on net yield. *In: Vistas in Seed Biology,* vol II (Singh T. & Trivedi P.C. eds) Printwell Publisher Jaipur, pp 61-81.

Sharma, M.P. and **Bhargava, S.** 1993. Potential of VAM fungus *G. fasciculatum* against the root-knot nematode *Meloidogyne incognita* on tomato. *Mycorrhiza News.* 4(4): 4-5.

Sikora, R.A., Schonbeck, F. 1975. Mycorrhiza (*Endogone mosseae*) on the population dynamics of the root-knot nematodes (*Meloidogyne incognita* and *M.hapla*). 8 th Int. Cong. *Pl. Prot.* 5 : 158-166.

Sikora, R.A. 1979. Predisposition of *Meloidogyne* infection by the endotropic mycorrhizal fungus *Glomus mosseae.* In: *Root-knot nematode Meloidogyne spp. Systematics, Biology and Control* (eds. F.Lamberti & C.Taylor), Academic Press, London. 399-404.

Singh, Y.P., Singh, R.S. and **Sitaramaiah, K.** 1990. Mechanism of resistance of mycorrhizal tomato against root-knot nematode, In: *Current Trends in Mycorrhizal Research*, Proc. National Conference on Mycorrhiza, ed. B.L. Jalali and H.Chand, Harayana Agricultural University, Hisar, India. 96-97.

Sitaramaiah, K. and **Sikora, R.A.** 1982. Effect of mycorrhizal fungus *Glomus fasciculatum* on the host-parasitic relationship of *Rotylenchus reniformis* in tomato, *Nematologia.* 28: 412-419.

Smith, G.S. 1987. Interaction of nematodes with mycorrhizal fungi. In : Vistas in Nematology (eds. J.A.Veech and D.W.Dickson). Soc. Nematologists. INC Hyattsville Maryland. 292-300.

Smith, G.S. and Kaplan, D.T. 1988. Influence of mycorrhizal fungus, phosphorus and burrowing nematode interactions on growth of rough lemon citrus seedlings, *J. Nematol,* 20: 539-544.

Smith, G.S., Hussey, R.S. and **Roncadori, R.W.** 1986a. Penetration and post infection development of *Meloidogyne incognita* on cotton as affected by *Glomus intraradices* and phosphorus. *J. Nematol.* 18: 429-435.

Smith, G.S., Roncadori, R.W. and **Hussey, R.S.** 1986b. Interaction of endomycorrhizal fungi, superphosphate and *Meloidogyne incognita* on cotton in microplot and field studies. *J. Nematol,* 18: 208-216.

Smith, G.S., Roncadori, R.W. and **Hussey, R.S.** 1986c. Penetration and post-infection development of *Meloidogyne incognita* as affected by *Glomus intraradices* and phosphorus, *J. Nematol.* 18: 429-435.

Smith, G.S. 1988. The role of phosphorus nutrition in interactions of vesicular-arbuscular mycorrhizal fungi with soilborne nematodes and fungi. *Phytopathology.* 78: 371-374.

Strobel, N.E., Hussey, R.S. and **Roncadori, R.S.** 1982. Interaction of vesicular-arbuscular fungi, *Meloidogyne incognita* and soil fertility on peach. *Phytopathology.* 72: 690-694.

Sundarababu, P., Sudha, S. and **Iyear, R.** 1995. Reaction of different subabul cultivars to root-knot nematode and the interaction of nematode and VA mycorrhiza on subabul. *Indian J. Nematol.* 25(1): 70-75.

Sundaram, H.D. and **Arangarasan, V.** 1996. Role of vesicular-arbuscular mycorrhizal fungi on the growth, nematode (*Meloidogyne incognita*) disease incidence and fruit yield of tomato Co.3 grown in loamy soils of Arcot District (Tamil Nadu). *In: Mycorrhizara: Biofertilizer for the future.* 100-103.

Suresh, C.K., and **Bagyaraj, D.J.** 1984. Interaction between vesicular-arbuscular mycorrhizae and a root-knot nematode and its effect on growth and chemical composition of tomato. *Nematol. Mediterr.* 12: 31-39.

Suresh, C.K., Bagyaraj, D.J. and **Reddy, D.D.R.** 1985. Effect of vesicular-arbuscular mycorrhizae on survival, penetration and development of root-knot nematode in tomato. *Plant and Soil.* 87: 305-308.

Taha, A.H.Y. and **Abdel-Kader, K.M.** 1990. The reciprocal effects of prior invasion of root-knot nematode or by endomycorrhiza on certain morphological and chemical characteristics of Egyptian clover plants, *Ann. Agricul. Sci.* (Cairo) 35(1): 521-532.

Thomson, K.M. and **Hussey, R.S.** 1981. Interactions of vesicular-arbuscular mycorrhiza and phosphorus on root-knot on tomato. *J. Nematol.* 23: 122-133.

Tinker, P.B. 1975. Effects of vesicular-arbuscular mycorrhizas on higher plants. *Symp. Soc. Exp. Biol.* 29: 325-349.

Tinker, P.B. and **Gildon, A.** 1983. *Mycorrhizal fungi and ion uptake.* In: *Metals and Micronutrients: uptake and utilization by plants.* D.A. Roble and W.S. Piperpoint (eds.) Academic Press, New York, 21–32.

Tisdall, J.M. 1991. Fungal hyphae and structural stability of soil. *Aust. J.Soil. Sci.* Res. 29: 729-743.

Trivedi, P.C. 1995. Biocontrol Potential of VAM fungi in the management of plant parasitic nematodes. Platinum Jubilee Volume *J. Indian Bot. Soc.* 74 A : 165-171.

Umesh, K.C., Krishnappa, K. and **Bagyaraj, D.J.** 1988. Interaction of burrowing nematode, *Radopholus similis* and VA mycorrhiza, *Glomus fasciculatum* in banana. *Indian J. Nematol.* 18:6-11.

Van Berkum, J.A. and **Seshadri, A.R.** 1970. Some important nematode problems in India. *X Inter Nematol. Sump. E.S.N. Pescara.* 1-143.

Wood, M. 1989. Underground allies of plants. *Agri. Res.* 37: 10-13.

Physico-Chemical and Biological Characterization of Ground Water Quality at Patancheru Industrial Area

I. Sreenu Babu, V. Venkat Ramana Kumar, M. Thirumala and S.K. Mahmood

Deptt. of Botany, Osmania University, Hyderabad - 500 007

ABSTRACT

Physico-chemical monitoring of Pashamailaram Village of Patancheru Industrial Area was carried out during the year 2001-2002. Ten industrial sites were selected on the basis of their importance. Effluents of Common Effuluent Treatment Plants (CETPL) and different industrial units which were being discharged into ground water. The analysis were carried out for the parameters namely, pH, DO, BOD, COD, TDS, Chlorides, Phosphates, Nitrates, Sulphates, Ca and Mg. Fungi and bacterial colonies were isolated and few fungi have been reported, among fungi *Geotrichum candidum* was dominating. The result obtained in the present investigation reveald that discharge of untreated industrial effluents and sewage had contributed considerable pollution in the ground water of different industrial localities, and is unsafe for consumption and also for irrigation, therefore needs serious attention.

KEY WORDS : Industrial effluents, Nala, Sewage, Bacteria and Fungi.

The effluents of Pantancheru Effulluent Treatment Limited (PETL) pass through the drain (Nala) sitatuated by the side of Industrial estate in Pashamailaram village of IDA Patanchuru. It is a well known fact that the common effluent treatment plant (CETPs) had failed to treat effectively, the effluent of the bulk drug industries, because of the huge quantity of toxic organic salts and inorganics they receive for which they were not designed. The present study was carried out to evaluate the physico-chemical and biological characteristics of ground water of IDA Patancheru.

Materials and Methods

The ground water samples were collected near different industrial localities as shown in the table(Table.1) within Patancheru (Pashamailaram) industrial area along with PETL drain (Nala).Total of 10 sampling points were selected during 2002-2003 on the basis of their importance. Samples for analysis were collected in the sterilized bottles using the standard procedure in accordance with the method. (APHA,1985,1995), i.e for temperature(thermometric), pH(potentiometric), DO and BOD (Azide modifiction)

COD (Dichromte reflux) TDS (Argentometric),Chloride (Gravimetric), sulphates, phosphates and nitrates (spectrophotometric), Ca as well as Mg (Titrimetric) methods were employed. Colony forming units (cfu) of fungi and bacteria was estimated by Waksman's serial dilution plate method (Waksman, 1922), potato sucrose agar for fungi luria Bertani Agar (tryptone=1g, yeast extract=0.5g, NaCl=0.5g, agar agar=2g and distilled water=100ml) for bacteria was utilized. Three replicates were maintained in each case. Fungi which were dominant during the study were identified up to species level (Ellis, 1971; 1976; Booth, 1971; Subramanian, 1971; Gilman, 1957; Mahmood *et al.*, 1989)

Results and Discussion

The physico-chemical parameters of the IDA Patancheru ground water samples, collected during 2002-2003 from 10 different points are given in Table-1. It was found that the maximum temperature of water was at the points F and H and minimum temperature was at point-C. The various chemical and biological reactions in water depend to a great extent on temperature. The observed values of temperature indicate that the water quality would be certainly affected by this parameter (Chaturvedi *et. al.*, 2003). Fungi and bacterial population, conductivity, pH and total dissolved solids are given in Table 1. A total of 22 fungal species belonging to 10 genera were isolated (data not shown).

The fungal species isolated can be categorized into the following groups according to Cooke (1960);

(i) ***Lymaphiles :*** These species are able to tolerate a pollution habitat as a livable one (organisms living in sewage habitat). Such species are *viz., Aspergillus niger, Fusarium oxysporum, Geotrichum candidum, Mucor* sp and *Pencillium* sp.

(ii) ***Lymaphobes :*** These species cannot tolerate pollution, although some of the species were found only once. Such species are *Alternaria alternata, Aspergillus flavipes, A. versicolor, Curvularia lunata, Cladosporium oxysporum, Penicillium expansum, Phoma humicola, Rhizopus nigricans, Trichoderma harzianum,* white, grey and brown sterile mycelium. The fungal population showed different behaviour in relation to important physico-chemical parameters:

PH of ground water varied from industry to industry. Maximum fungal population was recorded in Arabindo Pharma Ltd (58.12 CFU/ml) with lower pH (6.1). Reverse was true when the pH was high (8.8) and final population was low (6.13CFU/ml). As a consequence, the density of mycoflora showed negative significant correlation with Hydrogen on concentration as shown in Table 2 ($g = -0.863$, $P < 0.01$). Thus decomposition of organic matters by microbes including geofungi as well as bacteria were responsible for lowering the pH. It is well known fact that during summer and rainy season, when the microbial activity is supposed to be higher concentration of CO_2, Decaying vegetation and waste materials have acidifying effects in water (Bradshaw, *et. al.*, 1973).

Table 1 Physico-Chemical and Biological Characteristics of PETPL-IDA Patancheru

S.no	Ground Water locality	Fungi (cfu) x1K*	Bacteria (cfu) x50K*	Tempe-rature 0C	pH	Condu-ctivity um/cm	TDS mg/ml	Chlo-ride mg/ml	DO mg/l	BOD mg/l	COD mg/l	NO3 mg/l	PO4 mg/l	Mg mg/l	Ca mg/l
1	A	28.02	20.01	18.0	8.2	6380	3090	42	7.9	8.2	190	0.81	0.38	5.2	24.1
2	B	58.12	10.28	18.5	7.6	540	270	48	7.2	8.9	144	0.82	0.32	5.2	23.7
3	C	15.20	35.01	17.9	6.1	2250	1097	40	6.7	8.4	200	0.81	0.34	5.4	24.2
4	D	54.00	8.02	18.9	8.1	2800	1377	42	7.8	8.9	199	0.83	0.35	5.3	24.1
5	E	13.24	29.28	22.6	7.5	3200	1555	48	7.3	8.8	211	0.86	0.33	5.6	23.6
6	F	15.41	31.21	28.9	8.2	3930	1930	41	7.1	7.9	198	0.87	0.34	5.2	25.2
7	G	8.20	41.22	24.8	8.5	5990	2980	40	7.2	8.7	187	0.83	0,36	5.1	26.1
8	H	19.40	28.09	19.9	8.3	3290	1609	39	6.6	7.8	199	0.83	0.37	5.6	25.4
9	I	14.22	32.32	18.0	7.9	555	299	41	7.2	8.6	154	0.82	0.31	5.3	22.9
10	J	6.13	33.21	21.8	8.8	3900	985	30	7.9	8.9	180	0.88	0.32	5.7	26.2

A = Neuland Laboratories, B = Arabindo Pharma Ltd, C = Esnapur Village Area, D = Medicorp Industries, E = Agritech Ltd F = Pashamailaram village, G = Hyd, Chemils H = Ralchem, I = Akhil Pharma Ltd, J = Navapan Industries.* = An Average of Triplicate.0

Table 2 Correlation-Coefficient®value for the fungal population of Ground water with some parameter

S.No	Physico-Chemical Values	Correlation-Coefficient(r)	Regression Equation Y = c + mx
1	pH	− 0.862*	Y = 326.79 - 35.87x
2	TDS	+ 0.858*	Y = -19.21 + 0.057
3	Conductivity	+ 0.623**	Y = -28.80 + 0.0612x

* = Significant at 1%level ** = Significant at 5% level Y = Predicted No. of fungal Colonies.

X = Physico-Chemical Factor. C = Intercept with Y M=Slope Values.

The conductivity of drains varied through out the year. The highest fungal population was recorded when the conductivity also happened to be the maximum. Low fungal population was observed with minimum conductivity (Table 1). The higher density of ground water mycoflora was associated with greater conductivity and *vice-versa*. As a result, the population of ground water mycoflora showed significant positive correlation with the conductivity of sewer water (g = + 0.0603, P < 0.5). The pH of water varied from 6.1 to 8.8 showing slight alkalinity. It is known that pH of water does not cause any severe health hazard. TDS is an important parameter for drinking water and water to be used for other purposes. Beyond certain limit it imparts a peculiar taste to water and reduces its portability. TDS in water less than 1000 mg/l is classed as non-saline. D.O. is one of the most important parameters in assessing water quality and reflects the physical and biological processes prevailing in the water. A good water should have the solubility of oxygen 7.6 and 7.0 mg/l at 30°C and the permissible limit of this parameter suitable for drinking is 500 mg/l (WHO). All values of ground water exceeded this value (Kudesia 1985). Oxygen saturated water has pleasant taste. The DO of the Ganga river was in between 6.6 – 7.9. The maximum permissible value of COD is 10 mg/l for drinking water (De 1985). This parameter of all the ground water samples was beyond the limit. These high values indicated that ground water was rich either with respect to some dissolved organic compounds or oxidisable inorganic substances. The chloride and nitrate contents of ground water were small (Kumar *et. al.,* 2002). These amounts will not impart any taste to water. The amounts present do not exceed the maximum permissible limit i.e. 500 mg/l for drinking water prescribed (APHA, 1989) and sulphate has less effect on the taste of water compared to the presence of chloride. The desirable limit of sulphate in drinking water prescribed by ICMR is 200-400 mg/l. All the water samples collected have satisfied the drinking water quality so far presence of sulphate is concerned. Phosphate (0.31 – 0.32) were low in the ground water during study period. Certain fungal species viz., *Aspergillus niger, Geotrichum candidum, Mucor* sp and *Penicillium* sp were recorded frequently from the drain waters with maximum densities during the period of pollution. In this way some fungal species may be good indicator of water pollution (Cooke, 1954). The results of the present study indicate that the discharge of untreated industrial effluents and sewage contributed considerable pollution in the ground water at Pasha mailarum drain and serious steps should be taken to remove this pollution.

Acknowledgement

The authors are thankful to the scientist EPTRI –Hyderabad for their constant analytical help during the study and SKM acknowledges the financial assistance of UGC-SAP Botany, Osmania University.

References

APHA. 1989. Standard methods for the examination of water 17th editions, APHA Washington DC.

APHA (American Public Health Association). 1995 American Water Works Association and Water Pollution Control Federation. Standard Methods of　Examination of Water and Waste Water. 19th edition, New Yourk, U.S.A

Booth 1971. Methods in Microbiology. Academic Press, New York.

Bradshaw, J.S., Sunders, R.B., White, D.A., Barton, J.R., Fuhriman, D.K., Loveridge, E.L. and **Pratt, D.R.** 1973. Chemical response of Utah lake to nutrient inflow, *Jour. Wat. Pollut. Cont. Fed.* 45 (2) : 880-887.

Chaturvedi Samiksha, Dinesh Kumar and **R.V. Singh** 2003. Study on some physico-chemical characteristics of flowing water of Ganga River at Haridwar. *Res.J.Chem.Environ.* 7:78-79

Curtis, E.J.C. 1972. Sewage Fungus in Rivers in the U.K. *Water Pollution Control,* 71: 673-685.

Cooke, W.B. and **Kenneth, Aa. Bush** 1957. Activity of cellulose Decomposing fungi Isolated from Sewage, Polluted waters. *Sewage and Industrial Wastes,* 29 (2).

De A.K. 1985. The Saga of the Damodar River. *J. Indian Chem. Soc.,* 62, 1038

Ellis, M.B. 1971. Dematiaceous Hyphomycetes. C.M.I. Kew, Surrey, England.

Eillis, M.B. 1976. More Dematiaceous hyphomycetes. C.M.I., Kew, Surrey, England.

Gilman, J.C. 1957. A manual of soil fungi. IInd Ed., Oxford & IBH Publishing Co., New Delhi.

Kudesia, V.P. 1985. *Water Pollution.* Pragati Prakashan, Meerut

Mahmood, S.K., Renuka, B.R and **Rama Rao, P.** 1989. Effect of foliar sprays of rhiosphere microbes of *Alternanthera sessilis* R.Br.Geobios.16: 215-217..

Kumar, S. *et. al.,* 2002. "Nitrate a need for revision fro standards" Published in NEERI, JE & H. 4(4), Nagpur

Subramanian, C.V. 1971. Hyphomycetes. I.C.A.R., New Delhi Publication.

Waksmam, S.A. 1922, Methods for counting the no. of fungi in the soil. J. BACT. 7:339-341

53 | *Effect of Plant Extracts, Plant Products, Chemicals and Fungal Culture Filtrates on Four Pathogenic Fungi*

A. GOURINATH

Department of Botany, Osmania University, Hyderabad - 500007

ABSTRACT

Effect of plant extracts, pollen suspension, latex, gums, oils, homoeopathic drugs, chemicals, fungal culture filtrates and antifungal compounds have been tested on conidial germination and mycelial growth of four pathogenic fungi namely *Curvularia lunata* (Walker) Boedijn, *Cylindrocarpon lichenicola* (C. Massal) Hawksworth, *Fusarium solani* (Mart) Sacc. and *Myrothecium leucotrichum* (Peck) Tulloch. Among the plant parts tested, the extracts of root and flower have more inhibitory effect. The other plant products and chemicals had also shown more inhibitory effect on conidial germination and mycelial growth.

KEY WORDS : Plant extracts, chemicals, spore germination, growth.

Fungi are known to colonize and multiply on diversified habitats and known to live as saprophytes or parasites. In the part two decades, plant pathology has become an important aspect of science receiving increasing attention. Fungal species representing *Alternaria, Curvularia, Drechslera, Cercospora, Colletotrichum, Ascochyta, Cylindrocarpon, Fusarium, Myrothecium, Nigrospora* and many other such fungi have been reported as biotrophs on a wide variety of crop plants and forest plants besides causing pre and post harvest diseases on fruits and vegetables. Chemicals have been used to combat such diseases but this chemical technology has been plagued with several problems such as pesticides, lack of proper education to the farmer and development of resistance in many pathogens towards fungicides. In view of this in recent times, search has been started to invest and implement biological control methods and also to test the efficacy of plant extracts and plant products and possible means of controlling the diseases and also inhibiting the growth of plant pathogens. Therefore, an attempt has been made by the author to study the effect of 100 plant extracts, pollen suspension of 100 plants, latex from

19 plants, 15 gums, 10 oils, 10 homoeopathic drugs, 12 toxic chemicals, 10 culture filtrates and antifungal compounds such as lauricidin (Plus 41) JBC - 45, Lauricidin[R] lot 303 # 25A and Lauribic[R] 11 lot #10984 on the conidial germination and mycelial growth of four pathogenic fungi under laboratory conditions. The four pathogenic fungi viz., *Curvularia lunata* (Walker) Boedign, *Cylindrocarpon lichenicola* (C. Massal) Hawksworth causal agent of corm rot of Colcasia esculenta (L) Schott., *Fusarium solani* (Mart) Sacc., causal agent of Wilt in tomato and *Myrothecium leucotrichum* (Peck.) Tulloch., causal agent of seed rot disease in *Coriandrum sativum* (L). These pathogenic fungi were selected due to the fact that they were prevalent in this area and caused heavy damage to the concerned host plants.

Materials and Methods

The four pathogenic fungi, viz, *Curvularia lunata*, *Cylindrocarpon lichenicola*, *Fusarium solani* and *Myrothecium leucotrichum* were selected. Spore germination was studied by spore glass-slide germination test (American Phytopathological Society, 1943). The percentage inhibition of fungal spore germination was calculated.

To study the growth and sporulation of four pathogenic fungi, one ml. of the extract or chemical was incorporated into 24 ml. of Czapek's liquid medium taken in 100 ml. Erlenmeyer conical flasks. A uniform agar disc of the sporulating fungus was used for inoculating the flasks. The inoculated flasks (in triplicate) were incubated at $24 + 2^\circ$ C for 10 days. The dry weight and sporulations were recorded.

Results and Discussion

The existing literature of higher plants has demonstrated the occurrence of antifungal substances and are known to contain chemicals which help in resisting the pathogens. (Joshi and Keshwal, 1969; Lokendra singh, 1976; Khanna and Chandra, 1976; Garg; 1975; Agarwal, 1978; Vijaya Kumar *et. al.*, 1979; Saksena and Saksena, 1981). In India, some 200 to 300 plants have been screened against the conidial germination and mycelial dry weight of the test organisms. Nearly sixty plants have shown more than 60% inhibition of conidial germination. Similar results were obtained with reference to growth and sporulation of four pathogenic fungi. It is interesting to mention here that the plants such as, *Mangifera indica* L., *Annona squamosa* L., *Artbotrys odoratissisus* R.Br., *Nerium odorum* Soland, *Colocasia esculenta* (L) *Schott.*, *Caesalpinia pulcherrima* (L) *Swartz.*, *Cassia occidentalis* L., *Delonix regia* (Boger) Reg., *Cleome viscosa* L., *Commelina bengalensis* L., *Cosmos bipinnatus Cav.*, *Helianthus annus* L., *Tagetus erecta* L, *Zinnia elegans* Jacq., *Ipomea carnea Jacq.*, *Brassica nigra* (L) *Koch.*, *Raphanus sativus* L., *Cucumis sativus* L., *Lagenaria siceraria* (Molina) *Standley.*, *Acalypha indica* L., *Sorghum vulgare* L., *Triticum vulgare Vill.*, *Salvia plebeia* R.Br., *Abelmoschus esculentus* (L) *Moench.*, *Althea rosea* (L) Cav., *Hibiscus cannabinus* L., *Malvaviscus conzatti* Greenm., *Mimosa pudica* L., *Prosopis juliflora* (Sw) DC., *Eucalyptus lanceolatus* (L) Herit., *Psidium guajava* L., *Bouganvillea spectabilis* Wild., *Jasminum sambac* Ait., *Papaver somniferum* L., *Dolichos lablab* L., *Tephrosia purpurea* (L) Pers., *Hamelia patens* Jacq., *Ixora arborea* Roxb. ex Smith., *Citrus aurantium* (L)., *Citrus medica* L., *Citrus sinensis* (L) Osbeck., *Murraya koenigi* (L) *Spreng.*, *Ruta graveolens* L., *Antirrhinium majus* L., *Cestrum diurnum* L., *Cestrum nocturnum* L., *Datura metel* L., *Lycopersicon esculentum* L., *Petunia hybrida* Vilm., *Physalis minima* L., *Solanum xanthocarpum* S & W., *Carum stictocarpum* (Clarke) Wolff., *Daucas carota* L., *Foeniculum vulgare* Gaertn., *Clerodendron phlomidis* L., *Duranta erecta* L., and *Vitis vinifera* L. were not tested earlier for their antifungal activity. Hence these are considered as new additions of science of Chemotherapy. The fungitoxcity of plants is due to the presence of the antifungal principal and stimulation of test fungi by certain plant extracts may be due to the presence of stimulants. In the present study the extracts of root, stem, leaf, flower and fruit of 100 plants have been tested against the conidial germination of four pathogenic fungi. It has been shown that root extracts of nearly seventy plants inhibitory to more than 60%, while the flower extracts of 83 plants were exhibited 60% and above inhibition. The leaf extracts of Allium cepa L., *Allium sativum* L., *Mirabilis jalapa* L., *Papaver somniferum* L., *Ixora arborea Roxb.* ex Smith, *Citrus medica* L., *Cestrum nocturnum* L., *Datura metel* L., *Carum stictocarpum* (Clarke) Wolff. and *Foeniculum Vulgare* Gaertn, exhibited 100% inhibition. Further different plant parts of *Calotropis gigantea* (L). R.Br., *Datura metel* L., *Ocimum sanctum* L., *Ricinus communis* L., and *Tridax procumbens* L., have shown maximum antifungal activity as evidenced by poor values in mycelial dry weight and less sporulation of four test pathogens.

From the study of the effect of 100 plants on conidial germination of four pathogenic fungi and effect of five plants on growth and sporulation of the test fungi, it is possible to draw the following conclusions :

1. 60% of plants possess natural resistance against diseases causing agents as they have antifungal substances.

2. Among the plant parts tested, the extracts of root and flower were found to be more effective followed by leaf and others.

3. The exhaustive survey indicated that the active principles of antifungal activity can be isolated and their systematic activity can be studied on a variety of pathogenic fungi so as to find out a suitable plant chemical as possible chemotherapeutant besides predicting their possible use in the control of various plant diseases.

4. The fifty-nine plants which have not been worked out by earlier workers have been added as new chemotherapeutants.

Knowledge of pollen and fungal association and the effect of pollen suspension is very much meagre from Indian Sub-Continent. It is in this context that the author felt the necessity of screening the pollen of 100 plants. This study has given ample evidence that the pollen suspension of *Carica papaya* L., *Momordica charantia* L., *Mirabilis jalapa* L., *Papaver somniferum* L., *Zizyphus jujuba* Lamk., *Morinda tomentosa* Heyne. ex. Roth., *Citrus medica* L., *Cestrum diurnum* L., and *Solanum nigrum* L., showed 100% inhibition in conidial germination of four pathogenic fungi namely *Curvularia lunata, Cylindrocarpon lichenicola, Fusarium solani* and *Myrothecium leucotrichum.* 60% and above inhibition has been observed by the pollen suspension of *Mangifera indica* L., *Annona reticulata* L., *Artabotrys odoratissumus* R.Br., *Nerium odorum* Soland, *Vinca rosea* L., *Calotropis gigantea* L., R.Br., *Cryptostegia* (Bojer) Ref., *Cleome viscosa* L., *Gynandropsis pentaphylla* DC., *Helianthus esculentus* (L) Moench., *Abutilon indicum* G. Don. *Hibiscus cannabinus* L., *Prosopis juiliflora* (Sw) DC., *Cajanus cajan* L., *Crotolaria juncea* (L)., *Desmodium triflorum* (L) DC., *Tephrosia purpurea* (L) Pers., *Antigonon leptopus* HK & Arn., *Hamelia patens* Jacq., *Ixora arborea* Roxb. ex Smith., *Citrus sinensis* (L) Osbeck., *Murraya koenigii* (L) Spreng., *Capsicum annum* L., *Cestrum nocturnum* L., *Datura metel* L., *Coriandrum sativum* L., *Foeniculum vulgare* Gaertn., *Clerodendron phlomidis* L., *Duranta erecta* L., *Lantana camara* L. and *Verbena officinalis* (L). The growth and sporulation studies of four pathogenic fungi as affected by five pollen extracts showed that *Carica papaya* L. and *Citrus medica* L. were inhibitory. The present study clearly indicates that the pollen suspension which proved fungitoxic indicates their role in disease resistance.

Latex forms one of the important plant products and has been proved to be antifungal by some workers. There is no exhaustive survey of different latex samples of different plants with reference to their antifungal activity. Hence it has been proposed to study the effect of latex collected from 19 plants on the conidial germination and mycelial dry weight of four pathogenic fungi. Latex of *Plumeria rubra* L., *Tabernaemontana divaricata* (L) R.Br., *Calotropis gigantea* (L) R.Br., *Carica papaya* L., *Ipomoea carnea* Jacq., *Ficus bengalensis* L., *Ficus carica* L., *Ficus retusa* Hook f., *Euphorbia splendens* L., *Argemone mexicana* L., and *Achras sapota* L., has shown 100% inhibition on conidial germination of *Cylindrocarpon lichenicola* (C. Massal) Hawksworth, while *Calotropis gigantea* (L) R.Br., *Carica papaya* L., *Ipomoea carnea* Jacq., *Ficus bengalensis* L., *Ficus carica* L., *Ficus religiosa* L., *Ficus retusa* Hook f., *Ficus splendens* L. and *Achras sapota* L. have shown 100% inhibition on conidial germination of *Fusarium solani* (Mart) Sacc. *Calotropis gigantea* (L) R.Br., *Carica papaya* L., *Ipomoea carnea* Jacq. *Ficus bengalensis* L., *Ficus carica* L., *Euphorbia pulcherrima* Willd., *Euphorbia tircualli* L. and *Achras sapota* L., were inhibitory to the conidial germination of *Myrothecium leucotrichum* (Peck) Tulloch to an extent of 100%. On the whole, the mycelial growth has been reduced to a greater extent by the latex of *Plumeria rubra* L., *Tabernaemontana divaricata* (L) R.Br., *Calotropis gigantea* (L) R.Br., *Cryptostegia grandiflora* R.Br., *Carica papaya* L., *Ipomoea catnea* Jacq., *Artocarous integrifolia* Lam., *Ficus benghalensis* L., *Ficus carica* L., *Ficus religiosa* L., *Euphorbia pulcherrima* Willd., *Euphorbia splendens* L., *Manihot utilissimus* Willd., *Argemone mexicana* L. and *Achras sapota* L. It has been observed that latex of *Ficus retusa* Hook f. was stimulatory on the conidial germination of *Curvularia lunata*. Statistical data also conforms the similar views. Almost all plants were inhibitory statistically on *Curvularia lunata*, *Cylindrocarpon lichenicola*, *Fusarium solani* and *Myrothecium leucotrichum* at 5% level of significance. But the latex of *Artocarpus integrifolia* Lam., *Ficus bengalensis* L., *Ficus retusa* Hook f, *Euphorbia pulcherrima* Willd. and *Euphorbia tirculli* L. was least effective as the correlation was not significant.

From the literature, it is found that the gums form an important habitat for the colonization of fungi but no attempt has been made to study the efficacy of gum extracts as antifungal agents. In the present investigation, gum lextract of fourteen plants and commercial gum have been tested against conidial germination and mycelial dry weight of four pathogenic fungi. All the fifteen gums were antifungal to *Curvularia lunata* and *Fusarium solani,* while they were moderately effective on *Cylindrocarpon lichenicola* and *Myrothecium leucotrichum.*

There are some reports that the oils can be used as protectants against pre and post harvest diseases, anthracnose diseases of grape vines and some oils have been found to inhibit the mycelial growth. In the present investigation, ten oils have been screened against the conidial germination and mycelial growth of four pathogenic fungi. In the present study, clove oil and mustard oil have shown 60% and more inhibitory action. Of oils the three concentrations tested, 75% has shown more inhibitory activity than 50% and 25%. The present data clearly indicates that the oils can be used as fungitoxic agents.

In recent times, homoeopathic drugs are used against some pathogenic fungi. However, there is not much work about the antifungal activity of homoeopathic drugs. In view of this, the efficiency of ten homoeopathic drugs has been tested on conidial germination and mycelial growth of four pathogenic fungi. 200 potencies of Acid fluoric, Ranunculus, Sepia, Silicea and Tellurium were found to be inhibitory on *Curvularia lunata*. Similarly, except Tellurium all others were effective both at 30 and 200 potencies on *Cylindrocarpon lichenicola, Fusarium solani and Myrothecium leucotrichum*. However, Graphites, Mercurium Viv., Sepia, Silicea and Thuja were stimulatory to *Curvularia lunata*. Sepia and Tellurium were stimulatory to *Cylindrocarpon lichenicola*, while kali Bichrom and Thuja were least effective at 200 potency on *Fusarium solani*. *Curvulara lunata* has responded favourably in its growth and sporulation to Acid fluoric, Kali Bichrom, Sepia, Sulphur and Tellurium. Almost all homoeopathic drugs tested both at potency 30 and 200 were inhibitory to the mycelial dry weight of *Cylindrocarpon lichenicola, Fusarium solani* and *Myrothecium leucotrichum. Curvularia lunata* could sporulate well in all the homoeopathic drugs except in Mercurium viv., Ranunculus and Thuja, while the remaining pathogenic fungi did not show sporulation on the homoeopathic drugs tested. The statistical data of the interaction between homoeopathic drugs and the mycelial dry weight clearly indicates that all the homoeopathic drugs were effective at 5% level of significance.

Secondary metabolites of the fungi play an important role in being either beneficial or detrimental on the growth of other microorganisms. Though vast literature exists from different countries on this aspect, there is little information available from India.

Therefore, the effect of culture filtrates of ten fungi namely *Aspergillus flavus* Link., *Cladosporium cladosporioides* Fresen de vries., *Cunninghamella blakesleeana* Lendner., *Curvularia lunata* (Walker) Boedijn., *Cylindrocarpon lichenicola* (C . Massal) Hawksworth, *Drechslera halodes* Subramanian and Jain., *Fusarium solani* (Mart) Sacc., *Myrothecium leucotrichum* (Peck)l Tulloch., *Rhizopus nigricans* (Enrenberg) Gilman, J.C. and *Trichoderma viride* Pera, were studied on the conidial germination and mycelial growth of four pathogenic fungi. The data shows that the culture filtrates of *Drechslera halodes* completely inhibited the conidial germination of *Curvularia lunata*, while the conidial germination of *Cylindrocarpon lichenicola* was suppressed by the culture filtrates of *Aspergillus flavus, Drechslera halodes* and *Fusarium solani.* The culture filtrates of *Cylindrocarpon lichenicola* has inhibited the conidial germination of *Fusarium solani.* The conidial germination of *Myrothecium leucotrichum* was inhibited by *Cladosporium cladosporioides, Curvularia lunata, Cylindrocarpon lichenicola* and *Drechslera halodes.* It is interesting to note that the conidial germination of *Myrothecium leucotrichum* and *Fusarium solani* were stimulated by their own culture filtrates. The mycelial growth of all pathogenic fungi was affected adversely by the culture filtrates of all ten fungi tested.

A number of chemicals such as ephedrine, caffeine, coumarin, n-butyric acid, salicyclic acid, abscisic acid etc. have been screened by some workers in the recent past. In order to substantiate them as protectants nearly twelve chemicals have been tested on the conidial germination and mycelial growth of four pathogenic fungi. The chlorogenic acid, kojic acid and papanium were inhibitory to conidial germination of all the test fungi. The present data clearly indicates that chemicals like abscissic acid, caffein, chlorogenic acid, Kojic acid and papainum need to be applied against pathogenic fungi so as to formulate suitable control measures both *in vivo* and *in vitro*.

In recent time some antifungal agents which are generally regarded as safe have been tried on few fungi, by some workers. In the present study the antifungal activity of Lauricidin[R] (Plus 41) JBD - 45, Lauribic[R] 11 lot # 10984 land lauricidin[R] lot # 303-25A has been tested on four pathogenic fungi.

It is interesting to note that lauricidin[R] (Plus 41) JBD-45, Lauribic[R] 11 lot # 10984 and lauricidin[R] lot # 303-25A were totally inhibitory to the conidial germination of all the four test fungi at 0.5% concentration. The concentrations of 0.1% and 0.05% of Lauricidin[R] plus 41 JBD-45 and Lauribic[R] 11 lot # 10984 inhibited the conidial germination of all the four fungi by more than 60%. However, the Lauricidin[R] lot #303-25A completely inhibited the conidial germination of *Curvularia lunata, Fusarium solani* and *Myrothecium leucotrichum* at both 0.1% and 0.05% concentrations. The length of the germtubes was also effected by the three chemicals. 0.5% concentrations of Lauribic[R] 11 lot # 10984 and Lauricidin[R] lot#303-25A completely suppressed the mycelial growth of all the four test fungi. The growth of *Cylindrocarpon lichenicola, Fusarium solani and Myrothecium leucotrichum* was completely inhibited by Lauricidin[R] (Plus 41) JBD-45 at 0.5% concentrations, while *Curvularia lunata* has shown a little growth in all the three chemicals, lauricidin[R] lot # 303-25A has shown greater inhibitory activity on the growth of all the four test fungi. Sporulation was greatly affected in all concentrations of the three chemical compounds.

The scheme presented above throws some light on the biological activity of plant extracts, plant products and chemicals on some pathogenic fungi. Further it paves the way to identify such plants which can be used as controlling agents through its chemicals which may be less toxic and may not have residual effect. It is pertinent to mention here that the present data of plant extracts forms further addition in that chain of evidence obtained so far. The data with reference to antifungal activity of pollen suspension, latex, gums, oils, homoeopathic drugs, fungal culture filtrates, chemicals and antifungal compounds (GRAS) further strengthens the idea of using them as protectants in controlling the plant diseases besides adding more information to the less known field i.e., antifungal activity of plant products.

References

American Phytopathological Society. 1943. Committee on standardization to fungicidal tests. The slide germination method of evaluation protectant fungicides. *Phytopathology* 33: 627

Agrawal, P. 1978. Effect of root and bulb extracts of Alliun Spp. on fungal growth. *Trans Br. Mycol. Soc. 70:438-441.*

Garg, S.C. 1975. Antifungal activity of some essential oils. *Indian J. Pharma 36:46.*

Joshi, L.K. and **Keshwal R.L.** 1969. *Inhibition of Pythium debaryanum* Hesse by culture filtrates of *Penicillium notatum* and two species of *Aspergillus* from soil. JNKVV Res. J. *3: 93-94.*

Khanna, K.K. and **Chandra, S.** 1976. Effect of some homoeopathic drugs on the spore germination of four isolates of *Alternaria alternata. Indian Phytopath 29: 195-197.*

Lokendra Singh. 1976. Effect of certian chemicals on growth performance of two Aspergilli. *Acta Botanica Indica, 4: 71-73.*

Saxena, A.K and **Saksena, S.B.** 1981. Effect of Papaya latex on the germination on conidia of certain fungi. *Indian Phytopathol. 34: 505-506.*

Vijaya Kumar, C.S.K. and **Rao, A.S** 1979. Inhibitory effect of some phenolic and fungicidal compounds on the growth of Alternaria triticina. Indian J. Microbiology. 19: 243-245.

Species Diversity of Running Fresh Water Bodies of Uttara Kannada Region of Karnataka, India with Reference to Water Borne Conidial Fungi

Ch.Ramesh and S.Vijaykumar*

Deptt. of Botany, Karnatak University, Dharwad-580 003, Karnataka
*Associate Professor, Dept of Biotechnology, Institute of Technology
and Science, Madanapalle – 517 325.Chittoor dt (AP)

ABSTRACT

The mycoflora of fresh water ecosystem includes a diverse group of fungi, now known as water borne conidial fungi, which are characterized by their typical spore types. The present paper describes species diversity of running fresh water bodies of Uttara Kannada region of Karnataka with special reference to the water borne conidial fungi. All together 103 species belonging to 86 genera of these fungi have been recorded from different habitats viz. water, foam, submerged leaf litter and living plant roots. This is nearly 60% of the total known species of these fungi from India. Out of 103 species collected 43 of them are new records to the fungi of India.

KEYWORDS : Conidial fungi, water bodies, uttar kannada

The Uttara Kannada region and especially the ghat section are very rich in the plant material of varied types and are of interest to taxonomists, plant geographers, foresters etc. Saldanha (1984) has pointed out the importance of the flora of this region, since it is rich not only in angiospermic vegetation but also in Algae, Fungi, Bryophytes, and Pteridophytes. Kaveriappa and his school have worked on Aquatic Hyphomycetes from Dakshina Kannada region of Karnataka otherwise there is very little work done on water borne fungi of Karnataka and also there are reports from other parts of India (Bilgrami *et al.,* 1979, 1981 Kamat *et al.,* 1972: Sridhar *et al.,* 1992).

A detailed and comprehensive survey of water borne fungi from Uttara Kannada region is undertaken by the present worker due to their peculiar habitat and their role in the ecosystem. The results of the study are represented in this work.

Materials And Methods

The four streams viz: Kali, Panda, Bidaralla and Magod streams flowing in the western ghat forest region of Karnataka state, India were selected for the study. These streams were lined by thick vegetation including cultivated crops and the streambeds are mostly rocky and muddy. They carry a lot of organic materials including leaf litter on account of heavy rains during southwest monsoon.

Three sampling sites at a distance of 1-3 km along each stream were selected submerged leaves, foam and water samples were collected from three spots in each site at monthly intervals over a period of two years during 1991 – 1993 and analyzed as follows.

Table 1 Fungi Isolated from different substrastes

Fungi	Substrate
Pseudeurotium saltisporus (Saito & Minoura) Stolk	*Pongamia glabra*, wood
Thielaviopsis paradoxa (Dade) C. Moreau.	*Canabis sativa*. seeds
Humicola fuscoatra Traeen.	*Karselia sp*, Leaf litter
Bactrodesmiella masonii (Hughes) M.B. Ellis.	On wood
Helicomyces hyderabadensis Rao & Dev Rao.	*Cocos nucifera*, Litter
Monacrosporium phymatophagum (Drechsler) Subram.	*Typha angustaata*.
Fusariella hughesii Chabelska - Frydnam.	*Cannabis sativa*
Aspergillus fumigatus Fresenius	*Bursera delpechianua* wood.
Phoma capitulum- 1 Pawar, Mathur & Thirus.	*Marselia* sp, Leaf litter.
Phoma capitulum - II Pawar, Mathur & Thirus.	*Polygonust* sp, wood

1. Leaf litter

Submerged leaves of different kinds collected randomly from each spot were brought to the laboratory in the moist condition on polythene bags. They were washed several times in tap water and finally in distilled water to remove the extraneous sediments and invertebrates. The leaves were cut into small bits (5cm) and incubated (24±3°C) separately in Petridishes containing 20 ml distilled water. The water in the Petridishes was replaced once in two days to minimize the growth of bacteria and other organisms. The leaf bits were examined under low power microscope once in two days for 60 days to detect the colonies of water borne fungi appearing on them (Ingold 1975) The colonies when sporulated were identified and recorded on the basis of morphology of conidia.

2. Foam

Ten ml of foam was collected from each spot in clean plastic bottles, fixed in FAA (Mixture of 40% formaldehyde 10ml Glacial acetic acid, 5ml and 70% ethyl alcohol, 85ml) on the spot, and later examined under a low power microscope to detect the conidia of water borne fungi (Ingold, 1975). The conidia were identified and recorded.

3. Baiting

Two kinds of seeds were selected viz; Hemp and *Cicer* seeds as baites to isolate water borne fungi in the laboratory. The seeds were boiled for one hour, cooled and added to the Petriplates which contained a mixture of sample water and distilled water. Observations were made regularly for one month for the fungal growth.

4. Test wood blocks

Four different kinds of wood were selected viz; *Pinus roxburghii* Sarg, *Bursera delpechianum Poiss, Pongamia glabra* Vent and *Bambusa aurundinaceae* Willd. 4-5 cm long, 1-2 cm diameter blocks were sterilized for 30 min at 15 psi. A total of 48 blocks of each kind were submerged at each site (40 × 3) for each stream and monthly four blocks from each site were collected. All collections were brought to the laboratory in sterile condition. After thoroughly washing under tap water they were incubated in Petriplates with 20-30 ml sterilized distilled water as well as in moist chamber. Observations were recorded for six months.

The presence/absence of each fungus from the total fungi collected from each stream by the four methods was recorded. From this percentage occurrence of each fungus was calculated.

The water temperature, pH, rainfall of the stream were recorded at monthly intervals for two years. Water temperature was measured by suspending glass mercury thermometer in water for 5 min. The pH was measured by using a portable pH meter. The monthly rainfall data during the two years of study were obtained from meteorological observatory.

Results

Kali stream

Twenty four species of water borne fungi were recorded, out of which six species belonged to phycomycetes, eight species belonged to Ascomycetes and ten species from Fungi Imperfecti; all these forms were collected by three methods namely baiting, where as Ascomycetes on test wood block, Fungi Imperfecti were isolated on both test wood blocks and leaf litter.

Out of these *Dictyuchus monosporus, Zopfiella latipes, Savoryella lignicola, Diaymosphaeria dandeliensis, Helicosporium panacheum* and *Phaeoisaria sparsa* were recorded abundantly. All these organisms were recorded in all the methods. *Jahnula australiensis, Gliocladiopsis tenuis, Sporoschisma mirable* and *Stagnospora vitensis* were less abundant. The remaining sixteen species occurred rarely.

In Phycomycetes, *Dictyuchus monosporus* was the most predominant fungus and the frequency of occurrence of the rest of the fungi was low. In each case of Ascomycetes *Savoryella lignicola* was the predominant fungus, where as *Zopfiella lundquistii* and *Didymosphaeria dandeliensis* were less abundant. The frequency of occurrence of the other organisms was low.

In case of Fungi Imperfecti *Phaeoisaria sparsa* was the predominant fungus. *Helicosporium panacheum* was less abundant and the frequency of occurrence of other organisms was low (Table. 2)

Thirty two species of foam spora were recorded. Out of which *Triscelophorus accuminatus, Campylospora filicladia* and *Flabellospora verticillata* were the most abundant, while the occurrence of rest of the fungi were low (Table. 3, Fig. 1)

Table 2 Seasonal occurence of waterborne fungi from Kali stream

Fungus	Years								Number of occurences	Percentage Occurences
	1991-92				1992-93					
	Jun Aug	Sep Nov	Dec Feb	Mar May	Jun Aug	Sep Nov	Dec Feb	Mar May		
Phycomycetes										
Dictyuchus monosporus Leitgeb	3	-	-	-	-	-	-	-	3	6.38
Blastocladia ramosa Thaxter	-	1	-	-	-	-	-	-	1	2.12
Gonapodya polymorpha Thaxter	-	-	-	1	-	-	-	-	1	2.12
Gonapodya prolifera (Cornu) Seymour	-	-	-	1	-	-	-	-	1	2.12
Protoachlya oryzae Khulbe	1	-	-	-	-	-	-	-	1	2.12
Monoblepharis macrandra (Lagerheim) Woronin	-	-	-	1	-	-	-	-	1	2.12
Ascomycetes										
Savoryella lignicola Jones & Eaton	1	-	1	2	-	-	1	-	5	10.63
Didymosphaeria dandeliensis sp. nov.	-	-	1	2	-	-	-	-	3	6.38
Zopfiella latipes (Lundquist) Malloch & Cain	1	1	1	-	-	-	-	-	3	6.38
Jahnula australiensis Hyde.	1	1	-	-	-	-	-	-	2	4.25
Chaetomium medusarum Meyer & Lanneau	-	-	-	1	-	-	-	-	1	2.12
Leptosphaeria peruviana Spegazzini	1	-	-	-	-	-	-	-	1	2.12
Nectria dentifera Samuels	1	-	-	-	-	-	-	-	1	2.12
Pseudeurotium multisporus (saito & Mino) Stol.	-	-	-	-	-	-	1	-	1	2.12
Fungi Imperfecti										
Phaeoisaria sparsa Sutton	-	1	2	-	-	1	-	-	4	8.51
Helicosporium panaceum Moore	-	1	1	-	1	-	-	-	3	6.38
Gliocladiopsis tenuis (Bugnicourt Crou & Win)	2	-	-	-	-	-	-	-	2	4.25
Sporoschisma mirabile Berk & Br.	-	1	1	-	-	-	-	-	2	4.25
Stagnospora vitensis Unam.	-	1	1	-	-	-	-	-	2	4.25
Acrophialophora fusispora (Saksena) M.B. Ellis	1	-	-	-	-	-	-	-	1	2.12
Aspergillus fumigatus Fresenius	1	-	-	-	-	-	-	-	1	2.12
Denatophora necatrix Hartig	-	-	1	-	-	-	-	-	1	2.12
Drepunospora pannosa Berkeley & Curtis	1	-	-	-	-	-	-	-	1	2.12
Humicola fuscoatra Traaen	-	-	1	-	-	-	-	-	1	2.12
Total	16	8	10	8	1	1	2	0	47	

Table 3 Seasonal occurence of Foam spora of Kali stream

| Fungus | Years | | | | | | | | Number of occurences | Percentage Occurences |
| | 1991-92 | | | | 1992-93 | | | | | |
	Jun Aug	Sep Nov	Dec Feb	Mar May	Jun Aug	Sep Nov	Dec Feb	Mar May		
Triscelophorus accuminatus Nawawi	361	40	4	-	479	26	2	-	912	27.18
Campylospora filicladia Nawami	230	92	1	-	234	21	-	-	578	17.23
Flabellospora verticillata Alasoadura	185	29	-	-	122	3	-	-	339	10.1
Helicoayces colligatus Moore	10	306	-	-	13	-	-	-	329	9.8
Triscelophorus konajensis Sridhar & Kaver	100	6	-	-	137	-	-	-	243	7.2
Phalaqiospora constricta Nawawi	70	4	-	-	80	-	-	-	154	4.8
Isthrotricladia gonabakiensis Nawawi	70	6	-	-	75	-	-	-	151	4.5
Diplocladiella scalaroides Arnaud	39	28	-	-	54	-	-	-	121	3.6
Lateriranulsa uni-inflata Matsushima	68	-	-	-	40	-	-	-	108	3.2
Wiesneriomyces javanicus Koorders	29	19	-	-	26	-	-	-	74	2.2
Speiropsis pedatospora Tubaki	10	26	-	-	10	4	-	-	50	1.49
Flabellospora nultiradiata Nawawi	23	1	-	-	24	-	-	-	48	1.43
Tricladium anagulatum Ingold	10	31	-	-	7	-	-	-	48	1.43
Triscelophorus monosporus Ingold	37	-	-	-	10	-	-	-	47	1.4
Lunulospora curvula Ingold	1	16	1	-	5	4	1	3	31	0.92
Hycoleptodiscus lateralis Akron & Sutton	-	-	-	-	22	-	-	-	22	0.65
Anquillospora longissima Ingold	2	1	-	-	16	-	-	-	19	0.56
Clavatospora tentacula (Umplett) Nilsson	2	9	-	-	1	1	-	-	13	0.38
Beltrania rhombica Penzig	-	4	-	-	8	-	-	-	12	0.35
Tricladium splendens Ingold	6	-	-	-	6	-	-	-	12	0.35
Anquillospora curvula Iqbal	1	4	-	-	4	-	-	-	9	0.26
Canylospora chaetocladia Ranzoni	-	2	-	-	-	3	1	-	6	0.17
Tetraploa aristata Berk & Br	-	4	-	-	1	-	-	-	5	0.14
Hydronetrospora symmetrica Bonczol & Revay	1	-	-	-	3	-	-	-	4	0.11
Actinospora negalospora Ingold	-	1	-	-	-	-	1	-	2	0.05
Articulospora noni liforna Ranzoni	2	-	-	-	-	-	-	-	2	0.05
Speiropsis hyalospora Subram, & Lodha	-	-	-	-	2	-	-	-	2	0.05
Tripospermum canelopardus Ingold	-	-	-	-	2	-	-	-	2	0.05
Anquillompoma filiformis Greathead	-	-	-	-	1	-	-	-	1	0.02
Clavariana aquatica Nawami	-	1	-	-	-	-	-	-	1	0.02
Pestalotia Sp.	1	-	-	-	-	-	-	-	1	0.02
Tetracladiun setigerus (Grave) Grove Ingold	-	-	-	-	1	-	-	-	1	0.02
Total	1258	630	6	0	1383	62	5	3	3347	

The data on the number of species encountered in three methods, the average number of conidia present in the foam at quarterly intervals and the data on water temperature, rain fall and pH are represented graphically (Fig. 1).

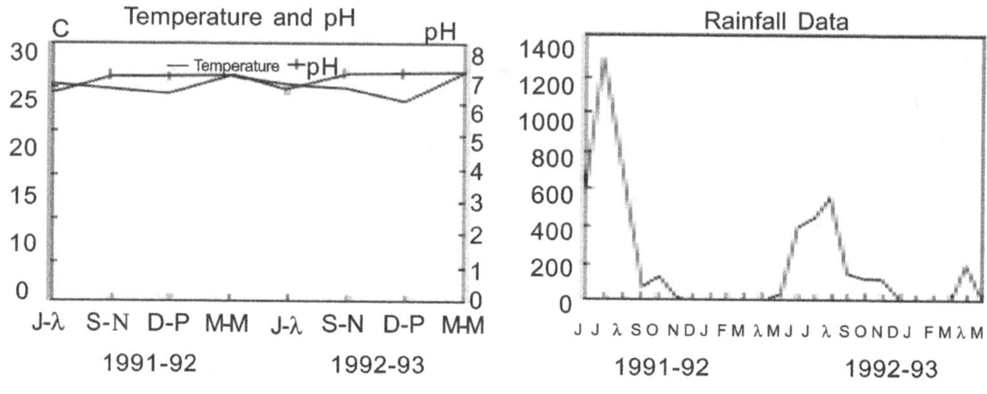

Fig 1 Kali stream

In all the methods the total number of the species encounter was the highest in the July – September months and it was lowest in the March – May months.

Panda stream

Twenty-eight species of water borne fungi were recorded. Out of which seven species were Phycomycetes, six species were Ascomycetes and fifteen species were Fungi Imperfecti. All these forms were encountered in three methods viz; Baiting, test wood blocks and leaf litter.

Out of these *Nowakowskiella profusa, Frondicola tunitricuspis, Savoryella lignicola, Sporoschisma mirabile* and *Septoria bambusae* were recorded abundantly. All these organisms were collected on all the methods. *Pseudoeurotium multisporum, Conriothyrium obiones, Phaeoisariopsis griseola* and *Verrucispora proteacearum* were less abundant. The remaining 19 species occurred rarely.

All the seven species belonging to Phycomycetes were isolated by baiting technique. *Nowakowskiella profusa* was most predominant fungus, while the frequency of other organisms is low.

All the six species of Ascomycetes were collected on test wood blocks out of which *Frondicola tunitricuspis* was very predominant fungus. *Savoryella lignicola* was less abundant while the frequency of occurrence of other organisms was low.

All the fifteen species of Fungi Imperfecti were collected either on test wood blocks or leaf litter. *Sporoschisma mirabile* was the most predominant fungus. *Septoria bambusae, Phaeoisariopsis griseola, Verrucispora proteacearum* and *Coniothyrium obiones* were less abundant. The frequency of occurrence of other organisms was low.(Table. 4)

Thirty species of foam spora were recorded out of which *Triscelophorus accuminatus* was most predominant and *T. konajensis* and *Lunulospora curvula* were less predominant while the occurrence of other fungi were low (Table. 5)

Table 4 Seasonal occurence of waterborne fungi from Panda stream

| Fungus | Years | | | | | | | | Number of occurences | Percentage Occurences |
| | 1991-92 | | | | 1992-93 | | | | | |
	Jun Aug	Sep Nov	Dec Feb	Mar May	Jun Aug	Sep Nov	Dec Feb	Mar May		
Phycomycetes										
Nowakowskiella profusa Karling	1	-	-	1	-	-	-	-	2	4.54
Alloamyces nacrogynus (Emerson) Emer & Wils	1	-	-	-	-	-	-	-	1	2.27
Gonapodya polymorpha Thaxter	-	-	-	1	-	-	-	-	1	2.27
Gonapodya prolifera (Cornu) Fischer	1	-	-	-	-	-	-	-	1	2.27
Achlya recemosa Hildebrand	1	-	-	-	-	-	-	-	1	2.27
Aphanomyces stellatus de Bary	1	-	-	-	-	-	-	-	1	2.27
Dictyuchus monosporus Leitgeb	-	1	-	-	-	-	-	-	1	2.27
Ascomycetes										
Frandicola tunitricuspis Hyde	-	-	2	2	-	-	-	-	4	9.09
Savoryella lignicola Jones & Eaton	-	-	1	1	-	1	-	-	3	6.81
Pseudeurotium multisporum (Saito & Mino) Stolk	1	1	-	-	-	-	-	-	2	4.54
Aniptodera chesapeakensis Shearer and Miller	-	-	1	-	-	-	-	-	1	2.27
Jahnula australiensis Hyde	-	1	-	-	-	-	-	-	1	2.27
Zopfiella latipes Shearer & Crane	-	-	-	1	-	-	-	-	1	2.27
Fungi Imperfecti										
Sporoschisma mirable Berk & Br	-	2	2	-	-	1	-	-	5	11.36
Septoria bambusae	-	-	2	1	-	-	-	-	3	6.81
Coniothyrium obiones Jaap	-	2	-	-	-	-	-	-	2	4.54
Phaeoisariopsis griseola (Sacc.) Ferraris	-	1	1	-	-	-	-	-	2	4.54
Verrucispora proteacearum Shaw & Alcorn	-	1	1	-	-	-	-	-	2	4.54
Acrenoniula sarcinellae (Pat & Har) Arn. ex Dei	-	1	-	-	-	-	-	-	1	2.27
Acrophialophora fusispora (saksena) M.B. Ellis										
Alveophona caballeroi Bausa Alcalde	-	1	-	-	-	-	-	-	1	2.27
Chaetospermum chaltosporum (Pat) Smith & Ramsb	1	-	-	-	-	-	-	-	1	2.27
Endophragmia elliptica (Berk & Br) M.B.Ellis	-	-	-	1	-	-	-	-	1	2.27
Gliocladiopsis tenuis (Bugnicourt) Crous & Win	1	-	-	-	-	-	-	-	1	2.27
Helicosporiun indicun Rao & Dev Rao	1	-	-	-	-	-	-	-	1	2.27
Leptodernella incarnata (Bres) Hohn	-	-	1	-	-	-	-	-	1	2.27
Phaeodactylium alpiniae (Sawada) M.B. Ellis	1	-	-	-	-	-	-	-	1	2.27
Tripospemum nyti (Lind) Hughes	-	-	1	-	-	-	-	-	1	2.27
Total	11	10	13	8	-	2	-	-	43	

Table 5 Seasonal occurence of Foam spora of Panda stream

Fungus	Years								Number of occurences	Percentage Occurences
	1991-92				1992-93					
	Jun-Aug-	Sep-Nov-	Dec-Feb-	Mar-May-	Jun-Aug-	Sep-Nov-	Dec-Feb-	Mar-May-		
Triscelophorus acuminatus Nawawi	69	19	8	-	86	8	1	-	191	54.72
Triscelophorus konajensis Sridhar & Kaveri	14	2	2	-	16	-	-	-	34	9.74
Lunulospora curvula Ingold	1	6	8	1	3	9	-	-	28	8.02
Campylospora filicladia Nawawi	11	-	-	-	4	-	-	-	15	4.29
Isthnotricladia gonbakiensis Nawawi	3	1	-	-	6	-	-	-	10	2.86
Tetraploa aristata Berk & Br.	-	4	1	-	2	-	-	-	7	2.00
Triscelophorus monosporus Ingold	5	2	-	-	-	-	-	-	7	2.00
Anguillospora longissima Ingold	-	-	-	-	6	-	-	-	6	1.7
Lateriranulosa uni-inflata Matsushima	6	-	-	-	-	-	-	-	6	1.7
Flabellospora vercillata Alasoadura	2	-	-	-	3	-	-	-	5	1.43
Helicomyces colligatus Moore	-	01	-	-	4	-	-	-	5	1.43
Flabellospora nultiradiata Nawawi	2	-	-	-	2	-	-	-	4	1.14
Helicomyces torquatus Lane and Shearner	-	-	-	-	4	-	-	-	4	1.14
Wiesnerioamyces javanicus Koorders	1	-	-	-	3	-	-	-	4	1.14
Phalangispora constricta Nawawi and Webster	1	-	-	-	1	1	-	-	3	0.85
Tricladium splendens Ingold	2	-	-	-	1	-	-	-	3	085
Beltrania rhonbica subram	-	-	1	-	1	-	-	-	2	0.57
Speiropsis hyalospora Subramand Lodha	-	1	-	-	-	1	-	-	2	0.57
Tricladium angulatun Ingold	1	-	-	-	1	-	-	-	2	0.57
Actinospora aegalospora Ingold	1	-	-	-	-	-	-	-	1	0.28
Anquillospora curvula Iqbal	-	1	-	-	-	-	-	-	1	0.28
Anquillospora gigantea Ranzoni	-	-	-	-	-	1	-	-	1	0.28
Articulospora tetracladia Ingold	-	1	-	-	-	-	-	-	1	0.28
Clavariana aquatica Nawawi	-	1	-	-	-	-	-	-	1	0.28
Diplocladiella scalaroides Arnaud	-	1	-	-	-	-	-	-	1	0.28
Mycoleptodiscus lateralis Alcorn and Sutton	-	-	-	-	1	-	-	-	1	0.28
Pestalotia sp.	-	1	-	-	-	-	-	-	1	0.28
Speiropsis pedatospora Tubaki	-	1	-	-	-	-	-	-	1	0.28
Tetracladiun setiqerum (Grovel) Inglod	1	-	-	-	-	-	-	-	1	0.28
Triposperum canelopardus Ingold (Gove)	-	-	1	-	-	-	-	-	1	0.28
Total	120	42	21	1	144	20	1	-	349	

The data on the number of species encountered in these methods, the average number of conidia present in the foam at the quarterly intervals and data on water temperature, rainfall and pH are represented graphically (Fig. 2)

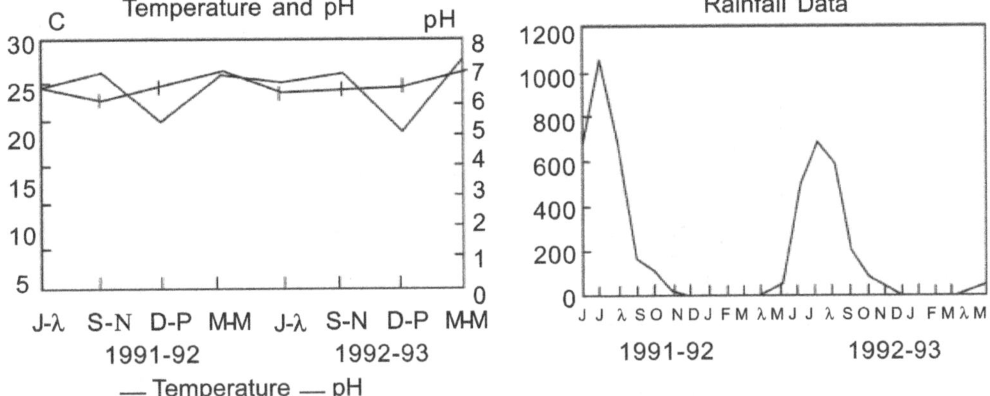

Fig. 2 Panda Stream

In all these methods the total number of species encountered in the three methods is less when compared to foam method. The maximum number of species occurred in foam during July – August and lowest in March – May season.

Bidaralla stream

Nine species of water borne fungi were recorded. Out of which three species belonged to Phycomycetes and six species Fungi Imperfecti. Ascomycetes could not be isolated due to natural disturbance as the test wood blocks washed away.

All the three species belonging to Phycomycetes were isolated by baiting technique. The frequency of occurrence of *Gonapodya polymorpha*, *G. prolifera*, and *Monoblepharsis macrandra* are common.

All the six species of fungi imperfecti were collected on leaf litter. *Phaeoisariopsis griseola* was most predominant fungus. *Helicomyces hyderabadensis* and *Verrucispora proteacearurn* were less abundant. The frequency of occurrence of other organisms was low (Table. 6)

Fourteen species of foam spora were recorded. Out of which *Triscelophorus accuminatus* was dominant, *Helicomyces torquatus*, and *Tripospermum camelopardus* were less abundant. While the occurrence of other fungi were low (Table. 7)

Table 6 Seasonal occurence of water borne fungi from bidaralla stream

Fungus	Years								Number of occurences	Percentage Occurences
	1991-92				1992-93					
	Jun Aug	Sep Nov	Dec Feb	Mar May	Jun Aug	Sep Nov	Dec Feb	Mar May		
	1	2	3	4	5	6	7	8	9	10
Phycomycetes										
Gonapodya polymorpha Thaxter	-	1	-	-	-	-	-	-	1	7.69
Gonapodya prolifera (Cornu) Fischer	-	-	1	-	-	-	-	-	1	7.69
Monoblepharis macrandra (Lagerheim) Woronin	-	1	-	-	-	-	-	-	1	7.69
Fungi Imperfecti										
Phaeoisariopsis griseola (sacc.) Ferraris	-	1	2	-	-	-	-	-	3	23.07
Helicomyces hyderabadensis Rao & Dev Rao	2	-	-	-	-	-	-	-	2	15.38
Verrucispora proteacearum Shaw and Alcorn	-	-	1	-	-	1	-	-	2	15.38
Drechslera spicifer Nelson	-	-	1	-	-	-	-	-	1	7.69
Phialosporostilbe setosa Bhat and Kendrick	1	-	-	-	-	-	-	-	1	7.69
Septoria bambusae sp. nov.	-	-	1	-	-	-	-	-	1	7.69
Total	3	3	6	-	-	1	-	-	13	

Table 7 Seasonal occurrence of Foam spora of bidaralla stream

Fungus	Years								Number of occurrences	Percentage occurrences
	1991-92				1992-93					
	Jun-Aug-	Sep-Nov-	Dec-Feb-	Mar-May-	Jun-Aug-	Sep-Nov-	Dec-Feb-	Mar-May-		
	1	2	3	4	5	6	7	8	9	10
Triscelophorus accuminatus Nawawi	18	3	2	-	23	-	4	-	50	2272
Heliocomyces torquatus Lane and S hearer	9	-	-	-	21	-	-	-	30	13.63
Triposperum comelopardus ingold	1	1	-	-	1	-	27	-	30	13.63
Anquillospora curvula Iqbal	8	-	1	-	11	-	-	-	20	9.09
Lunulospora curvula ingold	5	2	5	-	6	-	2	-	20	9.09
Triscelophorus konajensis Sridhar and Kaveri	7	2	-	-	11	-	-	-	20	9.09
Helicomyces colligatus Moore	6	1	-	-	9	-	-	-	16	7.27
Isthanotricladia qonbakiensis Nawawi	6	-	-	-	9	-	-	-	15	6.81
Tetraploa aristata Berk and Br.	1	4	1	-	1	-	-	1	8	3.63
Flabellospora verticillata Alasoadura	2	-	-	-	2	-	-	-	4	1.81
Wiesneriomyces javanicus Koorders	2	-	-	-	1	-	-	-	3	1.36
Phalangispora constricta Nawawi and Webster	1	-	-	-	1	-	-	-	2	0.90
Anquillospora filiformis Greathead	-	1	-	-	-	-	-	-	1	0.45
Mycoleptodiscus lateralis Alcorn and Sutton	-	-	-	-	1	-	-	-	1	0.45
Total	66	14	9	-	97	-	33	1	220	

The data on the number of species encountered in two methods, the average number of conidia present in the foam at quarterly intervals and the data on water temperature, rainfall and pH are represented graphically (Fig. 3)

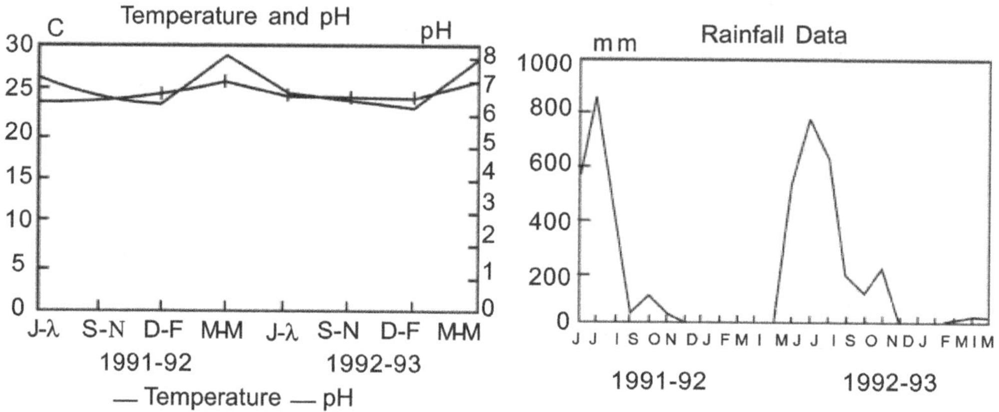

Fig. 3 Bidaralla Stream

The number of species encountered in the three methods is less when compared to the foam analysis. The maximum number of species occurred in foam during months of July – August in 91-92, 92-93 and lowest in December- February.

Magod stream

Nine species of water borne fungi were recorded. Out of which two species belong to Phycomycetes, two species belonged to Ascomycetes and rest five species belonged to Fungi Imperfecti. All these organisms were isolated by three methods viz; baiting technique, test wood blocks and leaf litter.

Both the species belonging to Phycomycetes were isolated by baiting technique. Both the species belonging to Ascomycetes were isolated by test wood blocks, *Savoryella lignicola* was the predominant fungus. The frequency of occurrence of *Nectria dentifera* was less abundant when compared to *Savoryella lignicola*.

Five species of Fungi imperfecti were recorded on leaf litter and test wood blocks. *Beltraniella odinae* was the most predominant fungus. The frequency of occurrence of the rest of the organisms was low (Table. 8)

Thirty species of foam spora were recorded. Out of which *Triscelophorus accuminatus* was predominant, while *Isthmotricladia gombakiensis* and *Triscelophorus konajensis* were dominant. While the occurrence of other fungi were low (Table. 9)

Table 8 Seasonal occurrence of water borne fungi from bidaralla stream

Fungus	Years 1991-92				1992-93				Number of occuences	Percentage Occurences
	Jun Aug	Sep Nov	Dec Feb	Mar May	Jun Aug	Sep Nov	Dec Feb	Mar May		
	1	2	3	4	5	6	7	8	9	10
Phycomycetes										
Gonapodya polymorpha Thaxter	1	-	-	-	-	-	-	-	1	7.69
Monoblepharis wacrandea (Lagerheim) Woroin	1	-	-	-	-	-	-	-	1	7.69
Ascomycetes										
Savoryella lignicola Jones and Eaton	-	-	3	-	-	-	-	-	3	23.07
Nectria dentifera Samuels	-	1	1	-	-	-	-	-	2	15.38
Hyphomycetes										
Beltraniella odinae Subram	-	-	1	-	1	-	-	-	2	15.38
Cryptophiale Mangiferae sp. nov	-	-	1	-	-	-	-	-	1	7.69
Denatophora necatrix Hartig	-	1	-	-	-	-	-	-	1	7.69
Nigrospora panici Zimm.	1	-	-	-	-	-	-	-	1	7.69
Verricispora proteacearum Shaw and Alcorn	-	-	1	-	-	-	-	-	1	7.69
Total	3	2	7	-	1	-	-	-	13	

Table 9 Seasonal occurrence of Foam spora of Magod stream

Fungus	Years								Number of occurences	Percentage Occurences
	1991-92				1992-93					
	Jun Aug	Sep Nov	Dec Feb	Mar May	Jun Aug	Sep Nov	Dec Feb	Mar May		
	1	2	3	4	5	6	7	8	9	10
Triscelophorus acuminatus Nawawi	164	27	3	-	184	9	4	-	391	28.71
Isthnotricladia qonbakiensis Nawawi	87	-	-	-	80	-	-	-	167	12.28
Triscelophorus konajensis Sridhar & Kaveri	52	6	5	-	50	-	-	-	113	8.31
Heliocoayces torquatus Lane & Shearer	36	-	-	-	36	4	-	-	76	5.59
Beltrania rhombica Penzig	32	-	-	-	34	-	-	-	66	4.85
Lunulospora curvua Ingold	1	9	22	-	4	2	24	-	62	4.56
Triscelophorus monosporus Ingold	29	1	-	-	28	1	-	-	59	4.34
Speiropsis pedatospora Tubaki	24	-	1	-	26	-	-	-	51	3.75
Phalagispora constricta Nawawi & Webstar	21	2	-	-	27	-	-	-	50	3.67
Flabellospora verticillata Alasoadura	26	-	-	-	20	1	1	-	48	3.53
Mycoleptodiscus lateralis Alcorn & Sutton	-	-	-	-	45	-	-	-	45	3.31]
Anquillospora longissima Ingold	15	2	1	-	18	-	1	-	37	2.72
Tetraploa aristata Berk & Br.	15	3	-	-	18	-	-	-	37	2.72
Flabellospora aultiradiata Nawawi	18	-	-	-	18	-	-	-	36	2.64
Helicomyces colligatus Moore	8	2	-	-	18	-	-	-	28	2.06
Diplocladiella scalariodes Arnaud	10	-	-	-	16	-	-	-	26	1.91
Tricladium splendens Ingold	7	-	-	-	7	-	-	-	14	1.03
Wiesneriomyces javancius Koorders	4	2	-	-	7	-	-	-	13	0.95
Campylospora filicladia Nawami	3	-	-	-	8	-	-	-	11	0.80
Lateriranulosa uni-inflata Matsushima	7	-	-	-	-	-	-	-	7	0.51
Tricladium anqulatea Ingold	2	-	-	-	4	-	-	-	6	0.44
Anquillospora gigantea Ranzoni	1	1	-	-	1	1	-	-	4	0.29
Actinospora negalospora Ingold	-	1	-	-	-	-	1	-	2	0.14
Campylospora chaetocladia Ranzoni	-	-	-	-	-	2	-	-	2	0.14
Lardospora appendiculata Nawawi	2	-	-	-	-	-	-	-	2	0.14
Pestalotia Sp.	-	-	-	-	2	-	-	-	2	0.14
Anquillospora curvula Iqbal	-	1	-	-	-	-	-	-	1	0.07
Articulospora moniliforna Ranzoni	1	-	-	-	-	-	-	-	1	0.07
Clavariana aquatica Nawawi	-	-	-	-	-	-	1	-	1	0.07
Speiropsis hyalospora Subram & Lodha	1	-	-	-	-	-	-	-	1	0.07
Total	566	57	32	-	651	21	32	-	1359	

The data on the number of species encountered in all the three methods, the average number of conidia present in the foam quarterly intervals and the data on water temperature, rainfall and pH are represented graphically. (Fig. 4).

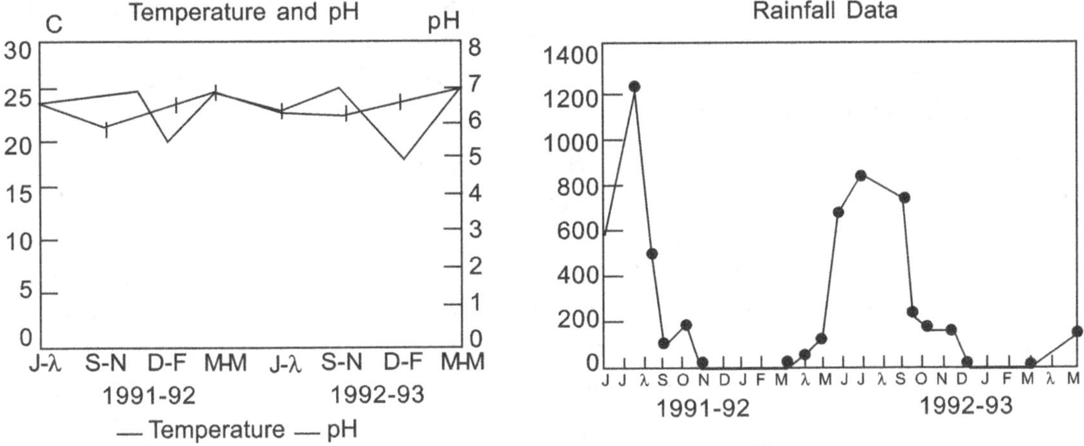

Fig. 4 Magod stream

The number of species encountered in the three methods was less when compared to foam method the maximum number of species encountered was the highest in July – August months and lowest in the months of December-February.

Comparison of occurrence of fungi in four areas

A total of 103 species belonging to 86 genera were recorded from the four areas during 1991 –1993. Twelve species were common to all streams viz; *Gonopodya polymorpha, Anguillospora curvula, Flabellospora verticillata, Helicomyces colligatus, Isthmotricladia gomabakiensis, Lateriramulosa uni-inflata, Mycoleptodiscus lateralis, Phalangispora constricta, Tetraploa aristata, Triscelophorus accuminatus, T. konajensis* and *Wiesneriomyces javanicus.* Fifty eight species were found in Kali stream, sixty species were found in Panda stream, thirty eight species in Magod stream and twenty three species in Bidaralla stream and Twenty three species from other bodies.

Discussion

Various Mycologists have studied the water borne fungi by isolating them from baiting methods, leaf litter, test wood blocks, foam and water (Iqbal and Webster 1973, Willoughby and Archer 1973; Barlocher and Kendrick 1974; Ingold 1975). Some investigators have attempted to trap conidia on submerged glass slides (Barlocher et al 1977; Muller Haeckal 1977). In the present study 103 species were recorded belonging to 86 genera in the four streams with four methods of analysis over a period of two years. Of the mycoflora recorded 12 species were common to all the four streams.

In the present investigation 103 species were recorded by one or the other method analysis in Kali, Panda, Bidaralla and Magod streams. Further all the mycoflora recorded in the streams were not found in all the samples while some species were recorded in most of the samples, almost throughout the year. Others were restricted to certain parts of the year. These observation do indicate that for correct assessment of the water borne fungi in a stream not only different methods of isolation have to be adapted but also samplings should be done in different times of a year.

Of the all the other organisms recorded in the present study *Gonapodya polymorpha, Anguillospora curvula, Flabellospora verticiliata, Helicomyces colligatus, Isthmotricladia gombakiensis, Laterirumulosa uni-inflata, Mycoleptodiscus lateralis, Phalangispora constricta, Tetraploa ariststa, Triscelophorus accuminatus, T. konajensis* and *Wiesneriomyces javanicus* were not only common to the four streams but also are found in all the seasons through out the two years of investigation. Hence, these may be considered as the most common water borne fungi of this region.

The occurrence of water borne fungi was also found to be affected by rainfall, water temperature and availability of leaf litter. The occurrence of water borne fungi have been reported earlier from temperate regions (Iqbal and Webster 1973; Muller Haeckel and Marvanova 1976, 1979 ; Lamore and Goos 1978; Del frate and Caretta 1983). Willoughby and Archer(1973) reported the occurrence of maximum number of species during wet period rather than dry period. Conway (1970) observed a considerable increase in the number of water borne fungi two weeks after the beginning of the leaf fall. The conidia of water borne Hyphomycetes in water were represented to more after the leaf deposition in the streams (Iqbal and Webster 1973; Muller Haecal and Maranova 1976, 1979; Delfrata and Caretta 1983). Lamore and Goos (1978) observed more number of species, Willoughby and Archer (1973) recorded maximum number of conidia in foam samples after heavy rainfall, Iqbal and Webster (1973b) correlated the abundance of conidia of *Clavarioopsis aquatica* and *Flagellospora curvula* coinciding with the intensive leaf deposition and rainfall on the occurrence of water borne Hyphomycetes, these observations were also obtained in the present investigation. In all the four streams the maximum number of species were recorded mostly from the collection of June-July (Monsoon) and November (Post- monsoon) and minimum in the May month (Summer) collections. The rainfall data on Dandeli, Londa, Yellapur and Sirsi regions during the two years of investigation shows in this region the south west monsoon started during April-May, reached maximum in the months of July, August and declined there after. The collection of water borne Hyphomycetes in the Bidaralla stream was also the lowest during summer months, where there were no rains. The leaf fall usually takes place during March – June in this region. The heavy rainfall during July – August carry the fallen leaves into the streams along with the terrestrial fungi and there by enrich the organic substrata and the mycoflora of the streams.

Wood – Eggenschwiler and Barlocher (1983) studied the waterborne fungi of 16 streams in France, Germany and Switzerland and correlated the species maxima with that of riparian vegetation of the streams. In the present investigation the number of species of water borne fungi was found to be more in Kali and Panda streams etc. due to heavy litter inflow, pH, temperature, rainfall and vegetation on the stream banks.

Thus abundance and varieties of plant litter in these Kali, Panda, Bidaralla and Magod streams are believed to be the reasons for recording of more number of species of water borne fungi in the four streams.

References

Barlocher, F. and Kendrick, B. 1974 Dynamics of fungal population on leaves in a stream. *J. Ecol.* 62: 761-791.

Barlocher, F. and Kendrick, B. 1977 Colonization of resin – coated slides by aquatic hyphomycetes . *Can. J. Bot.* 55: 1163-1168.

Bilgrami, M.S, Jamaluddin and Rizwi, M.A. 1979 *Fungi of India part-I* (List and references) 467, Today and tomorrow's printers and publishers, NewDelhi.

Bilgrami, M.S, Jamaluddin and Rizwi, M.A. 1981.*Fungi of India part-II* (List and references) 268 , Today and tomorrow's printers and publishers, NewDelhi.

Conway, K.E. 1970. The aquatic hyphomycetes of central Newyork. *Mycologia* 62: 516-530.

Delfrate, G. and Caretta, G. 1983. Aquatic hyphomycetes of a mountain streams in Valsesia (Piemonte). *Hydrobiologia*:102: 69-71.

Gonczol, J.1975. Ecological observations on the aquatic hyphomycetes of Hungary. *Acta. Bot. Acad. Sci. Hung.* 21: 243-264.

Ingold,C.T 1975. *An illustrated Guide to Aquatic and water borne Hyphomycetes (Fungi impefecti) with notes on their biology.* Fresh water Biological Assciation. Sci. Publ. no 30, Ferry House, Ambleside, Cumbria, England, 96.

Iqbal, S.H. and Webster, J. 1973a. The trapping of aquatic Hyphomycete spores by air bubbles. *Trans. Br. mycol. Soc.* 60: 37-48.

Iqbal, S.H. and Webster, J. 1973b. Aquatic Hyphomycetes spora of the river Exe and its tributaries. *Trans. Br. mycol. Soc.* 61: 331-346.

Kamat M.N; Patwardhan, P.G; Rao, V.G; and Sathe, A.V. 1972. Fungi of Maharashtra *M. P. K. V. Rahuri*, M.S Publ. No. 1, 124.

Lamore,B.J and Goos, R.D. 1978. Wood inhabiting fungi of a fresh water stream in Rhode island. *Mycologia.* 70: 1025-1034.

Muller Haeckel, A. 1977. Annual production of Hyphomycetes conidia in a subarctic stream. *Oikos.* 29: 396-397.

Muller Haeckel, A. and Marvanova, L. 1976. Konidien production und kolonisation Von susswasser Hyphomyzeten in kaltisjokk (Lappland). *Bot. Not.* 129 : 405-409.

Muller Haeckel,A and Marvanova, L. 1979a. Fresh water Hyphomycetes in Brackish and seawater. *Botanica Marina* 22: 421-424.

Muller Haeckel,A and Marvanova, L. 197a. Periodicity of aquatic Hyphomycetes in the subarchetic. *Trans. Br. mycol. Soc.*73: 109-116.

Saldanha Cecil, J. 1984. *Flora of Karnataka*, IBH, NewDelhi.

Sridhar, K.R., Chandrasekhar, K.R. and Kaveriappa K.M, 1992. Research on the Indian subcontinent in the ecology of aquatic Hyphomycetes ed. F. Barlocher. *Ecological studies* Vol.94, Springer - Verlag Heidelberg, 182-211.

Willoughby, L.G. and Archer, J.F. 1973. The fungal spora of fresh water stream and its colonization pattern on wood. *Freshwater Biol.* 3: 219-239.

Wood Eggenschwiler, S. and Barlocher, F. 1983. Aquatic Hyphomycetes in sixteen streams in France, Germany and Switzerland. *Trans. Br. mycol. Soc* 81: 371-379.

55 | *Screening of Ravenelia esculenta for Host-Parasite Interaction*

K. R. Gandhe[1], Rohini Chaubal[2] and N. R. Deshpande[2]

[1]Department of Botany, P. G. Research Centre, Modern College, Pune-411 005, India.

[2] T. R. Ingle Research Laboratory, Department of Chemistry, S. P. College, Pune-411 030, India.

ABSTRACT

Ravenelia esculenta Naras and Thirum, an edible rust, was analyzed chemically for host-parasite interaction. According to phytochemical tests, apparently there were no remarkable changes in the host plant even after infection. In both the healthy and infected material polar alkaloids, semi-polar and polar proteins; semi-polar and polar tannins were present. Percentage of volatile matter and sublimation product was found to be increased in the infected material. TLC of sublimation product from both the materials showed different compounds. Nicotine was present only in the infected material and was isolated quantitatively.

Keywords : *Ravenelia esculenta, Acacia eburnea, host-parasite interaction, Nicotine*

Introduction

Ravenelia esculenta Naras and Thirum on *Acacia eburnea* Willd (Narsimhan *et al.* 1961) was analyzed chemically as it is one of the important edible rusts. Adivasis consume this rust along with liquor. Due to the infection of this rust, the morphology of the host especially the structure of thorns, buds, flowers and fruits get completely changed. Healthy thorns are quite long, thin and pointed. Flowers are in typical head inflorescence and fruits are pod type. However, after infection especially due to invasion of aeciospores of *Ravenelia esculenta* the thorns, buds, flowers and fruits become succulent green and hypertrophoid with bizzare shapes. This may be due to the change in the levels of plant growth regulators which results in the disruption of physiology, metabolic activity and biochemistry of host (Isaac 1996).

In the present paper, mainly phytochemical tests in different solvents, extractive values, sublimation analysis along with thin layer chromatography (TLC) and isolation of nicotine was carried out.

Materials and methods

The healthy and infected material of *Acacia eburnea* willd was collected, shed dried and powdered for further analysis. Phytochemical tests were carried out. For these tests, the powder of healthy and infected material and their hexane, chloroform and ethanol extracts were tested as the solubility of different compounds is different in different solvents. Sublimation analysis was carried out in a sublimation unit. The sublimate was recovered and tested qualitatively by TLC. Isolation of Nicotine was carried out.

Results and Discussions

Healthy and infected material was tested for phytochemical tests especially for alkaloids, proteins, tannins and steroids. (Table No. 1).

Table 1 Phytochemical tests for infected material & healthy material

Test for	Test	A		B		C		D	
		I	H	I	H	I	H	I	H
Alkaloids	Material + picric acid →yellow ppt.	+ ve.	+ve.	−ve.	−ve.	−ve.	Little ppt.	+ ve.	+ve.
	Material + Wagner reagent → red ppt.	+ ve.	+ve.	Little ppt.	−ve.	+ ve.	Little ppt.	+ ve.	+ ve.
Proteins	Material + conc. HNO_3 → orange colour	+ ve.	+ ve.	−ve.	−ve.	Light yellow	Light yellow	+ ve.	+ ve.
Tannins	Material + NaOH heat, add FeCl3 → yellowish orange red ppt.	+ ve.	+ ve.	+ ve.	+ ve.	+ ve.	+ ve.	+ ve.	+ ve.
Steroids	Material + acetic acid → no change in colour	Light yellow	+ ve.	+ ve.	+ ve.	+ ve.	+ ve.	Light yellow	+ ve.

The results showed that qualitatively there were no remarkable changes in host plant even after infection, except for steroids for powdered material and ethanol extract. Test of

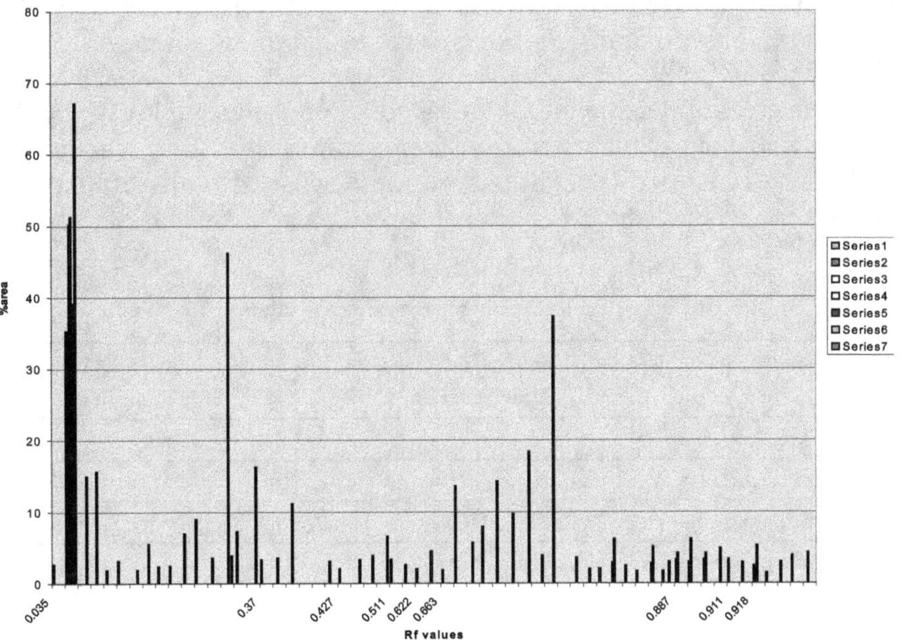

Figure 1. Extracts-Infected

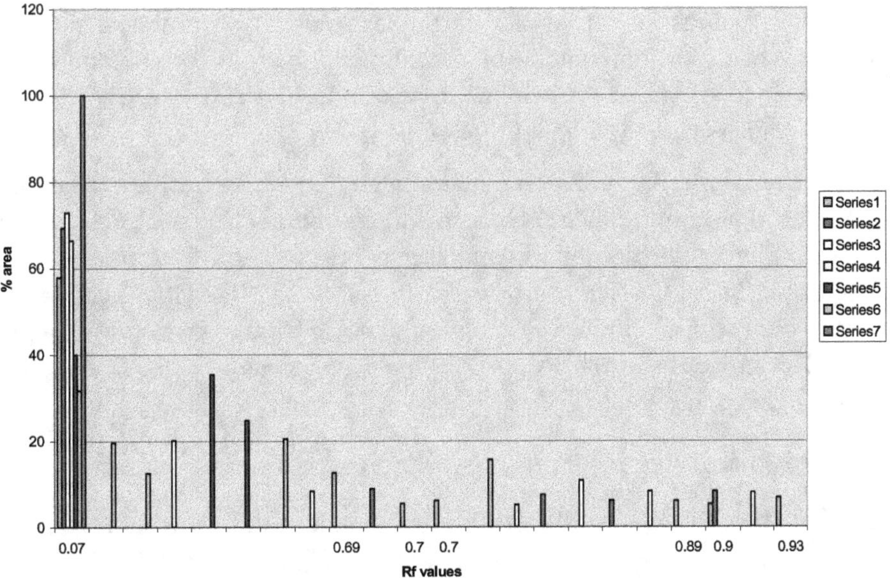

Figure 2. Extracts-Healthy

chloroform and ethanol extracts for proteins indicated the presence of semi polar and polar proteins. The positive tests for hexane, chloroform and ethanol for tannins indicated that non polar, semi polar and polar tannins were present in both the materials.

Cold solvent extraction :

Extractive values were determined by shaking the weighed sample (1.00 g) in different solvents (50 ml) for seven hrs at room temperature by using mechanical shaker. The material was kept overnight, filtered and washed with respective solvents. Removal of solvents gave the extractive values as shown in Table No.2.(3).

Table 2 Cold solvent extractive values

Solvent used	*A. eburnea* infected with *R. esculenta* (in %)	*A. eburnea* (in %)
Hexane	4.73	7.25
Benzene	0.48	1.29
Chloroform	8.69	7.92
Ethyl acetate	2.21	2.11
Acetone	6.01	6.52
Ethanol	7.83	7.15

The extractive values for semi-polar and polar solvents were almost similar for both the materials. While high percentage of the extractive values for Hexane and benzene indicated presence of more non-polar compounds like monoterpenes, diterpenes, triterpenes, steroids etc. in the healthy material.

Thin layer chromatographic and high performance thin layer chromatographic analysis:
Thin Layer Chromatography (TLC) of each solvent extract showed presence of various compounds. Quantitative estimation was carried out by High Performance Thin Layer Chromatography (HPTLC). The mobile phase used was 10% ethyl acetate in hexane. The extracts were spotted on Silica F_{254} plates by using Desaga Linomat. The plates were scanned at 275 nm densitometrically (Desaga Densitometer CD60) and the spectra obtained were compared.

Sublimation analysis :

The material (100 mg) was sublimed in a sublimation unit. The sublimation products were recovered in n-hexane, chloroform and ethanol. Thus three solutions of both healthy and infected material were obtained. The solutions were analyzed by TLC.

Table 3 HPTLC of solvent extracts

A	B	I		H		A	B	I		H	
		C	D	C	D			C	D	C	D
HEXANE	1.	0.035	2.7	0.074	57.9	Chloroform	1.	0.074	33.3	0.074	69.3
	2.	0.074	35.3	0.688	12.6		2.	0.113	1.9	0.695	8.9
	3.	0.143	3.2	0.696	5.4		3.	0.194	2.4	0.858	7.5
	4.	0.213	2.5	0.700	6.1		4.	0.238	7.0	0.866	6.1
	5.	0.288	9.0	0.886	6.0		5.	0.361	7.3	0.897	8.3
	6.	0.370	3.3	0.897	5.3		6.	0.525	2.6		
	7.	0.427	2.1	0.931	6.7		7.	0.635	4.5		
	8.	0.511	3.4				8.	0.700	8.0		
	9.	0.622	2.0				9.	0.859	2.1		
	10.	0.663	1.9				10.	0.866	2.9		
	11.	0.695	13.6				11.	0.872	2.5		
	12.	0.875	1.7				12.	0.877	2.8		
	13.	0.881	1.7				13.	0.887	4.2		
	14.	0.887	3.3				14.	0.898	6.2		
	15.	0.897	3.0				15.	0.906	3.4		
	16.	0.911	3.4				16.	0.916	2.9		
	17.	0.918	2.5				17.	0.929	3.0		
	18.	0.925	1.5				18.	0.937	3.1		
	19.	0.931	3.9								
Benzene	1.	0.075	50.2	0.074	72.9	Ethyl acetate	1.	0.075	51.3	0.074	66.4
	2.	0.304	3.6	0.865	10.8		2.	0.103	15.7	0.103	20.1
	3.	0.371	3.6	0.876	8.3		3.	0.154	5.5	0.368	8.4
	4.	0.418	3.1	0.918	8.0		4.	0.378	11.2	0.842	5.1
	5.	0.704	14.3				5.	0.697	5.7		
	6.	0.866	6.2				6.	0.842	3.6		
	7.	0.877	5.1				7.	0.858	2.0		
	8.	0.906	4.2				8.	0.907	4.9		
	9.	0.918	5.3								
	10.	0.942	4.3								
Acetone	1.	0.074	39.2	0.074	40.0	Dioxan	1.	0.075	24.3	0.074	31.7
	2.	0.150	1.9	0.359	35.3		2.	0.093	15.0	0.093	19.6
	3.	0.359	46.2	0.361	24.7		3.	0.366	16.3	0.096	12.5
	4.	0.716	9.8				4.	0.440	3.3	0.366	20.5
	5.	0.881	3.0				5.	0.483	3.9	0.73	15.7
							6.	0.743	37.3		
Ethanol	1.	0.074	67.2	0.074	100.0						
	2.	0.359	3.9								
	3.	0.493	6.6								
	4.	0.722	18.4								
	5.	0.725	3.9								

A = Solvent used for extraction, B = No. of peak, C = Rf value, D = Area (%), I = Infected material, H = Healthy material

Thin Layer Chromatographic analysis for the sublimation product of the two materials showed different spots indicating different compounds. The phytochemical tests of theses solutions were carried out for alkaloids, proteins, tannins and steroids. Table No. 4.

Table No. 4 Sublimation analysis

Sr. No.	Sample	Amount of sublimation product obtained (in mg)	No. of spots on TLC		
			Hexane	Chloroform	Ethanol
1.	A. eburnea infected with R. esculenta	9.20	5	7	5
2.	A. eburnea	8.20	6	6	5

The phytochemical tests for the sublimed product showed difference in alkaloid and steroid content of healthy and infected material for hexane extract and steroid content of ethanol extract (Table 5).

Table No. 5 Phytochemical tests for sublimation product

Tests	Hexane fraction		Chloroform fraction		Ethanol fraction	
	Infected	Healthy	Infected	Healthy	Infected	Healthy
Alkaloids	+ve	-ve	+ve	+ve	+ve	+ve
Proteins	+ve	+ve	+ve	+ve	+ve	+ve
Tannins	+ve	+ve	+ve	+ve	+ve	+ve
Steroids	-ve	+ve	-ve	-ve	+ve	-ve

Isolation of Nicotine

Healthy and infected material (10.0 g each) was taken in 250 ml beaker. HCL (20%, 100 ml) was added and kept overnight (for 20 hr). HCl extract was filtered and pH was adjusted to 12.5 with flakes of NaOH. The basic solution was steam distilled for 40 min. The distillate was made acidic to about pH 3 with solid oxalic acid and concentrated to get syrup. On cooling white crystals of nicotine oxalate (453.7 mg) were obtained from infected material and not from healthy material. The crystals were filtered and washed with cold water. Pure nicotine oxalate (277.0 mg) was taken in a 50 ml beaker, added excess of aqueous KOH and stirred. The KOH layer was extracted with diethyl ether. The organic phase was dried over anhydrous sodium sulphate and evaporated to get pure Nicotine (81.2 mg). TLC of ether layer showed single spot. The nicotine was confirmed by TLC and by comparing IR of nicotine oxalate crystals obtained above with that of oxalate of pure Nicotine. IR (V_{max}) 3860.1, 3854.2, 3849.2, 1718.5, 1666.4, 1636.5, 1573.7, 1559.9, 1503.6, 1473.8, 1441.4, 1404.9, 1301.5, 1255.5, 845.6, 817.7, 746.9, 721.5, 694.9, 683.0 cm^{-1}

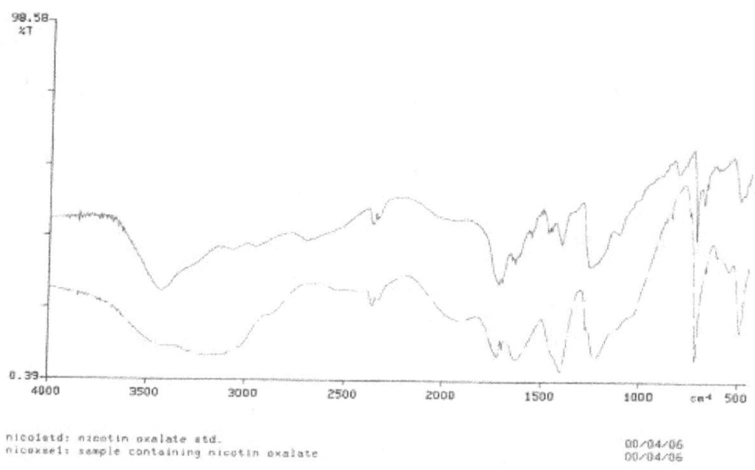

Figure 3.

Acknowledgements

The authors are grateful to the Principal, Modern College, Pune – 411005 and T. R. Ingle Laboratory and Principal, S. P. College, Pune- 411030 for providing the necessary facilities. The authors are also grateful to CSIR- New Delhi for financial assistance.

References

Narasimhan and ***Thirumalachar M.J.***(1961). Phytopathologische Zeitschrift 41 : 97-102

Isaac, Susan (1996). Fungal-Plant Interactions. Chapman and Hall, London, 418 pp.

Indian Pharmacopoeia (1996)

Sadasivam. S and ***A. Manickam*** (1992) Biochemical methods.

Nicotine-details Fluka, Switzerland

www.ingramcontent.com/pod-product-compliance
Lightning Source LLC
Chambersburg PA
CBHW060232120726

48009CB00004B/235